The Handbook of Health and Safety Practice

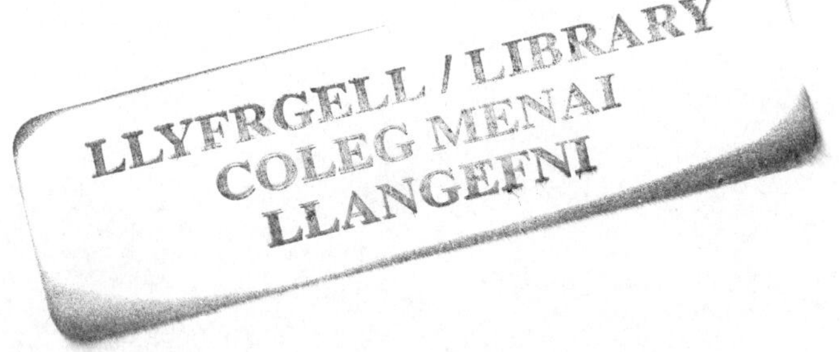

The Handbook of Health and Safety Practice

Sixth edition

Jeremy Stranks
MSc, FCIEH, FIOSH, RSP

PEARSON EDUCATION LIMITED

Head Office:
Edinburgh Gate
Harlow CM20 2JE
Tel: +44 (0)1279 623623
Fax: +44 (0)1279 431059

London Office:
128 Long Acre
London WC2E 9AN
Tel: +44 (0)20 7447 2000
Fax: +44 (0)20 7447 2170
Website: www.business-minds.com

First published in Great Britain in 1986
Sixth edition 2003

ISBN 0 273 66332 1

British Library Cataloguing in Publication Data
A CIP catalogue record for this book can be obtained from the British Library

10 9 8 7 6 5 4 3 2 1

Typeset by M Rules
Printed and bound in Great Britain by Biddles Ltd, Guildford & King's Lynn

The Publisher's policy is to use paper manufactured from sustainable forests.

Contents

Preface

The Handbook of Health and Safety Practice continues to be an important reference source for health and safety practitioners, enforcement officers and those studying for NEBOSH and other examinations in the field of occupational health and safety. This sixth edition has taken into account advances in legislation, in particular the Management of Health and Safety at Work Regulations 1999, the Working Time Regulations 1998, the Children (Protection at Work) Regulations 1998 and recent updates of other Regulations.

Particular attention has been given to the significance of quality systems in occupational health and safety and the revisions to the Approved Code of Practice and HSE Guidance dealing with the role of competent persons.

As with previous editions, I hope that all who use this book will find it helpful and informative in steering their way round the very comprehensive and broadly based subject.

Jeremy Stranks
January 2003

Acknowledgements

The author wishes to place on record his thanks to the following organisations and companies for their assistance in the preparation of *Health and Safety Practice*:

British Standards Institution
Draeger Safety
Dust Control Equipment Limited, Leicester
Doug Payne Kinetics Limited
National Vulcan Engineering Insurance Group Limited
Royal Society for the Prevention of Accidents
G.W. Sparrow & Sons Plc
Weller Engineering Limited

The law is stated at 1 January 2002

Extracts from British Standards are reproduced by permission of the British Standards Institution. Complete copies of the documents can be obtained from the British Standards Institution, Linford Wood, Milton Keynes MK14 6LE.

About the author

Jeremy Stranks has had over forty years experience in occupational health and safety, initially as an enforcement officer in local government and, more recently, as Chief Health and Safety Adviser to the Milk Marketing Board and its commercial subsidiary, Dairy Crest Limited. He is currently a consultant, trainer, author and expert witness in this field. He is the author of many books, including *Health and Safety Law* which is now in its fourth edition.

Principal Regulations

1972 Highly Flammable Liquids and Liquefied Petroleum Gases Regulations

1974 Industrial Tribunals (Improvement and Prohibition Notices Appeals) Regulations

1974 Agriculture (Tractor Cabs) Regulations

1975 Employers' Health and Safety Policy Statements (Exceptions) Regulations

1976 Fire Certificates (Special Premises) Regulations

1977 Safety Representatives and Safety Committees Regulations

1977 Health and Safety (Enforcing Authority) Regulations

1977 Offshore Installations (Life-Saving Appliances) Regulations

1978 Offshore Installations (Fire-Fighting Equipment) Regulations

1980 Safety Signs Regulations

1981 Health and Safety (First Aid) Regulations

1981 Health and Safety (Dangerous Pathogens) Regulations

1982 Notification of New Substances Regulations

1982 Notification of Installations Handling Hazardous Substances Regulations

1983 Asbestos (Licensing) Regulations

1985 Asbestos (Prohibition) Regulations

1985 Social Security (Industrial Injuries) (Prescribed Diseases) Regulations

1987 Control of Asbestos at Work Regulations

1989 Construction (Head Protection) Regulations

1989 Electricity at Work Regulations

1989 Noise at Work Regulations

1990 Health and Safety Training Regulations

1990 Dangerous Substances (Notification and Marking of Sites) Regulations

1990 Health and Safety (Training for Employment) Regulations

1992 Gas Appliances (Safety) Regulations

1992 Notification of Cooling Towers and Evaporative Condensers Regulations

1992 Workplace (Health, Safety and Welfare) Regulations

1992 Personal Protective Equipment at Work Regulations

1992 Manual Handling Operations Regulations

1992 Health and Safety (Display Screen Equipment) Regulations

1994 Chemicals (Hazard Information and Packaging for Supply) Regulations

1994 Carriage of Dangerous Goods by Road and Rail (Classification, Packaging and Labelling) Regulations

1994 Construction (Design and Management) Regulations

1994 Gas Safety (Installation and Use) Regulations

1995 Reporting of Injuries, Diseases and Dangerous Occurrences Regulations

1996 Construction (Health, Safety and Welfare) Regulations

1996 Gas Safety (Management) Regulations

1996 Health and Safety (Consultation with Employees) Regulations

1996 Health and Safety (Safety Signs and Signals) Regulations

1996	Carriage of Dangerous Goods (Classification, Packaging and Labelling) and the Use of Transportable Gas Receptacles
1997	Confined Spaces Regulations
1997	Fire Precautions (Workplace) Regulations
1998	Control of Lead at Work Regulations
1998	Lifting Operations and Lifting Equipment Regulations
1998	Provision and Use of Work Equipment Regulations
1998	Working Time Regulations
1998	Health and Safety (Information for Employees) Regulations
1998	Children (Protection at Work) Regulations
1998	Control of Lead at Work Regulations
1998	Health and Safety (Enforcing Authority) Regulations
1999	Control of Major Accidents Hazards Regulations
1999	Management of Health and Safety at Work Regulations
1999	Ionising Radiations Regulations
1999	Control of Substances Hazardous to Health Regulations
1999	Employers' Liability (Compulsory Insurance) Regulations
1999	Pressure Systems Safety Regulations

Principal Statutes

1954 Mines and Quarries Act
1956 Agriculture (Safety, Health and Welfare Provisions) Act
1957 Occupiers' Liability Act
1960 Radioactive Substances Act
1961 Factories Act
1963 Offices, Shops and Railway Premises Act
1969 Employers' Liability (Compulsory Insurance) Act
1969 Employers' Liability (Defective Equipment) Act
1971 Fire Precautions Act
1971 Mineral Workings (Offshore Installations) Act
1974 Health and Safety at Work etc., Act
1974 Control of Pollution Act
1975 Social Security Act
1976 Fatal Accidents Act
1977 Unfair Contract Terms Act
1978 Employment Protection (Consolidation) Act
1980 Employment Act
1982 Employment Act
1982 Administration of Justice Act
1982 Criminal Justice Act
1984 Occupiers' Liability Act
1985 Food and Environment Protection Act
1986 Social Security Act
1986 Safety at Sea Act
1986 Sex Discrimination Act
1987 Fire Safety and Safety of Places of Sport Act
1988 Environment and Safety Information Act
1989 Employment Act
1990 Environmental Protection Act
1995 Environment Act

Table of Cases

This Table includes, in alphabetical order, all relevant Cases. Where these are discussed in the text, the appropriate page number is given.

List of Abbreviations

ACOP	Approved Code of Practice
BS	British Standard
CHIPR	Chemicals (Hazard Information and Packaging for Supply) Regulations 1994
CLWR	Control of Lead at Work Regulations 1998
COMAH	Control of Major Accident Hazards Regulations 1999
COSHH	Control of Substances Hazardous to Health Regulations 1999
EMAS	Employment Medical Advisory Service
EPCA	Employment Protection (Consolidation) Act 1978
FA	Factories Act 1961
FPA	Fire Precautions Act 1971
FSSPSA	Fire Safety and Safety of Places of Sport Act 1987
HSC	Health and Safety Commission
HSE	Health and Safety Executive
HFL&LPGR	Highly Flammable Liquids and Liquefied Petroleum Gases Regulations 1972
HS(DSE)R	Health and Safety (Display Screen Equipment) Regulations 1992
HSWA	Health and Safety at Work etc. Act 1974
ILO	International Labour Organisation
LEV	Local exhaust ventilation
LOLER	Lifting Operations and Lifting Equipment Regulations 1998
MEL	Maximum exposure limit
MHOR	Manual Handling Operations Regulations 1992
MHSWR	Management of Health and Safety at Work Regulations 1999
OEL	Occupational exposure limit
OES	Occupational exposure standard
OLA	Occupiers Liability Act 1957
OSRPA	Offices, Shops and Railway Premises Act 1963
PPEWR	Personal Protective Equipment at Work Regulations 1992
PSSR	Pressure Systems Safety Regulations 2000
PUWER	Provision and Use of Work Equipment Regulations 1998
Reg	Regulation
RIDDOR	Reporting of Injuries, Diseases and Dangerous Occurrences Regulations 1995
Sch	Schedule
Sec	Section
SI	Statutory Instrument
SS(II)(PD)R	Social Security (Industrial Injuries) (Prescribed Diseases) Regulations 1985
TULRA	Trade Unions and Labour Relations Act 1974

UCTA	Unfair Contract Terms Act 1977
W(HSW)R	Workplace (Health, Safety and Welfare) Regulations 1992
WTR	Working Time Regulations 1998

Organisations/Publications

ACAS	Advisory, Conciliation and Arbitration Service
BS	British Standard
Cmnd	Command Paper
COIT	Central Office of the Industrial Tribunals
DSS	Department of Social Security
HSC	Health and Safety Commission
HSE	Health and Safety Executive
TSO	The Stationery Office
TUC	Trades Union Congress

Case citations

AC	Appeal Cases
AER	All England Reports
HSIB	Health & Safety Information Bulletin
IRLR	Industrial Relations Law Reports
KB	King's Bench
Macq	Macqueen (Scot) (ended 1865)
QB	Queen's Bench
SLT	Scottish Law Times
TLR	Times Law Reports
WLR	Weekly Law Reports

Legal terminology

J	Mr Justice, a junior judge, usually sitting in a court of first instance
LJ	Lord Justice, a senior judge, usually sitting in a court of appeal
plaintiff	the person presenting a claim in a civil action
defendant	the person against whom the claim is brought in a civil action
appellant	the person bringing an appeal in a civil action
respondent	the person against whom the appeal is brought in a civil action
tort	a species of civil action for injury or damage where the remedy or redress is an award of unliquidated damages
volenti non fit injuria	literally 'to one who is willing no harm is done' defence in a civil action.

SECTION A

Health and safety law

1 Origins of health and safety law

A

The law relating to health and safety at work has its origins in statute and common law going back to the early nineteenth century and the industrial revolution.

Nature of health and safety law

Statutes

The legislative history of occupational health and safety has been patchy. Indeed, until the Health and Safety at Work etc. Act (HSWA) was passed in 1974, thereby creating a new consultative approach to occupational health and safety, protective legislation consisted of a series of statutes which were aimed mainly at manufacturing industry. Passed on an ad hoc basis, they attempted to control safety and health hazards in specific work locations, such as factories, offices, shops, mines and quarries, etc., e.g. Factories Act 1961, Offices, Shops and Railway Premises Act 1963 and Mines and Quarries Act 1954. It should be appreciated, however, that this legislation is gradually being repealed and replaced by Regulations made under HSWA, e.g. Management of Health and Safety at Work Regulations 1999, Workplace (Health, Safety and Welfare) Regulations 1992.

Common law

Common law is the decisions of the courts which have been bound by the doctrine of precedent into a corpus of authoritative principles and rules. It is judge-made law and accounts for the greater part of the law of contract and tort, both of which – but particularly the latter (*see* Chapter 2) – have played a significant role in the development of civil liability relating to occupational health and safety. Common law is synonymous with case law and its rules and principles and doctrines are to be found in the law reports. Case law is a self-endorsing process, being perpetuated either by previous binding cases or by the interpretation of legislation. The main contribution of common law to the law relating to occupational health and safety is the body of rules developed in connection with the right of employees and their dependants to sue employers for damages for personal injury, disease or death at work, i.e. the civil liability of employers is to be found mainly in the rules of tort (*see* Chapter 2).

Judicial precedent

Judicial precedent is defined as 'a decision of a tribunal to which some authority is attached'. Precedents not only influence the development of law but are, in themselves, one of the material sources of the law.

To say that a case is a 'binding decision of precedent' means that the principle on which the decision was made will be binding in subsequent cases which are founded on similar facts.

A precedent may be authoritative or persuasive.

Authoritative precedents

These are decisions which judges are bound to follow. There is no choice in the matter. For instance, a lower court is bound by a previous decision of a higher court.

Persuasive precedents

These are decisions which are not binding on a court but to which a judge will attach some importance, for instance, decisions given by the superior courts in Commonwealth countries will be treated with respect in the English High Court.

The hierarchy of the courts

The doctrine force of precedent is based on the principle that the decisions of superior courts bind inferior courts. Thus:

1. The High Court is bound by a decision of the Court of Appeal, and the Court of Appeal is bound by a decision of the House of Lords, these courts being in the same hierarchy.
2. A High Court judge is not bound by a previous decision of the High Court. Such a decision enjoys only persuasive authority.

3. In civil matters, the Civil Division of the Court of Appeal will bind all inferior courts. It is bound by its own decisions and by those of the House of Lords.

   In criminal matters, the Criminal Division of the Court of Appeal will bind all inferior courts.

4. County Courts, Magistrates' Courts and Administrative Courts are bound, without question, by decisions of the superior courts.

Division into criminal and civil liability

Criminal liability

Criminal liability refers to the responsibilities under statute and the penalties which can be imposed by criminal courts, i.e. fines, imprisonment and remedial orders. The criminal courts in question are the magistrates' courts, which handle the bulk of health and safety offences (the less serious offences), and the Crown Courts, which deal with the more serious health and safety offences (*see* Chapter 4).

There are appeal procedures to the High Court, and beyond to the Court of Appeal and, assuming that leave is given, to the House of Lords. Appeals against decisions of magistrates' courts and Crown Courts on points of criminal law involving occupational health and safety are few. Employers and factory occupiers when found guilty tend to pay the fine and there the matter ends. After all, although not insignificant, a fine is generally much smaller than the civil remedy of an award of damages. Moreover, convictions for health and safety offences can be used as evidence of negligence in subsequent cases, in accordance with the Civil Evidence Act 1968, sec 11.

Criminal liability is either:

(a) statutory; or

(b) common law, e.g. murder, manslaughter.

As far as occupational health and safety law are concerned, criminal liability is exclusively of statutory origin (but see the remarks about the General Duties of HSWA on pages 25–32).

Statutory criminal liability – interpretation of penal duties

Penal duties in relation to occupational health and safety are strictly construed. This means that the courts only interpret words appearing in the statute or statutory instrument, and not words which do not appear. Examples of this occur often (and can sometimes have strange, if not irritating, commercial consequences). Here the chief way of 'rectifying' the law is by recourse to subsequent legislation, correcting the previous 'maverick' decision. The reversal by subsequent legislation of the House of Lords' decision in *John Summers & Sons Ltd* v. *Frost* [1955] 1 AER 870 provides a good illustration. Here the respondent was employed by the appellant as a maintenance fitter in a steelworks. His thumb was caught in a revolving grindstone, when grinding a piece of metal on a grinding machine. The machine had horizontal shafts at each end driven by an electric motor. The grindstones revolved downwards when the operator faced the machine and were enclosed to an extent commensurate with performing normal grinding operations (i.e. a small portion of each wheel remained exposed opposite its tool rest). It was held by the House of Lords that the grindstone was a 'dangerous part of machinery' for the purposes of the Factories Act 1937, sec 14(1). The legal view was that the machine was not 'securely fenced', as required, even though the consequence of 'securely fencing' would have been to make the machine commercially useless.

Types of statutory criminal liability

Statutory duties can be (a) absolute, (b) practicable and (c) reasonably practicable.

Absolute requirements: Where the risk of injury, etc., is inevitable if safety precautions are not taken, the statutory duty may well be absolute (i.e. liability is not referable to negligence, or failing to take practicable or reasonably practicable safety measures).

An example, perhaps the most important one in terms of factory safety, of an absolute statutory duty is the Factories Act 1961, sec 12(1). This requires that 'every flywheel directly connected to any prime mover and every moving part of any prime mover . . . shall be securely fenced, whether the flywheel or prime mover is situated in an engine house or not'.

If, in consequence, a machine becomes commercially impracticable or mechanically unusable when it has been securely fenced, this is irrelevant (*John Summers & Sons Ltd* v. *Frost* [1955] AC 746); alternatively, if the machine cannot be securely fenced, this is no defence (*Davies* v. *Thomas Owen & Co. Ltd* [1919] 2 KB 39).

Practicable requirements: A statutory obligation which has to be carried out so far as is practicable, and must be carried out if, in the light of current knowledge and invention, it is feasible. And this even though it may be difficult, inconvenient and costly (*Schwalb* v. *Fass (H) & Son* [1946] 175 LT 345).

Reasonably practicable requirements: A statutory duty which has to be carried out so far as is reasonably practicable, allowing for the balancing of costs and benefits. Thus, in the leading case of *Edwards* v. *National Coal Board* [1949] 1 AER 743, Asquith LJ said:

> 'Reasonably practicable' is the narrower term than 'physically possible', and seems to me to imply that a computation must be made by the owner in which the quantum or risk is placed on one scale and the sacrifice involved in the measures necessary for averting the risk (whether in money, time or trouble) is placed on the other, and that, if it be shown that there is a gross disproportion between them – the defendants discharge the onus on them. Moreover, this computation falls to be made by the owner at a point of time anterior to the accident.

Comparing 'practicable' and 'reasonably practicable' duties, Lord Reid in the leading case of *Marshall* v. *Gotham & Co.* [1954] 1 AER 937 observed, 'If a precaution is practicable it must be taken unless in the whole circumstances that would be unreasonable. And as men's lives may be at stake it should not lightly be held that to take a practicable precaution is unreasonable.'

To summarise a complex point, if a duty is required to be carried out 'so far as is reasonably practicable', the owner or occupier of the work premises, or any other person upon whom the duty is placed, can afford to run the risk of not instituting safety measures, procedures, etc., if the overall benefit, in terms of reducing accidents or improving health and safety conditions, is minimal compared with the cost and inconvenience of introducing the measures, etc.

Enforcement agencies

Criminal laws are enforced by state enforcement agencies, e.g. the police, trading standards departments and, in the case of occupational health and safety, HSE inspectors and environmental health officers. Breach of duties takes the form, on conviction, of a fine or imprisonment. Alternatives are also possible, e.g. remedial orders. By contrast with civil liability in tort, criminal liability cannot be insured against by either employers, directors or employees. It is, however, lawful to insure against court costs and the legal expenses of counsel in connection with a prosecution for health and safety offences. Many reputable insurers offer such cover.

Statutory civil liability

Civil liability refers to the 'penalty' which can be imposed by a civil court, i.e. the County Court; the High Court; the Court of Appeal (Civil Division); or the House of Lords. In general terms, the first two courts are courts of first instance (i.e. they try cases for the first time), whereas the latter two courts are courts of appeal. Such liability consists of awards of damages for injury, disease and/or death at work:

(a) in circumstances disclosing breach of common law and/or statutory duty (i.e. normally negligence) on the part of an employer/factory occupier (*see* Chapter 2 for an analysis of 'negligence', common to actions for both breach of common law and statutory duty); and
(b) arising out of and in the course of employment (*see* further Chapter 2).

Normally the employer is insured against this liability (*see* Chapter 2) and payment to the injured employee is made by the employer's insurer.

General and special damages

The damages which can be recovered following injury-causing accident/disease/death fall into two categories:

(a) *general damages* – which relate to losses incurred after the hearing of the action, viz. actual and probable (not merely 'possible') loss of future earnings following the accident;
(b) *special damages* – relating to quantifiable losses incurred before the hearing of the case, and consisting mainly of –
    (i) medical expenses incurred before the hearing of the case;
    (ii) loss of earnings incurred before the hearing of the case.

In the case of fatal injury, compensation for death negligently caused is payable under the Fatal Accidents Act 1976, sec 1; moreover, a fixed lump sum is payable, under the Administration of Justice Act 1982, sec 3, in respect of bereavement.

Parties to the action

Unlike criminal action, civil action is usually initiated by a private individual or by a company suing another private individual, company or local authority. This is known as a private interparty or adversarial action (or litigation). Where workplace injuries are concerned, such action is normally commenced by an injured worker, or, in the case of a fatality, by his dependants. If the worker is a member of a trade union, the union branch may initiate proceedings. Although brought against his employer or the factory occupier, the action is, very often, a claim on the employer's insurance company. If the employer is found legally liable (i.e. personally, or, more likely, vicariously negligent, *see* Chapter 2), the insurer will pay out to the injured worker.

Compulsory employers' liability insurance for negligence

Liability insurance cover is compulsory for most employers. Exceptions occur in the case of workplaces where the family and/or relatives of the employer are the only persons employed in the workplace, also the nationalised industries, local authorities, nuclear power installations and the police authorities, as well as certain other bodies financed out of public funds. Cover is provided in respect of injury, death or disease which is the result of legal liability on the part of the employer. Such legal liability may arise from (a) a breach of common law duty or (b) breach of statutory duty or (c) breach of both types of duty.

'Guarantee' of compensation: The existence of large funds of insurance monies, guaranteed by compulsory liability insurance legislation, has led to awards of damages being made in circumstances where the risk of injury (and certainly the risk of death) has not necessarily been reasonably foreseeable – the test necessary to establish liability for negligence (*see* Chapter 2). In consequence, the base of liability has gradually moved away from fault in the direction of strict liability.

Civil liability arising from statutory duties

Breach of statutory duties relating to occupational health and safety results principally in criminal liability. With the exception of the General Duties of HSWA, which can only give rise to civil liability at common law (*see above*), statutory duties can also create civil liability, i.e. breach can lead to prosecution (and conviction) in a criminal court *and* to an award of damages in a civil action. This point, relating to breach of safety provisions, was first established in *Groves* v. *Lord Wimborne* [1898] 2 QB 402, where a boy lost his arm when working at machinery in an ironworks and was later successful in his action for damages against the employer. Suing for breach of the fencing requirements of the FA is a standard practice today, and most injured employees sue their employers for breach of both statutory duty and common law duty, in what has become known as a double-barrelled action (*see* Chapter 2).

Breach and causation

Where there is a breach of statutory safety and/or health duty, before the injured workman can succeed in his action for damages, it is generally accepted that he must establish the three ingredients of negligence, viz. duty of care; breach of duty; and injury, disease or death caused substantially by the breach. The first two ingredients normally present no problems because, where a statutory instrument lays down a health and/or safety duty on an employer, this establishes the existence of a duty of care for the purposes of civil liability. But the third ingredient, causation, can all too often generate many problems for injured employees.

The law is that an injured employee, or a deceased employee's dependant, must show that statutory breach was the substantial cause of his injury, disease or death, i.e. he must show that the injury, disease or death was predominantly caused by breach rather than his own failure to act correctly and properly in the interests of his own health and safety. A leading case setting down the law on causation in relation to statutory health and safety duties is *Gardiner* v. *Motherwell Machinery & Scrap Co. Ltd* [1961] 3 AER 831 where Lord Reid said:

> When a man who has not previously suffered from a disease contracts that disease after being subject to conditions likely to cause it, and when he shows that it starts in a way typical of disease caused by such conditions, he establishes a prima facie presumption that his disease was caused by those conditions.

The distinctive nature of criminal and civil law in the field of occupational health and safety was succinctly elucidated by Lord Diplock in the leading case of *Boyle* v. *Kodak Ltd* [1969] 2 AER 439 when he said, 'When considering the civil liability engrafted by judicial decision upon the criminal liability imposed by statute, it is no good looking to the statute and seeing from it where the criminal liability would lie, for we are concerned only with civil liability. We must look to the cases.'

Time for bringing actions

All civil actions for personal injury, disease and death must be brought within three years of the injury (Limitation Act 1980, sec 11). Normally, where accidents causing injuries or fatality are concerned, this three-year limitation period causes no problems. But where certain (usually fatal) occupational diseases are concerned, manifesting themselves tardily over a 20- or 30-year period, the rule causes serious hardship. This is particularly true in the case of diseases such as occupational cancer, asbestosis and pneumoconiosis.

European Community Law

Under the Treaty of Rome, the Council of the European Committees can issue Directives. These usually provide for 'harmonisation' of the laws of Member States, including health and safety legislation. Representatives of the Member States meet to agree the content of the Draft Directives. Once agreed, Drafts are presented to the European Parliament for ratification.

Directives impose a duty on each Member State to:

(a) make Regulations to conform to the Directive; and

(b) enforce those Regulations.

Once a Directive is adopted, it is notified to Member States and takes effect following such notification. Directives incorporate a date by which their provisions must be implemented, where necessary by the enactment of domestic legislation.

Regulations are binding in their entirety and directly applicable in all Member States. This means that they are part of the domestic legislation of Member States and no further legislative measures are necessary. Regulations are the Community instruments which have the most direct impact on a wide range of duties and individual rights in Member States.

Other forms of Community law include decisions, recommendations and opinions, and agreements.

Decisions are made by either the European Commission or the Council of Ministers, being addressed to certain Member States or individuals. They require the person to whom they are addressed to take certain action or, alternatively, to refrain from certain activities, in order to comply with Community law.

Opinions and recommendations may also be issued by the Commission or the Council. Recommendations suggest a particular course of action in certain specific situations. Opinions, on the other hand, state a view on a situation or event in the Community or in a particular Member State.

Agreements between the Community and non-member countries, and between individual Member States, also form part of EEC law, but are limited in scope.

Stages in the development of an EC directive

Formal proposal by the European Commission

↓

Negotiations in Council Working Groups
(takes account of views on European Parliament)

↓

Council of Ministers agree a 'common position'

↓

Further negotiation

↓

Adoption by Member States

↓

Implementation

The court hierarchy in England and Wales

There are two distinct systems whereby the courts deal with criminal and civil actions respectively. Some courts have both criminal and civil jurisdiction, however.

1. Magistrates' Court

This is the lowest of the courts and deals mainly with criminal matters. Magistrates determine and sentence for many of the less serious offences. They also hold preliminary examinations into other offences to see if the prosecution can show a prima facie case on which the accused may be committed for trial at a higher court.

2. Crown Court

Serious criminal charges and some not so serious, but where the accused has the right to jury trial, are heard on indictment in the Crown Court. This court also hears appeals from magistrates' courts.

3. County Courts

These courts deal with civil matters only.

4. High Court of Justice

More important civil matters, because of the sums involved or legal complexity, will start in the High Court of Justice. The High Court has three divisions:

Queen's Bench – deals with contract and torts;

Chancery – deals with matters relating to, for instance, land, wills, partnerships and companies;

Family.

In addition, the Queen's Bench Division hears appeals on matters of law:

(a) from the magistrates' courts and from the Crown Court on a procedure called 'case stated', and

(b) from some tribunals, for example the finding of an industrial tribunal on an enforcement notice under the Health and Safety at Work Act.

It also has some supervisory functions over the lower courts and tribunals if they exceed their powers or fail to undertake their functions properly, or at all.

The High Court, the Crown Court and the Court of Appeal are known as the Supreme Court of Judicature.

5. The Court of Appeal

The Court of Appeal has two divisions:

(a) the Civil Division, which hears appeals from the county courts and the High Court;

(b) the Criminal Division, which hears appeals from the Crown Court.

6. The House of Lords

Further appeal, in practice on important matters of law only, lies to the House of Lords from the Court of Appeal and in restricted circumstances from the High Court.

7. European Court of Justice

This is the supreme law court, whose decisions on interpretation of European Community law are sacrosanct. Such decisions are enforceable through the network of courts and tribunals in all Member States.

Cases can only be brought before this court by organisations or individuals representing organisations.

Industrial tribunals and Health and Safety at Work

Industrial tribunals are empowered to hear complaints under the Health and Safety at Work Act 1974.

Composition

Each tribunal consists of a legally qualified chairman appointed by the Lord Chancellor and two lay members – one from management and one from a trade union, selected from panels maintained by the Department of Employment after nomination from employers' organisations and trade unions.

Representation

There is now a tendency to employ lawyers on both sides. However, a person is entitled to appear on his own behalf or have another person represent him. Legal aid is available but not actual representation. Employees with a grievance related to their employment, and who are members of a trade union, are normally represented by that union.

Decisions

When all three members of a tribunal are sitting, the majority view prevails.

Complaints on health and safety issues

Tribunals deal with the following health and safety issues:

(a) appeals against improvement and prohibition notices served by officers of the enforcement authorities;

(b) the provision by an employer of time off for the training of safety representatives;

(c) the failure by an employer to pay safety representatives for time off for carrying out their functions and for training;

(d) the failure of an employer to make a medical suspension payment;

(e) dismissal, actual or constructive, following a breach of health and safety law, regulation and/or term of employment contract.

Procedure

Procedures are laid down in the Industrial Tribunals (Rules of Procedure) Regulations 1985 and the Industrial Tribunals (Improvement and Prohibition Notices Appeals) Regulations 1974.

The standard procedure usually involves a pre-hearing and a full hearing.

1. *Pre-Hearing*

This takes the form below:

(a) The applicant i.e. person making the complaint, files an originating application setting out the nature of the claim and the redress sought.

(b) A copy of the application is sent to the employer (respondent) who submits a defence in another document, known as a 'notice of appearance'.

(c) Should a party wish to have further information about allegations made, he can request 'further particulars'.

(d) Should such a request be refused when such particulars are relevant to the case, the tribunal can order their discovery i.e. their production and disclosure.

(e) A date for the 'Pre-Hearing' of the complaint is set.

(f) Where a party proposes to refer to a document at the hearing, he should supply the other side with copies a week in advance of the hearing, as well as the chairman and two lay members; failure to do so could result in an adjournment to permit the other side time to refer to same.

(g) Persons present at all material times and occasions relating to the complaint should be available to give evidence as witnesses; where a witness is reluctant to appear, the tribunal can order him to do so.

(h) Both parties to the proceedings must be informed at least 14 days before the date fixed for the hearing, which consists of two parts –

    (i) the applicant (employee) presents the outline of his case and calls his evidence;

    (ii) the respondent (employer) calls his evidence and makes a closing speech.

2. *Full Hearing*

A Full Hearing is commonly held in public unless there is a good reason for not doing so e.g. national security.

The applicant makes an opening speech indicating his role in the organisation, the facts on which he bases his case and quoting any statutes and case law which may be relevant.

Witnesses are then called and examined under oath by the representatives of both parties. The relevant facts are then summarised in a closing speech by representatives of both parties, covering any questions raised by the other side or chairman or members of the tribunal.

Strict hearsay evidence rules do not apply in industrial tribunals, whereas first hand evidence of the applicant and witnesses carries much more weight.

After completion of the final submissions, the tribunal makes its oral decision normally on the day of the hearing. At a later date the decision is set out formally in summary form and sent to both parties.

Costs

Costs are not normally awarded to the successful party. For costs to be awarded it must be shown either:

(a) that one party acted 'frivolously or vexatiously' i.e. there never was a valid claim; or

(b) the proceedings were brought or conducted unreasonably e.g. failure to warn witnesses to attend, thereby resulting in a delay.

An order for costs is usually in keeping with the ability of a party to pay.

Expenses

Witnesses and other persons are paid an allowance by the tribunal for loss of pay and to cover the cost of attending the tribunal. Representatives, union officials, professional advisers, etc. are not entitled to be paid for attendance.

Review of tribunal decision

The decision of a tribunal can be challenged, or 'reviewed' on the following grounds:

(a) The decision discloses an error on the part of the tribunal.

(b) A party did not receive notice of proceedings.

(c) The decision was made in the absence of a party entitled to be heard.

(d) New evidence has come to light.

(e) The interests of justice require a review.

Appeals

Appeals are made against improvement and prohibition notices on the following grounds:

(a) the time limit for compliance; and/or

(b) substantive law involved.

Appeals on health and safety issues lie to the Divisional Court of the High Court. Like the Employment Appeals Tribunal, the latter can only interfere where the tribunal has erred in respect of the view of the law which it took.

Appeals from the Employment Appeals Tribunal and the High Court go to the Court of Appeal where litigants must be represented by counsel.

Criteria used by tribunals – appeals

The criteria which may influence tribunals in their hearing of appeals can include:

1. The kind of notice issued by the inspector

It would appear to be accepted by tribunals that the greater the risk to employees the more the likelihood of an inspector serving a Prohibition Notice. In an appeal following the service of an immediate Prohibition Notice, the employer has the difficult task of convincing a tribunal that the inspector is wrong in his opinion that there is a serious and imminent risk to health and safety.

2. The quality of communication between an employer and an inspector

If the employer has allowed correspondence to be unduly protracted, or has ignored advice from an inspector, a tribunal may take such delays into account in assessing the opportunities which have been available to an employer to rectify the situation and also how long employees have been exposed to risks.

3. Safety standards in common use throughout the industry

A tribunal may assess the extent to which it was reasonable for an employer to have been aware of, and to have implemented, industry-based standards in his workplace.

4. Time required and problems in making improvements

A tribunal will usually take into account the time and effort required to make, obtain and fit parts, such as machinery guards which have to be installed as a requirement of a notice. They will also make allowance for the technical difficulties of implementing a required improvement and the possibility of lay-offs of employees whose machines are required to be taken out of service as a requirement of a Prohibition Notice.

Improvement costs

Generally, tribunals do not make any allowance for the expense of making improvements. Neither have they been influenced by the frequency of accidents associated with a particular machine or process in question. Even if there has been no accident for many years, the authoritative statement of an inspector that there is a real risk will suffice as reason for rejecting an appeal.

2 Common law liability and workplace injuries

A

Common law is a law consisting of decisions of courts bound together by precedent into a body of authoritative rules. It is judge-made law and accounts for the greater part of contract and tort.

Contract law

A contract is an agreement between two parties and to be legally enforceable it requires certain basic features. It must be certain in its wording and consist of an offer made by one party which must be accepted unconditionally by the other. The great majority of contracts need not be in writing. There must, however, be *consideration* that flows from one party to the other. This is the legal ingredient that changes an informal agreement into a legally binding contract, for instance the exchange of goods for money, of work for wages, of a trip for the price of a ticket.

Two other essentials for a valid contract are that parties must intend to enter into a legally binding agreement, and both parties must have the legal capacity to make such a contract. Capacity to contract means that the parties must be sane, sober and over the age of 18 as a rule.

The law of tort

The rule of common law is that everyone owes a duty to everyone else to take reasonable care so as not to cause them foreseeable injury. Hence employers will be liable if they fail to take reasonable care to protect their workforces from foreseeable injury, disease and/or death. The basis of liability is negligence. The remedy or redress is unliquidated damages for injury, disease or death sustained at work.

A 'tort' is generally defined as 'a civil wrong'. It concerns the legal relationships between parties generally in the everyday course of their affairs, the duties owed by one to the other and the legal effect of a wrongful act of one party causing harm to the person, property, reputation or economic interests of another.

It is of universal application compared with the law of contract, which applies only to specific parties concerned.

Three separate branches of the law of tort are *trespass, nuisance* and *negligence,* the latter being the most important and applying in particular to the field of an employer's liability for accidental injury to his employee.

Negligence

Negligence could be defined as 'careless conduct injuring another'.

If an action for negligence is to be successful, three specific facts must be proved, namely:

(a) a duty of care is owed;

(b) a breach of that duty has occurred;

(c) injury, damage and/or loss has been sustained as a result of that breach.

{*Lochgelly Iron & Coal Co. Ltd.* v. *M'Mullen* [1934] AC 1}

Employers' duties at common law

The duties of employers at common law were identified by the House of Lords in the leading case of *Wilsons & Clyde Coal Co.* v. *English* [1938] AC 57 as follows. All employers must provide:

(a) a safe place of work, including safe access and egress;

(b) a safe system of work;

(c) safe plant and appliances;

(d) safe and competent fellow workers.

'I do not mean that employers warrant the adequacy of plant, or the competence of fellow-employees, or the propriety of the system of work. The obligation is fulfilled by the exercise of due care and skill' (per Lord Wright in *Wilsons & Clyde Coal Co.* v. *English*).

Safe place of work

This duty was established in *Brydon* v. *Stewart* (1855) 2 Macq 30. Here some miners had left the pit before the end of the shift. As they were on their way up the pitshaft, a stone fell from the side of the shaft, killing one of them. The cage in which they were travelling to the top was open. It was held that the mine owner was liable, since access to and from the pit was not safe.

A place of work can be outdoors. In *Bradford* v. *Robinson Rentals Ltd* [1967] 1 AER 267 an employee suffered frostbite after being required to drive 500 miles in a van without a heater and a leaking radiator in freezing conditions. He successfully sued his employer for damages, as frostbite was a reasonably foreseeable consequence of being required to drive a defective van. But it is only the foreseeable risk of injury for which provision has to be made (*Latimer* v. *AEC Ltd* [1953] 2 AER 449, a case which concerned sec 25 of the Factories Act 1937 (now sec 28(1) of the FA). This section requires that, so far as is reasonably practicable, floors, steps, stairs and passages, etc., should be kept free from obstructions and substances likely to cause persons to slip. During the course of an exceptionally heavy thunderstorm one weekend the factory had become flooded. The rainwater had become mixed with an oil liquid used for cooling machines and, as a result, a slippery film of oil was left on the factory floor. The factory floor was later treated with sawdust but, owing to the large area involved, the factory occupiers were unable to treat the whole floor. An employee slipped on an untreated part of the floor and was injured. It was held that the factory occupier was not in breach of the Factories Act 1937, sec 25(1). He had carried out all reasonably practicable measures to make the floor free from substances likely to cause persons to slip. (The result would be the same under the present requirement of the FA, sec 28(1), relating to floors.) At common law the factory occupier is not required to make provision for the unprecedented and freak hazard.

Safe system of work

This is the most extensive of the duties placed on an employer. The duty ranges from co-ordination of the various interrelated workplace departments to the general conditions of work, such as heating, lighting, ventilation, freedom from foreseeable risk of disease, safe working of lifts etc. In addition, it covers an effective system of supervision and training in the hazards of the job as well as instruction in the use of personal protective clothing and the safe use of appliances. This is in addition to any general or specific statutory duties placed on the employer, for instance by HSWA, sec 2.

Co-ordination of work departments

In *Wilsons & Clyde Coal Co.* v. *English* [1938] AC 57 an employee in a mine was injured whilst passing along one of the underground haulage roads at the end of the day shift at the same time as haulage plant was set in motion. His employer, the mine owner, was held liable at common law for having failed to provide a safe system of work. The haulage plant should have been stopped at the same time as the shift came to an end.

Effective supervision

This duty is well illustrated by the case of *Bux* v. *Slough Metals Ltd* [1974] 1 AER 262. Here the appellant, who was employed as a die-caster by the respondent, was engaged on piece-work in the die-casting foundry. His work consisted of melting ingots of aluminium alloy in a furnace, lifting out the molten metal with a ladle and pouring it into a die. Goggles had (eventually) been provided and the employee was instructed to wear them. He found, however, that they misted up, told his superintendent that they were unsuitable and stopped wearing them. He was not later exhorted to wear them by any member of management. He was blinded when molten metal splashed into his face. He sued his employer for damages for breach of (a) Reg 13(1)(*c*) of the Non-Ferrous Metals (Melting and Founding) Regulations 1962, in not having been provided with suitable goggles, and (b) the duty to 'police' a safe system of work through effective supervision at common law. It was held by the Court of Appeal that the employer had carried out his statutory duty but that nevertheless he was liable for breach of the duty of effective supervision at common law. 'It would be idle for the employers to urge that they have fulfilled their duty of reasonable care for the safety of their workmen by simply providing them with goggles and then stand by while the men did dangerous work with their eyes wholly unprotected' (per Edmund Davies LJ).

Safe plant and appliances

The duty of an employer is to 'provide' and 'maintain' safe plant and appliances for the job. The duty to 'provide' is satisfied where the employer makes available plant and appliances and safety equipment, which are readily to hand in an accessible place, and informs accordingly. In the absence of a specific statutory requirement to the contrary, it is not necessary for the employer to put the equipment into the physical possession of the worker, even less to 'wet nurse' experienced 'hands'. In *Woods* v. *Durable Suites Ltd* [1953] 2 AER 391 an employee, who was employed in the veneer department of a factory where he was engaged in the spreading of synthetic glue, suffered dermatitis. In order to warn of the risk of dermatitis, management had erected a notice outlining precautions to be taken to avoid dermatitis and instructing the workforce in the use of barrier cream, soap and washing facilities, all of which had been supplied. The appellant was an experienced workman and had been specifically instructed in the use of protective measures. He nevertheless chose to ignore them. In an action against his employer for damages at common law, it was held that the employer was not liable. He had done all that was reasonably expected of him.

Failure to provide plant for a job will involve the employer in liability. In *Lovell* v. *Blundells & Crompton & Co. Ltd* [1944] 2 AER 53 the plaintiff was employed by the defendant to overhaul the boiler tubes of a ship. Being unable to reach all the tubes himself, the employee found some planks and set up his own staging. The planking proved unsatisfactory and the employee was injured. It was held that the employer was liable. Similarly, if a correct quantity of plant for the job is not provided, the employer may well be liable. In *Machray* v. *Stewarts & Lloyds* [1964] 3 AER 716 the plaintiff was employed as a rigger. Whilst he was working 21 metres above ground a loop section of pipe, weighing over 500 kilograms, had to be fixed. In order to accomplish this it was necessary to move the loop section four metres. One of the established methods of doing this was by using chain block and tackle. The employee endeavoured to obtain two sets of chain block and tackle, but only one was available at the time. Because he had been told by the chargehand that the job was urgent, the employee used single chain block, in consequence of which the loop section swung out of hand, seriously injuring the plaintiff. He sued his employer for negligence at common law. During the course of giving judgment in favour of the injured workman, McNair J said:

> When I find a workman . . . adopting a course of conduct not for the sake of saving himself trouble, but in order to get on with his employer's business, and when I find that he has been prevented from doing the work in the way in which he would have preferred to do by the employer's breach in not providing him with proper tackle, I am very slow to put any blame on him.

Provision of safe plant and appliances includes the monitoring of safety performance accompanied by a regular defects reporting system. No matter how good the system of safety inspection of plant is in the factory, the employer could be liable if he has failed to institute a backup defects reporting procedure. In *Barkway* v. *South Wales Transport Co. Ltd* [1950] 1 AER 392 the appellant's husband was killed whilst travelling on a bus belonging to the respondent, as a result of a tyre burst. The vehicles were inspected and tested by the company regularly and thoroughly. Nevertheless, the company was held to be liable, because it did not require its drivers to report incidents which could have led to the discovery of hidden defects that might cause a tyre burst.

Duties owed only to employees

The common law duties of employers are owed only to employees. They are not owed to independent contractors and their employees who may be working on the premises. HSWA, sec 3(1), places a duty on an employer not to expose the workforce of an independent contractor to risks to health and safety. But breach of this duty only gives rise to criminal liability, HSWA, sec 47, preventing civil claims being brought in respect of breaches of the General Duties of HSWA. Enforcement of this duty is as much in the interests of the employer's workforce as that of the subcontractor's workforce.

The relationship of employer to employee is a complex one, and the following questions have to be answered in order to determine whether a workman is an employee:

(a) Does the person or company for whom the work is done control the way in which the work is done, e.g. specifying the work to be done, wages/salaries, etc., provision of social security stamps, holidays and hours of work, and the manner of work?
(b) Is a workman, although normally employed by an outside organisation, integrated even temporarily into the organisation of the enterprise in question?
(c) Does the workman trade in his own right as a self-employed person, providing his own insurance, assistants, tools and equipment? (And see generally the remarks made in *Ready Mixed Concrete (South East) Ltd* v. *Minister of Pensions and National Insurance* [1968] 1 AER 433.)

However, the fact that a person styles himself as self-employed does not mean that he cannot be an employee. In *Ferguson* v. *John Dawson & Partners (Contractors) Ltd* [1976] IRLR 346 it was held that a contract to hire labour-only subcontractors was a contract of employment. The labour-only subcontractor was, therefore, entitled to the protection of Regulation 28(1) of the Construction (Working Places) Regulation 1966, which required that the subcontractor who engaged him should provide a suitable guard rail. As a result of this not having been done, the subcontractor was held liable for injury suffered by the labour-only subcontractor (in spite of the fact that the latter regarded himself as self-employed).

Disclosure of information

Disclosure is directly connected with the legal concept of 'discovery and inspection' of documents. Discovery of documents is the procedure whereby a party discloses, to a court or to any another party, the relevant documents in the action that he has, or has had, in his possession, custody or power.

Documentary evidence plays an important part in nearly all civil cases. In an industrial injuries action, for example, the employers are likely to have internal accident reports, machinery maintenance records, records of complaints, etc. which it is very much in the plaintiff's interests to see while, on the other hand, the plaintiff may have documents relating to his medical condition, or to his earnings since the accident, or state benefits which he has received, all of which may be highly relevant to the qualification of his claim by the defendants.

Disclosure for purposes of civil liability proceedings

In civil cases, discovery may take place in two ways, under the Rules.

1. Discovery without order (Automatic Discovery)

Parties must, in any action commenced by writ, make discovery by exchanging lists within 14 days of the close of the pleadings. The lists are in prescribed form, i.e.:

(a) relevant documents, listed numerically, which the party has in his possession, custody or power and which he does not object to producing;
(b) relevant documents which he objects to producing;
(c) relevant documents which have been, but at the date of service of the list, are not in the possession, custody or power of the party in question.

A party serving a list of documents must also serve on his opponent a notice to inspect the documents in the list (other than those which he objects to producing) setting out a time within seven days and the place where the documents in (a) above may be inspected.

A right to inspect includes the right to take copies of the documents in question.

2. Discovery by order

In certain cases an order for discovery will be required.

The first case is where the action is one to which the automatic discovery rule does not apply or where any party has failed to comply with that rule. In this case an Application for an Order for Discovery must be made, should discovery be required.

The second case arises where a party is dissatisfied with the opponent's list. He may then apply for discovery of documents.

In either case an application must be made to the court for an order for discovery of the documents in question. The application must be accompanied by an affidavit setting out the grounds for the deponent's belief and identifying the documents of which discovery is required.

Irrespective of the above provisions, any party may, at any time, require inspection of any document referred to in any other party's pleadings or affidavits.

Where any party fails to comply with an order for discovery, the judge may order his statement of claim to be struck out, if he is a plaintiff, or his defence to be struck out and judgments entered against him, if a defendant. In a case of extreme and wilful disobedience, the offender may be committed for contempt of court.

The above procedures are detailed in the Administration of Justice Act 1970, whereby orders for discovery and inspection of documents against proposed parties may be made before the commencement of proceedings.

Under sec 34 of the Supreme Court Act 1981 there is, in addition, in actions for damages or death or personal injuries, power to order a person who is not a party to the proceedings, and who appears to be likely to have or has had in his possession, custody or power any documents which are relevant to an issue arising out of the action, to disclose whether these documents are in his possession, custody or power and to produce those documents which he has to the applicant. Such a power may be used, for example, by defendants to obtain a sight of a plaintiff's hospital notes, or by plaintiffs to obtain copies of manufacturers' instructions for using and maintaining machines, reports by employers to the Health and Safety Executive and documents of that type.

Documents privileged from production

The only circumstances in which a party may refuse to produce a document is where the document is one of a class which the law recognises as privileged. The privilege which attaches to documents must be carefully distinguished from privilege in the law of defamation. Thus a document which attaches qualified privilege for the purposes of defamation, such as the minutes of a company board meeting, is not necessarily privileged from discovery. The former is a rule of law, the latter a rule of evidence or procedure.

A party who wishes to claim privileges for a document must include it under those relevant documents which he objects to producing, together with a statement in the body of the list of the ground upon which he claims privilege for it. The principal classes of document which the law recognises as being privileged from production are documents which relate solely to the deponent's own case, incriminating documents, documents attracting legal professional privilege and documents whose production would be injurious to the public interest.

Disobeying safety orders

Where an employee disobeys a safety instruction, he can be prosecuted for breach of HSWA, sec 7 (*see* Chapter 3), as well as dismissed from his job (*see* Chapter 5). It is unlikely, however, that he will lose entitlement to damages, even though they may be reduced as a result of contributory negligence on his part. Moreover, if, whilst disobeying a safety instruction, an employee injuries either a co-employee or an innocent member of the public or an employee of an outside workforce, his employer will still be liable. In *Century Insurance Co. Ltd* v. *Northern Ireland Road Transport Board* [1942] AC 509 an employee of the respondent was employed to deliver petrol in tankers to garages. Whilst delivering petrol at a garage forecourt and whilst petrol was being transferred from the tanker to an underground tank at the garage, the employee decided to have a smoke. Having lit a match, he then threw it away, whilst still alight, and it landed by the underground tank. There was an explosion and considerable damage to persons and property. The Court of Appeal held that the respondent employer was liable, since the employee was doing what he was employed to do, namely deliver petrol, even though he was doing his job in a grossly negligent way (i.e. whilst smoking a cigarette).

More recently in *Rose* v. *Plenty* [1976] 1 AER 97 the respondent was employed as a milk roundsman by the Co-op. The latter had gone to great lengths to prevent employees continuing the practice of taking on young children and paying them for delivering bottles of milk and collecting empty bottles. Notices at depots urged that 'children must not in any circumstances be employed by you in the performance of your duties'. Contrary to this directive the respondent 'employed' a young boy. The latter was injured when he rode on the milk float with one leg dangling over the side so that he could jump off quickly to deliver bottles of milk. Evidence showed that the respondent drove negligently, with the result that the boy's leg was trapped and crushed between the wheel and the kerb. The trial judge ruled that the employee was 75 per cent to blame and the boy 25 per cent. On an appeal to the Court of Appeal to determine whether the employer was liable, it was held that he was. The fact was that the employee was doing what he was employed to do but was doing it negligently.

A case where an employee, whilst disobeying a safety instruction, suffered fatal injury and where damages were awarded to his widow was *Westwood* v. *The Post Office* [1973] 3 AER 184. The appellant's husband had been employed by the respondent. One day he went onto the flat roof of the telephone exchange building for his mid-morning break. The door to the roof was locked and so he entered the lift motor room, getting onto the roof through a window. The door to the lift motor room was normally locked but on this occasion it was slightly ajar. On the door was a prominent notice: 'Only the authorised attendant is permitted to enter'. The employee was not an authorised attendant. There was a floor in the lift motor room with a trap door which was not of sound construction, as required by sec 16 of the OSRPA. On his return the employee jumped from the roof onto the trap door, which gave way, and fell on a concrete floor and was killed. Though the Court of Appeal ruled that the employee was a trespasser and, therefore, not entitled to damages, the House of Lords took a different view. Giving judgment, Lord Kilbrandon said:

> The sole act of negligence giving rise to this accident was the respondents' breach of statutory duty. Any fault on the part of the deceased was a fault of disobedience, not a fault of negligence, because he had no reason to foresee that disregard of the order to keep out of the lift motor room would expose him to danger. It would indeed not have done so, had it not been that, unknown to him, the respondents were in breach of their duty to take care for the safety of those employed in the premises.

Vicarious liability

Once the doctrine of common employment was finally abolished by the Law Reform (Personal Injuries) Act 1948, vicarious liability became a reality. The importance of vicarious liability cannot be underlined too strongly, since it is generally on this ground that employers are held liable at common law. The substance of the doctrine is that if an employee, whilst acting in the course of his employment, negligently injures another employee or an employee of an outside contractor working on the premises, or even a member of the public, the employer will be liable for the injury. And since most accidents at work happen in this way, rather than as a result of the personal negligence of the employer, vicarious liability is the ground on which most claims for injury-causing accidents are successful.

Vicarious liability arises on the part of an employer simply because of his status as an employer. It must be insured against under the provision of the Employers' Liability (Compulsory Insurance) Act 1969 and employers cannot contract out of this liability, since it is prohibited by the Law Reform (Personal Injuries) Act 1948 and the Unfair Contract Terms Act 1977. The key to liability is that the accident causing injury/disease/death arises (a) out of and (b) in the course of employment. This does not normally include travelling to and from work, though it would be, as is the case with state benefit, if the mode of transport to and from work was within the employer's control, or provided by or in arrangement with him.

Notwithstanding vicarious liability, the employee can be sued instead of or in addition to the employer where he has been negligent (*see later Lister* v. *Romford Ice & Cold Storage Co. Ltd* [1957]). For obvious reasons this would be restricted to senior employees.

Breach of common law and statutory duties – the double-barrelled action

Because an injured employee is entitled to sue his employer for damages for injury resulting from a breach of (a) common law duty and (b) statutory duty, this has led to the emergence of the 'double-barrelled' action against employers; in such cases employees sue separately (though simultaneously) for breach of both duties on the part of the employer. This development can be traced back to the decision in *Kilgollan* v. *Cooke & Co. Ltd* [1956] 2 AER 294.

Employers' defences

There are two defences available to an employer sued for breach of common law, viz. (a) assumption of risk of injury (volenti non fit injuria) and (b) contributory negligence. The first is a complete defence and means that no damages will be payable; the second is a partial defence and means that the injured worker's damages will be reduced to the extent to which he is adjudged to blame for his injuries.

Volenti non fit injuria

This means 'to one who is willing no harm is done', and applies to the situation where an employee, being fully aware of the risks he is running in not complying with safety instructions, duties, etc., after being exhorted and supervised, and having received training and instruction in the dangers involved in not following safety procedures and statutory duties, suffers injury, disease and/or death as a result. Here it is open to an employer to argue that the employee agreed to run the risk of the injury, etc., involved. If this defence is successfully pleaded, the employer will be required to pay no damages.

This defence has not generally succeeded in actions by injured workmen, since there is a presumption that employment and the dangers sometimes inherent in it are not voluntarily accepted by workers; rather employment is an economic necessity. This has been the position since the decision of *Smith* v. *Baker & Sons* [1891] AC 325 where the appellant, who was employed to drill rocks, was injured by stones falling from a crane operated by a co-employee. Although the appellant knew of the risk he was running from the falling stones, it was held by the House of Lords that he had not agreed to run the risk of being injured.

An exception to the general position occurred in *ICI Ltd* v. *Shatwell* [1964] 2 AER 999. Here two brothers, the respondents, were employees of the appellant, being employed as skilled shotfirers in rock blasting, for which they were highly paid. A statutory duty was placed on the two employees personally (not on the employer) to take specific safety precautions when shotfiring was about to begin. The two employees had been thoroughly briefed in the dangers of the work and the risks involved, e.g. premature explosion. They knew of the statutory prohibition in question. Nevertheless, they decided to test although a cable was too short to reach the shelter, rather than wait a few minutes before a workmate could go and get another cable. One brother handed the other brother two wires and the latter applied them to the galvanometer terminals. An explosion occurred injuring both employees. On an appeal to the House of Lords (which reversed the decision of the Court of Appeal) it was held that the employer was not liable. The reasons given were as follows:

(a) the employer was not in breach of statutory duty; the breach of statutory duty was committed by the employees;
(b) the two employees were highly skilled operatives, part of a well-paid workforce elite and knew the dangers involved in taking short cuts where safety was concerned.

Contributory negligence

The Law Reform (Contributory Negligence) Act 1945, sec 1(1), provides that where injury is caused by the fault of two or more persons, liability or fault (and, in consequence, damages) must be apportioned in accordance with the extent to which both or more were to blame. If, therefore, following an employee's injury, it is established that both employer and employee were at fault (e.g. in breach of statutory requirements), when sued for damages, the employer can plead that the employee, through his negligence in failing to look after his own health and safety, should be held partly to blame and have his damages reduced to that extent. A classic case illustrating the operation of this rule (which is applied quite extensively in practice by judges) is *Uddin* v. *Associated Portland Cement Manufacturers Ltd* [1965] 2 AER 213. Here an employee, who was a machine minder, in that effort to retrieve a pigeon, leaned across a revolving shaft in a part of the factory in which he was not authorised to be, and was injured, losing an arm. His employment was confined to the cement-packing plant but his injury occurred in the dust-extracting plant. He sued his employer for breach of sec 14(1) of the Factories Act 1937. It was held that the employer was liable under that section, but that the employee's damages would be reduced by 80 per cent to account for his own contributory negligence.

Employers' liability

The Employers' Liability (Compulsory Insurance) Act 1969 deals with the duties of employers in terms of insuring themselves against claims which may be made by employees.

Sec 1 of the Act requires that every employer carrying on business in Great Britain shall insure and maintain insurance against liability for bodily injury or disease sustained by his employees, and arising out of and in the course of their employment in Great Britain in that business.

This insurance must be provided under one or more 'approved policies'. An approved policy is a policy of insurance not subject to any conditions or exceptions prohibited by regulations.

Cover

Cover is required in respect of liability to employees who either:

(a) are ordinarily resident in Great Britain; or

(b) though not ordinarily resident in Great Britain, are present in Great Britain in the course of employment here for a continuous period of not less than 14 days.

Extent of cover

The amount for which an employer is required to insure and maintain insurance is £2,000,000 in respect of claims relating to one or more of his employees, arising out of any one occurrence.

Issue and display of certificate of insurance

The insurer must issue the employer with a certificate of insurance, which has to be issued not later than 30 days after the date on which insurance was commenced or renewed. The certificate must be prominently displayed.

Penalties under the Act

1. Failure to insure or maintain insurance

Maximum penalty on conviction £1,000.

2. Failure to display a certificate of insurance

Maximum penalty on conviction £500.

Redress of employer's insurer

In the law of insurance there is a doctrine, known as subrogation, which entitles an insurer who has indemnified an insured (employer) in respect of his legal liability, to recover his losses as far as he can, and, in particular, to take over any action which his insured might have had against a third party. This principle has been applied to industrial injuries claims. In *Lister v. Romford Ice & Cold Storage Co. Ltd* [1957] 1 AER 125 the appellant was employed by the respondent as a lorry driver. His father was also employed by the same company. One day, whilst reversing the lorry, the appellant negligently injured his father, and his father was paid damages under the principle of vicarious liability (*see earlier*). The insurer who had paid the damages sought to recover that amount (£1,600) from the negligent employee, on the ground that it is an implied term of an employment contract that an employee carry out his employment with due care and skill (*see* Chapter 5). On an appeal to the House of Lords it was held that the employee had broken the duty of care and skill implied in his contract, and that as a consequence the insurer could recover the sum paid by way of damages.

Employees' duties at common law

Employees owe duties at common law towards employers, e.g. the duty to perform their work with reasonable care and skill. These are implied terms of the contract of employment and have, to a large extent, been superseded in importance by the law relating to unfair dismissal, as laid down in the Employment Protection (Consolidation) Act 1978 (EPCA). It is, therefore, better to consider them under that heading (*see* Chapter 5).

Occupiers' liability

Whilst health and safety law is largely concerned with the relationships between employers and employees, consideration must also be given to the duties of those people and organisations, such as private individuals, local authorities, organisations and companies, who occupy land and premises. Their land and premises are visited by people for a variety of purposes, such as to undertake work, provide goods and services, settle accounts, etc. The HSWA (sec 4), requires those people in *control* of premises to take reasonable care towards these other persons, and failure to comply with this duty can lead to prosecution and a fine on conviction.

Moreover, anyone who is injured while visiting or working on land or premises may be in a position to sue the occupier for damages, even though the injured person may not be their employee. Lord Gardner in the case of *Commissioner for Railways* v. *McDermott* [1967] 1 AC 169 explained the position thus:

> Occupation of premises is a ground of liability and is not a ground of exemption from liability. It is a ground of liability because it gives some control over and knowledge of the state of the premises, and it is natural and right that the occupier should have some degree of responsibility for the safety of persons entering his premises with his permission . . . there is 'proximity between the occupier and such persons and they are his neighbours'. Thus arises a duty of care.

Thus, occupiers' liability is a branch of civil law concerned with the duties of occupiers of premises to all those who may enter on to those premises. The legislation covering this area of civil liability is the Occupiers Liability Act 1957 and, specifically in the case of trespassers, the Occupiers Liability Act 1984.

Occupiers Liability Act 1957

Under the OLA 1957 an occupier owes a *common duty of care* to all lawful visitors. This common duty of care is defined as:

> a duty to take such care as in all the circumstances of the case is reasonable to see that the visitor will be reasonably safe in using the premises for the purposes for which he is invited or permitted by the occupier to be there.

Sec 1 of the OLA defines the duty owed by occupiers of premises to all persons lawfully on the premises in respect of 'dangers due to the state of the premises or to things done or omitted to be done on them'.

The OLA regulates the nature of the duty imposed in consequence of a person's occupation of premises. The duties are not personal duties but rather, are based on the occupation of premises, and extend to a person occupying, or having control over, any fixed or movable structure, including any vessel, vehicle or aircraft.

Visitors

Protection is afforded to all *lawful* visitors, whether they enter for the occupier's benefit, such as customers or clients, or for their own benefit, for instance, a police officer, though not to persons exercising a public or private right of way over premises.

Warning notices

Occupiers have a duty to erect notices warning visitors of imminent danger, such as an uncovered pit or obstruction. However, sec 2(4) states that a warning notice does not, in itself, absolve the occupier from liability, unless, in all the circumstances, it was sufficient to enable the visitor to be reasonably safe.

Furthermore, while an occupier, under the provisions of the OLA, could have excused his liability by displaying a suitable prominent and carefully worded notice, the chance of such avoidance is not permitted as a result of the Unfair Contract Terms Act 1977. This Act states that it is not permissible to exclude liability for death or injury due to negligence by a contract or by a notice, including a notice displayed in accordance with sec 2(4) of the OLA.

Trespassers

A trespasser is defined in common law as a person who:

(a) goes on premises without invitation or permission;

(b) although invited or permitted to be on premises, goes to a part of the premises to which the invitation or permission does not extend;

(c) remains on premises after the invitation or permission to be there has expired;

(d) deposits goods on premises when not authorised to do so.

Occupiers Liability Act 1984

Sec 1 of this Act imposes a duty on an occupier in respect of trespassers, namely persons who may have a lawful authority to be in the vicinity or not, who may be at risk of injury on the occupier's premises. This duty can be discharged by issuing some form of warning such as the display of hazard warning notices, but such warnings must be very explicit. For example, it is insufficient to display a notice that merely states EYE HAZARD where there may be a risk to visitors from welding operations carried out. A suitable notice in such circumstances might read:

> **RISK OF EYE INJURIES FROM WELDING OPERATIONS**
>
> **NO PERSON IS ALLOWED TO ENTER THIS AREA**
> **UNLESS WEARING APPROVED EYE PROTECTION**

It is not good enough, however, merely to display a notice. The requirements of notices must be actively enforced by management.

Generally, the displaying of a notice, the clarity, legibility and explicitness of such a notice, and evidence of regularly reminding people of the message outlined in the notice, may count to a certain extent as part of a defence when sued for injury by a simple trespasser under the OLA 1984.

Individual risk taking

Under the OLA 1984, there is no duty on the part of occupiers to persons who willingly accept risks (sec 2(5)). Further, the fact that an occupier has taken precautions to prevent persons going into his premises or onto his land, where some form of danger may exist, does not mean that the occupier has reason to believe that someone would be likely to come into the vicinity of the danger, thereby owing a duty to the trespasser under the OLA 1984.

Children

Children, generally, from a legal viewpoint have always been deemed to be less responsible than adults. The OLA 1957 is quite specific on this matter. Sec 2(3)(a) requires an occupier to be prepared for children to be less careful than adults. Where, for instance, there is something, or a situation, on the premises that is a 'lure or attraction' to a child, such as a pond, an old motor car, a derelict building or scaffolding, this can constitute a 'trap' as far as a child is concerned. Should a child be injured as a result of this trap, the occupier could then be liable. Much will depend upon the location of the premises, for instance, whether or not it is close to houses or a school or is in an isolated location, such as a farmyard deep in the countryside but, in all cases, occupiers must consider the potential for child trespassers and take appropriate precautions.

Contractors and their employees

The relationship between occupiers and contractors has always been a tenuous one. Sec 2(3)(b) of the OLA 1957 states that an occupier may expect that a person, in exercising their calling, such as a window cleaner, bricklayer or painter, will appreciate and guard against any risks ordinarily incident to that calling, for instance the risk of falling, so far as the occupier gives them leave to do so. This means that the risks associated with the system of work on a third party's premises are the responsibility of the contractor's employer, not the occupier. (It should be appreciated, however, that while the above may be the case at civil law, the situation at criminal law, namely the duties of employers towards non-employees under sec 3 of the HSWA, is quite different.)

Where work is being carried out on a premises by a contractor, the occupier is not liable if he:

(a) took care to select a competent contractor; and

(b) satisfied himself that the work was being properly done by the contractor (sec 2(4)(b)).

However, in many cases, an occupier may not be competent or knowledgeable enough to ascertain whether or not the work is being 'properly done'. For instance, an occupier may feel that an unsafe system of work adopted by a contractor's employee, such as cleaning the external window surfaces to fourth floor offices without using any form of access equipment, such as a suspended scaffold or a safety line, is standard practice amongst window cleaners! In such cases, the occupier might need to be advised by a surveyor, architect or health and safety consultant in order to be satisfied that the work is being done properly and safely.

This relationship between occupiers and contractors has been substantially modified through the Construction (Design and Management) Regulations 1994.

3 Health and safety at work – the principal legal requirements

The principal legal requirements relating to occupational health and safety are covered in the Health and Safety at Work etc., Act 1974 (HSWA) and the Management of Health and Safety at Work Regulations 1999.

Health and Safety at Work etc., Act 1974 (HSWA)

This Act implemented the majority of the proposals of the Robens Report. More exactly:

(a) it established the Health and Safety Commission (HSC);

(b) it established the Health and Safety Executive (HSE);

(c) it conferred accident prevention powers on HSE inspectors (*see* Chapter 4);

(d) it placed broad general duties on employers, employees and manufacturers of industrial products as well as on the self-employed and occupiers of buildings where people work (*see below*);

(e) it provided for participation of management and workers in the identification and monitoring of workplace hazards by requiring the appointment of union safety representatives from the workforce to carry out periodic inspections of the workplace, and by requiring that changes and improvements in occupational health and safety be discussed at meetings of safety committees consisting of management and union representation. These requirements were laid down in the Safety Representatives and Safety Committees Regulations 1977 (SRSCR).

Health and safety regulations

One of the principal functions of the Health and Safety Commission is to propose regulations on health and safety at work to the Secretary of State for Employment. Health and safety regulations can do one of the following:

(a) Repeal or modify any of the existing 'relevant statutory provisions' (for the meaning of this expression, *see* Chapter 4).

(b) Exclude or modify any of the 'existing statutory provisions'. An example of the power of modification is the Employers' Health and Safety Policy Statements (Exception) Regulations 1975, which exempts employers employing fewer than five employees from complying with HSWA, sec 2(3), and having to issue a written company safety policy, and revise it as often as necessary.

(c) Make a specified authority responsible for the enforcement of any of the 'relevant statutory provisions'. An example of the exercise of this power is the enforcement of some of the 'relevant statutory provisions' by HSE and others by local authorities (*see* Chapter 4) (HSWA, sec 15(3)).

Issue and approval of codes of practice

The HSC is empowered to approve and issue codes of practice for the purpose of providing guidance on health and safety duties laid down in statute or regulations. A code of practice can be drawn up by HSC or some other body. In every case, however, any relevant government department, or other body, must be consulted beforehand and the approval of the Secretary of State must be obtained (HSWA, sec 16). Any code of practice approved in this way is an 'approved code of practice' (HSWA, sec 16(7)). (For the effect of codes of practice in civil and criminal proceedings, *see* Chapter 4.)

Health and Safety at Work etc., Act – the General Duties

HSWA places broad General Duties upon five separate classes of persons:

(a) employers;
(b) employees;
(c) manufacturers and suppliers of industrial products;
(d) the self-employed;
(e) occupiers of buildings in which persons work, other than one's own employees.

Duties of employers

It is the duty of every employer, so far as is reasonably practicable, to ensure the health, safety and welfare at work of all his employees (HSWA, sec 2(1)). More particularly, this includes:

(a) the provision and maintenance of plant and systems of work that are safe and without health risks (HSWA, sec 2(2)(*a*));
(b) arrangements for ensuring safety and absence of health risks in connection with the use, handling, storage and transport of articles and substances (HSWA, sec 2(2)(*b*));
(c) the provision of such information, instruction, training and supervision as is necessary to ensure the health and safety at work of employees (HSWA, sec 2(2)(*c*));
(d) the maintenance of any place of work under the employer's control in a condition that is safe and without health risks, including means of access and egress (HSWA, sec 2(2)(*d*));
(e) the provision and maintenance of a working environment for employees that is safe and free from health risks, with adequate facilities and arrangements for employees' welfare (HSWA, sec 2(2)(*e*)).

Health and safety policy

Employers must prepare and, as often as is necessary, revise a written Statement of Health and Safety Policy (HSWA, sec 2(3)). There is an exception to this requirement in the case of employers employing at any time fewer than five employees (Employer's Health and Safety Policy Statements (Exception) Regulations 1975). (For further information relating to Health and Safety Policies, *see* Chapter 6.)

Legal importance of Health and Safety Policy statements

Requirements detailed in the Statement of Health and Safety Policy (and preferably published in works rules, handbooks or codes of practice relating to specific tasks) probably do not constitute terms of a contract of employment. Nevertheless the rule book may be construed (and, indeed, has been so construed) as constituting reasonable orders which an employer is entitled to give to an employee. Failure on the part of an employee to obey such rules would amount to a breach of contract of employment. The matter was summarised as follows by Roskill LJ (referring to the employers' rule book) in *Secretary of State* v. *ASLEF* [1972] 2 AER 949:

> It was not suggested that strictly speaking this formed part of the contract of employment as such. But every employer is entitled within the terms and scope of the relevant contract of employment to give instructions to his employees and every employee is correspondingly bound to accept instruction properly and lawfully so given. The rule book seems to me to constitute instructions given by the employer to the employee in accordance with that general legal right.

The importance of the law relating to the contractual rights and duties of employers and employees has been largely superseded by the law relating to unfair dismissal, as contained in the Employment Protection (Consolidation) Act 1978 (EPCA). The present position would, therefore, seem to be that failure to carry out an order based on a rule or procedure in a Statement of Health and Safety Policy (assuming that the employee was well aware of it) could lead to the dismissal of the employee. Or, if an employer was in breach, to the constructive dismissal of an employee (*see* Chapter 5). This view is supported by the fact that it was always an implied term of an employment contract at common law that an employee would carry out his duties honestly and with due care and skill (*Lister* v. *Romford Ice & Cold Storage Co. Ltd* [1957] 1 AER 125). A case confirming this approach is *Martin* v. *Yorkshire Imperial Metals Ltd*, COIT No. 709/147, Case No. 32793/77, where an employee was held to have been fairly dismissed for having removed a guard from a machine so that he might get his work done more quickly.

Failure to bring the contents of a Statement of Health and Safety Policy to the attention of employees, albeit even on a departmental basis, can have serious consequences, as illustrated in the case of *Armour* v. *Skeen* 1977 SLT 71. Here two painters employed under a direct labour contract by Strathclyde Regional Council fell from scaffolding erected for the work. One was killed. Both the Regional Council and its Director of Roads were prosecuted for having failed to comply with HSWA, sec 2(1) and (3) and sec 37(1) – the last section concerning offences committed by a body corporate and individual functional directors. Although the council had issued a statement of safety policy, as required by sec 2(3), which had been circulated to the various departments, instructing directors and heads of departments to ensure safe conditions of work, as well as an 'Advisory Bulletin' instructing each department to issue a safety statement specifically related to the work being carried out, including training and instruction in safe practices, the director of roads had not brought this to the attention of his employees and persons carrying out contracts for the council. He was fined £125 and his appeal against conviction under section 37(1) was turned down. Moreover, the council was found guilty of breach of HSWA, sec 2(1), and fined £500. (For more about sec 37, *see later* in this chapter.)

Duty to provide information to employees

Under sec 2(2)(*c*) of HSWA employers have a duty to 'provide such information, instruction, training and supervision as is necessary to ensure, so far as is reasonably practicable, the health and safety at work of their employees'. Similar provisions apply in the case of non-employees in situations where the way an employer conducts his undertaking might affect the health and safety of non-employees, e.g. contractors and sub-contractors. Furthermore, employers are required, under the SRSCR, to disclose information to trade union-appointed safety representatives which is necessary for them to carry out their functions (Reg 7(2)). This wide requirement relating to the giving of information, which is detailed further in the Approved Code of Practice accompanying the Regulations, relates not only to technical information and consultants' reports, but also to information concerning the employer's future plans and proposed changes insofar as they affect the health, safety and welfare at work of their employees.

Under sec 28(8) of HSWA inspectors are obliged to supply safety representatives with technical information, i.e. factual information obtained during their visits, such as the results of any noise measurements, tests, monitoring activities, together with notices of prosecution, copies of correspondence and copies of any improvement or prohibition notices served on their employer.

The Health and Safety Information for Employees Regulations 1998

These Regulations require information relating to health, safety and welfare to be furnished to employees by means of posters or leaflets in the form approved and published for the purposes of the Regulations by the

HSE. Copies of the form of poster or leaflets approved in this way may be obtained from TSO.

The Regulations require the name and address of the enforcing authority and the address of the Employment Medical Advisory Service to be written in the appropriate space on the poster; and where the leaflet is given the same information should be specified in a written notice accompanying it.

The Regulations provide for the issue of Certificates of Exemption by the HSE and a defence for contravention of the Regulations. They repeal, revoke and modify various enactments and instruments relating to the provision of information to employees. The Regulations do not apply to the Master and crew of a sea-going ship.

Duty of consultation and establishment of safety committees

Employers are required, in the interests of making and maintaining arrangements to enable them and their employees to co-operate in promoting and developing measures to ensure the health and safety at work of employees and in monitoring the effectiveness of such measures:

(a) to consult with workplace representatives ('safety representatives', whose powers and functions are set out in the SRSCR – *see* Chapter 5);

(b) to set up a safety committee, if requested to do so by two or more safety representatives, to monitor measures taken to ensure the health and safety of employees (HSWA, sec 2(6)).

Duties of employees

Employees, whilst at work, must:

(a) take reasonable care for their own health and safety, and that of other persons (including members of the public) who may foreseeably be affected by their acts or omissions at work;

(b) co-operate with their employer so far as is necessary for him to comply with any duty or requirement under any of the 'relevant statutory provisions' (for the meaning of this expression, *see* Chapter 4) (HSWA, sec 7(*a*) and (*b*));

(c) not intentionally or recklessly interfere with or misuse anything provided for the purpose of health and safety at work in furtherance of a statutory requirement (HSWA, sec 8).

In addition, employees owe employers a contractual duty to carry out their work with care and skill. Failure to do this can lead to dismissal as being in breach of a term of their contract of employment (*see* Chapter 5).

As for the definition of 'employee' and the legal problems involved in determining the existence of an employer/employee relationship, *see* Chapter 2.

Consequences of breach of duties in HSWA by employers and employees

Breach by employer

An employer who is in breach of his duties under HSWA, sec 2, can be:

(a) prosecuted by an HSE inspector, and, if found guilty, made to pay a fine, or possibly sent to prison (*see* Chapter 4);

(b) sued for negligence at common law if the breach causes injury to an employee. An action for breach of statutory duty does not lie for breach of the General Duties of HSWA (HSWA, sec 47) (*see* Chapter 2);

(c) sued for constructive dismissal before an industrial tribunal by an employee if the breach constitutes a fundamental term of the contract of employment (*see* Chapter 5).

Breach by employee

An employee who is in breach of his duties under HSWA can:

(a) be prosecuted by an HSE inspector and made to pay a fine, if convicted;

(b) if his breach causes injury to himself or to other employees or to a member of the public, involving his employer in vicarious liability (*see* Chapter 2), though, where the injury is to himself, his damages will be reduced on the ground of his contributory negligence;

(c) be dismissed from his employment for being in breach of the term of his contract of employment that he will carry out his work with proper care and skill. If the employer has gone through the necessary procedures set out in the ACAS Code, 'Disciplinary Practice and Procedures in Industry', an industrial tribunal may well declare such dismissal fair (*see* Chapter 5).

Duties of employers to persons other than their employees

Every employer must conduct his undertaking in such a way as to ensure, so far as is reasonably practicable, that persons not in his employment are not exposed to risks to health or safety (HSWA, sec 3(1)).

Duties of occupiers of premises to persons other than their employees

Health and safety duties are placed on persons in control of premises for the benefit of persons working in such premises, but who are not their employees, or who use plant and substances made available to them for their use there (HSWA, sec 4(1)). An example of the operation of this section would be factory or other work premises owned and leased out by a local authority. Here it is the duty of the person or persons or body in control to take steps to ensure, so far as is reasonably practicable, that (*a*) the premises, (*b*) the means of access and egress, and (*c*) the plant and substances provided for use there, are safe and without health risks (HSWA, sec 4(2)). The protection under this section extends to visitors who are:

(a) workers, e.g. employees of a contract window-cleaning firm engaged to clean windows on a local authority owned or privately owned building;

(b) non-working visitors, who use plant and equipment on premises provided for their use, e.g. visitors to a launderette or school children visiting a public swimming pool.

Meaning of control

The duty laid down in HSWA, sec 4, is associated with occupation of a building rather than ownership, though it would extend to owner-occupiers. HSWA places responsibility for health and safety upon a person or persons having control 'to any extent' (HSWA, sec 4(2)), and this duty extends to both working and non-working visitors.

In law, control is a synonym for occupation, which includes management of a building. Today occupation or control of a building or buildings is often shared between owner and manager or owner and occupier. Indeed, the idea of single exclusive control based on physical occupation is practically redundant. This much emerged from the decision of the House of Lords in *Wheat* v. *E. Lacon & Co. Ltd* [1965] 2 AER 700, where, following a fatal injury to a visitor at a public house, both owner, the brewery, and the occupier, the licensee, would have been liable (under the Occupiers' Liability Act 1957), had it not been for a technicality of law, the court taking the view that, since the owners could control the use to which the licensee put the premises, they were 'to some extent' in control.

The position under HSWA, sec 4, will, therefore, be as in *Wheat* v. *Lacon* regarding occupation of buildings, unless that position has been modified by contract or lease. Where a contract or lease expressly or by implication places responsibility for:

(a) maintenance and/or repair of premises, or

(b) health and safety obligations in connection with plant and substances in premises,

upon a particular person or body, that person or body is deemed to be in control (HSWA, sec 4(3)).

Duties of self-employed persons

HSWA places responsibilities upon self-employed persons. This is the first time that such general statutory duties (as distinct from common law duties, which are applicable to everyone, *see* Chapter 2) have been placed on the self-employed. This is a measure largely designed to remedy certain 'practices' in the construction industry, where many self-employed persons are engaged, if only on a temporary basis (but *see also* the case of *Ferguson* v. *John Dawson Partners (Contractors) Ltd* in Chapter 2).

Every self-employed person must conduct his undertaking in such a way as to ensure that persons not in his employment are not exposed to health and safety risks (HSWA, sec 3(2)).

Duties of designers, manufacturers and suppliers of industrial products

For the first time in the history of occupational health and safety legislation, general duties have been placed on designers, manufacturers and suppliers of articles and substances for use at work. HSWA, sec 6, requires that, so far as is reasonably practicable, designers, manufacturers (including submanufacturers), importers and suppliers of articles and substances for use at work:

(a) ensure that they are designed and constructed so as to be safe and without health risks (HSWA, sec 6(1)(*a*), (2) and (4));

(b) carry out or arrange for the carrying out of testing, research and examination which may be necessary to comply with the duty in section 6(1)(*a*) (*above*) (HSWA, sec 6(1)(*b*) and (5));

(c) take steps which are necessary to secure that adequate information is available about the use for which the product has been designed and tested, and about any conditions necessary to ensure that, when put to that use, the product will be safe and without health risks (HSWA, sec 6(1)(*c*) and (4)(*c*)).

Special responsibilities of designers and manufacturers

The duty to research and test industrial products rests with designers and manufacturers (*see* sec 6(1), (2), (4) and (5)). The requirements applicable to importers, distributors and suppliers of industrial products are rather less onerous. It is their duty to obtain information, on an ongoing basis, from designers and manufacturers of industrial products, so as to be able to advise their client-users how to use safely products which they buy or hire. Provision of data sheets by designers, manufacturers, etc., is not necessarily enough for these purposes. Importers and suppliers are not required to repeat any research and testing already carried out (HSWA, sec 6(6)).

Duties of installers of industrial equipment

Installers and erectors of equipment and machinery for use at work must, so far as is reasonably practicable, ensure that no health and safety hazards arise from the method of installation or erection (HSWA, sec 6(3)).

The effect of the Consumer Protection Act 1987

Sec 6 of HSWA places specific duties on the designers, manufacturers, importers and suppliers of articles and substances for use at work. These duties were amended and extended through sec 36 and Schedule 3 of the Consumer Protection Act thus.

1. Articles for use at work

(a) To ensure, so far as is reasonably practicable, that the article is so designed and constructed that it will be safe and without risks to health at all times when it is being set, used, cleaned or maintained by a person at work.

(b) To carry out or arrange for the carrying out of such testing and examination as may be necessary for the performance of the duty imposed on him by the preceding paragraph.

(c) To take such steps as are necessary to secure that persons supplied by that person with the article are provided with adequate information about the use for which the article is designed or has been tested and about any conditions necessary to ensure that it will be safe and without risks to health at all such times as are mentioned in paragraph (a) above when it is being dismantled or disposed of.

(d) To take such steps as are necessary to secure, so far as is reasonably practicable, that persons so supplied are provided with all such revisions of information provided by them by virtue of the preceding paragraph as are necessary by reason of it becoming known that anything gives rise to a serious risk to health or safety.

The amendment also incorporates specific provisions relating to the manufacturers, designers, importers or suppliers of articles of fairground equipment.

2. *Substances*

(a) To ensure, so far as is reasonably practicable, that the substance will be safe and without risks to health at all times when it is being used, handled, processed, stored or transported by a person at work or in premises to which sec 4 applies.

(b) To carry out or arrange for the carrying out of such testing and examination as may be necessary for the performance of the duty imposed on him by the preceding paragraph.

(c) To take such steps as are necessary to secure that persons supplied by that person with the substance are provided with adequate information about any risks to health or safety to which the inherent properties of the substance may give rise, about the results of any relevant tests which have been carried out on or in connection with the substance and about any conditions necessary to ensure that the substance will be safe and without risks to health at all such times as are mentioned in paragraph (a) above and when the substance is being disposed of.

(d) To take such steps as are necessary to secure, so far as is reasonably practicable, that persons so supplied are provided with all such revisions of information provided by them by virtue of the preceding paragraph as are necessary by reason of it becoming known that anything gives rise to a serious risk to health or safety.

Definition of articles and substances

'Article for use at work' means:

(a) any plant designed for use or operation (whether exclusively or not) by persons at work, or who erect or install any article of fairground equipment;

(b) any article designed for use as a component in any such plant or equipment.

'Substance' means any natural or artificial substance (including micro-organisms) intended for use (whether exclusively or not) by persons at work.

Duties of companies, corporations and their officials

HSWA makes provision for offences committed by companies, corporations and local authorities, as well as by officials. Most directors and other corporate officials are employees, no matter how senior, and so can be dismissed for health and safety offences like any other employee (*see* Chapter 5), in addition to any prosecution which might be brought against them by HSE inspectors. Moreover, the more senior the official, the more severe the penalty imposed by the magistrates is likely to be.

Offences committed by companies and local authorities

Where there has been a breach of one of the 'relevant statutory provisions' (e.g. HSWA, OSRPA and FA) on the part of a body corporate (i.e. limited company or local authority) and the offence can be proved:

(a) to have been committed with the consent or connivance of; or

(b) to be attributable to any neglect on the part of,

any director, manager, secretary or other similar officer of the body corporate, he, as well as the body corporate, can be found guilty and punished accordingly (HSWA, sec 37(1)). Breach of this section has the following consequences:

(a) Where an offence is committed through neglect by a board of directors or the council of a local authority, the company itself or the local authority can be prosecuted as well as the directors or council members individually who may have been to blame.

(b) Where an individual functional director or executive of a local authority is guilty of an offence, he can be prosecuted as well as the company or local authority.

(c) A company or local authority can be prosecuted even though the act or omission resulting in the offence was committed by a junior official or executive or even a visitor to the company or local authority premises. (This last point is made clear in HSWA, sec 36(1) (*see below*).) Thus an offence committed by a training officer or safety officer of a local authority or company can involve the council of the local authority or the board of directors of the company in liability under sec 37(1).

Offences committed by other 'corporate' persons

Offences committed by junior corporate officials e.g. training officers, safety officers, etc., are provided for in sec 36(1). Thus:

> Where the commission by any person of an offence under any of the relevant statutory provisions is due to the act or default of some other person, that other person shall be guilty of the offence, and a person may be charged with and convicted of the offence . . . whether or not proceedings are taken against the first-mentioned person.

Sec 36(1) can be used not only against junior members of management but also against visitors to workplaces. Case law seems to show that 'other person' refers to persons lower down the corporate tree than those mentioned in sec 37(1), e.g. safety officers, training officers. This is the effect of the decision (involving the Trade Descriptions Act 1968) in *Tesco Stores Ltd* v. *Nattrass* [1971] 2 AER 127.

Breach of sec 36(1) is not a defence for a company or local authority prosecuted under sec 37(1). This is made clear in sec 36(1) itself. (This was a defence under the Trade Descriptions Act 1968 where the company could show fault on the part of an employee who was not himself of sufficiently senior corporate status to 'commit' the company.) It is thought, however, that breach of sec 36(1) would entitle an HSE inspector to use discretion in deciding whether to proceed against the body corporate under sec 37(1).

'Manslaughter' charge

In December 1989 a company director was found guilty, for the first time in the history of health and safety legislation, of manslaughter. The manslaughter charges were brought following the inquest into an employee's death in the company's factory in May 1988. In this case the coroner considered that the case might be one of unlawful killing and referred it to the Crown Prosecution Service who asked the police to investigate. All the prosecutions were conducted by the Crown Prosecution Service. The director concerned pleaded guilty to the charge and received a 12-month sentence suspended for two years.

A similar situation arose in 1987 involving a custodial sentence when a company director was sentenced to 18 months' imprisonment, suspended for two years. Here the director was found guilty under sec 37(1) of HSWA 1974 on the grounds that offences by one of his companies engaged in asbestos removal were committed with his consent or connivance or were attributable to an employee's death.

Work experience schemes

Students and children attending school who take part in work experience schemes were brought under the general protection of health and safety legislation. The Health and Safety (Training for Employment) Regulations 1990 require that all those receiving training or work experience from an employer in the workplace are deemed to be employees for the purposes of health and safety legislation. These Regulations extend the Health and Safety Training Regulations 1990 which give health and safety protection to pupils and students on work experience schemes. The extension covers school-age pupils on work experience and college students on sandwich courses. The Regulations also re-enact the protection already provided for participants on government training schemes.

Management of Health and Safety at Work Regulations 1999

Introduction

These regulations are, perhaps, the most significant regulations in current health and safety legislation. They are accompanied by an ACOP and HSE Guidance. The strict duties on employers, because of their wide-ranging general nature, overlap with many duties under existing regulations, such as the Control of Substances Hazardous to Health (COSHH) Regulations 1999. Where duties overlap, compliance with the duty in the more specific regulation will normally be sufficient to comply with the corresponding duty under the MHSWR. However, where the duties in these regulations go beyond those in the more specific regulations, additional measures will be needed to comply fully with the MHSWR. The principal requirements of the Regulations are dealt with below.

Reg 1 – Interpretation

The 1996 Act means the Employment Rights Act 1996.

The assessment – means, in the case of an employer, the assessment made by him in accordance with Reg 3.

Child:

(a) As respects England and Wales, means a person who is not over compulsory school age, construed in accordance with sec 8 of the Education Act 1996.

(b) As respects Scotland, means a person who is not over compulsory school age, construed in accordance with sec 31 of the Education (Scotland) Act 1980

Employment business – means a business (whether or not carried on with a view to profit and whether or not carried on in conjunction with any other business) which supplies persons (other than seafarers) who are employed in it to work for and under the control of other persons in any capacity

Fixed-term contract of employment – means a contract of employment for a specific term which is fixed in advance or which can be ascertained in advance by reference to some relevant circumstance

Given birth – means delivering a live child or, after 24 weeks of pregnancy, a stillborn child.

New or expectant mother means an employee who is pregnant, who has given birth within the previous six months, or who is breastfeeding

The preventive and protective measures – means the measures which have been identified by the employer or by the self-employed person in consequence of the assessment as the measures he needs to take to comply with the requirements and prohibitions imposed upon him by or under the relevant statutory provisions and Part II of the Fire Precautions (Workplace) Regulations 1997.

Young person – means any person who has not attained the age of 18 years.

Reg 2 – Disapplication of these Regulations

These regulations shall not apply to or in relation to the master or crew of a sea-going ship or to the employer of such persons in respect of the normal shipboard activities of a ship's crew under the direction of the master.

Regs 3(4), (5), 10(2) and 19 shall not apply to occasional work or short-term work involving:

(a) domestic service in a private household; or

(b) work regulated as not being harmful, damaging or dangerous to young people in a family undertaking.

Reg 3 – Risk assessment

1. Every employer shall make a suitable and sufficient assessment of:

(a) the risks to the health and safety of his employees to which they are exposed whilst at work;

(b) the risks to the health and safety of persons not in his employment arising out of or in connection with the conduct by him of his undertaking, for the purpose of identifying the measures he needs to take to comply with the requirements and prohibitions imposed upon him by or under the relevant statutory provisions and by Part II of the Fire Precautions (Workplace) Regulations 1997.

2. Similar provisions as in paragraph 1 above apply in the case of self-employed persons (excluding the provisions of Part II of the Fire Precautions (Workplace) Regulations 1997).

3. Any assessment shall be reviewed by the employer or self-employed person who made it if:

(a) there is reason to suspect it is no longer valid; or

(b) there has been a significant change in the matters to which it relates,

and where as a result of any such review changes to an assessment are required, the employer or self-employed person shall make them.

4. An employer shall not employ a young person unless he has, in relation to risks to the health and safety of young persons, made or reviewed an assessment in accordance with paragraphs 1 and 5.

5. In making or reviewing the assessment, an employer who employs or is to employ a young person shall take particular account of:

(a) the inexperience, lack of awareness of risks and immaturity of young persons;

(b) the fitting-out and layout of the workplace and the workstations;

(c) the nature, degree and duration of exposure to physical, biological and chemical agents;

(d) the form, range and use of work equipment and the way in which it is handled;

(e) the organisation of processes and activities;

(f) the extent of the health and safety training provided to young persons; and

(g) risks from processes, agents and work listed in the Annex to the Council Directive 94/33/EC on the protection of young people at work.

6. Where the employer employs five or more employees, he shall record:

(a) the significant findings of the assessment;

(b) any group of his employees identified by it as being especially at risk.

Comment

The duty on employers to undertake 'a suitable and sufficient risk assessment' is a principal feature of all modern protective legislation. The practical application of risk assessment is covered in the ACOP.

Reg 4 – Principles of prevention to be applied

Where an employer implements any preventive and protective measures he shall do so on the basis of the principles specified in Schedule 1 to these regulations.

Reg 5 – Health and safety arrangements

1. Every employer shall make and give effect to such arrangements as are appropriate, having regard to the nature of his activities and the size of his undertaking, for the effective planning, organisation, control, monitoring and review of the preventive and protective measures.

2. Where the employer employs five or more employees, he shall record these arrangements.

Comment

There is a need here to consider the systems necessary to ensure effective management of health and safety procedures and requirements. Such systems should be integrated with other management systems, e.g. those for financial, human resources, production, engineering, purchasing and other areas of management activity. The principal elements of management practice i.e. planning, organising, controlling, monitoring and reviewing of the 'preventive and protective measures' must be taken into account. The HSE publication HS(G()65 *Successful health and safety management* provides guidance on the installation of management systems.

Reg 6 – Health surveillance

Every employer shall ensure that his employees are provided with such health surveillance as is appropriate having regard to the *risks to their health* and safety which are identified by the assessment.

Comment

Health surveillance may already be required in order to comply with existing legislation, such as the COSHH Regulations.

However, health surveillance may also be necessary where the risk assessment under the MHSWR indicates:

(a) there is an identifiable disease or adverse health condition related to the work concerned;

(b) valid techniques are available to detect indications of the disease or condition e.g. certain forms of biological monitoring;

(c) there is a reasonable likelihood that the disease or condition may occur under the particular conditions of work;

(d) surveillance is likely to further the protection of the health of the employees to be covered.

The principal objective of any health surveillance activity is to detect adverse health effects at an early stage, thereby enabling further harm to be prevented.

Reg 7 – Health and safety assistance

1. Every employer shall, subject to paragraphs 6 and 7, appoint one or more competent persons to assist him in undertaking the measures he needs to take to comply with the requirements and prohibitions imposed upon him by or under the relevant statutory provisions and by Part II of the Fire Precautions (Workplace) Regulations 1997.

2. Where an employer appoints persons in accordance with paragraph 1, he shall make arrangements for ensuring adequate co-operation between them.

3. The employer shall ensure that the number of persons appointed under paragraph 1, the time available for them to fulfil their functions and the means at their disposal are adequate having regard to the size of his undertaking, the risks to which his employees are exposed and the distribution of those risks throughout the undertaking.

4. The employer shall ensure that:

(a) any person appointed by him in accordance with paragraph 1 who is not in his employment –

(i) is informed of the factors known by him to affect, or suspected by him of affecting, the health and safety of any other person who may be affected by the conduct of his undertaking;

(ii) has access to the information referred to in Reg 10; and

(b) any person appointed by him in accordance with paragraph 1 is given such information about any person working in his undertaking who is –

(i) employed by him under a fixed-term contract of employment; or

(ii) employed in an employment business,

as is necessary to enable that person properly to carry out the function specified in the paragraph.

5. A person shall be regarded as competent for the purposes of paragraphs 1 and 8 where he has sufficient training and experience or knowledge and other qualities properly to undertake the measures referred to in paragraph 1.

6. Paragraph 1 shall not apply to a self-employed employer who is not in partnership with any other person where he has sufficient training and experience or knowledge and other qualities properly to undertake the measures referred to in that paragraph himself.

7. Paragraph 1 shall not apply to individuals who are employers and who are together carrying on business in partnership where at least one of the individuals concerned has sufficient training and experience or knowledge and other qualities:

(a) properly to undertake the measures he needs to take to comply with the requirements and prohibitions imposed upon him by or under the relevant statutory provisions;

(b) properly to assist his fellow partners in undertaking the measures they need to take to comply with the requirements and prohibitions imposed upon them by or under the relevant statutory provisions.

8. Where there is a competent person in the employer's employment, that person shall be appointed for the purposes of paragraph 1 in preference to a competent person not in his employment.

Comment

The concept of 'competent persons' is not new to health and safety legislation, the appointment of such persons being required under, for instance, the Construction (Health, Safety and Welfare) Regulations 1996 and the Noise at Work Regulations 1989.

The degree of competence of the competent person for the purposes of these regulations will depend on the risks identified in the risk assessment.

Whilst there is no specific emphasis on the qualifications of such persons, broadly a competent person should have such skill, knowledge and experience as to enable him to identify defects and understand the implications of those defects. The depth of training, accountabilities and responsibilities, authority and level of reportability within the management system of the competent person is, therefore, important. The competent person should not be placed so far down the management system that his recommendations carry no weight.

Reg 8 – Procedures for serious and imminent danger and for danger areas

1. Every employer shall:

(a) establish and where necessary give effect to appropriate procedures to be followed in the event of serious and imminent danger to persons at work in his undertaking;

(b) nominate a sufficient number of competent persons to implement those procedures insofar as they relate to the evacuation from premises of persons at work in his undertaking;

(c) ensure that none of his employees has access to any area occupied by him to which it is necessary to restrict access on grounds of health and safety unless the employee concerned has received adequate health and safety instruction.

2. Without prejudice to the generality of paragraph l(a), the procedures referred to in that sub-paragraph shall:

(a) so far as is practicable, require any persons at work who are exposed to serious and imminent danger to be informed of the nature of the hazard and of the steps to be taken to protect them from it;

(b) enable the persons concerned (if necessary by taking appropriate steps in the absence of guidance or instruction and in the light of their knowledge and the technical means at their disposal) to stop work and immediately proceed to a place of safety in the event of their being exposed to serious, imminent or unavoidable danger;

(c) save in exceptional cases for reasons duly substantiated (which cases and reasons shall be specified in those procedures), require the persons concerned to be prevented from resuming work in any situation where there is still a serious and imminent danger.

3. A person shall be regarded as competent for the purposes of paragraph

1(b) where he has sufficient training and experience or knowledge and other qualities to enable him properly to implement the evacuation procedures referred to in that sub-paragraph.

Comment

Employers must establish procedures to be followed by employees and others:

(a) in situations which present serious and imminent danger; together with

(b) the circumstances where they should stop work and move to a place of safety.

Fundamentally, the requirement is for an organisation's emergency plans/procedures to cover foreseeable high risk situations, such as fire, explosion, etc., and to ensure training of staff in these procedures.

Reg 9 – Contacts with external services

Every employer shall ensure that any necessary contacts with external services are arranged, particularly as regards first aid, emergency medical care and rescue work.

Reg 10 – Information for employees

1. Every employer shall provide his employees with comprehensible and relevant information on:

   (a) the risks to their health and safety identified by the assessment;

   (b) the preventive and protective measures;

   (c) the procedures referred to in Reg 8(1)(a) and the measures referred to in Reg 4(2)(a) of the Fire Precautions (Workplace) Regulations 1997;

   (d) the identity of those persons nominated by him in accordance with Reg 8(1)(b) and Reg 4(2)(b) of the Fire Precautions (Workplace) Regulations 1997;

   (e) the risks notified to him in accordance with Reg II(I)(c).

2. Every employer shall, before employing a *child,* provide a parent of the child with comprehensible and relevant information on:

   (a) the risks to his health and safety identified by the assessment;

   (b) the preventive and protective measures;

   (c) the risks notified in accordance with Reg II(I)(c).

Comment

The significant feature of this regulation is that information provided to employees must be 'comprehensible', i.e. written in such a way as to be easily understood by the people to which it is addressed, and relevant to the circumstances of their work and the hazards which could arise.

On this basis, the mode of presentation of such information should take account of their level of training, knowledge and experience. There may also be a need to consider people with language difficulties or with disabilities which may impede their receipt of information. Information can be provided in whatever form is most suitable in the circumstances, so long as it is comprehensible, e.g. staff handbook, posters.

Specific provisions apply in the case of the employment of children.

Reg 11 – Co-operation and co-ordination

1. Where two or more employers share a workplace (whether on a temporary or permanent basis) each such employer shall:

(a) co-operate with the other employers concerned so far as is necessary to enable them to comply with the requirements and prohibitions imposed upon them by or under the relevant statutory provisions and by Part II of the Fire Precautions (Workplace) Regulations 1997;

(b) (taking into account the nature of his activities) take all reasonable steps to co-ordinate the measures he takes to comply with the requirements and prohibitions imposed upon him by or under the relevant statutory provisions and Part II of the Fire Precautions (Workplace) Regulations 1997 with the measures which the other employers concerned are taking to comply with the requirements and prohibitions imposed upon them by that legislation;

(c) take all reasonable steps to inform the other employers concerned of the risks to their employees' health and safety arising out of or in connection with the conduct by him of his undertaking.

2. Paragraph 1 (except as it refers to Part II of the Fire Precautions (Workplace) Regulations 1997) shall apply to employers sharing a workplace with self-employed persons and to self-employed persons sharing a workplace with other self-employed persons as it applies to employers sharing a workplace with other employers; and the references in that paragraph to employers and the reference in the said paragraph to their employees shall be construed accordingly.

Comment

This regulation makes provision for employers jointly occupying a work site e.g. a construction site, trading estate or office block, to co-operate and co-ordinate their health and safety activities, particularly where there are risks common to everyone, irrespective of their work activity. Fields of co-operation and co-ordination include health and safety training, safety monitoring procedures, emergency arrangements and welfare amenity provisions, In some cases, the appointment of a health and safety co-ordinator should be considered.

Reg 12 – Persons working in host employers' or self-employed persons' undertakings

1. Every employer and every self-employed person shall ensure that the employer of any employees from an outside undertaking who are working in his undertaking is provided with comprehensible information on:

(a) the risks to those employees' health and safety arising out of or in connection with the conduct by that first-mentioned employer or by that self-employed person of his undertaking;

(b) the measures taken by that first-mentioned employer or by that self-employed person in compliance with the requirements and prohibitions imposed upon him by or under the relevant statutory provisions and by Part II of the Fire Precautions (Workplace) Regulation 1997 insofar as the said requirements and prohibitions relate to those employees.

2. Paragraph 1 (except insofar as it refers to Part II of the Fire Precautions (Workplace) Regulations 1997) shall apply to a self-employed person who is working in the undertaking of an employer or a self-employed person as it applies to employees from an outside undertaking who are working therein; and the reference in that paragraph to the employer of any employees from an outside undertaking who are working in the undertaking of an employer or a self-employed person and the references in the said paragraph to employees from an outside undertaking who are working in the undertaking of an employer or a self-employed person shall be construed accordingly.

3. Every employer shall ensure that every person working in his undertaking who is not his employee and every self-employed person (not being an employer) shall ensure that any person working in his undertaking is provided with appropriate instructions and comprehensible information regarding any risks to that person's health and safety which arise out of the conduct by that employer or self-employed person of his undertaking.

4. Every employer shall:

(a) ensure that the employer of any employees from an outside undertaking who are working in his undertaking is provided with sufficient information to enable that second-mentioned employer to identify any person nominated by that first-mentioned employer in accordance with Reg 8(1)(b) to implement evacuation procedures as far as those employees are concerned;

(b) take all reasonable steps to ensure that any employees from an outside undertaking who are working in his undertaking receive sufficient information to enable them to identify any person nominated by him in accordance with Reg 8(1)(b) to implement evacuation procedures as far as they are concerned.

5. Paragraph 4 shall apply to a self-employed person who is working in an employer's undertaking as it applies to employees from an outside undertaking who are working therein; and the reference in that paragraph to the employer of any employees from an outside undertaking who are working in an employer's undertaking and the references in the said paragraph to employees from an outside undertaking who are working in an employer's undertaking shall be construed accordingly.

Comment

This regulation applies to a wide range of work activities where employees fundamentally undertake work at other people's premises, e.g. contract cleaners, maintenance staff and building workers. The host employer must provide these persons with adequate information and instructions regarding the risks to their health and safety and the precautions necessary.

Reg 13 – Capabilities and training

1. Every employer shall, in entrusting tasks to his employees, take into account their capabilities as regards health and safety.

2. Every employer shall ensure that his employees are provided with adequate health and safety training:

(a) on their being recruited into the employer's undertaking;

(b) on their being exposed to new or increased risks because of –

(i) their being transferred or given a change of responsibilities within the employer's undertaking;

(ii) the introduction of new work equipment into or a change respecting work equipment already in use within the employer's undertaking;

(iii) the introduction of new technology into the employer's undertaking; or

(iv) the introduction of a new system of work into or a change respecting a system of work already in use within the employer's undertaking.

3. The training referred to in paragraph 2 shall:

(a) be repeated periodically where appropriate;

(b) be adapted to take account of new or changed risks to the health and safety of the employees concerned;

(c) take place during working hours.

Comment

This regulation introduces, for the first time in health and safety legislation, a human factors approach to ensuring appropriate levels of health and safety provision. When allocating work to employees, employers must ensure that the demands of the job do not exceed the employees' capabilities to carry out the work without risk to themselves and others. There is a need, therefore, for employers to consider both the physical and mental capabilities of employees before allocating tasks, together with their level of knowledge, training and experience. Training, in particular, should reflect this aspect of human capability. Circumstances where there is an absolute duty on employers to provide health and safety training are also specified, e.g. on induction.

Reg 14 – Employees' duties

1. Every employee shall use any machinery, equipment, dangerous substance, transport equipment, means of production or safety device provided to him by his employer in accordance both with any training in the use of the equipment concerned which has been received by him and the instructions respecting that use which have been provided to him by the said employer in compliance with the requirements and prohibitions imposed upon that employer by or under the relevant statutory provisions.

2. Every employee shall inform his employer or any other employee of that employer with specific responsibility for health and safety of his fellow employees:

(a) of any work situation which a person with the first-mentioned employee's training and instruction would reasonably consider represented a serious and immediate danger to health and safety;

(b) of any matter which a person with the first-mentioned employee's training and instruction would reasonably consider represented a shortcoming in the employer's protection arrangements for health and safety, insofar as that situation or matter either affects the health and safety of that first-mentioned employee or arises out of or in connection with his own activities at work, and has not previously been reported to his employer or to any other employee of that employer in accordance with this paragraph.

Comment

This regulation reinforces and expands the duties of employers towards employees under sec 7 of the HSWA. In the light of these requirements, employers should instal some form of hazard reporting system, whereby employees can report hazards to their employer or appointed competent person. Such a system should ensure a prompt response where hazards are identified, with the competent person signing off the hazard report when the hazard has been eliminated or controlled.

Reg 15 – Temporary workers

1. Every employer shall provide any person whom he has employed under a fixed-term contract of employment with comprehensible information on:

(a) any special occupational qualifications or skills required to be held by that employee if he is to carry out his work safely;

(b) any health surveillance required to be provided to that employee by or under any of the relevant statutory provisions, and shall provide the said information before the employee concerned commences his duties.

2. Every employer and every self-employed person shall provide any person employed in an employment business who is to carry out work in his undertaking with similar comprehensible information as indicated in 1(a) and (b) above.

3. Every employer and every self-employed person shall ensure that every person carrying on an employment business whose employees are to carry out work in his undertaking is provided with comprehensible information on:

   (a) any special occupational qualifications or skills required to be held by those employees if they are to carry out their work safely; and

   (b) the specific features of the jobs to be filled by those employees (insofar as those features are likely to affect their health and safety); and the person carrying on the employment business concerned shall ensure that the information so provided is given to the said employees.

Comment

This regulation supplements previous regulations requiring the provision of information with additional requirements on temporary workers, i.e. those employed on fixed duration contracts and those employed in employment businesses, but working under the control of a user company. The use of temporary workers will also have to be notified to the competent person.

Reg 16 – Risk assessment in respect of new or expectant mothers

1. Where:

   (a) the persons working in an undertaking include women of child-bearing age;

   (b) the work is of a kind which could involve risk, by reasons of her condition, to the health and safety of a new or expectant mother, or to that of her baby, from any processes or working conditions, or physical, biological or chemical agents, including those specified in Annexes I and II of Council Directive 92/85/EEC on the introduction of measures to encourage improvements in the safety and health at work of pregnant workers and workers who have recently given birth or are breastfeeding,

   the assessment required by Reg 3(1) shall also include an assessment of such risk.

2. Where, in the case of an individual employee, the taking of any other action the employer is required to take under the relevant statutory provisions would not avoid the risk referred to in paragraph 1 the

employer shall, if it is reasonable to do so, and would avoid such risks, alter her working conditions or hours of work.

3. If it is not reasonable to alter the working conditions or hours of work, or if it would not avoid such risk, the employer shall, subject to sec 67 of the 1996 Act suspend the employee from work so long as is necessary to avoid such risk.

4. In paragraphs 1 to 3 references to risk, in relation to risk from any infectious or contagious disease, are references to a level of risk at work which is in addition to the level to which a new or expectant mother may be expected to be exposed outside the workplace.

Reg 17 – Certificate from registered medical practitioner in respect of new or expectant mothers

Where

(a) a new or expectant mother works at night;

(b) a certificate from a registered medical practitioner or a registered midwife shows that it is necessary for her health or safety that she should not be at work for any period of such work identified in the certificate,

the employer shall, subject to sec 67 of the 1996 Act, suspend her from work for so long as is necessary for her health or safety.

Reg 18 – Notification by new or expectant mothers

1. Nothing in paragraph 2 or 3 of Reg 16 shall require the employer to take any action in relation to an employee until she has *notified* the employer in writing that she is pregnant, has given birth within the previous six months, or is breastfeeding.

2. Nothing in paragraphs 2 or 3 of Reg 16 or in Reg 17 shall require the employer to maintain action taken in relation to an employee:

(a) in a case –

(i) to which Reg 16(2) or (3) relates;

(ii) where the employee has notified her employer that she is pregnant, where she has failed within a reasonable time of being requested to do so in writing by her employer, to produce for the employer's inspection a certificate from a registered medical practitioner or registered midwife showing that she is pregnant;

(b) once the employer knows that she is no longer a new or expectant mother;

(c) if the employer cannot establish whether she remains a new or expectant mother.

Reg 19 – Protection of young persons

1. Every employer shall ensure that young persons employed by him are protected at work from any risks to their health or safety which are a consequence of their lack of experience, or absence of awareness of existing or potential risks of the fact that young persons have not yet fully matured.

2. Subject to paragraph 3, no employer shall employ a young person for work:

(a) which is beyond his physical or psychological capacity;

(b) involving harmful exposure to agents which are toxic or carcinogenic, cause heritable genetic damage or harm to the unborn child or which in any other way chronically affect human health;

(c) involving harmful exposure to radiation;

(d) involving the risk of accidents which it may reasonably be assumed cannot be recognised or avoided by young persons owing to their insufficient attention to safety or lack of experience or training; or

(e) in which there is a risk to health from:

(i) extreme cold or heat;

(ii) noise; or

(iii) vibration,

and in determining whether work will involve harm or risks for the purposes of this paragraph, regard shall be had to the results of the assessment.

3. Nothing in paragraph 2 shall prevent the employment of a young person who is no longer a child for work:

(a) where it is necessary for his training;

(b) where the young person will be supervised by a competent person; and

(c) where any risk will be reduced to the lowest level that is reasonably practicable.

4. The provisions contained within this regulation are without prejudice to:
   (a) the provisions contained elsewhere in these regulations;
   (b) any prohibition or restriction, arising otherwise than by this regulation, on the employment of any person.

Reg 20 – Exemption certificates

These regulations may, by certificate of the Secretary of State for Defence, exclude members of the forces generally.

Reg 21 – Provisions as to liability

Nothing in the relevant statutory provisions shall operate so as to afford an employer a defence in any criminal proceedings for a contravention of those provisions by reason of any act or default of:

(a) any employee of his; or
(b) a person appointed by him under Reg 7.

Reg 22 – Exclusion of civil liability

1. Breach of a duty imposed by these Regulations shall not confer a right of action in any civil proceedings. (Similar provisions apply in the case of secs 2–6 of HSWA.)
2. Paragraph 1 shall not apply to any duty imposed by these regulations on an employer:
   (a) to the extent that it relates to risk referred to in Reg 16(1 to an employee; or
   (b) which is contained in Reg 19.

Schedule 1

General principles of prevention

(This Schedule specifies the general principles of prevention set out in Article 6(2) of the Council Directive 89/391/EEC.)

(a) avoiding risks;
(b) evaluating the risks that cannot be avoided;

(c) combating the risks at source;

(d) adapting the work to the individual, especially as regards the design of workplaces, the choice of work equipment and the choice of working and production methods, with a view, in particular, to alleviating monotonous work and work at a predetermined work rate and to reducing their effect on health;

(e) adapting to technical progress;

(f) replacing the dangerous by the non-dangerous or the less dangerous;

(g) developing a coherent overall prevention policy which covers technology, organisation of work, working conditions, social relationships and the influence of factors relating to the working environment;

(h) giving collective protective measures priority over individual protective measures;

(i) giving appropriate instructions to employees.

Footnote to the MHSWR

It must be appreciated that the vast majority of the duties specified in these regulations are of an absolute nature, compared with the HSWA and other regulations, where the duties are qualified by the term 'so far as is reasonably practicable', a lesser level of duty.

The 1999 version of the regulations brought in specific provisions with regard to young persons, children and pregnant workers, first aid, emergency medical care and rescue work, together with specifying general principles of accident and ill-health prevention. The regulations revoke:

the Management of Health and Safety at Work Regulations 1992

the Management of Health and Safety at Work (Amendment) Regulations 1994

the Health and Safety (Young Persons) Regulations 1997

Part III of the Fire Precautions (Workplace) Regulations 1997.

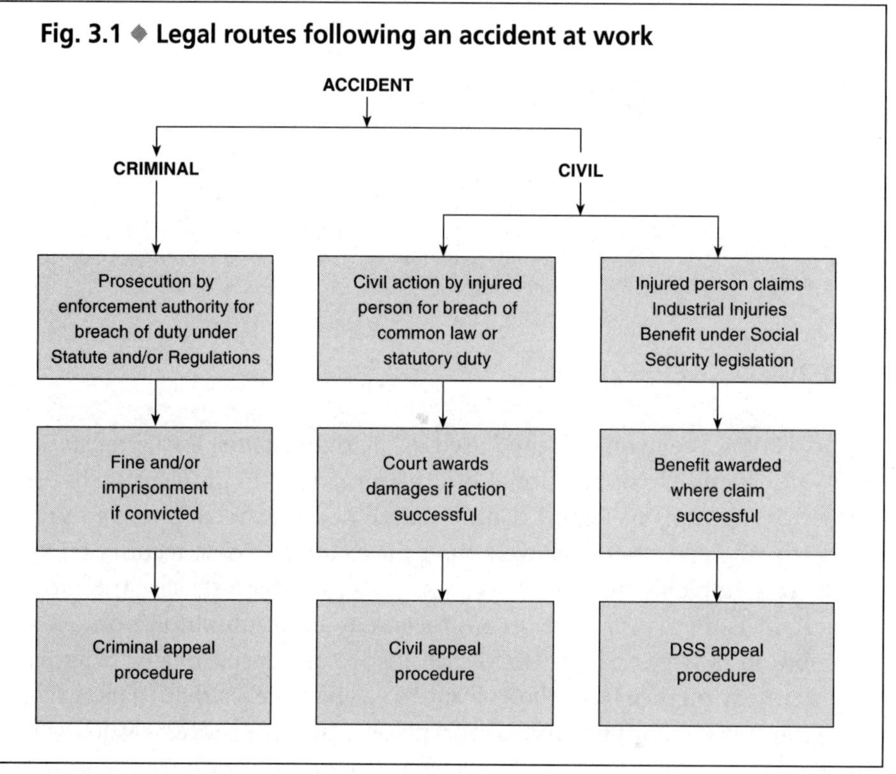
Fig. 3.1 ◆ Legal routes following an accident at work
ACCIDENT
CRIMINAL
CIVIL
Prosecution by enforcement authority for breach of duty under Statute and/or Regulations
Civil action by injured person for breach of common law or statutory duty
Injured person claims Industrial Injuries Benefit under Social Security legislation
Fine and/or imprisonment if convicted
Court awards damages if action successful
Benefit awarded where claim successful
Criminal appeal procedure
Civil appeal procedure
DSS appeal procedure

4 Enforcement

Introduction

Before HSWA, the main 'weapon' used in the fight against breach of statutory duty relating to occupational health and safety was prosecution by an inspector. The Robens Report (Cmnd 5034), however, recommended that inspectors be given wider powers than prosecution. These recommendations were implemented in HSWA, secs 21–6. Of these powers the most important is the power to serve improvement and prohibition notices. A prohibition notice can be served by an inspector, irrespective of whether any statutory offence has been committed, if he believes that there is risk of serious personal injury from a workplace activity or process. Moreover, inspectors can use these executive powers to enforce not only the provisions of HSWA but also pre-HSWA protective legislation, e.g. the Offices, Shops and Railway Premises Act 1963 (OSRPA) and the Mines and Quarries Act 1954 (i.e. 'relevant statutory provisions').

Despite these powers conferred on inspectors, it should not be assumed that prosecution is a thing of the past. On the contrary, prosecution is still an important 'weapon', though of ancillary importance to the improvement or prohibition notice. Employers and employees, whether junior, middle or senior management including the managing director and individual functional directors, who commit breaches of 'relevant statutory provisions' could be prosecuted instead of being served with an improvement or prohibition notice. This applies also to corporate bodies.

'Relevant statutory provisions'

Enforcement powers and procedures apply to all the 'relevant statutory provisions'. This expression refers to the following statutory requirements:

(a) all the provisions of HSWA, part I (i.e. secs 1–54);

(b) health and safety regulations passed pursuant to HSWA (e.g. the Notification of New Substances Regulations 1982);

(c) 'existing statutory provisions' (i.e. all enactments specified in HSWA, Schedule 1, including any regulations made under them insofar as they continue in force) (HSWA, sec 53(1)).

(A full list of the 'relevant statutory provisions' is given in the Appendix to this chapter.)

The enforcing authorities

Generally, the enforcing authority is the HSE, except the following cases:

(a) Where Regulations stipulate that the local authority is the enforcing authority. The Regulations which so stipulate are the Health and Safety (Enforcing Authority) Regulations 1977, as amended by SI 1980 No. 1744 (referred to as the 'Enforcing Authority Regulations').

(b) Where one of the 'relevant statutory provisions' stipulates that some other body is responsible for enforcement (HSWA, sec 18(1), (7)(*a*)).

There are draft proposals to transfer more enforcing powers to local authorities and environmental health officers.

Where HSE is the enforcing authority

HSE is the enforcing authority for:

(a) premises occupied by, or under the control of, a local authority or the Crown, e.g. colleges of further education;

(b) the following activities carried on in any premises (whether or not they are the main activity) –

  (i) construction work;

  (ii) installation, maintenance or repair of gas, water or electricity systems;

  (iii) areas of responsibility specifically excluded from local authorities (Enforcing Authority Regulations, Regs 4(1) and (2)).

These last (iii) refer to:

(a) premises controlled or occupied by a railway undertaking;
(b) sale or storage of water and sewage, or gas;
(c) wholesale distribution of flammable, toxic or certain other dangerous substances, including petrol;
(d) the following premises where consumer services are carried on –
    (i) dry cleaning;
    (ii) radio and television repairs;
    (iii) the maintenance or repair of motor vehicles (Enforcing Authority Regulations, Reg 3, 1 Sch);
(e) premises occupied or under the control of a fire authority or police authority or the UK Atomic Energy Authority; or
(f) circumstances where responsibility for licensing and enforcement are vested under an 'existing statutory provision' in a county council or a harbour authority (Enforcing Authority Regulations, Regs 4(1), (3)).

Where the local authority is the enforcing authority

The local authority is the enforcing authority in the following situations, where the *main* activity is:

(a) sale or storage of goods for retail or wholesale distribution (other than those mentioned in (*d*)(iii) above);
(b) office activities;
(c) catering services;
(d) provision of residential accommodation;
(e) shop premises providing consumer services except dry cleaning, radio and television repairs, etc. (mentioned in (*d*) above); or
(f) dry cleaning in launderettes (Enforcing Authority Regulations 1977, Reg 3, 1 Sch).

Thus local authorities have enforcement powers in respect of offices, shops, public houses, and restaurants and hotels.

Transfer of responsibility

Enforcement of any of the 'relevant statutory provisions' may be transferred by agreement from HSE to local authorities, and vice versa. Moreover, the Health and Safety Commission has power to order such transfer without agreement (Enforcing Authority Regulations, Reg 5).

Enforcement powers of HSE inspectors and local authority officers

The two main enforcement powers of inspectors are:

(a) service of improvement and prohibition notices;

(b) prosecution.

Improvement notices and prohibition notices

Improvement notices

An inspector may serve an improvement notice if he is of the opinion that a person:

(a) is contravening one or more 'relevant statutory provisions';

(b) has contravened one or more of those provisions in circumstances that make it likely that the contravention will continue or be repeated (HSWA, sec 21).

The period stipulated in the notice within which the requirement must be carried out must not be less than 21 days – the time allowed for lodging an appeal with an industrial tribunal (HSWA, sec 21).

Prohibition notices

If an inspector is of the opinion that an industrial activity involves or will involve a risk of serious personal injury, he may serve a prohibition notice (HSWA, sec 22(2)). Prohibition notices differ from improvement notices as follows:

(a) in the case of a prohibition notice it is not necessary that an inspector believe that a provision of HSWA or any other statutory provision is being or has been contravened;

(b) prohibition notices are served in anticipation of danger.

Improvement notices and prohibition notices compared

Unlike an improvement notice, where time is allowed in which to put the offending matter right, a prohibition notice can take immediate effect. 'A direction . . . shall take immediate effect if the inspector is of the opinion . . . that the risk of serious personal injury is . . . or will be imminent' (in any other case it will have effect at the end of a period stipulated in the notice) (HSWA, sec 22(4)). Thus a prohibition notice which is a direction to stop the work activity in question, rather than put the offending matter right, can take effect immediately it is issued; alternatively (and not infrequently) it may allow time to implement certain modifications (it is then known as a 'deferred prohibition notice'). Both immediate and deferred prohibition notices will usually refer to a schedule of work which the inspector will require to be carried out.

Effect of non-compliance with a notice

If, after the time specified in the notice has expired, or, in the event of an appeal, after the expiry of any additional time allowed for compliance by the tribunal, an applicant does not comply with the notice, he can be prosecuted. If convicted by the Crown Court of contravening a prohibition notice, he may be imprisoned (HSWA, sec 33(1)(*g*), (4)(*d*)).

Appeals against improvement notices and prohibition notices

A person against whom an improvement notice or prohibition notice is served may appeal to an industrial tribunal within 21 days from the date of service of the notice. The tribunal may extend this time where it is satisfied, on application made in writing (either before or after the expiry of 21 days), that it was not reasonably practicable for the appeal to be brought within the 21-day period. On appeal the tribunal may either affirm or cancel the notice, and if it affirms it, may do so with modifications, in the form of additions, omissions or amendments (HSWA, secs 24(2), 82(1)(*c*)); Industrial Tribunals (Improvement and Prohibition Notices Appeals) Regulations 1974, Reg 3 and Sch 2.

Effect of an appeal

Where an appeal is lodged against an improvement notice, that automatically suspends operation of that notice. But a prohibition notice will continue to apply, unless there is a direction to the contrary from the tribunal. More specifically:

(a) in the case of an improvement notice, the bringing of an appeal has the effect of suspending operation of the notice;

(b) in the case of a prohibition notice, the lodging of an appeal only suspends operation of the notice –

  (i) if the tribunal so directs, on the application of the applicant;

  (ii) the suspension is then effective from the date when the tribunal so directs (HSWA, sec 24(3)).

Prosecution for breach of the 'relevant statutory provisions'

Prosecution often follows non-compliance with an improvement notice or prohibition notice. Alternatively, inspectors may simply prosecute without serving notices. Prosecutions normally take place before the magistrates but HSWA makes provision for prosecution on indictment (HSWA, sec 33(3)(*b*)). The test of prosecution on indictment is gravity of the particular offence.

Burden of proof in criminal cases

Throughout criminal law the general burden of proving guilt is on the prosecution to show that the accused committed the offence (*Woolmington* v. *DPP* [1935] AC 462). While not detracting from this general principle, HSWA makes the burden of the prosecution considerably easier than it otherwise would have been. Thus HSWA, sec 40, provides that it is incumbent upon the accused to show either (a) it was not practicable; *or* (b) it was not reasonably practicable to do more than was, in fact, done; *or* (c) there was no better practicable means than were, in fact, used to satisfy the statutory requirement.

If the accused cannot discharge this duty, the case will be regarded as proved against him. In this connection the effect of an approved code of practice may be crucial.

Effect of an approved code of practice

A provision in an approved code of practice is not in itself a statutory duty and so cannot of itself be enforced by HSE inspectors. Nor of itself can it give rise to civil liability (HSWA, sec 17(1)). Nevertheless, failure on the part of a person to whom its provisions apply to comply with it can have serious practical consequences, since it brings into operation a presumption that he was in breach of the statutory requirement(s). This applies only in the case of criminal liability (HSWA, sec 17(2)).

Principal offences and penalties

Health and safety offences are of three kinds:

(a) summary offences (i.e. triable by magistrates);

(b) offences triable summarily and on indictment;

(c) offences triable only on indictment.

By far the greatest number of health and safety offences fall into categories (a) and (b).

Summary offences and offences triable either way

The principal health and safety offences which are triable (a) summarily or (b) either summarily or on indictment are the following:

(a) Failure to carry out one or more of the General Duties of HSWA, secs 2–7.

(b) Contravening either of the following sections –

  (i) HSWA, sec 8 – intentionally or recklessly interfering with anything provided for safety;

  (ii) HSWA, sec 9 – levying payment for anything which an employer must by law provide in the interests of health and safety (i.e. personal protective clothing).

(c) Contravening any health and safety regulations.

(d) Contravening a requirement imposed under HSWA, sec 14, or under regulations passed pursuant to it (power of HSC to order an investigation).

(e) Contravening a requirement imposed by an inspector.

(f) Preventing or attempting to prevent a person from appearing before an inspector (e.g. a safety representative), or from answering his questions.

(g) Intentionally obstructing an inspector.

(h) Intentionally or recklessly making false statements, where the statement is made –

(i) to comply with a requirement to furnish information; or

(ii) to obtain the issue of a document.

(i) Intentionally making a false entry in a register, book, notice, etc., which is required to be kept.

(j) Falsely pretending to be an inspector (HSWA, sec 33(1)).

Summary trial or trial on indictment

Many of these offences are triable either way, i.e. summarily before the magistrates, or on indictment before the Crown Court. The offences mentioned in (*d*) to (*g*) as well as (*i*) are, however, only triable summarily. And in the case of (*e*) it is only an offence consisting of contravening a requirement imposed by an inspector under section 20 which is triable summarily; contravention of a requirement imposed under section 25, i.e. power to seize and destroy dangerous articles and substances, is triable either way.

Offences triable either way are usually tried summarily before the magistrates. The Offshore Safety Act 1992 amended the HSWA and allows magistrates to impose fines of up to £20,000 per offence for breaches of sections 2 to 6 of HSWA, and for breach of an improvement notice or prohibition notice, and of £5000 for other offences. Where, following a conviction, a person continues to contravene an improvement notice or a prohibition notice, he is liable to a maximum fine of £500 for every day the offence or contravention continues. (This Act also widened the range of health and safety offences for which the higher courts can impose prison sentences.)

However, offenders could be tried on indictment, for instance, when the magistrates feel that the case is too serious for them to try and, accordingly, requires the imposition of a penalty greater than they can impose, under their powers under HSWA. This could well be the situation where the commission of the offence involved serious threat to life, or where the offender was a persistent one, or the offence(s) revealed flagrant disregard of health and safety precautions. If such a case goes to the Crown Court, that court can impose an unlimited fine. (HSWA, sec 33(3)(*b*)(ii))

Offences triable on indictment only

The following are more serious offences and are triable on indictment only:

(a) contravening any of the 'relevant statutory provisions' by doing something unlicensed which requires a licence from HSE;

(b) contravening a term or condition or restriction attached to such licence;

(c) acquiring or attempting to acquire, possessing or using an explosive article or substance in contravention of the 'relevant statutory provisions';

(d) contravening a requirement or prohibition imposed by a prohibition notice under section 33(1)(*g*);

(e) wrongful use or disclosure of information, contrary to section 33(1)(*j*) (HSWA, sec 33(4)).

Penalties for indictable offences

Offences under section 33(4) are punishable on conviction in the Crown Court by:

(a) a maximum of 2 years' imprisonment; or

(b) an unlimited fine; or

(c) up to 2 years' imprisonment and an unlimited fine (HSWA, sec 33(3)(*b*)(i)).

As regards other indictable offences (i.e. those triable either way (*see above*)), the Crown Court can impose an unlimited fine (HSWA, sec 33(3)(*b*)(ii)).

Appendix: the 'relevant statutory provisions'

Explosives Act 1875	The whole Act except sec 30 to 32, 80 and 116 to 121
Alkali, etc., Works Regulation Act 1906	The whole Act
Anthrax Prevention Act 1919	The whole Act
Employment of Women, Young Persons and Children Act 1920	The whole Act
Celluloid and Cinematograph Film Act 1922	The whole Act
Explosives Act 1923	The whole Act
Petroleum (Consolidation) Act 1928	The whole Act
Hours of Employment (Conventions) Act 1936	The whole Act except sec 5
Petroleum (Transfer of Licences) Act 1936	The whole Act
Hydrogen Cyanide (Fumigation) Act 1937	The whole Act
Ministry of Fuel and Power Act 1945	Section 1(1) so far as it relates to maintaining and improving the

	safety, health and welfare of persons employed in or about mines and quarries in Great Britain
Coal Industry Nationalisation Act 1946	Sections 42(1) and (2)
Radioactive Substances Act 1948	Section 5(1)(*a*)
Fireworks Act 1951	Sections 4 and 7
Agriculture (Poisonous Substances) Act 1952	The whole Act
Emergency Laws (Miscellaneous Provisions) Act 1953	Section 3
Agriculture (Safety, Health and Welfare Provisions) Act 1956	The whole Act
Factories Act 1961	The whole Act except sec 135
Public Health Act 1961	Sec 73
Offices, Shops and Railway Premises Act 1963	The whole Act
Nuclear Installations Act 1965	Sections 1, 3 to 6, 22 and 24, Schedule 2
Mines Management Act 1971	The whole Act
Employment Medical Advisory Service Act 1972	The whole Act except secs 1 and 6 and Schedule 1

A

Appendix: provisions of the Health and Safety at Work etc., Act 1974

What powers have Health and Safety Inspectors?

- to enter premises; make investigations; take samples/photos; ask questions, etc. (Sec 20)
- to serve *Improvement Notices* on a person:
  - where there is breach of Regulations;
  - minimum time limit of 21 days;
  - if Employer appeals to Industrial Tribunal, Notice void, pending Appeal. (Sec 21)
- to serve *Prohibition Notices* on persons controlling activities:
  - where there is risk of serious personal injury;
  - with immediate effect (or deferred where necessary);
  - if Employer appeals, Notice stays on pending Appeal. (Sec 22)
- to *seize and destroy articles/substances* that may cause serious personal injury (Section 25)
- to give *information to Safety Representatives* under Sec 28(8):
  (a) factual information on hazards, etc.
  (b) what he's asked the employer to do, e.g. Notices.

What are the penalties for breaking the Act?

■ at magistrates' courts	– maximum fine of £20,000 per contravention
■ at jury trials (indictment)	– unlimited fine and/or
	– up to 2 years in prison for some offences

listed in Section 33(4)

- after conviction, continuing contravention of an Improvement or Prohibition Notice, a maximum of £500 for every day that the offence is continued or the contravention continues.

Does the Act apply to Crown property?

Yes – Part 1, including General Duties 'binds the Crown'. (Sec 48.1)

Can the Act be enforced against the Crown?

No – at least not through the Courts, nor through enforcing Improvement and Prohibition Notices (Section 48.1)

Can employees of the Crown be prosecuted?

Yes – Sec 48(2), 36(2) and Sec 37 allow prosecution against Crown employees where they are at fault.

Appendix: summary of features of 'notices'

Notice	Circumstances: Contravention of a 'relevant statutory provision'	Circumstances: Risk involved	When notice takes effect	Effect of appeal (section 24*)	Person on whom notice is served
Improvement (section 21)	Must have been one and it is likely that it will be continued or repeated	No risk specified and covers cases where there is no risk	When specified but not earlier than 21 days after issue	Suspends the notice until appeal is determined or the appeal is withdrawn	Person contravening provision
Immediate Prohibition Notice (section 22)	Not necessary	Where there is or will be immediate risk of serious personal injury†	Immediate	No suspension unless the Industrial Tribunal rules otherwise	Person under whose control the activity is carried on or by whom it is carried on
Deferred Prohibition Notice (section 22)	Not necessary	Risk of serious personal injury not imminent	At the end of the period specified in the notice	As in 'immediate notice'	As in 'immediate notice'

* Appeals against notices are dealt with in regulations made under section 24 as follows:
The Industrial Tribunals (Improvement and Prohibition Notices Appeals) Regulations 1974, (SI 1974 No 1925) – for notices served in England and Wales and The Industrial Tribunals (Improvement and Prohibition Notices Appeals) (Scotland) Regulations 1974, (SI 1974 No 1926) – for notices served in Scotland.
† 'Personal injury' includes any disease and any impairment of a person's physical or mental condition (section 53)

Fig. 4.1 ◆ Specimen Improvement Notice under the Health and Safety at Work Act

Health and Safety at Work, etc, Act 1974 Sections 21, 23 and 24

IMPROVEMENT NOTICE

Name and address (See Section 46)	To____________________
(a) Delete as necessary	(a) Trading as____________________
(b) Inspector's full name	I (b)____________________
(c) Inspector's official designation	one of (c) ____________________
(d) Official address	of (d) ____________________ __________ Telephone Coventry 83 33 33 __________ hereby give you notice that I am of the opinion that at
(e) Location of premises or place and activity	(e)____________________ you, as (a) an employer/a self-employed person/a person wholly or partly in control of the premises, or:
(f) Other specified activity	(f) ____________________ (a) are contravening/have contravened in circumstances that make it likely that the contravention will continue or be repeated/or:
(g) Provisions contravened	(g)____________________ The reasons for my said opinion are: ____________________ and I hereby require you to remedy the said contraventions, or as the case may be, the matters occasioned by them by: (h)__________ (a) in the manner stated in the attached schedule which forms part of this notice. Signature____________________Date__________ Being an Inspector appointed by an instrument in writing made pursuant to Section 19 of the said Act and entitled to issue this notice. (a) An Improvement Notice is also being served on__________ of____________________
P.S. 95059	related to the matters contained in this notice.

Fig. 4.2 ◆ Specimen Prohibition Notice under the Health and Safety at Work Act

Health and Safety at Work, etc. Act 1974, Sections 22, 23 and 24

PROHIBITION NOTICE

Name and address (See Section 46)	To____________________
(a) Delete as necessary	(a) Trading as____________________
(b) Inspector's full name	I (b)____________________
(c) Inspector's official designation	one of (c)____________________
(d) Official address	of (d)____________________
	__________Telephone Coventry 83 33 33__________
	hereby give you notice that I am of the opinion that the following activities, namely:____________________
(e) Location of activity	which are (a) being carried out by you/about to be carried out by you/under your control at (e)____________________
	involve, or will involve (a) a risk/an imminent risk, of serious personal injury. I am further of the opinion that the said matters involve contravention of the following statutory provisions:____________________
	because____________________
	and I hereby direct that the said activities shall not be carried on by you or under your control (a) immediately/after
(f) Date	(f)____________________
	unless the said contraventions and matters included in the schedule, which forms part of this notice, have been remedied.
	Signature____________________Date__________
	being an Inspector appointed by an instrument in writing made pursuant to Section 19 of the said Act and entitled to issue this notice.

P.S. 95058

5 Health and safety and industrial relations law

Need for disciplinary procedure

Established, mutually agreed, in-house disciplinary procedure and practice are essential. This should contain a clearly identifiable appeals procedure. Employees should be permitted to give an explanation of their conduct and management should carry out a thorough investigation of matters leading to the alleged offence (*Henry* v. *Vauxhall Motors Ltd*, COIT No. 664/85, Case No. 25290/77). Moreover, in addition to an in-house procedure, embodying the ACAS Code, a clearly communicable formal system should be instituted. Thus the contract of employment and supporting job description must identify requirements and responsibilities of all employees and management. This can be done by cross-referencing to job description in the contract. Moreover, the company health and safety policy statement should specify the statutory duties under HSWA and other legislation placed on employees and management.

Relevance of company Health and Safety Policy Statement to the contract of employment

Statutory duties under HSWA and other legislation, as well as Regulations made under HSWA and other legislation, should be itemised in the company Health and Safety Policy Statement – itself a requirement of HSWA, sec 2(3). Such duties would then qualify as contractual terms, breach of which could lead to dismissal, though not necessarily instant dismissal, since not all breaches would amount to gross misconduct leading to instant dismissal. The Employment Protection (Consolidation) Act 1978 (ECPA), sec 1(4), requires that the written particulars of terms of employment, which all full-time employees are entitled to receive by the thirteenth week of employment, include reference to disciplinary rules, procedures, etc., or a document specifying such rules, procedures etc. However, sec 1(5) expressly provides that these requirements do not apply to rules, disciplinary decisions and procedures in connection with health and safety at work.

Works rules

A

Not all works rules will be based on or referable to statutory duties and/or regulations. This may well be the position in respect of self-evident risks taken at work, e.g. skylarking, smoking, drinking, etc., which are not in themselves specific health and safety offences. Most companies wishing to specify in their works rules the 'penalties' for such conduct would do well to bear in mind, however, that, unless such conduct:

(a) constitutes a serious risk of injury to the employee or fellow employees; or

(b) breaches specific health and safety duties and/or regulations; or

(c) is persistent or repeated in contravention of earlier warnings,

it is not likely to be regarded as an implied term of an employment contract but rather as a reasonable order or instruction which an employer is entitled to give to an employee. The practical consequence is, however, probably the same as if it constituted a term of an employment contract. Therefore, an employee who disobeyed such an instruction would be in breach of his general duty to co-operate with his employer and so, after institution of the necessary procedures laid down in the ACAS Code, liable to be dismissed (*see Secretary of State* v. *ASLEF* [1972] 2 AER 949 mentioned in Chapter 3).

General offences

Smoking

There are certain regulations which prohibit smoking either completely or in certain circumstances. Breach of such regulations could lead to instant dismissal, assuming that the employee knew that he was committing an offence and was aware of the risk involved. The specific regulations prohibiting smoking at work are:

(a) Explosives Act 1875, sec 10.
(b) Celluloid (Manufacture, etc.) Regulations 1921, Reg 6.
(c) Manufacture of Cinematograph Film Regulations 1928, Reg 10.
(d) Magnesium (Grinding of Castings and Other Articles) Special Regulations 1946, Reg 13.
(e) Pottery (Health and Welfare) Special Regulations 1950, Reg 14.
(f) Factories (Testing of Aircraft Engines and Accessories) Special Regulations 1952, Reg 20.
(g) Petroleum Spirit (Conveyance by Road) Regulations 1957.
(h) Highly Flammable Liquids and Liquefied Petroleum Gases Regulations 1972, Reg 14.
(i) Organic Peroxides (Conveyance by Road) Regulations 1973.
(j) Control of Lead at Work Regulations 1980, Reg 10.

In other cases smoking in itself, even though a risk may be involved (unless, of course, the risk of danger is high, as, say, in the case of a fire risk), would not normally be enough to justify instant dismissal. In *Bendall* v. *Paine & Betteridge* [1973] IRLR 44 an employee who had smoked for 15 years was told periodically not to do so, as there was a fire risk. One day he was summarily dismissed for smoking. His dismissal was held to be unfair (under the provisions of the Industrial Relations Act 1971, the Act then in force governing unfair dismissal). The dismissal was unfair because previously, although the employee had been warned about smoking, he was not made aware that he was risking instant dismissal. The applicant should have been given a final warning in writing advising him of the likelihood of dismissal, before he was actually dismissed.

Drinking

Like smoking, drinking in itself is probably not sufficient to justify instant dismissal (*McGibbon* v. *Gillespie Building Co. Ltd* [1973] IRLR 105), but if drunkenness constitutes a safety hazard, then the necessary disciplinary procedures should be instituted immediately. If this attracts instant dismissal, the employee should be made aware of this on entering employment, the penalties being spelt out to him in the works rules and the contract of employment. This was made clear in *Abercrombie* v. *Alexander Thomson & Son* [1973] IRLR 326, where the applicant was found drunk whilst operating a crane. Management decided to dismiss him, but this was not done until two weeks later. In the interim period the applicant was permitted to carry on working. It was held that the delay in dismissing him made the dismissal unfair.

Other 'dismissible' offences

Other conduct which could give rise to dismissal, instant or otherwise, is:

(a) sleeping while employed;

(b) vandalism of company property;

(c) theft of company property;

(d) failure to attend in-plant or residential training courses;

(e) failure to observe hygiene standards in the food industry, e.g. smoking or failing to wash regularly.

General duty on employer to maintain discipline

Because of his status as an employer, he is vicariously liable for negligent injury caused by an employee to a co-employee or to a member of an outside workforce or the general public (*see* Chapter 2). Nor can he contract out of this liability. For this reason employers must, through line management, ensure that employees are protected from the negligent conduct of other employees. This gives the employer the right (indeed, imposes upon him the duty) to dismiss, either after due warnings, or instantly in the case of gross misconduct, an employee who constitutes a danger to co-employees. 'If a fellow workman . . . by his habitual conduct is likely to prove a source of danger to his fellow employees, a duty lies fairly and squarely on the employer to remove the source of danger' (per Streatfield J in *Hudson* v. *Ridge Manufacturing Co. Ltd* [1957] 2 AER 229).

Gross misconduct causing health and safety risks

Gross misconduct, involving health and safety risks, is 'punishable' with instant dismissal. The offending employee is not entitled to a warning, oral or in writing, nor does the employer have to abide by the ACAS procedures. What constitutes gross misconduct will normally be specified, for the benefit of employees, in the works rules, the company health and safety policy statement as well as in individual contracts of employment. Normally, however, only the following matters would be construed as serious enough to constitute gross misconduct:

(a) Serious and/or flagrant disregard of or disobedience to health and safety duties, regulations and rules, involving risk of injury to life or property damage. Thus in *Ashworth* v. *Needham & Sons Ltd* 1977COIT No. 681/78, Case No. 20161/77, an employee who erected a collapsible fence around a cellar, instead of steel or concrete plates, was held to have been fairly dismissed.
(b) Persistent or repeated health and safety offences, or hygiene offences.
(c) Failure to observe safety procedures on the part of a person engaged on an intrinsically dangerous operation. The categories of such occupations are not exhaustive but the following would come within this category: an airline pilot (*Taylor* v. *Alidair Ltd* [1978] IRLR 82 per Lord Denning MR), a nuclear scientist, the driver of a petrol or chemical container – in all these cases, with one split-second mistake a disaster could ensue.

Law relating to unfair dismissal

The law relating to unfair dismissal was introduced by the Industrial Relations Act 1971 and is currently contained in the Employment Protection (Consolidation) Act 1978 (EPCA), secs 54–81, as amended by the Employment Acts 1980 and 1982. Under this Act an employee can present a claim to an industrial tribunal, and if the complaint is upheld the employer will be ordered to compensate the ex-employee for loss of earnings, or alternatively reinstate or re-engage him.

The following conditions govern applications for unfair dismissal. The applicant is only entitled to present a complaint if:

(a) he worked for the employer as an employee and not as an independent contractor;
(b) he has either been dismissed by the employer (actual dismissal) or has resigned in circumstances amounting to constructive dismissal;
(c) he has in general worked for the employer for at least two years (as from 1 June 1985); however, if an employee is dismissed on medical grounds in compliance with any law or regulation or code of practice providing for health and safety at work (e.g. Control of Lead at Work Regulations 1980), the qualifying period for presenting a claim is only one month (EPCA, sec 64(2); Employment Act 1982, 2 Sch 5).

Actual and constructive dismissal

Dismissal is of two kinds: actual and constructive.

Actual dismissal

Here the employer terminates the contract of employment.

Constructive dismissal

Here the employee voluntarily resigns because of the employer's conduct in breaking a fundamental term of the contract of employment.

Reason for dismissal

In order to determine whether a dismissal was fair or unfair, the employer must identify the reason (or the main reason) for the dismissal and at the same time establish that the reason was one of those cited in EPCA, sec 57(2), or some other 'substantial' reason justifying the dismissal (EPCA, sec 57(1)). For these purposes, reasons that might well be relevant to health and safety at work are those:

(a) related to the capability or qualifications of the employee; or

(b) related to the conduct of the employee (EPCA, sec 57(2)).

Once an employer has established the reason for dismissal, it must then be shown that he acted reasonably in treating the reason as sufficient cause for dismissing the employee. This is a matter for the tribunal to decide. Here the tribunal must have regard to the size and administrative resources of the employer's undertaking when determining the reasonableness of the dismissal (EPCA, sec 57(3), and the Employment Act 1980, sec 6). Arguably, a small concern would be treated more sympathetically than a larger one with greater management resources to draw on.

Relevance of ACAS Code

In order to determine whether an employer has acted reasonably in dismissing an employee, regard should be had to the question whether the employer complied with the provisions of the ACAS Code relating to disciplinary procedure and practice, except in cases of gross misconduct, where instant dismissal is justified (*see above*). Failure to follow procedures outlined in the ACAS Code means, in all probability, that the employer would be construed as having acted unreasonably.

Duty of employer to consult with safety representatives

It is the duty of every unionised employer to consult with union-appointed safety representatives so as to make and maintain arrangements which will enable the employer and his workforce to co-operate in both promoting and developing health and safety at work measures and procedures, and monitoring their effectiveness (HSWA, sec 2(6)). Moreover, every such employer must, if requested to do so by two or more safety representatives, establish a safety committee whose purpose is to monitor health and safety at work (HSWA, sec 2(7), and the Safety Representatives and Safety Committees Regulations 1977, Reg 9(1) (SRSCR)).

Definition of a safety representative

A safety representative is an employee appointed by his trade union to represent the workforce in consultations with the employer on all matters affecting health and safety at work, and to carry out periodic inspections of the workplace for hazards. A safety representative appointed by an employer to fulfil the same functions is not a safety representative for present purposes and would not enjoy the powers and immunities given by the SRSCR. In short, he would not be recognised by law.

Role of safety representatives

Safety representatives exist in order to:

(a) represent their fellow employees at the workplace;
(b) carry out workplace investigations and inspections;
(c) handle information (e.g. information which an inspector might have disclosed in furtherance of his duty under HSWA, sec 28(8) (*see below*)).

None of the functions of a safety representative confers legal duties upon him (SRSCR, Reg 4(1)).

Disclosure of information to safety representatives by employer

Employers are required to disclose to safety representatives information necessary in order to enable them to carry out their functions (SRSCR, Reg 7(2)). The full list of information is to be found in paragraph 6 of the Approved Code of Practice on Safety Representatives and Safety Committees, and an employer who failed to comply with this code would be deemed to be in breach of the law (*see* Chapter 4). The sort of information referred to is technical information and consultants' reports, information concerning the employer's future plans and any proposed changes insofar as they might affect health, safety and welfare at work.

Disclosure of information to safety representatives by an inspector

HSE inspectors are required by HSWA, sec 28(8), to supply safety representatives with technical information obtained during their visits (e.g. measurements, testing the results of sampling and monitoring), notices of prosecution, copies of correspondence and copies of any improvement notice or prohibition order served upon the employer. An inspector has no discretion in this matter; it is mandatory. During discussions with safety representatives an inspector must inform them what action he proposes to take as a result of his visit to the workplace.

Employer's duty to consult and provide facilities and assistance

Reg 17 of and the Schedule to the Management of Health and Safety at Work Regulations 1992 (MHSWR) modified the Safety Representatives and Safety Committees Regulations 1977 specifying circumstances and situations where every employer shall consult with safety representatives as follows:

1. Without prejudice to the generality of sec 2(6) of HSWA, every employer shall consult safety representatives in good time with regard to –

   (a) the introduction of any measure at the workplace which may substantially affect the health and safety of the employees the safety representatives concerned represent;

   (b) his arrangements for appointing or, as the case may be, nominating persons in accordance with Reg 6(1) of the MHSWR 1992;

   (c) any health and safety information he is required to provide to the employees and safety representatives concerned represent by or under the relevant statutory provisions;

   (d) the planning and organisation of any health and safety training he is required to provide to the employees the safety representatives concerned represent by or under the relevant statutory provisions;

(e) the health and safety consequences for the employees the safety representatives concerned represent of the introduction (including the planning thereof) of new technologies into the workplace.

2. Without prejudice to Regs 5 and 6 of these Regulations, every employer shall provide such facilities and assistance as safety representatives may reasonably require for the purpose of carrying out their functions under sec 2(4) of the 1974 Act and under these Regulations.

Safety committees

In prescribed cases every employer must establish a safety committee, if requested to do so by two or more safety representatives (SRSCR, Reg 9(1)). The purpose of the safety committee is to monitor health and safety at work measures and procedures (HSWA, sec 2(7)).

When setting up a safety committee the employer must:

(a) consult with both –
   (i) the safety representatives who made the request;
   (ii) the representatives of a recognised trade union or unions whose members work in any workplace where it is proposed that the committee will function;

(b) post a notice, stating the composition of the committee and the workplace or workplaces to be covered by it, in a place where it can be easily read by employees;

(c) establish the committee within three months after the request for it was made (SRSCR, Reg 9(2)).

Health and Safety (Consultation with Employees) Regulations 1996

These Regulations came into operation on 1 October 1996 and brought in changes to the law with regard to the health and safety consultation process between employers and employees.

Under the Safety Representatives and Safety Committees Regulations 1997, employers must consult safety representatives appointed by any trade unions they recognise.

Under the Health and Safety (Consultation with Employees) Regulations 1996 employers must consult any employees who are not covered by the Safety Representatives and Safety Committees Regulations. This may be by direct consultation with employees or through representatives elected by the employees they are to represent.

HSE Guidance

HSE Guidance accompanying the Regulations details:

(a) which employees must be involved;

(b) the information they must be provided with;

(c) procedures for the election of representatives of employee safety;

(d) the training, time off and facilities they must be provided with;

(e) their functions in office.

Running a safety committee

As with any committee, it is essential that the constitution of a safety committee be in written form. The following aspects should be considered when establishing and running a safety committee.

Objectives

To monitor and review the general working arrangements for health and safety.

To act as a focus for joint consultation between employer and employees in the prevention of accidents, incidents and occupational ill-health.

Composition

The composition of the committee should be determined by local management, but should normally include equal representation of management and employees, ensuring all functional groups are represented.

Other persons may be co-opted to attend specific meetings e.g. health and safety adviser, company engineer.

Election of committee members

The following officers should be elected for a period of one year:

the Chairman

the Deputy Chairman

the Secretary

Nominations for these posts should be submitted by a committee member to the Secretary for inclusion in the agenda of the final meeting in each yearly period. Members elected to office may be re-nominated or re-elected to serve for further terms.

Election should be by ballot and should take place at the last meeting in each yearly period.

Frequency of meetings

Meetings should be held on a quarterly basis or according to local needs. In exceptional circumstances, extraordinary meetings may be held by agreement of the Chairman.

Agenda and minutes

The agenda should be circulated to all members at least one week before each Committee meeting. The agenda should include:

1. **Apologies for absence**
   Members unable to attend a meeting should notify the Secretary and make arrangements for a deputy to attend on their behalf.
2. **Minutes of the previous meeting**
   Minutes of the meeting should be circulated as widely as possible and without delay. All members of the Committee, senior managers, supervisors and trade union representatives should receive personal copies. Additional copies should be posted on notice boards.
3. **Matters arising**
   The minutes of each meeting should incorporate an Action Column in which persons identified as having future action to take, as a result of the Committee's decisions, are named.

   The named person should submit a written report to the Secretary, which should be read out at the meeting and included in the minutes.
4. **New items**
   Items for inclusion in the agenda should be submitted to the Secretary in writing, at least seven days before the meeting. The person requesting the item for inclusion in the agenda should state in writing what action has already been taken through the normal channels of

communication. The Chairman will not normally accept items that have not been pursued through the normal channels of communication prior to submission to the Secretary.

Items requested for inclusion after the publication of the agenda should be dealt with, at the discretion of the Chairman, as emergency items.

5. Safety Adviser's report

The Safety Adviser should submit a written report to the Committee, copies of which should be issued to each member at least two days prior to the meeting and attached to the minutes.

The Safety Adviser's report should include, for example:

(a) a description of all reportable injuries, diseases and dangerous occurrences that have occurred since the last meeting, together with details of remedial action taken;

(b) details of any new health and safety legislation directly or indirectly affecting the organisation, together with details of any action that may be necessary;

(c) information on the outcome of any safety monitoring activities undertaken during the month e.g. safety inspections of specific areas;

(d) any other matters which, in the opinion of the Secretary and himself, need a decision from the Committee.

6. Date, time and place of the next meeting

Women and young people

The Employment Act 1989 amends the legal requirements on a range of issues, including the removal of many of the restrictions on the employment of women and young people for reasons of health and safety.

The Employment Act 1989 narrows the exemption which makes an act of sex discrimination lawful if it is necessary to comply with a statutory requirement, but retains the exemption where the act is necessary for the special protection of women because of their childbearing capacity. It removes most of the restrictions on the employment of women and young persons, except those which are justifiable on health and safety grounds.

HSE leaflet IND(G)83(L) 'Working hours' describes the change from the prescriptive regulatory regime, which dates back to the early nineteenth century exploitation of women and young persons, to the present relaxed position where, as the leaflet says, 'in general, hours of work are a matter for agreement between employers and employees or their representatives'. Generally, the law does not specify the hours which people over school-leaving age can work. However, the following points need consideration with regard to the employment of young persons:

(a) exceptions exist for drivers of goods and passenger vehicles;
(b) shop assistants are entitled to meal breaks and a half-day off each week;
(c) it is illegal to employ children under the age of 13 years in any kind of work in an industrial activity or undertaking;
(d) some children can be employed under strict limitations as set out in sec 18 of the Children and Young Persons Act 1933 and in local authority by-laws generally administered by Local Education Authorities;
(e) employers have a general duty under HSWA to ensure the safety, health and welfare of employees so far as is reasonably practicable; this duty implies no excessive hours or unsuitable shift patterns likely to lead to accidents or ill-health, together with reasonable rest periods;
(f) adequate supervision, refreshment facilities, days off and training arrangements.

Working Time Regulations 1998

These regulations implemented the EC Directive on Working Practices or 'Working Time' Directive and apply to all areas of employment with the exception of air, rail, road, sea, inland waterway and lake transport, sea fishing, other work at sea and the activities of doctors in training.

Fundamentally, the regulations apply to atypical workers, that is those not in normal daytime employment, together with shift workers, part-time workers and night workers.

Principal requirements

These requirements:

(a) entitle workers to a minimum daily rest period of 11 consecutive hours in each 24-hour period;
(b) provide for workers to receive a rest break where the working day is longer than six hours, the duration of which is to be determined by 'collective agreements or agreements between the two sides of industry or, failing that, by national legislation';

(c) require workers to receive a minimum weekly rest period of 35 hours in each 7-day period, although this period may be reduced to 24 hours 'if objective, technical or work organisation considerations so justify';
(d) limit the average working time during each 7-day period to 48 hours;
(e) require that workers receive four weeks annual paid leave (Article 7);
(f) restrict the normal hours of work of night workers to an average of 8 hours in any 24-hour period, and impose an absolute limit of 8 hours where the work involves special hazards or heavy physical or mental strain;
(g) state that night workers should be entitled to a free health assessment, which may be provided through the national health system, prior to their assignment and at regular intervals thereafter;
(h) that night workers, suffering from ill-health recognised as being connected with the fact that they perform such work are, whenever possible, transferred to day work to which they are suited;
(i) state that night and shift workers must be provided with safety and health protection arrangements that are equivalent to those applying to other workers, and that are available at all times;
(j) require employers to take account of the general principles of adapting work to the worker, with a view in particular to alleviating monotonous work and work at a predetermined work rate, depending upon the type of activity, and of safety and health requirements, especially as regards to breaks during working time.

Children (Protection at Work) Regulations 1998

These regulations amend the Children and Young Persons Acts 1933 and 1963 in order to implement, with regard to children, the EC Directive on the protection of Young People at Work (94/33/EC). The principal features of these regulations are:

(a) the minimum age at which a child may be employed in any work, other than as an employee of his parent or guardian in light agricultural or horticultural work on an occasional basis, is 14 years;
(b) anything other than light work is prohibited; 'light work' is work which does not jeopardise a child's safety, health, development, attendance at school or participation in work experience;

(c) the employment of children over the age of 13 years in categories of light work specified in local authority byelaws is permitted;

(d) the hours which a child over the age of 14 years may work, and the rest periods which are required, are specified; a child must have at least one two-week period in his school holidays free from any employment;

(e) the 1933 Act is amended to extend the prohibition against a child going abroad for the purposes of performing for profit without a local authority licence, to further cover a child going abroad for the purpose of taking part in sport or working as a model in circumstances where payment is made;

(f) where children take part in public performances, existing requirements for a local authority licence are extended to require such a licence to be obtained before a child may take part in sport or work as a model in circumstances where payment is made either to the child or to someone

Head protection on construction sites

The Employment Act 1989 exempts Sikhs from the requirements of the Construction (Head Protection) Regulations 1989 to wear safety helmets on construction sites when wearing a turban, whether at work or otherwise. Associated requirements related to or connected with the wearing, provision or maintenance of safety helmets imposed on any other person, such as the employer, no longer apply under the Act.

Current trends in health and safety legislation

1. All modern health and safety legislation is largely driven by European Directives. For instance:

   (a) the Directive 'on the health and safety of workers at work' was implemented in the UK as the Management of Health and Safety at Work Regulations 1992; and

   (b) the Directive 'on temporary and mobile construction sites' was implemented in the UK as the Construction (Design and Management) Regulations 1994.

2. Regulations produced since 1992 do not, in most cases, stand on their own. They must be read in conjunction with the general duties imposed on employers under the Management of Health and Safety at Work Regulations 1992, in particular the duties relating to:

(a) risk assessment;

(b) the operation and maintenance of safety management systems;

(c) the appointment of competent persons;

(d) establishment and implementation of emergency procedures;

(e) provision of information to employees which is comprehensible and relevant;

(f) co-operation, communication and co-ordination between employers in shared workplaces, e.g. construction sites, office blocks;

(g) provision of comprehensible health and safety information to employees from an outside undertaking;

(h) assessment of human capability prior to allocating tasks;

(i) provision of health and safety information, training and instruction.

3. Duties imposed on employers tend to be largely of an absolute nature, as opposed to qualified duties, such as 'so far as is reasonably practicable', as with the Health and Safety at Work etc., Act 1974.

4. Risk assessment, taking into account the 'relevant statutory provisions', is the starting point of all health and safety management systems.

5. Most modern health and safety legislation requires some form of documentation, such as risk assessments and planned preventive maintenance systems, and the maintenance of records.

SECTION B

Health and safety management

6 Safety management and policy

B

Introduction

Under the Management of Health and Safety at Work Regulations 1999 (MHSWR), considerable emphasis must be placed by organisations on the actual management of health and safety procedures and systems. The HSE publication *Successful Health and Safety Management* (HS(G)65) provides guidance on this issue, including the need for effective health and safety policies, organising for health and safety, planning for health and safety, measuring performance and auditing and reviewing performance.

Management principles

What is management? One definition is: *The effective use of resources in the pursuit of organisational goals*. 'Effective' implies achieving a balance between the risk of being in business and the cost of eliminating or reducing those risks.

Management entails leadership, authority and co-ordination of resources, together with:

(a) planning and organisation;
(b) co-ordination and control;
(c) communication;
(d) selection and placement of subordinates;
(e) training and development of subordinates;
(f) accountability;
(g) responsibility.

Management resources include:

(a) people;
(b) land and buildings; and
(c) capital; together with
(d) time;
(e) management skills in co-ordinating the use of resources.

Health and safety management

Health and safety management is no different from other forms of management. It covers:

(a) the management of the health and safety operation at national and local level – planning, organising, controlling, objective setting, establishing accountability and the setting of policy;
(b) measurement of health and safety performance on the part of individuals and specific locations;
(c) motivating managers to improve standards of health and safety performance.

Decision-making is a very important feature of the management process (*see* Fig 6.1)

Management performance

Management is concerned with people at all levels of the organisation and human behaviour, in particular personal factors such as attitude, perception, motivation, personality, learning and training. Communication brings these various behavioural factors together.

The management of health and safety is also concerned with organisational structures, the climate for change within an organisation, individual roles within the organisation and the problem of stress which takes many forms. Several questions must be asked:

(a) Are managers good at managing health and safety?
(b) What is their attitude to health and safety? Is it of a pro-active or reactive nature?
(c) How do their attitudes affect the development of health and safety systems?

(d) Is health and safety a management tool, or a problem thrust on to them by the enforcement agencies?

Fig. 6.1 ◆ Decision-making in the management process

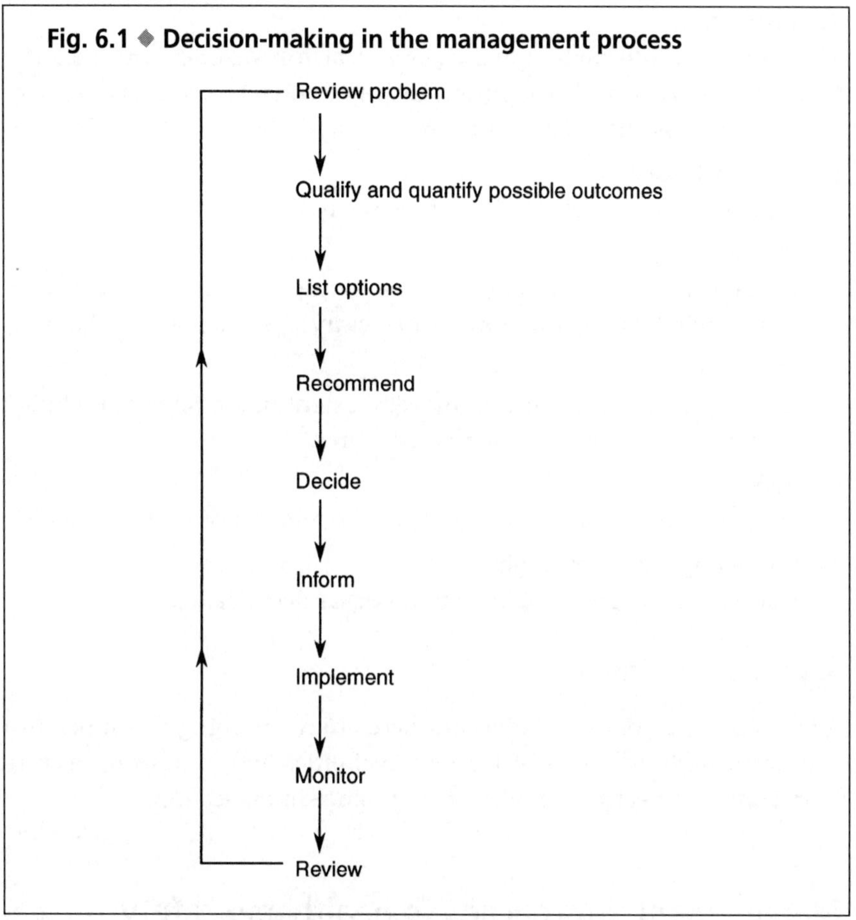

B

The role of the supervisor

The supervisor has a significant role in the management health and safety. But what are the most important features of an effective supervisor? These can be summarised thus:

1. **Introduction**
   Getting to know the employees in his charge and to be known by them.
2. **Instruction**
   Passing on information and theory in a clear manner with regard to safe systems of work, the correct use of personal protective equipment, accident reporting procedures, etc.
3. **Demonstration**
   Actually showing how a job is done safely.
4. **Practice**
   Making reasonable allowance for employees to become proficient in tasks, including any precautions necessary to ensure safe working.
5. **Monitoring**
   Observing and measuring employees' extent of proficiency, including compliance with formal safety procedures.
6. **Reporting**
   Making a fair evaluation of employees' performance for management.
7. **Correcting and encouraging**
   Correcting and encouraging employees as necessary.

Supervisor training

All the above factors should be considered in the training of a supervisor, and particularly in terms of his responsibilities and duties for ensuring sound levels of health and safety performance in his section.

Management approaches to health and safety

Approaches to health and safety vary considerably at all levels of management. They can be categorised thus:

1. **Legalistic** – 'Comply with the law, but no more!'
2. **Socio-humanitarian** – Considering the human resources aspects; people are important.
3. **Financial-economic** – All accidents and ill health cost money. Most organisations are good at calculating the costs of health and safety, e.g. machinery guarding, health and safety training, etc. but are not good at calculating the losses associated with accidents, sickness and

generally poor health and safety performance. These losses tend to get absorbed in the operating costs of the business.

With the greater emphasis on the human factors approach to health and safety, there is a need to actually identify those organisational characteristics which influence safety-related behaviour. These include:

(a) the need to produce a positive climate in which health and safety is seen by both management and employees as being fundamental to the organisation's day-to-day operations, i.e. they must create a positive safety culture;
(b) the need to ensure that policies and systems which are devised for the control or risk from the organisation's operations take proper account of human capabilities and fallibilities;
(c) commitment to the achievement of progressively higher standards which is shown at the top of the organisation and cascaded through successive levels of same;
(d) demonstration by senior management of their active involvement, thereby galvanising managers throughout the organisation into action;
(e) leadership, whereby an environment is created which encourages safe behaviour.

Health and safety management in practice

Whilst health and safety management covers many areas, there are a number of aspects which are significant, namely:

(a) the company Statement of Health and Safety Policy;
(b) procedures for health and safety monitoring and performance measurement;
(c) clear identification of the objectives and standards which must be measurable;
(d) the system for improving knowledge, attitudes and motivation and for increasing individual awareness of health and safety issues, responsibilities and accountabilities;
(e) procedures for eliminating potential hazards from plant, machinery, substances and working practices through the design and operation of safe systems of work and other forms of hazard control;
(f) measures taken by management to ensure legal compliance.

Key elements of successful health and safety management

The key elements of successful health and safety management are summarised in the diagram below (Fig. 6.2).

Fig. 6.2 ◆ Key elements of successful health and safety management

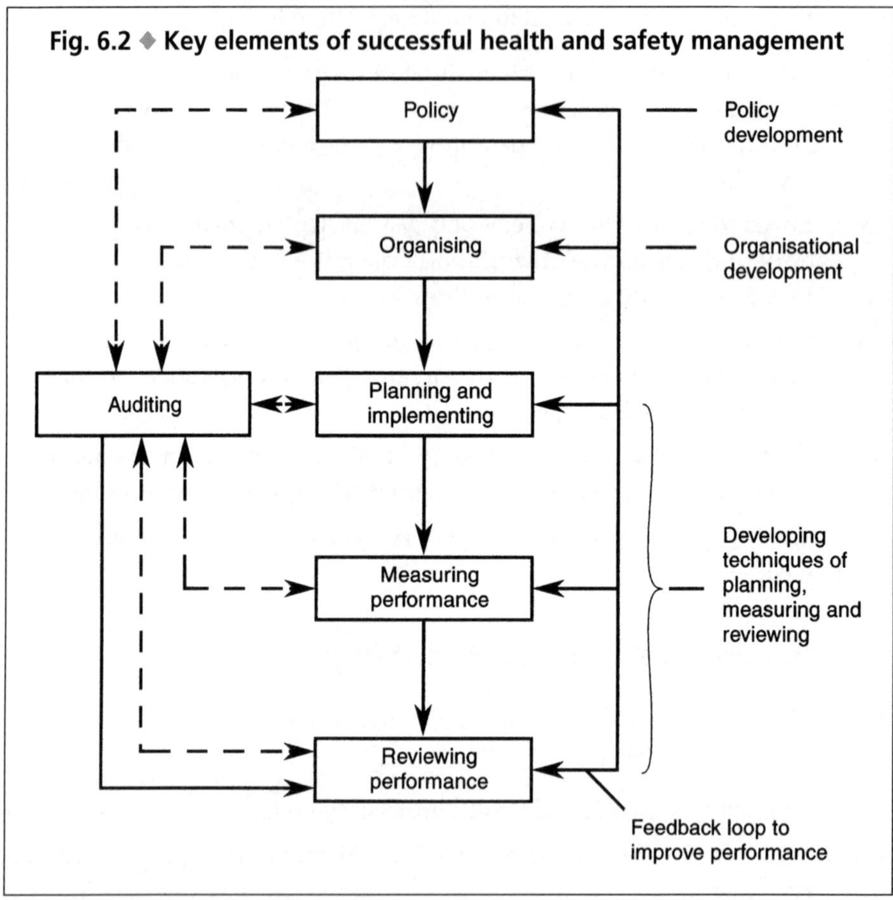

Quality management and health and safety

BS 8800: 1996 guide to occupational health and safety management systems

Introduction

BS 8800 offers an organisation the opportunity to review and revise its current occupational health and safety arrangements against a standard that has been developed by industry, commerce, insurers, regulators, trade unions and occupational health and safety practitioners.

The standard offers all the essential elements required to implement an effective occupational health and safety management system. It is equally applicable to small organisations as it is to large complex organisations.

The aims of the standard are to 'improve the occupational health and safety performance of organisations by providing guidance on how management of occupational health and safety may be integrated with the management of other aspects of the business performance in order to:

(a) minimise risk to employees and others;
(b) improve business performance;
(c) assist organisations to establish a responsible image in the workplace'.

B

The benefits

Whilst organisations recognise that they have a duty to manage health and safety, their senior managers may also believe the management of same is a financial burden that gives very little positive return, like VAT and PAYE. This belief frequently results in a lack of, or limited, commitment to health and safety by these managers.

Conversely, those organisations that do subscribe with commitment usually find enormous benefit. Apart from the obvious direct effect on staff morale, they find a positive contribution to the bottom line of their operational costs. The following benefits can be gained:

(a) improved commitment from staff (and positive support from trade unions);
(b) reduction in staff absenteeism;
(c) improved production output, through reductions in downtime from incidents;
(d) reduction in insurance premiums;
(e) improved customer confidence;
(f) reductions in claims against the organisation;
(g) a reduction in adverse publicity.

Achieving progress

The way forward for successful health and safety management is to involve everyone in the organisation, using a proactive approach to identify hazards and to control those risks that are not tolerable. This ensures that those employees at risk are aware of the risks that they face and of the need for the control measures.

In order to achieve positive benefits, health and safety management should be an integral feature of the undertaking contributing to the success of the organisation. It can be an effective vehicle for efficiency and effectiveness, encouraging employees to suggest improvements in working practices. In an ideal environment, health and safety is an agenda item alongside production, services, etc. at any senior management review of the undertaking, rather than an inconvenient add-on item.

Status review of the health and safety management system

In any review of an organisation's current health and safety management system, BS 8800 recommends the following four headings:

1. Requirements of relevant legislation dealing with occupational health and safety management issues.
2. Existing guidance on occupational health and safety management within the organisation.
3. Best practice and performance in the organisation's employment sector and other appropriate sectors (e.g. from relevant HSC's industry advisory committees and trade association guidelines).
4. Efficiency and effectiveness of existing resources devoted to occupational health and safety management.

Effective policies

BS 8800 identifies nine key areas that should be addressed in a policy, each of which allows visible objectives and targets to be set, thus:

1. Recognising that occupational health and safety is an integral part of its business performance.
2. Achieving a high level of occupational health and safety performance, with compliance to legal requirements as the minimum and continual cost effective improvement in performance.
3. Provision of adequate and appropriate resources to implement the policy.
4. The publishing and setting of occupational health and safety objectives, even if only by internal notification.
5. Placing the management of occupational health and safety as a prime responsibility of line management, from most senior executive to the first-line supervisory level.

6. Ensuring understanding, implementation and maintenance of the policy statement at all levels in the organisation.
7. Employee involvement and consultation to gain commitment to the policy and its implementation.
8. Periodic review of the policy, the management system and audit of compliance to policy.
9. Ensuring that employees at all levels receive appropriate training and are competent to carry out their duties and responsibilities.

The models

There are two recommended approaches depending on the organisational needs of the business and with the objective that such an approach will be integrated into the total management system.

BS 8800

BS 8800 offers two well-known approaches, one based on *Successful Health and Safety Management* HS(G)65 and one on ISO 14001.

ISO 14001 is compatible with the environmental standard

1. **An approach based on HS(G)65 – Successful Health and Safety Management**

2. **An approach based on ISO 14001: The Plan–Do–Check–Act cycle (Fig. 6.3).**

The structure of this approach is such that interfacing or integration with environmental management is relatively straightforward.

A pro-active approach to health and safety management

For any new system to succeed, there must be commitment at the highest management levels, with the plan for implementing such a system underwritten by the ruling body of the organisation, e.g. Board of Directors. Planning improvements in health and safety performance should take a proactive approach identifying the risks and the immediate, short,

Fig. 6.3 ◆ ISO 1400. The plan – Do – check – act cycle

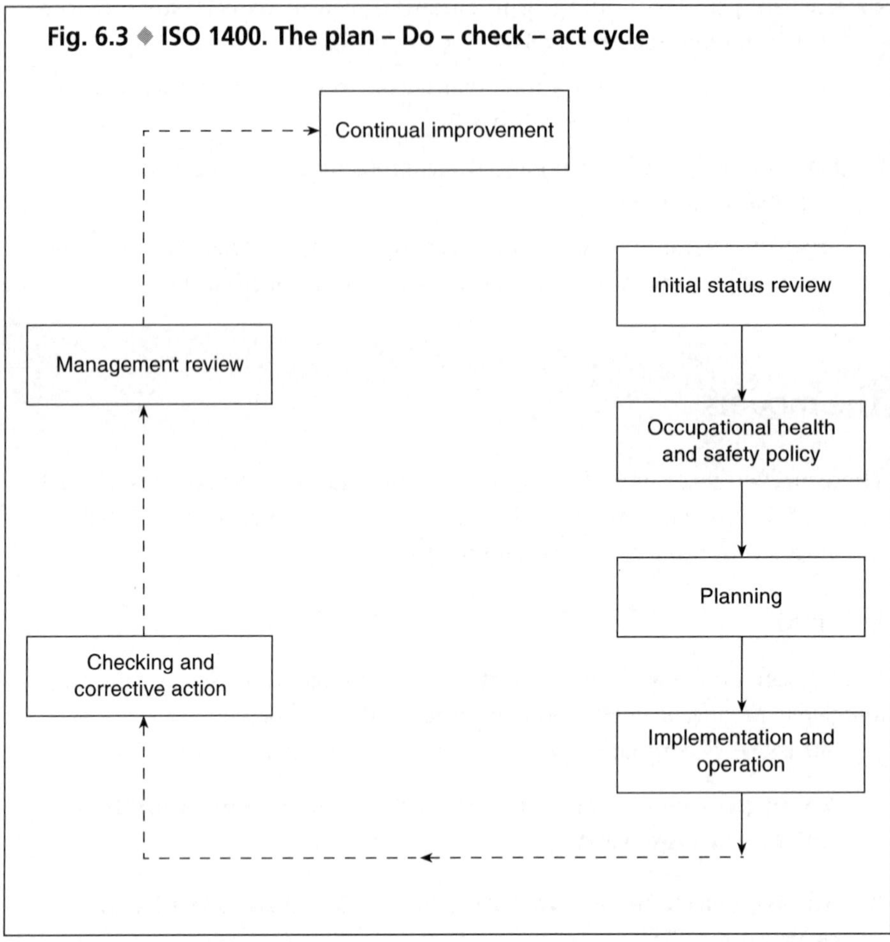

medium and long-term actions required, as opposed to the commonly encountered reactive approach based on the analysis of accident causes and the preparation of statistical information. Any proactive approach should:

(a) specify the actions necessary and the criteria for assessing satisfactory implementation of same;

(b) set deadlines for completion of the actions necessary;

(c) identify the persons responsible for ensuring these actions are taken.

The standard specifies a staged approach for developing and implementing a plan, incorporating key stages (see Fig. 6.4).

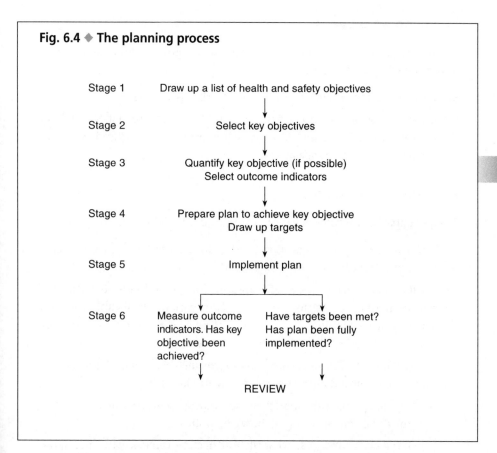

Fig. 6.4 ◆ The planning process

Stage 1	Draw up a list of health and safety objectives
	↓
Stage 2	Select key objectives
	↓
Stage 3	Quantify key objective (if possible) Select outcome indicators
	↓
Stage 4	Prepare plan to achieve key objective Draw up targets
	↓
Stage 5	Implement plan
	↓
Stage 6	Measure outcome indicators. Has key objective been achieved? ↓ / Have targets been met? Has plan been fully implemented? ↓
	REVIEW

The key objectives for Stage 1 are identified from the initial or status review and by undertaking risk assessments and identifying the legal requirements to which organisation subscribes. This stage identifies the long-term objectives of the organisation.

Key objectives dealing with the most significant risks should be identified. These objectives should be both measurable and achievable and, where possible, their implementation should involve employees directly in order to promote the programme and gain their commitment.

Note: Fig. 6.4 covers both the planning and implementation stages to indicate the complete process. Planning involves Stages 1 to 4.

Involving employees

In order to gain commitment to the process, employees can be involved in a number of ways, such as:

(a) training certain members to undertake risk assessments;

(b) undertaking safety tours of the workplace to identify aspects of non-conformance, e.g. failure to wear personal protective equipment or use specified protective measures.

Planning requirements

The principal requirements for planning under BS 8800 are:

(a) *Risk assessment:* the organisation should undertake risk assessments, including the identification of hazards.

(b) *Legal and other requirements:* the organisation should identify the legal requirements, in addition to the risk assessment, applicable to it and also any other requirements to which it subscribes which are applicable to occupational health and safety management.

(c) *Occupational health and safety management:* the organisation should make arrangements to cover the following key areas –

   (i) overall plans and objectives, including personnel and resources, for the organisation to achieve its policy;

   (ii) have or have access to sufficient occupational health and safety knowledge, skills and experience to manage its activities safely and in accordance with legal requirements;

   (iii) operational plans to implement arrangements to control risks identified and to meet the requirements identified;

   (iv) planning for operational control activities;

   (v) planning for performance measurement, corrective action, audits and management reviews;

   (vi) implementing corrective actions shown to be necessary.

Risk assessment

The MHSWR impose an absolute duty on an employer to undertake a 'suitable and sufficient assessment' of the risks to his employees and other persons who may be affected by the conduct by him of his undertaking for the purpose of identifying the measures he needs to take to comply with the requirements and prohibitions imposed by or under the relevant statutory provisions.

Having carried out the risk assessment, an employer must introduce effective systems to ensure the effective planning, organisation, control, monitoring and review of the preventive and protective measures arising from the risk assessment.

Employees should be involved in the risk assessment process as, in most cases, they will be aware of the hazards arising from work activities.

The risk assessment process

Risk assessment fundamentally takes place in a series of stages (Fig. 6.5).

Note: **Tolerable risk** means that the risk has been reduced to the lowest level that is reasonably practicable.

Fig. 6.5 ◆ Risk assessment

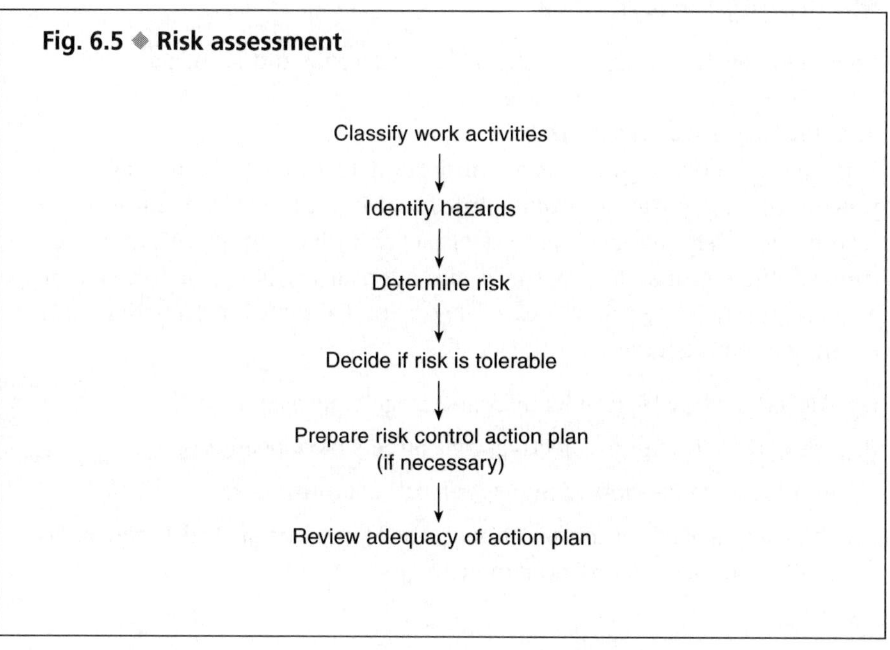

Effective implementation and operation

There must be total commitment from the board and senior management downwards for effective implementation and operation. Many organisations successfully ensure this by the appointment of a director or executive with responsibility for occupational health and safety and with particular responsibility for the regular review of performance across the organisation. This may be accompanied by a mission statement or other document outlining the commitment of the organisation to the continual improvement of health and safety.

In addition to ensuring that everyone recognises their individual and collective responsibilities, BS 8800 specifies a number of key issues which must be addressed:

(a) structure and responsibility;

(b) training, awareness and competence;

(c) communication;

(d) occupational health and safety management system documentation;

(e) document control;

(f) operational control;

(g) emergency preparedness and response.

Monitoring performance

This involves the key areas of checking, correcting and auditing.

1. Checking and correcting

Performance measurement is an important feature of the management process and a key way to provide information on the effectiveness of the health and safety management system. Qualitative and quantitative measures should be considered with the aim of ensuring that proactive systems are recognised as the prime means of control. Examples of proactive monitoring data include:

(a) the extent to which objectives and targets have been set;

(b) the extent to which objectives and targets have been met;

(c) employee perception of management's commitment;

(d) the adequacy of communication of the Statement of Health and Safety Policy and other 'core' documents;

(e) the extent of compliance with the 'relevant statutory provisions' and voluntary codes, standards and guidance;
(f) the number of risk assessments carried versus those actually required;
(g) the extent of compliance with risk controls;
(h) the time taken to implement actions on complaints or recommendations;
(i) the frequency of inspections, audits and other forms of monitoring.

BS 8800 identifies three other key areas to be addressed apart from monitoring and measurement, namely:

(a) corrective action;
(b) records;
(c) audit.

2. Auditing

There must be an effective audit system which entails a critical appraisal of all the elements of the management system. This is best undertaken through a form of ongoing audit programme. The process given in BS 8800 for establishing an audit programme is shown in Fig. 6.6.

The audit process leads to the following questions:

(a) Is the organisation's overall health and safety management system capable of achieving the required standards of health and safety performance?
(b) Is the organisation fulfilling all its obligations with regard to health and safety?
(c) What are the strengths and weaknesses of the health and safety management system?
(d) Is the organisation (or part of it) actually doing and achieving what it claims to do?

Health and safety audits should 'be conducted by persons who are competent and as independent as possible from the activity that is being audited, but may be drawn from within the organisation'. The auditing process should include:

(a) structured interviews being carried out with key personnel and others to determine that sound procedures are in place, understood and followed;

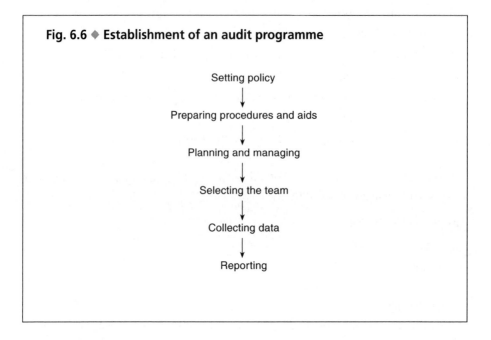

Fig. 6.6 ◆ Establishment of an audit programme

Setting policy
↓
Preparing procedures and aids
↓
Planning and managing
↓
Selecting the team
↓
Collecting data
↓
Reporting

(b) relevant documentation being examined, such as Statements of Health and Safety Policy, risk assessments, written instructions and procedures, health and safety manuals, etc.;

(c) inspections being undertaken to confirm that the documented procedures and any statements made;

(d) the analysis and interpretation of data such as accidents, near misses, etc.

Reviewing progress

A periodic management review is an essential component of the management system. It differs significantly from the audit but draws on the audit, amongst other indicators, to determine the robustness of the health and safety management system. BS 8800 suggests that reviews should consider:

(a) the overall performance of the occupational health and safety management system;

(b) the performance of the individual elements of the system;

(c) the findings of audits;

(d) internal and external factors, such as changes in organisational structure, legislation pending, the introduction of new technology, etc.

and identify what action is necessary to remedy any deficiencies.

Examples of information which should be considered

1. Changes or amendments to legislation.
2. The impact of emerging legislation, such as EC Directives.
3. Codes of Practice.
4. Inspections and/or warnings from equipment and material suppliers.
5. Proposed changes in the organisation.
6. Changes in equipment, plant, buildings, infrastructure, etc.
7. Benchmarking information from similar organisations.
8. Risk assessment reviews.
9. Reports from other management areas such as production, quality, etc.
10. Accident/incident statistic trends

Integration with other systems

1. Quality management systems

There may be some difficulty in integrating BS 8800 with the ISO series of management systems, e.g. ISO 9001/2/3. This is largely due to the fact that such systems are procedurally orientated, whereas BS 8800 and ISO 14001 are risk based. This risk-based approach requires the organisation to identify a series of organisational activities that need to be controlled and interlinked for effective management of those areas. Both BS 8800 and ISO 14001 have a dominant requirement that there is a continual improvement, whereas this is not a strong requirement of ISO 9001/2/3. Where it may be desirable to align current systems, the following areas should be considered:

1. **Policy generation**
   Initial review
2. **Planning**
   Management responsibility
   Resource allocation
3. **Implementation and operation**
   Operational control
   Document control

Records
Training, competence and awareness
Communication and reporting

4. **Measurement, and preventive and corrective action**
   Monitoring and evaluating performance
   Corrective action
   Internal audit
5. **Management review**

2. Environmental management systems

Integrating with ISO 14001 is relatively simple as both this system and BS 8800 share a common structure. It is vital that these two systems interface closely to ensure that there is no conflict in the decision-making process.

Establishing a safety culture – the principles involved

With the greater emphasis on health and safety management implied in the MHSWR, attention should be paid by managers to the establishment and development of the correct safety culture within the organisation.

Both the HSE and the CBI have provided guidance on this issue. The main principles involved, which involve the establishment of a safety culture, accepted and observed generally, are:

(a) the acceptance of responsibility at and from the top, exercised through a clear chain of command, seen to be actual and felt through the organisation;
(b) a conviction that high standards are achievable through proper management;
(c) setting and monitoring of relevant objectives/targets, based upon satisfactory internal information systems;
(d) systematic identification and assessment of hazards and the devising and exercise of preventive systems which are subject to audit and review; in such approaches, particular attention is given to the investigation of error;
(e) immediate rectification of deficiencies;
(f) promotion and reward of enthusiasm and good results.

Rimington, J.R. (1989), *The Onshore Safety Regime, HSE Director General's Submission to the Piper Alpha Inquiry*, December 1989.

Developing a safety culture – essential features

(Excerpt from *Developing a Safety Culture* – CBI (1991))
Several features can be identified from the study which are essential to a sound safety culture. A company wishing to improve its performance will need to judge its existing practices against them.

1. Leadership and commitment from the top which is genuine and visible. This is the most important feature.
2. Acceptance that it is a long-term strategy which requires sustained effort and interest.
3. A policy statement of high expectations and conveying a sense of optimism about what is possible supported by adequate codes of practice and safety standards.
4. Health and safety should be treated as other corporate aims, and properly resourced.
5. It must be a line management responsibility.
6. 'Ownership' of health and safety must permeate at all levels of the workforce. This requires employee involvement, training and communication.
7. Realistic and achievable targets should be set and performance measured against them.
8. Incidents should be thoroughly investigated.
9. Consistency of behaviour against agreed standards should be achieved by auditing and good safety behaviour should be a condition of employment.
10. Deficiencies revealed by an investigation or audit should be remedied promptly.
11. Management must receive adequate and up-to-date information to be able to assess performance.

The statement of health and safety policy

The Health and Safety at Work etc., Act 1974, requires employers to prepare and, as often as may be appropriate, revise their written statement of general policy with respect to the health and safety at work of their employees, together with the organisation and arrangements in force at the time for carrying out the policy, and to bring the statement, and any revisions, to the notice of all employees. However, employers employing fewer than five employees are excepted from this requirement (Employers' Health and Safety Policy Statements (Exception) Regulations 1975) (*see* Chapter 3). To this end the Health and Safety Executive (HSE) produced useful guidance on the format and utilisation of health and safety policy statements in industry and commerce. A policy should comprise three parts:

Part 1: **General Statement of Intent** (Specifying Objectives)
This part should outline the company's overall philosophy in relation to the management of health and safety.

Part 2: **Organisation** (People and their Duties)
This part should clearly show who is responsible to whom and for what – i.e. the chain of command. (A management structure diagram is of use in this regard.) Also, this part should demonstrate how accountabilities are fixed, how policy implementation is to be monitored, how safety committees and safety representatives are to function, and how individual job descriptions should reflect health and safety responsibilities and associated accountabilities.

Part 3: **Arrangements** (Systems and Procedures)
This part should detail the practical arrangements in force to assist in overall policy implementation. These include safety training, safety monitoring, accident reporting/investigation, safe systems of work, permit to work systems, noise/environmental control, utilisation of protective equipment, fire safety, liaison with contractors, machine guarding, emergency procedures and medical/welfare considerations. In practice, this part of the policy may be very lengthy and is often best presented in the form of a manual with appropriate cross-references made to it in Parts 1 and 2.

Each policy should be signed by the managing director/chief executive, and should be dated to facilitate revision. The policy should be brought to the attention of *all* employees, i.e. both existing and new, through training sessions, inserts in pay packets, display on notice boards, individual issue and by incorporation in employee handbooks. Any policy revision should also be brought to the attention of all employees.

The HSE publication *Effective Policies for Safety and Health* suggests that a policy statement should incorporate the following elements:

(a) A written safety policy which states the basic objectives and is supplemented by more detailed rules and procedures to cater for specific hazards.

(b) Definition of both the duties and the extent of responsibility of specified line management levels for safety and health, with identification made at the highest level of an individual with overall responsibility.

(c) Clear definition of the function of the safety officer and his relationship to line management made clear.

(d) The system for monitoring safety performance and publishing information about such performance.

(e) An identification and analysis of hazards, together with the precautions necessary on the part of staff, visitors, contractors, etc.

(f) An information system which will be sufficient to produce an identification of needs and can be used as an indicator of the effectiveness of the policy. (The amount of information required by such a system will depend on these needs, bearing in mind that the cost of obtaining the information should be realistically related to the expected benefit.)

(g) A training policy for all management and staff levels.

(h) A commitment to consultation on health and safety and to a positive form of worker involvement.

The HSE further suggests that it is not sufficient to publish a policy, however comprehensive it may be, unless the policy is translated into effective action at all levels within the organisation. Success, in terms of an acknowledged and continuing commitment at all management levels, is more likely where the policy is underwritten by the Board. They need also to make adequate financial provision for bringing the policy into effect. The requirement to monitor performance at a high level in the organisation is of particular importance. A successfully operated policy ought to result in:

(a) high standards of compliance with environmental requirements;

(b) high standards of compliance with the physical requirements, i.e. in regard to plant, machinery and process operations;

(c) carefully designed and observed safe systems of work;

d) high standards of cleanliness and housekeeping;
(e) maintenance of safe means of access and egress;
(f) control of toxic, corrosive and flammable substances, where such control is relevant;
(g) effective consultation for safety and positive participation by work people;
(h) training programmes which are well developed and run on a continuing basis for all levels.

The primary aim of the policy must be the motivation and involvement of the workforce. This is best achieved through a demonstrated commitment and concern on the part of management. In this context, first line supervision is of primary importance and more recognition should be given to its role. This means that particular attention should be paid to training in hazard identification and to the support of first line supervision in actions needed and taken. The needs of workers relate to protective equipment and devices, to training relative to their jobs and the hazards involved, and to job descriptions. The importance of their contributions can be stressed in terms of the reporting of defects, in inspection and audit of the workplace and in the correct use of rules and procedures relating to safe systems of work.

Health and safety organisation: policy implementation

In order to ensure effective policy implementation, the organisation of health and safety management within a company should:

(a) clearly and unambiguously show, in written and pictogram format, the unbroken and logical delegation of duties through line management operating where risks arise and accidents happen;
(b) identify key personnel, by title rather than name, to be held accountable to senior management for ensuring that detailed arrangements (systems and procedures) for safe working are developed, utilised and maintained;
(c) define the roles of both line and functional management by the use of specific job descriptions;

d) provide adequate support through relevant functional management, i.e. safety advisers, medical advisers, designers, chemists, engineers, etc;

(e) nominate competent persons to measure and monitor overall safety performance;

(f) provide the means/authority to deal with failures to meet the requirements of the policy;

(g) fix management's accountability for health and safety in a similar manner to other management functions;

(h) ensure that the organisation unambiguously indicates to each individual exactly what he must do to fulfil his role;

(i) ensure that the organisation makes it known, in terms of both time and money, what resources are available for health and safety. The individual must be certain of the extent to which he is actually supported by the policy and the organisation needed to fulfil it.

If the above elements are incorporated into the organisation framework of a company, then the efficiency by which the policy is implemented will be greatly enhanced.

Health and safety professionals

Occupational health and safety is a multi-disciplinary area of study, entailing an understanding of many subject areas, such as law, toxicology, human factors, engineering, risk management, occupational hygiene and fire protection, each of which is an area of study in its own right. Health and safety professionals, therefore, must meet a high level of academic attainment in order to be in a position to advise employers on these issues. Effective occupational safety and health management requires fully trained, competent and experienced professionals capable of taking on a range of responsibilities from risk assessment to advising on health and safety policy.

The Institution of Occupational Safety and Health (IOSH) is the principal professional institution in the United Kingdom for health and safety practitioners. As such, this Institution represents over 18,000 individual members working across the full spectrum of industry and commerce – from multinationals to small consultancies.

IOSH is the focus for important matters affecting the profession. It is consulted by government departments for the views of its members on draft legislation, approved codes of practice and guidance notes, and is represented on the committees of national and international standards-making bodies. In addition, IOSH periodically publishes policy statements on key topics and issues. Close relationships are maintained with organisations with common interests throughout the world.

The Institution works to promote excellence in the discipline and practice of occupational safety and health and helps its members working in this specialist sphere of work by:

(a) granting corporate membership as a mark of technical ad professional competence and the attainment of experience;
(b) providing facilities to maintain and enhance professional skills and knowledge.

Register of Safety Practitioners

IOSH'S Register of Safety Practitioners (RSP) provides a means of identifying and measuring practical ability in addition to professional occupational safety and health knowledge. It provides employers, clients and regulatory bodies with an accepted standard of competence and capability for general safety practitioners. Entrance to the Register is only open to Corporate Members of IOSH.

Qualifications and experience

Membership of IOSH is at both Corporate and Non-corporate level. Entrance at Corporate level depends upon a combination of academic qualifications, experience and achievement. It is open to those holding an appropriate qualification coupled with a minimum of three years' professional experience. Appropriate qualifications are:

(a) an accredited degree or diploma in occupational safety and health or a related discipline;
(b) the National Examination Board in Occupational Safety and Health (NEBOSH) Diploma; or
(c) Level 4 of the Vocational Qualifications for Occupational Health and Safety Practice.

People whose roles include some health and safety responsibilities, for example, assisting more highly qualified occupational safety and health professionals, or dealing with routine matters in low risk sectors, join IOSH at Associate level. An Associate must hold a NEBOSH Certificate or equivalent.

Affiliate level is designed for those who have an active interest in occupational safety and health, but who are not eligible to join at other grades of membership.

Joint consultation in health and safety

B

This section examines the complex role of management, workers and health and safety specialists in the development and subsequent implementation of health and safety policies through joint consultation.

In its report, the Robens Committee stated that real progress in the promotion of safety and health at work was impossible without the full co-operation and commitment of all employees. The committee felt that if workpeople were to accept their full share of responsibility, they must first be able to participate in the making and monitoring of arrangements for safety and health at work. These recommendations were incorporated in the HSWA more specifically secs 2(4), (6) and (7), and the SRSCR (*see* Chapter 5).

The principal aim, therefore, of joint consultation on, or participation in, health and safety matters is to involve those people at the 'sharp end' who are actually having the accidents and who are, therefore, well placed to identify risks and to assist in the prevention of further accidents. Hence line management, supervisors and workers should be actively encouraged to work together with health and safety specialists by participating in safety audits and safety committees. In this way, they will be given a say in developing a more realistic approach to accident prevention. More specifically, the HSC booklet *Safety Representatives and Safety Committees* gives excellent guidance on the interaction between the participants in the joint consultation process.

The role of the health and safety specialist in joint consultation is generally that of a catalyst. Primarily, his role is advisory, leaving executive decisions to be made by line management/safety committees, based on advice given to them. In fulfilling this role the health and safety specialist should work closely with management and workers and their representatives, with the object of ensuring a safe and healthy workplace in line with the organisation's health and safety policy.

Information, instruction and training requirements

There are extensive duties under current legislation, such as Health and Safety at Work Act, the Management of Health and Safety at Work Regulations and the Control of Substances Hazardous to Health (COSHH) Regulations 1994, for employers to provide their employees and others, such as the employees of contractors, with appropriate information, instruction and training.

Information and instructions

Information and instructions are frequently incorporated in an organisation's safety manuals and staff handbooks, which may incorporate the various rules for safe working.

Inadequate and/or ineffective codes of practice, rules and instructions to operators contribute to the causes of accidents and ill-health at work. In some cases, rules and instructions may be ambiguous, badly worded or simply not available to the people to whom they are directed.

Health and safety manuals

Codes of practice are ideally incorporated in a company Health and Safety Manual which provides guidance to managers, in particular, and staff on various matters, such as:

(a) accident reporting, recording and investigation procedures;

(b) hazardous substances in use;

(c) the selection and use of personal protective equipment.

These codes of practice should be seen as an extension of the organisation's Statement of Health and Safety Policy and should be referred to in same.

Company safety rules

Rules relating to safe working procedures and practices are commonly covered by a staff Health and Safety Handbook which is issued to all staff at the induction training stage. The handbook should incorporate, for example:

(a) basic 'DOs' and 'DO NOTs' with regard to safe working;

(b) details of safe systems of work;

(c) information on hazardous substances in use;

(d) an indication of individual responsibilities of all grades of staff;

(e) procedures for reporting accidents, incidents, near misses, ill-health and sickness;

(f) the role of the safety adviser and trade union/staff safety representative;

(g) fire precautions, including action to be taken in the event of fire;

(h) safe manual handling procedures.

It should also incorporate any other matters identified as the outcome of risk assessments and the various forms of safety monitoring.

Provision of information to employees

All employers have to provide such information, instruction, training and supervision as is necessary to ensure, so far as is reasonably practicable, the health and safety at work of their employees (HSWA, sec 2(2)(*c*)). In addition to this general duty there are specific Regulations, e.g. the SRSCR, Reg 7, the Health and Safety (First Aid) Regulations 1981, Reg 4, which require employers to provide relevant information to employees, together with the specific requirements of the Health and Safety Information for Employees Regulations 1989 (*see* Chapter 3). The MHSWR impose more specific requirements (*see* Chapter 3.).

Contracts of employment/employee handbooks

Although a contract of employment need not be in writing, all employees are entitled to receive, after 13 weeks of employment, details in writing of the main terms of their contract (EPCA, sec 1(4)(*a*)). Surprisingly, section 1(5) states that this statutory requirement does not include details relating to health and safety at work. It is, however, a common law requirement that employers take reasonable care for the health and safety of their employees and, therefore, such a term is implied in all contracts of employment (*Lister* v. *Romford Ice & Cold Storage Co. Ltd* [1957] 1 AER 125). To this end employee handbooks are a useful vehicle for disseminating information to the workforce relating to working conditions, health and safety requirements and Regulations as well as codes of practice. Moreover, induction training sessions will assist in this regard, as relevant information can be promptly brought to the attention of all new employees.

Measurement of safety performance

The measurement of safety performance is a complex evaluation and should take into account both positive (accident prevention) and negative (accident statistics) features. Positive features include aspects such as safety audits and inspections, risk identification, evaluation and control, loss control profiling, and fault-tree analysis (*see* Chapter 8). Negative features to be incorporated include data collected via the use of accident records (*see* Chapter 7), i.e. accident statistics. Accident statistics may be used to monitor the overall effectiveness of the implementation of a health and safety policy, but a clearer picture will be obtained if other indicators, such as the results of a safety audit, are used in conjunction with statistical data.

Within the UK a number of standard indices are used. These are:

(a) $\text{Frequency Rate} = \dfrac{\text{Total number of accidents}}{\text{Total number of man-hours worked}} \times 100{,}000$

(b) $\text{Incidence Rate} = \dfrac{\text{Total number of accidents}}{\text{Number of persons employed}} \times 1000$

(c) $\text{Severity Rate} = \dfrac{\text{Total number of days lost}}{\text{Total number of man-hours worked}} \times 1{,}000$

(d) $\text{Mean Duration Rate} = \dfrac{\text{Total number of days lost}}{\text{Total number of accidents}}$

(e) $\text{Duration Rate} = \dfrac{\text{Number of man-hours worked}}{\text{Total number of accidents}}$

For the statistics to be of use, the limitations of these indices should be realised. For example, the indices should not be used for comparing one factory or department with another, because of variables such as differing work patterns, differing degrees of risk and different levels of training. Ideally, the indices should only be used to compare the performance of the same factory/department over similar periods of time, i.e. to compare this year with last year.

Accident data – the problems with their use as a measure of performance

Studies by Amis and Booth (1991) questioned the significance and relevance of accident statistics as a measure of health and safety performance. Their conclusions were as follows:

B

1. They measure failure, not success.
2. They are difficult to use in staff appraisal.
3. They are subject to random fluctuations; there should not be enough accidents to carry out a statistical evaluation. Is safety fully controlled if, by chance, there are no accidents over a period?
4. They reflect the success, or otherwise, of safety measures taken some time ago. There is a time delay to judge the effectiveness of new measures.
5. They do not measure the incidence of occupational diseases where there is a prolonged latent period.
6. They measure injury severity, not necessarily the potential seriousness of the accident. Strictly, they do not even do this. Time off work as a result of injury may not correlate well with true injury severity. Data may be affected by known variations in the propensity of people to take time off work for sickness in different parts of the country.
7. They may under-report (or over-report) injuries, and may vary as a result of subtle differences in reporting criteria.
8. They are particularly limited for assessing the future risk of high consequence, low probability accidents. (A fatal accident rate based on data from single fatalities, may not be a good predictor of risk of multiple fatal emergencies.)

The crucial point is that counting numbers of accidents provides incomplete, untimely, and possibly misleading, answers to the questions:

- Are we implementing fully our safety plan?
- Is it the right plan?

Where management has not drawn up a safety plan, counting accidents is the only measure of safety performance available, apart, of course, from auditing compliance with statutory hardware requirements. This is a reason safety management via accident rate comparison is attractive to the less competent and committed employer.

Amis, R.H. and Booth, R.T. (1991) *Monitoring Health and Safety Management*, IOSH Conference Paper 1991.

Sources of information

Occupational health and safety covers an extremely broad field of knowledge – law, engineering, toxicology, behavioural science, ergonomics, management, etc. Sources of information range from Acts of Parliament to Hazard Data Sheets provided by the manufacturers and suppliers of hazardous substances, from EEC Directives to published British Standards. In order to keep abreast of the law and technology, the health and safety practitioner must be aware of the various sources of information, both general and specific, which are available. The most common sources of information are outlined below.

1. Acts of Parliament (Statutes)

Statutes outline the general principles to be applied in the field of occupational health, safety and welfare. An Act of Parliament commences its life as a Bill and follows a specific procedure in both the House of Commons and the House of Lords before receiving the Royal Assent, i.e. First Reading, Second Reading, Committee Stage, Report Stage and Third or Final Reading.

The principal statute is the Health and Safety at Work etc., Act 1974. Whilst a statute specifies general provisions it generally gives the Minister or Secretary or State power to make Regulations.

2. Regulations

Regulations give substance to the provisions detailed in statutes and contain the specific details. Draft Regulations may be submitted to the Secretary of State by the Health and Safety Commission. They may also be produced as a requirement of an EEC Directive. When the Regulations have passed through the consultation stage, for instance with the Health and Safety Commission and other interested parties, they are then laid on the table of the House. It is usual for Regulations to come into force after forty days unless a negative resolution is passed by either House.

At present there are over four hundred Regulations in force made under the relevant statutory provisions. Typical examples are the Health and Safety (First Aid) Regulations 1981, the Control of Substances Hazardous to

Health (COSHH) Regulations 1999 and the Construction (Health, Safety and Welfare) Regulations 1996. Regulations may be of general application, such as the Safety Signs Regulations 1980, or specific to certain industries or risks. In most cases, a breach of the Regulations implies a breach of the more general requirements outlined in the statute under which the Regulations are made.

3. Codes of practice

Increasingly codes of practice now accompany Regulations as a means of providing information and guidance on the practical interpretation of the same. Typical examples are the Approved Codes of Practice issued by the Health and Safety Commission 'Control of substances hazardous to health' and 'Control of carcinogenic substances' which accompany the COSHH Regulations. In this case the codes of practice are 'approved' by the Secretary of State following consultation with interested parties. However, not all codes of practice are approved in this way and there has been a tendency on the part of the regulatory authorities to limit the number of Approved Codes of Practice.

Codes of practice do not have legal status. However, in criminal proceedings they may be used as evidence of what is reasonably practicable.

4. Health and Safety Executive (HSE)

The Health and Safety Executive publishes a wide range of Guidance Notes under the following subject headings – Medical, Environmental Hygiene, Chemical Safety, Plant and Machinery, and General. An extremely helpful source of information, Guidance Notes, whilst having no legal status, give general and specific advice on practical implementation of statutory and other requirements. Typical examples are Guidance Note EH40 *Occupational exposure limits*, which should be read in conjunction with the COSHH Regulations and various Approved Codes of Practice accompanying the Regulations, and Guidance Note MS23 *Health aspects of job placement and rehabilitation*.

Memoranda of Guidance are also issued by the HSE, such as the *Memorandum of guidance on the Electricity at Work Regulations 1989*.

The HSE produces an annual Publications Catalogue, available from TSO, which lists all relevant health and safety literature by subject, with key word cross-referencing, by title of legislation, by title of guidance literature/reports, by title of approved factory forms and by film/video title. This organisation also produces a 'publications in series' list which identifies all series of publications produced by the HSC and the HSE, such as guidance notes, guidance booklets, codes of practice, occasional papers, research papers and toxicity reviews.

5. Annual Report of the Health and Safety Commission

This is largely a financial report to the Secretary of State for the particular year in question. However, it contains details of the year's activities and short addenda from the various Inspectorates and can be used to identify trends in legislation and enforcement priorities.

6. Annual Report of the Health and Safety Executive

This Report is divided into a number of sections dealing with separate areas on an industry basis. Comments are made about the performance of various industries during the period under review and the Report incorporates statistical information dealing with contraventions of health and safety legislation, enforcement action taken and fines imposed in the Courts. As with the HSC Annual Report, this Report gives a broad indication of trends in the development of health and safety enforcement.

7. European Directives

Increasingly European Directives will have a significant effect on UK health and safety legislation. Under the Treaty of Rome, the Council of the European Committees can issue Directives. Directives usually provide for harmonisation of the laws of Member States, including those covering health and safety. Representatives of the Member States meet to agree the contents of Draft Directives and, once agreed, the Drafts are presented for ratification to the European Parliament. Directives impose a duty on each Member State to:

(a) make Regulations to conform to the Directive;

(b) ensure the enforcement of such Regulations.

Stages in the development of an EC Directive are:

(a) formal proposal by the European Commission;

(b) negotiations in Council working groups, which take account of the views of the European Parliament;

(c) agreement of a 'common position' by the Council of Ministers;
(d) further negotiation;
(e) adoption by the Member States;
(f) implementation of the Directive, whereby individual Member States prepare Regulations, bring them into operation and arrange for enforcement.

8. Consultative documents

Such documents are issued by the HSC or HSE as part of the consultative procedure prior to recommending the framing of Regulations by the Minister or Secretary of State. As with draft Directives, the proposed legislation is set out and explanations given. Interested parties are invited to comment on the contents of the document by a specified date. Consultative documents give the best indication of future legislative intentions and standards.

9. Specific reports

Government departments frequently commission specific reports on matters of general concern or particularly following a disaster. The report on the investigation into the Clapham Junction railway accident, commissioned by the Secretary of State for Transport under the Regulation of Railway Act 1871, is a typical example. Similar reports of significance in health and safety include the reports into the accident at Markham Colliery, Derbyshire, in 1973 and that at Brent Cross in 1964.

Other government departments issue reports giving guidance on topics within their particular areas of activity, such as the Safety in Mines Research Establishment and the Building Research Establishment.

10. British Standards

The British Standards Institution produces Safety Standards and Codes. These standards cover a wide range of health and safety-related topics, such as BS 6842: 1987 *Guide to measurement and evaluation of human exposure to vibration transmitted to the hand* and BS 7018: 1988 *Guide to the use of electrical insulating materials containing asbestos*.

Standards and Codes of Practice are produced through committees formed to deal with a specific subject. Standards contain details relating to the construction and materials incorporated in an item, for instance, and, where necessary, prescribe methods of testing to establish compliance.

Codes, on the other hand, tend to deal with safe working practices and systems of work.

British Standards and Codes are of particular significance in the field of health and safety. Whilst they have no legal status, they can be interpreted by courts as being the authoritative guidance on a particular topic.

11. Hazard data sheets

These documents incorporate information relating to the safe use of potentially hazardous substances and are issued to users by the manufacturers or suppliers of such substances. They are an important form of information when undertaking health risk assessments required under the COSHH Regulations.

12. Books, periodicals, films and videos, microfilm systems, computer programs

There are many textbooks and monthly publications available which give varying degrees of information on safety, health and welfare. Computerised information systems have been developed in the last decade, together with microfilm or microfiche systems on a range of topics, particularly following the introduction of the COSHH Regulations. A wide range of textbooks and periodicals are held by public libraries.

13. Existing written information

This may take the form of Statements of Health and Safety Policy (*see* Chapter 3), specific company policies, e.g. on the use and storage of dangerous substances, current agreements with trade unions, company rules, regulations and codes of practice, and job safety instructions.

14. Interviews and discussions

These are a useful way of obtaining subjective information, perceptions, ideas and feelings, and for identifying informal roles and relationships, group norms, attitudes, levels of knowledge and skills; the Health and Safety Committee is a valuable information source.

15. Direct observation

This is the actual observation of work being carried out, which identifies interrelationships, hazards, dangerous practices and occurrences, and risk situations.

16. Work study techniques

Included here are the results of activity sampling, surveys, method study, work measurement and process flows.

17. Personal experience

People have their own experience of particular jobs and the hazards their jobs present. Individual experience of the operation of plant and machinery, systems of work and practices used is a valuable source of information. The personal experience of accident victims, particularly in the events leading to their particular accident, is a useful form of feedback in preventing further accidents of the same type.

18. Job descriptions

A job description should incorporate health and safety elements. It should take account of the physical and mental requirements and limitations of certain jobs and past modifications to the job description as a result of experience. Representations from operators and safety representatives on safety aspects of their jobs should always be taken into account.

19. Manufacturers' and suppliers' information and instructions

Information about the safe use of articles and substances used at work which is provided under HSWA, sec 6, and operating instructions for machinery and plant should be sufficiently comprehensive to enable a judgement to be made on their safe use.

20. Accident, illness and absence statistics

Statistical information on past accidents, cases of occupational ill-health and time lost as a result of same may identify unsatisfactory trends in operating procedures, unsafe systems of work, the use of hazardous substances or poor environmental conditions in the workplace. In many cases such statistics may identify areas of poor management supervision and control. These factors must be taken into account in the design of safe systems of work and in health risk assessments under the COSHH Regulations.

21. Task analysis

Information produced from the analysis of tasks, such as mental and physical requirements, manual operations, skills required, influences on behaviour, learning methods and specific hazards, must be taken into account as an important source of information, particularly when undertaking Job Safety Analysis, which is a development of Task Analysis (*see* Chapter 13 'Safe systems of work').

Sources of professional advice on health and safety issues

A wide range of sources of advice on health and safety issues is available to management and health and safety practitioners. These include:

1. HSE public enquiry points. There are two HSE public enquiry points based on the HSE's Library and Information Services at London, Sheffield and Bootle, Lancashire. Free leaflets and advice may be obtained from the Area Offices of the HSE.
2. Local authorities. Most local authority environmental health departments provide advice on health and safety issues in those matters which come within their jurisdiction.
3. Public utilities (gas, electricity and water authorities).
4. British Standards Institute.
5. Universities and colleges of further education.
6. National safety organisations e.g. Royal Society for the Prevention of Accidents (RoSPA) and the British Safety Council.
7. National Radiological Protection Board (NRPB).
8. Consultant organisations.

The provision of information

The sources of information on occupational health and safety are many and varied as indicated previously in this chapter. Most of the examples quoted are, however, of a textual or written nature.

The process of giving and receiving information, however, takes a particular pathway commencing with the information source through the information provider to the actual information format for the information user. The actual format can vary substantially, thus:

(a) *Textual* – books, letters, diagrams, charts, etc.
(b) *Verbal* – conversation face-to-face, lectures, etc.
(c) *Information technology* – computer printouts, on-line systems.
(d) *Visual* – photographs, films, videos.
(e) *Numerical* – statistical information, bar charts, pie charts.

Other means of information transfer can be of an auditory and tactile nature.

Documentation requirements

Under current health and safety legislation, considerable emphasis is placed on the documentation of health and safety procedures and systems.

The following are some of the more common documents which are required to be produced and maintained:

1. Statement of Health and Safety Policy (Health and Safety at Work etc., Act 1974).

2. Risk assessments in respect of:

   (a) workplaces (Management of Health and Safety at Work Regulations 1999 and Workplace (Health, Safety and Welfare) Regulations 1992);
   (b) work activities (Management of Health and Safety at Work Regulations 1999 and Workplace (Health, Safety and Welfare) Regulations 1992);
   (c) work equipment (Provision and Use of Work Equipment Regulations 1998);
   (d) personal protective equipment (Personal Protective Equipment at Work Regulations 1992);
   (e) manual handling operations (Manual Handling Operations Regulations 1992);
   (f) display screen equipment (Health and Safety (Display Screen Equipment) Regulations 1992);
   (g) substances hazardous to health (Control of Substances Hazardous to Health Regulations 1999);

   (h) noise levels in excess of 85 dBA (Noise at Work Regulations 1989).
3. Safe systems of work, including permits to work and method statements.
4. Pre-tender stage health and safety plan and construction phase health and safety plan (Construction (Design and Management) Regulations 1994).
5. Planned preventive maintenance schedules (Workplace (Health, Safety and Welfare), Regulations 1992 and Provision and Use of Work Equipment Regulations 1992.
6. Cleaning schedules (Workplace (Health, Safety and Welfare), Regulations 1992.
7. Written schemes of examination (Pressure Systems Safety Regulations 2000).

Information retrieval systems

Such systems provide information under a combination of formats, e.g. information technology, textual and visual. They all, however, incorporate the basic pathway mentioned above and shown in Fig. 6.7.

Safety budgets

The economic argument for health and safety is strongly based on the fact that accidents cost money, in terms of both insured and uninsured costs, as well as economic sanctions such as improvement notices, prohibition notices, prosecutions and fines in the courts. (A fuller discussion on the costs of accidents is presented in Chapter 11.) Moreover, the use of accident costs to facilitate budgeting for safety and accident prevention enables a degree of financial accountability to be built into the overall budgetary control system.

Indeed, the Robens Report suggested that better knowledge of accident costs could contribute towards more informed decision making. The report advocated the displaying of accident costs on the balance sheet, so encouraging management to apply the same effort and technique to accident reduction as is customarily applied to other facets of the business. Accident prevention then becomes a part of the standard economic activity of the company.

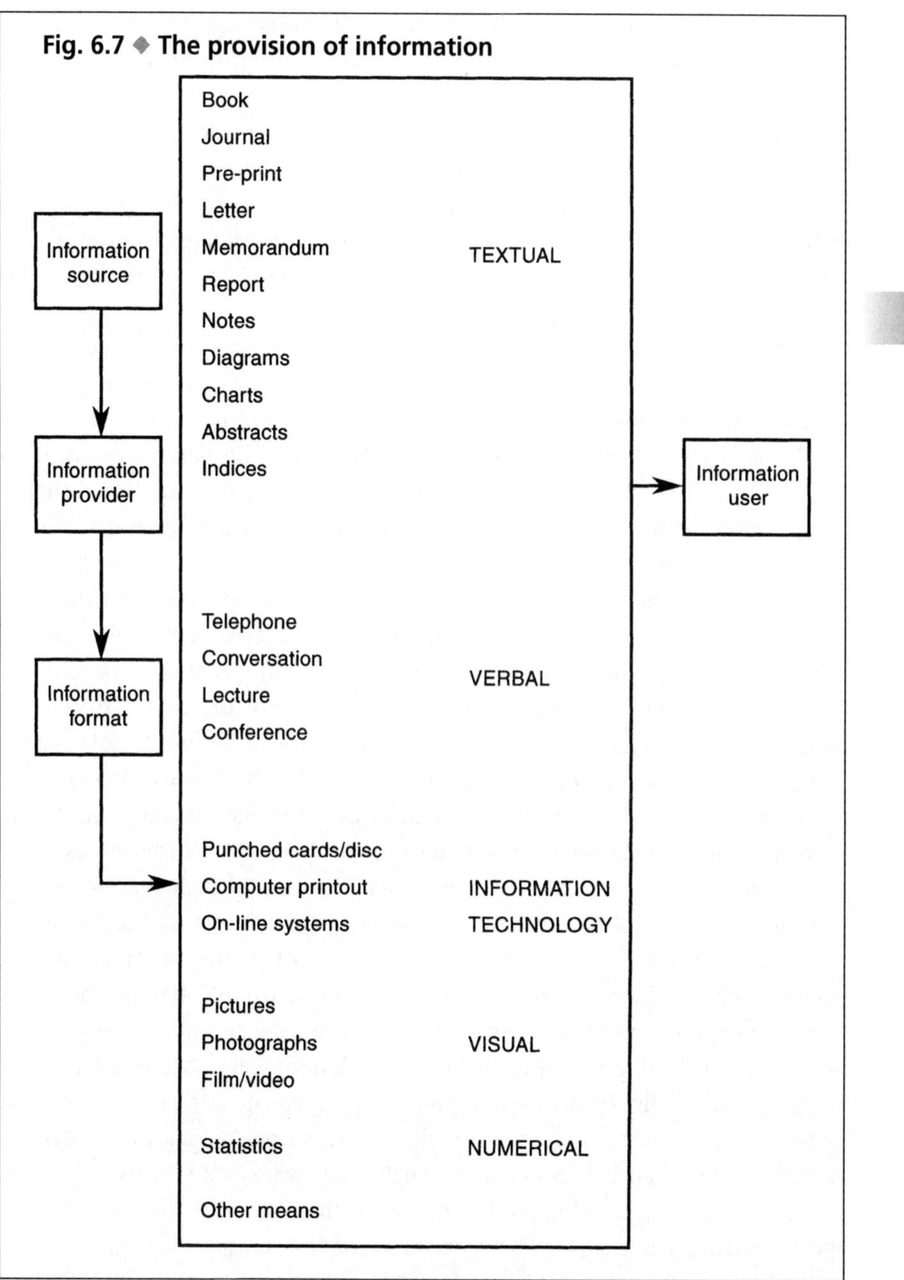

Fig. 6.7 ◆ The provision of information

The present system of accident prevention operating throughout most of industry does not attempt to make line management financially accountable for accidents and uninsured losses, and very little use is made of economic arguments in stimulating management interest in accident prevention. Any arguments put forward rely mainly on legal and humanitarian considerations which, in some instances, fail to convince management that there is a need for accident prevention, at least beyond compliance with statutory duties.

The use of budgetary control introduces economic accountability into the field of accident prevention. Such a measure involves the reorganisation of the existing accounting procedures in most companies in order to overcome the lack of accountability for accidents.

When an accident occurs within a factory department, the cost of the accident is usually absorbed into the running costs of the factory as a whole, and will not be itemised on the departmental balance sheet. Nor will many of the indirect costs be paid for from the departmental manager's budget. Furthermore, the direct cost, i.e. insurance premiums such as employer's liability, will generally be paid from a central fund, usually administered by Head Office.

However, when a safety adviser or factory inspector recommends safety measures such as guarding for machinery, the cost is usually charged against the departmental manager's budget, though it is very unlikely that it would be itemised as an accident prevention cost. Thus under accounting systems employed in many companies, accident costs are not charged to the departmental manager's budget, whereas accident prevention is charged. Hence the departmental manager has no economic motivation to undertake any accident prevention, rather the reverse.

A positive economic motivating factor for encouraging accident prevention may be introduced by transposing these items in the budgetary system. For each accident, injury or incidence of damage that occurs in a department, a charge is made against that department. Any accident prevention expenditure that is required within the department is financed from a central fund subject to approval by the risk manager or safety adviser. The result, as far as the departmental manager is concerned, is that it is costly to have accidents but not to prevent them. Thus line management becomes accountable for accidents occurring within their areas of control.

At the end of the financial year, a realistic allowance for accidents will be set within the new budget as a target of the manager to achieve. This allowance forms an integral part of the management plan as with budgetary control in other areas. However, the number, and hence the cost, of accidents budgeted for should be less than that of the previous year so that reduction of accidents becomes part of the management plan. This reorganised system helps to bring about the necessary economic accountability and makes full use of the knowledge and data obtained in establishing what accidents are costing the company in financial terms.

Once the costs of accidents have been established (*see* Chapter 11), the reorganised budgetary system can be implemented. The charges to be made against the departmental manager's budget can then be calculated and allocated on a monthly basis. The departmental manager receives a monthly report giving information on the costs of accidents and accident prevention expenditure. This enables him to plan any action necessary to maintain or improve the level of safety within his department. It also facilitates decision making in connection with the allocation of scarce accident prevention resources. Any deficiencies in the current safety programme are highlighted in cost terms rather than by a frequency rate, a measure of safety questioned by both management and academics!

On their own, the legal and humanitarian arguments for accident prevention may not be sufficient to achieve a reduction in accidents and other losses, e.g. damage to plant. The addition of economic accountability, through safety budgeting, greatly assists in the overall aim of minimising losses resulting from accidents.

Safety propaganda

Safety propaganda – publications, posters, slide/tape programmes, films, etc. – is utilised in both formal and informal training situations with the intention of modifying behaviour through attitude change. In order to decide which aspect of safety propaganda is best suited to the needs of the organisation, it is useful to note the following. People tend to remember:

10 per cent of what they read.

20 per cent of what they hear.

30 per cent of what they see.

50 per cent of what they see and hear.

70 per cent of what they say in conversation.

90 per cent of what they say as they do a thing.

It follows, therefore, that the relative usefulness of safety propaganda will depend to a large extent on the method of communication utilised. For example, people will tend to remember 50 per cent of a film, slide/tape programme or video as against 90 per cent if they describe a task as they do it under supervision.

Generally, safety propaganda may be communicated through a variety of media:

(a) Publications; programmed learning (*read*) – 10 per cent.
(b) Taped commentaries; lectures (*hear*) – 20 per cent.
(c) Slides; posters; overhead transparencies (*see*) – 30 per cent.
(d) Films; slide/tape programmes; videos (*see and hear*) – 50 per cent.
(e) Discussion groups (*say in conversation*) – 70 per cent.
(f) On the job training; simulation exercises; role playing (*say as they do a thing*) – 90 per cent.

Pirani and Reynolds (1976) examined the use of five different methods of communication, using different facets of safety propaganda designed to persuade employees to utilise protective equipment. Their results, in rank order form best (highest percentage improvement in utilisation) to worst (little or no improvement), were as follows:

(a) role playing;
(b) films;
(c) posters;
(d) discussion;
(e) discipline.

These results tend to confirm that, for safety propaganda to be effective, active participation is required in order to create the necessary changes in behaviour.

Safety signs in the workplace

The Safety Signs Regulations 1980 require that any sign displayed in the workplace must comply with the specifications of signs contained in BS 5378: Part 1: 1980 *Safety Signs and Colours – Specifications for Colour and Design*. A safety sign is defined as 'a sign that gives a message about health or safety by a combination of geometric form, safety colour and symbol or text, or both'. Apart from certain exceptions, e.g. those regulating road, rail or air traffic, coal mines, etc., the Regulations apply to any sign giving health and safety information or instruction to persons at work. The duty to comply with the Regulations rests with the employer, self-employed person, or the person having control of the place of work.

Classification of safety signs

There are four basic categories of safety signs, thus:

1. Prohibition signs

These signs indicate that certain behaviour is prohibited or must stop immediately, e.g. smoking in a 'No Smoking' area. The signs are recognised by a red circle with a cross bar running top left to bottom right on a white background. Any symbol is reproduced in black within the circle. The colour red (Red for Danger) implies STOP or DO NOT DO.

2. Warning signs

These are signs which give warning or notice of a hazard. The signs are black outlined triangles filled in by the safety colour – yellow. The symbol or text is in black. The combination of black and yellow identifies the need for CAUTION.

3. Mandatory signs

These signs indicate that a specific course of action is required, e.g. EYE PROTECTION MUST BE WORN. The safety colour is blue with the symbol or text in white. The sign is circular in shape.

4. Safe condition signs

These signs provide information about safe conditions. The signs are rectangular or square in shape, coloured green with white text or symbol. The safety colour green is associated with GO.

Altogether, there are 8 Prohibition, 19 Warning, 14 Mandatory and 7 Safe Condition symbols. Where additional information is required, supplementary text may be used in conjunction with the relevant symbol, provided that it is apart and does not interfere with the symbol. The text should be incorporated in an oblong or square box of the same colour as the sign with the text in the relevant contrasting colour, or white with black text.

Fire safety signs

Such signs are specified in BS 5499, which indicates the characteristics of signs for fire fighting equipment, fire precautions and means of escape in case of fire (cf BS 5499 Part 1: 1978 *Fire Safety Signs, Notices and Graphic Symbols*).

Table 6.1 ◆ Safety signs (BS 5378: Part 1: 1980 *Safety Signs and Colours – Specification for Colour and Design*)

Meaning or Purpose	*Safety Colour*	*Examples of Use*	*Contrasting Colour (if required)*	*Symbol*
Stop Prohibition	Red	STOP signs; prohibition signs; identification of emergency shutdown devices	White	Black
Caution Risk of danger	Yellow	Warning signs, e.g. electric current on, harmful vapours, obstacle ahead, scaffold incomplete, asbestos	Black	Black
Mandatory action	Blue	Obligation to wear personal protective equipment, e.g. eye protection; report damage immediately; keep out; switch off machine when not in use	White	White
Safe condition	Green	Identification of first aid posts, safety showers, fire exits	White	White

The Standard uses the basic framework concerning safety colours and design adopted by BS 5378 (*see* Table 6.1).

Danger identification

The danger identification sign consists of yellow and black diagonal stripes and is used to identify the perimeter of a hazard or hazardous area. Many sign manufacturers now produce a continuous ‘tiger striping’ sign in the form of adhesive tape which can be used in temporary situations where hazards may exist.

Examples of classified safety signs

Fig. 6.8 ◆ Safety signs

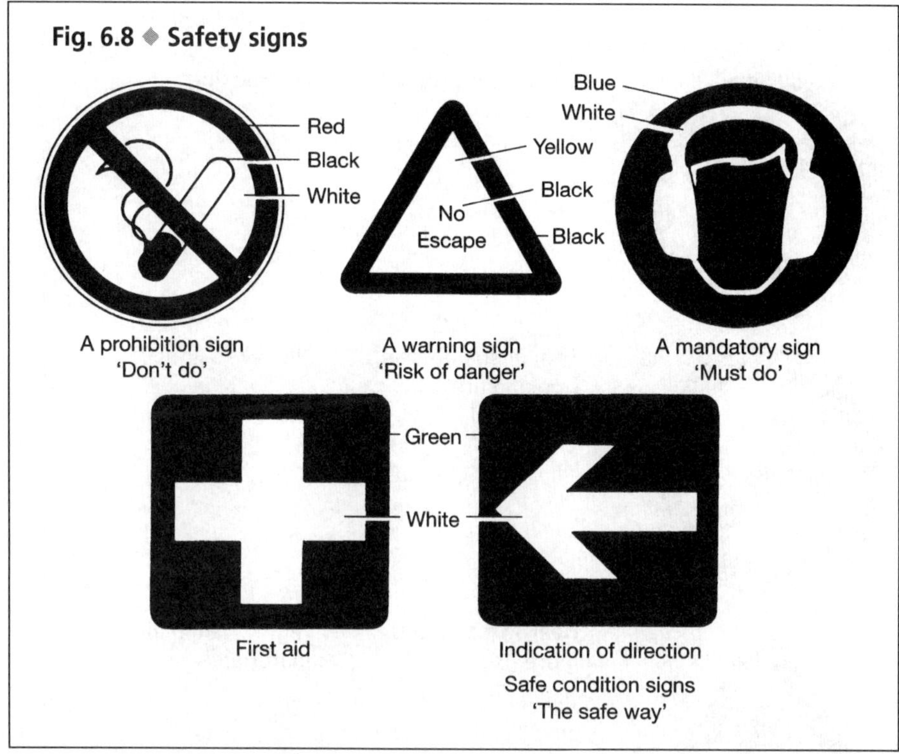

B

Health and Safety (Safety Signs and Signals) Regulations 1996

These regulations implement the European Safety Signs Directive 92/58/EEC on the provision and use of safety signs at work, and apply to all workplaces, including offshore installations.

The purpose of the Directive was to encourage the standardisation of safety signs throughout the member states so that safety signs, wherever they are seen, have the same meaning.

The regulations cover various means of communicating health and safety information, including the use of illuminated signs, hand and acoustic signals (e.g. fire alarms), spoken communication and the marking of pipework containing dangerous substances.

These requirements are in addition to traditional signboards such as prohibition and warning signs. Fire safety signs e.g. signs for fire exit and fire-fighting equipment are also covered. The signboards specified in the

Fig. 6.9 ◆ Fire safety signs

Symbol	*Meaning*	*Standard wording*
	Smoking is prohibited	No smoking
	Fire, open lights and smoking are prohibited	No naked light
	Nearby material is a fire risk	Highly flammable material
	Nearby material is liable to explode	Explosive material
		Fire exit or Emergency exit or Exit

Source: BS S499 Part 1 1978 *Fire Safety signs, notices and graphic symbols*

regulations are covered BS 5378: Parts 1 and 3: 1980 Safety signs and colours.

Principal requirements

Employers must use a safety sign where a risk cannot be adequately avoided or controlled by other means. Where a safety sign would not help to reduce that risk, or where the risk is not significant, a sign is not required. The regulations promote a general move towards symbol-based signs. They extend the term 'safety sign' to include hand signals, pipeline marking, acoustic signals and illuminated signs.

The number of safety symbols is increased, and new colour meanings are introduced. The regulations require, where necessary, the use of road traffic signs within workplaces to regulate road traffic. Employers are required to:

(a) maintain the safety signs which are provided by them;

(b) explain unfamiliar signs to their employees and tell them what they need to do when they see a safety sign.

The regulations apply to all places and activities where people are employed, but exclude signs and labels used in connection with the supply of substances, products and equipment or the transport of dangerous goods.

Employers are required to mark pipework containing dangerous substances, for example, by identifying and marking pipework at sampling and discharge points. The same symbols or pictograms need to be shown as those commonly seen on containers of dangerous substances, but using the triangular-shaped warning signs.

Although the regulations specify a code of hand signals for mechanical handling and directing vehicles, they permit other equivalent codes to be used such as BS 6736: 1986 Code of practice for hand signalling use in agricultural operations, and BS 7121: Part 1: 1989 Code of practice for safe use of cranes.

Dangerous locations, for example, where people may slip, fall from heights, or where there is low headroom, and traffic routes may need to be marked to meet requirements under the Workplace (Health, Safety and Welfare) Regulations 1992.

Although these regulations require stores and areas containing significant quantities of dangerous substances to be identified by the appropriate warning sign, i.e. the same signs as are used for marking pipework, they will mainly impact on smaller stores. This is because the majority of sites on which 25 tonnes or more of dangerous substances are stored can be expected to be marked in accordance with the Dangerous Substances (Notification and Marking of Sites) Regulations 1990. These have similar marking requirements for storage of most dangerous substances.

Stores need not be marked if:

(a) they hold very small quantities;

(b) the labels on the containers can be seen clearly from outside the store.

Fire safety signs

In general, the regulations do not require any changes where existing fire safety signs containing symbols comply with BS 5499: Part 1: 1990 Fire safety signs, notices and graphic symbols, perhaps in order to comply with the requirements of a fire certificate. This is because the signs in BS 5499, although different in detail to those specified in the regulations, follow the same basic pattern and are, therefore, considered to comply with the regulations.

Fire warning systems

Where evacuation from buildings is needed, the regulations require the fire alarm signal to be continuous. Fire alarms conforming to BS 5839: Part 1: 1988 Fire detection and alarm systems for buildings do not need changing, nor do other acceptable means such as manually operated sounders, such as rotary gongs or handbells.

Existing signs

In the case of fire safety signs, where employers decide that a previously acceptable sign is not of a type referred to in the regulations, they have until 24 December 1998 to replace it. All other signs must meet the requirements of the regulations.

Fig. 6.10 ◆ Fire exit signs

B

7 Investigation, reporting and recording of accidents

There has been much research over the last century into the actual causes of accidents. These causes are amplified in Chapter 12. In the majority of cases, two factors are directly relevant:

(a) the objective danger, e.g. the inadequately fenced machine, unguarded floor opening, defective ladder;

(b) the subjective perception of risk of the potential accident victim.

Cause of accident and cause of injury

People investigating accidents are sometimes unable to distinguish between the actual causes of an accident and the cause of an injury. These are two distinct factors, which can best be demonstrated by the Cause–Result Accident Sequence (*see* Fig. 7.1).

Fundamentally, all accidents involve one or more events which lead to the accident and resulting injury. There are both direct and indirect causes of accidents and direct and indirect *results* of accidents. The direct causes are generally associated with unsafe acts and/or omissions and unsafe conditions. The indirect causes are, however, frequently overlooked. These may be people related, in terms of knowledge and skill deficiencies on the part of operators, or be caused through poor safety specification and design of work equipment. The accident, e.g. a slip, trip or fall, is the cause of the injury and the direct result of the accident. The indirect results of the accident, again frequently overlooked, can be significant for both the injured person and the organisation respectively in terms of loss of earnings and earning capacity and, for the organisation, poor reputation as a result of adverse media publicity.

Fig. 7.1 ◆ The cause–result accident sequence

INDIRECT CAUSES	DIRECT CAUSES	*THE ACCIDENT*	*DIRECT RESULTS*	*INDIRECT RESULTS*
PERSONAL FACTOR Definition: Any condition or characteristic of a man that causes or influences him to act unsafely. 1) Knowledge and skill deficiencies (a) Lack of hazard awareness (b) Lack of job knowledge (c) Lack of job skill 2) Conflicting motivations (a) Saving time and effort (b) Avoiding discomfort (c) Attracting attention (d) Asserting independence (e) Seeking group approval (f) Expressing resentment 3) Physical and mental incapacities	*UNSAFE ACT* Definition: Any act that deviates from a generally recognised safe way of doing a job and increases the likelihood of an accident. *BASIC TYPES* 1) Operating without authority 2) Failure to make secure 3) Operating at unsafe speed 4) Failure to warn or signal 5) Nullifying safety devices 6) Using defective equipment 7) Using equipment unsafely 8) Taking unsafe position 9) Repairing or servicing moving or energized equipment 10) Riding hazardous equipment 11) Horseplay 12) Failure to use protection	Definition: An unexpected occurrence that interrupts work and usually takes this form of an abrupt contact. *BASIC TYPES* 1) Struck by 2) Contact by 3) Struck against 4) Contact with 5) Caught in 6) Caught on 7) Caught between 8) Fall to different level 9) Fall on same level 10) Exposure 11) Overexertion/ strain	Definition: The immediate results of an accident. *BASIC TYPES* 1) 'No results' or near miss 2) Minor injury 3) Major injury 4) Property damage	Definition: The consequences for all concerned that flow from the direct result of accidents. *FOR THE INJURED* 1) Loss of earnings 2) Disrupted family life 3) Disrupted personal life 4) And other consequences *FOR THE COMPANY* 1) Injury costs 2) Production loss costs 3) Property damage costs 4) Lowered employee morale 5) Poor reputation 6) Poor customer relations 7) Lost supervisor time 8) Product damage costs
SOURCE CAUSES Definition: Any circumstances that may cause or contribute to the development of an unsafe condition. *MAJOR SOURCES* 1) Production employees 2) Maintenance employees 3) Design and Engineering 4) Purchasing practices 5) Normal wear through use 6) Abnormal wear and tear 7) Lack of preventive maintenance 8) Outside contractors	*UNSAFE CONDITIONS* Definition: Any environmental condition that may cause or contribute to an accident. *BASIC TYPES* 1) Inadequate guards and safety devices 2) Inadequate warning systems 3) Fire and explosion hazards 4) Unexpected movement hazards 5) Poor housekeeping 6) Protruding hazards 7) Congestion, close clearance 8) Hazardous atmospheric conditions 9) Hazardous placement or storage 10) Unsafe equipment defects 11) Inadequate illumination, noise 12) Hazardous personal attire			

The cause–result accident sequence

The cause–result accident sequence can be summarised as in Fig. 7.2. In order to understand the various components of this sequence, it is necessary to analyse each individually.

Fig. 7.2 ◆ Cause–result accident sequence

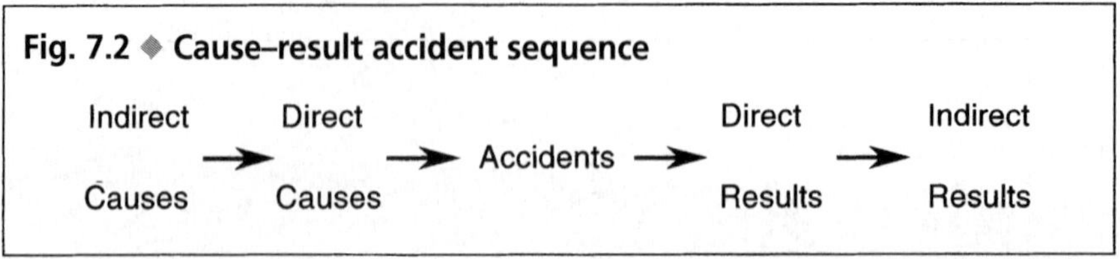

Indirect causes

These are generally associated with two particular aspects, viz. people and the environment, which can be termed 'personal factors' and 'source causes'.

Personal factors

A personal factor is defined as 'any characteristic or condition of a person that causes or influences that person to act unsafely'. Personal factors include:

(a) knowledge and skill deficiencies –
  (i) lack of hazard awareness;
  (ii) lack of job knowledge;
  (iii) lack of job skill;
  (iv) lack of adequate instruction;
(b) conflicting motivations –
  (i) saving time and effort:
  (ii) avoiding discomfort;
  (iii) attracting attention;
  (iv) asserting independence;
  (v) seeking group approval;
  (vi) expressing resentment;
(c) physical and mental incapacities.

Source causes

Source causes are any circumstances that may cause or contribute to the development of an unsafe condition. Major sources include:

(a) Management, e.g. failure by management and supervision to inform, instruct, maintain and lead in safety procedures.

(b) Production employees, e.g. failure to follow a specific safety procedure.

(c) Maintenance employees, e.g. inadequate or careless maintenance of plant.

(d) Design and engineering, e.g. the provision of insufficient or ineffective safety devices on machinery.

(e) Purchasing practices, e.g. failure to write safety requirements into specifications.

(f) Normal wear through use resulting in hazardous failure.

(g) Abnormal wear and tear.

(h) Lack of preventive maintenance.

(i) The presence or activities of third parties including outside contractors, e.g. unsafe working practices by contractors working on the premises.

Direct causes

Accidents are attributed to:

(a) unsafe acts and/or omissions;

(b) unsafe conditions.

Unsafe acts and/or omissions

An unsafe act is any act that deviates from a generally recognised safe way of doing a job and increases the likelihood of an accident. The following are types of unsafe act:

(a) operating without authority, e.g. unauthorised use of potentially dangerous machinery;

(b) operating at unsafe speed, e.g. internal factory transport;

(c) nullifying safety devices, e.g. operators defeating safety devices on machines;

(d) use of defective equipment;

(e) using equipment unsafely, e.g. home-made 'lash-ups';

(f) taking an unsafe position, e.g. working on a fragile roof without crawl boards;
(g) attempting repair or servicing of moving or energised equipment;
(h) riding hazardous equipment, e.g. conveyor belts;
(i) horseplay.

Omissions or failures on the part of management and workers can result in accidents. Typical omissions are:

(a) failure to make secure, e.g. the load on a hand truck;
(b) failure to warn or signal, e.g. fork lift truck drivers working in congested areas;
(c) failure to use personal protective equipment. (Various statutes and regulations require employees and other persons to wear personal protective equipment when engaged in certain work or processes.)

Unsafe conditions

Unsafe conditions are fundamentally associated with the quality of the working environment and may be defined as 'any environmental conditions that may cause or contribute to an accident or occupational illness/disease'. The basic types include:

(a) inadequate guards and safety devices;
(b) inadequate warning systems;
(c) fire and explosion hazards;
(d) unexpected movement hazards, e.g. conveyors;
(e) poor housekeeping and cleaning activities;
(f) protruding objects, e.g. nails in scaffold boards;
(g) congestion;
(h) hazardous atmospheric conditions, e.g. during welding operations;
(i) hazardous placement or storage (various statutory requirements regulate the placement and/or storage of dangerous substances, e.g. the Highly Flammable Liquids and Liquefied Petroleum Gases Regulations 1972);
(j) unsafe equipment and defective equipment (by definition 'a part of machinery is dangerous if it is a possible cause of injury to anybody acting in a way in which a human being may be reasonably expected to act in circumstances which may be reasonably expected to occur', per du Parcq J in *Walker* v. *Bletchley-Flettons Ltd* [1937] 1 AER 170);

(k) inadequate or unsuitable illumination;

(l) noise;

(m) hazardous personal attire, e.g. wearing loose jewellery or clothing.

The accident

Direct results

One or more of the following are the direct (or immediate) results of an accident:

(a) 'no results' or near misses;

(b) minor injury;

(c) major injury or death;

(d) property damage.

Indirect results

Indirect results refer to the consequences flowing from the direct results of accidents. They include:

(a) for the injured person –
  - (i) loss of earnings and of earning capacity;
  - (ii) disrupted family life;
  - (iii) disrupted personal life; and
  - (iv) other consequences, e.g. continued pain and suffering.

(b) for the company –
  - (i) injury costs;
  - (ii) production loss costs;
  - (iii) property damage costs;
  - (iv) reduction in employee morale;
  - (v) poor reputation as a result of adverse media publicity;
  - (vi) poor customer relations following adverse media publicity;
  - (vii) lost time – supervisors, workers and others;
  - (viii) product damage costs;
  - (ix) first aid and medical costs;
  - (x) increased employer's liability premiums;

(xi) legal costs, e.g. fines imposed by courts, legal representation fees;

(xii) cost of changes in practice arising from prosecution or as a result of enforcement action, e.g. prohibition order or improvement notice;

(xiii) training costs, e.g. retraining of injured employee, training of replacement labour.

(*See further* Chapter 11.)

Reporting of injuries, diseases and dangerous occurrences

The Reporting of Injuries, Diseases and Dangerous Occurrences Regulations 1995 (RIDDOR) cover the requirement to report certain categories of injury and disease sustained by people at work, together with certain specified dangerous occurrences and gas incidents, to the relevant enforcing authority. The relevant authority in most cases is the HSE. In the case of premises covered by the OSRPA, injuries and diseases must be reported to the local authority, in most cases the environmental health officer (*see* Table 7.1, p. 153).

Notification and reporting of injuries and dangerous occurrences (Reg 3)

The Regulations require the 'responsible person' to *notify* the relevant enforcing authority 'by quickest practicable means' (i.e. telephone or fax) and subsequently make a *report* within ten days on the approved form:

(a) the death of any person as a result of an accident arising out of or in connection with work;

(b) any person at work suffering a specified major injury (see definition p.160) as a result of an accident arising out of or in connection with work;

(c) any person who is not at work suffering an injury as a result of an accident arising out of or in connection with work and where that person is taken from the site of the accident to a hospital for treatment in respect of that injury;

(d) any person who is not at work suffering a major injury as a result of an accident arising out of or in connection with work at a hospital; or

(e) where there is a dangerous occurrence.

The responsible person must also report as soon as practicable, and in any event within ten days of the accident, using the approved form, any situation where a person at work is incapacitated for work of a kind which he might reasonably be expected to do, either under his contract of employment, or, if there is no such contract, in the normal course of his work, for more than three consecutive days (excluding the day of the accident but including any days which would not have been working days) because of an injury resulting from an accident arising out of or in connection with work. (Note: The injured person may not necessarily be away from work but, perhaps, undertaking light duties.)

Reporting of the death of an employee (Reg 4)

Where an employee, as a result of an accident at work, has suffered a reportable injury which is a cause of his death within one year of the date of that accident, the employer must inform the relevant enforcing authority in writing of the death as soon as it comes to his knowledge, whether or not the accident has been reported.

Reporting of cases of disease (Reg 5)

Where:

(a) a person at work suffers from any of the occupational diseases specified in column 1 of Part I of Schedule 3 and his work involves one of the activities specified in the corresponding entry in column 2 of that Part; or

(b) a person at an offshore workplace suffers from any of the diseases specified in Part II of Schedule 3,

the responsible person shall forthwith send a report thereof to the relevant enforcing authority on the approved form, unless he forthwith makes a report thereof to the HSE by some other means so approved.

The above requirement applies only if:

(a) in the case of an employee, the responsible person has received a written statement prepared by a registered medical practitioner diagnosing the disease as one of those specified in Schedule 3; or

(b) in the case of a self-employed person, that person has been informed, by a registered medical practitioner, that he is suffering from a disease so specified.

Reporting of gas incidents (Reg 6)

Whenever a conveyor of flammable gas through a fixed pipe distribution system, or a filler, importer or supplier (other than by means of retail trade) of a refillable container containing liquefied petroleum gas receives notification of any death or any major injury which has arisen out of or in connection with the gas distributed, filled, imported or supplied, as the case may be, by that person, he must forthwith notify the HSE of the incident and within 14 days send a report on the approved form.

Whenever an employer or self-employed person who is a member of a class of person approved by the HSE for the purposes of paragraph (3) of the Gas Safety (Installation and Use) Regulations 1994 (i.e. a CORGI registered gas installation business) has in his possession sufficient information for it to be reasonable for him to decide that a gas fitting or any flue or ventilation used in connection with that fitting, by reason of its design, construction, manner of installation, modification or servicing, is or has been likely to cause death, or any major injury by reason of:

(a) accidental leakage of gas;

(b) inadequate combustion of gas; or

(c) inadequate removal of the products of combustion of gas,

he must within 14 days send a report of it to the HSE on the approved form, unless he has previously reported such information.

Reports must be made on:

Form 2508 – Report of an injury or dangerous occurrence

Form 2508A – Report of a case of disease

Form 2508G – Report of a gas incident

Records

A record must be kept by the responsible person of all reportable injuries, diseases and dangerous occurrences.

In the case of reportable injuries and dangerous occurrences, the record must contain in each case:

(a) the date and time of the accident or dangerous occurrence;

(b) in the case of an accident suffered by a person at work, the following particulars of that person –

    (i) full name;

    (ii) occupation;

(iii) nature of injury;

(c) in the case of an accident suffered by a person not at work, the following particulars of that person (unless they are not known and it is not reasonably practicable to ascertain them) –

(i) full name;

(ii) status (for example 'passenger', 'customer', 'visitor' or 'bystander');

(iii) nature of injury;

(d) place where the accident or dangerous occurrence happened;

(e) a brief description of the circumstances in which the accident or dangerous occurrence happened;

(f) the date on which the event was first reported to the relevant enforcement authority;

(g) the method by which the event was reported.

In the case of diseases specified in Schedule 3 and reportable under regulation 5, the following information should be recorded:

(a) date of diagnosis of the disease;

(b) name of the person affected;

(c) occupation of the person affected;

(d) name or nature of the disease;

(e) the date on which the disease was first reported to the relevant enforcing authority;

(f) the method by which the disease was reported.

The system for recording accidents and diseases is not specified in RIDDOR. An employer can simply retain photocopies of reports in a file or, in the case of accidents, utilise the BI510 Accident Book, identifying those accidents reportable under RIDDOR. Alternatively, details may be stored on a computer, provided that details from the computer file can be retrieved and printed out readily when required.

Notifiable and reportable major injuries

These are listed in Schedule 1 to RIDDOR thus:

1. Any fracture, other than to the fingers, thumbs or toes.

2. Any amputation.

3. Dislocation of the shoulder, hip, knee or spine.

4. Loss of sight (whether temporary or permanent).

5. A chemical or hot metal burn to the eye or any penetrating injury to the eye.
6. Any injury resulting from electric shock or electrical burn (including any electrical burn caused by arcing or arcing products) leading to unconsciousness or requiring resuscitation or admittance to hospital for more than 24 hours.
7. Any other injury:
   (a) leading to hypothermia, heat-induced illness or to unconsciousness;
   (b) requiring resuscitation; or
   (c) requiring admittance to hospital for more than 24 hours.
8. Loss of consciousness caused by asphyxia or by exposure to a harmful substance or biological agent.
9. Either of the following conditions which result from the absorption of any substance by inhalation, ingestion or through the skin:
   (a) acute illness requiring medical treatment; or
   (b) loss of consciousness.
10. Acute illness which requires medical treatment where there is reason to believe that this resulted from exposure to a biological agent or its toxins or infected material.

Dangerous occurrences

A dangerous occurrence is not what many people loosely describe as a 'near miss'. Dangerous occurrences under RIDDOR are listed in Schedule 2. This Schedule classifies dangerous occurrences into five groups, namely:

1. General

This group includes those dangerous occurrences involving lifting machinery, etc., pressure systems, freight containers, overhead electric lines, electrical short circuit, explosives, biological agents, malfunction of radiation generators, etc., breathing apparatus, diving operations, collapse of scaffolding, train collisions, wells, pipelines or pipeline works, fairground equipment and carriage of dangerous substances by road.

This group also includes dangerous occurrences which are reportable except in relation to offshore workplaces, namely collapse of a building or structure, explosion or fire, escape of flammable substances and escape of substances.

2. Dangerous occurrences which are reportable in relation to mines

Listed in this group are dangerous occurrences involving fire or ignition of gas, escape of gas, failure of plant or equipment, breathing apparatus, injury by explosion of blasting material, use of emergency escape apparatus, inrush of gas or water, insecure tip, locomotives and falls of ground.

3. Dangerous occurrences which are reportable in relation to quarries

This group includes sinking of craft, explosive-related injuries, projection of substances outside a quarry, misfires, insecure tips, movement of slopes or faces and explosions or fires in vehicles or plant.

4. Dangerous occurrences which are reportable in respect of relevant transport systems

A 'relevant transport system' is defined as meaning a railway, tramway, trolley vehicle system or guided transport system. Reportable dangerous occurrences are accidents to passenger trains, accidents not involving passenger trains, accidents involving any kind of train, accidents and incidents at level crossings, accidents involving the permanent way and other works on or connected with a relevant transport system, accidents involving failure of the works on or connected with a relevant transport system, incidents of serious congestion and incidents of signals passed without authority.

5. Dangerous occurrences which are reportable in respect of an offshore workplace

This group includes release of petroleum hydrocarbon, fire or explosion, release or escape of dangerous substances, collapses of an offshore installation or part thereof, certain occurrences having the potential to cause death, e.g. the failure of equipment required to maintain a floating offshore installation on station, collisions, subsidence or collapse of seabed, loss of stability or buoyancy, evacuation situations and falls into water, i.e. where a person falls more than two metres into water.

Note: It should be appreciated that dangerous occurrences are notifiable and reportable to the HSE irrespective of whether death and/or major injury has resulted from same.

Table 7.1 Summary of reporting requirements

Summary of reporting requirements

Reportable event	*Person affected*	*Responsible person*
1. Special cases		
All reportable events in mines		The mine manager
All reportable events in quarries or in closed mine or quarry tips		The owner
All reportable events at offshore installations, except cases of disease reportable under regulation 5		The owner, in respect of a mobile installation, or the operator in respect of a fixed installation (under these Regulations the responsibility extends to reporting incidents at subsea installations, except tied back wells and adjacent pipeline)
All reportable events at diving installations, except cases of disease reportable under regulation 5		The diving contractor
2. Injuries and disease		
Death, major injury, over 3-day injury, or case of disease (including cases of disease connected with diving operations and work at an offshore installation)	of an employee at work	That person's employer
	of a self-employed person at work in premises under the control of someone else	The person in control of the premises * at the time of the event * in connection with the carrying on of any trade, business or undertaking
Major injury, over 3-day injury, or case of disease	of a self-employed person at work in premises under their control	The self-employed person or someone acting on their behalf

Death, or injury requiring removal to a hospital for treatment (or major injury occurring at a hospital)	of a person who is not at work (but is affected by the work of someone else), e.g. a member of the public, student, a resident of a nursing home	The person in control of the premises where, or in connection with the work going on at which, the accident causing the injury happened: * at the time of the event; and * in connection with their carrying on any trade, business or undertaking
3. Dangerous occurrences One of the dangerous occurrences listed in Schedule 2 to the Regulations, except * where they occur at workplaces covered by part 1 of this table (i.e. at mines, quarries, closed mine or quarry tips, offshore installations or connected with diving operations): or * those covered below		The person in control of the premises where, or in connection with the work going on at which, the dangerous occurrence happened: * at the time the dangerous occurrence happened; and * in connection with their carrying on trade, business or undertaking
A dangerous occurrence at a well		The concession owner (the person having the right to exploit or explore mineral resources and store and recover gas in any area, if the well is used or is to be used to exercise that right) or the person appointed by the concession owner to organise or supervise any operation carried out by the well
A dangerous occurrence at a pipeline, but not dangerous occurrence connected with pipeline works		The owner of the pipeline
A dangerous occurrence involving a dangerous substance being conveyed by road		The operator of the vehicle

B

Reportable diseases

Reportable diseases are those listed in Schedule 3 of RIDDOR. There are 47 diseases and they are listed under three groups:

1. **Conditions due to physical agents and the physical demands of work** e.g. malignant disease of the bones due to ionising radiation, decompression illness and subcutaneous cellulitis of the hand (beat hand)
2. **Infections due to biological agents** e.g. anthrax, brucellosis, leptospirosis, Q fever and tetanus
3. **Conditions due to substances** e.g. poisoning by, for instance, carbon disulphide, ethylene oxide and methyl bromide, cancer of a bronchus or lung, bladder cancer, acne (through work involving exposure to mineral oil, tar, pitch or arsenic) and pneumoconiosis (excluding asbestosis).

Accident investigation

There are very good reasons for the effective and thorough investigation of accidents:

(a) On a purely humanitarian basis, no one likes to see people killed or injured.

(b) The accident may have resulted from a breach of statute by the organisation, the accident victim, the manufacturers and/or suppliers of articles and substances used at work, or other persons, e.g. contractors, with the possibility of civil proceedings being instituted by the injured party against his employer.

(c) The accident may be reportable to the enforcing authority under the RIDDOR.

(d) the accident may result in lost production.

(e) From a management viewpoint, a serious accident, particularly a fatal one, can have a long-term detrimental effect on the morale of the workforce.

(f) There may be damage to plant and equipment, resulting in the need for repair or replacement, with possible delays in replacement.

(g) In most cases, there will be a need for immediate remedial action in order to prevent recurrence of the accident.

Humanitarian reasons apart, all accidents represent losses to the organisation, both directly and indirectly (*see* Chapter 11). There are, therefore, legal and financial reasons for investigating accidents to identify causes and produce strategies to prevent recurrences. Above all, the purpose of accident investigation is *not* to apportion blame or fault, though this may inevitably emerge from investigation.

Which accidents should be investigated?

It may be impracticable to investigate every accident, but the following factors are helpful in determining which accidents should be investigated as a priority:

(a) the type of accident, e.g. fall from a height, chemical handling, machinery;

(b) the form and severity of injury, or the potential for severe injury and/or damage;

(c) whether the accident indicates the continuation of a particular trend in accident experience;

(d) the extent of involvement of articles and substances used at work, e.g. machinery, plant, dangerous substances, and the ensuing damage or loss;

(e) the possibility of a breach of the law;

(f) whether the accident is by law reportable to the enforcing authorities;

(g) whether the accident should be reported to the insurance company as it could result in a claim being made.

Moreover, since an occupational disease or condition, sometimes referred to as a 'slow accident', can have the same effects on an individual as an accident, the above factors should be considered when investigating the possible causes of occupational diseases, such as non-infective dermatitis or noise-induced hearing loss.

Practical accident investigation

In any accident situation, particularly a fatal accident, one resulting in major injury, or a scheduled dangerous occurrence, speed of action is essential. This is particularly true when it comes to interviewing the injured person and any witnesses. The following procedure is recommended:

(a) Establish the facts as quickly and completely as possible about –

(i) the general environment;

(ii) the particular plant, machinery, practice or system of work involved;

(iii) the sequence of events leading to the accident.

(b) Use an instant camera to take photographs of the accident scene.

(c) Draw sketches and take measurements with a view to producing a scale drawing of the accident scene.

(d) List the names of all witnesses, i.e. those who saw, heard, felt or smelt anything; interview them thoroughly in the presence of a third party, if necessary, and take full statements. Do not prompt or lead the witnesses.

(e) Evaluate the facts, and individual witnesses' versions of same, as to accuracy, reliability and relevance.

(f) Endeavour to arrive at conclusions as to the *cause* of the accident on the basis of the relevant facts.

(g) Examine closely any contradictory evidence. Never dismiss a fact that does not fit in with the rest. *Find out more.*

(h) Learn fully about the system of work involved. Every accident occurs within the context of a work system. Consider the personnel involved in terms of their ages, training, experience and level of supervision, and the nature of the work, e.g. routine, sporadic or incidental.

(i) In certain cases, it may be necessary for plant and equipment to be examined by a specialist, e.g. consultant engineer.

(j) Produce a report for the responsible manager emphasising the causes and remedies to prevent a recurrence, including any changes necessary.

(k) In complex and serious cases, consider the establishment of an investigating committee comprising managers, supervisors, technical specialists and trade union representatives.

The investigators

The investigation of accidents is not the sole prerogative of the employer or his representative. Clearly, the employer, or his health and safety specialist, must endeavour to ascertain the cause of the accident with a view to preventing recurrence and improving the work situation. (The decision of the House of Lords in *Waugh* v. *British Railways Board* [1979] 2 AER 1169 has established that a safety officer's detailed report after an injury-causing accident and/or fatality at work is only privileged from 'discovery' by the plaintiff or his representatives if its dominant purpose is the preparation of the employer's defence in subsequent litigation.) A number of other organisations, however, have a direct interest in the investigation of accidents, in particular the following.

Trade union safety representatives

Such persons have a right to undertake an inspection of that part of the workplace concerned in any reportable or notifiable accident as far as is necessary for determining the cause of the accident. They have a duty to notify the employer of their intention to undertake such an inspection. (*See* Chapter 5 for the specific powers of safety representatives.)

Insurance company surveyor

Whether or not a claim has been made by the injured person, most insurance companies insist on inspection rights, particularly where the accident may identify trends and/or there is potential for a claim being submitted later. This is advisable also from a legal viewpoint, as the facts themselves may disclose evidence of negligence (res ipsa loquitur).

Legal representative

In the event of a civil action for negligence, the legal representative of the injured person, as well as the company's legal representative, would expect reasonable access to ascertain the cause of the accident. They may both need to interview witnesses.

Enforcing authority

HSE inspectors and local authority environmental health officers have powers of entry for the purpose of investigating accidents to ascertain whether there has been a breach of statute (*see* Chapter 4).

The outcome of accident investigation

Whether accident investigation is carried out by individuals or by a committee, it is necessary, once the cause has been identified, to submit recommendations to management with a view to preventing a recurrence. The organisation of 'feedback' is crucial in large organisations, especially those which occupy more than one site or premises. An effective investigation should result in one or more of the following recommendations being made:

(a) the issuing of specific instructions by management, perhaps attached to the Statement of Health and Safety Policy, regarding systems of work, the need for more effective guarding of machinery, etc.;

(b) the establishment of a working party or group to undertake further investigation, perhaps in conjunction with a safety committee and/or safety representatives;

(c) the preparation and issue of specific codes of practice or guidance notes dealing with the procedures necessary to minimise a particular risk;

(d) the identification of specific training needs for groups of individuals – e.g. managers, foremen, supervisors, machinery operators, drivers – and the implementation of a training programme designed to meet these needs;

(e) the formal analysis of the job or system in question, perhaps using Job Safety Analysis techniques, to identify specific skill and safety components of the job;

(f) identification of the need for further information relating to articles and substances used at work, e.g. equipment, chemical substances;

(g) identification of the need for better environmental control, e.g. noise reduction at source or improved lighting;

(h) general employee involvement in health and safety issues, e.g. the establishment of a health and safety committee (this may be a legal requirement – *see* Chapter 5);

(i) identification of the specific responsibilities of groups with regard to safe working practices.

Above all, a system of monitoring should be implemented to ensure that the lessons which have been learned from the accident are put into practice or incorporated in future systems of work, that supervisors have been trained to monitor such systems, and that procedures and operating systems have been produced for all grades of staff (*see* further Chapter 8).

Near misses

A chapter on accident reporting and investigation would not be complete without reference to the question of near-miss reporting, recording and investigation. A near miss could be defined as 'an unplanned and unforeseeable event that could have resulted, but did not result in human injury, property damage or other form of loss'. Studies by Frank Bird, the American exponent of Total Loss Control, showed a correlation between serious or disabling injuries, minor injuries, i.e. those requiring first aid treatment only, accidents involving damage to property and plant, and accidents with no injury or damage, i.e. the near miss types of accident. The outcome of this study can be depicted by Bird's Triangle (Fig. 7.3).

Fig. 7.3 ◆ Bird's Triangle

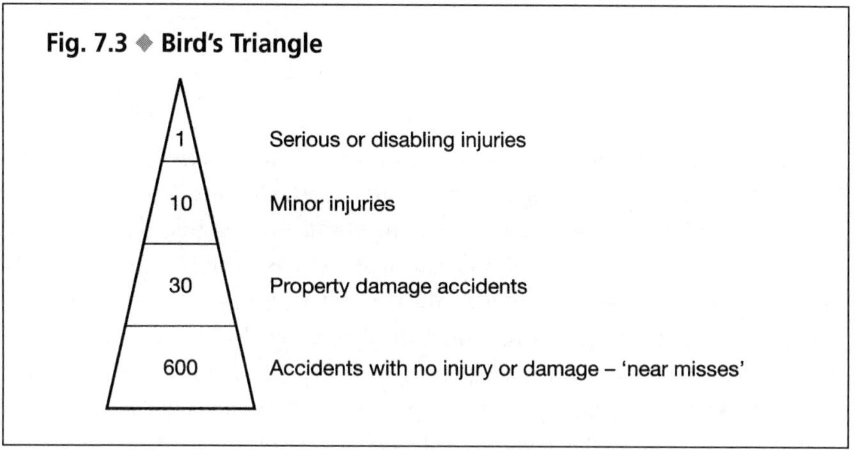

On this basis, it is essential that all near misses be reported and action taken in terms of investigation, recording and analysis, so that appropriate remedial action can be taken on the philosophy that 'Yesterday's near miss could be tomorrow's serious accident'.

Incident recall is a system that can be used to gain information about near-miss accidents. In this system a random sample of employees is interviewed with a view to obtaining information about non-compliance with safe systems of work, unsafe conditions, unsafe working practices and near-miss situations. Each person is asked to discuss, on a purely confidential basis, examples of the above in which he was involved, had knowledge or actually witnessed. From this information, remedial action can then be taken before further similar situations result in accidents involving injury and property damage or both.

8 Health and safety monitoring

Health and safety monitoring refers to all of those activities which are concerned with ensuring good standards of health and safety management. In order to assess and evaluate standards of health and safety performance, and to bring about improvements which will reduce the potential for accidents, it is essential to operate one or more forms of health and safety monitoring. There may be a case for analysing or evaluating certain tasks from a health and safety viewpoint, or a need to motivate staff, for instance, to use personal protective equipment or to adhere to a specifically designed safe system of work. All these activities come within the scope of 'health and safety monitoring' and, as such, are closely related to the 'arrangements' for ensuring a safe and healthy workplace which must be outlined in the Statement of Health and Safety Policy (*see* Chapter 3).

The various forms of health and safety monitoring are outlined below. The adoption of a specific technique will depend on the hazards present, the number of people exposed to the hazard, the levels of training received, the standards of supervision and control and the climate within the organisation for bringing about improvement.

Safety audits

A safety audit is the systematic measurement and validation of an organisation's management of its health and safety programme against a series of specific and attainable standards (RoSPA). Fundamentally, an audit subjects each area of an organisation's activities to a systematic critical examination with the object of minimising injury and loss. It should commence with a full audit of the complete programme, in order to identify the strengths and weaknesses. This is followed by recommendations leading to achievable targets being set prior to each subsequent audit. Safety auditing is, therefore, an ongoing process aimed at ensuring effective health and safety management. Every component of the total system is included, e.g. management policy, attitudes, training, features of processes, personal protection needs, emergency procedures, etc.

A specimen safety audit check-list appears at the end of this chapter.

Health and safety surveys

A survey is a detailed examination of a number of critical areas of operation or an in-depth study of the whole health and safety operation of premises. For instance, a survey might examine in depth health and safety management and administration, environmental factors, occupational health and hygiene provisions, the diverse field of safety and accident prevention as it affects the premises, and the current system for health and safety training of staff, from the Board of Directors to shopfloor workers. Contractors and other groups who may enter the premises on a casual or infrequent basis might also be covered. Much will depend on the inherent risks, the number of employees, the location of the premises in relation to urban population, the potential for pollution incidents and the range of products manufactured.

Reporting procedure

Following a survey, a report is published. In most cases, this report is purely critical and produced on an observation and recommendation basis. Recommendations would be phased according to the degree of risk, current legal requirements and the cost of eliminating or reducing the risks. A survey report is mainly concerned with risks and the system for bringing about a gradual upgrading of standards, e.g. in welfare amenity provisions. Recommendations could include the eventual replacement of old and dilapidated buildings which may be totally unsuitable for the work processes being undertaken within. The purpose of the health and safety report is to present management with a phased programme of health and safety improvement covering, say, a five-year period. Recommendations would be phased as follows.

Phase 1: These recommendations apply to situations where there is a serious risk and/or direct breach of the law, and to situations where relatively minor improvements can be made without capital expenditure, e.g. redecoration, repairs to floors. Phase 1 items would also include recommendations relating to health and safety management and administration, such as updating the Statement of Health and Safety Policy, the publication of specific objectives for the health and safety committee or the establishment of a company code of practice dealing with, perhaps, the supply and use of personal protective equipment. It would be normal to expect implementation of phase 1 recommendations within six months of the presentation of the report.

Phase 2: These recommendations would normally be implemented within two years. They would include items which do not require large-scale expenditure but which are essential for the improvement of working conditions and safety standards, e.g. improvements to lighting in working areas, the installation of specific exhaust ventilation to processes or improvements in access to parts of the premises.

Phase 3: Recommendations in this category are those for which long-term planning is needed, together with approval for capital expenditure, such as for the provision of new amenities, replacement of outdated plant and equipment, provision of improved chemical storage facilities and fire protection measures. In many cases, such improvements can be incorporated in the long-term development plans for workplaces.

The health and safety survey thus covers the whole field of the working environment, from plant and machinery to systems of safe working and the establishment of training programmes. For the impact of the report to continue, regular monitoring of progress in its implementation is needed by the individual undertaking the survey and by company management, supported by progress reports to senior management.

Safety inspections

Many people are involved in safety inspections, such as company health and safety specialists who may visit on a regular basis, plant safety officers, local management, trade union safety representatives, inspectors of the enforcing authorities and insurance company surveyors. Each of these individuals has different motives for undertaking an inspection. The enforcement agencies carry out inspections to identify possible breaches of the law and the causes of a notified accident. In-company health and safety specialists inspect to identify hazards, to ensure that specific procedures designed to promote safe working are being operated, and to protect their employers from liability. The safety representative carries out inspections to protect members of his trade union from hazards.

Generally, however, a safety inspection is a scheduled inspection of premises or part of same by personnel within that organisation, possibly accompanied by an external specialist. The inspection may examine maintenance standards, employee involvement, working practices and housekeeping levels, and check that work is undertaken in accordance with company codes of practice or procedures. This form of monitoring tends to be a general examination of the situation at a specific point in time rather than the in-depth and broad approach taken with health and safety surveys. Whatever system exists for safety inspection, it is vital that the objectives are clearly defined and made known to the workforce.

Safety tours

This is an unscheduled examination of a work area, carried out by a manager, possibly accompanied by health and safety committee members, to ensure that, for instance, standards of housekeeping are at an acceptable level, fire protection measures are being observed and maintained, or personal protective equipment is being used correctly. Safety tours tend to be of limited value unless related to decisions made by local management and/or health and safety committee members. To be effective, it is essential that deficiencies noted during the tour are remedied immediately.

Contact schemes

The objective of a contact scheme is to provide added incentive and facilities to middle and junior management to persuade them to impart information and/or instructions on health and safety matters to the workers under their control. Many supervisors, for instance, are conscious of the fact that they are not instructors or trainers. Even short talks with workers are not within their scope, quite apart from the fact that such talks tend to disrupt production. If, however, a supervisor tours a section and talks to each person separately he can usually communicate various ideas through this one-to-one relationship. The contact scheme enables him to do this and keep a record of the extent of instruction received by each worker.

A simple system involves the use of a piece of squared paper. Topics for discussion are entered in the left-hand column and names of operators in the row at the top of the page. Subjects to be covered may include the use of personal protection equipment, housekeeping, manual handling techniques or the use of hand tools. The supervisor then prepares his programme for informal discussions and, over a period, ensures that each worker is contacted, a tick being placed in the appropriate square after each contact session (Table 8.1).

Table 8.1 ◆ Safety contact scheme

	Operator						
	A	B	C	D	E	F	G
1. Fire protection procedures 2. Safe systems of work for: (a) (b) (c) (d) 3. Personal protective equipment 4. Hand tools 5. Machinery and plant: (a) (b) (c) 6. Manual handling							

Safety sampling

This technique is designed to measure, by random sampling, the accident potential in a specific workshop or process by identifying safety defects or omissions. The area to be sampled is divided into sections and an observer appointed to each section. A prescribed route through the area is planned and observers follow their itinerary in the time allowed, about 15 minutes. During the sampling period they note safety defects on a safety sampling sheet, with a limited number of points to be observed, e.g. housekeeping, eye protection being worn, correct handling procedures. Other aspects for observation may include obstructed fire exits, environmental factors such as lighting and ventilation, faulty hand tools and damaged guards to machinery. The staff undertaking the inspections should be trained in the technique and have a broad knowledge of procedures and processes carried out. The results of the sampling activity are collated and presented in graphical form by a specific manager, e.g. safety officer. The system monitors the effectiveness of the overall safety programme (Table 8.2).

Table 8.2 ◆ Safety sampling exercise

			Area		
			A	B	C	D
1.	Housekeeping/cleaning	(Max 10)				
2.	Personal protection	(Max 10)				
3.	Machinery	(Max 10)				
4.	Chemical storage	(Max 5)				
5.	Chemical handling	(Max 5)				
6.	Manual handling	(Max 5)				
7.	Fire protection	(Max 10)				
8.	Structural hazards	(Max 10)				
9.	Internal transport e.g. FLTs	(Max 5)				
10.	Access equipment	(Max 5)				
11.	First Aid boxes	(Max 5)				
12.	Hand tools	(Max 10)				
13.	Internal storage (racking systems)	(Max 5)				
14.	Structural safety	(Max 10)				
15.	Temperature	(Max 5)				
16.	Lighting	(Max 5)				
17.	Ventilation	(Max 5)				
18.	Noise	(Max 5)				
19.	Dust and fumes	(Max 5)				
20.	Welfare amenities	(Max 10)				
		Total (Max 140)				

Hazard and operability studies

These studies incorporate the application of formal critical examination to the process and engineering intentions regarding new facilities, e.g. new production processes. Their aim is to assess the hazard potential arising from incorrect operation of each item of equipment and the consequential effects on the facility as a whole. Remedial action is then usually possible at a very early stage of the project with maximum effectiveness and minimum costs. These studies examine the potential for error amongst operators, with regard to new plant and machinery in particular, on the basis that the perpetuation of errors in design can result in accidents, plant stoppage and loss of production. While unsuspected hazards may be revealed by hazard and operability studies, the use of a formal check-list in engineering design departments helps to ensure that account is taken of accumulated experience, knowledge of the technology and best practice in the initial design.

Damage control

The theory of damage control or damage costing is that non-injury accidents are as important as injury accidents. The elimination of non-injury accidents will, in many cases, remove other forms of accident. For example, a pallet stack falls and may just miss a man standing close by. No injury results and an accident is not recorded, but there may be damage to wooden pallets, the building fabric, plant and equipment. However, the next time a pallet stack falls over, someone could be seriously injured or killed. The fact is recorded. The elimination of the cause of the first pallet stack falling – e.g. bad stacking or the use of defective pallets or stacking on an uneven floor – might have prevented the injury when the second pallet stack fell.

Damage control aims to ensure a safe working environment and calls for keen observation and co-operation by staff who see or experience a condition which may lead to an accident. It relies heavily on an effective inspection and reporting system for all damage and defects and a programme of preventive maintenance. Moreover, evidence of damage to property and plant is often an indication of poor safety performance.

Job Safety Analysis

Just as productivity can benefit from work study or job analysis, health and safety benefits from Job Safety Analysis. Furthermore, the two are intimately connected. The work study engineer should not ignore safety and the safety specialist should not ignore productivity considerations.

Job Safety Analysis, whether undertaken as part of work study or not, can do much to eliminate the hazards of a job. The analysis isolates each single operation, examines the individual hazards, and indicates remedies to reduce the hazards. It involves the examination of plant and work processes, systems of work, including permit to work systems, influences on behaviour, the qualifications and training required for the job, and the degree of instruction, supervision and control necessary (*see* Chapter 13).

Permits to work

Permits to work are essential where work with a foreseeably high hazard content must be undertaken with a commensurate need for numerous precautions, e.g. entry into confined spaces (*see* Chapter 13).

Safety incentive schemes

Opinions vary considerably as to the desirability or value of incentives to encourage workers to consider their own safety. However, the principal philosophy behind safety incentive schemes is exactly the same as that behind other incentive schemes for production, sales or marketing, i.e. to motivate people to perform better. The ways of securing motivation are based on:

(a) identifying targets, for which a reward can be given if the target is reached;
(b) making the reward meaningful and desirable to the people concerned.

Essentially these schemes are concerned with two aspects of human behaviour and performance, viz. motivation and attitude. People should be motivated to perform or behave in a particular way and, in many cases, attitude changes should be brought about so that they will behave more safely, thereby bringing about a reduction in accidents. The most successful safety incentive schemes are those which are linked to a programme of safety inspections or safety audits of the workplace. To be effective, inspections should be undertaken on an irregular basis, otherwise there is the danger that workers will become 'conditioned' to the system and not behave as they normally would. The most common system is that of awarding points for areas of health and safety performance, e.g. systems of work, procedures, housekeeping. Inspections should be carried out by a mixed team – e.g. safety specialist, personnel officer, trade union safety representative and the line manager concerned – in order to maintain a degree of fairness. The establishment of the correct targets is important in such schemes. Everyone must fully understand the scheme, and there must be total support from senior and line management if the scheme is to work. Such schemes as the RoSPA Safe Driving Awards for commercial vehicle drivers do much to develop a personal and collective pride in the job and, generally, these schemes are more successful when the final result or reward is achieved by collective effort.

Many existing safety incentive schemes, however, have serious deficiencies. If the rewards are directly related to the number of accidents in a department, this can reduce the incidence of accident reporting in order to attain the target. They can alter the threshold at which accidents are reported and, in many cases, injured persons may be discouraged from obtaining first aid treatment if all treatments are reported as part of the scheme. It does not affect serious accidents which must be reported anyway. One of the principal criticisms of safety incentive schemes is that they tend to be short-lived. There have been accusations of gimmickry, 'nine-day wonders' and lack of organisation and control. They can also shift responsibility from management to worker for the general control of health and safety and, over a relatively short period of time, serious situations can develop over accidents that have not been reported. People may even disclaim knowledge of an accident ever taking place, due to the desire to earn the rewards offered in the scheme.

Generally, however, safety incentive schemes have much to offer in terms of improved operator behaviour and performance, and this should result in a reduction in accidents. These schemes are probably most effective where people are restricted to one area of activity, e.g. driving commercial vehicles or fork lift trucks, and the operation of machinery or well-controlled processes where it is relatively simple to measure and demonstrate safety performance. They do, however, need regular stimulation and rejuvenation, and must have management and trade union support. Additionally, they should be supported adequately by safety propaganda.

Hazard reporting systems

An important adjunct to all forms of safety monitoring is the operation of an effective hazard reporting system. Under such a system, any person –

manager, employee, visitor or contractor – should have access to a standard Hazard Report form which he or she can complete when concerned about a particular hazard. Figure 8.1 shows a typical Hazard Report form.

Once the hazard has been formally reported, the departmental manager or supervisor, in conjunction with the safety officer, should consider the hazard and decide the action necessary, including priority action according to the risk involved. In certain cases financial approval may be necessary and the form incorporates a section for certification of such approval. When action has been completed the Report is signed off by the responsible manager, following a check by the safety officer.

Hazard Reporting Systems, when operated effectively, go a long way in reassuring workers of the organisation's intentions so far as health and safety are concerned. They also provide useful feedback in the design of future safe systems of work, identification of training needs and in generally preventing accidents and incidents.

Conclusion

The above monitoring systems are commonly used. Their success depends upon the expertise available, the inherent risks and the system for remedying deficiencies. No matter which monitoring system is operated, there is clearly a need to operate some form of system with a view to preventing accidents and occupational disease, rather than undertaking the 'fire-fighting exercise' characteristic of so many situations following an accident.

Fig. 8.1 ◆ Hazard report form

HAZARD REPORT

1. **Report** (to be completed by person identifying hazard).
Date .... Time.......... Department.................
Reported to: Verbal .... Written .... (Names)
Description of Hazard (including location, plant, machinery etc.) ....
....
....
Signature .... Position ....

2. **Action** (to be completed by Departmental Manager/Supervisor)
*Hazard Verified YES/NO Date .... Time...............
....
Remedial Action (including changes in system of work) ....
....
Action to be taken by: Name .... Signature ....
+*Priority Rating: 1 2 3 4 5 Estimated Cost ....
Completion: Date .... Time ....
Interim Precautions ....
....
Signature .... (Dept. Manager)

3. **Financial Approval** (to be completed by / /Unit Manager or his assistant where cost exceeds departmental authority).
The expenditure necessary to complete the above work is approved.
Signature .... (Manager/Assistant Manager) Date ....

4. **Completion** The remedial action described above is complete.
Actual cost ....
Date .... Date ....
Signatures .... (persons completing work)
....(Dept. Head)

5. **Safety Officer's Check** I have checked completion of the above and confirm that the hazard has been eliminated.
Signature .... (Safety Officer) Date/Time ....

*Delete as appropriate
+Priority Ratings 1(immediate) 2(48 hours) 3(1 week) 4 (1 month) 5 (3 months)

Appendix: specimen safety audit check-list

Documentation	Yes	No
1. Are management aware of all health and safety legislation applying to their workplace? Is this legislation available to management and employees?		
2. Have all Approved Codes of Practice, HSE Guidance Notes and internal codes of practice been studied by management with a view to ensuring compliance?		
3. Does the existing Statement of Health and Safety Policy meet current conditions in the workplace? Is there a named manager with overall responsibility for health and safety? Are the 'organisation and arrangements' to implement the Health and Safety Policy still adequate? Have the hazards and precautions necessary on the part of staff and other persons been identified and recorded? Have individual responsibilities for health and safety been clearly detailed in the Statement?		
4. Do all job descriptions adequately describe individual health and safety responsibilities and accountabilities?		
5. Do written safe systems of work exist for all potentially hazardous operations? Is Permit to Work documentation available?		
6. Has a suitable and sufficient assessment of the risks to staff and other persons been made, recorded and brought to the attention of staff and other persons? Have other risk assessments in respect of: (a) substances hazardous to health; (b) risks to hearing; (c) work equipment; (d) personal protective equipment; (e) manual handling operations; and (f) display screen equipment, been made, recorded and brought to the attention of staff and other persons?		
7. Is the fire certificate available and up-to-date? Is there a record of inspections of the means of escape in the event of fire, fire appliances, fire alarms, warning notices, fire and smoke detection equipment?		

Documentation	Yes	No
8. Is there a record of inspection and maintenance of work equipment, including guards and safety devices? Are all examination and test certificates available, e.g. lifting appliances and pressure systems?		
9. Are all necessary licences available, e.g. to store petroleum spirit?		
10. Are workplace health and safety rules and procedures available, promoted and enforced? Have these rules and procedures been documented in a way which is comprehensible to staff and others, e.g. Health and Safety Handbook? Are disciplinary procedures for unsafe behaviour clearly documented and known to staff and other persons?		
11. Is a formally written emergency procedure available?		
12. Is documentation available for the recording of injuries, near misses, damage only accidents, diseases and dangerous occurrences?		
13. Are health and safety training records maintained?		
14. Are there documented procedures for regulating the activities of contractors, visitors and other persons working on the site?		
15. Is hazard reporting documentation available to staff and other persons?		
16. Is there a documented planned maintenance system?		
17. Are there written cleaning schedules?		
Health and safety systems		
1. Have competent persons been appointed to: (a) co-ordinate health and safety measures; and (b) implement the emergency procedure? Have these persons been adequately trained on the basis of identified and assessed risks? Are the role, function, responsibility and accountability of competent persons clearly identified?		
2. Are there arrangements for specific forms of safety monitoring, e.g. safety inspections, safety sampling? Is a system in operation for measuring and monitoring individual management performance on health and safety issues?		

Documentation		Yes	No
3.	Are systems established for the formal investigation of accidents, ill-health, near misses and dangerous occurrences?		
	Do investigation procedures produce results which can be used to prevent future incidents?		
	Are the causes of accidents, ill-health, near misses and dangerous occurrences analysed in terms of failure of established safe systems of work?		
4.	Is a hazard reporting system in operation?		
5.	Is a system for controlling damage to structural items machinery, vehicles, etc. in operation?		
6.	Is the system for joint consultation with trade union safety representatives and staff effective?		
	Are the role, constitution and objectives of the Health and Safety Committee clearly identified?		
	Are the procedures for appointing or electing Committee members and trade union safety representatives clearly identified?		
	Are the available facilities, including training arrangements, known to committee members and trade union safety representatives?		
7.	Are the capabilities of employees as regards health and safety taken into account when entrusting them with tasks?		
8.	Is the provision of first aid arrangements adequate?		
	Are first aid personnel adequately trained and retrained?		
9.	Are the procedures covering sickness absence known to staff?		
	Is there a procedure for controlling sickness absence?		
	Are managers aware of the current sickness absence rate?		
10.	Do current arrangements ensure that health and safety implications are considered at the design stage of projects?		
11.	Is there a formally-established annual health and safety budget?		
Prevention and control procedures			
1.	Are formal inspections of machinery, plant, hand tools, access equipment, electrical equipment, storage		

B

Documentation		Yes	No
	equipment, warning systems, first aid boxes, resuscitation equipment, welfare amenity areas, etc. undertaken?		
	Are machinery guards and safety devices examined on a regular basis?		
2.	Is a Permit to Work system operated where there is a high degree of foreseeable risk?		
3.	Are fire and emergency procedures practised on a regular basis?		
	Where specific fire hazards have been identified, are they catered for in the current fire protection arrangements?		
	Are all items of fire protection equipment and alarms tested, examined and maintained on a regular basis?		
	Are all fire exits and escape routes marked, kept free from obstruction and operational?		
	Are all fire appliances correctly labelled, sited and maintained?		
4.	Is a planned maintenance system in operation?		
5.	Are the requirements of cleaning schedules monitored?		
	Is housekeeping of a high standard, e.g. material storage, waste disposal, removal of spillages?		
	Are all gangways, stairways, fire exits, access and egress points to the workplace maintained and kept clear?		
6.	Is environmental monitoring of temperature, lighting, ventilation, humidity, radiation, noise and vibration undertaken on a regular basis?		
7.	Is health surveillance of persons exposed to assessed health risks undertaken on a regular basis?		
8.	Is monitoring of personal exposure to assessed health risks undertaken on a regular basis?		
9.	Are local exhaust ventilation systems examined, tested and maintained on a regular basis?		
10.	Are arrangements for the storage and handling of substances hazardous to health adequate?		
	Are all substances hazardous to health identified and correctly labelled, including transfer containers?		
11.	Is the appropriate personal protective equipment available?		

Documentation	Yes	No
Is the personal protective equipment worn or used by staff consistently when exposed to risks?		
Are storage facilities for items of personal protective equipment provided?		
12. Are welfare amenity provisions, i.e. sanitation, hand washing, showers and clothing storage arrangements, adequate?		
Do welfare amenity provisions promote appropriate levels of personal hygiene?		
Information, instruction, training and supervision		
1. Is the information provided by manufacturers and suppliers of articles and substances for use at work adequate?		
Do employees and other persons have access to this information?		
2. Is the means of promoting health and safety adequate?		
Is effective use made of safety propaganda, e.g. posters?		
3. Do current safety signs meet the requirements of the Safety Signs Regulations 1980 and Health and Safety (Safety Signs and Signals) Regulations 1996?		
Are safety signs adequate in terms of the assessed risks?		
4. Are fire instructions prominently displayed?		
5. Are hazard warning systems adequate?		
6. Are the individual training needs of staff and other persons assessed on a regular basis?		
7. Is staff health and safety training undertaken:		
(a) at the induction stage;		
(b) on their being exposed to new or increased risks because of:		
(i) transfer or change in responsibilities;		
(ii) the introduction of new work equipment or a change respecting existing work equipment;		
(iii) the introduction of new technology;		
(iv) the introduction of a new system of work or change in an existing system of work.		
Is the above training:		
(a) repeated periodically;		
(b) adapted to take account of new or changed risks; and		

Documentation	Yes	No
(c) carried out during working hours?		
8. Is specific training carried out regularly for first aid staff, fork lift truck drivers, crane drivers and others exposed to specific risks?		
Are selected staff trained in the correct use of fire appliances?		
Final question		
Are you satisfied that your organisation is as safe and healthy as you can reasonably make it, or that you know what action must be taken to achieve that state?		

9 Major incidents and emergency procedures

B

A major incident is one that may:

(a) affect several departments within a premises;

(b) endanger the surrounding communities;

(c) be classed as a fatal or major injury accident or scheduled dangerous occurrence under RIDDOR; or

(d) result in adverse publicity for an organisation with ensuing loss of public confidence and market place image.

Such incidents may be precipitated by a malfunction of the normal operating procedures, by the intervention of some outside agency, such as a crashed aircraft, a severe electrical storm, flooding, a rail disaster, an act of arson or sabotage, or as a result of terrorist activities. Bomb threats and product interference or contamination threats should be classed as major incident situations.

A major incident could result from a scheduled dangerous occurrence, such as a pressure vessel explosion or collapse of a crane. Alternatively, it could be a food poisoning outbreak associated with a food manufacturer's products or a number of cases of food poisoning amongst staff due to unsound hygiene practices in catering operations. Discovery of asbestos in an existing premises could be classed as a major incident, as could a notified case of AIDS amongst the workforce.

In any strategy aimed at reducing the potential for major incidents, organisations should be asking the question 'What is the worst possible situation from a health and safety viewpoint that could arise?'. Much will depend upon the nature of the business, the number of employees, the dangerous substances used, the age of the buildings and their method of construction, their vulnerability in the market-place in terms of industrial sabotage, proximity to the local population and the layout of the premises or site.

In considering the types of emergency that might occur, account must be taken of the areas likely to be affected. The interdependence and proximity of plant, buildings, including storage areas, and exit points from the site should be considered, Particular attention should be paid to the effect of wind direction and strength, and the effect these may have on the spread of fire and toxic materials. Furthermore, the civil authorities will need information about areas outside the premises which could be affected by toxic fumes, dusts or gases. An emergency plan, therefore, is essential to cover this type of situation.

Emergency procedures

Reg 8 of the Management of Health and Safety at Work Regulations 1999 requires that every employer shall:

(a) establish and where necessary give effect to appropriate procedures to be followed in the event of serious and imminent danger to persons at work in his undertaking;

(b) nominate a sufficient number of competent persons to implement those procedures insofar as they relate to the evacuation from the premises of persons at work in his undertaking;

(c) ensure that none of his employees has access to any area occupied by him to which it is necessary to restrict access on the grounds of health and safety unless the employee concerned has received adequate health and safety instruction.

Principal stages

A properly conceived emergency procedure or plan will take account of four phases or stages of an emergency.

Phase 1 – Preliminary action

This refers to:

(a) the preparation of a plan, tailored to meet the special requirements of the site, products and surroundings, including –

  (i) a list of all key telephone numbers;

  (ii) the system for the provision of emergency lighting, e.g. hand lamps and torches;

  (iii) designation of exit routes;

  (iv) a plan of the site layout identifying hydrant points and the location of shut-off valves to energy supplies, e.g. gas;
  (v) notes on specific hazards on site for use by the emergency services;

(b) the familiarisation of every employee with the details of the plan, including the position of essential equipment;
(c) the training of key personnel involved; competent persons;
(d) the initiation of a programme of inspection of potentially hazardous areas, testing of warning systems and evacuation procedures;
(e) stipulating specific periods at which the plan is to be re-examined and updated.

Phase 2 – Action when emergency is imminent

There may be a warning of the emergency, in which case this period should be used to assemble key personnel, to review the standing arrangements in order to consider whether changes are necessary, to give advance warning to external authorities, and to test all systems connected with the emergency scheme.

Phase 3 – Action during emergency

If phase 1 has been properly carried out, and phase 2, where applicable, phase 3 proceeds according to plan. However, it is likely that unexpected variations in a predicted emergency will take place. The decision-making personnel, selected beforehand for this purpose, must be able to make precise and rapid judgements and see that proper action follows the decision made.

Phase 4 – Ending the emergency

There must be a procedure for declaring plant, systems and specific areas safe, together with an early reoccupation of buildings where possible.

Implementation of procedure

Implementation of an emergency procedure involves the following.

Liaison with external authorities and other companies

The closest contact must be maintained with the fire, police and health authorities, together with the Health and Safety Executive and local authority. A mutual aid scheme involving neighbouring premises is best undertaken at this stage. A major emergency may also involve a failure in the supply of gas, electricity, water and/or telephone communications.

Discussions with the appropriate authority will help to determine priorities in re-establishing supply.

Emergency controller

A senior manager, with a thorough knowledge of all processes and their associated hazards, should be nominated Emergency Controller, and a deputy appointed to cover absence, however brief this may be. Out of normal working hours, the senior member of management on site should take initial control until relieved by the Emergency Controller.

Emergency control centre

A sound communication system is essential if a major emergency is to be handled effectively. A control centre should be established and equipped with means of receiving information from the forward control and assembly points, transmitting calls for assistance to external authorities, calling in essential personnel and transmitting information and instructions to personnel within the premises. Alternative means of communication must be available in the event of the main system being rendered inoperative, e.g. field telephones. A fall-back control centre may be necessary in certain situations such as a rapidly escalating fire.

Initiating the procedure

The special procedure for handling major emergencies must only be initiated when such an emergency is known to exist. A limited number of designated senior managers should be assigned the responsibility for deciding if a major emergency exists or is imminent. Only these persons should have authority to implement the procedure.

Notification to local authorities

Notification can be achieved by a predetermined short message, transmitted via an emergency line or by the British Telecom lines. The warning message should mention routes to the premises which may become impassable. Alternative routes can then be used.

Call out of competent persons

A list of competent persons required in the event of a major emergency should be drawn up, together with their internal and home telephone numbers and addresses. The list should be available in control centres and constantly updated.

Immediate action on site

Any emergency would be dealt with by action by supervisors and operators designed to close down and make safe those parts which are affected or likely to be affected. Preservation of human life and the protection of property are of prime importance, and injured persons should be conveyed to hospital with the least possible delay. This may require temporary facilities at points in a safe area accessible to ambulances.

Evacuation

Complete evacuation of non-essential personnel immediately the alarm is sounded is usually advisable, though it may not be necessary or advisable in large factories. In either situation, however, an evacuation alarm system should be installed and made known to all employees, for the purposes of evacuating the premises. Evacuation should be immediately followed by a roll call at a prescribed assembly point to ensure its success.

Access to records

Because relatives of injured and/or deceased employees will have to be informed by the police, each control centre should keep a list of names and addresses of all employees.

External communication

It must be possible to transmit urgent calls from the factory telephone exchange without delay. If this is out of order, alternative means of communication should be available.

Public relations

As a major incident will attract the attention of the media, it is essential to make arrangements for official releases of information to the press and other news services. This is best achieved through a specifically designated Public Relations Officer. Other employees should be instructed not to release information, but to refer any inquiries to the PRO, who should keep a record of any media inquiries dealt with during the emergency.

Catering and temporary shelter

Emergency teams will need refreshment and temporary shelter if the incident is of long duration. Where facilities on the premises cannot be used, it may be possible for the local authority or neighbouring companies to provide facilities.

Contingency arrangements

A contingency plan should be drawn up covering arrangements for repairs to buildings, drying out and temporary water-proofing, replacement of raw materials, alternative storage and transport arrangements.

Training

It is difficult to predict the way people will react in a major emergency situation, but knowledge of the correct procedure will increase the probability of the situation being resolved safely. Training exercises should include the participation of outside services, such as the fire brigade, ambulance service and police. Where mutual aid schemes with neighbouring organisations exist, all possible participants should take part in any form of training exercise.

Familiarisation of competent persons and of all staff concerned with the procedure, together with training exercises at regular intervals, will help reduce the risk of fatal and serious injuries following an emergency. For this reason, the emergency procedure should form an integral part of the company Statement of Health and Safety Policy (*see* Chapter 3).

Bomb threats

Any bomb threat, received by telephone, letter or other means of communication, should be treated extremely seriously, and a formal company procedure should be written to cover such incidents, including the responsibilities of named individuals for ensuring evacuation, dealing with the civil authorities and agreeing re-entry to buildings.

Immediately following notification of a bomb on the premises, the senior manager on site should be notified immediately by telephone. Departmental managers and supervisors should, in turn, be notified by cascading telephone calls with a view to a controlled evacuation of the building. Emergency services, i.e. police, ambulance and fire, should be notified at the same time by ringing 999.

The fire alarm should then be sounded, and staff from individual departments assembled at their specified assembly points. A check should be made to ensure that everyone has been evacuated.

Following an investigation and search of the building by the police, the senior police officer will advise local management on reoccupation of the building or not. No one should be allowed entry into the building until final clearance has been given by the senior police officer concerned.

Fatal and major injury accidents

The statutory procedure under RIDDOR is outlined in Chapter 7. Given the possibility of the occurrence of a fatal or major injury accident, it is important that organisations have a clearly established procedure, including the allocation of responsibility for specific tasks, following the accident. Generally, this procedure can be subdivided into two aspects, the notification sequence and the post-accident procedure.

Notification sequence

On receipt of a report of a fatal accident or one causing major injury, a senior manager should immediately contact by telephone the ambulance service, the factory doctor where appropriate and, in the case of a fatal accident, the police. In the case of a fatal accident, the manager, in consultation with the police, should arrange to visit the next of kin to advise that person of the accident.

Following this, the senior manager or safety specialist should inform, by telephone, the following:

(a) the divisional manager, where appropriate;

(b) HSE or local authority environmental health officer, as appropriate;

(c) the head of the central health and safety department, where appropriate;

(d) the personnel manager for the organisation;

(e) the insurance company.

Post-accident procedure

The area of an accident, including machinery, plant and equipment, should be immediately sealed off by the use of ropes or barriers. Unauthorised access to the scene of the accident must be prevented and the accident area not disturbed, other than to render the situation safe, until permission is given by the officer of the enforcing authority. The actual position of the corpse should be marked out on the floor or other surface prior to removal. (For the position regarding admissibility of evidence for litigation purposes following a fatality, *see* Chapter 5.)

At this stage it is essential that:

(a) photographs of the scene of the accident are taken using either an instant camera or standard camera, before any disturbance takes place;

(b) witnesses to the accident are identified;

(c) the trade union shop steward and/or safety representative, where appropriate, are informed (*see* Chapter 5 for the powers of safety representatives);

(d) consideration is given to the establishment of a formal investigating committee;

(e) a person trained in the taking of statements is briefed.

Statements from witnesses and other persons should be taken as soon as possible by the nominated person or manager whilst the accident is still fresh in the minds of witnesses. Witnesses should be cautioned and interviewed separately and not allowed to discuss the accident together prior to the taking of statements. A shorthand typist should be made available if possible, and witnesses' statements should be read out to them at the completion of the process. All witness statements should be signed and dated.

In the event of involvement with the media (press, radio, television), a bare statement of the facts should be given as follows:

(a) an accident has taken place which is the subject of investigation;

(b) the person involved has been taken to hospital;

(c) a further statement may be made later.

A note should be made of all contacts with the media and a senior manager or a public relations officer (PRO) should be delegated to handle any future contact. Extreme care must be taken to avoid any admission of liability and the making of unnecessary assumptions by the media, e.g. no mention of a 'fatality' should be made at this stage.

Summary

The above procedure is only one way of tackling a fatal or major accident situation. Speed is of the utmost significance, particularly in taking photographs, preparation of sketches and interviewing witnesses. Facts surrounding the accident must be established quickly and accurately, bearing in mind that the accident will be the subject of specific investigation by enforcement agencies, insurance officials, representatives of the deceased or injured person, trade union safety representatives and staff of the Department of Social Security.

10 Health and safety training and communication

B

Training is defined by the Department of Employment as 'the systematic development of attitude, knowledge and skill patterns required by an individual to perform adequately a given task or job'. Under HSWA, it is a general requirement applying to all levels of management and the shop floor. Clearly, safety officers, competent persons and trade union safety representatives require specialised health and safety training.

Systematic training

The term 'systematic' immediately distinguishes such training from the traditional apprenticeship consisting of 'sitting by Nellie', i.e. learning through listening and observation. Systematic training makes full use of skills available in training staff. It attracts recruits, achieves the target of an experienced operator's skill in one half to one third of the traditional time, and creates confidence in the minds of trainees. It guarantees better safety performance and morale, and results in greater earnings and productivity, ease of mind, a sense of security and contentment at work. Furthermore, it excludes misfits and diminishes unrest, whilst facilitating the understanding and acceptance of change. Systematic training implies the following:

(a) the presence of a trained and competent instructor working with suitable trainees;

(b) defined training objectives;

(c) a content of knowledge broken down into sequential units which can be readily assimilated;

(d) a content of skills analysed by elements;

(e) a clear and orderly training programme;

(f) an appropriate place in which to learn;

(g) suitable equipment and visual aids;

(h) sufficient time to attain a desired standard of knowledge and competence, with frequent testing to ensure trainees understand and know what has to be learned.

A well-designed training programme can be truly educational. First, it is possible to place the specific job knowledge or skill in a wider context, relating it to affiliated topics, to what is already known by the trainee, and to possible future trends and developments. Second, any programme of technical instruction must be supplemented by other courses which enable an operator to relate to his entire sociopolitical environment, and answer his questions about the structure and meaning of society, the destiny of Man, the meaning of law and the lessons of history.

Education, on the other hand, is a lot more than the giving and receiving of information. As Maritian said, it is a 'human awakening'. The acid test of both education and training is the degree and quality of skill which they finally engender – conceptual skills, numerical skills, verbal skills, social skills, administrative and managerial skills.

The training process

Any process of systematic training must take place in a number of clearly defined stages. These stages are outlined below.

Identification of training needs

A training need exists when the optimum solution to an organisation's problem is through some form of training. For instance, a training need may exist where there is damage to stored goods and buildings through fork lift truck activity, or where a particular procedure, such as a permit to work system, is about to be introduced. For training to be effective, it must be integrated with the selection, placement and promotion policies of the organisation. Selection, however, must ensure that the trainees are capable of learning what is to be taught. Promotion must ensure that the promoted person is fully aware of his responsibilities in terms of health and safety at work.

Training needs should be identified with regard to the retraining, or the reinforcement of training, of existing personnel, and the induction training of new recruits. The identification should show:

(a) what kind of training is needed, e.g. safety procedure, duties and responsibilities, general safety training for supervisors;
(b) when such training will be needed, i.e. immediate, medium-term and long-term training requirements;
(c) for how many people training is needed;
(d) the standard of performance to be attained by the trainees.

Development of the training plan and programme

Training programmes must be co-ordinated with the personnel needs of the organisation. The first step in the development of the training programme is the definition of training objectives. Objectives or aims may be designed by job specification in the case of new training, or by detailed Task Analysis in the case of existing jobs. Safety training objectives may be assessed through the process of Job Safety Analysis (*see* Chapter 13).

Two important factors which need to be considered carefully in the development of a training programme are:

(a) what has to be taught – theoretical and practical areas;
(b) how it can best be taught.

What has to be taught will be determined by a number of factors, e.g. legal, production, health and safety, and industrial relations requirements. A comprehensive job specification for the trainees is relevant in clarifying such matters.

How is the new knowledge and/or skill to be imparted? Is it to be by a process of 'sitting by Nellie' or through a planned training programme? There are great disadvantages to the former, such as:

(a) the trainee acquiring bad habits, particularly in relation to safe working and procedures;
(b) Nellie not being able to explain what she is doing or not knowing the best or safest way of doing the job;
(c) the problem of the skilled operator being dissimilar to those of the trainee;
(d) the sizes of the learning stages being too large for the trainee;
(e) personality clashes developing between trainer and trainee who may be unsuited.

'Sitting by Nellie', however, does have some advantages:

(a) such a system can fit into a planned training programme, particularly if Nellie is a trained trainer and aware of the programme requirements;

(b) it involves a one-to-one relationship, which is a good training relationship, particularly where detailed safety requirements or procedures must be explained in stages.

Implementation of the training programme

It is important that, at this stage, a distinction is made between 'learning' and 'training'.

Learning goes on all the time. People are continually learning from each other and through their own experiences. It is the process by which manpower becomes and remains effective. If learning does not occur, an organisation will stagnate and soon cease to exist.

Training is simply a way to help people to learn. It may impart skills and/or alter a person's basic characteristics. To be effective, training must be planned and organised, i.e. systematic training.

In the implementation of the training programme, there are three factors which must be considered.

Organising the training

This covers the provision of training staff and equipment, and the training, certification and appointment of trained trainers. The facilities for running the training must be organised in terms of training rooms provided with blackboards, projectors and other equipment. The general environment of the training room with regard to temperature, lighting, ventilation, arrangement of tables and chairs, location of toilets and washing facilities, together with the arrangements for meals and refreshment periods, must also be considered.

Undertaking the training

With many training activities, such as the training of fork lift truck drivers, only trained trainers should be used. There must be a correct mix of both active and passive learning systems, and a system for monitoring the effectiveness of the trainers.

Recording the results

It is necessary to record improvements in operator performance and productivity, safety awareness and other factors included among the training objectives. Feedback systems must be incorporated with a view to measuring the effectiveness of the training by observation, questionnaire or the examination results of trainees.

Evaluation of the results

There are two questions that need to be answered at this stage:

(a) Have the training objectives been met?

(b) If they have been met, could they have been met more effectively?

Operator training in industry requires an appraisal of those skills needed to perform the task satisfactorily and safely. It is normal, therefore, to incorporate the results of such appraisals in the basic training objectives. A further objective is to bring about long-term changes in attitude on the part of trainees linked with performance in the job. Any decision, therefore, as to whether training objectives have been met cannot be taken immediately the trainee returns to work or after a short period of time. It may be several months, or even years, before a valid evaluation can be made, and only after continuous assessment of the trainee. A significant factor in training effectiveness is the extent of the transfer of training achieved by the trainee. 'Transfer of training' means the ability to carry over the knowledge acquired in one task to another task.

The answer to the second question can only be achieved through feedback from staff monitoring the performance of trainees, and the trainees themselves. Feedback can usefully be employed in setting objectives for further training, in the revision of training content and in the analysis of training needs of other groups.

A model illustrating the training process with regard to health and safety is shown in Fig. 10.1.

Fig. 10.1 ◆ The training process: a model for health and safety

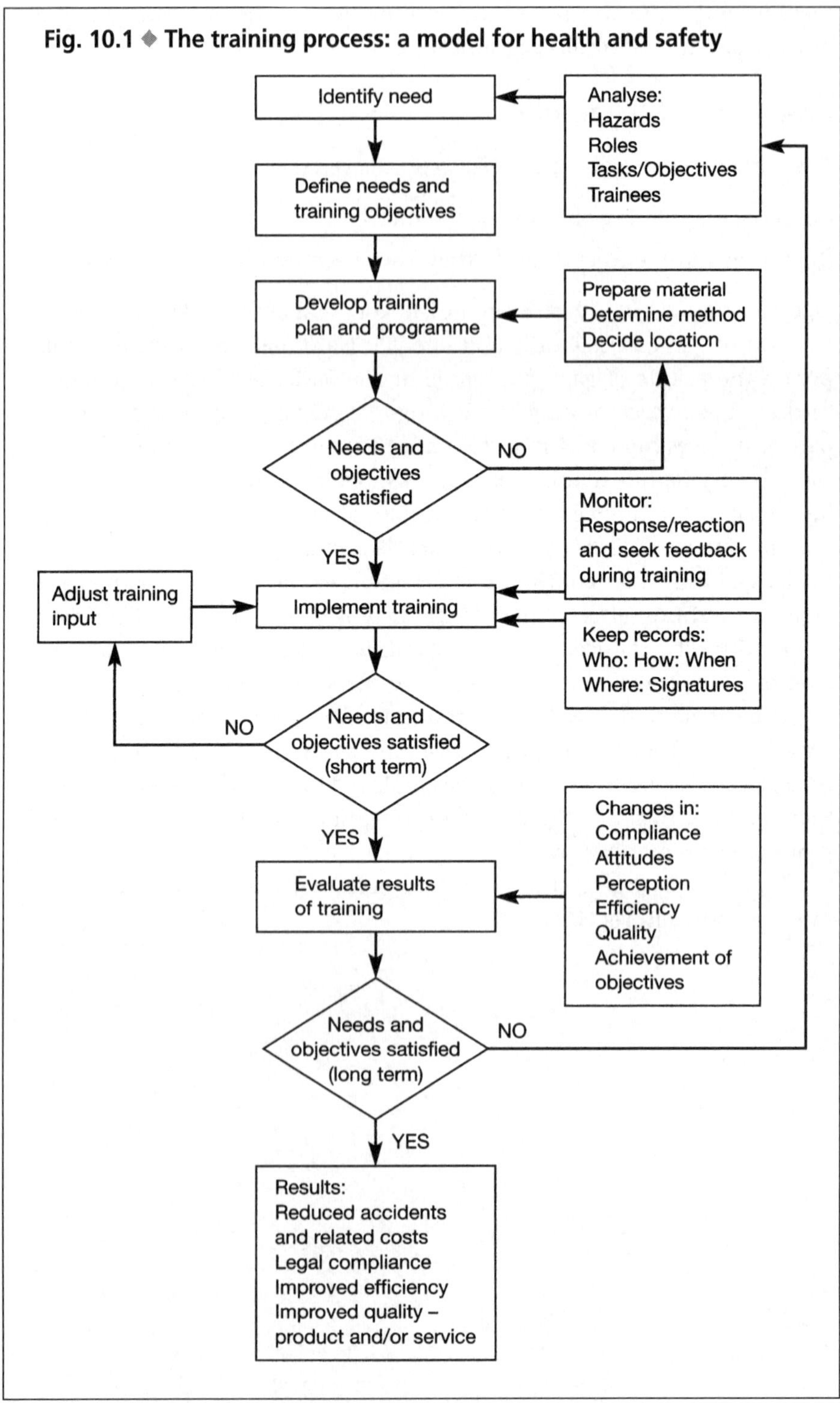

Induction training

All staff, irrespective of status within an organisation, should receive induction training, some of which will be off-the-job. Typical topics for inclusion in induction training are as follows:

1. The organisation's Statement of Health and Safety Policy and the individual responsibilities of all concerned.
2. Procedures for reporting hazards, accidents, near misses and ill-health.
3. Details of the hazards specific to the job, which may be associated with machinery, hazardous substances or energy sources used, the operating instructions and precautions necessary, together with formally-written safe systems of work and emergency procedures.
4. Procedures to follow in the event of fire – means of escape, assembly areas, the use of fire appliances – together with procedures to follow when the fire alarm sounds.
5. Safety monitoring procedures currently in operation and the systems for measurement of health and safety performance throughout the organisation.
6. Current welfare arrangements – sanitation, hand washing, showers, clothing storage, first aid and arrangements for taking meals.
7. Sources of health and safety information available.
8. The role and function of the company health and safety specialist, safety representatives and the health and safety committee.
9. The correct use of personal protective equipment, where provided.

Orientation training

This form of training should be undertaken where people change their job and/or work location. Topics for inclusion include:

1. Details of hazards specific to the job. (*See* 3 above.)
2. Procedures to follow in the event of fire. (*See* 4 above.)
3. Safety monitoring systems in operation. (*See* 5 above.)

The actual extent of orientation training will vary according to the perceived needs of the individual and the hazards to which he is exposed.

Management of Health and Safety at Work Regulations 1999

Reg 13 of the MHSWR stipulates circumstances where there is a duty on employers to train employees.

The importance of health and safety training

The identification of training needs, particularly in relation to health and safety, is an important task in any organisation. It has the basic objectives of:

(a) determining the content of the required training;

(b) indicating the best method of undertaking the training;

(c) highlighting the problem of motivating the organisation to implement the recommended training methods or to use the training provided;

(d) revealing the ultimate problem of motivating the organisation to apply the training once it has been given.

Health and safety training should emphasise certain basic themes summarised below:

(a) The importance of distinguishing between (i) accident and injury, and (ii) prevention and protection.

(b) The link between safety performance and operational efficiency, and that between accident prevention management and company management as a whole, must be recognised.

(c) Training must be regarded as the creation of learning situations, not simply by systematic instruction, but by better design of organisational structures and the adoption of appropriate management styles. Organisations are viewed as learning systems. This applies in all situations, including health and safety.

(d) The concept of training must be extended to developing people's full potential so that organisations can satisfy human needs by effective utilisation of manpower. It has been said that accidents 'downgrade the system'. In order to ensure effective utilisation of manpower, there is clearly a need for regular health and safety training of all staff, from the Board room downwards.

(e) Health and safety training should be given at each stage of a person's career, e.g. at induction stage, following the introduction of new processes, substances, systems of work and legislation, on promotion to inform of individual responsibilities, and on a refresher basis to keep individuals up to date on new legal requirements, procedures, systems, including emergency procedures, and new technology.

Communication

Many forms of communication operate within organisations. That which is officially inspired is often referred to as 'formal' communication, while communication which is unofficial, unplanned and spontaneous can be classified as 'informal' communication.

A communication system can transmit information up, down and sideways within an organisation on a one-way or two-way basis.

When communication is one-way there is no opportunity to receive a reaction or response from the receiver of the message, but in two-way communication the receiver can provide a response and is encouraged to do so.

What is communicated?

Communication is the transfer of information, ideas and emotions between one individual or group of individuals and another. The basic function is to convey meanings.

Functions of communication

1. *Instrumental* – to achieve or obtain something.
2. *Control* – to get someone to behave in a particular way.
3. *Information* – to find out or explain something.
4. *Expression* – to express feelings or put yourself over in a particular way.
5. *Social contact* – to enjoy another person's company.
6. *Alleviation of anxiety* – to sort out a problem, ease a worry.
7. *Stimulation* – because you find it interesting.
8. *Role-related* – because the situation required it.

Other factors in communication

Communication should not only be treated in terms of the verbal content of what is being said. It should include all the other things being communicated, for instance the variation in social status of the participants, their emotional attitudes, the way each tries to dominate or manipulate by using non-verbal signs, such as eye gaze, body signals, etc., the social context of the communication and each individual's perception of the other.

How is it communicated?

1. The way the words are spoken.
2. The accompanying non-verbal information.
3. Facial expression, gestures and posture.
4. The expectation of participants.
5. The context in which the transaction occurs.

Interpersonal communications consist of a complex fabric of interacting 'cues' or signals of moving kinds – sequence of words coloured by voice, tone, pitch, stress or rhythm, and movement of eyes, hands and body.

Communication almost always requires the use of a 'code' of some kind, e.g. language, non-verbal cues, emotional states, relationships, understandings.

Language

There are three kinds of meaning attached to words:

(a) *Denotive* – the key features that distinguish it;

(b) *Connotive* – varies according to experience, association and context;

(c) *Indexical* – provides an indication of the nature of the speaker.

Understanding

This is a critical factor in the communication process. Vital information with respect to hazards, for instance, should be repeated at intervals. The FIDO principle is worth noting. It says that learning by communication is enhanced by:

Frequency
Intensity
Duration
Over again

Functions of non-verbal communication

1. It can give support to verbal communication thus:
   (a) gestures can add to or emphasise words;
   (b) terminal glances help with speech synchronisation;
   (c) tone of voice and facial expression indicate the mood in which remarks are intended to be taken;
   (d) feedback on how others are responding to what is being said is obtained by non-verbal devices.
2. It can replace speech where speech is not possible.
3. It can perform ritualistic functions in everyday life and can communicate complex messages in greeting and farewell ceremonies.
4. It can express feelings we have about others, e.g. like or dislike.
5. It can express what condition we are in or feelings we have about others – happiness, anger, anxiety – although we may attempt to control them.
6. It can be used to convey how we would like other people to see us, e.g. by the way in which we present ourselves for public scrutiny.

Aspects of non-verbal communication

1. *Visual aspects:*
   (a) involuntary (non-controllable) – blushing, pallor, perspiration;
   (b) physical appearance – gestures, facial expression, gaze;
   (c) posture – static, movement and change of posture;
   (d) orientation; and
   (e) proximity.
2. *Tactile aspects:*
   (a) aggression – hitting and striking;
   (b) caressing and stroking;
   (c) guiding;
   (d) greeting (often formalised).
3. *Olfactory aspects:*

   Associated with how people smell.
4. *Auditory aspects:*

   Associated with how people hear things and their non-verbal responses.

Functions of verbal communication

This incorporates a number of features:

1. *Reflexes* – e.g. coughing.
2. *Non-verbal noises* – e.g. grunts; they need non-verbal accompaniment to clarify their meanings.
3. *Voice quality* – e.g. accent; for categorisation of the individual and the group to which he belongs.
4. *Linguistic* – verbal (choice of words); grammar and style;

   Paralinguistic – timing, speech, rhythms, tone of voice.

Barriers to communication

A number of pitfalls hinder effective communication:

1. The communicator is unable to think clearly.
2. There may be problems or difficulties in encoding the message.
3. Transmission of the message can be interrupted by noise.
4. Selectivity in reception, interpretation and retention of information is exercised.
5. The receiver of the message is too quick to jump to conclusions, or he becomes defensive.
6. An unsuitable environment acts as an impediment to good communication.
7. A misunderstanding of feedback from the receiver of the message arises, or there is no perceptible reaction from the receiver.
8. Rumour fills the gap in the formal communication system, and is normally associated with the 'grapevine'.

Communications failures

Associated with:

(a) time defects;

(b) spacial segregation;

(c) work groups who fail to co-ordinate activities;

(d) conflict between the staff organisation, specialists and line management.

Successful communication on health and safety issues

Communication of the right kind has a vital part to play in health and safety as a participative process. What is the right kind? The following forms of communication are significant.

1. Safety propaganda

The use of posters, films, exhibitions and other forms of repeating a specific message. Safety posters should be used to reinforce current health and safety themes, e.g. the use of eye protection, correct manual techniques, and should be changed on a regular basis. Videos should be used as part of scheduled training activities.

2. Safety incentive schemes

Various forms of planned motivation directed at rewarding good safety behaviour on the basis of formally agreed objectives and criteria have proved successful. Safety incentive schemes should *not* be based on a reduction in accident rates, however, as this can reduce or restrict accident reporting by employees.

3. Effective speech

Health and safety communication should incorporate sincerity, authority, confidence, accuracy and humour. There may be a need for training of speakers and trainers in these aspects.

4. Management leadership and example

This is perhaps the strongest form of non-verbal communication, and has a direct effect on attitudes to health and safety held by operators. There should be:

(a) a desire to attain a common goal;
(b) insight into ever-changing situations;
(c) alertness to the needs and motives of others;
(d) an ability to bear responsibility;
(e) competence in initiating and planning action;
(f) social interaction aimed at promoting the health and safety objectives of the organisation;
(g) communication – upwards and downwards;
(h) clear identification by senior management with the health and safety promotional activities undertaken by the organisation;
(j) management setting an example at all times on health and safety-related procedures.

Written communication

This may be of a formal or informal nature. It may take the form of business letters, memoranda and reports.

1. *Business letters*

Letters are sent to external organisations either giving or requesting information. As a formal means of communication, they may also be sent internally conveying information or to elicit information. Business letters must be laid out in a logical and concise manner.

2. *Memoranda*

Generally, memoranda comprise a few paragraphs and are an informal means of communication between members of an organisation. They should be simple and to the point.

3. *Reports*

A report is defined as 'a written record of activities based on authoritative sources, written by a qualified person and directed towards a predetermined group'. Reports tend to be of an impersonal nature. They state the facts or findings of the author of the report, e.g. following an accident investigation, make recommendations and, in some cases, seek approval for, say, expenditure, certain actions to be taken or not taken.

Reports provide information, formulate opinions and are directed at assisting people to make decisions.

Reports should follow a logical sequence. They should be written with clarity and may be accompanied by diagrams, tables and photographs. The principal objective of a report is to enable the reader to reach a conclusion as to future action in certain situations. A report should generally terminate with a recommendation or series of recommendations.

Communication on health and safety issues

Increasingly supervisors and line managers are required to prepare and undertake short training sessions on health and safety issues for their staff (*see Fig. 10.2*).

The following matters need consideration if such activities are to be successful and get the appropriate messages over to staff.

1. A list of topics to be covered should be developed, followed by the formulation of a specific programme.
2. Sessions should last no longer than 30 minutes.

3. Extensive use should be made of visual aids – films, videos, slides, flip charts, etc.
4. Topics should, as far as possible, be of direct relevance to the group.
5. Participation should be encouraged with a view to identifying possible misunderstandings or concerns that people may have. This is particularly important when introducing a new safe system of work or operating procedure.
6. Topics should be presented in a relatively simple fashion in the language that operators can understand. The use of complicated technical, legal or scientific terminology should be avoided, unless an explanation of the terms is incorporated in the session.
7. Consideration must be given to eliminating any boredom, loss of interest or adverse response from participants. On this basis, talks should be given on as friendly a basis as possible and in a relatively informal atmosphere. Many people respond adversely to the formal classroom situation commonly encountered in staff training exercises.

Fig. 10.2 ◆ The presentation and the presenter: important points

1. The presenter	
Appearance	Gestures and mannerisms
Body movement	Eye contact
Firmness	Empathy
Distracting habits	Pitch, Pace and Pause
Timing	Use of notes and visual aids
Punctuality	Convey – Conviction
SMILE!!	Confidence
	Sincerity
2. The presentation	
Preparation and rehearsal	Construction – beginning, middle and end
Content	
Word pictures	Vocabulary
3. The environment	
Room arrangements – layout, seating, tables	Temperature, lighting, ventilation
4. Support material	
Visuals – seen and understood; interest	Equipment – working order; spares available
Handouts	

11 The cost of accidents and ill-health

The term *accident* refers to personal injury and 'loss' (*see* Chapter 12). Given that a minimum level of profitability is essential for a business, loss prevention is a key factor.

All accidents, whether they result in personal injury, property and plant damage and/or interruption of business, represent losses to an organisation. The aim of this chapter is to consider the range of losses which can be incurred and how such losses can be quantified, and to explain how loss identification can assist in the establishment of priorities where the improvement of health and safety standards is concerned. Although senior management will complain of the 'burden of health and safety at work', it is the escalating costs of accidents which need closer scrutiny by management.

Health and safety – a multiple approach

The reasons for preventing accidents are threefold – legislative, humanitarian and economic:

1. The legislative approach is based on the enforcement of statutory provisions contained in Acts, Regulations, orders and approved codes of practice.
2. The humanitarian or social approach is based on the premise that no one likes to think that his error or omission has been the principal cause of an accident. No two people respond identically in their assessment or quantification of hazards. An approach based on subjective criteria of a situation is, therefore, of very little value.
3. The economic or financial approach presupposes that accidents and occupational diseases represent financial losses to the employing organisation. The approach aims to persuade managers that money can be saved and profits increased if accidents are reduced.

It should be appreciated that, in the prevention of accidents, all three approaches must be considered together.

The direct costs of accidents

Direct costs, sometimes referred to as 'insured costs', are largely concerned with a company's liabilities as employer and occupier. Direct costs are covered by premiums paid to an insurance company to provide cover against claims made by injured parties. The premiums paid are determined by the claims history of the organisation and the risks involved in the business. (For the legal liabilities of employers and occupiers, *see* Chapter 2.) Other direct costs of accidents are claims by insured persons and users of products manufactured by the organisation, which are settled either in or out of court, together with fines imposed by courts for breaches of health and safety law. (For the law relating to claims and penalties connected with health and safety at work, *see* Chapter 2.) Also included in this category might be substantial legal defence costs.

Direct costs are relatively simple to calculate annually and they may be included in the operating costs of the organisation. The cost of claims and fines, and court costs, are more difficult to predict and quantify.

The indirect costs of accidents

Indirect costs of accidents are many and varied and are often difficult to predict. Some indirect costs may be included, and thus hidden, in other costs, e.g. labour costs, production costs and administration costs. It is common for indirect costs to be ignored owing to the difficulty of separating them from other costs. Some indirect costs, however, are simple to quantify. These are outlined below.

Treatment costs

These can be divided as follows:

(a) *First aid* – the cost of first aid materials, including any re-treatment or re-dressing of injuries.

(b) *Transport* – the cost of transport for the injured person to home, doctor or hospital.

(c) *Hospital* – any charges made by a hospital for emergency or casualty treatment.

(d) *Other costs* – within these costs can be included attendance charges made by a local doctor, provision of specific medical materials or treatment not provided under the National Health Service, or the services of a specialist after referral by a doctor.

Lost time costs

Lost time costs include the cost of a person being away from the job for which he is normally paid. They are incurred by the injured person, management, first aid staff and others. These costs are assessed on the basis of time (number of hours) away from the job multiplied by the hourly rate for the job.

Production costs

These can be subdivided as follows:

(a) *Lost production* – production losses as a result of an accident.

(b) *Extra staff payments* – payment of overtime rates following hold-up in the normal production cycle after an accident.

(c) *Damage costs* – through damage to property, plant, vehicles, raw materials and finished products as a result of an accident; these costs should be calculated on the basis of –

   (i) replacement and/or repair of materials and plant;

   (ii) labour costs involved in replacement/repair.

(d) *Training and supervision costs* – where replacement labour is introduced to maintain production, costs should be calculated on the basis of –

   (i) actual man-hours for replacement labour;

   (ii) specific training costs for replacement labour, i.e. time of trainers and trainees together with other training costs.

Investigation costs

A number of people may be involved in the investigation of an accident – management, safety officer, shop-floor personnel, safety representative, together with the injured person. Investigation costs should be based on the total man-hours involved, with the hourly rates for individuals taken into account.

Miscellaneous costs

Miscellaneous costs may include ex-gratia payments made by the organisation, perhaps to a widow, the replacement costs of personal items belonging to the injured person and others, incidental costs incurred by witnesses, etc.

Costs to the state

These are difficult to quantify but are potentially enormous. They include the costs of medical treatment provided by the National Health Service, including hospitalisation, attendance of medical staff and provision of ambulance services, together with payment of injury/occupational illness benefit under the Social Security Act.

Costs to the injured person(s)

This chapter has previously considered the cost of accidents to the employer and, to a lesser extent, the state. The personal costs (apart from pain and suffering) also need consideration. They can include:

(a) possible loss of earnings, dependent on terms and conditions of employment, for the period of absence from work;
(b) possible total loss of earning capacity due to physical and/or mental injury and resultant inability to do the original job;
(c) possible reduced earning capacity for the reasons given in (*b*) above;
(d) possible legal costs in pursuing claims for injury where these are payable by a trade union;
(e) possible legal costs, both criminal and civil, where the injured person has been found guilty of reckless or negligent behaviour.

An example of a typical accident costing form is shown in Fig. 11.1.

Managing sickness absence

Many organisations pay great attention to accidents at work and the cost, both direct and indirect, of these accidents. On the other hand, the management of absence is commonly neglected on the basis that the problem is insoluble and outside the control of management. However, absence from work on the part of staff and others can be a substantial cost to an organisation. It can take a number of forms:

Fig. 11.1 ◈ Accident costing form

Details of accident Unit

Date. . . . . Time . . . . Place of accident. . . . .

Injured person
Name in full . . . .
Address . . . .
Occupation . . . . Age . . . .
Length of service . . . . Sex. . . . .
Accident details . . . .
Injury details . . . .

Accident costs **Direct costs**				£	p
1.	% of occupier's liability premium		These figures to be inserted by the insurance company		
2.	% of increased premiums payable				
3.	Claims				
4.	Fines and damages awarded in court				
5.	Court and legal representation costs				
Indirect costs					
6.	Treatment	(a)	First aid		
		(b)	Transport		
		(c)	Hospital		
		(d)	Others		
7.	Lost time	(a)	Injured person		
		(b)	Management		
		(c)	Supervisor(s)		
		(d)	First aiders		
		(e)	Others		
8.	Production	(a)	Lost production		
		(b)	Overtime payments		
		(c)	Damage to plant, vehicles, etc.		
		(d)	Training/supervision of replacement labour		
9.	Investigation	(a)	Management		
		(b)	Safety officer		
		(c)	Others, e.g. safety reps		
		(d)	Liaison with enforcement authority		
10.	Other costs	(a)	Ex-gratia payment to injured person		
		(b)	Replacement of personal items of		
			(i) injured person		
			(ii) other persons		
		(c)	Other miscellaneous costs		
			TOTAL COSTS		

(a) certificated or uncertificated sick leave, which may be of a long-term or short-term nature;

(b) other forms of authorised absence, e.g. to attend training courses;

(c) unauthorised absence and lateness.

Approximately 200 million working days are lost each year in the UK due to absence. This equates to 10–11 days absence per person per year, or a national absence rate of 4.6 per cent. Many organisations may be totally unaware of their absence rate, whilst others may indicate absence rates way in excess of the average 4.6 per cent. Whatever the absence rate may be, sickness absence, in particular, represents a substantial continuing loss. It is essential, therefore, that the whole area of sickness absence be successfully managed.

The legal position

Informal studies indicate that approximately 75 per cent of the sickness absence in some organisations is taken by 25 per cent of the workforce. A substantial part of this absence is accounted for by people taking frequent periods of short-term sickness absence (one or two days per month) which tends to go unnoticed. Long-term sickness absence, on the other hand, does tend to be noticed due, in many cases, to the need to recruit temporary labour or to require staff to work longer hours in order to cover that period of absence. However, it is only when attendance records are checked that the true scale of sickness absence is identified.

The legal issues are concerned with dismissal resulting from absence associated with:

(a) failure to produce sick notes;

(b) frequent and short unconnected periods of absence; and/or

(c) prolonged or continuous absence.

In the case of (a) and (b) above, it is essential for the organisation to go through a series of stages in order to counter a claim by an employee for unfair dismissal. In most cases an employee has a contract of employment, which would indicate a statement of terms and conditions of employment. He would be provided with details of any sick pay and the procedure for reporting sickness absence. An employee may be in breach of his contract of employment through either non-compliance with the scheme, e.g. failure to produce a sick note, or actual compliance with the scheme but with continuing frequent, short unconnected absences from work. In such cases, it would be appropriate for the company to withhold payment with a view to investigating whether there has been a case of misconduct in these situations. (Procedures for dealing with such an investigation and its outcome are featured in many national agreements between employers and trade unions.)

Before an employee can be dismissed on the basis of absence, it must be established that the employer has behaved reasonably in dealing with the matter. For instance, he may have been interviewed by management or an occupational health nurse with a view to ascertaining the causes of his absencee, e.g. a sick wife or children. Where it can be established that there are no real grounds for his previously poor attendance record, and the matter has been brought to his attention both verbally and in writing, it would be reasonable for the employer to dismiss the employee.

Prolonged or continuous absence, on the other hand, hinges around 'capability' rather than 'misconduct'. 'Capability' is assessed by reference to skill, aptitude, health or any other physical or mental quality. In many cases, perhaps through illness or injury, an employee may no longer have the capacity to undertake the work for which he is employed.

For a fair dismissal decision to be reached, particularly where an employee complains of unfair dismissal to an Employment Appeals Tribunal, the employer should be able to answer satisfactorily the following questions:

1. Were offers of alternative employment considered?
2. Was the employee consulted sufficiently in terms of his problems, prospects and the possibility of dismissal?
3. Was the employee warned that further sickness could result in the possibility of dismissal?
4. Has there been a chance recently for the employee to comment on his health or capability to work?
5. Was an independent medical opinion obtained?
6. Has the employer sought medical advice on the employee's condition?
7. Has the employer fully investigated all relevant matters relating to the dismissal decision?
8. How long, apart from illness, would the employment be likely to last?
9. Can the employer wait any longer for the employee to return? (Here a balance must be struck between the position of the employee, the interests of the organisation and the need to be fair.)

10. How important is it to replace the employee concerned?

11. Has the employee been consulted in the final step of the procedure?

Other questions which need consideration by an employer include:

1. Were the terms of the employee's contract of employment, including sick pay provisions, fulfilled?
2. Was the nature of the employment such that it constituted a key position?
3. What is the nature of the illness/condition?
4. How long has it continued?
5. What are the prospects of recovery?
6. What is the total period of employment?

If there is no adequate improvement in the attendance record, it is likely that, in most cases, the employer would be justified in treating the persistent absences as sufficient reason for dismissing the employee. At the completion of this deliberation stage, the employee must be advised of management's decision and the decision confirmed in writing. Where an employee complains to an Employment Appeals Tribunal, an employer must provide the following information:

(a) evidence of a fair review of the employee's attendance record and the reasons for that record;

(b) evidence of appropriate warnings being given, after the employee concerned has been given an opportunity to make representations.

The unquantifiable costs

There are some indirect costs which it is difficult to measure. These costs could conceivably amount to a further 25 per cent increase in actual calculated indirect costs. They may include:

(a) reduced output and performance owing to loss of workforce morale; this is particularly evident following a fatal accident and can affect some people for many months afterwards;

(b) reduced sales as a result of adverse media publicity culminating in customer alienation;

(c) trade union action or reaction;

(d) increased attention from enforcement agencies.

In total loss control terminology, 'accidents downgrade the system'. They cause losses which increase costs and reduce profits. This must be one of the most cogent motives for improving safety performance.

SECTION C

Accident prevention

12 Principles of accident prevention

What is an accident?

C

A number of definitions have been put forward over the last 30 years and they are quoted below:

(a) 'An unforeseeable event, often resulting in injury.'

(b) 'An unplanned and uncontrolled event which has led to or could have caused injury to persons, damage to plant or other loss' (RoSPA).

(c) 'An unexpected, unplanned event in a sequence of events that occurs through a combination of causes. It results in physical harm (injury or disease) to an individual, damage to property, business interruption or any combination of these' (Department of Occupational and Environmental Health, University of Aston in Birmingham).

(d) 'An accident is an undesired event that results in physical harm to a person or damage to property. It is usually the result of a contact with a source of energy (i.e. kinetic, electrical, chemical, thermal, ionising radiation, non-ionising radiation, etc.) above the threshold limit of the body or structure' (Frank Bird, Executive Director, American Institute of Loss Control).

(e) 'A management error; the result of errors or omissions on the part of management.'

A number of significant factors emerge from these definitions. Generally accidents are:

(a) unforeseeable, as far as the accident victim is concerned;

(b) unplanned;

(c) unintended;

(d) unexpected.

In most cases there is a sequence of events leading to the accident and there may be a number of contributory causes. An accident may result in injury, damage to property and plant, stoppage of or interference with the work process and resultant interruption of business, or any combination of these results.

An analysis of accidents and their causes requires consideration of the events leading up to a specific accident, 'the pre-accident situation'. It is also vital to know what to do after the accident, first, to minimise the effects of injuries, and second, to prevent a recurrence.

The pre-accident situation

There are two specific aspects to accidents, viz. risk, and how people perceive or recognise the risk. The element of risk may be associated with processes, machinery, access to buildings, inadequate supervision and control, or use of unsafe materials. How people perceive risk is a feature of their past experiences, training and general attitude to safety. Summarised simply, therefore, accidents are concerned with:

(a) the objective danger at a particular point in time associated with a specific machine, process, system of work or substance used at work;

(b) the subjective perception of risk on the part of the individual, which varies according to the features mentioned above, and also according to behavioural factors such as their level of arousal prior to the risk arising, motivation, memory, personality and attitude.

Accident prevention strategies should thus be directed at, first, bringing about a reduction in the objective danger in the workplace, and second, increasing the perception of risk on the part of individual workers. This is brought about, in the first case, by the use of 'safe place' strategies, and in the second case, by 'safe person' strategies, which are outlined below.

'Safe place' strategies

Safe place strategies feature in much of the occupational health and safety legislation that has been enacted over the last century, in particular the Health and Safety at Work etc., Act 1974. Their principal aim is to bring about a reduction in the objective danger at the workplace through measures such as machinery guarding, improvements in the working environment to reduce the risk of occupational disease, or the design of safe systems of work. They may be classified as follows.

Safe premises

This relates to the general structural requirements of industrial and commercial premises, such as stability of buildings, soundness of floors and the load-bearing capacity of beams. Environmental working conditions, such as levels of lighting and ventilation, feature in this classification, together with the system for assessment of environmental stressors and standards used in the evaluation of stress. (Legal requirements concerning floors, lighting, ventilation and temperature control are to be found in the Workplace (Health, Safety and Welfare) Regulations 1992.)

Safe plant

A wide range of plant and machinery, their power sources, location and use are relevant here. The safety aspects of individual processes, procedures for vetting new machinery and plant, and systems for maintenance and cleaning must be considered.

Safe processes

All factors contributing to the operation of a specific process must be considered, e.g. plant and machinery, raw materials, procedures for loading and unloading, the ergonomic aspects of machine operation, dangerous substances used in the process, and the operation of internal factory transport such as fork lift trucks.

Safe materials

Relevant here are the safety aspects of chemical substances, radioactive substances, raw materials of all types and specific hazards associated with the handling of materials. Adequate information on their correct use, storage and disposal must be provided. (There is a general requirement under HSWA, sec 6, to provide materials which are safe for use at work. More specifically, however, dangerous substances, supplied in the ordinary course of industrial use, or conveyed in bulk by road, must be packaged and labelled in accordance with current packaging and labelling requirements.)

Safe systems of work

The design and implementation of safe systems of work is a key feature of any safe place strategy. It incorporates planning, involvement of workers, training, and designing out hazards which may have existed with previous systems of work (*see* Chapter 13).

Safe access to work

This refers to safe access both to the factory or workplace from the road outside and to the working position, which may be several hundred metres up in the air, as with construction workers, or several miles below the Earth's surface in the case of miners. Consideration must be given, therefore, to matters such as factory approach roads, yards, work at high level, the use of portable and fixed access equipment – e.g. ladders and lifts – and the shoring of underground workings. (The legal requirements relating to access to workplaces are outlined in specific Regulations, e.g. the Construction (Working Places) Regulations 1966.)

Adequate supervision

In all organisations, there must be adequate safety supervision directed by senior management through supervisory management to the shop floor. The responsibilities of directors, senior managers, departmental managers, supervisors, specialists – such as health and safety practitioners – and shop-floor workers must be clearly identified in writing in the Statement of Health and Safety Policy. Moreover, since many workers have a written form of job description, this should incorporate their various duties and responsibilities related to health and safety. Failure to observe these contractual duties can lead to disciplinary action including, in cases of flagrant abuse, summary dismissal. Supervisory management has a key role in ensuring the maintenance of a safe place of work. They are the 'linkmen' between management and shop floor and, therefore, have the most immediate contact with the workforce. The duties of supervisors should be clearly identified and they should receive sufficient training and management support to discharge these duties effectively. (Failure to have properly trained supervisors can have serious legal consequences. First, there will be a criminal liability on the part of the employer for breach of HSWA, sec 2(2)(*c*) and, second, there may be a civil liability if injury is suffered by an employee as a result of lack of or inadequacy of supervision (*Bux* v. *Slough Metals Ltd* [1974] 1 AER 262).)

Competent and trained personnel

The duty to train all staff levels in safe systems of work is laid down in HSWA, sec 2(2)(*c*). Every employee will need some form of health and safety training. This should be satisfied through induction training, on-the-job training, and training in specialised aspects, such as the operation of permit to work systems or driver training. Moreover, persons who are designated 'competent persons', in order to comply with specific requirements – e.g. Electricity at Work Regulations 1989 – will also require specific training, as will the various categories of first aiders under the Health and Safety (First Aid) Regulations 1981. A well-trained labour force is a safe labour force and in organisations which undertake safety training accident costs (*see* Chapter 11) tend to be lower.

'Safe person' strategies

These strategies are concerned with protecting the individual in specific situations where a Safe Place strategy may not be wholly appropriate or possible to implement. They depend upon the individual conforming to certain prescribed standards, e.g. wearing personal protective equipment. Safe person strategies may be classified as follows.

Care of the vulnerable

In any work situation inevitably there will be some people who are more vulnerable than others to certain specific risks, e.g. where workers may be exposed to toxic substances, to small levels of radiation or to dangerous metals such as lead. Typical examples of 'vulnerable' groups are young persons who, through their lack of experience, may be unaware of hazards; pregnant women, where there may be specific danger to the foetus; and disabled persons, who may be limited in specific tasks which they are able to undertake. In a number of cases there may be a need for the medical surveillance of such persons. In particular, there are certain occupations which require periodic compulsory medical examination. These include persons employed with lead, in certain processes in chemical works, in chromium-plating processes and as divers.

Personal hygiene

The potential for occupational skin conditions caused by contact with certain substances – e.g. solvents, glues, adhesive and the wide range of skin sensitisers – needs consideration. There is also the risk of ingestion of chemical substances as a result of contamination of food and drink and their containers. Personal hygiene is very much a matter of individual upbringing. For instance, washing the hands after using the toilet is a basic hygiene drill instilled in most people by their parents and teachers, but it is amazing how many people fail to carry out this basic drill.

In order to promote good standards of personal hygiene, it is vital that the organisation provides adequate wash-basins, showers, hot and cold water, nail brushes, soap and towels. Soap is best provided in the liquid form from a wall-mounted dispenser, and may incorporate a bactericide, particularly where workers are at risk of suffering minor hand and arm abrasions. Drying facilities in the form of disposable paper towels or wallmounted cabinets dispensing non-returnable roller towels are recommended as opposed to simple roller towels and washable hand towels.

Personal protective equipment

Personal protective equipment is considered in Chapter 24. Personal protective clothing should be selected carefully for reasons of hygiene and workers should be made aware of the limitation of such equipment.

Careful conduct for the safety of the individual and others

Dangerous behaviour and 'horseplay' at work have always existed to some extent. Examples of dangerous behaviour are the wilful removal of machine guards; fighting; smoking in designated 'no smoking' areas; the dangerous driving of vehicles; abuse of equipment, gases and flammable substances; failure to wear personal protective clothing and equipment, such as eye protection or safety helmets, where the need arises; and the playing of practical jokes which expose workers to risk of injury. HSWA places a duty on every 'employee' while at work to take reasonable care for the health and safety of himself and of other persons who may be affected by his acts or omissions at work.

Caution towards danger

All workers and management should appreciate the risks in the workplace, and these risks should be clearly identified in the Statement of Health and Safety Policy, together with the precautions to be taken by workers to protect themselves from such risks.

Post-accident strategies

Accidents are usually unforeseeable (*see* above). This may be attributable to insufficient research into hazards: the 'it can't happen to me' philosophy, which is particularly prominent among skilled workers: human error, failure on the part of management or manufacturers of articles and substances to provide information; or simple mechanical failures in plant and machinery. Thus, whilst there must be an emphasis on accident prevention through the use of pre-accident strategies, provision must be made, by means of post-accident strategies, to deal with the aftermath of accidents and to learn from mistakes. Post-accident strategies can be classified in the following way.

Disaster/emergency planning

It is important, particularly in any large industrial or commercial enterprise, that a disaster or emergency plan be developed. Depending upon the inherent risks, the plan should be tailored to take account of the procedure in the event of a large fire, an explosion, a sudden release of toxic liquids, gases or fumes, building collapse or a major vehicular traffic accident on the premises (*see* Chapter 9).

Ameliorative strategies

These strategies are concerned mainly with minimising the effects of injuries as quickly and effectively as possible following an accident. They will include the provision and maintenance of first aid services, procedures for the rapid hospitalisation of injured persons and, possibly, a scheme for rehabilitation following major injury (*see* Chapter 29).

Feedback strategies

There is much to be learned following an accident. Investigation must be directed at identifying the causes of the accident and not just the effects. By the study of past accidents, feedback strategies attempt to prevent their recurrence. The correct use and interpretation of statistical information is vital here.

The causes of accidents

No chapter dealing with strategies in accident prevention would be complete without an examination of the causes of accidents. In major injury or multiple accidents, there may be up to twenty factors which contribute to the accident. In disasters such as the Flixborough incident the contributory causes are even greater in number. Whatever the accident situation, almost inevitably there will be some form of human involvement, however. People are unpredictable, they make mistakes, their attitude to safety varies and they forget things. Whilst the 'human factor' features strongly in the majority of accidents, it is only part of the process, and many other factors are relevant. A resumé of the principal causes of accidents is given below.

Design and/or layout of the working area

Bad design and layout of working areas, in many cases producing congestion and overcrowding, is one of the principal contributory causes of accidents. Ideally, work processes should follow a sequential flow with the raw materials coming in at one end of the building and the finished products going out at the other end. However, most manufacturing tends to be something of a compromise. Unless a factory is purpose-built to accommodate a particular process, the process will have been fitted into an existing building with either too much or too little space. Safety considerations must, therefore, be taken into account at the design stage of projects, particularly where existing buildings are to be modified for a new process.

Structural features

The floor is the most important structural feature in any work situation, and every year many workers are injured through slips, trips and falls on defective floors, badly drained floors or slippery floor finishes. Floor finishes should be specified according to the needs of the process and the people who have to operate the process. To facilitate cleaning, floors should be non-slip, resistant to oils, solvents, fats and other spillages, and be maintained in a good state of repair. Other structurally important features are staircases, elevated working platforms and teagle openings, all of which need adequate safeguards to prevent workers falling from one level to another.

Environmental features

Poor standards of temperature, lighting, ventilation and noise control characterise many accidents. They may be associated with people failing to see where they are going, failing to hear warning signals due to high ambient sound pressure levels or becoming lethargic as a result of high temperature and humidity levels.

Mechanical or materials failure

Machinery and plant breakdown frequently occur due to a failure to undertake regular preventive maintenance. Many accidents are associated with these breakdowns, particularly where operators carry out their own temporary maintenance to keep the line running. Maintenance schedules indicating the item of plant or machinery, the system of maintenance, the frequency of maintenance and the responsibility for ensuring that the maintenance is undertaken should be produced and implemented. Raw materials should be inspected on a regular basis to ensure soundness and suitability for use on machinery and plant.

Inadequate machinery guarding

With many machines there is a risk of contact with the moving parts, which can result in various forms of injury (*see* Chapter 32).

Bad housekeeping

Bad housekeeping is found to be the cause of many accidents. This may take the form of failure to clear spillages; articles left on the floor or in gangways, where they can be stepped on or tripped over; inadequate systems for storage of refuse; or a simple failure to adopt the principle of 'clean as you go'. Cleaning and housekeeping tasks are unpopular with workers, because they are considered 'menial', and with management, who often consider them to be 'unproductive'. Therefore, in many cases there is an unacceptably poor level of housekeeping. To combat this problem, cleaning schedules should be introduced and enforced (*see* Chapter 17). Staff whose specific occupation is cleaning should be employed on a permanent basis. They should use, and be trained in the use of, modern mechanically operated cleaning equipment.

Inadequate supervision and control

The duty of all levels of management to supervise and control safety at work is well established in law (*see* Chapters 2 and 4). Many accidents are a result of a failure of management to do this.

Inadequate training

Management have a duty to train staff in the general and specific areas of health and safety at work. As with supervision and control, the level of training received by workers varies greatly. A clearly defined health and safety training policy should be included in every Statement of Health and Safety Policy, and the training needs of individual groups of workers identified. Poorly trained workers, or workers who have received no health and safety training, are a danger to themselves and others, particularly in high-risk situations. There is a need to establish the principle that a safe worker is an efficient one.

Deficiencies in personal protective systems

Every year many accidents are associated with deficiencies in personal protection. There may be a failure by management to provide the protection; alternatively, workers may refuse to wear or use the protection because it is uncomfortable, hinders their working routine or, in the case of eye protection, reduces their vision owing to misting. There is a need, therefore, to assess individual requirements by consulting the workers. This approach is established in case law, much of which has concerned the suitability of eye protection. In certain cases, the wrong type of personal protection may be provided, e.g. respiratory and hearing protection, with the result that the worker does not receive adequate protection. Moreover, most types of personal protection need regular testing for efficiency and the detection of faults. This presupposes efficient supervision and control with, if necessary, the implementation of disciplinary procedures for persistent offenders.

Inadequate or ineffective rules and instructions

The occupier has a duty to explain the hazards to workers and the precautions necessary to ensure safe working (*see* Chapter 3). Most organisations undertake this activity and use internal codes of practice, working instructions, operating manuals or other forms of documentation. In many cases, rules and instructions are ambiguous, badly worded or simply not available to the people to whom they are directed. Alternatively, where consultation with the workforce has not taken place, workers may consider the instructions so unworkable or unreasonable that they are disregarded. Whatever the cause of this situation, there is clearly a need for training and briefing of operators in any new set of rules or procedures which may be introduced. Moreover, the procedures to be adopted when there are breaches of rules or instruction by workers must be fully understood. This requires vigilance on the part of supervisors and regular reinforcement training of operators.

Physical disability

Whilst not a significant factor in the causes of accidents, there are many people who, because of physical limitations or disabilities, may be prevented from undertaking certain tasks or operations. There is a clearcut need to assess regularly the physical limitations of disabled persons to ensure that there has not been a deterioration in their ability to perform certain tasks safely. (Such factors have been documented in case law established at industrial tribunals. Many cases concerned with unfair dismissal complaints have underlined the fact that if, through physical or mental disability or both, an employee is a source of danger to himself and/or other employees, the employer has a duty to dismiss him or, at least, transfer him to other, less dangerous work.)

Poor ergonomic design

There is much evidence to support the view that the poor ergonomic design of items such as controls and displays on plant and machinery, and of working positions such as those in crane drivers' and lorry drivers' cabs, has in the past been a contributory factor in accidents. Poor ergonomic design results in stress on the operator in terms of general fatigue, eye strain, discomfort, muscular strain and cramps. This may, in turn, lead to loss of interest, reduced arousal level, reduced perceptual powers and altered patterns of thought. All these effects can contribute to the 'human error' aspects of accidents.

Risk assessment

Under Reg 3 of the MHSWR employers and the self-employed must assess the risks to anyone who might be affected by their working activities. Assessments must be revised if they cease to be valid, and employers with five or more employees must make their assessments in writing and record any group of employees identified to be at high risk. The findings of the risk assessment will govern the way that employers comply with most of their subsequent duties. This concept of requiring an employer to assess the risks is very much in line with similar requirements under recent health and safety legislation. For instance, employers have to make health risk assessments under the COSHH Regulations and hearing risk assessments under the Noise at Work Regulations.

The Approved Code of Practice (ACOP), which accompanies the Regulations, states 'A risk assessment should usually involve identifying the hazards present in any undertaking (whether arising from work activities or from other factors, e.g. the layout of the premises) and then evaluating the extent of the risks involved, taking into account whatever precautions are already being taken'.

Risk assessment generally takes place in three clearly-defined stages:

(a) identification of all the hazards;

(b) evaluation of the risks;

(c) implementation of measures to eliminate or control the risks.

The ACOP states:

> (a) a *hazard* is something with the potential to cause harm (this can include substances or machines, methods of work and other aspects of work organisation);
>
> (b) *risk* expresses the likelihood that the harm from a particular hazard is realised;
>
> (c) the *extent* of the risk covers the population which might be affected by a risk, i.e. the number of people who might be exposed and the consequences for them.

Risk therefore reflects both the likelihood that harm will occur and its severity.

The risk assessment process

Fundamentally, risk assessment takes into account a number of factors, in particular the probability or likelihood that an accident or incident could be caused as a result of a particular activity, the severity of the outcome of that accident or incident in terms of injury, damage or loss and the number of people affected, and the frequency of exposure to risk. A risk assessment must:

(a) identify all the hazards;

(b) evaluate the risks arising from such hazards;

(c) record the significant findings where more than five persons are employed;

(d) identify any specific group of employees or individuals who are especially at risk;

(e) identify others who may be specifically at risk, e.g. members of the public, visitors;

(f) evaluate current control procedures, including the provision of information, instruction and training;

(g) assess the probability of an accident or incident occurring as a result of uncontrolled risk;

(h) record any circumstances arising from the assessment where there is a potential for serious or imminent danger;

(j) specify information requirements for employees, including precautionary measures and emergency arrangements;

(k) provide an action plan giving information on the implementation of additional controls, in order of priority, and with an appropriate timescale for such implementation

Identification of hazards

In many cases, hazards can be simply identified. In rather more complex processes or activities the use of certain techniques, such as job analysis and job safety analysis, may be necessary.

Quantifying the risks

A quantitative approach to the assessment of risk is commonly used through use of 'risk ratings', taking into account the factors of probability, severity and frequency, each of which can be measured on a scale from 1 to 10. A risk rating is calculated thus:

Risk Rating = Probability (P) × Severity (S) × Frequency (F)

which gives a range of risk ratings between 1 and 1000. Standard Probability, Severity and Frequency Tables are used. (*See* Tables 12.2–4.)

The urgency or priority of action in respect of a particular risk can be evaluated as shown in Table 12.1.

Table 12.1 ◆ Priority of action in respect of risk

Below 200	No immediate action necessary, but keep under review
200–400	Action within next year
400–600	Action within next three months
600–800	Action within next month
800–1000	Immediate action/possible prohibition of use

Table 12.2 ◆ Probability index

Probability index	*Descriptive phase*
10	Inevitable
9	Almost certain
8	Very likely
7	Probable
6	More than even chance
5	Even chance
4	Less than even chance
3	Improbable
2	Very improbable
1	Almost impossible

Table 12.3 ◆ Severity index

Severity index	*Descriptive phrase*
10	Death
9	Permanent total incapacity
8	Permanent severe incapacity
7	Permanent slight incapacity
6	Absent from work for more than three weeks with subsequent recurring incapacity
5	Absent from work for more than three weeks but with subsequent complete recovery
4	Absent from work for more than three days but less than three weeks with subsequent complete recovery
3	Absent from work for less than three days with complete recovery
2	Minor injury with no lost time and complete recovery
1	No human injury expected

Table 12.4 ◆ Frequency index

Frequency	*Descriptive phrase*
10	Hazard permanently present
9	Hazard arises every 30 seconds
8	Hazard arises every minute
7	Hazard arises every 30 minutes
6	Hazard arises every hour
5	Hazard arises every shift
4	Hazard arises once a week
3	Hazard arises once a month
2	Hazard arises every year
1	Hazard arises every five years

C

The assessment/evaluation of hazards using the above technique is best undertaken on a team basis, the team to comprise a senior manager, supervisor, safety adviser and trade union safety representative. It is essential that the senior manager and/or safety adviser have sufficient authority within the organisation to instigate action, particularly where risk ratings are assessed at over the 800 mark. The system should be backed up by formal documentation, e.g. a Risk Rating Manual, which identifies each risk, indicates the assessed risk rating, identifies the manager responsible for dealing with same, the action required and a time specification for completion of action. Records should be maintained by the safety adviser of all assessed risks and action taken.

The HSE approach to risk assessment

In their publication *Five Steps to Risk Assessment* the HSE recommend the following procedure for undertaking a risk assessment as shown in Fig. 12.1.

Fig. 12.1 ◆ HSE risk assessment procedure

Assessment of risk for:

Company name ....

Company address ....

....

.... Post code ....

Assessment undertaken (date) ....

Signed ....

Assessment review date ....

Hazard

Look only for hazards which you could reasonably expect to result in significant harm under the conditions in your workplace. Use the following examples as a guide:

- Slipping/tripping hazards
- Electricity
- Dust and/or fume
- Work at heights
- Ejection of material from machines
- Pressure systems
- Vehicles
- Fire
- Chemicals
- Moving parts of machinery
- Manual handling
- Noise
- Poor lighting
- Low temperature

List hazards here:

....

....

....

....

....

....

Who might be harmed?

There is no need to list individuals by name. Just think about groups of people doing similar work or who may be affected, eg –

- Office staff
- Maintenance personnel
- Contractors
- People sharing your workplace
- Operators
- Cleaners
- Members of the public

Pay particular attention to:

- Staff with disabilities
- Visitors
- Inexperienced staff
- Lone workers

They may be more vulnerable.

Fig. 12.1 ◆ continued

List groups of people who are especially at risk from the significant hazards which you have identified.

. . . .

. . . .

. . . .

. . . .

. . . .

Is the risk adequately controlled?
Have you already taken precautions against the risks from the hazards you listed? For example, have you provided:

- adequate information, instruction and training?
- adequate systems of procedures?

Do the precautions:

- meet the standards set by a legal requirement?
- comply with a recognised industry standard?
- represent good practice?
- reduce risk as far as reasonably practicable?

If so, then the risks are adequately controlled, but you need to indicate the precautions you have in place. You may refer to procedures, manuals, company rules, etc. giving this information.

List existing controls here or note where the information may be found:

. . . .

. . . .

. . . .

. . . .

. . . .

What further action is necessary to control the risk?
What more could you reasonably do for those risks which you found were not adequately controlled?

You will need to give priority to those risks which affect large numbers of people and/or could result in serious harm. Apply the principles below when taking further action, if possible in the following order:

- Remove the risk completely
- Try a less risky option
- Prevent access to the hazard
- Organise work to reduce exposure to the hazard
- Issue personal protective equipment
- Provide welfare facilities eg washing facilities, first aid

Fig. 12.1 ◆ continued

List the risks which are not adequately controlled and the action you will take where it is reasonably practicable to do more. You are entitled to take cost into account, unless the risk is high.

. . . .

. . . .

. . . .

. . . .

HSE (1994): *Five Steps to Risk Assessment*: HSE Enquiry Points.

A 'suitable and sufficient' risk assessment

A suitable and sufficient risk assessment should:

(a) identify the significant risks arising out of work;

(b) enable the employer to identify and prioritise the measures that need to be taken to comply with the relevant statutory provisions;

(c) be appropriate to the nature of the work and such that it remains valid for a reasonable period of time.

In particular, a risk assessment should:

(a) ensure that all relevant hazards are addressed;

(b) address what actually happens in the workplace or during the work activity;

(c) ensure that all groups of employees and others who might be affected are considered;

(d) identify groups of workers who might be particularly at risk;

(e) take account of existing preventive or precautionary measures.

The level of detail in a risk assessment should be broadly proportionate to the risk.

Prevention/control of risks

The third stage of the risk assessment process is defining the preventive or control measures to be installed, e.g. prohibition of a particular practice, replacement of a hazardous substance with a safer alternative, machinery guarding, implementation of a safe system of work, enclosing the hazard, provision of information, instruction and training, etc.

Managing the risks

It must be appreciated that risk assessment is not a precise science. What may be a 'risky' operation for one person may be 'a matter of common sense' to another. It is necessary, therefore, to quantify the risks in a way which is understandable and credible to all concerned.

Once the risks have been assessed and the controls installed, then they must be maintained by a number of techniques, including safety monitoring activities, planned maintenance procedures, operation of cleaning schedules, environmental monitoring, testing and examination of certain items of plant and equipment and occupational health surveys.

13 Safe systems of work

A safe system of work is defined as 'the integration of personnel, articles and substances in a suitable environment and workplace to produce and maintain an acceptable standard of safety. Due consideration must also be given to foreseeable emergencies and the provisions of adequate rescue facilities'. HSWA requires the provision and maintenance of plant and systems of work that are, so far as is reasonably practicable, safe and without risks to health (*see* Chapter 3).

Components of a safe system of work

1. **People**

   People are the human assets of the enterprise. A safe system of work should incorporate safe behaviour, sound knowledge, skills, both mental and physical, willingness to conform to the system, motivation, resistance to pressure to behave unsafely and job experience.

2. **Machinery, plant and equipment**

   Sound design and safety specification of plant, machinery and equipment, including consideration of ergonomic factors, together with efficient and planned maintenance, are important features of a safe system of work (*see* Chapter 32).

3. **Materials**

   Materials must be safe during processing and as finished products, meet quality assurance standards, and be safe for disposal as waste products and by-products of manufacture.

4. **Environment**

   The fourth aspect includes control of temperature, lighting and ventilation, dust, fumes, vapours, radiation, chemical and biological hazards, and provision of safe access and egress, sound levels of welfare amenity provision and safe levels of noise and vibration.

5. **Place of work**

   This may be a factory workshop, construction site or office. The place of work should be safe in terms of its construction, means of fire protection, including means of escape in the event of fire, and layout.

Requirements for a safe system of work

The principal requirement is planning the safest way to combine people and plant so that the work can be undertaken in a specific area. This includes:

(a) a layout which allows for safe access to and egress from the working area and plant within, and adequate space between machines and operating plant;

(b) a correct sequence of operations with materials and products conveyed mechanically, wherever appropriate, to and from work positions;

(c) analysis of tasks, including Job Safety Analysis (*see below*), and the provision of clear job instructions;

(d) identification of safe procedures, both routine and emergency, including requirements that –

  (i) the authority for starting and stopping machines is clearly allocated and obvious;

  (ii) clear instructions are given to those allowed to lubricate or carry out maintenance work, including the circumstances under which this work may be done;

  (iii) adequate arrangements are made for removal of materials, components, scrap, trimmings, swarf and dirt from plant and the immediate floor area;

  (iv) preventive maintenance schedules incorporate safety checks;

  (v) there is a firm commitment to cleaning and housekeeping procedures;

(e) provision of a safe and healthy working environment, in particular –

(i) illumination levels which prevent glare and sharp contrasts between light and shadow;

(ii) heating and ventilation systems which avoid extremes of temperature and humidity and which allow circulation of fresh air to all parts of working areas;

(iii) ambient noise levels kept to within the limits imposed by current legal requirements, otherwise hearing protection may be necessary;

(iv) localised exhaust ventilation at work stations where dusts, fumes, gases or vapours are emitted from the work process.

Information sources

The importance of information sources in the design of safe systems of work cannot be overstated. The various sources of information are outlined in Chapter 6 'Safety management and policy'.

Job Safety Analysis

Job Safety Analysis is a technique which identifies all accident prevention measures appropriate to a particular job or area of work activity, and the behavioural factors which most significantly influence whether or not these measures are taken. The approach is both diagnostic and descriptive.

This analysis reflects the contribution which should be made by all personnel – managers, supervisors, safety representatives, health and safety specialists, engineers, contractors – in the creation of an overall 'safety climate' within which the individual receives the maximum support for his own accident prevention role. Hence it is possible to create an integrated approach to accident prevention through analysis which ensures that all functions are involved in a co-operative effort. Having recognised the remedial measures necessary, all functions can then become committed to their adoption. This would include design modification, methods improvement and machinery guarding. The training specialist would extract the relevant parts from the Job Safety Analysis for incorporation in his training analysis. In this way subsequent training programmes would make provision for all functions to be able to fulfil their accident prevention roles. If the training analysis has already been prepared, it can often be used as a basis for Job Safety Analysis.

Job Safety Analysis can be:

(a) job based, e.g. machinery operators, fork lift truck drivers; or
(b) activity based, e.g. manual handling activities, window cleaning, work at heights.

In all circumstances, however, it is normal to undertake Job Safety Analysis in two stages: Initial Job Safety Analysis and Total Job Safety Analysis.

Initial job safety analysis

The following information is required for effective analysis:

(a) job title;
(b) department or section;
(c) job operations, i.e. a stage-by-stage breakdown of the physical and mental tasks required in the job;
(d) machinery and equipment used;
(e) materials used, i.e. raw materials and finished products;
(f) protection needed. e.g. machinery guarding, personal protective equipment;
(g) the hazards that may be encountered;
(h) degree of risk involved;
(i) work organisation, including the responsibilities of supervisor and operator, current safety requirements and procedures;
(j) specific tasks – Task Analysis would split the job into various stages, e.g. setting up, feeding, controlling, unloading, machine maintenance and housekeeping aspects.

An example of this initial stage of the process is shown in Fig. 13.1.

Total Job Safety Analysis

From the basic information provided in the Initial Job Safety Analysis, Total Job Safety Analysis proceeds to an examination of the following:

(a) operations;
(b) hazards;
(c) skills required –
  (i) knowledge;
  (ii) behaviour;

Fig. 13.1 ◆ Initial job safety analysis

INITIAL JOB SAFETY ANALYSIS	
Job Title	Filling machine operator
Department	Bottling Department
Purpose	Filling mineral water bottles using automatic filler and capper

Machinery and Equipment
Rotary filler incorporating capping press, empty bottle conveyor, marshalling table and filled bottle conveyor; hand tools – spanners and steel lever.

Materials
Empty bottles; mineral water

Protection Clothing – One piece overall, apron and safety wellingtons, full face protection, hair enclosed in combined drill cap with snood attachment, heavy-duty gloves and ear defenders.

Machinery – 2 m high polycarbonate enclosure to filler with interlocked gates at access points; interlocked guard to capping press; fixed guards to conveyor end adjacent to filler.

Intrinsic Hazards
1. Hand contact with moving filler and capping press, and when changing cap supply.
2. Cuts to hands from removal of broken glass from filler.
3. Entanglement in in-running nip to conveyor.
4. Flying glass splinters due to occasional bottle explosions.
5. Broken glass on floor – risk of foot injuries.
6. Noise – risk of occupational deafness.
7. Injuries from slips and falls on wet floor.

Degree of Risk
Assume that four machines of this type are in regular use in the same department. The operator with least experience has been doing this job for eight months, others having up to 20 years' experience. The principal risks are in the removal of broken glass from the filler and falls due to wet floors.

Work Organisation
Assume further that a bonus system is operated on weekly output and work is mainly repetitive. After setting up, a large order could run for two or three days. On the other hand, a series of small orders are occasionally processed, which requires re-setting the machine after every four-minute cycle, when much more time would be spent setting than running the machine. Moreover, operator and assistant are interchangeable, but the operator makes the adjustment to the flow control device to ensure even and smooth running. Little maintenance work is required as the machine is very reliable. Typical maintenance includes replacing the driving belt and adjusting the brake – straightforward operations occupying a short period of time.

Specific Tasks
1. Setting up machine.
2. Regulating flow control device.
3. Controlling input of bottles to machine.
4. Removing broken bottles and caps, together with broken glass present.
5. Housekeeping activities, including removal of broken bottles from conveyor and liquid spillages.

(d) external influences on behaviour –
   (i) nature of influences, e.g. noise;
   (ii) source of influence, e.g. machinery;
   (iii) activities, e.g. loading procedure;
(e) learning method.

A standard form, which covers the above principal elements of the analysis, is used for this exercise. Four examples of Job Safety Analysis are shown in Table 13.1 (page 248).

Finally, one of the principal duties of the employer under HSWA is identifying the hazards and the precautions needed by staff, this information to be itemised in the Statement of Health and Safety Policy (*see* Chapter 3). Job Safety Analysis can be undertaken by groups of workers or as a joint exercise by the health and safety committee, the final analysis being incorporated in codes of practice or documentation which is referred to in the statement.

The management of safe systems of work

The actual management of safe systems of work takes place in a number of clearly defined stages.

1. Identification of need

Identification of the need for a formally-written safe system of work may arise as a result of a series of similar accidents, e.g. hand accidents, following a reportable dangerous occurrence, at the introduction of new machinery or process, or through routine observation of jobs and work activities. Poor standards of supervision and control could also be the cause for formal documentation of safe systems of work with a view to identifying hazards and the individual responsibilities of supervisors and operators.

2. Design of the system

This takes place through the use of Job Safety Analysis, using the sources of information available (*see* Chapter 6).

Table 13.1 ◆ Total job safety analysis related to task analysis

Task	*Hazards*	*Skills*	*Influences on behaviour*	*Learning method*
1. Sharpening a knife using a steel.	Cutting hand or fingers.	Co-ordination of movement of knife and steel.	Sharpness of knife. Condition of floor. Condition of knife. Space limitations. Other people present.	Demonstration of technique. Repetitive practice until speed increases.
2. Use of bench grinder to sharpen a chisel.	Sparks and metal particles in eyes and face. Hand contact with wheel. Entanglement of sleeve or tie in wheel. Flying fragments from possible wheel burst.	Co-ordination of cutting edge of chisel with rotating wheel. Correct use of tool rest.	Speed and condition of wheel. Condition of bench and floor. Space limitations. Illumination. Condition of chisel. Other people present.	Demonstration of method of operation and use. Repetitive practice at machine using guard. Inspection of tools and wheel – action on defects.
3. Dispensing strong chemical compounds from bulk.	Burns to eyes, face, hands and rest of body. Inhalation of fumes.	Correct use of tap, drum cradle, drip tray and container, and protective clothing.	Strength of chemicals and fumes. Corrosive effects. Illumination and ventilation. Arrangement of system. Space limitations. Condition of floor.	Demonstration of correct dispensing method. Practice using apron, gloves and full face protection. Recognition of hazards from spillage, splashing and fumes.

Table 13.1 ◆ continued

Task	*Hazards*	*Skills*	*Influences on behaviour*	*Learning method*
4. Working on fragile roof.	Falling through roof. Sliding down and falling off edge of roof. Algae on roof causing slips and falls. Cuts from roof sheets while handling. Eye and hand injuries while cutting and shaping roof sheets. Inhalation of dust while cutting. Electrical and mechanical hazards from use of drill, saw and cutting disc.	Correct position of access ladder, staging and crawl boards. Correct working position on crawl board. Knowledge of strong and weak points of roof. Correct handling and elevation of roof sheets. Correct placing, lapping, drilling and bolting of roof sheets. Correct use of safety harness. Knowledge of wind effects while handling roof sheets.	Weather conditions, e.g. wind speed. Height of roof. Existing condition of roof. Illumination. Spacing and location of structural roof members. Footwear worn. Provision and use of safety harness.	Demonstration of correct placing and use of ladder and crawl boards, erection and use of access system. Practice placing, lapping, drilling and bolting roof sheets. Recognition of dangerous conditions on existing roofs.

3. Documentation

Documentation of the safe system may entail reference to further information sources, such as British Standards, HSE Guidance Notes, manufacturers' information or hazard data sheets, etc. The documentation should be simple with the safe system outlined in a series of stages. There should be extensive consultation with operators at this stage, together with safety representatives and shop stewards, where appointed, with a view to reassuring operators of any changes necessary in working practices and obtaining their commitment to operating the safe system of work.

4. Instruction and training

Theoretical and practical instruction and training of supervisors and operators should take place prior to implementation of the system.

5. Operation of the system

Initial operation of the safe system of work should be closely monitored to ensure compliance with safe operation, to make small modifications where necessary and to obtain feedback from operators as to their willingness or reluctance to operate the safe system of work.

6. Review

After a stated period, the safe system of work should be reviewed with a view to assessing its effectiveness in terms of preventing accidents. In certain cases, there may be a need for revision of the documentation, reissue of the safe system of work and retraining of all concerned.

7. Maintenance of the system

Management and supervision should ensure that, once introduced and finalised, the safe systems of work are followed at all times. Failure to conform to an established safe system of work, perhaps incorporated in the company safe systems manual, should be regarded as a failure in performance. Persistent failure to conform should, therefore, become a matter for disciplinary action at all levels of the organisation.

Permit to work systems

A permit to work system is a formal safety control system designed to prevent accidental injury to personnel, damage to plant, premises and product, particularly when work with a foreseeably high hazard content is undertaken and the precautions required are numerous and complex. The permit to work is essentially a document which sets out the work to be done and the precautions to be taken. It predetermines a safe drill and is a clear record that all foreseeable hazards have been considered and that all precautions are defined and taken in the correct sequence. It does not, in itself, make the job safe, but is dependent for its effectiveness on specified persons carrying it out conscientiously and with a high degree of supervision, control and training of staff.

Permit to work systems have largely grown out of the requirements laid down in the FA. Certain sections of that Act require particular precautions to be taken, e.g. in respect of work involving dangerous substances and fumes (sec 30), where there may be lack of oxygen, the presence of inflammable or explosive dusts, vapours, gases and other substances, or where staff may need to enter confined spaces.

HSWA, secs 2(1) and 2(2) places a duty on the employer to ensure the health and safety at work of all of his employees and, in particular, to provide safe systems of work together with adequate supervision.

Requirements of the system

The system must be formal, but simple to operate, so as to ensure the commitment of those who operate and who are affected by it. Permits to work will involve the engineering function and production and service departments. In some cases, engineering staff may not be involved. In the operation of a permit to work system, the following principles must be observed:

(a) The permit must provide concise and accurate information about who is to do the work, the time span over which the permit is valid, specific work to be undertaken and precautions.

(b) The work instruction in the permit must be considered the principal instruction and, until it is cancelled, this instruction overrides all other instructions.

(c) No one must, in any circumstances, work at a place or on apparatus not indicated as safe by the permit.

(d) No one must undertake any work whatsoever which is not described in the permit to work. In the event of a change in the work programme, the permit must be amended, or cancelled and a new permit issued.

(e) Only the originator, or another person taking over responsibility, may amend or cancel the permit. Anyone taking over a permit, either as a matter of routine or in an emergency, must familiarise himself with and assume full responsibility for the work until he has formally handed the permit back to the originator, or the work is completed.

(f) Anyone accepting a permit is, from that moment, responsible for the safe conduct of the work within the limits of the permit. Above all, he must not allow himself to be persuaded to disregard its conditions.

(g) There must be effective liaison with controllers of other plant or work areas whose activities could be affected by the permitted work.

(h) When work has to be undertaken on part of a site, or on specific plant or equipment, the boundary or limits of the work area must be clearly marked or defined.

(i) Contractors undertaking specific tasks must be included in the permit to work system, including any briefing prior to commencement. Observance of safety rules and procedures, including the use of permits to work, should be a condition of contract. Training may be necessary for contractors' staff together with the provision of advice and assistance in certain cases.

The system in operation

The decision to issue a permit to work depends upon foreseeable hazards. A permit to work system should be operated for the following activities:

(a) entry into confined spaces, closed vessels and vessels containing agitators or other moving parts;

(b) work involving the breaking of pipe lines or the opening of plant containing steam, ammonia, chlorine, other hazardous chemicals and hot substances, or vapours, gases or liquids under pressure;

(c) work on certain electrical systems;

(d) welding and cutting operations in areas other than workshops;

(e) work in isolated locations, locations with difficult access or at high level;

(f) work in the vicinity of, or requiring the use of, highly flammable, explosive or toxic substances;

(g) work which may cause atmospheric pollution of the workplace;
(h) certain work involving commissioning, particularly pressure testing;
(i) certain fumigation activities using dangerous substances in gaseous form, e.g. methyl bromide;
(j) certain work involving ionising radiations;
(k) work involving contractors in any of the above activities on or about the premises.

Furthermore, it is necessary to assess the degree of risk to which personnel, contractors' staff, possibly members of the public, property and product are exposed in respect of:

(a) the type of work undertaken;
(b) the working method used;
(c) the location of the work;
(d) any articles and substances used which may affect, or be affected by, the work.

Operation of permit to work systems

The operation of a system should take place in a number of clearly defined stages as outlined below.

Assessment

This is the most important stage and should be undertaken by an authorised person, appointed in writing by senior management, who is experienced in the work and, where specialist plant is concerned, is familiar with the relevant process, chemistry and engineering. The person appointed must be allowed sufficient time to examine each part of the operation and personally check each stage of the action necessary in the task. Assessment should consider the work to be done, the methods by which the work can be done, and the hazards inherent in the plant in relation to the task. The ultimate objective of the assessment is to determine the steps which should be taken to make the job safe and the precautions which should be adopted during the actual working.

Withdrawal from service

Before plant is prepared for actual work or entry, it should be withdrawn from service and clearly designated by notices and/or fencing as a permit to work area, so that there is no chance of personnel not covered by the permit opening valves or activating machinery whilst others are inside or working on the plant. After withdrawal has been completed, the person in charge of the process must sign to that effect on the permit. This entry on the permit should also state that all operators concerned have been advised of the situation. Warning notices, as appropriate, should be displayed.

Isolation

After withdrawing the plant from service, it should be physically isolated by barriers displaying warning notices, and mechanically or electrically isolated. In certain cases, it is appropriate to indicate the stages in the isolation procedure in an attachment to the permit though, in general, the permit to work form should be specifically designed to take the authorised person through a series of logical precautions included on the form itself and designed to ensure that he does not fail to consider each and every such precaution. A declaration that the plant has been isolated should then be entered on the permit. It may be necessary also to undertake atmospheric testing, particularly if personnel are to enter a confined space.

Cancellation of permit to work

When scheduled operations have been completed, the permit to work certificate should be cancelled and returned to the originator, who should ensure that all the work has been completed satisfactorily. This person should sign the declaration that all personnel and equipment have been removed from the plant.

Return to service

The plant should now be returned to service. The person responsible for the plant should first check that the permit has been properly cancelled, and then make the final entry on the certificate accepting responsibility for the plant.

Training

Training in the operation of the system is necessary for management, supervisory staff, engineering staff and operators, including contractors' staff, who may operate a permit to work system.

Administrative procedures for permits to work

The permit to work must be raised, by the senior person responsible for carrying out the work, before the work, or the phase of the work requiring the permit, is commenced. The authority for issuing permits to work should be limited to specified appointed persons. When senior management is considering the appointment of 'authorised persons' for the purposes of signing permits to work, the following factors should be taken into account in relation to the person:

(a) age and experience;

(b) training and academic qualifications;

(c) knowledge of the actual plant or process, etc., involved;

(d) status and ability to control the permit to work operation.

Only a limited supply of permits to work should be available at any one time.

Work involving permits should be carefully planned to cause the least possible interference with, or interruption of, working processes. One benefit of planning is that time is made available to gain specialist advice or refer to written information sources to ensure adequate knowledge of the situation, the hazards involved and preventive methods needed. In any emergency situation, where a permit may be required quickly, the degree of risk may be much greater; hence the need to remain calm and consider the above factors before preparing a permit. The actual preparation of a permit must, in many cases, be classed as a 'team job'. It may require the knowledge and experience of the engineer, chemist, safety specialist, production manager and other specialists from both inside and outside the organisation. In some cases, it may be necessary to issue a permit in the middle of the night or on a Sunday afternoon, when there may be limited staff available. It is at such times that people may take short cuts in the procedure or fail to consider all the implications at the assessment stage of the operation. Where numerous permits are required, a system of early application may be needed, e.g. a request for a permit is made to the authorised person 24 hours before it is required, thereby giving him time to consider and arrange any necessary precautions, such as atmospheric testing.

All potential permit to work situations should be considered for individual locations for identified types of work, e.g. welding in confined spaces. Checklists should be prepared, particularly for isolation procedures, to ensure a uniform approach.

Documentation of the permit to work system

The permit to work should be printed in triplicate, self-carbonned and serial numbered, perhaps with different coloured pages for the original, first copy and second copy. These should be distributed by the originator as follows:

(a) The original should go to the person undertaking the work, and possibly posted at the place of work.
(b) The first copy should be given to the person responsible for the department or area in which the work is to be carried out.
(c) The second copy should be retained by the originator.

On completion of the work and final clearance of the permit, all copies should be returned to the originator for destruction, except for the second copy which should be kept for record purposes for a period of not less than two years.

Typical examples of permits to work are shown in Figs 13.2–13.6 at the end of this chapter.

Work in confined spaces

What is a confined space?

A 'confined space' is defined as a place which is substantially, though not always entirely, enclosed and where there is a risk that anyone who may enter the space could be:

(a) injured due to fire or explosion;
(b) overcome by gas, fumes, vapour, or the lack of oxygen;
(c) drowned;
(d) buried under free flowing solids, such as grain; or
(e) overcome due to high temperature.

Whilst some confined spaces are readily identifiable, for example, closed tanks, vessels and sewers, others, such as open-topped tanks, vats, silos, freight containers, ship holds, wells, deep trenches and closed unventilated rooms, may be less obvious but equally dangerous.

Risk assessment

The employer must undertake a risk assessment to determine the measures necessary to comply with the 'relevant statutory provisions' i.e. the Health and Safety at Work etc. Act 1974 and regulations made under same. This assessment will help to identify the measures that will be needed to enable the work to be undertaken without entering the confined space and the measures to ensure a safe system of work. Such a risk assessment should include consideration of the need for:

(a) a Permit to Work system;
(b) atmospheric monitoring;
(c) respiratory protective equipment and personal protective equipment;
(d) equipment for safe access to and egress from the confined space;
(e) suitable and sufficient emergency rescue arrangements.

Permit to Work systems

A Permit to Work system will usually be required where there is a significant risk in entering and working in the confined space. The system should be used as a formal procedure for ensuring all the elements of the safe system of work are in place before anyone is allowed to enter or work in the confined space. However, a Permit to Work system would not normally be needed if:

(a) the risks are low and can easily be controlled;
(b) the system of work is simple;
(c) it is known that other work activities that are being carried out will not affect safe working.

Requirements for a safe system of work

The precautions required to be included in the safe system of work will depend upon the nature of the confined space and the associated risks. The main elements of a safe system of work are:

1. **Supervision**
   An adequate level of supervision should be ensured. However, where the confined space work involves a very low risk, the employer might simply instruct the employee on the work to be undertaken and periodically check that all is well.
2. **Competence**
   Previous experience of working in similar confined spaces and the training received must be taken into account when selecting operators.
3. **Communications**
   Adequate communication is required which allows communication inside and outside the confined space, and to summon assistance in the event of an emergency.

4. **Atmospheric testing/monitoring**

   The atmosphere within the confined space may require testing for hazardous fumes, gases or vapours or to check the concentration of oxygen prior to entry to the space. Continued monitoring will usually be required to ensure that there is no change in the atmosphere.

5. **Gas purging**

   If the risk assessment identified the presence or possible presence of flammable or toxic gases there may be a need for purging of the gases.

6. **Ventilation**

   Some confined spaces will require mechanical ventilation to provide operators with sufficient fresh air.

7. **Personal and respiratory protective equipment**

   Where necessary, suitable equipment must be provided and used.

8. **Residue removal**

   Appropriate measures must be taken to remove any hazardous residues produced.

9. **Isolation of gases etc.**

   The flow of gases, liquids and other materials which could present a risk to operators working in the confined space must be isolated.

10. **Isolation of electrical and mechanical equipment**

    Power supplies to mechanically- or electrically-operated equipment inside the confined space must be isolated.

11. **Suitability of equipment**

    All work equipment provided for use should be suitable for the purpose.

12. **Gas cylinders and engines**

    Petrol-fuelled internal combustion engines should never be used in confined spaces. Portable gas cylinders are usually inappropriate, but if their use cannot be avoided, adequate ventilation should be provided.

13. **Gas supply**

    The use of pipes and hoses for conveying oxygen or flammable gases into a confined space should be controlled to minimise the risks, and at the end of every working period the supply valves securely closed and pipes withdrawn.

14. **Access and egress**

    Safe access and egress must be provided along with adequately sized openings.

15. Fire prevention
Where there is a risk of a flammable atmosphere all sources of ignition, including smoking and static discharges, must be avoided. Combustible materials should not be stored in spaces.

16. Lighting
Any lighting provided must be suitable for use in the confined space, including intrinsically safe lighting where necessary.

17. Emergencies
Suitable emergency rescue arrangements, with the necessary equipment, must be in place before anyone enters or starts work.

18. Time
In some circumstances there may be a need to limit the time period personnel are allowed to work in the space.

Confined Spaces Regulations 1997

Work in confined spaces has always been a high-risk activity and a major source of deaths in construction activities, chemical processing operations, agriculture and the public utilities, such as water undertakings.

These regulations require employers to:

(a) avoid entry to confined spaces, for example, by doing the work from outside;

(b) follow a safe system of work, e.g. a Permit-to-Work system, if entry to a confined space is unavoidable;

(c) put in place adequate emergency arrangements before work starts, which will also safeguard rescuers.

The regulations are accompanied by an ACOP and HSE Guidance.

Lone workers

What is a lone worker?

A 'lone worker' or 'solitary worker' is defined as anyone who works alone out of contact with other persons. A company representative may, for example, be classified as a lone worker even though he may be in contact with other people, such as customers, sub-contractors or suppliers.

General duties of employers

Whilst there is no specific legal prohibition on lone working, the employer must, under the Management of Health and Safety at Work Regulations 1999, plan, organise, control, monitor and review the activities of lone workers to ensure that they are not subjected to any more significant risks than other employees who may work together.

There is also a general duty on employers, under the Health and Safety at Work etc. Act 1974 to ensure, so far as is reasonably practicable, a safe system of work, safe access to and safe egress from the workplace.

Lone working arrangements

In the design of systems of work involving lone working, the following factors should be considered:

(a) Careful selection of operators who are fit, competent and reliable.

(b) The need to undertake a suitable and sufficient risk assessment of the lone working activity, which must be kept under review, together with regular monitoring of individual performance.

(c) The operators concerned must be provided with such information, instruction and training so that they are quite clear as to all the significant foreseeable risks which may arise and the measures they must take to ensure their own safety and the safety of other persons.

(d) A formally established safe system of work, such as a Permit-to-Work system, which incorporates detailed emergency procedures, must always be operated.

(e) Suitable and sufficient communication must be maintained, such as a radio or telephone-based buddy system, central control or electronic monitoring incorporating non-body movement indication/panic alarm and radio/satellite location, appropriate to the environment in which operators may be working.

(f) There must be adequate recognition of the more serious consequences for lone workers of fatigue and stress whilst travelling or undertaking their particular duties.

Competent persons

One way of ensuring the operation of a safe system of work is by the designation and employment of specifically trained operators who appreciate the risks involved, or by the use of external specialists. This may be for undertaking certain inspections, examination and testing of work equipment, or for undertaking certain activities where there may be a high degree of foreseeable risk.

The expression 'competent person' occurs frequently in construction safety law. For example, under the Construction (Health, Safety and Welfare) Regulations 1996 certain inspections, examinations, operations and supervisory duties must be undertaken by competent persons.

It should be noted that the term 'competent person' is not generally defined in law except in the Electricity at Work Regulations 1989 and the Pressure Systems Safety Regulations 2000. Therefore the onus is on the employer to decide whether persons are competent to undertake these duties. An employer might do this by reference to the person's training, qualifications and experience. Broadly, a competent person should have practical and theoretical knowledge as well as sufficient experience of the particular machinery, plant or procedure involved as will enable him to identify defects or weaknesses during plant or machinery examinations, and to assess their importance in relation to the strength and function of that plant and machinery (*Brazier* v. *Skipton Rock Company Limited* (1962) 1 AER 955).

Duties and functions of competent persons

Lifting Operations and Lifting Equipment Regulations (LOLER) 1998

The duties of competent persons under these regulations with respect to the 'thorough examination' and 'inspection' of lifting equipment, together with the making of reports, are extensive.

The regulations require employers to ensure the following thorough examinations and inspections are undertaken by competent persons:

1. In the case of lifting equipment for lifting people, and where there are no suitable devices to prevent the risk of a carrier falling, inspection of the rope or chain to that carrier every working day (Reg 5).
2. The planning of lifting operations involving lifting equipment (Reg 8).

3. Thorough examination for any defect in lifting equipment before being put into service for the first time unless either:

(a) the lifting equipment has not been used before;

(b) in the case of lifting equipment for which an EC declaration of conformity could or (in the case of a declaration under the Lifts Regulations 1997) should have been drawn up, the employer has received such declaration made not more than 12 months before the lifting equipment is put into service; or if obtained from the undertaking of another person, it is accompanied by the appropriate physical evidence i.e. that the last thorough examination required to be carried out under Reg 9 has been carried out (Reg 9).

4. Where the safety of lifting equipment depends upon the installation conditions, thorough examination:

(a) after installation and before being put into service for the first time;

(b) after assembly and before being put into service at a new site or in a new location (Reg 9).

5. Where lifting equipment is exposed to conditions causing deterioration which is liable to result in dangerous situations:

(a) thorough examination –

(i) in the case of lifting equipment for lifting persons or an accessory for lifting, at least every six months;

(ii) in the case of other lifting equipment, at least every 12 months; or

(iii) in either case, in accordance with an examination scheme; and

(iv) each time that exceptional circumstances which are liable to jeopardise the safety of the lifting equipment have occurred; and

(b) if appropriate for the purpose, inspection by a competent person at suitable intervals between thorough examinations, to ensure that health and safety conditions are maintained and that any deterioration can be detected and remedied in good time (Reg 9).

6. Following a thorough examination for an employer of lifting equipment under Reg 9, a competent person must:

(a) notify the employer forthwith of any defect in the lifting equipment which in his opinion is or could become a danger to persons;

- (b) as soon as is practicable make a report of the thorough examination in writing authenticated by him or on his behalf by signature or equally secure means and containing the information specified in Schedule 1 to –
    - (i) the employer;
    - (ii) any person from whom the lifting equipment has been hired or leased;
- (c) where there is in his opinion a defect in the lifting equipment involving an existing or imminent risk of serious personal injury send a copy of the report as soon as is practicable to the relevant enforcing authority (Reg 10).

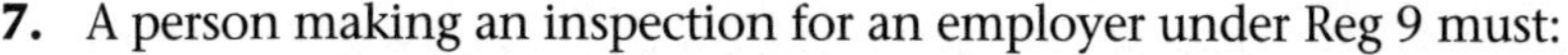

7. A person making an inspection for an employer under Reg 9 must:

- (a) notify the employer forthwith of any defect in the lifting equipment which in his opinion is or could become a danger to persons;
- (b) as soon as is practicable make a record of the inspection in writing (Reg 10).

Provision and Use of Work Equipment Regulations (PUWER) 1998

These regulations make reference to both 'inspection' and 'thorough inspection' of work equipment.

In the case of the inspection of work equipment under paragraph 1 or 2 of Reg 6, this means:

- (a) such visual or more rigorous inspection by a competent person as is appropriate for the purpose described in that paragraph;
- (b) where it is appropriate to carry out testing for the purpose, includes testing the nature and extent of which are appropriate for the purpose.

Under Reg 32, which deals with the thorough examination of power presses, guards and protection devices, 'thorough inspection' means:

- (a) a thorough examination by a competent person;
- (b) includes testing the nature and extent of which are appropriate for the purpose described in the paragraph.

See further Regs 6 and 32 of PUWER 1998.

Noise at Work Regulations 1989

A competent person must make a noise assessment which is adequate for the purpose of:

(a) identifying employees' noise exposure;

(b) providing the employer with appropriate information so as to enable him to facilitate compliance with his duties.

Pressure Systems Safety Regulations 2000

Owners and users of pressure systems to have a Written Scheme of Examination drawn up by a competent person for the examination of the system at specified intervals. Competence is based on the type of work undertaken i.e. minor, intermediate or major systems:

(a) advises user on the scope of the Written Scheme of Examination;

(b) draws up or certifies Schemes of Examination;

(c) undertakes examinations under the Scheme.

A competent person must be sufficiently independent from the interests of all other functions to ensure adequate segregation of accountabilities.

Electricity at Work Regulations 1989

No person must carry out a work activity where technical knowledge or experience is necessary to prevent danger or injury, unless he has such knowledge or is under the appropriate degree of supervision. (Whilst the term does not appear, competence is implied.)

Construction (Design and Management) Regulations 1994

Competence must be taken into account by:

(a) a client when appointing a planning supervisor;

(b) any person when arranging for a designer to prepare a design;

(c) any person when arranging for a contractor to carry out or manage construction work.

Construction (Health, Safety and Welfare) Regulations 1996

Competent persons must be appointed for:

1. Supervision of:

  (a) the installation or erection of any scaffold and any substantial addition or alteration to a scaffold;

  (b) the installation or erection of any personal suspension equipment or any means of arresting falls;

(c) erection or dismantling of any buttress, temporary support or temporary structure used to support a permanent structure;

(d) demolition or dismantling of any structure, or any part of any structure, being demolition or dismantling which gives rise to a risk of danger to any person;

(e) installation, alteration or dismantling of any support for an excavation;

(f) construction, installation, alteration or dismantling of a cofferdam or caisson;

(g) the safe transport of any person conveyed by water to or from any place of work.

2. Inspection of places of work as specified in Schedule 7 to the Regulations.

Management of Health and Safety at Work Regulations 1999

These Regulations bring in important new provisions relating to the appointment and mode of operation of 'competent persons', both generally to ensure the employer is complying with legal requirements (Reg 6), and specifically in connection with 'procedures for serious and imminent danger and for danger areas' (Reg 7).

Reg 7 – Health and safety assistance

1. Every employer shall, subject to paragraphs 6 and 7, appoint one or more competent persons to assist him in undertaking the measures he needs to take to comply with the requirements and prohibitions imposed upon him by or under the relevant statutory provisions.

2. Where an employer appoints persons in accordance with paragraph 1, he shall make arrangements for ensuring adequate co-operation between them.

3. The employer shall ensure that the number of persons appointed under paragraph 1, the time available for them to fulfil their functions and the means at their disposal are adequate having regard to the size of his undertaking, the risk to which his employees are exposed and the distribution of those risks throughout the undertaking.

4. The employer shall ensure that:

(a) any person appointed by him in accordance with paragraph 1 who is not in his employment –

C

    (i) is informed of the factors known by him to affect, or suspected by him of affecting, the health and safety of any other person who may be affected by the conduct of his undertaking;
    (ii) has access to the information referred to in Reg 8;
  (b) any person appointed by him in accordance with paragraph 1 is given such information about any person working in his undertaking who is –
    (i) employed by him under a fixed-term contract of employment; or
    (ii) employed in an employment business,

  as is necessary to enable that person to properly carry out the function specified in that paragraph.

5. A person shall be regarded as competent for the purposes of paragraph 1 where he has sufficient training and experience or knowledge and other qualities to enable him properly to assist in undertaking the measures referred to in that paragraph.

6. Paragraph (1) shall not apply to a self-employed employer who is not in partnership with any other person where he has sufficient training and experience or knowledge and other qualities properly to undertake the measures referred to in that paragraph himself.

7. Paragraph (1) above shall not apply to individuals who are employers and who are together carrying on business in partnership where at least one of the individuals concerned has sufficient training and experience or knowledge and other qualities:

  (a) properly to undertake the measures he needs to undertake to comply with the requirements and prohibitions imposed upon him by or under the relevant statutory provisions;
  (b) properly to assist his fellow partners in undertaking the measures they need to take to comply with the requirements imposed upon them or under the relevant statutory provisions.

The ACOP accompanying the regulations makes the following points:

1. Employers are solely responsible for ensuring that those they appoint to assist them with health and safety measures are competent to carry out the tasks they are assigned and are given adequate information and support. In making decisions on who to appoint, employers themselves need to know and understand the work involved, the principles of risk assessment and prevention, and current legislation and health and safety standards. Employers should ensure that anyone they appoint is capable of applying the above to whatever task they are assigned.

2. Employers must have access to competent help in applying the provisions of health and safety law, including these Regulations. In particular they need competent help in devising and applying protective measures, unless they are competent to undertake the measures without assistance. Appointment of competent people for this purpose should be included among the health and safety arrangements recorded under Reg 5(2). Employers are required by the Safety Representatives and Safety Committees Regulations 1977 to consult safety representatives in good time on arrangements for the appointment of competent assistance.

The HSE Guidance goes further on this matter:

1. When seeking competent assistance employers should look to appoint one or more of their employees, with the necessary means, or themselves, to provide the health and safety assistance required. If there is no relevant competent worker in the organisation or the level of competence is insufficient to assist the employer in complying with health and safety law, the employer should enlist an external service or person. In some circumstances a combination of internal and external competence might be appropriate, recognising the limitations of the internal competence. Some regulations contain specific requirements for obtaining advice from competent people to assist in complying with legal duties. For example, the Ionising Radiations Regulations requires the appointment of a radiation protection adviser in many circumstances, where work involves ionising radiations.
2. Employers who appoint doctors, nurses or other health professionals to advise them of the effects of work on employee health, or to carry out certain procedures, for example health surveillance, should first check that such providers can offer evidence of a sufficient level or expertise or training in occupational health. Registers of competent practitioners are maintained by several professional bodies, and are often valuable.
3. Competence in the sense that it is used in these Regulations does not necessarily depend on the possession of particular skills or qualifications. Simple situations may require only the following:

   (a) an understanding of relevant current best practice;

(b) an awareness of the limitations of one's own experience and knowledge;

(c) the ability to supplement existing experience and knowledge, when necessary by obtaining external help and advice.

4. More complicated situations will require the competent assistant to have a higher level of knowledge and experience. More complex or highly technical situations will call for specific applied knowledge and skills which can be offered by appropriately qualified specialists. Employers are advised to check the appropriate health and safety qualifications (some of which may be competence based and/or industry specific), or membership of a professional body or similar organisation (at an appropriate level and in an appropriate part of health and safety), to satisfy themselves that the assistant they appoint has a sufficiently high level of competence. Competence-based qualifications accredited by the Qualifications and Curriculum Authority and the Scottish Qualifications Authority may also provide a guide.

Reg 8 – Procedures for serious and imminent danger and for danger areas

Reg 8 of the MHSWR is concerned with emergency procedures and the appointment by employers of competent persons with sufficient training and experience or knowledge and other qualities to enable them properly to implement such procedures where appropriate. The principal duties on employers are:

(a) to establish and give effect to, where appropriate, procedures to be followed in the event of serious and imminent danger;

(b) to nominate a sufficient number of competent persons to implement these procedures as they relate to the evacuation of persons at work from the premises;

(c) to prevent access to any area for health and safety purposes unless the employee concerned has received adequate health and safety instruction.

The 'procedures' referred to in (a) above shall:

(a) so far as is practicable, require persons exposed to serious and imminent danger to be informed of the nature of the hazards and the precautions necessary;

(b) enable the person concerned to stop work immediately and proceed to a place of safety;

(c) save in exceptional cases for reasons duly substantiated, require the persons concerned to be prevented from resuming work in any situation where there is still a serious and imminent danger.

Fig. 13.2 ◆ Permit to work – plant

PERMIT TO WORK CERTIFICATE SERIAL NO: **1414**

LOCATION: **ORIGINATOR:** **DATE:**

PART A
Valid from . . . . (time) to . . . . (time) on . . . . (date)
Issued by . . . . to . . . .
This permit is issued for the following work . . . .
in . . . . department/area/section.

PART B – PRECAUTIONS	YES/NO	N/A	SIGNATURE
1 The above plant has been removed from service and persons under my supervision have been informed.			
2 The above plant has been isolated from all sources of: (a) ingress of dangerous fumes, flammable and toxic substances; (b) electrical and mechanical power; (c) heat, steam and/or hot water.			
3 The above plant has been freed of dangerous substances.			
4 Atmospheric tests have been carried out and the atmosphere is safe.			
5 The area is roped off or otherwise segregated from adjacent areas.			
6 The appropriate danger/caution notices have been displayed.			
7 The following additional safety precautions have been taken: (a) the use of safety belt and life line; (b) the use of goggles and/or gloves; (c) the use of flameproof lamps; (d) the use of fresh air/self-contained breathing apparatus; (e) prohibition on naked lights/ sources of ignition; (f) . . . . (g) . . . . (h)			

PART C – DECLARATION
I hereby declare that the operations detailed in Parts A and B have been completed and that the above particulars are correct.

Signed . . . . Date . . . . Time . . . .

PART D – RECEIPT/ACCEPTANCE OF CERTIFICATE
I have read and understand this certificate and will undertake to work in accordance with the conditions in it.

Signed . . . . Date . . . . Time . . . .

Fig. 13.2 ◆ continued

PART E – COMPLETION OF WORK
The work has been completed and all persons under my supervision, materials and equipment have been withdrawn.
Signed . . . . Date . . . . Time . . . .

PART F – REQUEST FOR EXTENSION
The work has NOT been completed and permission to continue is requested.
Signed . . . . Date . . . . Time . . . .

PART G – EXTENSION
I have re-examined the plant detailed above and confirm that the certificate may be extended to expire at . . . . (time).
Further precautions . . . .
Signed . . . . Date . . . . Time . . . .

PART H – CANCELLATION OF PERMIT
I hereby declare this Permit to Work cancelled and that all precautionary measures have been withdrawn.
Signed . . . . Date . . . . Time . . . .

PART I – RETURN TO SERVICE
I accept the above plant back into service.
Signed . . . . Date . . . . Time . . . .

PART J – REMARKS, SPECIAL CONDITIONS AND EXTRA INFORMATION

. . . .

. . . .

. . . .

. . . .

. . . .

. . . .

. . . .

. . . .

. . . .

C

Fig. 13.3 ◆ Permit to work – electrical apparatus

*Delete as appropriate throughout form

PERMIT TO WORK ON/TEST* ESSENTIALLY LIVE ELECTRICAL APPARATUS. PERMIT No **Z 7201**

1. DETAILS OF APPARATUS AND WORK/TEST* TO BE CARRIED OUT
   .... Contract No ....
   .... Location ....

2. REASON WHY WORK/TEST* CANNOT BE CARRIED OUT WITH APPARATUS ISOLATED
   ....

3. PERSON(S) INVOLVED IN WORK/TEST*
   *(only Competent Persons may work on/test live apparatus at or above 55 volts AC/DC)

COMPETENT PERSON	OBSERVER
Name and Initials	Name and Initials ....
Name and Initials	Name and Initials ....

4. PRECAUTIONS AND AUTHORISATION

4.1 SAFETY EQUIPMENT The following safety equipment will be provided and used through the work/test*

...... PAIRS INSULATING RUBBER GLOVES	– MAX SAFE VOLTAGE ....
...... PAIRS INSULATING RUBBER BOOTS	– MAX SAFE VOLTAGE ....
...... INSULATING RUBBER MATS	– MAX SAFE VOLTAGE ....
...... INSULATING PUSH BARS	– MAX SAFE VOLTAGE ....

OTHER TOOLS AND EQUIPMENT. (State type and when to be used) ....
....

4.2 ADJACENT LIVE EQUIPMENT The following precautions are to be taken to ensure that the persons named in Para 3 cannot come into contact with adjacent live equipment
....
....

4.3 DANGER AND CAUTION NOTICES have been posted at ....
....

4.4 ATMOSPHERIC CONDITIONS. The following precautions are to be taken to avoid danger from wet or humid conditions ....
....

Fig. 13.3 ◆ continued

4.5 VALIDATION PERIOD: .... This permit is effective
From .... hrs Date ....
To .... Date ....

4.6 AUTHORISATION
I declare that all precautions specified in Para 4 are in force and that the work/test* described in Para 1 may now begin.
Signed .... Manager Time .... hrs
Date ....

---

5. ACCEPTANCE BY COMPETENT PERSON(S) AND OBSERVER(S)

5.1 I acknowledge receipt of the Top (Yellow) Copy of this permit, and understand/will use the safety precautions listed in Para 4.
I will work only under the surveillance of the Observer, and will return this permit to the Manager when work is complete.
Signed .... Competent Person(s) Time .... hrs
Date ....

5.2 I acknowledge receipt of the first (Pink) Copy of this permit and will monitor the safe progress of the Competent Person. I have been instructed what to do in case of emergency, and will return the permit to the Manager when the work/test* is complete.
Signed ....Observer(s) Time .... hrs
Date ....

---

6. CLEARANCE
I hereby declare that the work/test* described in Para 1 is complete. The apparatus is safe and tools/gear have been withdrawn.
Signed .... Competent Person
Date .... Time ....

---

7. CANCELLATION
I hereby declare this permit cancelled. I have received the Yellow and Pink copies of the permit back respectively from the Competent Person and the Observer. These copies have been destroyed.
Signed .... Manager
Date .... Time ....

NOTES
(I) Top Copy (Yellow) and First Copy (Pink) to be issued by Manager respectively to Competent Person and Observer and retained by them during the work/test. Both copies to be returned to Manager for destruction on completion of work/test.
(II) Where the Manager is also the Competent Person he should issue the Top (Yellow) Copy to himself as a check on correct procedure.

Fig. 13.4 ◆ Permit to work – confined spaces

DANGEROUS CONFINED SPACE WORK PERMIT

Permit Serial No ....0028....
Date of Issue ....

1.	DETAILS AND LOCATION OF WORK TO BE CARRIED OUT Brief description of confined space ______ Permit Request No. ______			
2.	THIS PERMIT IS VALID FROM ______ hrs. Date______ to ______ hrs. Date ______			
3.	DESCRIPTION OF POSSIBLE HAZARDS WITHIN CONFINED SPACE (Quote actual gases, liquids, dusts, vapours, chemicals, oxygen deficiency/enrichment involved, and quote where possible LEL & UEL):			
4.	PRECAUTIONS	YES	NO	COMMENT
	Valve(s) isolated			
	Spade(s) fitted			
	Inert gas purged			
	No smoking/naked lights			
	Total disconnection			
	Closed/open steamed			
	Mech. through vent.			
	Drained free of liquid			
	Explosion-proof electrics			
	Lighting			
	Isolated machinery			
	'Don't Touch' labels			
	Breathing app. to be worn (specify type)			
	Observer/rescue staff outside space			
	Ops. to work in pairs			
	Lifeline to be worn			
	Standby resus. equipment			

Fig. 13.4 ◆ continued

Constant auto alarm monitor			
Ops. trained in hazards & precautions			
Other necessary precautions			

5. ATMOSPHERIC TEST RESULTS
Person carrying out tests (print name) ________ TITLE ________ DATE ________ TIME ______
Toxic gas results ________ ppm. (breathing app. to be worn at 50% of OES or above.)
Flammable gas test result ________ % explosive limit. (Entry not permitted if reading exceeds 0%LEL)
Oxygen sufficiency/deficiency test result ________ (Entry not permitted below 20% or above 20.8%).
Dust/fibre count result ________ (Breathing app. to be worn at 50% OES or above)

6. FURTHER SPECIAL CONDITIONS AND PRECAUTIONS:

7. AUTHORISATION: I have personally checked the above conditions & consider it safe to carry out this work.
Competent Person (print name) ________ TITLE ________
Signature ________ DATE ________ TIME ________

8. ACKNOWLEDGEMENT: I understand the hazards of this work and the precautions to be taken. These have also been fully explained to the operatives carrying out this work, and I consider them competent to do it safely. I will return my copy of this permit to the Manager when the work has been safely completed.
Performing Supervisor (print name) ________ TITLE ________
Signature ________ DATE ________ TIME ________

9. TIME EXTENSION: Subject to the following further precautions
(mark N/A if none is required) the expiry time of this permit is extended from ________ hrs. Date ________
to ________ hrs. Date ________ Signed ________ (Manager) Date ________

10. CANCELLATION
10.1 I have completed the work detailed in this permit, and have restored the location to a safe and orderly condition. I have returned my copy (yellow) and the display copy (pink) of this permit to the Manager.
Signed ________ (Performing Supervisor) Date ________
10.2 I accept that the work has been safely completed. The top (yellow) and 1st (pink) copy of this permit have been destroyed.
Signed ________ (Manager) Date ________ Time ________

NOTES: PERMIT IS AUTOMATICALLY SUSPENDED UPON SOUNDING OF EMERGENCY ALARMS, INSTRUCTIONS VIA PUBLIC ADDRESS SYSTEM, ETC. CHECK WITH MANAGER BEFORE RECOMMENCING WORK.

Fig. 13.5 ◆ Permit to work – ionising radiations

IONISING RADIATIONS WORK PERMIT

Permit Serial No ....0002....
Date of Issue ....

1.	DETAILS AND LOCATION OF WORK TO BE CARRIED OUT Permit Request No. ______			
2.	THIS PERMIT IS VALID FROM ______ hrs. Date ______ to ______ hrs. Date ______			
3.	DETAILS OF RADIATION SOURCE: (a) X-ray apparatus (make/type) ______ Max tube voltage ______ kV (b) Sealed source type ______ Strength ______ Curies Identification data ______			
4.	PRECAUTIONS	YES	NO	COMMENT
	Rad. area fenced			
	Caution notices posted			
	Warning lights, etc., positioned			
	Rad. area boundaries monitored			Min. reading = Max reading =
	Scatter shielding			
	Rad. meters checked/calibrated			
	Audio warning signal			
	Remote handling equipment checked			
	Sealed source container checked			
	Sealed source storage arrangements			
	Notification of HSE			Date
	Other special precautions			
	Name & initials of overseeing Competent Person			

Fig. 13.5 ◆ continued

5. DETAILS OF CLASSIFIED WORKERS CARRYING OUT WORK:
Name & initials ______ Name & initials ______ Name & initials ______
Name & initials ______ Name & initials ______ Name & initials ______

6. AUTHORISATION: I have personally checked the above conditions and consider it safe to carry out this work.
Competent Person (print name) ______ TITLE ______
Signature ______ DATE ______ TIME ______

7. ACKNOWLEDGEMENT: I understand the hazards of this work and the precautions to be taken. These have also been explained to the above Classified Workers, who will wear film badges/personal dosemeters throughout the work. I will return my copy of this permit to the Manager when the work has been safely completed.
Performing Supervisor (Competent Person) ______ (print name) TITLE ______
Signature ______ DATE ______ TIME ______

8. TIME EXTENSION: Subject to the following further precautions
(mark N/A if none is required) the expiry time of this permit is extended from ______ hrs. Date ______
to ______ hrs. Date ______ Signed ______ (Manager) Date ______

9. CANCELLATION
9.1 I have completed the work detailed in this permit, and have restored the location to a safe and orderly condition. I have returned my copy (yellow) and the display copy (pink) of this permit to the Manager.
Signed ______ (Performing Supervisor) Date ______
9.2 I accept that the work has been safely completed. The top (yellow) and 1st (pink) copy of this permit have been destroyed.
Signed ______ (Manager) Date ______ Time ______

NOTES:
PERMIT IS AUTOMATICALLY SUSPENDED UPON SOUNDING OF EMERGENCY ALARMS, INSTRUCTIONS VIA PUBLIC ADDRESS SYSTEM, ETC. CHECK WITH AUTHORISED PERSON BEFORE RECOMMENCING WORK.

Fig. 13.6 ◆ Permit to work – confined space

PERMIT TO WORK CERTIFICATE

PLANT DETAILS (Location, identifying number, etc.)		ACCEPTANCE OF CERTIFICATE	I have read and understand this certificate and will undertake to work in accordance with the conditions in it Signed Date Time
WORK TO BE DONE			
WITHDRAWAL FROM SERVICE	The above plant has been removed from service and persons under my supervision have been informed Signed Date Time	COMPLETION OF WORK	The work has been completed and all persons under my supervision, materials and equipment withdrawn Signed Date Time
ISOLATION	The above plant has been isolated from all sources of ingress of dangerous fumes etc. Signed The above plant has been isolated from all sources of electrical and mechanical power Signed The above plant has been isolated from all sources of heat Signed Date Time	REQUEST FOR EXTENSION	The work has not been completed and permission to continue is requested Signed Date Time

Fig. 13.6 ◆ continued

<table>
<tr><td>CLEANING AND PURGING</td><td>The above plant has been freed of dangerous materials
Material(s): Method(s):
Signed
Date Time</td><td rowspan="2">EXTENSION</td><td rowspan="2">I have re-examined the plant detailed above and confirm that the certificate may be extended to expire at:
Further precautions:
Signed
Date Time</td></tr>
<tr><td>TESTING</td><td>Contaminants tested Results
Signed
Date Time</td></tr>
<tr><td colspan="2" rowspan="2">I CERTIFY THAT I HAVE PERSONALLY EXAMINED THE PLANT DETAILED ABOVE AND SATISFIED MYSELF THAT THE ABOVE PARTICULARS ARE CORRECT
*(1) THE PLANT IS SAFE FOR ENTRY WITHOUT BREATHING APPARATUS
(2) BREATHING APPARATUS MUST BE WORN
Other precautions necessary:
Time of expiry of certificate:
*Delete (1) or (2)
Signed
Date Time</td><td colspan="2">THIS PERMIT TO WORK IS NOW CANCELLED. A NEW PERMIT WILL BE REQUIRED IF WORK IS TO CONTINUE
Signed
Date Time</td></tr>
<tr><td>RETURN TO SERVICE</td><td>I accept the above plant back into service
Signed
Date Time</td></tr>
</table>

SECTION D

The working environment

14 The organisation of the working environment

D

The term 'environment' is of French origin (*les environs* – the neighbourhood; that which surrounds us: the surroundings). The working environment embraces structural aspects of workplaces and the problems of stress brought about by poor standards of environment, e.g. extremes of temperature, lighting and ventilation, the presence of dust, fumes and gases, noise. Thus, occupational health and safety law – e.g. Health and Safety at Work etc., Act 1974 Workplace (Health, Safety and Welfare) Regulations 1992 – enshrines the employer's duty to provide a sound and healthy working environment, both in relation to the actual organisation of the environment and in the control of environmental stressors.

Environmental conditions have a direct effect on the way people behave at work, on the degree of risk of occupational disease and injury, and on morale, management/worker relations, labour turnover and profitability. This chapter examines the factors which are important in the organisation of the working environment, at the same time making recommendations for those involved in planning the environment and the enforcement of legal provisions relating to it.

Location of workplaces

Particularly with large undertakings, considerations should be given to the locality, density of surrounding buildings, availability of vehicle parking areas, access for employees and transport, including employees' own vehicles, and fire, ambulance and police vehicles. The need for a sound traffic control system, which does not expose pedestrians to risk of injury, must further be considered, together with the vulnerability of the general public to involvement in major accidents or diseases, e.g. from leaking gases or fumes, such as ammonia, explosions and/or large fires, contamination from toxic, corrosive or carcinogenic chemical substances, or heavy transport entering and leaving the premises.

Layout of workplaces

The term 'layout' refers to the space available for those employees working within a particular room or area, and the situation of plant, equipment, machinery, furnishings and stored goods in relation to the operative and the tasks performed. An efficient layout should make a material contribution to preventing or reducing overcrowding, minimising the physical and mental effort required to perform the operation, expediting the process in an orderly and sequential flow, and ensuring maximum safety and hygiene standards throughout. Moreover, the premises should be large enough and designed to allow for an orderly sequence of work without undue crossing of lanes and gangways, and without unnecessary manhandling of materials, but with easy movement between one part of the premises and another.

General requirements for workplaces

Under Reg 5 of the Workplace (Health, Safety and Welfare) Regulations 1992 (W(HSW)R), there is a general duty on employers to maintain the workplace, and equipment, devices and systems, thus:

1. The workplace and the equipment, devices and systems to which this regulation applies shall be maintained, (including cleaning as appropriate) in *an efficient state, in efficient working order and in good repair.*
2. Where appropriate, the equipment, devices and systems shall be subject to a suitable system of maintenance.
3. The equipment, devices and systems to which this regulation applies are:
   (a) equipment and devices a fault in which is liable to result in failure to comply with any of these Regulations;
   (b) mechanical ventilation systems provided pursuant to Reg 6 (whether or not they include equipment or devices with sub-paragraph (a) of this paragraph).

There is, therefore, a duty to instal and implement planned maintenance procedures which cover structural safety aspects of the workplace, in addition to those covering equipment, devices and systems.

Planned maintenance systems

Such systems should be produced on a formal written basis. They should incorporate the following elements:

(a) clear identification of the area, constructional item, item of equipment, device or system which requires maintenance;

(b) the inspection/testing and maintenance procedure to be followed;

(c) the frequency of inspection/testing and maintenance;

(d) clear identification of the individual with responsibility for ensuring the above procedure is followed;

(e) any general and specific precautions necessary, e.g. the operation of a Permit to Work system, segregation of the area.

Overcrowding

Reg 10 of the W(HSW)R requires that 'every room where persons work shall have sufficient floor area, height and unoccupied space for the purposes of health, safety and welfare.' Workplaces subject to the Factories Act 1961 should comply with Part 1 of Schedule 1 of the Regulations, as follows –

Part 1 – Space

1. No room in the workplace shall be so overcrowded as to cause risk to the health or safety of persons at work in it.

2. Without prejudice to the generality of para 1 the number of persons employed at a time in any workroom shall not be such that the amount of cubic space allowed for each is less than 11 cubic metres.

3. In calculating for the purposes of this Part the amount of cubic space in any room, no space more than 4.2 metres from the floor shall be taken into account and, where a room contains a gallery, the gallery shall be treated for the purposes of this Schedule as if it were partitioned off from the remainder of the room and formed a separate room.

Structural safety in relation to the worker

Floors and traffic routes

Specific provisions relating to the safety of floors and traffic routes in workplaces are dealt with in Reg 12 of the above Regulations. Floors and surfaces of traffic routes must be of suitable construction, free from dangerous holes, slopes and uneven and slippery surfaces, and provided with effective means of drainage where necessary. So far as is reasonably practicable, every floor and the surface of every traffic route shall be kept free from obstructions and from any article or substance which may cause a person to slip, trip or fall.

Floors should be of sound construction, free from obstruction and sudden changes in level, and of non-slip finish. Where safety levels, production or the storage of goods are materially assisted, storage areas should be clearly marked by the use of yellow or white lines. 'No Go' areas should be cross-hatched with yellow lines. All openings in floors or significant differences in floor level should be fenced. Attention should be paid to ensure that floor loading does not produce structural instability. (*See Greaves & Co. (Contractors) Ltd* v. *Baynham Meikle & Partners* [1975] 1 WLR 1905. This case concerned a badly designed floor which collapsed when heavy mobile machinery was installed. As a consequence, the structural engineering consultancy involved was held liable in negligence.)

Where a wet process is carried out, e.g. slaughterhouses, or where frequent floor washing is necessary, the floor should be laid to a fall to a drain. Floor channels incorporating metal gratings or covers can sometimes be used as an alternative, but this method may create hygiene risks.

Stairs, ladders and catwalks

Suitable and sufficient handrails and, if appropriate, guards must be provided on all traffic routes which are staircases except in circumstances in which a handrail cannot be provided without obstructing the traffic route. In the case of very wide staircases, further handrails may be necessary in addition to those at the sides. If necessary the space between the handrail and the treads should be filled in, or an intermediate rail fitted. Fixed vertical ladders and catwalks, including bridges to them, should be securely fixed. Where practicable back rings should be fitted to vertical ladders from a height of two metres upwards and spaced at one metre intervals. Catwalks and bridges should be adequately fenced by means of one metre high guard rails, 500 mm high intermediate rails and toe boards.

External areas, traffic routes and approach roads

Reg 17 of the W(HSW)R deals with the organisation and safety features of traffic routes, with the general requirement that every workplace shall be organised in such a way that pedestrians and vehicles can circulate in a safe manner. A 'traffic route' is defined as meaning a route for pedestrian traffic, vehicles or both and includes any stairs, staircase, fixed ladder, doorway, gateway, loading bay or ramp (Reg 2).

Traffic routes must be suitable for the persons or vehicles using same, sufficient in number, in suitable positions and of sufficient size. Particular precautions must be taken to prevent pedestrians or vehicles causing danger to persons near that route, ensure there is sufficient separation of traffic routes for vehicles from doors, gates and pedestrian routes, and where pedestrians and vehicles use the same traffic routes, ensure there is sufficient separation between them. Traffic routes must be suitably indicated where necessary for reasons of health or safety.

To facilitate access to and egress from the premises by people and vehicles, external areas should have impervious and even surfaces and be adequately drained to a stormwater drain. The provision of water supply points and hoses, for washing down yards and approaches, is recommended.

Windows, doors, gates, and walls, etc.

Specific provisions relating to structural items such as windows, doors, gates, walls, skylights and ventilators are incorporated in the W(HSW)R thus.

Every window or other transparent or translucent surface in a wall or partition and every transparent or translucent surface in a door or gate shall, where necessary for reasons of health and safety:

(a) be of safety material or be protected against breakage;

(b) be appropriately marked or incorporate feature so as, in either case, to make it apparent (Reg 14).

No window, skylight or ventilator which is capable of being opened shall be likely to be opened, closed or adjusted in a manner which exposes any person performing such operation to a risk to his health or safety. No window, skylight or ventilator shall be in a position which is likely to expose any person in the workplace to a risk to his health or safety (Reg 15).

All windows and skylights in a workplace shall be of a design or so constructed that they may be cleaned safely. In considering whether a window or skylight is safe, account may be taken of equipment used in conjunction with the window or skylight or of devices fitted to the building (Reg 16).

Doors and gates shall be suitably constructed (including fitted with any necessary safety devices) (Reg 17). Specific safety provisions apply to sliding doors/gates, upward opening doors/gates, powered doors and doors/gates which are capable of being opened by being pushed from either side.

Workplace glazing

The revision to the ACOP to the regulations qualifies the above requirements as follows.

In assessing whether it is necessary, for reasons of health or safety, for transparent or translucent surfaces in doors, gates, walls and partitions to be of safety material or be adequately protected against breakage, particular attention should be paid to the following cases:

(a) in doors and gates, and door and gate side panels, where any part of the transparent or translucent surface is at shoulder height or below;

(b) in windows, walls and partitions, where any part of the transparent or translucent surface is at waist level or below, except in glasshouses where people are likely to be aware of the presence of glazing and avoid contact with it.

This paragraph does not apply to narrow panes up to 250 mm wide measured between glazing beads.

Hospitals and nursing homes

Following a number of fatal accidents involving falls by patients from windows in hospitals and nursing homes, NHS Estates issued the following guidance to NHS Trust Managers and owners of registered nursing homes in 1989.

This guidance states that a restricted window opening of not more than 100 mm (four inches) is recommended for general use where windows are within easy reach of patients and is essential where windows are accessible to children. This standard is frequently used in geriatric, maternity and psychiatric areas.

Safety glazing should be considered where necessary for reason of health and safety. Those most at risk are the elderly, especially if they are confused, demented or anxious, and those suffering from mental illness. Disorientation and suicidal tendencies are foreseeable conditions for these types of patients.

It is important that the risk assessment considers the needs of the types of patients likely to be resident in, or visiting, the hospital or home and looks carefully at all situations which could give rise to a risk. This assessment must then be acted upon and appropriate precautions introduced and maintained.

Guidance referred to is Department of Health/Welsh Office Technical Memorandum No. 55 *Windows* in the Building Components series. (HMSO)

Walls

Interior walls have a contribution to make to illuminance levels, colour schemes, maintenance of physical cleanliness, sound insulation and the prevention of fire spread. They should be substantial, durable, smooth, easily cleaned and reflect light. The use of hollow partitions and the practice of battening out of walls are not recommended.

Ceilings and inner roof surfaces

The ceilings or inner roof surfaces of workrooms should assist in the maintenance of satisfactory illuminance levels, heat insulation, sound insulation and physical cleanliness. The height of ceilings should be a minimum of 2.4 metres. If false or suspended ceilings are incorporated in the structure, provision should be made for safe access to the space created above the ceiling for maintenance and cleaning purposes.

Escalators and moving walkways

Reg 19 of the W(SHW)R makes special provisions for escalators and moving walkways. In both cases, they shall:

(a) function safely;

(b) be equipped with any necessary safety devices;

(c) be fitted with one or more emergency stop control which is easily identifiable and readily accessible.

In all the above cases, further detail is incorporated in the ACOP accompanying the Regulations.

Colour

Colour is an important factor in the maintenance of a sound working environment, and influences:

(a) the extent to which the creation of a congenial environment is achieved;

(b) the amount of visual assistance afforded to employees by –

   (i) general and specific illuminance levels;

   (ii) drawing attention to specific parts of the workplace, e.g. fire escape routes;

   (iii) the control of glare;

(c) general safety performance, e.g. colour coding of safety signs and symbols as per the Safety Signs Regulations 1980, the use of 'tiger striping' to identify particular hazards.

Waste disposal

No waste material or refuse should be allowed to accumulate within a working area (W(HSW)R, Reg 9), and an adequate supply of containers should be provided at convenient points, together with an external storage and disposal area. Specifically, flammable wastes should be separated and stored in closed metal containers.

Good housekeeping requires that refuse and waste materials be removed not less than once daily and stored in a suitable enclosure. The use of refuse compactors in conjunction with an industrial side-loading waste container is recommended from a fire protection and hygiene viewpoint.

Traffic management

Many industrial accidents involve vehicles – fork lift truck, lorries, tractors, vans and cars. In most undertakings there is a need, therefore, for an efficient traffic control system. This assists in reducing internal traffic accidents and improves standards of security, particularly where employees and visitors are permitted to bring vehicles on to the premises. Effective traffic management embraces the following.

Segregation

There should be adequate segregation of pedestrians from vehicular traffic (W(HSW)R, Reg 17), of incoming from outgoing traffic, and of general parking areas from loading and unloading areas. In order to achieve this segregation, parking areas for commercial vehicles and cars must be clearly identified, properly marked out to indicate parking spaces, with directional signs showing the correct flow of traffic into, around and out of these areas. There is often a case for instigating a one-way system around the external parts of the premises, with separate entrance and exit to the highway. Road markings and signs should comply with the Road Traffic Regulation Act 1984 and BS 5378: *Safety Signs and Colours*, e.g. road junction marking, 'No Entry' signs.

Vehicular traffic control

The system for traffic control should be effective and regularly reviewed. This should include designation of 'No Go' areas by yellow crosshatched floor and road marking where there is extensive fork lift truck activity or in loading bays, or where pedestrians may be using a particular entrance to the premises. Moreover, speed control is most important in any system of traffic control. Apart from the standard speed limit signs (10 mph, 15 mph) there may be a need for the installation of 'sleeping policemen' or speed ramps, which have the effect of slowing down vehicles to an acceptable speed.

Pedestrian movement

Where there is extensive pedestrian movement at specific times of the day, it may be appropriate to install pedestrian crossings, with or without barriers, to ensure safe entry to and egress from the premises. The use of convex mirrors is recommended at obscured junctions or where vision of drivers coming from either direction is impaired.

Control over drivers

Above all, there must be a system to control drivers whilst on the premises. Constant speeding, unsafe parking and bad driving should result in disciplinary action. In the case of fork lift and other types of trucks, there is usually a case for a 'permit to drive' system, so that only those drivers who have passed an appropriate test are granted authority to drive trucks. Training undertaken by RoSPA is considered suitable (*see* Chapter 38). Fork lift truck 'cowboys' should be subject to disciplinary action, including the cancellation of the permit to drive in extreme cases.

Environmental control

Lighting in external roadways, parking areas, loading points and pedestrian walkways must be sufficient to allow for safe vehicle movement. An illuminance level of 50 lux is recommended in these areas, together with spot lighting where specific hazards may exist or where there is a need to ensure handling at night. This is consistent with current best practice.

Transport

In an analysis of transport fatal accidents, undertaken by HSE (*Transport Kills*, 1982), the following specific operations were involved:

Reversing vehicles	20 per cent
Vehicles under repair/maintenance	15 per cent
Loading/unloading	12 per cent
Vehicles overturning	10 per cent
Lift trucks overturning	7 per cent

Fatalities involving reversing vehicles when analysed by vehicle type reveal:

Heavy goods vehicles	41 per cent
Lift trucks	21 per cent
Mobile plant	13 per cent

Fatalities involving loading/unloading operations when analysed by vehicle type reveal:

Articulated lorries	40 per cent
Non-articulated lorries	28 per cent
Tipper lorries	12 per cent

Finally, it is interesting to note the occupations of the persons killed by transport:

Drivers/plant operators	29 per cent
Factory employees	14 per cent
Fitters/mechanics	12 per cent

An analysis of the causative factors which resulted in the fatal accidents reveals:

Safe system of work not provided	34 per cent
Inadequate training, information, instruction	23 per cent
Human error	15 per cent
Failure to follow safe system of work	14 per cent
Lack of maintenance/faulty vehicle	14 per cent
Poor management organisation	11 per cent

These figures serve to illustrate where action needs to be taken to improve transport operations.

Transport safety audit

Internal transport systems can involve heavy goods vehicles, private and company cars, fork lift trucks and commercial and other vehicles driven on to the premises by visitors and persons delivering goods. Accidents to staff and visitors, together with members to the public, may arise if transport safety arrangements are poor. As a means of identifying strong and weak areas of performance in the transport safety field, the Transport Safety Audit should be undertaken (Table 14.1). As with any form of safety audit, a plan indicating short-term, medium-term and long-term objectives to improve health and safety performance should be produced and implemented according to the agreed time schedule.

D

Table 14.1 ◆ Transport safety audit

TRANSPORT SAFETY AUDIT
In undertaking this audit, any response other than a positive (yes) response indicates a need for some form of action by local management.

	Yes	No
1. Organisation, systems and training		
Have all health and safety aspects of the transport operation been assessed?		
Have the organisation and arrangements for securing sound health and safety standards been incorporated in the company Statement of Health and Safety Policy?		
Has a trained person been appointed to deal with transport safety?		
Have safe systems of work been established?		
Are activities monitored to ensure safe systems of work are followed?		
Are all drivers adequately trained, examined and re-examined?		
Is there a formal company authorisation system for all drivers?		
Have all personnel been trained, informed and instructed about safe working practices where transport is involved?		
Is supervision adequate and effective?		
2. External roadways and manoeuvring areas		
Are they of adequate dimensions?		
Are they of sound construction?		

Table 14.1 ◆ continued

	Yes	No
Are they well maintained, adequately drained and well illuminated?		
Are road surfaces scarified when smooth?		
Are road surfaces gritted when slippery due to weather conditions?		
Are road surfaces and vehicle areas kept free of refuse and obstructions?		
Are there suitable and sufficient road markings, warning signs and speed limit notices?		
Is there a one-way system in operation?		
Is there provision for vehicles to reverse where necessary?		
Are there adequate pedestrian walkways and crossings?		
Are barriers installed by exit doors adjacent to roadways?		
Is there a separate vehicle parking area?		
Is storage located well away from vehicle movement areas?		
Is the road system suitable for internal factory transport, e.g. fork lift trucks?		
3. Internal transport arrangements		
Are internal traffic routes demarcated and separated from pedestrian routes?		
Are there separate internal and external doors for trucks and pedestrians?		
Are vision panels fitted?		
Are convex mirrors located at blind corners?		
Do trucks use a satisfactory warning system?		
4. Vehicles		
Is there a vehicle defect/fault reporting system?		
Are there regular vehicle inspections to ensure safety standards are being maintained?		
Are keys kept secure when vehicles and mobile plant are not in use?		
Is suitable access provided to elevated working places or vehicles?		
Are tractors and lift trucks equipped with protection to prevent the driver being struck by falling objects or thrown from his cab in the event of overturning?		
Are all dangerous parts of vehicles, e.g. power take-offs, adequately fenced?		
Are there fittings for earthing vehicles with highly flammable loads?		

Table 14.1 ◆ continued

	Yes	No
Are all loads correctly labelled? Are vehicles suitable for use in all areas of operation? Where passengers ride on vehicles, is a safe riding position provided? **5. Loading and unloading** Are loading and unloading positions separated from other vehicular activities? Do pedestrian walkways afford protection to people not involved in loading and unloading? Are precautions available in the event of specific hazards arising, e.g. flammable liquid spillages? Is there a yard manager to supervise operations, control movement of vehicles, and who can act as a banksman during reversing operations? Is the yard manager adequately trained in the use of signals and is cover provided during his absence? In the case of loading docks, does the layout prevent trucks falling off the edge of the dock or from colliding with other trucks or stored goods? Are mechanical hazards created by dock levellers adequately controlled? Are methods of loading and unloading assessed to ensure maximum safety provision? Are loads stable, properly secured and sheeted? Is there a pallet inspection scheme? **6. Vehicle maintenance and repair** Are the safety arrangements for the draining and repair of fuel tanks adequate? Are the arrangements for tyre repair and inflation adequate? Are there arrangements to ensure brakes are applied and wheels chocked? Is safe access provided to elevated working positions? Is portable electrical equipment of low voltage type and adequately earthed? Is the control over movement of vehicles in the workshop adequate? Are engines run with the brakes on and in neutral gear? Are raised bodies always propped? Are jacks and axle stands used correctly?		

15 Temperature, lighting and ventilation

Any discussion on the working environment would be incomplete without consideration being given to temperature, lighting and ventilation. They are significant in the maintenance of comfort. Furthermore, poor standards of provision and control may result in stress on the part of the operator accompanied by gradual deterioration in health and, in some cases, the risk of heat stroke, heat stress, eyesight deficiencies and fume fevers.

Comfort

Comfort is a subjective assessment of the conditions in which an individual works, sleeps, relaxes, travels, etc. Sensations of comfort vary with a person's state of health, vitality and age. Despite the fact that comfort is a personal state, a degree of unanimity is usually found whenever a group of people are asked to assess a given atmospheric condition. Research indicates that there are four factors which are chiefly responsible for the production of the sensation of thermal comfort, namely air temperature, radiated heat, humidity and the amount of air movement. There are limits to the ability of the human body to adapt to achieve this state of comfort. Few people, for example, would choose to live in a greenhouse or a house without windows all the year round, yet in industry workers may be expected to do just that! The 'indoor climate' is of prime importance. Failure to recognise this fact can result in poor standards of performance and efficiency, discontent, increased labour turnover, increased accident rates and absenteeism. (*See* later in this chapter.)

Temperature

In order to understand why it is necessary to control the thermal environment, the important process of body temperature regulation, or 'thermoregulation', should be understood.

The chemical process for generating heat by food conversion is an important feature of a person's metabolism. Food is a source of energy and the body converts approximately 20 per cent of this energy to mechanical energy, the remaining 80 per cent being utilised as heat.

A healthy person has a body temperature 36.9°C which is kept remarkably constant, largely by the body continually varying the flow rate of blood. When body temperature increases, blood flows to the skin and dissipates its heat through the skin surface by thermal exchange. Conversely, when body heat is low, heat is conserved in the deep tissues to maintain what is commonly known as the 'core temperature'. If the air temperature is too high and the differential between the skin and the surrounding air is small, insufficient heat is lost through normal exchange. As a result, the body overheats and the sweat glands are activated. In very hot conditions as much as one litre of body fluid can be lost each hour. This reduces the body fluid level and salt deficiencies may be created.

Stress conditions

Although reference is made to heat stroke in Chapter 19, it is appropriate here to consider stress conditions associated with extremes of temperature.

Heat stress

Under most industrial working conditions operators tend to be self-limiting in their thermoregulatory control. The average worker will tend to withdraw from a hot environment or heat source before he becomes liable to heat stroke. Whilst there are no specific heat exposure limits in the UK, threshold limits for permissible heat exposure (indices of thermal stress) have been established in the USA. The most commonly used index of thermal stress is that based on physiological observations and related to wet bulb globe temperature shown with a whirling hygrometer. This index was developed for use in the desert under wartime conditions and is now incorporated in the American Conference of Governmental Industrial Hygienists' (1980) *Threshold Limit Values for Chemical Substances Physical Agents in the Workroom* as recommended practice. The following equations are used to calculate the wet bulb globe temperature (WBGT) values:

Outdoor work with a solar load – WBGT = 0.7WB + 0.2GT + 0.1DB

Indoor work, or outdoor work with no solar load – WBGT = 0.7WB + 0.3GT

where WB = natural wet bulb temperature
DB = dry bulb temperature
GT = globe thermometer temperature

The natural wet bulb temperature is that recorded from the sling hygrometer without any rotation of the sling. After calculating the wet bulb globe temperature, the number in degrees Celsius is compared with recommended limits of work: rest schedules using either a table or a graph (*see* Fig 15.1 and Table 15.1.) Workers should not be permitted to continue their work when their core temperature reaches 38°C.

Fig. 15.1 ◆ WBGT values plotted against rate of work for various work: rest schedules

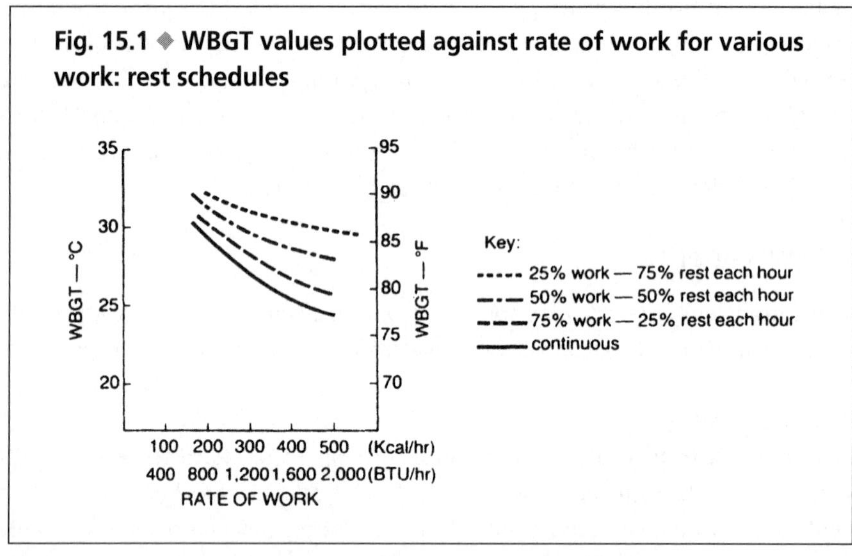

Table 15.1 ◆ Maximum permissible wet bulb globe temperature readings

	Work load		
Work: rest schedule (per hour)	*Light*	*Moderate*	*Heavy*
Continuous work	30.0°C	26.7°C	25.0°C
75% work, 25% rest	30.6°C	28.0°C	25.9°C
50% work, 50% rest	31.4°C	29.4°C	27.9°C
25% work, 75% rest	32.2°C	31.1°C	30.0°C

Cold stress

When outdoor clothing is worn, the limit of tolerance is as shown in Table 15.2. Whilst special protective clothing would be necessary for, say, cold store work, and work at –40°C, some cooling is inevitable. Great attention should be paid in this case to protecting the extremities, i.e. hands, feet and head.

Table 15.2 ◆ Limit of cold stress tolerance

Air temperature	*Time*
– 12°C	6 hours
– 23°C	4 hours
– 40°C	1½ hours
– 57°C	½ hour

Ideal comfort conditions

Air temperature

This is the most important factor and depends upon the type of work being undertaken, e.g. relatively sedentary work such as office work, light work and heavy (manual) work. For each of these classifications, other thermal conditions notwithstanding, there is either an optimum air temperature or a fairly wide air temperature range (comfort range) within which the majority of workers will not feel any discomfort. It is apparent from the range of activities in Table 15.3, that as the activity increases so the optimum ambient temperature reduces. In any of these categories the air temperature should be reduced when operators are exposed to radiant heat.

Radiant heat

Optimum temperature	18.3°C
Comfort range	16.6 to 20°C

It is important to regulate the worker's exposure to radiated heat. There should not be a temperature gradient between head and feet of more than 3°C. In fact, any temperature gradient should be negative, i.e. cool head, warm feet.

Table 15.3 ◆ Optimum working temperatures

Sedentary/office work	
Comfort range	19.4 to 22.8°C
Light work	
Optimum temperature	18.3°C
Comfort range	15.5 to 20°C
Heavy work	
Comfort range	12.8 to 15.6°C

Relative humidity

Relative humidity is defined as 'the actual amount of moisture present in air expressed as a percentage of that which would produce saturation'. It is generally accepted that relative humidity should be between 30 and 70 per cent. If the relative humidity is too low, a feeling of discomfort is produced due to the drying of the throat and nasal passages. Conversely, high relative humidity produces a feeling of stuffiness, and reduces the rate at which body moisture evaporates (sweat), thereby reducing the efficiency of the body's thermoregulatory system.

Air movement

Air movement is an important factor in the consideration of comfort conditions. Movement of air is just perceptible at about nine metres per minute and complaints of draughts will be received when it exceeds 30 metres per minute. Below 6 metres per minute, a room could be considered airless. The sensation of air movement is directly related to air temperature and skin sensitivity. If the air is cool, even slight draughts are detectable, whereas if the air temperature is controlled then the draughts, although still present, may not be detected. Air movement assists in cooling and causes distress if excessive.

Temperature control in the workplace

In order to reduce stress associated with inadequate temperature control, a number of aspects need consideration at the design stage of projects and in the assessment of an existing thermal environment.

Reg 7 of the W(HSW)R covers the legal requirements relating to temperature in indoor workplaces. During working hours, the temperature in all workplaces inside buildings shall be reasonable. (*See* Table 15.3, Optimum working temperatures.) A method of heating or cooling shall not be used which results in the escape into the workplace of fumes, gas or vapour of such character and to such extent that they are likely to be injurious or offensive to any person. A sufficient number of thermometers shall be provided to enable persons at work to determine the temperature in any workplace inside a building.

Central heating and other systems should operate independently of any hot water or steam installation, so that central heating can be reduced or turned off in summer months. Individual control of radiators and other heating appliances, together with installation of, where appropriate, thermostat control switches, should be provided where control of central heating is by electrical means.

Heating appliances with exposed electrical elements and heaters producing combustion gases which vent to the workplace should not be used. The use of privately owned heating appliances, particularly those of the open radiant type, should be strictly controlled or prohibited in view of the fire risk associated with them. In offices, wall-mounted convector heaters should be provided if it is necessary to supplement existing central heating during cold weathers. Portable free-standing appliances, which are easily knocked over, create a greater fire hazard.

Lighting

Two specific aspects of health and safety are relevant in relation to lighting:

(a) a gradual deterioration in an individual's visual acuity and performance;
(b) the increased likelihood of accident brought about by a worker's failing or incorrect perception.

The purpose of this section on lighting, therefore, is to consider lighting deficiencies which can result in the inability of people to perceive danger and cause deterioration in visual performance. In turn, this involves an examination of lighting in terms of:

(a) the quantity and quality of light required for a given task as well as the relationship of lighting to the general environment of the workplace;
(b) the basis for lighting design and specific applications, as in the case of visual display units (VDUs).

Present legal requirements

Legal requirements relating to lighting of workplaces are dealt with in Reg 8 of the W(HSW)R. Every workplace shall have suitable and sufficient lighting which, so far as is reasonably practicable, shall be by natural light. Furthermore, suitable and sufficient emergency lighting shall be provided and maintained in any room in circumstances in which persons are specially exposed to danger in the event of failure of artificial lighting.

The provisions relating to emergency lighting in the Regulations are mainly directed at situations where sudden loss of light would present a serious risk, for example, if process plant needs to be shut down under manual control or a potentially hazardous process needs to be made safe, and this cannot be done safely without lighting.

Emergency lighting should be powered by a source independent from that of normal lighting. It should be immediately effective in the event of failure of the normal lighting, without need for action by anyone. It should provide sufficient lighting to enable persons at work to take any action necessary to ensure their, and others', health and safety (ACOP).

Quantitative aspects of lighting

The quantity of light flowing from a source such as a light bulb or fluorescent light (luminaire) is the luminous flux or light flow, which is generally termed 'illuminance'. The units of measurement of luminous flux were formerly foot candles or lumens per square foot, but more recently the unit has become the lux, which is the metric unit of measurement. Thus:

Foot candles	=	lumens per square foot
Lux	=	lumens per square metre
10.76 lux	=	1 lumen per square metre
1 lux	=	0.093 lumens per square metre

On this basis a conversion factor of 10 or 11 is used for converting from lumens per square foot to lux, i.e. 20 lumens per square foot = 200 lux.

The lux, therefore, is the unit of illuminance (not 'illumination'), and the value in lux, measured with a standard photometer or light meter, is an indication of the quantity of light present at a particular point.

Lighting standards are detailed in HSE Guidance Note HS(G)38 *Lighting at work*. The Guidance Note distinguishes between 'average illuminances' and 'minimum measured illuminances' according to the general activity undertaken and the type of location of work undertaken (*see* Table 15.4). The ratio between working areas and adjacent areas is also featured in the Guidance Note (*see* Table 15.5 on page 304.)

The Guidance Note recommends that where there is conflict between the recommended average illuminance shown in Table 15.4 and the maximum ratios of illuminance in Table 15.5, the higher value should be taken as the appropriate average illuminance.

Table 15.4 ◆ Average illuminances and minimum measured illuminances for different types of work

General Activity	*Typical Locations/ Types of Work*	*Average Illuminance Lux (Lx)*	*Minimum Measured Illuminance Lux (Lx)*
Movement of people, machines and vehicles (1)	Lorry parks, corridors, circulation routes	20	5
Movement of people, machines and vehicles in hazardous areas; rough work not requiring any perception of detail (1)	Construction site clearance, excavation and soil work, docks, loading bays, bottling and canning plants	50	20
Work requiring limited perception of detail (2)	Kitchens, factories assembling large components, potteries	100	50
Work requiring perception of detail (2)	Offices, sheet metal work, bookbinding	200	100
Work requiring perception of fine detail (2)	Drawing offices, factories assembling electronic components, textile production	500	200

Notes

1. Only safety has been considered, because no perception of detail is needed and visual fatigue is unlikely. However, where it is necessary to see detail, to recognise a hazard or where error in performing the task could put someone else at risk, for safety purposes as well as to avoid visual fatigue, the figure should be increased to that for work requiring the perception of detail.
2. The purpose is to avoid visual fatigue: the illuminances will be adequate for safety purposes.

Table 15.5 ◆ Maximum ratios of illuminance for adjacent areas

Situations to which recommendation applies	*Typical location*	*Maximum ratio of illuminances*		
		Working area		*Adjacent area*
Where each task is individually lit and the area around the task is lit to a lower illuminance	Local lighting in an office	5	:	1
Where two working areas are adjacent, but one is lit to a lower illuminance than the other	Localised lighting in a works store	5	:	1
Where two working areas are lit to different illuminances but are separated by a barrier and there is frequent movement between them	A storage area inside a factory and a loading bay outside	10	:	1

Qualitative aspects of lighting

The concept of average illuminances in the Guidance Note relates only to the quantity of light, and in the design or assessment of lighting installations consideration must be given to the qualitative aspects. Factors which contribute to the quality of lighting include the presence or absence of glare in its various forms, the degree of brightness, the distribution of light, diffusion, colour rendition, contrast effects and the system for lighting maintenance.

Glare

This is the effect of light which causes discomfort or impaired vision, and is experienced when parts of the visual field are excessively bright compared with the general surroundings. This usually occurs when the light source is directly in line with the visual task or when light is reflected off a given surface or subject. Glare is experienced in three different forms:

(a) Disability glare is the visually disabling effect caused by bright bare lamps directly in the line of sight. The resulting impaired vision (dazzle) may be hazardous if experienced when working in high-risk processes, at heights or when driving. It is seldom experienced in workplaces because most bright lamps, e.g. filament and mercury vapour, are usually partly surrounded by some form of fitting.

(b) Discomfort glare is caused mainly by too much contrast of brightness between an object and its background, and is associated with poor lighting design. It causes visual discomfort without necessarily impairing the ability to see detail, but over a period can cause eye strain, headaches and fatigue. Discomfort glare can be reduced by –

  (i) careful design of shades which screen the lamp;

  (ii) keeping luminaires as high as practicable;

  (iii) maintaining luminaires parallel to the main direction of lighting.

(c) Reflected glare is the reflection of bright light sources on shiny or wet work surfaces such as glass or plated metal, which can almost entirely conceal the detail in or behind the object which is glinting. Care is necessary in the use of light sources of low brightness and in the arrangement of the geometry of the installation, so that there is no glint at the particular viewing position.

Distribution

The distribution of light, or the way in which light is spread, is important in lighting design. Poor lighting distribution may result in the formation of shadowed areas which can create dangerous situations, particularly at night. For good general lighting, regularly spaced luminaires are used to give evenly distributed illuminance. This evenness of illuminance depends upon the ratio between the height of the luminaire above the working position and the spacing of fittings.

Colour rendition

This refers to the appearance of an object under a given light source, compared to its colour under a reference illuminant, e.g. natural light. Colour rendition enables the colour appearance to be correctly perceived. The colour-rendering properties of light fitments should not clash with those of natural light, and should be equally effective at night when there is no daylight contribution to the total illumination of the workplace.

Brightness

Brightness or, more correctly, 'luminosity', is essentially a subjective sensation and cannot be measured. It is possible, however, to consider a brightness ratio, which is the ratio of apparent luminosity between a task object and its surroundings. To achieve the recommended brightness ratio, the reflectance of all surfaces in the workplace should be carefully maintained and consideration given to reflectance values in the design of interiors. Given a task illuminance factor of 1, the effective reflectance values should be:

Ceilings	0.6
Walls	0.3 to 0.8
Floors	0.2 to 0.3

Diffusion

This is the projection of light in many directions with no directional predominance. The directional effects of light are just as important as the quantity of light, however, as the directional flow of light can often determine the density of shadows, which may affect safety. Diffused lighting can soften the output from a particular source and so limit the amount of glare that may be encountered from bare fittings.

Stroboscopic effects

All lamps that operate from an alternating current electricity supply produce oscillations in light output. When the magnitude of the oscillations is great and their frequency is a multiple or sub-multiple of the frequency of movement of machinery, that machinery will appear to be stationary or moving in a different manner. This is called the 'stroboscopic effect'. It is not common with modern lighting systems but where it does occur it can be dangerous, so appropriate action should be taken to avoid it. Possible remedial measures include:

(a) supplying adjacent rows of lighting fittings from different phases of the electricity supply;
(b) providing a high frequency supply;
(c) washing out the effect with local lighting which has much less variation in light output, e.g. tungsten lamp; and/or
(d) use high frequency control gear if applicable.

Lighting maintenance

A well-organised maintenance programme is necessary for permanently good illumination to be achieved. The programme should incorporate regular cleaning and replacement of lamp fittings as a basic consideration, together with regular assessment of illuminance levels with a standard photometer at predetermined points. Furthermore, the actual function of the lighting provided should be reviewed in line with changes that may be made in production, storage or office arrangements. To facilitate safe lamp cleaning and replacement, high-level luminaires should be fitted with raising and lowering gear, so that this work can be undertaken at floor level.

Emergency lighting

When the normal lighting installation fails, emergency lighting can be used to provide either standby or escape lighting.

Standby lighting

Standby lighting enables essential work to continue and the illuminance needed depends upon the nature of the work. It may be between 5 and 100 per cent of the illuminance produced by the normal installation. The illuminance recommendations in the HSE Guidance Note *Lighting at work* can be taken as a guide.

Escape lighting

Escape lighting enables a building to be evacuated safely. The illuminances required are given in BS 5266 Part 1: *Code of Practice for emergency lighting of premises other than cinemas and certain other specified premises used for entertainment*. This Code of Practice recommends that escape lighting should reach the required illuminance within five seconds of the failure of the main lighting system, although if the occupants are familiar with the building this time may be increased to 15 seconds. Battery-powered escape lighting is usually designed to operate for between one and three hours according to the size of the building and the likely problems of evacuation. Escape lighting installations powered by a generator will operate for as long as the generator runs, which should at least match the operating times of battery-powered installations.

The design of lighting

In the design of lighting installations, many factors need consideration. These may include the following:

(a) *General lighting requirements*. Illuminance levels for the principal operations within the premises and for specific parts of the premises, both internally and externally, should be considered. Both average and minimum measured illuminance levels should be specified, bearing in mind the current HSE Guidance Note recommendations and also the effects of lighting on worker performance and his potential for fatigue. Specific vision defects amongst workers and the hazards associated with incorrect or faulty perception must further be considered.

(b) *Availability of natural lighting.* Whilst natural lighting is the best form of illuminance, it must often be considered a secondary option in lighting design owing to its unreliability.

(c) *Specific areas and processes.* The type of lighting and lighting needs of specific areas and/or processes should be considered, e.g. access points, corridors, fine assembly work, catering, internal traffic lanes.

(d) *Colour rendition aspects.* These aspects need consideration to ensure correct perception of colour under both natural and artificial lighting. The positioning of safety signs and notices should take account of this factor.

(e) *Glare.* The potential for glare should be considered in the positioning of machinery, and for particular surfaces adjacent to machinery. The potential for dazzle should also be taken into account – extraneous light sources may produce it.

(f) *Structural aspects.* The effects of structural items such as screens, pillars and plant should be taken into account. Such items can obstruct light flow and reduce illuminance levels.

(g) *Atmospheric influences.* The presence of steam, fumes and mists should be considered, and illuminance levels upgraded where these influences may bring about a reduction in general lighting.

(h) *Lamp and window cleaning.* The need for frequent cleaning, replacement and maintenance of lamps will depend upon the types of process being undertaken in the area. Suitable access equipment must be provided, e.g. raising and lowering gear to permit lamp cleaning at floor level. Window cleaning should be incorporated in the cleaning schedule for the premises (*see* Chapter 17).

(i) *High-risk areas.* The potential for fire and explosion in certain areas may indicate the need for sealed light fittings, e.g. petroleum installations.

(j) E*mergency lighting.* An emergency lighting system, using an independently operated emergency generator, battery-operated systems or hand lamps, is necessary in most activities. These systems should operate at least to between 5 and 100 per cent of usual operating illuminance.

(k) *Energy considerations.* The need for the use of time-switches or photoelectric switching devices should be considered.

Finally, the physiological effects of poor lighting, such as reduced acuity and performance, and the psychological effect on perception and attitudes of workers, must be considered. Lighting design has a direct effect on worker performance and accident potential, as do other environmental factors, such as temperature and ventilation.

Ventilation

Ventilation, i.e. the movement of air through a building, may be by natural and/or mechanical means. Natural ventilation is generally taken to mean ventilation produced without the aid of mechanically induced draught, implying the movement of air through permanent openings in the fabric of a building, e.g. windows, doors, airbricks, flues, etc. Reg 6 of the W(HSW)R requires that effective and suitable ventilation shall be made to ensure that every enclosed workplace is ventilated by a sufficient quantity of fresh or purified air. Any plant used for complying with the above requirement shall include an effective device to give visible or audible warning of any failure of the plant where necessary for reasons of health or safety. This regulation shall not apply to any confined space in a workplace subject to the provisions of sec 30 of the Factories Act 1961. This process, concerned with providing sufficient air to breathe for occupants and, to some extent, regulating temperature, is known as '*comfort ventilation*'. Comfort ventilation is directly related to the number of air changes per hour according to the external ambient temperature and the actual rate of air movement.

The principal features of a ventilation system are, therefore:

(a) the provision and maintenance of the circulation of fresh air in every occupied part of a workplace;

(b) the rendering harmless of all potentially injurious airborne contaminants, e.g. dusts, fumes, vapours and gases.

Ventilation systems, whether natural or artificial, should be designed assuming a maximum air temperature of 32.2°C (90°F) and a minimum air temperature of 0°C (32°F) and should operate, from a comfort viewpoint, to give the numbers of air changes per hour, summer and winter, shown in Table 15.6. The following points should be considered in the design ventilation systems:

1. The production of an atmosphere which is cool rather than hot, dry rather than damp, moving rather than still, with relative humidity between 40 and 70 per cent.

2. Incoming air should be drawn from a clean source or should be filtered.
3. Every room, passage and staircase should be separately ventilated and door openings should not be included in ventilation calculations.
4. Heat should be removed as close to the source of emission as possible.
5. Inputs should be sited to give a flow of air from operating positions towards heat sources and thence to extracts.
6. Fresh air intakes on roofs should stand at least 700 mm clear of the roof surface to avoid picking up heated air.
7. The total extract volume should be 80 per cent of the input volume to allow for a positive plenum, i.e. keeping the building under pressure.

Sec 63 of the FA, on the other hand, requires that in every factory all practicable measures be taken where there is given off, in connection with a process carried on there, dust or fumes or other impurities which are likely to be injurious or offensive to employees, or any substantial quantity of dust of any kind. This implies the need for effective control of airborne contaminants. This duty is greatly reinforced under the COSHH Regulations, in particular the provision and maintenance of local exhaust ventilation (LEV) systems. (*See further* Chapter 26 'Prevention and control strategies in occupational hygiene'.)

Table 15.6 ◆ Air changes

Location	*Summer*	*Winter*
Offices	6	4
Corridors	4	2
Amenity areas	6	4
Storage areas	2	2
Production areas, with heat-producing plant	20	20
Production areas (assembly, finishing work)	6	6
Workshops	6	4

16 Welfare amenity provision

Welfare amenities include arrangements for sanitation (water closets, urinals), washing (wash-basins and showers), drinking water, storage for clothing, including protective clothing, taking meals (canteens, mess-rooms) and the provision of seats for certain work. First aid provision is also relevant here (*see* Chapter 28).

D

Sanitation and washing arrangements

Three factors are important, namely:

(a) the number of fitments in relation to the total number of employees, e.g. water closets, urinals, wash-basins, shower units;

(b) the arrangement of fitments in a correctly designed amenity area (*see later*), which should normally incorporate facilities for storing clothing;

(c) the relative ease of maintenance and cleaning of the facilities.

General requirements

Requirements for amenities are laid down in the Workplace (Health, Safety and Welfare) Regulations 1992 (WHSWR), with specific detail incorporated in the accompanying HSC Approved Code of Practice (ACOP) and Guidance. It should be appreciated that the majority of the requirements relating to the provision and maintenance of welfare amenity provisions under the Regulations are of an absolute nature.

Sanitary conveniences (Reg 20)

Reg 20(1) requires that suitable and sufficient sanitary conveniences shall be provided at readily accessible places. Conveniences shall not be suitable unless:

(a) the rooms containing them are adequately ventilated and lit;

(b) they and the rooms containing them are kept in a clean and orderly condition;

(c) separate rooms containing conveniences are provided for men and women, except where and so far as each convenience is in a separate room the door of which can be secured from inside. (Reg 20(2))

In existing workplaces, compliance with Part II of Schedule 1 to the Regulations shall be sufficient compliance with the requirement in paragraph 1.

Schedule 1 – provisions applicable to factories which are not new workplaces, extensions or conversions

Part II – number of sanitary conveniences

4. In workplaces where females work, there shall be at least one suitable water closet for use by females only for every 25 females.

5. In workplaces where males work, there shall be at least one suitable water closet for use by males only for every 25 males.

6. In calculating the number of males or females who work in any workplace for the purposes of this Part of this Schedule, any number not itself divisible by 25 without fraction or remainder shall be treated as the next number higher than it which is so divisible.

Washing facilities (Reg 21)

Reg 21(1) of the WSHWR states that suitable and sufficient washing facilities, including showers if required by the nature of the work or for health reasons, shall be provided at readily accessible places. Washing facilities shall not be suitable unless:

(a) they are provided in the immediate vicinity of every sanitary convenience, whether or not provided elsewhere as well;

(b) they are provided in the vicinity of any changing rooms required by the regulations, whether or not provided elsewhere as well;

(c) they include a supply of clean hot and cold, or warm, water (which shall be running water so far as is practicable);

(d) they include soap or other suitable means of cleaning;

(e) they include towels or other suitable means of drying;

(f) the rooms containing them are sufficiently ventilated and lit;

(g) they and the rooms containing them are kept in a clean and orderly condition and are properly maintained;

(h) separate facilities are provided for men and women, except where and so far as they are provided in a room the door of which is capable of being secured from inside and the facilities in each room are intended to be used by only one person at a time (Reg 21(2)).

Paragraph 2(h) above shall not apply to facilities which are used for washing the hands, forearms and face only.

Minimum numbers of facilities

The ACOP amplifies the requirements of Regs 20 and 21 by means of two Tables. Table 16.1 shows the minimum number of sanitary conveniences and washing stations which should be provided. The number of people at work shown in column 1 refers to the maximum number likely to be in the workplace at any one time. Where separate sanitary accommodation is provided for a group of workers, for example men, women, office workers or manual workers, a separate calculation should be made for each group.

D

Table 16.1 ◆ Water closet and wash station provision

1 *Number of people at work*	*2* *Number of water closets*	*3* *Number of wash stations*
1–5	1	1
6–25	2	2
26–50	3	3
51–75	4	4
76–100	5	5

In the case of sanitary accommodation used only by men, Table 16.2 may be followed if desired, as an alternative to column 2 of Table 16.1. A urinal may be either an individual urinal or a section of urinal space which is at least 600 mm long.

The ACOP further recommends that an additional water closet, and one additional washing station, should be provided for every 25 people above 100 (or fraction of 25). In the case of water closets used only by men, an additional water closet for every 50 men (or fraction of 50) above 100 is sufficient provided at least an equal number of additional urinals are provided.

Table 16.2 ◆ Provision of water closets and urinals for men

1 *Number of men at work*	*2* *Number of water closets*	*3* *Number of urinals*
1–5	1	1
16–30	2	1
31–45	2	2
46–60	3	2
61–75	3	3
76–90	4	3
91–100	4	4

Where work activities result in heavy soiling of the face, hands and forearms, the number of washing stations should be increased to one for every 10 people at work (or fraction of 10) up to 50 people; and one extra for every additional 20 people (or fraction of 20).

Where facilities provided for workers are also used by members of the public the number of conveniences and washing stations specified above should be increased as necessary to ensure that workers can use the facilities without undue delay.

The ACOP also makes specific recommendations covering remote workplaces and temporary work sites.

Ventilation, cleanliness and lighting

Any room containing a sanitary convenience shall be well ventilated, so that offensive odours do not linger. Measures should also be taken to prevent odours entering other rooms. This may best be achieved by, for example, providing a ventilated area between the room containing the convenience and the other room. Alternatively it may be possible to achieve it by mechanical ventilation or, if the room containing the convenience is well sealed from the workroom and has a door with an automatic closer, by good natural ventilation. However, no room containing a sanitary convenience should communicate directly with a room where food is processed, prepared or eaten (ACOP).

Arrangements should be made to ensure that rooms containing sanitary conveniences or washing facilities are kept clean. The frequency and thoroughness of cleaning should be adequate for this purpose. The surfaces of the internal walls and floors of the facilities should normally have a surface which permits wet cleaning, for example, ceramic tiling or a plastic coated surface. The rooms should be well lit; this will also facilitate cleaning to the necessary standard and give workers confidence in the cleanliness of the facilities. Responsibility for cleaning should be clearly established, particularly where facilities are shared by more than one workplace (ACOP).

Drinking water (Reg 22)

Reg 22(1) requires that an adequate supply of wholesome drinking water shall be provided for all persons at work in the workplace. Every supply of drinking water required by paragraph 1 above shall:

(a) be readily accessible at suitable places;

(b) be conspicuously marked by an appropriate sign where necessary for reasons of health or safety (Reg 22(2)).

Where a supply of drinking water is required by Reg 22(1), there shall also be provided a sufficient number of suitable cups or other drinking vessels unless the supply of drinking water is in a jet from which persons can drink easily (Reg 23(3)).

The ACOP makes specific recommendations relating to adequacy of water supplies, the prevention of contamination of taps, the provision of disposal or non-disposable cups, with facilities for washing the latter and marking of supplies where there is a risk of people drinking from non-drinkable water supplies. The Guidance accompanying the ACOP recommends marking of supplies which could become grossly contaminated.

Accommodation for clothing (Reg 23)

Reg 23 of the WSHWR deals with accommodation for clothing and Reg 24, facilities for changing clothing. Suitable and sufficient accommodation shall be provided:

(a) for any person at work's own clothing which is not worn during working hours;

(b) for special clothing which is worn by any person at work but which is not taken home (Reg 23(1)).

Without prejudice to the generality of paragraph 1 the accommodation mentioned in that paragraph shall not be suitable unless:

(a) where facilities to change clothing are required, it provides suitable security for clothes not so worn;

(b) where necessary to avoid risks to health or damage to the clothing, it includes separate accommodation for clothing worn at work and for other clothing;

(c) so far as is reasonably practicable, it allows or includes facilities for drying clothing;

(d) it is in a suitable location (Reg 23(2)).

The ACOP to the regulations recommends that accommodation for work clothing and workers' own personal clothing should enable it to hang in a clean, warm, dry, well-ventilated place where it can dry out during the course of a working day if necessary. If the workroom is unsuitable for this purpose, then accommodation should be provided in another convenient place. The accommodation should consist of, as a minimum, a separate hook or peg for each worker.

Where facilities to change clothing are required by Reg 24, effective measures should be taken to ensure the security of clothing. This may be achieved, for example, by providing a lockable locker for each worker (ACOP).

Where work clothing (including personal protective equipment) which is not taken home becomes dirty, damp or contaminated due to the work it should be accommodated separately from the worker's own clothing. Where work clothing becomes wet, the facilities should enable it to be dried by the beginning of the following work period unless other dry clothing is provided (ACOP).

It should be noted that civil action lies against a factory occupier in the event of personal clothing being stolen (*McCarthy* v. *Daily Mirror Newspapers Ltd* [1949] 1 AER 801).

Facilities for changing clothing (Reg 24)

Suitable and sufficient facilities shall be provided for any person at work in the workplace to change clothing in all cases where:

(a) the person has to wear special clothing for the purpose of work;

(b) the person cannot, for reasons of health or propriety, change in another room (Reg 24(1)).

Without prejudice to the generality of paragraph 1, the facilities mentioned in that paragraph shall not be suitable unless they include separate facilities for, or separate use of facilities by, men and women where necessary for reasons of propriety (Reg 24(2)).

The ACOP recommends that a changing room should be provided for workers who change into special work clothing and where they remove more than outer clothing. Changing rooms should also be provided where necessary to prevent workers' own clothing being contaminated by a harmful substance. Changing facilities should be readily accessible from workrooms and eating facilities, if provided. They should be provided with adequate seating and should contain, or communicate directly with, clothing accommodation and showers or baths if provided. They should be constructed and arranged to ensure privacy of the user. Furthermore, the facilities should be large enough to enable the maximum number of persons at work expected to use them at any one time to do so without overcrowding or unreasonable delay. Account should be taken of starting and finishing times and the time available to use the facilities.

Facilities for rest and to eat meals (Reg 25)

Suitable and sufficient rest facilities shall be provided at readily accessible places (Reg 25(1)). Rest facilities provided by virtue of Reg 25(1) shall:

(a) where necessary, for reasons of health or safety include, in the case of a new workplace, extension or conversion, rest facilities provided in one or more rest rooms, or, in other cases, in rest rooms or rest areas;

(b) include suitable facilities to eat meals where food eaten in the workplace would otherwise be likely to become contaminated (Reg 25(2)).

Rest rooms and rest areas shall include suitable arrangements to protect non-smokers from discomfort caused by tobacco smoke (Reg 25(3)).

Suitable facilities shall be provided for any person at work who is a pregnant woman or nursing mother to rest (Reg 25(4)).

Suitable and sufficient facilities shall be provided for persons at work to eat meals where meals are regularly eaten in the workplace (Reg 25(5)).

The ACOP makes extensive recommendations on the question of rest facilities and facilities for taking meals in the workplace. These include:

(a) where workers have to stand to carry out their work, suitable seats should be provided for their use if the type of work gives them an opportunity to sit from time to time;

(b) suitable seats should be provided for workers for use during breaks;

(c) rest areas or rooms should be large enough, and have sufficient seats with backrests and tables, for the number of workers likely to use them at any one time;

(d) where workers frequently have to leave their work area, and wait until they can return, there should be a suitable rest area where they can wait;

(e) where workers regularly eat meals at work suitable and sufficient facilities should be provided for the purpose;

(f) seats in working areas can be counted as eating facilities provided they are in a sufficiently clean place and there is a suitable surface on which to place food;

(g) eating facilities should include a facility for preparing or obtaining a hot drink, such as an electric kettle, vending machine or a canteen;

(h) workers who work during hours or at places where hot food cannot be obtained in, or reasonably near to, the workplace should be provided with the means for heating their own food;

(j) eating facilities should be kept clean to a suitable hygiene standard;

(k) canteens or restaurants may be used as rest facilities, provided there is no obligation to purchase food in order to use them;

(l) good hygiene standards should be maintained in those parts of rest facilities used for eating or preparing food and drinks.

Amenity areas

Since current requirements relating to welfare amenities lay down *minimum* standards, prudent employers will endeavour to provide better standards for their employees. In factories and labour-intensive operations it is relevant to consider the provision of an amenity area. An amenity area includes all the general welfare provisions such as washing and sanitation, clothing storage, first aid and even catering, incorporated as one purpose-built unit, separate from the main activity, but with reasonable access to it. Much will depend upon existing layout, the needs of workers who may be separated from the main activity, e.g. engineers, and the degree of shift work. Sometimes smaller 'satellite' amenity areas, as distinct from one large central unit, may be more appropriate.

The design of amenity areas

The following factors should be taken into account in the design and use of amenity areas.

Sanitation

(a) The surfaces of the floors, walls, ceilings, doors and fittings should be capable of being readily cleaned and maintained.

(b) There should be total separation of sexes, except where fewer than three persons of one sex or the other are employed.

(c) There should be an intervening ventilated space between any sanitation area and a workroom, office, foodroom or store.

(d) Facilities for the disposal of sanitary dressings should be provided in female staff sanitation areas, e.g. incinerator, comminuter or chemical method.

(e) Adequate lighting and ventilation, by both natural and artificial means, should be provided.

(f) Walls, doors and fittings should be of vandal-proof design.

Hand washing and showers

(a) The above provision relating to separation of the sexes, surfaces, lighting, ventilation and the use of vandal-proof fittings also apply here.

(b) Wash-basins and showers, where installed, should have adequate supplies of hot and cold water or of hot water at a suitably controlled temperature.

(c) Supplies of soap, clean towels and nail brushes should be provided. The use of wall-mounted liquid soap dispensers and disposable paper towels – or the continuous non-returnable type of roller towel cabinet – are recommended as opposed to tablet soap and the simple form of returnable roller towel. In food preparation and manufacture, the use of a bactericidal hand cleanser is recommended. Lidded containers should be installed for the storage of soiled paper towels.

Clothing storage

(a) The above provisions relating to separation of the sexes, surfaces, lighting, ventilation and the use of vandal-proof fittings apply here too.

(b) Suitable seats, preferably of the wall-mounted type, should be provided to permit easy changing of footwear.

(c) There must be effective means for drying clothing.

(d) Use of purpose-manufactured clothing storage units, as opposed to simple wall hooks, is recommended. Such units incorporate a fixed rail, with clothes hangers fixed to the rail, and banks of small personal lockers at each end of the unit for storage of personal items, such as handbags. This arrangement allows for effective drying of clothing, assisted by either wall-mounted fan heaters or floor-mounted tubular steel heaters.

Drinking water

(a) Provision of fountains, as opposed to the use of taps and cups, is recommended.

(b) Drinking water installations should be located outside any sanitation area.

Layout

(a) In activities requiring high standards of personal hygiene, such as food preparation and manufacture, layout should ensure that staff have separate access to clothing storage facilities, sanitation, hand cleansing and shower facilities, and are not required to pass through the working area in outdoor clothes.

(b) Where large numbers of staff use an amenity area, layout should ensure the operation of a sequential flow from outdoor clothing storage area to hand-washing and sanitation area and thence to protective clothing storage area, before entering the working area. The reverse process should take place on completion of work.

(c) The environmental health officer should be consulted at the design stage of new amenity provisions.

17 Cleaning and hygiene

Hygiene has been defined as 'the science of health' or 'rules for health'. In factories and other workplaces it is concerned with the promotion of good health through the maintenance of satisfactory sanitary conditions and catering activities, and the prevention of contamination, in particular by pest infestation, which can create insanitary conditions and contaminate or damage food, raw materials, manufactured goods and the structure of the premises.

D

Legal requirements

Workplace legislation has always recognised the need for satisfactory levels of factory cleanliness with a view to the prevention of occupational ill-health associated with insanitary working conditions.

Reg 9 of the W(SHW)R deals with cleanliness and waste materials in the workplace. Reg 9 requires that every workplace and the furniture, furnishings and fittings therein shall be kept sufficiently clean. The surfaces of the floor, walls and ceiling of all workplaces inside buildings shall be capable of being kept sufficiently cleaned. So far as is reasonably practicable, waste materials shall not be allowed to accumulate except in suitable receptacles.

Maintenance of sanitary conditions

The operation of satisfactory cleaning and housekeeping procedures is a prerequisite in the prevention of accidents and occupational ill-health. Indeed, poor housekeeping is a contributory factor in a large proportion of accidents.

Management of the cleaning operation is best undertaken by the use of a cleaning schedule or plan, produced after a hygiene survey of the premises, supported by frequent inspections to ensure effective implementation.

Cleaning schedules/plans

Schedules are most correctly produced in tabular form, identifying the following:

(a) what is to be cleaned, e.g. item of plant, area, room, surface;
(b) location;
(c) the nature and extent of soiling;
(d) frequency of cleaning necessary;
(e) method and materials to be used;
(f) responsibility for ensuring satisfactory completion of the cleaning task;
(g) monitoring procedure;
(h) any precautions necessary, e.g. use of chemical-based cleaning agents.

Schedules should be reviewed annually, or more frequently if standards deteriorate.

A typical schedule for food-manufacturing premises or a large catering establishment is shown in Table 17.1. (This can be modified to suit other types of premises.)

The identification of management responsibility for implementation of schedules is the most important factor in ensuring sound levels of factory hygiene. Cleaning schedules must be linked with preventive maintenance systems, and cleaning staff must be trained in the correct and safe use of cleaning preparations.

Cleaning preparations

The type of preparation used will depend on the nature of the soiling. Cleaning preparations may be classified as shown in Table 17.1.

Acids

These are used to remove hard water scale and deposits on urinals formed by a combination of scale and urine salts. Commonly an inhibited hydrochloric acid preparation is used, although sulphamic acid preparations are less corrosive.

Table 17.1 ◆ Cleaning schedule for food-manufacturing or catering establishment

Item	*Frequency*	*Responsibility*	*Equipment*	*Materials*	*Method*	*Special precautions*
Yard	Daily		Hose and broom		Sweep up all refuse. Hose and brush down yard surfaces	Keep drain gullies clear
Floor Stores Food preparation Kitchen Wash-up Servery Dining room Cloakroom Lavatory Washroom	Daily		Vacuum cleaner Suction polisher Polisher Floor scrubber Floor drier Sponge mop Plastic bucket	Detergent/ sanitiser Warm water	Vacuum clean Wash floor Dry floor Polish (where appropriate)	
Walls All rooms as above	Fortnightly		Vacuum cleaner (with attachment)		Vacuum clean	Ventilation fans should be included in the cleaning, but the electric supply to the fan must be switched off at the main
	Monthly		Mechanical wall washer or sponge and plastic bucket	Detergent/ sanitiser Warm water	Wash down	Include canopies over cooking equipment
Ceilings	Monthly		Vacuum cleaner (with attachment)		Vacuum clean	
	6-monthly		Sponge and plastic bucket	Detergent/ sanitiser Warm water	Wash ceilings	

Table 17.1 ◆ continued

Item	*Frequency*	*Responsibility*	*Equipment*	*Materials*	*Method*	*Special precautions*
Windows Interior Exterior	Fortnightly Monthly		Chamois leather Plastic bucket	Water	Wash	Safe working conditions must be ensured. Ladders must be sound and firmly placed
Electric fittings	Monthly		Drying cloth or vacuum cleaner			The electricity supply should be switched off at the main
WCs Pans	Daily		Nylon brush	Approved cleanser	Sprinkle cleanser around bowl at end of each day. Brush entire bowl surface following morning. Flush pan, holding brush under flush water	When an approved cleanser is used, no other cleansing agent must be mixed with it, or used at the same time
Seat Flushing handle Door Furniture	Daily		Disposable cloth Plastic bucket	Detergent/ sanitiser	Thoroughly wash the door, door furniture, flushing handle and pedestal seat. Disposal of the cloth via the WC pan	
Washbasins, Showers	Daily		Sponge	Approved cleanser	Thoroughly wash the basin	

Alkalis

Caustic soda (sodium hydroxide) preparations will break down fat, grease and carbon deposits, but will seriously affect metals such as aluminium. Alkalis are neutralised by acids and in the presence of acids are ineffective as cleaning agents.

Detergents

These are mildly alkaline. Most detergents are manufactured in the form of washing powders and liquids for washing dishes and utensils. Detergents incorporate 'wetting agents' to increase surface contact, and added to acids, for example, provide a far greater degree of penetration, thereby ensuring quicker removal of soil.

Solvents

Solvent-based products are used to soften fats and greases. They may incorporate a detergent. Solvents incorporating methylene chloride.

NOTE. It should be appreciated that the majority of cleaning preparations come within the scope of the COSHH Regulations (*see* Chapters 18 and 39).

SECTION E

Occupational health and hygiene

18 Toxicology and health

Toxicology

Toxicology is the quantitative study of the body's responses to toxic substances. In order to understand the subject and interpret toxicological data and information it is important to know the meaning of expressions and definitions commonly used.

E

Definitions

(a) **Toxicity** is the ability of a chemical substance to produce injury once it reaches a susceptible site in or on the body. The effects may be acute or chronic, local or systemic.

(b) **Acute effect** is a rapidly produced effect following a single exposure to an offending agent.

(c) **Chronic effect** is produced as a result of prolonged exposure or repeated exposures of long duration. Concentration of the offending agent may be low in both cases. One single prolonged exposure can result in chronic effects, however.

(d) **Sub-acute effect** generally implies a reduced form of acute effect.

(e) **Progressive chronic effect** continues to develop after exposure ceases.

(f) **Local effect** is usually confined to the initial point of contact. The site may be the skin, mucous membranes of the eyes, nose or throat, liver, bladder, etc.

(g) **Systemic effects** occur in parts of the body other than at the point of initial contact, and are associated with a particular body system, e.g. respiratory system, central nervous system.

(h) **Dose** is the level of environmental contamination multiplied by the length of time (duration) of exposure to the contaminant.

(i) **Minimum lethal dose**, in experimental toxicology, is the minimum quantity of toxic substance per unit body of experimental animal which will have a fatal effect. It is expressed in milligrams per kilogram of body weight.

(j) **LD$_{50}$** is a more commonly used term for the amount of toxic material which will kill 50 per cent of the test animal population of an experimental group (lethal dose 50 per cent kill). This figure generally forms the basis for comparison between different chemical compounds.

(k) **LC$_{50}$** is the lethal concentration of a toxic substance in air which will kill 50 per cent of the test animal population of an experimental group (lethal concentration 50 per cent kill).

(l) **Toxic hazard** is a measure of the likelihood of toxic effects occurring.

(m) **Total inhalable dust** is the fraction of airborne material which enters the nose and mouth during breathing and is therefore available for deposition in the respiratory tract.

(n) **Respirable dust** is intended to simulate the fraction which penetrates to the gas exchange region of the lung.

(o) **Percutaneous absorption** is absorption through the skin, and can result from local contamination, for example from a splash on the skin or clothing, or in certain cases from exposure to high atmospheric concentrations of vapour.

Toxic substances

Routes of entry

Inhalation

Inhalation of toxic substances in the form of dust, fume, gas, vapour or mist, accounts for the majority of deaths and illnesses associated with toxic substances. The results may be acute (immediate) as in the case of many gassing accidents, e.g. chlorine, carbon monoxide (at high concentrations), hydrogen sulphide and nitric oxide; or chronic (prolonged and cumulative) as with exposure to, for example, chlorinated hydrocarbons, lead compounds, dusts which produce pneumoconiosis, mists and fogs, such as paint spray and oil mists, and fume, notably that from welding operations.

Absorption (pervasion)

The skin, if intact, is proof against most but not all inputs. There are certain substances and micro-organisms which are capable of passing straight through the intact skin into underlying tissue or even into the blood stream, without apparently causing any change in the skin itself (percutaneous effect). The resistance of the skin to external irritants varies with age, sex, race, colour and, to a certain extent, diet. Absorption, as a route of entry, is normally associated with occupational dermatitis, the causes of which may be broadly divided into two groups.

Primary irritants: These substances will cause dermatitis at the site of the contact if permitted to act for sufficient length of time in sufficient concentrations, e.g. strong acids, strong alkalis and solvents.

E

Secondary cutaneous sensitisers: These substances do not necessarily cause skin changes, but effect a specific sensitisation of the skin. If further contact occurs after an interval of approximately seven or more days, a dermatitis will develop at the site of the second contact. Examples of secondary sensitisers are some rubber additives, nickel, certain wood dusts and proteolytic enzymes.

Ingestion

Certain substances are carried into the intestine from which some will pass into the body by pervasion through the intestinal wall. Like the lung, the intestine behaves as a selective filter which keeps out many, but not all, harmful agents presented to it.

Injection, inoculation and implantation

A forceful breach of the skin, perhaps as a result of injury, can carry harmful substances through the skin barrier.

Target organs and target systems

Many dangerous substances, when taken into the body, are deposited in and act upon a particular body organ (target organ) and/or a particular body system (target system).

The most common target organs are the lungs, liver, bladder, skin and brain, and the most common target systems, the central nervous system, circulatory system, reproductive system and the urino-genital system.

Typical examples

Substance	Target organ/system
Beta-naphthylamine	Bladder
Solvents e.g. trichlorethylene	Liver
Asbestos and other particulates	Lungs
Lead, mercury, cadmium and other heavy metals; most organic solvents	Blood/circulatory system
Mercury and derivatives	Central nervous system

Biological indicators

The biological samples where indicators of absorption into the body may be determined consist of:

(a) blood, urine, saliva, sweat, faeces;

(b) hair, nails, etc;

(c) expired air.

Dose/response relationship

A basic principle of occupational disease prevention is the concept of threshold limits of exposure or dose, which normal people can tolerate without long-term or short-term damage to their health. For many chemical substances found in common industrial use, it is possible to discern a link between dose and the body's response, a characteristic known as the 'dose/response relationship'.

With many dusts, for instance, the body's reaction or response is directly proportional to the dose received over a period of time: the greater the dose, the more serious the resulting condition, and vice versa (*see* Fig. 18.1(*a*)). In the case of other substances, the dose/response curve remains at a level of no response at a point greater than zero on the dose axis, and this point of cut-off identifies the threshold dose (*see* Fig. 18.1(*b*)).

Fig. 18.1 ◆ (a) Response directly proportional to the dose received – dusts; (b) Identification of threshold dose – gases, etc.

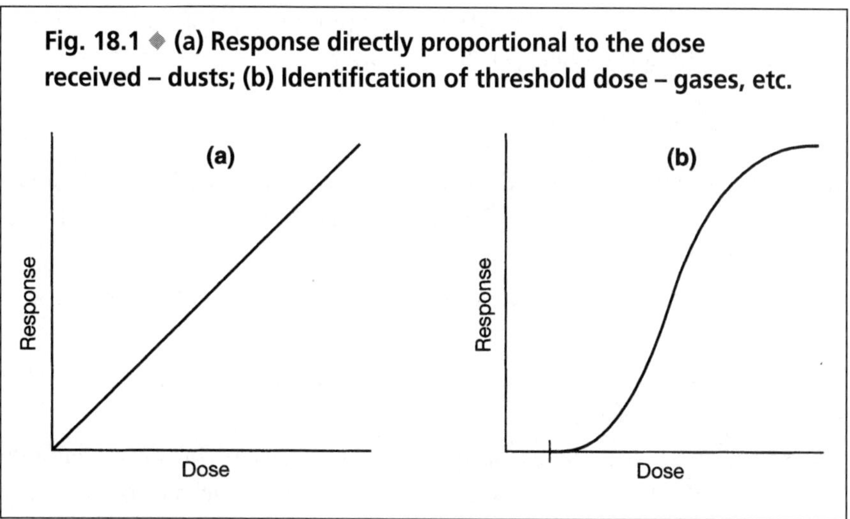

E

Occupational exposure limits

Threshold limit values were values published by the American Conference of Governmental Industrial Hygienists in the USA (where, in effect, they are enforced through state inspectors as if they were minimum statutory standards) and adopted for use in the UK by the HSE. They refer to airborne concentrations of substances and represent conditions under which it is believed that nearly all workers may be repeatedly exposed, day after day, without adverse effects. Threshold limit values refer to time-weighted average concentrations for a seven- or eight-hour work day and a 40-hour working week. They have now been replaced by occupational exposure limits.

HSE Guidance Note EH 40 *Occupational exposure limits* gives details of occupational exposure limits (OELs) which should be used for the purposes of determining the adequacy of control of exposure by inhalation to substances hazardous to health. These limits form part of the requirements of the COSHH Regulations 1999.

A **substance hazardous to health** is defined in Reg 2 as meaning any substance (including any preparation) which is:

(a) a substance which is listed in Part I of the approved supply list as dangerous for supply within the meaning of the Chemicals (Hazard Information and Packaging for Supply) Regulations 1994 and for which an indication of danger specified for the substance in Part V of that list is very toxic, toxic, harmful corrosive or irritant;

(b) a substance for which the HSC has approved a maximum exposure limit or occupational exposure standard;

(c) a biological agent;

(d) dust of any kind, except dust which is a substance within the first two paragraphs above, when present at a concentration in air equal to or greater than –

  (i) 10 mg/cu.m, as a time-weighted average over an 8-hour period of total inhalable dust; or

  (ii) 4 mg/cu.m, as a time-weighted average over an 8-hour period of respirable dust;

(e) a substance, not being a substance mentioned in the sub-paragraphs above, which creates a hazard to the health of any person which is comparable with the hazards created substances mentioned in those sub-paragraphs.

The advice given in Guidance Note EH 40 should be taken in the context of the requirements of the COSHH Regulations, especially Reg 6 (assessment), Reg 7 (control of exposure), Regs 8 and 9 (use and maintenance of control measures) and Reg 10 (monitoring of exposure). Additional guidance may be found in the COSHH General Approved Code of Practice.

Legal requirements

Reg 2(1) of COSHH states that:

> 'the **maximum exposure limit**' for a substance hazardous to health means the maximum exposure limit for that substance set out in Schedule 1 in relation to the reference period specified therein when calculated by a method approved by the Health and Safety Commission (HSC);
>
> the occupational exposure standard for a substance hazardous to health means a standard approved by the HSC for that substance in relation to a specified reference period when calculated by a method approved by the HSC.

Reg 7(4) of COSHH requires that where there is exposure to a substance for which an MEL is specified in Schedule 1, the control of exposure, so far as inhalation of that substance is concerned, shall only be treated as being adequate if the level of exposure is reduced so far as is reasonably practicable and in any case below the MEL.

Reg 7(5) of COSHH requires that, without prejudice to the generality of Reg 7(1), where there is exposure to a substance for which an OES (Occupational exposure standard) – *see* page 361 has been approved, the control of exposure shall, so far as the inhalation of that substance is concerned, be treated as being adequate if:

(a) the OES is not exceeded; or

(b) where the OES is exceeded, the employer identifies the reasons for the standard being exceeded and takes appropriate action to remedy the situation as soon as is reasonably practicable.

Units of measurement

E

The lists of OELs given in the Guidance Note, unless otherwise stated, relate to personal exposure to substances hazardous to health in the air of the workplace. Concentrations of gases and vapours in air are usually expressed as parts per million (ppm), a measure of concentration by *volume*, as well as in milligrams per cubic metre of air (mg m^{-3}), a measure of concentration by *mass*. Concentrations of airborne particles (fume, dust, etc.) are usually expressed in milligrams per cubic metre, with the exception of mineral fibres, which are expressed as fibres per millilitre of air.

Maximum exposure limits (MELs)

MELs are listed in both Schedule 1 of the COSHH Regulations and Table 1 of Guidance Note EH 40.

An MEL is the maximum concentration of an airborne substance, averaged over a reference period, to which employees may be exposed by inhalation under any circumstances and is specified, together with the appropriate reference period, in Schedule 1 of the COSHH Regulations.

Reg 7(4) of COSHH, when read in conjunction with Reg 16, imposes a duty on the employer to take *all reasonable precautions* and to exercise *all due diligence* to achieve these requirements.

In the case of substances with an eight-hour long-term reference period, unless the assessment carried out in accordance with Reg 6 shows that the level of exposure is most unlikely ever to exceed the MEL, to comply with this duty the employer should undertake a programme of monitoring in accordance with Reg 10 so that he can show, if it is the case, that the MEL is not normally exceeded, that is, that an occasional result above the MEL is without real significance and is not indicative of a failure to maintain adequate control.

Some substances mentioned in Schedule 1 of the COSHH Regulations have been assigned short-term MELs, e.g. a ten-minute reference period. These substances give rise to acute effects and the purpose of limits of this kind is to render insignificant the risks to health resulting from brief exposure to the substance. For this reason short-term exposure limits should never be exceeded.

In determining the extent to which it is reasonably practicable to reduce exposure further below the MEL, as required by Reg 7(4), the nature of the risk presented by the substance in question should be weighed against the cost and the effort involved in taking measures to reduce the risk.

Occupational exposure standards (OESs)

An OES is the concentration of an airborne substance, averaged over a reference period, at which, according to current knowledge, there is no evidence that it is likely to be injurious to employees if they are exposed by inhalation, day after day to that concentration and which is specified in a list approved by the HSC.

OESs are approved by the HSC following a consideration of the often limited available scientific data by the Working Group on the Assessment of Toxic Chemicals (WATCH).

For a substance which has been assigned an OES, exposure by inhalation should be reduced to that standard. However, if exposure by inhalation exceeds the OES, then control will still be deemed to be adequate provided that the employer has identified why the OES has been exceeded, and is taking appropriate steps to comply with the OES as soon as is reasonably practicable. In such a case, the employer's objective must be to reduce exposure to the OES, but the final achievement of this objective may take some time. Factors which need to be considered in determining the urgency of the necessary action include the extent and cost of the required measures in relation to the nature and degree of exposure involved.

Long-term and short-term exposure limits

Substances hazardous to health may cause adverse effects, e.g. irritation of the skin, eyes and lungs, narcosis or even death after short-term exposure, or via long-term exposure through accumulation of substances in the body or through the gradual development of increased risk of disease with each contact. It is important to control exposure so as to avoid both short-term and long-term effects. Two types of exposure limit are therefore listed in Guidance Note EH 40.

The *long-term exposure* limit is concerned with the total intake over long periods and is therefore appropriate for protecting against the effects of long-term exposure.

The *short-term exposure* limit is aimed primarily at avoiding the acute effects, or at least reducing the risk of the occurrence. Specific short-term exposure limits are listed for those substances for which there is evidence of a risk of acute effects occurring as a result of brief exposures.

For those substances for which no short-term exposure limit is listed, it is recommended that a figure of three times the long-term exposure limit averaged over a 15-minute period be used as a guideline for controlling exposure to short-term excursions.

Both the long-term and short-term exposure limits are expressed as airborne concentrations averaged over a specified reference time. The period for the long-term limit is normally eight hours; when a different period is used this is stated. The averaging period for the short-term exposure limit is normally 15 minutes, such a limit applying to any 15-minute period throughout the working shift.

'Skin' annotation

Certain substances listed in Guidance Note EH 40 carry the 'Skin' annotation (Sk). This implies that the substance can be absorbed through the skin. This fact is important when undertaking health risk assessments under the COSHH Regulations.

Examples of MELs and OESs

Table 18.1 gives some examples of substances listed in Guidance Note EH 40.

Guidance Note EH 40 and the COSHH Regulations

Guidance Note EH 40 is an extremely significant document in the interpretation and implementation of the COSHH Regulations. Along with the Approved Codes of Practice (ACOPS) and other documentation issued with the Regulations, regular reference to the Guidance Note should be made in activities directed at securing compliance with Regs 6 to 12 of the Regulations. Further guidance is given on the COSHH Regulations in Chapter 39 'Dangerous substances'.

Table 18.1 ◆ Hazardous substances listed in Guidance Note EH 40

	Formula	*Long-term exposure limit*		*Short-term exposure limit*		*Note*
		ppm	mg/m^{-3}	ppm	mg/m^{-3}	
Maximum exposure limits						
Acrylonitrile	$CH_2{=}CHCN$	2	4	–	–	Skin
Carbon disulphide	CS_2	10	30	–	–	Skin
Isocyanates		–	0.02	–	0.07	
Trichlorethylene	$CCl_2{=}CHCl$	100	535	150	802	Skin
Occupational exposure standards						
Ammonia	NH_3	25	18	35	27	
Sulphur dioxide	SO_2	2	5	5	13	
Carbon monoxide	CO	50	55	300	330	
Disulphur decafluoride	S_2F_{10}	0.025	0.25	0.075	0.75	
Mercury & compounds (except mercury alkyls)	Hg	–	0.05	–	0.15	

Lead and asbestos

Lead and asbestos are excluded from the requirement of the COSHH Regulations because legislation covering them already exists, i.e. Control of Lead at Work Regulations 1998 and Control of Asbestos at Work Regulations 1987.

Lead

The Control of Lead at Work Regulations 1998 define lead as meaning lead (including lead alkyls, lead alloys, any compounds of lead and lead as a constituent of any substance or material) which is liable to be inhaled, ingested or otherwise absorbed by persons except where it is given off from the exhaust system of a vehicle on a road within the meaning of section 192 of the Road Traffic Act 1988.

These regulations:

(a) lay down occupational exposure limits for lead and lead alkyls;

(b) introduce –

  (i) blood-lead action levels;

  (ii) blood-lead suspension levels and urinary lead suspension levels for women of reproductive capacity and young persons and other employees;

(c) impose a prohibition in respect of women of reproductive capacity and young persons in specified activities only;

(d) require an employer to carry out an assessment as to whether the exposure of any employee to lead is liable to be significant;

(e) require an employer to ensure that only persons responsible for undertaking necessary work are permitted into an area where a significant increase in exposure to lead is likely to occur as a result of a failure of a control measure;

(f) impose requirements concerning the examination and testing of engineering controls and respiratory protective equipment and the keeping of personal protective equipment;

(g) impose sampling procedures in respect of air monitoring;

(h) impose requirements in relation to medical surveillance;

(i) require that information given to employees by employers includes the results of air monitoring and health surveillance and its significance;

(j) require the keeping of records in respect of examination and testing of control measures, air monitoring and health surveillance for specified periods.

The ACOP to the former regulations is still relevant.

Asbestos

The Control of Asbestos at Work Regulations 1987 give statutory backing to the existing Control Limits and supplement them with a new ten-minute Control Limit. The Regulations introduced a measure of exposure, the 'Action Level', which determines whether or not certain Regulations apply in a given case. HSC has approved a method of measurement for the purposes of checking dust levels against the Control Limits or Action Levels. For comparison with the four-hour Control Limit the approved method is identical to the European Reference Method, but variations are introduced when measurements are used for other purposes.

Details of the Control Limits and measurement techniques for asbestos are set out in HSE Guidance Note EH 10 *Asbestos exposure limits and measurement of airborne dust concentrations*. This Note gives guidance, in particular, on the use of airborne fibre measurements for comparison with exposure limits, to check the effectiveness of enclosures or other control measures, or for site clearance when work is finished. MDHS 39/2 *Asbestos fibres in air: light microscope methods for use with the Control of Asbestos at Work Regulations* describes in detail the measurement methods for airborne fibre, whilst other HSE guidance on asbestos is listed in Guidance Note EH 10.

Toxicological assessment

Toxicological assessment refers to the collection, assembly and evaluation of data on a potentially toxic substance and the conditions of its use, in order to determine the danger to human health, systems for controlling the danger, the detection and treatment of over-exposure and, where such information is insufficient, the need for further investigation. It is closely related to the duties of manufacturers and importers of substances used at work, under HSWA, sec 6, to ensure the safety of their products, to undertake testing and examination, and to provide adequate information for the user to ensure safe working (*see* Chapter 3). These duties are further extended in the Notification of New Substances Regulations 1982 and the approved codes of practice.

In assessing toxic hazards, the following basic information is required:

(a) the name of the substance, including any synonyms;
(b) a physical or chemical description of the substance;
(c) information on potential exposure situations;
(d) details of exposure limits;
(e) general toxicological aspects, such as –
  (i) the route of entry into the body;
  (ii) the mode of action in or on the body;
  (iii) signs and symptoms;
  (iv) diagnostic tests;
  (v) treatment;
  (vi) the disability potential.

Moreover, under the Notification of New Substances Regulations 1982 specific duties are placed on the manufacturers and importers of new substances with regard to testing and notification, announcement procedures, submission of information to the 'competent authority', the form, content and time of notification and procedures for further notification.

Protection against toxic substances

The COSHH Regulations place a duty on employers to either prevent or control exposure to substances hazardous to health. Details of the various strategies for preventing or controlling exposure are covered in Chapter 26 'Prevention and control strategies in occupational hygiene'.

Biological monitoring

This can be defined as a regular measuring activity where selected validated indicators of the uptake of toxic substances are determined in order to prevent health impairment. Biological monitoring could well feature in the health or medical surveillance of persons exposed to hazardous substances under the COSHH Regulations. It is undertaken through the determination of the effects certain substances produce on biological samples of exposed individuals, and these determinations are used as biological indicators.

Dose-effect relationship

The evaluation of the relationship between the dose of an offending agent and the effect on the body of that specific dose is based on analysis of the degree of association existing between an indicator of dose, e.g. blood, urine, saliva, faeces, and an indicator of effect on the body, e.g. loss of consciousness, lacrimation, respiratory response. The study of the dose-effect relationship identifies at which concentration of toxic substance the indicator of effect exceeds the values currently accepted as 'normal'.

Dose-response relationship

This term is discussed earlier in this chapter. Since not all individuals in a group react in the same manner to exposure to a contaminant, it is necessary to study how the group responds by evaluating the appearance of the effect compared to the internal dose. This is what is meant by 'response', which is the percentage of subjects in the group who show a specific quantitative variation of an indicator of each dose level.

Biological indicators

The biological samples where the indicators may be determined consist of:

(a) blood, urine, saliva, sweat, faeces;

(b) hair, nails;

(c) expired air.

With biological monitoring, information can be obtained that would not otherwise be available:

(a) on the evaluation of absorption and/or exposure over a prolonged period of time;

(b) on the amount of substance absorbed as a result of movements within the working environment or of accidental causes, which often cannot be checked;

(c) on the amount absorbed by the organism via various routes of entry;

(d) on the evaluation of overall exposure, as the sum of different sources of contamination, which may also exist outside the working environment;

(e) on the amount absorbed by the subject, taken as an individual, as related not only to his workplace, but taking into account climatic factors, specific physical resistance, age, sex, individual genetic characteristics, etc.;

(f) on whether the subject has been exposed to a risk which could not be proven in any other way and, in some cases, when.

The indicators of internal dose can be further divided into:

(a) *true indicators of dose*, i.e. capable of indicating the quantity of the substance at the sites where it exerts its effect;

(b) *indicators of exposure*, which can provide an indirect estimate of the degree of exposure, since the levels of substances in the biological samples closely correlate with levels of environmental pollution;

(c) *indicators of accumulation* that can provide an evaluation of the concentration of the substance in organs and/or tissues from which the substance, once deposited, is slowly released.

Biological exposure indices (BEIs)

BEIs are set by the American Conference of Governmental Industrial Hygienists (ACGIH) to reflect the average body fluid concentration of a toxic substance or its metabolite found in workers exposed at the equivalent Threshold Limit Value (TLV).

Biological tolerance values

These values are established in Germany and are defined as 'the maximum permissible quantity of a chemical compound or its metabolites or any deviation from norm of biological parameters induced by these substances in exposed humans'. According to current knowledge, these conditions do not impair the health of employees, even if the exposure is repeated and of long duration. The German values are thus health-based values and are set as acceptable upper limits, whereas the ACGIH levels are only for the guidance of occupational health professionals. In a group of workers exposed at a given TLV there will be a typical biological spread of the concentration of toxic substances or its metabolites in biological fluids.

The Biological Exposure Index (BEI) is, therefore, not an acceptable upper limit and should only be considered as a guide to the occupational physician or hygienist. In fact, all it may indicate is the average concentration to be expected in a worker exposed at the TLV, and it is not directly a health-based limit.

Detection by biological monitoring

Environmental monitoring and control of the absorption of some toxic materials into the body is not possible without monitoring the total uptake of the material in the body. Biological monitoring is a valuable additional tool available to occupational health professionals working in this area.

Furthermore, environmental monitoring of breathing zone air may not be a good guide to the levels of contaminant absorbed by the body in the long term, because of the variations in:

(a) breathing rate (work rate);

(b) metabolic exit rate;

(c) absorption rate in the lungs.

Analysis of the following substances (Table 18.2) can be used to gauge atmospheric exposure levels.

Table 18.2 ◆ Hazardous substances that can be revealed by medical analysis

Substance	*Technique*
Benzene	Phenol in urine; benzene in breath
Inorganic lead	Lead in blood/urine; coproporphyrin III in urine
Elemental mercury/inorganic mercury	Mercury in urine; protein in urine
Methyl mercury	In faeces
Arsenic	In urine, hair and nails
Cadmium	In blood and urine
Trichlorethylene	In urine as trichloracetic acid
Organo-phosphorus compounds	Cholinesterase in blood/urine; nerve conduction velocity; electromyography

19 Occupational diseases and conditions

Occupational diseases are those diseases contracted as a result of a particular employment. This chapter considers the wide range of occupational diseases and conditions, their causes and symptoms. Many occupational diseases are prescribed, in which case benefit is payable under the Social Security Act 1975.

Prescribed diseases

A disease may be prescribed if:

(a) it ought to be treated, having regard to its causes and incidence and other relevant considerations, as a risk of occupation and not as a risk common to all persons;

(b) it is such that, in the absence of special circumstances, the attribution of particular cases to the nature of the employment can be established with reasonable certainty (Social Security Act 1975, sec 76(2)).

Current law relating to prescribed occupational diseases is to be found in the Social Security (Industrial Injuries) (Prescribed Diseases) Regulations 1985, SI 967 (SS(II)(PD)R).

Reportable diseases

A reportable disease is a disease listed in Schedule 2 column 1 of RIDDOR and related to a particular kind of employment listed in Schedule 2 column 2 of same. Where such cases are diagnosed by a medical practitioner, an employer must report same to the enforcement authority on Form 2058A within seven days of notification.

In certain cases, an occupational disease or condition may be both reportable under RIDDOR and prescribed for the purposes of industrial injuries benefit.

Classification of the causes of occupational disease

Physical causes

Examples include:

(a) *Heat* – heat cataract, heat stroke (prescribed disease A2).

(b) *Lighting* – miner's nystagmus (prescribed disease A9).

(c) *Noise* – noise-induced hearing loss (occupational deafness) (prescribed disease A10).

(d) *Vibration* – vibration-induced white finger (prescribed disease A11).

(e) *Radiation* – radiation sickness (at ionising wavelengths), burns, arc eye.

(f) *Dust* – silicosis, coal worker's pneumoconiosis (prescribed disease D1).

(g) *Pressure* – decompression sickness (prescribed disease A3).

Chemical causes

Examples include:

(a) *Acids and alkalis* – dermatitis (non-infective dermatitis is prescribed disease D5).

(b) *Metals* – lead and mercury poisoning (prescribed diseases C1 and C5).

(c) *Non-metals* – arsenic and phosphorus poisoning (prescribed diseases C4 and C3).

(d) *Gases* – Carbon monoxide poisoning, arsine poisoning (prescribed disease C4).

(e) *Organic compounds* – occupational cancers, e.g. bladder cancer (prescribed disease C23).

(f) *Dusts* – mercury poisoning (prescribed disease C5).

Biological causes

Examples include:

(a) *Animal-borne* – anthrax, brucellosis, glanders fever (prescribed diseases B1, B2 and B7).

(b) *Human-borne* – viral hepatitis (prescribed disease B8).

(c) *Vegetable-borne* – aspergillosis (farmer's lung) (prescribed disease B6).

Ergonomic causes

Examples include:

(a) *Job movements* – cramp (in relation to handwriting or typewriting) (prescribed disease A4).

(b) *Friction and pressure* – bursitis, cellulitis, i.e. beat hand, traumatic inflammation of the tendons or associated tendon sheaths of the hand or forearm, i.e. tenosynovitis (prescribed diseases A5 and A8).

The physical causes

Heat

Heat cataract

Cataracts of the eye, caused by excessive exposure to heat and microwaves, have been common in many industries, e.g. glass blowing, chain making and others requiring the operation of furnaces. Continuous exposure to radiant heat results in the opacity of the lens of the eye. Such radiations, it is thought, disturb the nutrition of the lens and cause localised coagulation of the protein. (Heat cataract is included in prescribed disease A2.)

Heat stroke

Heat stroke is occasionally encountered in workers in hot processes. The symptoms are due to a defect in thermoregulation – the ability of the body to vary its temperature according to external factors. The onset is usually abrupt: the patient falls unconscious and could have a temperature of 40.6°C or more. Emergency treatment is aimed at reducing the body temperature to 40.0°C within one hour by all possible means, thus minimising the risk of damage to the central nervous system. (Heat stroke is included in prescribed disease A2.)

Heat cramps

Such cramps may be encountered by workers in the heat treatment of metals, e.g. forging or casting, or as a result of heat from microwave radiation. Most cases occur during summer months. Cramps take the form of pain in the muscles beginning in the calves and spreading to the arms and abdomen. The pains are of an intermittent nature, occurring with increasing severity every few minutes. Generally, taking a drink containing common salt (saline) is the only treatment necessary, and it is normal in many industries to have stocks of salt tablets available.

Lighting

Headaches, vertigo, insomnia and 'eye strain' (visual fatigue) are common in many work situations. They are associated with poor lighting and the level of visual performance of the operator (*see* Chapter 15). There is one occupational disease prescribed in relation to lighting (or the lack of it), i.e. miner's nystagmus (prescribed disease A9).

Miner's nystagmus

This disease is associated with poor lighting conditions in underground working operations, and is a complex psychological malady. It is thought to be primarily the result of poor lighting. However, exposure to toxic gas, the adoption of unusually awkward working postures or the onset of a state of anxiety have often been precipitating factors. The disease is associated with the more or less rhythmic oscillation of the eyeballs often coupled with persistent headaches, vertigo and insomnia. It may be accompanied by contraction of the fields of vision, poor visual acuity and photophobia, nervous symptoms and tremor, as well as nuchal rigidity (a condition of the brain) with a characteristic posture in walking. Generally, complete recovery takes place after cessation of underground work and a return to exposure to good illuminance levels.

Noise

The most common condition associated with exposure to noise is occupational deafness. Under the provisions of the SS(II)(PI)R the condition is described as follows: 'Substantial sensorineural hearing loss amounting to at least 50 dB in each ear, being due in the case of at least one ear to occupational noise, and being the average of pure tone losses measured by audiometry over the 1, 2 and 3 KHz frequencies.'

Noise may affect hearing in three ways:

1. Temporary threshold shift is the short-term effect (i.e. a temporary reduction in hearing acuity) which may follow exposure to noise. The condition is reversible. The effect depends upon individual susceptibility.

2. Permanent threshold shift takes place where the limit of tolerance is exceeded in terms of time, level of noise and individual susceptibility. Recovery from permanent threshold shift will not proceed to completion, but will effectively cease at some particular point in time after the end of the exposure. ('Persistent threshold shift' is used to denote the degree of hearing impairment remaining after at least 40 hours.) The term 'permanent' is reserved for conditions which may be reasonably supposed to have no possibility of further recovery. Some recovery of hearing may be found after two days away from noise, and even longer.
3. Acoustic trauma is quite a different condition from occupational deafness (noise-induced hearing loss). It involves sudden aural damage resulting from short-term intense exposure or even from one single exposure. Explosive pressure rises are often responsible, such as exposure to gunfire, major explosions or even fireworks.

For most steady types of industrial noise, intensity and duration of exposure are the principal factors in the degree of noise-induced hearing loss. Hearing ability also deteriorates with age (presbyacusis), and it is sometimes difficult to distinguish between the effects of noise and normal age deterioration in hearing. Research by the UK Medical Research Council and the National Physical Laboratory has shown that the risk of noise-induced hearing loss can be related to the total amount of noise energy that is taken in by the ears over a working lifetime.

Symptoms of noise-induced hearing loss

Mild form of noise-induced hearing loss: There is sometimes difficulty in conversing with people, the wrong answers may be given occasionally, and speech on television and radio seems indistinct. Moreover, there is difficulty in hearing normal domestic sounds, such as a clock ticking.

Severe form of noise-induced hearing loss: With a severe degree of deafness, there is difficulty in conversing, even when face to face with people, as well as hearing what is said at public meetings, unless sitting right at the front. Generally, people seem to be speaking indistinctly, even on radio and television, and there is an inability to hear the normal sounds of home and street. It is often impossible to tell the direction from which a sound is coming, and to assess the distance from the sound. (This last mentioned feature is a contributory factor in accidents.) In most severe cases, there is a sensation of whistling or ringing in the ears (tinnitus). This condition can give rise to liability at common law in an action against an employer, even though the resultant deafness is quite insignificant (*O'Shea* v. *Kimberly-Clark Ltd*, reported in *The Guardian*, 8 October 1982). Moreover, it has recently been decided that it is legitimate to apportion an employer's liability for an employee's occupational deafness according to the length of time the employer can be shown to have been in breach of his duty of care (*Thompson, Gray, Nicholson* v. *Smiths Ship Repairers (North Shields) Ltd* [1984] IRLR 93–116). This is contrary to the earlier decision in *Heslop* v. *Metalock (Great Britain) Ltd, The Observer*, 29 November 1981.

In order to understand the mechanism of noise-induced hearing loss, see the description below.

Hygiene standards for noise-induced hearing loss

Standards are based on noise-induced deafness, age-based deafness, and speech range at the frequencies of interest, i.e. 500Hz, 1kHz, 2kHz and 3kHz. The various standards are as follows:

E

(a) National Insurance (Industrial Injuries) Commission: 50dB hearing loss averaged through 1, 2 and 3KHz in the ear which hears best (SS(II)(PD)R, A10 Part 1, Schedule 1).

(b) American Academy of Ophthalmologists and Otolaryngologists (AAOO): 25 dB hearing loss averaged through 0.5, 1 and 2 kHz in both ears.

(c) British Association of Ophthalmologists (BAO): 40 dB hearing loss averaged through 1, 2 and 3kHz in both ears.

Cause of noise-induced hearing loss

In order to ascertain how noise-induced hearing loss takes place, it is necessary to understand the physiology of the human hearing system. The ear is composed of three specific parts, the outer, middle and inner ears (*see* Fig. 19.1). The outer ear comprises the pinna, with the auditory canal (meatus) leading to the ear-drum (tympanic membrane). The middle ear is a chamber containing three linked bones or ossicles, the malleus (hammer), incus (anvil) and stapes (stirrup). The function of the ossicles is to transform the vibration caused by sound waves impinging on the tympanic membrane into mechanical movements. These are transmitted to the fluid of the inner ear by the stapes bone, which fits into one of the two holes connecting the middle and inner ears, the fenestra ovalis (oval window).

The inner ear contains the important organ of hearing, the cochlea, which comprises a coiled fluid-filled tube, similar in appearance to a snail's shell. The cochlea has a basilar membrane, an auditory nerve and frequency-responsive hair cells incorporated along its coil. The hair cells run along the length of the basilar membrane. A sound entering the cochlea causes vibration of fluid resulting in the hair cells also being vibrated. At the beginning of the cochlean coil the hair cells detect the lower frequencies whilst progressively higher frequencies are detected by the hair cells located towards the centre of the coil. The hair cells effectively turn mechanical energy into electrical impulses which they send to the brain by means of the auditory nerve.

Fig. 19.1 ◆ The human hearing system

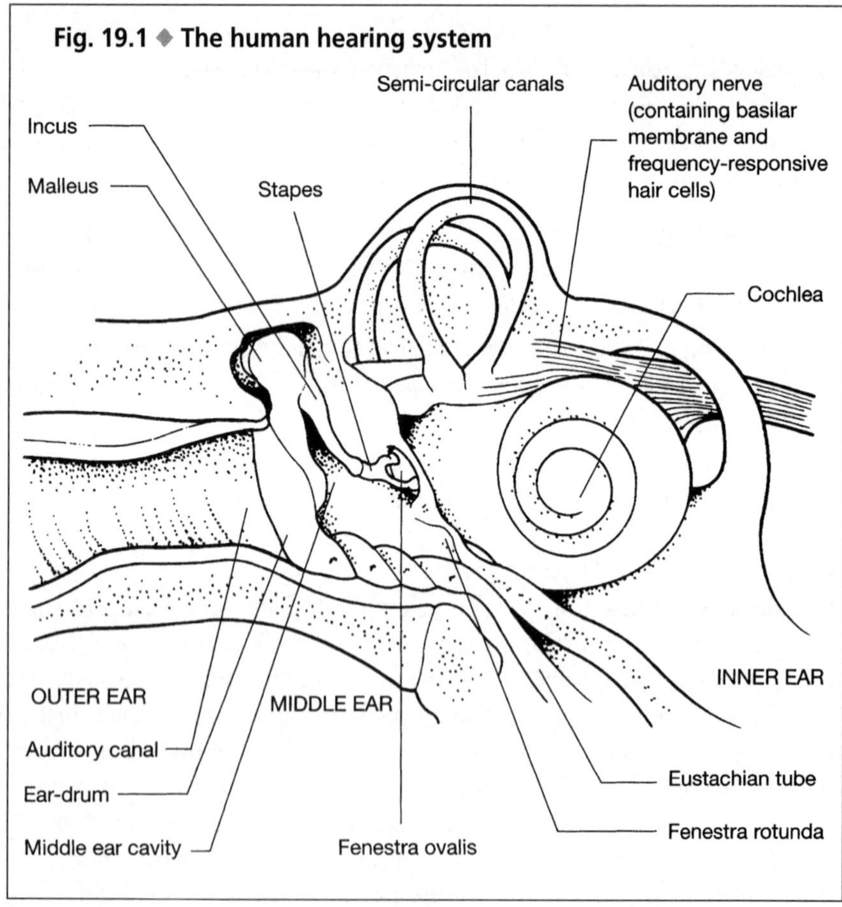

When the ear-drum is vibrated, this vibration is amplified by the ossicles and transmitted to the cochlea. The cochlea transforms the vibrations into nerve impulses which are sent to the brain via the auditory nerve. If the ear-drum is vibrated between 20 and 20,000 times per second a sound will be heard. Normally the hair cells to the cochlea respond well to vibration, and they normally recover after periods of excessive vibration from noise. However, noise can so fatigue one of these cells that it no longer revives and actually dies; thus the row of cells gradually thins out. In cases of noise-induced hearing loss, the projecting cells can be compared with a cornfield that has been subjected to storm or people walking through it. Some of the hair cells are flattened and many will be broken off at their base. This is the effect of excessive noise on the fine hair cells of the cochlea, resulting in varying degrees of noise-induced hearing loss, and variable ability to hear sound at different frequencies.

Measuring hearing loss – audiometry

Hearing loss is measured by audiometry, the most widely used technique being pure tone audiometry. This involves the subject sitting in a soundproof booth and listening through earphones to a series of pure tone sounds. Each sound is gradually increased in intensity until the subject can hear it, whereupon the subject presses a button to indicate that he has perceived that sound. In this way, the hearing threshold, or lowest level at which the sound can be heard, is established over a range of frequencies. At the end of the audiometric test, an audiogram is produced which records the hearing of the subject over these frequencies. An audiogram is produced for each ear.

The audiogram is then compared to an audiogram which, notionally, would be produced by perfect hearing. No one has perfect hearing and the audiogram for a subject with normal hearing is shown in Fig. 19.2(*a*). In 'normal' hearing the audiograms for both ears will be virtually identical. An audiogram is used to assess the degree of hearing loss across the frequencies of interest, which are particularly those at which normal speech takes place, i.e. 0.5, 1 and 2 kHz. In Fig. 19.2(*b*) there is substantial hearing loss at 3, 4 and 6 kHz, but not predominantly at the frequencies of interest. This would be described as a moderate degree of hearing damage.

Vibration

The principal condition related to vibration is 'vibration-induced white finger' (VWF), which is associated with the use of vibrating hand tools, such as compressed air pneumatic hammers, electrically operated rotary tools and chain-saws. Prolonged exposure to local vibration causes this condition. VWF is a prescribed occupational disease. People suffering from the similar condition, Raynaud's syndrome, are likely to have the condition dramatically worsened by exposure to vibration.

Fig. 19.2 ◆ (a) Audiogram for a subject with normal hearing; (b) Audiogram for a subject with a moderate degree of hearing damage.

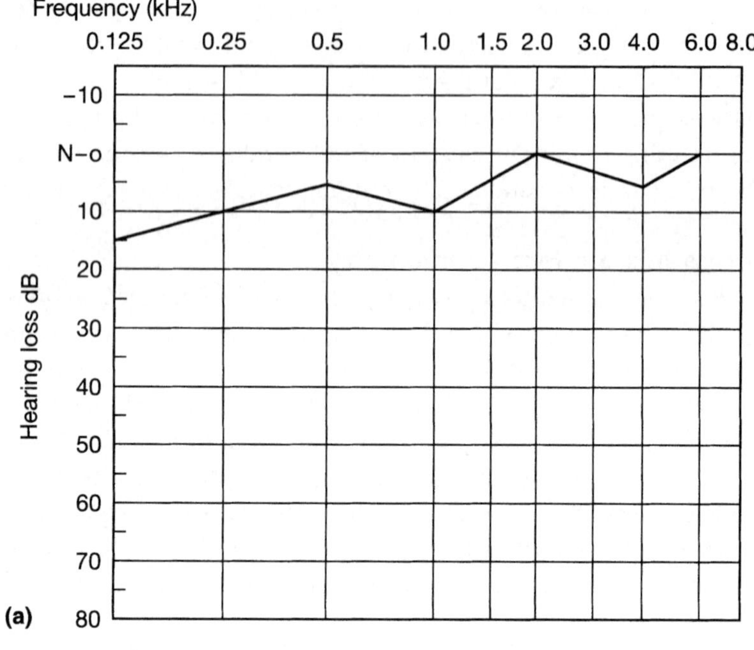

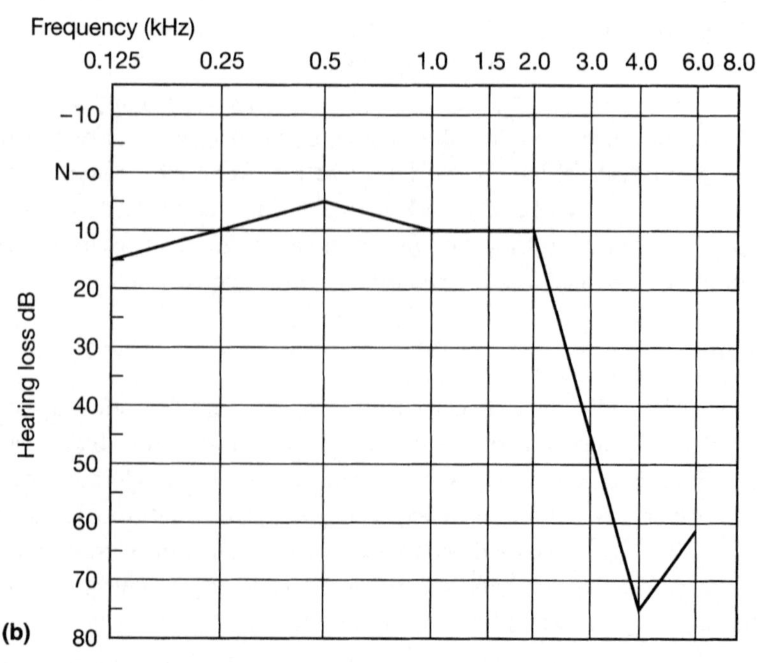

Vibration-induced white finger

The condition is defined as episodic blanching, occurring throughout the year, affecting the middle or proximal phalanges or in the case of a thumb the proximal phalanx, of

(a) in the case of a person with five fingers (including thumb) on one hand, any three of those fingers; or

(b) in the case of a person with only four such fingers, any two of those fingers; or

(c) in the case of a person with less than four such fingers, any one of those fingers or . . . the one remaining finger. (Social Security (Industrial Injuries) (Prescribed Diseases) Regulations 1985, Sch 4.)

Raynaud's syndrome: This may be caused by a number of conditions unassociated with vibration. In fact, research indicates that vibration is not the primary cause. It was originally seen as a manifestation of Raynaud's disease (or 'constitutional white finger') which is thought to be of a hereditary nature. The symptoms may also be caused by diseases of the connective tissue, such as rheumatoid arthritis, trauma, thrombotic conditions, blood disorders and certain neurological diseases. In order to put a specific label on a characteristic set of symptoms caused by exposure to vibration, the Industrial Injuries Advisory Committee has favoured the term 'vibration-induced white finger'.

Symptoms of VWF: The first signs of VWF, which often pass unnoticed or are not attributed to vibration, are mild tingling and numbness of the fingers, similar to 'pins and needles'. Later, the tips of the fingers which are most exposed to vibration become blanched, typically early in the morning or in cold weather. On further exposure, the affected area increases, sometimes to the base of the fingers, sensitivity during attacks is reduced and the characteristic reddening of the areas affected marks the end of an attack causing severe pain. Prolonged and intense exposure may cause further advancement of the condition with the fingers sometimes taking on a blue-black appearance. In severe cases, gangrene and necrosis (death of living tissue) have been reported. Table 19.1 shows a classification of the severity of VWF developed by Taylor and Pelmear in 1975.

There is a latent or symptom-free period from the commencement of regular exposure to the onset of ill-effects, the length of which is thought to be related to the intensity of the vibration. In people subjected to very high intensity levels, stages 2 and 3 may be reached within a few months.

Usually the condition progresses more slowly, and a typical latent period is around five years. In general, the shorter the latent period, the more severe the condition will become as exposure increases.

Table 19.1 ◆ Stages of vibration-induced white finger

Stage	*Condition of digits*	*Work and social interference*
0	No blanching of digits	No complaints
0_T	Intermittent tingling	No interference with activities
0_N	Intermittent numbness	No interference with activities
1	Blanching of one or more finger tips with or without tingling and numbness	No interference with activities
2	Blanching of one or more fingers with numbness; usually confined to winter	Slight interference with home and social activities. No interference at work
3	Extensive blanching; frequent episodes, summer as well as winter restriction	Definite interference at work, at home and with social activities;
4	Extensive blanching; most fingers; frequent episodes, summer and winter	Occupation changed to avoid further vibration exposures because of severity of signs and symptoms

Note: Complications are not considered in this grading
Source: Taylor and Pelmear, 1975

Other vibration-related conditions

These include:

(a) *Osteoarthritis of the arm joints:* this condition is encountered most commonly in the elbow joints, but also in the wrist and shoulder joints.

(b) *Injury to the soft tissues of the hand:* this is mainly injury to the palm of the hand (Dupuytren's contracture) and bursitis; atrophy (shrinkage or wasting) of the palmar muscles and injury to the ulnar nerve are less common.

(c) *Decalcification of the carpus:* this condition in the main bone at the base of the hand has been noted in the hands of workers using pneumatic tools, but it does not deteriorate.

Vibratory hand tools

With hand tools the energy level at particular frequencies is significant in the prevention of VWF. Percussive action tools in the range 2,000–3,000 'beats per minute', equal to a frequency range of 33–50Hz, are the worst. With rotary tools, the range 40–125Hz is common and promotes similar damage. Many vibratory hand tools are used in industry, including pneumatic tools (for riveting, caulking, fettling, rock drilling and hammering), combustion engine operated tools (for chain-sawing, drilling, vehicle operation, the operation of flex-driven machinery) and electrically powered tools (for grinding, concrete levelling, swaging, drilling and burring).

Reduction of vibration injuries

Vibration is best reduced either by redesign of the tool or by introducing more automation to isolate the operator's hand from the source of vibration. Alternatively, it may be possible to introduce a shock-absorbing mechanism between the vibration source and the handles of the tool. If this is impracticable, the following procedures should be carried out:

(a) Pre-employment health screening of potentially exposed workers to assess susceptibility to VWF or manifestation of Raynaud's syndrome.

(b) Ensure body warmth before the start of work in order to achieve good circulation to the extremities.

(c) Ensuring that the temperature in the workplace maintains body warmth throughout the working day.

(d) In the case of outside workers, the supply of wind-resistant clothing and gloves together with replacements for wet clothing.

(e) Minimising smoking because of its effect on circulation.

(f) Where workers are reaching or have already reached the irreversible stage, i.e. stage 4, preclusion from further exposure.

(g) Regular health examinations.

(h) Redesign of portable hand tools at reduced frequencies.

(i) Mechanisation of grinding methods.

(j) Provision and use of cotton gloves with rubber inserts or padded with absorbent material.

(k) Use of specific working methods to reduce exposure time, such as job rotation.

(l) Proper maintenance of tools, such as sharpening of cutters, tuning of engines and renewal of vibration isolators.

(m) Training in correct work techniques to minimise exposure.

Whole body vibration

Whilst VWF tends to be the principal problem associated with vibration, the effects of vibration on the body generally should not be ignored. Whole body vibration, associated, for instance, with driving heavy lorries long distances, can cause blurred vision, loss of balance and loss of concentration, the latter being a causative factor in accidents. In 1974 the International Standards Organisation published recommendations concerned with vibration and the human body (ISO 2631–1974). The recommendations cover cases where the human body is subjected to vibration on one of three supporting surfaces, i.e. the feet of a standing person, the buttocks of a person whilst sitting down and the areas supporting a lying person. Three severity criteria are specified:

(a) a boundary of reduced comfort, applying to fields such as passenger transportation;

(b) a boundary of fatigue-decreased efficiency that is relevant to drivers of vehicles and to certain machine operators;

(c) an exposure limit boundary, which indicates danger to health.

Research shows that in the longitudinal direction, i.e. head to feet, the human body is most sensitive to vibration in the frequency range 4–8Hz, whilst in the transverse direction, i.e. finger tip to finger tip, the body is most sensitive in the range 1–2Hz (*see* Chapter 21 and 'Work-related upper limb disorders').

Radiation

Electromagnetic radiations are those radiations by which energy is transmitted without the necessity of material medium. The spectrum of electromagnetic radiation extends from the very short X or gamma rays to long radio waves and includes X-rays, ultraviolet, visible light, infrared and short radio waves. Ionising particles can be the corpuscular products of atomic disintegration and the result of high-energy electron beams. The bodily effects of radiation across the electromagnetic spectrum are developed in Chapter 23.

Dust

The group of lung diseases of a chronic fibrotic nature due to the inhalation of dust are generally classified as pneumoconiosis (*see* Chapter 22).

Pneumoconiosis

This disease is defined by the International Labour Organisation (ILO) as 'the accumulation of dust in the lungs and the tissue reactions to its presence'. It is divided by the ILO into the collagenous and noncollagenous forms. (Collagen is a protein-based substance which forms the principal component of connective tissue. Its molecules are assembled like three-strand ropes. The collagen diseases or connective tissue diseases have as their common factor a disorganisation of collagen strands. In all collagen diseases there is inflammation without infection.) Noncollagenous pneumoconiosis is caused by non-fibrogenic dust and has the following characteristics: intact alveolar architecture, minimal supporting tissue reaction and potentially reversible effects. Collagenous pneumoconiosis, on the other hand, may be caused by fibrogenic dusts or an altered tissue response to a non-fibrogenic dust. It has the following characteristics: authenticated damage to alveolar architecture, appreciable supporting tissue reaction and permanent (irreversible) scarring of the lung. The following types of collagenous pneumoconiosis may occur following inhalation of dust: anthracosis (coal dust), silicosis (free silica particles in gold, tin, zinc, iron and coal mining, sand blasting, metal grinding, slate quarrying, granite, sandstone and pottery work), siderosis (iron particles), lithosis (stone particles), asbestosis (asbestos) and byssinosis (cotton). Specific aspects of these diseases are outlined below.

Coal worker's pneumoconiosis: This disease takes two forms, simple and complicated. Simple coal worker's pneumoconiosis is a relatively harmless condition. There are nodular lesions with little fibrosis. Emphysema, the abnormally distended condition of the lungs, is slight and there is little distortion of the lung architecture. This condition does not progress in the absence of further dust exposure. Neither does it regress, however.

Complicated coal worker's pneumoconiosis (progressive massive fibrosis) is a different matter, however. Dust collections are embedded in the diseased areas of fibrous tissue, and there is considerable distortion of lung architecture and elasticity producing interference with lung function. There is still considerable argument as to whether the simple form progresses to the complicated form or whether they are two separate disease entities.

Asbestosis: This is a fibrotic condition of the lung, resulting in scarring and thickening of the lung tissue, which may occur after many years of exposure to high concentrations of asbestos dust. The disease may manifest itself some years after occupational exposure has terminated. The risk of contracting asbestosis appears to be related to the duration and level of exposure to asbestos dust. The characteristic symptoms are a progressive breathlessness and unproductive cough. Lung damage takes the form of a diffuse fibrosis or scarring throughout the lungs accompanied by emphysema and collagenous thickening of the pleural lining. The skin may have a bluish discoloration due to cyanosis and sputum may contain asbestos bodies. Other features include the presence of pleural placques (small patches attached to the pleura and peritoneum), asbestos bodies in the lung tissue and asbestos warts on the hand where the fibres penetrate the skin. A characteristic feature is also finger clubbing, i.e. thickening of the fingers. Evidence so far suggests that asbestosis usually has a long induction period (10 to 20 years) and there is abundant evidence that those who smoke suffer both a greatly enhanced risk of contracting the disease and the prospect of a much worse overall lung condition. There is increasing confidence among researchers that a dose-response relationship exists between the amount of dust inhaled and the emergence of asbestosis.

Asbestosis, which is included in occupational disease D8, is prescribed in relation to

(a) the working or handling of asbestos or any admixture of asbestos; or
(b) the manufacture or repair of asbestos textiles or other articles containing or composed of asbestos; or
(c) the cleaning of any machinery or plant used in any of the foregoing operations and of any chambers, fixtures and appliances for the collection of asbestos dust; or
(d) substantial exposure to the dust arising from any of the foregoing operations (SS(II)(PD)R).

Research indicates that around 50 per cent of asbestosis sufferers develop lung cancer and/or cancer of the bronchus, and that smoking and asbestos act synergistically, thereby producing a far greater risk (some say more than 20 times) of bronchogenic lung cancer.

Mesothelioma, also included in occupational disease D8, is prescribed in relation to the same occupations as asbestosis. Moreover, bilateral diffuse pleural thickening is a prescribed occupational is disease (D9) in relation to the same occupations as asbestosis.

Work activities involving asbestos are now regulated by the Control of

Asbestos at Work Regulations 1987, together with the ACOP 'Work with asbestos insulation and asbestos coating'.

Silicosis

This is a condition resulting in fibrosis of the lung. Nodular lesions are formed which ultimately destroy the lung structure. It is caused by the inhalation of respirable-sized particles of free silica. There is a strong predisposition to tuberculosis as a result of contracting silicosis. Silica takes a number of forms in both the crystalline and the amorphous state:

(a) crystalline – tridymite, cristobalite, quartz;

(b) morphous (after heating) – vitreous silica, diatomite, silica fume and dried silica gel.

Sources of silicosis are:

(a) potteries, tile making – the drying out process produces silica dust;

(b) masonry industry – granite polishing, cutting, chipping, quarrying;

(c) furnaces – cutting of refractory bricks, stripping of furnaces;

(d) ceramics – manufacture of insulators;

(e) mining – coal-face working, coal washing;

(f) steel foundries – foundry sand, parting powders;

(g) sand-blasting processes – use of sandstone wheels.

Byssinosis

This disease is predominantly associated with the textile industry. It is a chronic respiratory disease which progresses to bronchitis and emphysema. No physical change is noted in the lung before bronchitis develops. It is characterised by tightness of the chest and increased breathlessness and leads to respiratory disability. Typical is the 'Monday feeling' or 'Monday fever' which occurs when the individual has been away from cotton dust for some days and then returns.

Byssinosis is a response to a substance or substances found in cotton. It appears to occur either as a result of a direct action on the bronchi or broncheoles or through a histamine release mechanism. The severity of the pattern of symptoms is related to the actual tasks carried out by the individual. Textile manufacturing follows some ten or eleven different stages. 'Ginning', the coarse sorting of cotton balls from bracts, leaves and soil, is the first stage and it takes place overseas. Bales are opened by machine after which 'carding' takes place. Here the cotton is separated by a carding engine producing 'slub', a coarse, loosely wound rope of cotton fibres. 'Spinning' results in the strands becoming thinner as the fibres are pulled out. The strand is strengthened by 'twisting' and 'doubling' where more than one thread may be brought together. Workers involved in the early stages of the process, i.e. ginning, opening, carding and spinning, tend to contract byssinosis more than workers involved in the later stages, e.g. twisting, doubling, winding, beaming and weaving. Symptoms are as follows:

Grade 0 – No chest tightness or difficulty in breathing.

Grade ½ – Occasional chest tightness or difficulty in breathing on the first day of a working week.

Grade 1 – Chest tightness or difficulty in breathing on the first day of every working week.

Grade 2 – Chest tightness or difficulty in breathing on the first and other days of every working week.

Grade 3 – Chest tightness or difficulty in breathing on the first and other days of every working week, accompanied by permanent incapacity with diminished effort tolerance or reduced ventilatory capacity.

It is necessary to distinguish between the term 'byssinosis' as defined above and that used for the purpose of assessing benefit under the SS(11)(PD)R. For the latter purpose, the existence of Grade 2 or more would normally be expected in the absence of other complications.

Normally the condition does not progress after removal from exposure, except when emphysema has developed or chronic bronchitis is present. In the early stages the symptoms may remit or disappear.

Pressure

The most commonly encountered condition associated with exposure to pressure is the decompression sickness syndrome (prescribed disease A3).

Decompression sickness

Work in compressed air is undertaken during civil engineering excavations in water-bearing strata or under water, as well as in occupations involving diving. The work is effected either by single divers using diving suits or by a group of workers in a caisson or a diving bell. A caisson consists of a working chamber and shaft communicating at the surface with an air-lock. This, in turn, communicates with the outer air. Caissons are pressurised with cool compressed air from a pipeline. Within the caisson proper, in which the men work, and in the shaft, the air pressure must be equal to the pressure exerted by the water outside, and so must be raised in proportion to the depth at which the work is being carried out, i.e. raised approximately 1 atmosphere for each 10 metres of depth (*see* Diving Operations at Work Regulations 1981).

Decompression sickness is associated with the release within the blood and tissues, when the air pressure is reduced, of gases driven into solution when the air pressure was higher. The gases concerned are oxygen, carbon dioxide and nitrogen. Oxygen is removed very rapidly by reabsorption and carbon dioxide by exhalation from the lungs. Nitrogen, however, is relatively insoluble in the body fluids and collects as minute bubbles of gas which coalesce to form emboli. Nitrogen is, however, five or six times as soluble in fats and lipoids as in the body fluids, and tissues such as the nervous system and bone marrow hold proportionately more of the gas than others, releasing it in bulk when the pressure is dropped.

The symptoms of decompression sickness usually appear within the first few hours following decompression, but may not develop for 12 hours or longer. They depend, in general, on the location of the emboli formed, and may therefore simulate many other diseases. The most common symptom is pain in the limbs which when mild is known as the 'niggles' and when severe, the 'bends'. Generalised itching, vertigo, nausea, vomiting, epigastric pain and dyspnoea ('the chokes') may also occur at this stage. In some cases there is involvement of the central nervous system, and symptoms such as paralysis of the skeletal muscles or of the bladder may occur. Emboli in the blood vessels of the lungs, brain or heart may be fatal.

Electricity

Electric shock

This is the principal hazard and has been the cause of many fatalities. A number of factors affect the severity of electric shock. The actual harm depends on the current flowing through the body and particularly the heart. This is affected, however, by voltage, the resistance of the skin and internal organs (dry skin – high; organs – low), the type of current (a.c. or d.c.), the current pathway through the body, the duration of the current flow and the surface area of contact. The concept of 'let-go currents' is important here. *Let-go current* is defined as 'the maximum d.c. current a person can tolerate when holding an electrode and still let go using muscles stimulated by the shock'. Average let-go currents are 16.0mA for men and 10.5mA for women. A current of only 10mA passing through the heart can have a serious effect on it depending on the duration of the flow. At higher currents, ventricular fibrillation (heart 'flutter') can set in, until at 100mA cessation of the heart beat altogether is the likely result.

Other effects of electricity

Other bodily effects of electricity include burns, 'arc eye' and broken bones through muscular spasm.

The chemical causes

The effects on the body of exposure to chemical substances are many and varied. They can, however, be split into a number of well-defined areas to include the dermatoses, the various forms of poisoning, occupational cancers, the dust-borne diseases and gassing situations or gassing accidents.

Dermatoses

This group includes the range of skin conditions referred to as 'eczema', an inflammation of the skin. Dermatitis is by far the most common occupational disease, and prevention is largely aimed at or associated with improvements in personal hygiene. Only non-infective dermatitis is classified as a prescribed disease (D5), in relation to exposure to dust, liquid, vapour or other skin irritant (SS(II)(PD)R).

Most cases of dermatitis are either:

(a *endogenous* – controlled by factors within the person, and mainly a matter of medical concern; or

(b) *exogenous* – controlled by factors from outside the person, which is very common and preventable.

Within endogenous dermatitis, withdrawal of the affected person from any future contact with the offending factor or agent may be the only solution. Most cases of occupational dermatitis are, however, exogenous in nature, being associated with contact with a specific factor or agent. Agents which produce the typical skin lesions can be classified thus:

(a) mechanical factors – friction, pressure and trauma;

(b) physical factors – heat, cold, electricity, sunlight, radiation;

(c) chemical agents – organic and inorganic substances;

(d) plants and their products, resins and lacquers;

(e) other biological agents such as insects and mites.

Chemical agents are by far the greatest cause of occupational skin disorders, but the other factors mentioned above will be briefly commented upon.

Mechanical factors

These give rise to cuts, abrasions and skin lesions which may become secondarily infected.

Physical factors

Heat may affect the skin, causing excessive perspiration which softens the protective horny layers. When this is combined with frictional stress, a heat rash can develop. This is common in people working on hot processes, e.g. metal workers, furnacemen. Cold injuries to the skin include chilblains and, in extreme cases, frostbite. The secondary effects of burns at work should also be considered, together with possible exposure to ionising radiations.

Chemical agents

The main defences of the skin against external irritants are the cornified layer of the epidermis (*see* Fig. 19.3) and the glandular secretions (sebum). The epidermis may, to some small extent, withstand the action of fairly strong acids, but it is particularly damaged by acids and sulphides. The glandular secretions normally form a fatty and slightly acid protective coating against water soluble irritants, but maceration of the epidermis, by friction, heat and excessive sweating may lead to an inflammatory response to external irritants. Similarly, the openings of the sebaceous glands and hair follicles provide entry for irritants, particularly when the irritant is fat soluble, and inflammatory changes may result. Oil folliculitis and chloracne are examples of such changes. The resistance of the skin to external irritants varies with age, sex, race, colour, diet and state of health.

Plants and their products

Certain plants can cause dermatitis by virtue of the chemical compounds contained within them, e.g. members of the *Liliaceae* and *Primulaceae*. There may be some degree of risk to nurserymen, gardeners and florists. Dermatitis from certain woods is common amongst carpenters, wood machinists and cabinet makers, the sawdust, sap, polishings or oil in the wood being the principal causative agents. Such woods usually produce a 'sensitivity' and a worker may become affected several days after beginning

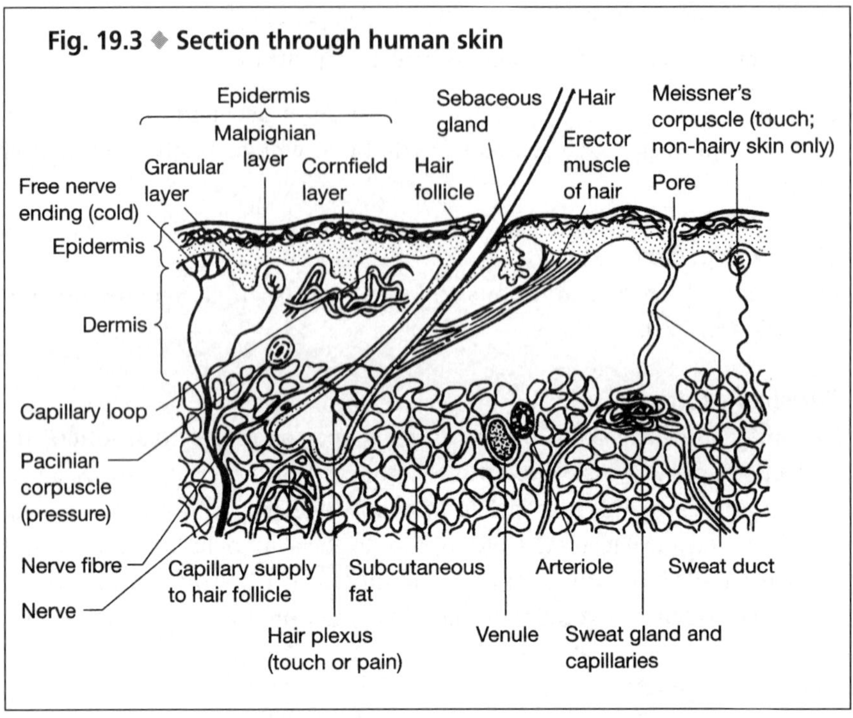

Fig. 19.3 ◆ Section through human skin

work with a new wood. Most of those affected become desensitised, but a small minority may remain sensitised, and will need medical treatment for the rash formed on the hands and arm. Where an employer either actually knew or ought to have known of the harmful properties of wood dust, an action against him will lie at the suit of the affected employee (*Ebbs* v. *James Whitson & Co.* Ltd [1952] 2 QB 877, where the action was unsuccessful).

In some cases more symptoms may be produced. For instance, *Gonioma kamassi*, known as Kamassi boxwood or African boxwood, causes systemic effects in susceptible individuals, owing to the liberation of an alkaloid which has effects similar to curare, i.e. languor, mental dullness and influenza-like symptoms. Skin disorders associated with occupations involving the manufacture of articles from *Gonioma kamassi* are prescribed occupational disease C16.

Other biological agents

Grain itch, barley itch, grocer's itch and copra itch are varieties of dermatitis caused by mites amongst workers involved in agriculture, in the handling of cargoes of grain or in agricultural mills. Scabies, commonly associated with bad hygiene in camps and similar settings, can be contracted by veterinary practitioners and animal breeders and handlers.

The causes of occupational dermatitis

The causes may be broadly divided into two groups of substances:

(a) Primary irritants will cause dermatitis at the site of contact if permitted to act for a sufficient length of time in sufficient concentration, e.g. solvents, strong acids and alkalis. Some agents, such as fibreglass, can also cause dermatitis by straight mechanical (abrasive) action.

(b) Secondary sensitisers – e.g. plants, rubber, nickel and many chemical compounds – do not necessarily cause skin changes on first contact, but effect a specific sensitisation. If further contact occurs after seven days or more, dermatitis will develop at the site of the second contact.

A primary irritant may also be sensitiser in that an initial exposure may so condition the skin that subsequent exposures result in dermatitis.

Most cases of occupational dermatitis go through a number of distinct stages – redness of the skin, swelling, blistering, cracking, scaling and crusting. The principal causes are the use of poor or totally unsuitable hand-cleansing agents, such as paint stripper, paraffin, petrol and proprietary substances which defat the skin, thereby removing the normal defences provided by the cornified layer of the epidermis and the glandular secretions of the skin. Unrefined neat mineral cutting oils and certain coal tar derivatives have promoted dermatitis and can also cause skin cancer in the longer term. The backs of hands are invariably worse affected than the palms, due to the greater degree of softness. Dermatitis is also sometimes the first indication of an unsatisfactory degree of exposure to toxic materials.

Occupational cancers

A tumour or neoplasm (new growth) consists of a mass of cells which have undergone some fundamental and irreversible change in their physiology and structure which leads to a continuous and unrestrained proliferation. The word ‘tumour’ strictly means ‘lump’, but is often used to describe a solid neoplasm. Any cell type within the body can give rise to a neoplasm, although this occurs more frequently with some cell types than with others. Cells which do not undergo regeneration or replacement, nerve cells or voluntary muscle cells, are the ones least likely to give rise to tumours. The rate of growth of neoplasms varies greatly. Some may take many years to develop, whilst others may show an increase within a few days and extend far beyond their point of origin. The variation in growth behaviour and rate forms the basis of a broad classification of tumours into the benign and malignant classes. Benign tumours are those which grow slowly and less expansively, and unless they occur in some vital site, e.g. the brain, or interfere with an important organ, are well tolerated and do not necessarily interfere with a person's well-being or shorten his life. They are composed of well-differentiated and mature types of cells. The cells of the neoplasm resemble the cells of the original tissue. Malignant tumours, on the other hand, grow more rapidly, will infiltrate and extend into normal tissue and structures and, unless effectively treated, interfere with health and eventually cause death. They are often composed of more embryonic (primitive) or poorly differentiated cell types, resembling less the cells of origin. They spread by means of secondary deposits or 'metastases' to other sites in the body. (*See* comparison in Table 19.2.)

Table 19.2 ◆ Characteristics of tumours

Benign	*Malignant*
Remain localised	Can form secondary bodies – metastases
Slow growth	Rapid growth
Often encapsulated	Encapsulation rare and incomplete
Very similar to cells origin	Less similar and, in some cases, totally undifferentiated
Cells of uniform size and appearance	Cells and nuclei vary in size and structure – nuclei often hyperchromatic
Degenerative changes relatively uncommon	Degenerative changes relatively common

There are three main ways in which tumours spread to form metastases – by the lymphatic system, through the bloodstream or through transcoelomic spread, i.e. from the pleural or peritoneal cavities.

Some cancer-promoting agents or 'carcinogens' have a 10–40 year latent period. This is the time between first exposure and the actual diagnosis of the tumour. This could mean that a new chemical substance introduced to industry, and which is later identified as a bladder carcinogen, may result in a tumour after, say, 12 years following the initial exposure (induction period). Even if the use of this compound were immediately prohibited, there would still be the problem of workers exposed during the 12-year period and there could be chemically induced tumours recorded up to 40 years after the last recorded exposure. This was the situation with β-naphthylamine. Although exposure was prohibited many years ago, bladder cancer cases are still coming to light.

Classic examples of the carcinogens discovered in the last 65 years are shown in Table 19.3.

Table 19.3 ◆ Carcinogenic hazards discovered

Occupation	*Agent*	*Site*
Chimney sweeps Distillers of brown coal Makers of patent fuels Manufacturers of coal gas Road workers, boat builders and others exposed to tar and pitch Cotton mule spinners	Combustion products of coal, shale oil (polycylic hydrocarbons)	Scrotum and other parts of skin; bronchus
Dye manufacturers, rubber workers, manufacturers of coal gas	α- and β-naphthylamine, benzidine and methylene bis-ortho-chloroaniline (MBA)	Bladder
Radiologists, radiographers	Ionising radiation and X-rays	Skin
Chemical workers	4-aminodiphenol	Bladder
PVC manufacturers	Vinyl chloride monomer	Liver
Haematite miners	Radon	Bronchus
Asbestos workers, insulation workers, dock workers	Asbestos	Bronchus, pleura and peritoneum
Chromate workers	Chrome ore and pigments	Bronchus
Workers with glues and varnishes	Benzene	Marrow (myeloid and erythro-leukaemia)

Chemical poisoning

The effects on the body of exposure to chemical compounds are many and varied. Some of the more common chemical substances and their effects on humans are detailed below.

Lead

Lead is a microconstituent of many foodstuffs, and is normally found in human tissues and fluids in small amounts, although it is not thought to be an essential element. It may enter the body by inhalation of the fume or dust, by ingestion of contaminated food or through pervasion of the unbroken skin in the form of organic compounds such as tetraethyl lead. Exposure to lead or a lead compound is a prescribed occupational disease (C1).

When absorbed, lead rapidly becomes widely distributed. (Its degree of toxicity depends largely on the ratio between the rate of absorption and the rate of excretion.) When absorption is slow and continuous over a long period of time, the storage factor is significant. Lead is a cumulative poison which, in cases of chronic absorption, is deposited in the calcareous portion of the bones as an insoluble and harmless triple phosphate. In certain circumstances, lead may be released from the bones into the blood stream, producing symptoms similar to those of acute poisoning.

Many processes may result in lead poisoning, the most common being lead smelting, melting and burning; vitreous enamelling on glass and metal; the glazing of pottery; the manufacture of lead compounds, such as red and white lead, and lead colours; the manufacture of lead accumulators; shipbuilding and ship breaking; painting; plumbing and soldering operations, and the manufacture of rubber. (All these processes are controlled by specific Regulations, including the Control of Lead at Work Regulations 1980, together with the approved code of practice.)

Lead poisoning occurs in two distinct forms, namely by inorganic lead, and by lead in the organic form, mainly tetraethyl lead and tetramethyl lead, tetraethyl lead being the more common. In acute cases, usually caused by exposure to fumes, the initial symptoms are a sweetish taste in the mouth, especially on smoking, with anorexia, nausea, vomiting and headache, sometimes persistent constipation and intermittent colic. Acute poisoning is often fatal, although exposure may have been of only a few days' duration. In severe cases the effect on the nervous system is startling – restlessness, talkativeness, excitement, muscular twitchings accompanied by insomnia, delusions, hallucinations and even acute and violent mania. It is typically accompanied by a fall in blood pressure and body temperature.

In more chronic cases the classical symptoms and signs are headache, pallor, a blue line around the gums, anaemia, palsy (drop wrist) and encephalopathy, a form of mental disorder characterised by mental dullness, inability to concentrate, faulty memory, tremors, deafness, convulsion and coma. The great majority of lead-poisoning cases are, in fact, occasioned by inhalation of fumes containing lead due to heating the metal, for various reasons, above a temperature of 500°C.

Tetraethyl lead is volatile at room temperature. It is readily inhaled and, on absorption, exercises a focal effect on the central nervous system.

Mercury

Mercury and its compounds enter the body in dangerous amounts via the alimentary tract, through unbroken skin or mucous membranes, or by inhalation of the vapour or dust. Poisoning occurs commonly in the following occupations:

(a) mercury mining and the recovery of the metal from the ore;
(b) use of metallic mercury in the manufacture of thermometers, barometers and electric meters;
(c) associated mercury distillation processes;
(d) manufacture of salts of mercury;
(e) manufacture and use of organic compounds of mercury as fungicides in seed dressing;
(f) manufacture and use of disinfectants;
(g) use as a carroting agent in the treatment of animal skins in the manufacture of fur felts.

Mercury poisoning is prescribed occupational disease C5.

Mercury is used in the form of metallic mercury and inorganic mercury compounds. Generally, mercury poisoning results from exposure to metallic mercury or the dust of its compounds. Chronic poisoning produces the characteristic tremors which are noticeable from a simple handwriting test, and drowsiness by day coupled with insomnia at night. Often there are symptoms similar to paranoia or persecution complex, which were very common in the hatting industry a century ago. (The Mad Hatter in *Alice in Wonderland* was not a figment of Lewis Carroll's imagination, but like real people commonly found in the hatting trade!)

There are important differences in the effects of exposure to metallic mercury and the different forms of mercurial compound. With metallic mercury, its oxides and inorganic salts, early symptoms include nausea, frequent headaches, tiredness and chronic diarrhoea. The characteristic features are stomatitis (inflammation of the mucous membranes of the mouth), muscular tremors and psychic disturbances. Effects on the mouth may vary from a mere metallic taste to salivation, bleeding of the gums, ulceration and loosening of the teeth. Muscular tremors appear early, often starting in the fingers and spreading to the tongue, lips, eyes and lower limbs. The psychic disturbance of mercurial erethism manifests itself in abnormal shyness and loss of confidence, coupled with irritability, vague fears and depression, often leading to loss of memory, hallucinations and deterioration of intellect. Metallic mercury has a dangerously high vapour pressure and tends to release dangerous levels of fume at normal room temperature. Most cases of inorganic mercury poisoning occur due to inhalation of the vapour in such circumstances.

In poisoning by methyl and ethyl (alkyl) organo-mercury compounds, the symptoms are more pronounced. Early stages may be no more than tiredness, followed by sensations of tingling or numbness in the fingers and toes. In the later stages, however, loss of co-ordination of movement (ataxia), tremors, difficulty in speaking clearly (dysarthria) and constriction of the visual field, in extreme cases amounting to 'tunnel vision', may occur.

Chromium

Chromium is a hard, steel-grey, brittle metal. It forms two series of salts, the trivalent and the hexavalent, the latter being of great industrial importance, for instance, in chromium plating. It also forms an acid oxide, chromium trioxide (chromic acid), from which chromates and dichromates are formed.

Chromium comes within the group of external irritants responsible for occupational dermatitis. The common symptom with chromium, however, is the formation of chromium ulcers or chromium holes on the hands and forearms due to direct contact. These are also common on the eyelids or, if chromium is inhaled, in the nostrils, and a hole can be formed right through the nasal septum. Chromic ulceration is included in prescribed occupational disease D5.

Chromic ulceration of the skin may be caused by chromic acid, the alkali chromates and dichromates, and zinc chromate. The penetration of these substances through a minute break in the skin may cause a raised hard lump which breaks down at the centre revealing a deep ulcer with rounded and thickened edges and a slough-covered base. (Slough is a portion of dead tissue cast off from living tissue.) The risk of contracting this form of occupational dermatitis is associated with the following uses of chromium:

(a) as metallic chromium in the formation of alloys; and

(b) as chromium compounds in chromium plating and anodising, in metal treatment processes, as tanning agents in the leather industry, in the impregnation of timber for preservation, as a constituent of anti-corrosion paints, as sensitisers in the photographic industry and in the manufacture of dyestuffs.

The benzene family

Included in this family are benzene (benzole), nitrobenzene and aniline. Benzene is produced from coal tar distillation and as a by-product of petrochemical processes. Until recently it was the starting point for many processes in the chemical industry, and was used in the manufacture of paints, plastics, lacquers, rubber and adhesives. Benzene has virtually been banned from industry and replaced by the safer toluene, but may still be found in research situations, where its use is strictly controlled. Toluene and xylene are structurally similar to benzene, but much less toxic than benzene in its pure state. The commercial forms, Toluol and Xylol, however, may contain significant concentrations of benzene and are thus rendered much more hazardous, and should be regarded as potential carcinogens. Benzene poisoning is prescribed occupational disease C7.

In acute poisoning cases where there may have been accidental exposure, the early symptoms include euphoria, giddiness, headache and vomiting. Unconsciousness and death from respiratory failure can follow. Chronic benzene poisoning has its principal effect on the bone marrow, leading to leukaemia, which can be delayed for some years after cessation of exposure. There are usually no symptoms in the early stages, but if any *are* present they tend to be vague and non-specific. The first symptoms are tiredness, mild gastro-intestinal disturbance and giddiness, followed by haemorrhages from mucous membranes and the development of skin rashes. Anaemia is commonly encountered and, in serious cases, leukaemia may develop.

Xylene and toluene are metabolised in a similar way to benzene. They have narcotic properties and produce symptoms ranging from drowsiness, fatigue and headache, to unconsciousness and death. Toluene is less powerful a narcotic than benzene.

Nitrobenzene is used in perfumery and in the manufacture of aniline, and is readily absorbed through the skin. Aniline is used in the rubber industry, in dyeing and in resin extraction. Contact may be through inhalation of aniline-contaminated dust or aniline fumes, together with absorption through the skin.

Trichlorethylene and other chlorinated hydrocarbons

Within this group are included carbon tetrachloride, methyl chloride, tetrachloroethane, methyl bromide and perchloroethylene, all solvents with varying degrees of toxicity.

Trichlorethylene, commonly known as 'Trike', 'Triklone' and 'Trilene', has widespread use, either in the pure state as an anaesthetic or mixed with other solvents. Its main uses are as a degreasing agent for metals, as a dry cleaning agent and as a refrigerant. Inhalation can produce drowsiness, giddiness, unconsciousness and death. If the fumes are drawn through the lighted tip of a cigarette, the danger is greatly increased due to the formation of acidic products, including phosgene.

Carbon tetrachloride has similar properties to trichlorethylene. It is used as a fat and rubber solvent and in dry cleaning, but principally as a refrigerant, where it is used in the manufacture of 'Freon' (trade name of a group of ICI fluorocarbons). As with other solvents, the main hazard is from the vapour. This solvent should only be used in a well-ventilated room.

Carbon tetrachloride has a narcotic effect which produces unconsciousness. This may be preceded by signs of central nervous disturbance. Nowadays, severe cases are rare; the symptoms are predominantly those of liver or renal damage, the renal symptoms being more evident. Initially the patient will complain of persistent headaches, nausea and vomiting, colic and diarrhoea, and hepatic tenderness. Renal and hepatic damage appear after a variable latent period.

Tetrachloroethane is one of the most toxic chlorinated hydrocarbons. It is between eight and ten times more toxic than carbon tetrachloride and four times as toxic as chloroform, with a smell similar to that of chloroform. The vapour constitutes the main hazard, and it should only be used where the ventilation is adequate. Tetrachloroethane principally affects the liver and central nervous system. The effects on the liver, or hepatic syndrome, occur in four specific stages, commencing with persistent headache, lassitude, anorexia and vomiting. There may be an unpleasant taste in the mouth, together with stomach pain. Then follows the jaundice stage, closely followed by the typical toxaemic stage, with enlargement of the liver and deepening jaundice. Vomiting is severe and there may be signs of liver damage. There may be delirium, stupor, rashes and oedema. In the final stage of the disease the patient may develop ascites – a swelling of the abdomen due to fluid exuded from the blood vessels – but he may die before reaching this stage. Tetrachloroethane poisoning is prescribed occupational disease C10.

The neurological symptoms of tetrachloroethane poisoning include numbness and tingling of the fingers and toes. There may be tremor and twitching of the face muscles. At a later stage, power in the hands and feet is lost and this weakness can spread to the rest of the body.

Tetrachloroethane has many industrial uses as a general purpose solvent, and as an intermediary in the manufacture of tetrachlorethylene and trichlorethylene.

Methyl bromide is an extremely volatile liquid in gaseous form at a temperature above 4.5°C. It is used as a refrigerant, insecticide and fumigant. Past occupational exposures have largely been associated with leakage of the gas from pipework and storage vessels where, because of its high volatility, even a minute crack or pin hole will permit large quantities of the gas to be liberated. Methyl bromide poisoning is prescribed occupational disease C12.

Whilst methyl bromide is highly toxic, there is a latent period of up to 48 hours before bodily symptoms are apparent. After the latent period initial effects are irritation of the respiratory tract, followed by nausea, vomiting, headache, cough, watering of the eyes and abdominal pain. Vision becomes blurred, then double vision occurs. At this stage an affected person will appear drunk, with a staggering walk, slurred speech and loss of balance. In severe cases, pulmonary oedema, an excess of tissue fluid, may be present, together with convulsions. The symptoms are directly related to the length of exposure and the concentration of the gas.

Direct splashing on to the skin of methyl bromide results initially in a tingling sensation followed by burning. The skin becomes inflamed and, after several hours, small vesicles (blisters) appear. The vesicles become distended with straw-coloured fluid but, if punctured, they do not refill. Healing takes place after a few days. If the degree of exposure is insufficient to produce vesicles, a form of dry eczema may be produced.

Methyl chloride is less toxic than methyl bromide, but has a greater narcotic action. It is used as a refrigerant, in the manufacture of chloroform and in the dyestuffs industry. Typical poisoning symptoms are similar to those of methyl bromide, but there are several important differences. Ocular symptoms are more frequent a day or so following exposure, but symptoms such as headache and dizziness follow immediately upon exposure. Anorexia and vomiting are common within 24 hours. Generally, mild cases recover, but this may take six to nine months, depending on the duration of exposure and the concentration of the gas.

Isocyanates

A range of isocyanates is used in industry, principally in the manufacture of urethane foams and resins, for instance toluene di-isocyanate (TDI) and methylene bisphenyl di-isocyanate (MDI). TDI is an extremely volatile compound and should only be used under closely controlled conditions in an enclosed system or with well-controlled ventilation. The vapour can be evolved during the manufacture of foams, and also during cutting of the finished product with a hot wire. A dust hazard may arise from the use of 1:5 napthalene di-isocyanate (NDI) which is sometimes used in powder form. In the latter case there is a potential hazard from the spray when NDI in solid form is being sprayed with polyurethane lacquers. This practice is common in the furniture and allied industries. The spraying of urethane foams is a particularly hazardous operation which requires a high degree of environmental and personal protection.

TDI is the most volatile of the isocyanates and one of the most toxic. Symptoms are usually reduced when the patient is removed from contact, but a severe respiratory reaction, as with MDI, may follow on second or subsequent exposures, even if the exposure is to extremely low concentrations of the vapour. Vapour pressure, i.e. the percentage gas by volume in air, is particularly important in controlling the hazards associated with isocyanates. Vapour pressure increases very rapidly with temperature (*see* Fig. 19.4).

Fig. 19.4 ◆ The relationship between vapour pressure and temperature

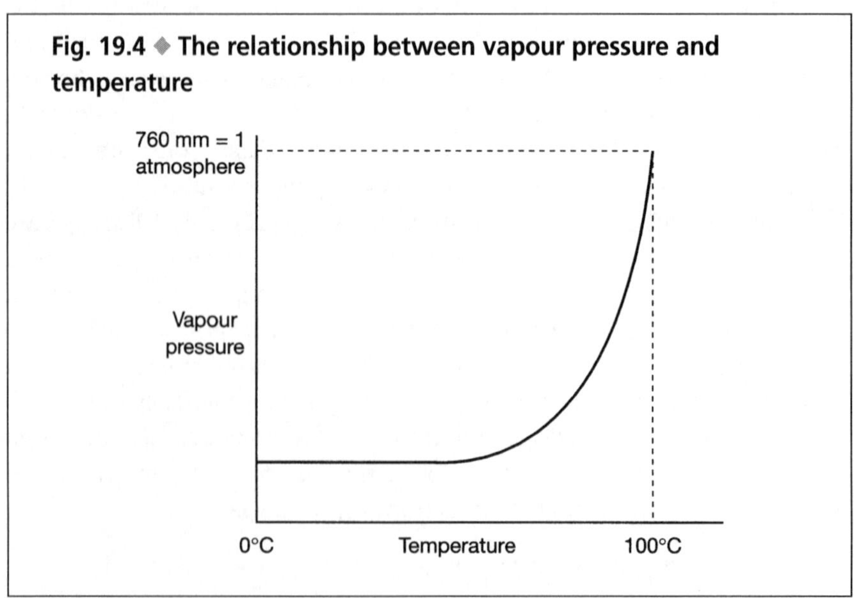

The degree of volatility of a vapour affects the degree of inhalation and therefore the risk associated with a particular substance. The vapour pressure of a liquid at room temperature can be estimated from the boiling point, i.e. the temperature at which the vapour pressure equals the pressure of the atmosphere; the lower the boiling point, the greater the vapour pressure at room temperature. (7.6 mm mercury vapour pressure, 10,000 ppm or 1 per cent concentration is the concentration at which most solvents will produce narcosis. All solvents have a greater vapour pressure than 7.6 mm of mercury.) TDI has a vapour pressure at 25°C of 0.025 mm of mercury, and is very volatile. MDI has a vapour pressure of 0.00009 mm of mercury.

Most isocyanates can produce varying degrees of dermatitis, and exposure can result in skin sensitisation effects in rare cases. The main consideration, however, is that isocyanates are a potent primary irritant to the respiratory tract and, in some cases, cause dramatic sensitisation, i.e. asthmatic effects once an individual has become sensitised. Eye splashes may cause severe chemical conjunctivitis. Handling in open vessels should, therefore, be prohibited, any transference being undertaken in fully enclosed systems. Spillages should be cleared up immediately and decontaminants should be readily available, e.g. 5 per cent ammonia in sawdust. Health supervision should include pre-employment and routine periodic health examinations for all operators and other staff coming into contact with isocyanates. Asthma due to exposure to (*inter alia*) isocyanates is prescribed occupational disease D7(*a*).

E

Alcohols

Any of the simple alcohols, e.g. amyl, ethyl, butyl, methyl and propyl alcohols, will produce intoxication if an excess of vapour is inhaled. In rare cases, contact can cause temporary or even permanent blindness. Where an alcohol is combined with chlorine, a highly poisonous solvent can be formed, e.g. ethylene chlorohydrin, a solvent used in lacquers. Where poisoning has occurred, the symptoms of vomiting, headache, shortage of breath and unconsciousness may precede death, with evidence of damage to internal organs. In most cases, poisoning takes place through inhalation or absorption through the skin.

Acetates

The principal acetates used in industry are ethyl, methyl and amyl acetate. They are closely related to solvents and are highly flammable. There is little toxic risk, exposure typically causing running eyes and smarting.

Ethers

Diethylether is a commonly used industrial solvent. Toxic effects are usually limited to loss of consciousness and, in most cases of exposure, a good recovery is made. Diethyl ether is highly flammable. Diethylene dioxide, or 'Dioxan', another member of the ether family, is both poisonous and highly flammable. It may be inhaled as a vapour or absorbed through the skin.

Carbon disulphide

This is an extremely toxic solvent with one of the lowest flashpoints. It can be ignited by the heat from a radiator or an electric light bulb. It is used extensively in the production of rayon and paper. As with other solvents, bodily effects rest on the degree and length of exposure. Symptoms include headache, fainting, vomiting, breathlessness, hallucinations, degrees of blindness and mental disorders. Death commonly follows cases of severe exposure.

Carbon monoxide

This gas is produced as a result of combustion. It causes death by poisoning due to its ability to deprive the body tissues of oxygen. Its propensity to combine with the haemoglobin in the blood is at least 200 times greater than that of oxygen, and the victim of carbon monoxide gassing dies from asphyxiation.

Cyanides

The use of cyanide-based substances has been reduced considerably in the last two decades. Cyanides are used in heat treatment and electroplating processes. Cyanide is renowned for the fact that it kills swiftly through inhalation of cyanide fumes, ingestion of cyanide-contaminated food or absorption through the skin. Contact between cyanides and other substances, notably acids, causes the evolution of hydrocyanic acid which is lethal.

Arsenic

Arsenic poisoning in industry occurs in two distinct forms, with totally different symptoms. The first form, an acute and serious type, arises from inhalation of arseniuretted hydrogen (arsine) gas. The second, which is the more common and chronic type, results from absorption into the body, usually by inhalation over a long period, of dusts of arsenical compounds.

Arsine is produced when nascent hydrogen is accidentally released in the presence of arsenic. This type of accident can occur wherever dilute sulphuric or hydrochloric acids are used in the processing of ores or residues, while 'pickling' (acid cleaning of metals), while clearing acid tanks of sludge, or in the manufacture of electric accumulators. Arsine can be produced accidentally in laboratories. Solid arsenical compounds are encountered in smelting and other industrial processes. More commonly, they are used as weed killers and insecticides, in the preservation of hides, skins and furs, and in glass making.

With arsenic poisoning, 25–30 per cent of all cases die in a relatively short period of time. Mild cases, after a latent period of several hours to a day or more, show varying degrees of nausea, headache, shivering, exhaustion, giddiness, stomach pain and vomiting, all of which are of sudden onset. In more severe cases, the latent period may be reduced to six hours or less, after which haemoglobinuria appears. Within 24 hours, jaundice develops followed by anaemia and severe kidney damage. If treatment is not rapid, the patient drifts into a 'typhoid' state with death from anuria.

Poisoning from solid compounds is usually through inhalation of the dust. There may also be local irritation of the mucous membranes of the nose and mouth together with skin inflammation and ulceration, resulting in localised dermatitis, conjunctivitis and ulceration of the nasal septum.

Phosphorus

This element occurs as either red phosphorus or yellow (white) phosphorus. Red phosphorus is relatively non-toxic, being used in the manufacture of safety matches, and as a starting point for other preparations. Yellow phosphorus is a waxy solid which ignites spontaneously on contact with air, emitting poisonous fumes. The handling of yellow phosphorus can cause severe burns. Yellow phosphorus was once widely used in the manufacture of matches, but the practice was banned in 1906 owing to the necrotic condition of the jaw bone or mandible ('phossy jaw') produced in workers dipping matches. Yellow phosphorus has limited industrial use as a rodenticide and in the production of non-ferrous alloys. The chronic poisoning, typified by phossy jaw, is now rarely encountered, but there is a need to consider the results of acute poisoning. This follows the ingestion of yellow phosphorus, where the symptoms are delayed. The main symptoms are abdominal pain, vomiting, depression and general weakness followed, after an interval of days or weeks, by toxic jaundice and possibly haemorrhages of the mucous membranes. In fatal cases, the principal cause of death is atrophy, i.e. wastage and shrinking, of the liver.

Phosphine, a gas with the odour of decaying fish, may be evolved during the preparation and use of calcium phosphide, in the manufacture of acetylene from impure calcium carbide and when zinc phosphide, a grain fumigant constituent, is accidentally wetted. Quenching metal alloys with water may also produce the gas, as may the manufacture of certain forms of graphite.

Symptoms of phosphine poisoning include abdominal pain, nausea and vomiting. Ataxia (loss of muscle co-ordination), convulsions, coma and death may follow within 24 hours. In milder cases, the gas may produce some degree of respiratory irritation but recovery could be complete. Chronic poisoning, with resulting effects on the central nervous system, may occur in persons regularly exposed to low concentrations of phosphine.

Organo-phosphorus compounds are being used increasingly as insecticides. The principal route of entry is through the skin with minor local irritation, but entry may also be through inhalation and ingestion. The effect of these chemical substances is to inhibit the action of the enzyme cholinesterase present in red blood cells and motor nerve end plates. The action is cumulative and a toxic concentration may be built up by repeated slight exposure. Early symptoms of organo-phosphorus poisoning are non-specific, but may include anorexia and nausea (the latter characteristically increased by taking food and smoking), giddiness, drowsiness, diarrhoea and fatigue. Within a few hours, muscular twitchings, cramps, incontinence, coma, convulsions, paralyses and signs of pulmonary oedema may develop. The severity of symptoms is directly related to the dose.

The biological causes

Several occupational diseases are transmissible from animals to humans. These are the 'zoonoses' and include such diseases as anthrax, leptospirosis, orf (contagious pustular dermatitis), glanders fever and brucellosis. Whilst the incidence of such diseases is low, there is always some degree of risk to anyone working with animals, especially veterinary surgeons, meat inspectors, pet shop workers, people working in zoos, artificial inseminators and farmers. Anthrax, glanders, leptospirosis and brucellosis are all prescribed occupational diseases (B1, B7, B3 and B2).

Anthrax

This is a disease which may occur in humans and certain animals, e.g. cattle and sheep, as a result of infection by *Bacillus anthracis*, a spore-forming organism which, although killed by boiling for ten minutes, may survive for years in the soil and in animal remains. Cattle are the main source of the infection, and infection in humans may occur through contact with fresh infective material containing the bacillus; people at risk include agricultural workers, veterinary surgeons, knackers, slaughtermen, and those working with dried animal products such as hides, skins, hair, wool, hooves, bone-meal and contaminated implements. Infection in humans may be of the cutaneous type (malignant pustule) or internal, e.g. the pulmonary form (wool sorter's disease). The disease manifests itself in almost every case as a grave toxaemia, with headache, shivering, muscle and joint pains, nausea, vomiting and collapse, together with additional symptoms depending on the site and type of infection.

Malignant pustule is the more common form of anthrax. Infection takes place through cuts and abrasions on the skin. After an incubation period of one to four days an irritant pimple develops. The pimple rapidly enlarges and breaks down with a black necrotic centre. The lesion may be ringed with small vesicles and inflammatory swelling. Local lymph nodes may be slightly enlarged. In 90 per cent of cases the pustule is situated on some exposed part of the body such as the face or neck, and in such cases the intense oedema may be fatal.

Internal anthrax takes place through ingestion or inhalation of the bacilli. In these cases, even more than in external cases, the general intense toxaemia, with sudden vertigo, somnolence, dyspnoea (difficulty in breathing), croup and marked prostration, is prevalent, and death may ensue. In typical pulmonary cases there is widespread congestion and oedema or an atypical pneumonia with frothy blood-stained sputum. If untreated, death occurs from septicaemia in the first few days.

Glanders fever

Glanders fever or 'farcy' is a disease of horses, mules and donkeys. The infecting organism is *Bacillus mallei* or *Pfeifferella mallei.* Infection in humans is now rare, but is always caused by contact with an infected animal. The disease occurs in both acute and chronic forms. In the acute form there is an incubation period of two to three days before general malaise is experienced, together with headaches, anorexia and joint pains. The site of infection becomes ulcerated and there is marked lymphangitis (inflammation of the lymph vessels). Nodular abscesses form along the lymphatic vessels and these break down to form painful ulcers. There is a marked fever, highest between the sixth and twelfth days, after which time eruptions appear on the face and on the nasal, palatal and pharyngeal mucosae. The lesions typically begin as patches which eventually enlarge and form pustules. The pustules ulcerate with the destruction of bone and cartilage or produce a thick blood-stained purulent discharge. A form of arthritis may also occur with the development of abscesses in the muscles.

The chronic form is more rare than the acute form, but is again characterised by the formation of abscesses, which break down to form painful ulcers. The lungs may be involved in terms of pneumonia, pleural effusion, lung abscesses and empyema (a collection of pus in a natural body cavity, e.g. in the space between the lung and outer wall of the chest). The disease runs a long course and an acute phase may supervene at any time.

Leptospira ictero-haemorrhagica

This disease is also known as 'leptospiral jaundice', 'spirochaetal jaundice', 'spirochaetosis icterohaemorrhagica' and 'Weil's disease'. It is a feverish condition caused by the organism *Leptospira icterohae-morrhagica*, commonly found in rats, which are the source of human infection. Infection may be due to ingestion of food or water contaminated with the urine of infected rats; alternatively, it may enter the skin or through the mucous membranes of the eyes, nose and mouth. The disease sometimes occurs amongst men who work in rat-infested locations such as mines, slaughterhouses and fish docks.

After an incubation period of 6–12 days there is an abrupt onset of high fever, rigors, headache, muscular pain and vomiting, accompanied by prostration. At this time, the leptospires multiply in the blood and may be carried to and affect any organ. Conjunctival haemorrhages are common together with a body rash, often accompanied by petechial (pinpoint sized) haemorrhages in the skin. There may be mild liver damage and jaundice is common two to five days after onset of the fever. There is usually a steady improvement after the second week of the illness and mild cases recover completely without specific treatment. Fatalities are rare.

Brucellosis

Brucellosis in humans in caused by contact with infected animals. Three species of the organism account for most human disease. These species show an affinity for particular animal hosts, so that *Brucella abortus* is found in cattle, *Brucella melitensis* in sheep and goats, and *Brucella suis* in pigs. The disease may be contracted by persons working in slaughterhouses or among those handling meat, meat products or the byproducts and waste from slaughtering. Veterinary surgeons and meat inspectors are an outstanding high-risk group.

The routes of infection can be through inhalation, ingestion and direct contact with infected material, e.g. the uterus of an infected animal, direct contact being the most important. In the latter case, this occurs usually through handling the placenta or foetal parts during the delivery of a calf or in post-mortem examinations. The organism gains access through cuts and abrasions in the skin or through the mucous membranes, including the conjunctivae.

Brucellosis takes two forms, the acute attack and the chronic condition. In acute cases, onset may be gradual with non-specific signs such as headache, joint pains, fever, insomnia and low back pain, or it may be abrupt with fever, rigors and prostration. Usually the disease subsides within two weeks and the patient makes a complete recovery. Some patients will continue, however, to have intermittent bouts of fever, back pain, a feeling of lethargy and depression which may last for several months or years.

Chronic brucellosis has all the symptoms of an acute attack, i.e. lassitude, malaise, joint pains and prolonged depression. There is not always a history of an acute attack and, in many cases, the occupation of the patient may be the only clue in diagnosis, e.g. a stockman on a farm. In chronic brucellosis there may be complications including endocarditis (inflammation of the heart lining) and spondylitis (inflammation of the vertebrae).

Q fever

This is an infection caused by an organism, *Rickettsia burneti*. The infection is found most frequently in farm workers who contract the disease from sheep and cows by the inhalation of infected dust or by drinking infected raw milk. Veterinary surgeons, meat inspectors and abattoir workers are particularly high-risk groups in this case. The symptoms are very similar to those of influenza and it is common for cases of Q fever to be diagnosed as such. Typically, the illness begins with fever accompanied by shivering, sweating and backache, inflammation of the throat and suffused conjunctivae. In many cases, the patient has an unproductive cough, photophobia and muscular pains.

Orf (contagious pustular dermatitis)

Orf is a viral infection of sheep and goats which is transmitted occasionally to abattoir workers and animal handlers. The disease takes the form of a mild skin rash occurring at the site of infection. Clinical signs appear 4–12 days after infection, with the development of a red macule (a spot level with the surface of the surrounding skin) or papule (a raised spot on the surface of the skin). This enlarges until it becomes 1–4 cm in diameter containing first clear fluid and then pus. There may be some local tenderness and lymphadenitis (inflammation of the lymph nodes), and the lesion is sometimes painful. Healing is usually complete within four to six weeks.

Viral hepatitis

Hepatitis (inflammation of the liver) is most commonly ascribed to various infections. Hepatitis B (serum hepatitis) occurs more frequently amongst members of the medical and allied professions than among the general public, the risk being greatest among those who handle blood or blood products, and who work in renal dialysis units. The symptoms of the disease include malaise, myalgia (muscle pain), headache, nausea, vomiting, anorexia, abdominal pain and pruritis (itching). The patient becomes jaundiced and the liver is enlarged. Generally the disease runs a mild course, although some cases may turn to chronic hepatitis.

Hepatitis A, on the other hand, is a form of epidemic jaundice spread through human contact or through contaminated food and water supplies. Viral hepatitis is prescribed occupational disease B8.

Aspergillosis (farmer's lung)

Exposure to the dust of mouldy hay or other mouldy vegetable produce can result in pulmonary disease. It is characterised, along with many other similar conditions such as mushroom picker's lung and malt worker's lung, by an influenza-like illness, during which the person feels generally unwell, has pain in the limbs and is feverish. The patient will also have a dry cough and dyspnoea. Farmer's lung is one form of extrinsic allergic alveolitis, an inflammatory condition of the lung tissue associated with hypersensitivity to the spores of mouldy hay. It is usually a transitory condition where the symptoms abate after three to four days. It is prescribed occupational disease B6.

Legionnaire's disease

This disease is caused by the inhalation of aerosols containing a specific bacterium, *Legionella pneumophila*. *Legionella* bacteria are widely distributed in the environment and occur in at least ten different forms. They are commonly encountered in water cooling systems, rivers, streams, ponds, lakes and in the soil. Most reported cases occur in the 40 to 70 years age group.

Initial symptoms of the disease include high fever, chills, headache and muscle pain, and there is an incubation period which may range from two to ten days, but usually three to six days. After a short period a dry cough develops and most patients suffer difficulty with breathing. About one-third of patients also develop diarrhoea or vomiting and about 50 per cent of patients may become confused or delirious. The disease may not always be severe and mild cases may be recognised which would probably have escaped detection except for the increased awareness of this disease amongst doctors and managers.

Conditions that affect the proliferation of *Legionella* bacteria include:

(a) the presence of sludge, scale, rust, algae and organic particulates which, although the ecology of *Legionella* is not fully understood, are thought to provide nutrients for growth;
(b) water temperatures in the range 20°C to 45°C which favours growth.

Legionella is frequently found in many recirculating and hot water systems, particularly large complex systems, such as those incorporated in multi-storey office blocks, factories and hospitals. Particular sites for bacterial growth are air conditioning systems, cooling towers, water standing in ductwork and condensate trays, humidifiers, hot and cold water storage tanks, calorifiers, pipework and plant.

Regular sampling of water in these types of installation is recommended, together with regular disinfection of the system. Further guidance is available in HSE Guidance Note EH 48 *Legionnaire's disease* and publications by local water authorities and environmental health departments.

The ergonomic causes

A number of occupational conditions are associated with repetitive job movements, e.g. cramp, or with friction and pressure on limbs and joints.

Cramp

This disability, known as 'writer's cramp', 'twister's cramp', 'occupational cramp' or 'craft palsy', is characterised by attacks of spasm, tremor and pain in the hand or forearm caused by attempts to perform a familiar act involving frequently repeated muscular action. Muscular co-ordination necessary for the performance of the repetitive movements breaks down and the continuation of the movements becomes impossible. The causative factors in this condition are unknown, but may be attributed to a combination of physical fatigue of muscles and nerves and an underlying psychoneurosis. It is prescribed occupational disease A4.

Beat hand

Referred to as 'subcutaneous cellulitis of the hand', this condition or disability, prescribed occupational disease A5, is the result primarily of the bruising of the skin and the underlying tissues and the implantation there, by friction or pressure, of 'dirt' and particles. The condition is liable to follow frequent jarring of the hand in the use of pick and shovel, and is more likely to occur in wet conditions. It is principally found in the hand of people unaccustomed to manual labour or who have been away from such activity for a long time. When accompanied by local infection, it may become acutely disabling. This condition is encountered in the palm of the hand and takes the form of, first, an acute inflammation, followed in many cases by a suppurative condition, i.e. broken skin and the presence of pus, due to infection.

Beat knee

Officially described as 'bursitis or subcutaneous cellulitis arising at or about the knee due to severe or prolonged external friction or pressure at or about the knee', this condition, prescribed occupational disease A6, is similar in aetiology to beat hand. It occurs in those unaccustomed to working in a kneeling position or on returning to such work after a prolonged absence, and is more likely to occur if the skin is wet and sodden. Repeated or lengthy pressure, together with regular pivoting on the knee, as in the case of roof tilers or carpet fitters who persistently kneel, is a potential cause.

Cellulitis of the skin generally proceeds to the suppuration stage and may involve the bursa of the knee. In bursitis, the enlargement of the knee joint may be due to acute effusion (leakage of fluid into a body cavity) or to infection of a chronic enlargement. Depending on the severity of the condition, incapacity may last only a few weeks or surgery may be necessary to remedy the condition.

Beat elbow

This condition is similar in aetiology to beat hand and beat knee, but with the elbow a single, although perhaps sustained, injury during work is more easily identified as the cause. Here again there are the classical signs of acute inflammation. The elbow is swollen and painful, signs of deep inflammation set in, and the swelling rapidly extends down the back of the forearm. The prognosis, as with other 'beat' conditions, depends on the degree of severity of the condition.

Work-related upper limb disorders

Work-related upper limb disorders caused by repetitive strain injuries (RSI) were first defined in the medical literature by Bernardo Ramazzini, the Italian father of occupational medicine, in the early eighteenth century. The International Labour Organisation recognised RSI as an occupational disease in 1960 as a condition caused by forceful, frequent, twisting and repetitive movements.

E

Repetitive strain injury covers some well-known conditions such as tennis elbow, flexor tenosynovitis and carpal tunnel syndrome. It is usually caused or aggravated by work, and is associated with repetitive and over-forceful movement, excessive workloads, inadequate rest periods and sustained or constrained postures, resulting in pain or soreness due to the inflammatory conditions of muscles and the synovial lining of the tendon sheath (see Fig. 19.5).

Fig. 19.5 ◆ Sites of RSI

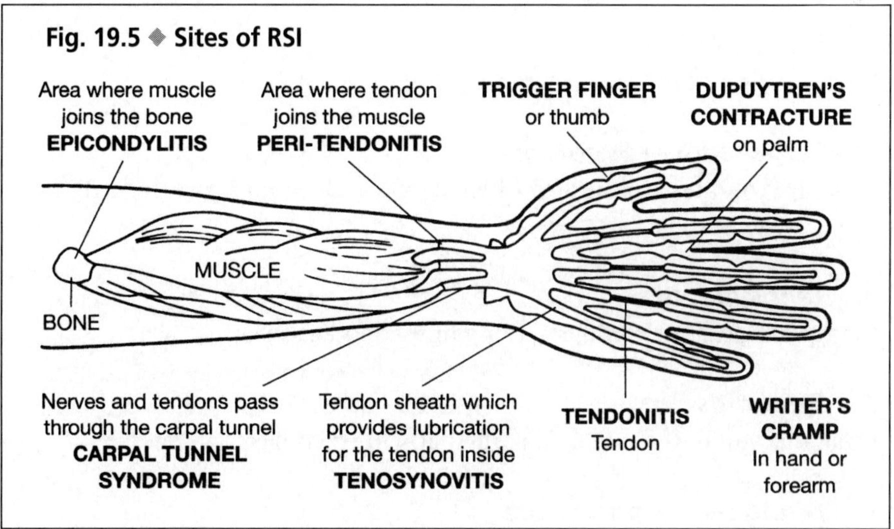

Present approaches to treatment are largely effective, provided the condition is treated in its early stages. Tenosynovitis has been a prescribed industrial disease since 1975, and the HSE have proposed changing the name of the condition to 'work-related upper limb disorder' on the grounds that the disorder does not always result from repetition or strain, and is not always a visible injury.

Many people, including assembly workers, supermarket checkout assistants and keyboard operators, are affected by RSI at some point in their lives.

Clinical signs and symptoms

These include local aching pain, tenderness, swelling and crepitus (a grating sensation in the joint) aggravated by pressure or movement. Tenosynovitis, affecting the hand or forearm, is the second most common prescribed industrial disease, the most common being dermatitis. True tenosynovitis, where inflammation of the synovial lining of the tendon sheath is evident, is rare and potentially serious. The more common and benign form is peritendinitis crepitans, which is associated with inflammation of the muscle-tendon joint that often extends well into the muscle.

Forms of RSI

1. *Epicondylitis*

Inflammation of the area where a muscle joins a bone.

2. *Peritendinitis*

Inflammation of the area where a tendon joins a muscle.

3. *Carpal Tunnel Syndrome*

A painful condition in the area where nerves and tendons pass through the carpal bone in the hand.

4. *Tenosynovitis*

Inflammation of the synovial lining of the tendon sheath.

5. *Tendinitis*

Inflammation of the tendons, particularly in the fingers.

6. *Dupuytrens Contracture*

A condition affecting the palm of the hand, where it is impossible to straighten the hand and fingers.

7. Writer's Cramp

Cramps in the hand, forearm and fingers.

Prevention of RSI

Injury can be prevented by:

(a) improved design of working areas, e.g. position of keyboard and VDU screens, heights of workbenches and chairs;

(b) adjustments of workloads and rest periods;

(c) provision of special tools;

(d) health surveillance aimed at detecting early stages of the disorder; and

(e) better training and supervision.

If untreated, RSI can be seriously disabling.

20 Occupational health practice

Occupational health is essentially a branch of preventive medicine which examines the relationship between work and health and the effects of work on the worker. Occupational health practitioners include the occupational health nurse, occupational physician, occupational hygienist, the health and safety specialist and the trained first aider, all of whom have a specific contribution to make in the provision and maintenance of healthy conditions at work. The principal areas of occupational health practice are outlined below.

Placing people in suitable work

As industry becomes more sophisticated it is of vital significance that workers should be physically and mentally suited for the tasks they are required to undertake. The pre-employment medical examination for restricted groups of people has been common practice for many years, but over the last quarter of a century the more prudent employer has extended this form of examination to all grades of staff. In many cases the examination is undertaken by a registered medical practitioner paid on a retainer basis, or by an appointed factory doctor. However, in recent years the concept of health screening by a trained occupational health nurse has found favour with many organisations, and such a pre-employment health screen for prospective employees is a standard feature of their recruitment policies. This is particularly appropriate in the food and catering industries where not only the health of the worker is important but the potential for his contaminating the product must be given considerable prominence.

Pre-employment screening activities now include not only an assessment of general fitness for the job but specific aspects of it such as vision screening of drivers, VDU operators and people engaged in fine assembly work, the assessment of disability levels where heavy work is involved, certain tests for suitability as food handlers and aptitude testing for a wide range of tasks.

Health surveillance

Reg 5 of the MHSWR requires that every employer shall ensure that his employees are provided with such health surveillance as is appropriate having regard to the risks to their health and safety which are identified by the (risk) assessment (required under Reg 3).

Health surveillance concentrates on two main groups of workers:

(a) those at risk of developing further ill-health or disability by virtue of their present state of health, e.g. people exposed to excessive noise levels;

(b) those actually or potentially at risk by virtue of the type of work they undertake during their employment, e.g. radiation workers.

Health surveillance of such groups usually takes the form of ongoing health examinations at predetermined intervals of, say, 6 months or 12 months according to the degree of risk involved. Such a system allows for early detection of evidence of occupational disease and for its early treatment. Under the COSHH Regulations, health surveillance may be necessary where employees are exposed to, or are liable to be exposed to, a substance hazardous to health.

Providing a treatment service

This activity has for many years been the principal function of some occupational health services. However, with the greater emphasis on prevention, there has been a tendency to reduce the importance of this activity. Nevertheless, the efficient and speedy treatment of injuries, acute poisonings and minor ailments is important because it prevents complications and aids rehabilitation. Such a service does have an important role to play in keeping people at work, thereby reducing lost time associated with attendance at casualty departments or doctors' surgeries.

Yet another important feature of a treatment service is that of detecting trends in accidents and injuries, with a view to improving preventive measures, and assisting injured persons, through counselling, in their rehabilitation after an accident. A joint approach between occupational health practitioners and safety practitioners can be effective here.

Primary and secondary monitoring

Primary monitoring is concerned largely with the clinical observation of sick people who may seek treatment or advice on their condition. Such observation will identify new risks which were previously not considered. For instance, there may be a sudden increase in the number of workers reporting signs of dermatitis, which could subsequently, through investigation, indicate the total unsuitability of a new adhesive or similar solvent-based product being used for the first time.

Secondary monitoring, on the other hand, is directed at controlling the hazards to health which have already been recognised. Audiometry is a classic form of secondary monitoring whereby the hearing levels of workers are tested on a six-monthly or annual basis to assess whether there has been any further hearing loss due to exposure to noise. Similar secondary monitoring may be carried out for workers using vibratory hand tools in order to assess early stages of vibration-induced white finger.

Avoiding potential risks

This is an important feature of occupational health practice with the principal emphasis on prevention, in preference to treatment, for a known condition. The occupational health practitioner can make a significant contribution to the planning and design of work layouts, and to considering the ergonomic aspects of jobs and the potential for fatigue amongst workers. The effects of shift working, long hours of work and the physical and mental effects of repetitive tasks would be taken into account in any assessment of risks involved.

Supervision of vulnerable groups

There is no doubt that certain groups are more vulnerable to accidents and occupational disease than others. Included in this grouping of 'vulnerable' workers are young persons, the aged, the disabled and people generally who may have long periods of health-related absence. Special attention must be given to such persons in terms of counselling on a wide range of matters, assistance with rehabilitation in the workplace and, possibly, assistance in the reorganisation of their tasks to remove harmful factors. Routine health examinations to assess their continuing fitness for work should be a standard feature here.

Monitoring for early evidence of non-occupational disease

Many industries are associated with specific occupational diseases. For instance, the pottery industry has long been associated with silicosis, the mining industry with coal worker's pneumoconiosis and the cotton industry with byssinosis. Whilst improvements in environmental working conditions have greatly reduced the incidence of such diseases, routine monitoring of workers not exposed to such conditions is an important feature of occupational health practice. Here the principal objective is that of controlling diseases prevalent in industrial populations with a view to their eventual eradication. Such monitoring also makes a great contribution to the control of the stress-related diseases and conditions such as mental illness and heart disease.

Counselling

Counselling, carried out by a trained occupational physician or occupational health nurse, is, perhaps, the most significant component of occupational health practice. This may take two forms, viz. counselling on health-related matters and counselling on personal, social and emotional problems. There is no doubt that, in the second case, many people would benefit from a counselling session with an occupational health nurse. Most people, at some time in their lives, have social and emotional problems. In many cases, for a variety of reasons, they are unable or unwilling to consult their spouse or their family or general practitioner and, over a period of time, develop a high state of stress. This results in an inability to concentrate for long periods, fatigue, frustration and absence from work. It may be associated with an inability to cope with problems or, perhaps, a feeling of injustice brought about by certain events. The availability of a sympathetic ear, independent of organisational controls, can assist the individual to come to terms with such problems more easily.

Health education

This is a particularly broad area of occupational health practice. It is primarily concerned with the education of employees towards healthier modes of living, but can also include training of management and staff in their respective responsibilities for health and safety at work, in healthy working techniques and in the avoidance of health hazards. In the food industry, where the purity of the product is of utmost significance, it can include the training of production staff and catering staff in food hygiene. It can incorporate feedback from other areas of occupational health practice, such as the reasons for certain aspects of health surveillance or the reinforcement of the need for the wearing of personal protective equipment.

Clearly, any health education, as with other areas of education, must be related to the health risks present and must be directed at bringing about an improvement in attitudes in individual areas of health care.

First aid and emergency services

Included in this area are the supervision of first aid facilities and ancillary equipment such as emergency showers, eye wash stations and emergency breathing apparatus, together with the preparation of contingency plans to cover major disasters such as fire, explosion or gassing accidents. This would entail the training of first aiders, rescue staff and key members of the management team in preparation for such disasters. (*See* Chapter 28 – First aid.)

Welfare amenity provision

Occupational health practice can include procedures for advising management on legal requirements for sanitation, hand-washing facilities, showers, arrangements for storing and drying clothing, and the provision of drinking water. Routine surveillance of such installations and other amenities such as kitchens, canteens, rest rooms and day nurseries feature strongly in the maintenance of sound health standards. (*See* Chapter 16 – Welfare amenity provision.)

Environmental control and occupational hygiene

Control of the working environment and the environment outside the workplace are important components of occupational health and hygiene practice. The employer must provide a safe working environment by recognition, measurement, evaluation and control of long-term health hazards. He must also ensure that he does not expose people living in the vicinity of the workplace to health risks or public health nuisances from pollution of the air, land, water, drainage system, watercourse or surrounding land.

E

Liaison

Staff of occupational health services liaise with a wide range of enforcement officers, such as medical and nursing advisers of the Employment Medical Advisory Service (EMAS) Factories Inspectors, environmental health officers, planning officers and staff of the Area Health Authority.

The importance of the relationship between members of the occupational health team and general medical practitioners must not be overlooked in planning and implementing any programme of health supervision and care. It is important that the general practitioner, who has primary responsibility for the health of individual workers registered with him, is kept informed of any health matters of significance and is involved in the care of the patient while at work. Similarly, the occupational physician or nurse should always be involved in cases where management receive a communication from a general practitioner about the health of an individual employee.

Health records

The maintenance of suitable records relating to the health state of individual employees features significantly in occupational health practice. The purpose of such records is to:

(a) assist occupational health staff to provide efficient health surveillance, emergency attention, health care and continuity of such care;
(b) enable staff to undertake epidemiological studies to identify general health and safety problems and trends arising amongst employees and to identify problem areas and specific risks;
(c) establish, maintain and keep up-to-date written information relating to people, hazards and current monitoring activities;
(d) facilitate assessment of problems, decision making, recommendations and the writing of reports.

The following records on individual employees, although not required by law, are desirable in an occupational health department:

(a) initial and subsequent health questionnaire, interview, examination and screening test results;
(b) relevant medical and occupational history, smoking habits, disabilities and handicaps;
(c) attendance in the department for first aid, treatment, re-treatment, general health care and counselling;
(d) injuries resulting from occupational and non-occupational accidents;
(e) illness occurring at work or on the way to or from work;
(f) sickness absences;
(g) occupational conditions and diseases;
(h) care and treatment provided;
(i) advice given, recommendations and work limitations imposed;
(j) referrals made to other specialists or agencies;
(k) correspondence relating to the health of employees;
(l) dispersal of cases following emergencies and treatment;
(m) communications between occupational health staff and others, including written reports.

The following information should be included in occupational health records:

(a) Personal identification details. Personal records are necessary for identifying and tracing individual employees and groups of employees exposed to particular risks. Identification details which are of particular value are –
    (i) National Health Service number;

(ii) National Insurance number;

(iii) surname and forenames (maiden name, where applicable);

(iv) sex;

(v) date of birth, country of birth and place of birth;

(vi) usual address and date of taking up residence there.

Some of these items are useful for tracing individuals who are no longer employed by the organisation.

(b) Job history. Before he commences work with a new employer, an occupational history should be taken from the prospective employee. Details of the occupations in the current employment should appear on the individual record, including transfers to alternative work with dates and duration in each job.

Other records which should be maintained include accident records, the results of work area visits, a daily attendance record of employees visiting the occupational health department and information relating to such matters as potential health hazards, drugs, medical equipment and departmental procedures.

Reg 11 of the COSHH Regulations requires that health surveillance be provided for persons who are, or are liable to be, exposed to a substance hazardous to health. Where such cases arise, the employer must ensure that a health record, containing particulars approved by the HSE, in respect of each of his employees so exposed or liable to be exposed, be made and maintained. The record, or a copy thereof, must be kept in suitable form for at least 30 years from the date of the last entry made in it. Where such an employer ceases to trade, he must forthwith notify the HSE in writing and offer these records to the HSE.

Specific aspects of occupational health practice

In addition to the areas detailed in the earlier part of this chapter, occupational health practice includes a number of specific areas which have come into prominence, largely as a result of medical research, over the last 20 years. Included in this group are, for instance, the problem of drug taking, hearing and eyesight defects, the relationship of social habits to work – e.g. smoking and the taking of alcohol – and the relationship of physical defects to the safety of the employee. These various aspects are discussed below.

Drug addiction

Here we must consider the problem of addiction to drugs such as opium, cocaine, morphine, heroin, etc., which is common amongst certain age groups and ethnic groups. Addiction, in its broadest sense, implies that the individual has developed a need for the particular drug in order to stay both physically and mentally normal. Once access to the drug is prevented or removed, certain physical and/or mental symptoms become apparent in the addict. With the increased publicity that has been given to the problem of drug addiction, most people would be aware of the weakness and depression of the cocaine taker, the persistent diarrhoea of the morphine addict or the excessive excitement of the marijuana smoker. It is possible to become addicted to the strangest drugs, for instance to chloroform and ether, the benzedrene in nasal inhalers, and opium in the form of chlorodyne in certain cough medicines.

In most cases addiction to a particular drug will bring about some changes in behaviour or bouts of abnormal behaviour and, whilst the health of the individual addict needs careful attention from the occupational health practitioner and his doctor, attention must also be given to the safety of individuals with whom the addict may come into contact whilst at work. Drug addicts do represent a serious threat to safety and, therefore, must be carefully controlled in terms of the tasks they undertake. Health surveillance and primary monitoring of such persons on a regular basis feature strongly in good occupational health practice.

Hearing and eyesight defects

It is a fact of life that as people get older so their ability to see and hear reduces. This is part of the normal ageing process, so that where workers are exposed to high noise levels or need to undertake close visual tasks a form of secondary monitoring is necessary. In the first case, this may be undertaken by annual audiometric testing and the comparison of audiograms from previous tests carried out for the individual. Such examination should indicate the current level of hearing ability and identify whether there is a need for, perhaps, a change of job to a less noisy part of the factory, increased emphasis on the wearing of hearing protection or further assistance from a medical specialist. The main objective is, of course, to prevent further deterioration in hearing whether this be associated purely with ageing (presbyacusis) or exposure to excessive noise (sociocusis) or both.

Vision screening now features prominently in both pre-employment screening and secondary monitoring activities. Here again the problem of ageing must be taken into account, together with the visual demands of certain tasks. With advancing age the human eye gradually loses its ability to adapt for near and/or distance vision, so that frequently objects tend to be held further from the eyes in order to bring them into focus, unless corrective spectacles are worn.

One indicator of deteriorating focusing ability is the distance of the 'near-point', which is the shortest distance at which an object can be brought into sharp focus. Conversely, the 'far point' is the furthest distance at which an object can be focused.

Colour-blindness in men is another difficulty which increases with age. One in ten young adults has some degree of colour-blindness but many more lose some power of colour differentiation as they get older.

As with defects in hearing, defects in visual acuity and performance can be a cause of or contributory factor in accidents. Specialist groups to whom particular attention should be paid include drivers of all types of vehicle, including fork lift trucks; crane drivers; machinery operators; VDU operators; laboratory staff, who may use optical equipment such as microscopes; and most clerical workers.

Smoking and alcoholism

The relationship of cigarette smoking in particular to various forms of cancer is now well established. Many experts would argue that smoking is not a true form of addiction due to the fact that many people give up smoking quite easily without the usual symptoms of true addiction such as trembling, loss of appetite or a high excitement level. However, there is no doubt that smoking has a direct effect on a high proportion of people in terms of reduced lung function and an increased potential for lung conditions such as bronchitis. The synergistic effect of smoking and, say, asbestos, producing a vastly increased risk of lung cancer, should receive careful consideration.

Alcoholism, on the other hand, is a true addiction, and the alcoholic must be encouraged to obtain medical help and advice. There is no doubt that the abuse of alcohol leads to broken homes, broken marriages, lost jobs, a certain amount of crime and unhappiness generally for all those who may come into contact with the alcoholic, together with varying degrees of physical and mental disease. On the other hand, it is a fact that many people can consume very large quantities of alcohol throughout a long life without showing any apparent ill-effects whatever, and that in most cases alcoholism is a symptom rather than a disease in itself.

The general, although by no means universally accepted, belief today is that the physical diseases brought about by the excessive consumption of alcohol are the result of its indirect effect in producing malnutrition rather than its direct toxic one. The repeated consumption of strong spirits, especially on an empty stomach, can lead to chronic gastritis, and possible inflammation of the intestines which interferes with the absorption of food substances, notably those in the vitamin B group. This, in turn, damages the nerve cells causing alcoholic neuritis, injury to the brain cells leading to certain forms of insanity and, in some cases, cirrhosis of the liver.

The alcoholic is not necessarily the person who becomes obviously drunk on frequent occasions but is more commonly the man or woman who drinks steadily throughout the day, often without any immediate effect being apparent to others. Later, however, symptoms which are partly due to physical effects, partly to the underlying neurosis which is at the root of the trouble in most cases, and partly social, begin to show themselves. The individual eats less and drinks more, often begins the day with vomiting or nausea which necessitates taking the first drink before he can face the public, his appearance tends to become bloated and the eyes are often red and congested. His work suffers, he forgets to keep appointments and he becomes indifferent to his social responsibilities. His craving for drink becomes insatiable, and when he is unable to get it he becomes shaky, irritable and tense. Since he is ashamed of his condition, he tries to hide it and often, instead of drinking openly, hides his bottles about the house and perhaps his office. His emotions are less controlled and he gets angry or tearful readily, tells facile lies, and a minor illness or cessation of the supply may lead to an attack of 'DTs' (delirium tremens).

In severe cases the alcoholic may die from cirrhosis of the liver; or an attack of pneumonia or some other infection, not generally fatal to healthy people, may be so in his case. No matter how alcoholism manifests itself, the alcoholic needs help, particularly if his condition is prejudicing the safety of his fellow workers. In most cases, this implies complete abstention for a period of time under controlled conditions away from the normal temptations of the home and the workplace, perhaps psychotherapy to assess any psychological causes of the condition, and the general building up of impaired physical health.

It is in cases of alcoholism that the occupational health practitioner can be of considerable support and assistance in bringing about the gradual rehabilitation necessary, perhaps through advising on the various social and therapeutic treatments available. The occupational health practitioner is also trained in the early detection of cases of alcohol abuse and, through counselling and routine surveillance, can prevent the situation from deteriorating further.

Many organisations have now prepared and issued statements of policy on both smoking at work and alcohol at work, with a view to improving standards of employee health and reducing sickness absence associated with smoking and alcohol consumption in the workplace and during working hours.

Conclusion

This chapter has endeavoured to cover the very broad field of occupational health practice. It should be appreciated that only the more general aspects of this discipline have been mentioned, and that many more aspects could be added. Fundamentally, as stated at the beginning of the chapter, occupational health is concerned with prevention of accidents and ill-health at work, the treatment, in certain cases, of injuries and illness sustained at work, and the promotion, maintenance and restoration of health. Many people are directly involved in the promotion of sound occupational health standards but, as with safety, everyone at work must consider his own health and potential health hazards and, where possible, take preventive action. The training of staff in avoiding hazards to health is, therefore, vitally important if the toll of occupational disease and conditions affecting health is to be reduced.

21 Noise and vibration

Sound and noise

Sound

Sound is defined as 'any pressure variation in air, water or some other medium that the human ear can detect'. Sound, within the physical sense, is a vibration of particles in a gas, liquid or solid.

Noise

Noise is generally defined as 'unwanted sound'. It is a problem to humans for many reasons. First, environmentally it can be a nuisance, resulting in disturbance and loss of enjoyment of life, loss of sleep and fatigue. Its nuisance effect, whether as noise from a factory, motorway or discothèque, varies from person to person. Second, it can distract attention and concentration, mask audible warning signals or interfere with work, thereby becoming a causative factor in accidents. Finally, exposure to excessive noise can result in hearing impairment. However, provided the exposure period is of sufficient duration, even 'wanted sound', such as loud music, can lead to hearing impairment.

The nature of sound

Sound is a series of pressure waves or fluctuations impinging on the eardrum (sound waves). The sounds of everyday life are composed of a mixture of many simple sound waves. Sound is generated from any energy source which sets up rapid pressure variations in the surrounding air. The rate at which variations occur ('frequency' or 'pitch') is expressed in hertz (Hz) (cycles per second, i.e. the number of complete air waves passing a fixed point per second). The normal human ear is sensitive to frequencies between about 20 and 20,000 Hz, being particularly sensitive in the range 2,000 to 6,000 Hz (with maximum sensitivity at 4,000 Hz), and is progressively less sensitive at higher and lower frequencies. This fact is very important when measuring sound, since two sounds of equal intensity, but of different frequency, may appear, subjectively, to be of different loudness.

Characteristics of sound waves

Sound may be 'pure tone', that is of one frequency only, such as the sound produced by a tuning fork (*see* Fig. 21.1(*a*)). Some industrial noise is of this type, but most is highly complex, with components distributed over a wide range of frequencies. Noise of this type is referred to as 'broad band' (*see* Fig. 21.1(*b*)), common examples being noise produced by looms, an air jet or printing presses.

E

Fig. 21.1 ◆ (a) Pure tone sound (b) Broad band noise (c) Impulse noise

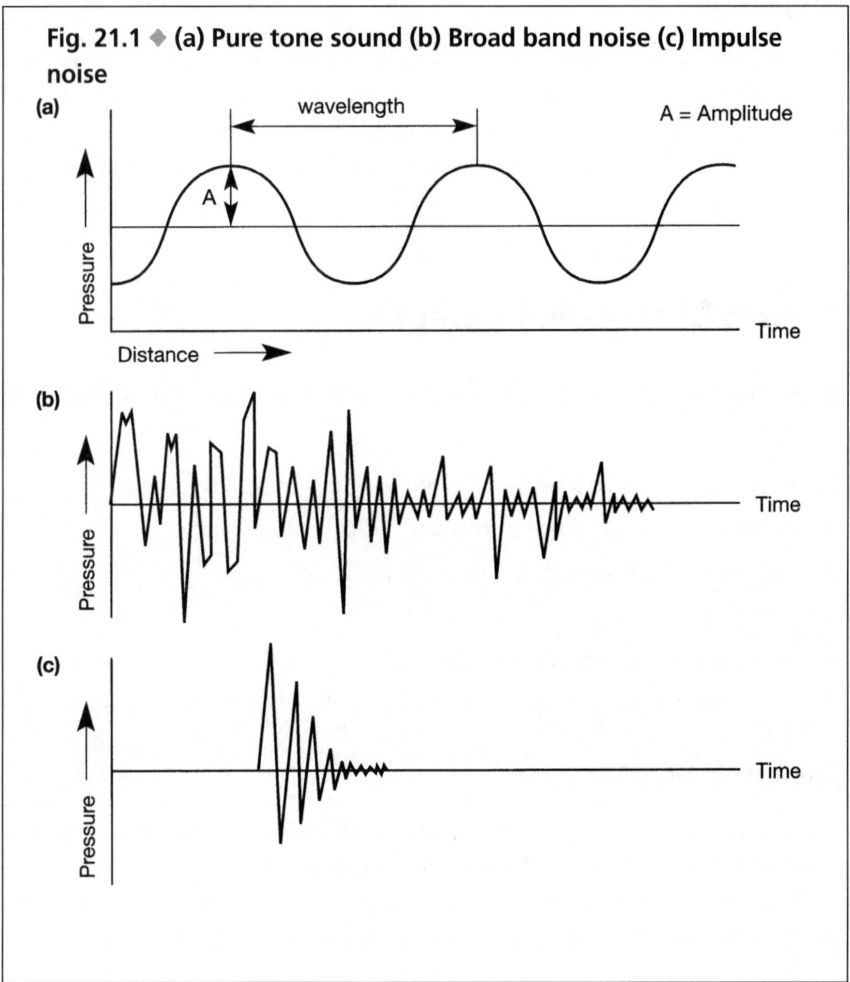

Industrial noise is often produced by impact between metal parts. If there are many impacts per second, as in a riveting machine, the noise produced is usually treated as broad band noise, but if the noise is produced by widely spaced impacts, as from a drop hammer or cartridge-operated hand tool, then the noise produced is termed 'impulse noise' (*see* Fig. 21.1(*c*)). Impulse noise can present special difficulties in measurement and in assessing the risk to hearing.

Wavelength

This is the physical distance in air from one peak of a sound wave to the next. It equals the ratio of the speed of sound in the medium to the fundamental frequency.

Amplitude

The amplitude of a sound wave is the maximum displacement of a particle from its rest position. For practical purposes, this means the sound wave size and, in turn, therefore, the amount of sound energy involved. The amplitude of a sound wave determines loudness, although the two are not directly related.

Sources of noise and vibration

Noise and vibration in the working environment can include the following:

(a) noise produced as a result of vibration in machines;
(b) noise taking a structure-borne pathway;
(c) radiation of structural vibration into the air;
(d) turbulence created by air or gas flow;
(e) noise taking an airborne pathway;
(f) noise produced from vibratory hand tools, e.g. chain-saws.

The effect on hearing

All the above sources of noise result in vibration of the ear-drum. This vibration is amplified by the ossicles (malleus, incus and stapes) in the middle ear, and transmitted to the cochlea. In the cochlea, the vibrations are transformed into nerve impulses which are sent to the brain. It is in this way that a sound is heard. (*See* Chapter 19, which deals with noise-induced hearing loss.)

Important aspects of sound and noise

Sound intensity

Sound intensity describes the particular power of a sound or the level of sound energy with which it confronts the ear. Intensity thus describes the rate of flow of sound energy. High-intensity sound has more energy than low-intensity sound.

Sound pressure level

Sound intensity is difficult to measure directly, but the passage of sound energy through air is accompanied by fluctuations in atmospheric pressure. These fluctuations can be measured and related to the amount of sound energy that is flowing. Therefore, it is usual to measure sound pressure level, which is a measurement of the magnitude of the air pressure variations or fluctuations which make up sound. The root mean square value of the pressure variations is used and expressed in decibels (dB).

Frequency

Frequency is the number of complete pressure variations passing a fixed point per second. It is measured in hertz (Hz), i.e. 1 Hz = 1 cycle per second; 1 kHz = 1,000 cycles per second. The frequency of a sound gives it its distinguishing character. For instance, high-frequency sound, such as a train whistle, will sound high-pitched, whereas a low-frequency sound, such as that from a double bass, will sound low-pitched. The more rapidly the vibrations occur, the higher is the frequency and vice versa.

Pitch

This is the subjective quality of a sound which determines its position in the musical scale. It is determined by frequency.

Tone

The tone and quality of a note depend upon the particular overtones or harmonics which are sounding together with the fundamental note.

Loudness

The loudness of a sound depends upon its intensity and the amplitude of the sound waves involved. However, since the human ear is less sensitive to high and low frequencies, it also depends upon frequency and the subjective perception of sound by human beings. (*See also* 'phons' on page 380.)

Basic theory of noise measurement

A sound pressure level meter measures sound intensity on a comparative basis. The range of intensities to which the ear responds, however, is enormous, from the threshold of hearing to the threshold of pain. For example, at 1,000 Hz the threshold of pain is 100,000,000,000,000 (10^{14}) times more intense than the threshold of hearing, where sound is just discernible. It is clearly difficult to express such ratios on a simple arithmetic scale, so a logarithmic scale is used. The ratio would therefore be expressed as

$$\log_{10} \frac{10^{14}}{1} \text{ or 14, rather than}$$

$$\frac{10^{14}}{1}$$

The unit used is the bel. Thus 1 bel is $\log_{10}10^1$ (a tenfold change in intensity), 2 bel is $\log_{10}10^2$ (a hundred-fold change in intensity) and so on. The bel, however, is a very large unit, so it is further split into tenths, called decibels (dB); 1 bel equals 10 decibels. For example, 10 $\log_{10}10^{14}$ equals 140 dB. Thus 1 decibel equals a change of intensity of 1.26 times, since $10^{1/10}$ is 1.26 (or 1.26^{10} is 10). Also, a change of intensity of 3 dB = 1.26^3 = 2, so that doubling the intensity of a sound gives an increase of 3 dB.

If there are two sounds of intensities I_1 and I_2 and they differ by n dB, then

$$n = 10 \log_{10} \frac{I_1}{I_2}$$

It is normal practice to relate intensity to a standard reference level, so that

$$n = 10 \log_{10} \frac{I_1}{I_0}$$

and I_0 is taken as 10^{-12} watts per square metre.

However, as intensity is proportional to pressure squared,

$$n = 10 \log_{10} \frac{P^2}{P_0^2}$$

$$= 10 \log_{10} \left[\frac{P}{P_0} \right]^2$$

$$\text{or } n = 20 \log \frac{P}{P_0} \; dB$$

where P is the standard reference level of 2×10^{-5} newtons per square metre (pascals) and n is sound pressure level in dB. Pressure is the easiest quantity to measure, hence the use of dB sound pressure level. The standard reference level of 2×10^{-5} N/m^2 is chosen since it is the average threshold of audibility at 1,000 Hz (i.e. it is 0 dB).

NOTE. Under the SI system, sound pressure is expressed in pascals. A pascal is a unit of pressure corresponding to a force of one newton acting uniformly upon an area of 1 square metre. Hence 1 Pa = 1 N/m^2.

The use of a logarithmic scale in sound measurement has a further advantage, because the evaluation of intensities is simplified by the replacement of multiplication with addition and of division with subtraction. Furthermore, the response of the ear tends to follow a logarithmic scale.

The addition of decibels is carried out on a ratio basis, rather than an arithmetic one, and Table 21.1 may be used to simplify the procedure. To add two sound pressure levels, take the difference between the two levels and add the corresponding figure in the right-hand column to the higher sound pressure level.

Table 21.1 ◆ Addition of decibels

Difference (dB)	*Add to higher (dB)*
0.0–0.5	3.0
1.0–1.5	2.5
2.0–3.0	2.0
3.5–4.5	1.5
5.0–7.0	1.0
7.5–12.0	0.5
Over 12.0	0.0

Other units used in noise measurement

Phons

The human ear does not respond equally to all frequencies. Sounds of different frequency at a constant sound pressure level do not evoke equal loudness sensations. This phenomenon is linear with neither amplitude nor frequency, and 'loudness level' is measured in phons, the sound being compared again to a standard reference signal of 1,000 Hz. The loudness level in phons of any sound is taken as that which is subjectively as loud as a 1,000 Hz tone of known level. 0 phon is 0 dB at 1,000 Hz. 50 phon is the loudness of any tone which is as loud as a 1,000 Hz tone of 50 dB. This can be demonstrated by equal loudness curves for pure tones shown in Fig. 21.2. Maximum sensitivity occurs between 1 and 5 kHz. The curves are obtained by finding the sound levels at different frequencies which seem equally loud to the listener in comparison with a reference sound at 1 kHz.

This is a linear unit of loudness on a scale designed to give scale numbers approximately proportional to loudness. The scale is precisely defined by its relation to the phon scale.

Octave bands and octave band analysis

It is possible to make a single measurement of the overall sound pressure of the entire range of audible frequencies but this measurement, if taken in linear decibels, is of limited use since the ear is more sensitive to some frequencies than others. Use of the 'A' weighted decibel scale (*see below*)

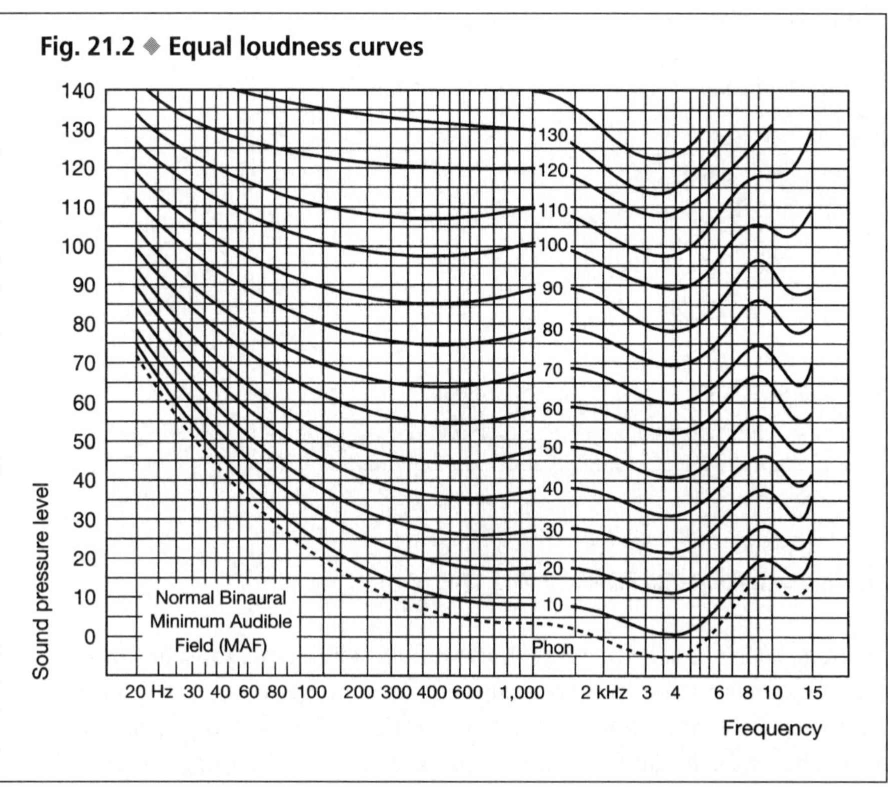

Fig. 21.2 ◆ Equal loudness curves

provides a reasonable means of assessing likely risk to hearing but a knowledge of the way in which the sound is distributed throughout the frequency spectrum provides a much more accurate picture. This can be obtained by dividing the noise into octave bands and measuring the sound pressure level at the centre frequency of each band. (An octave represents a doubling of frequency, so that the range 90–180 Hz is one octave, as in the range 1,400–2,800 Hz.)

The octave bands are usually identified by their geometric centre frequencies. For example, the geometric centre frequency of the octave 90–180 Hz is approximately 125 Hz. The standard range of octave bands has the geometric centre frequencies shown in Table 21.2. Octave band analysis is used for assessing risk of noise-induced hearing loss and in the specification of certain forms and types of hearing protection. It is also used in the diagnosis of machinery noise and in the selection of noise attenuation methods.

Table 21.2 ◆ Standard range of octave bands

Limits of band (Hz)	*Geometric centre frequency (Hz)*
45–90	63
90–180	125
180–355	250
355–710	500
710–1,400	1,000
1,400–2,800	2,000
2,800–5,600	4,000
5,600–11,200	8,000

The sound pressure level meter

A sound pressure level meter is an instrument which measures linear sound pressure level in the human audiofrequency range unless provided with and set to various weighting networks. The 'A' weighted network gives objective measurements of sound pressure level in accordance with the manner of response of the human ear. A typical mode of operation is shown in Fig. 21.3.

Fig. 21.3 ◆ Measurement of sound pressure level in accordance with the manner of response of the human ear

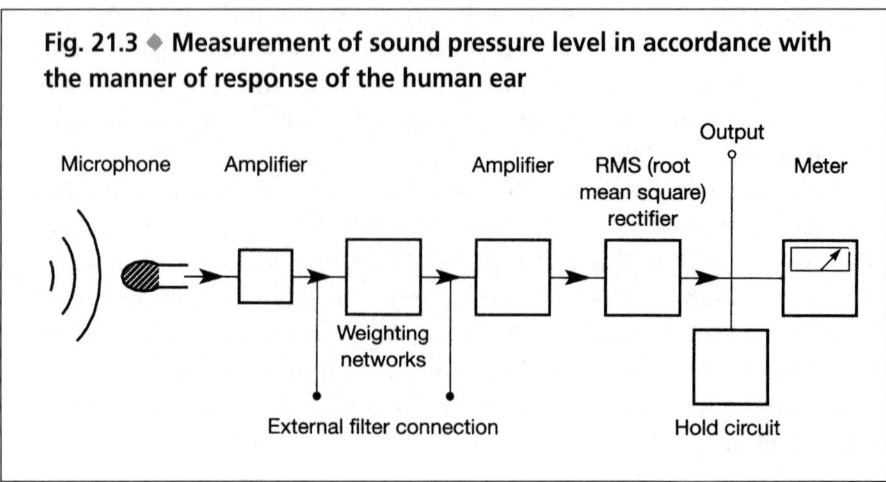

The microphone senses the air pressure fluctuations and converts mechanical vibration to an electrical signal containing amplitude and frequency components. The amplifier increases the weak signal from the microphone and incorporates gain adjustment, which enables the instrument to cope with the very wide range of pressure amplitudes which the ear can sense. The sound signal is also available as an output socket so that it may be fed to external instruments such as recorders or noise dosemeters.

Since an accurate response from the sound level meter is necessary, provision is made to calibrate it for accurate results. This is best done by the use of a portable acoustic calibrator placed directly over the microphone. The calibrator is basically a miniature audible signal generator giving a precisely defined sound pressure level to which the sound level meter can be calibrated. Electronic oscillators are most commonly used.

When the sound level fluctuates, the meter needle should follow these variations. However, if the level fluctuates too rapidly, the meter needle may move so erratically that it is impossible to obtain a meaningful reading. For this reason, two meter response characteristics are used:

(a) Fast: this gives a fast-reacting indicator response which enables the user to follow and measure noise levels which are not fluctuating too rapidly.

(b) Slow: this gives a damped response and helps average out meter fluctuations which would otherwise be impossible to read.

Weighting networks

The sound level meter incorporates electrical circuits known as weighting networks. These provide for various sensitivities to sounds of different frequencies, the original object being to simulate the response characteristics of the human ear at different frequencies. These weighting networks are known as A, B, C or D weighted decibel scale operating conditions and can be selected on a sound level meter.

A scale

The A scale is normally used for industrial noise measurement. This scale makes the instrument more sensitive to the middle range of frequencies, and less sensitive to high and low frequencies, and is the one which most closely approximates to the response of the human ear. Measurements of the sound pressure level using this scale are designated dBA.

B scale

This scale was intended for the measurement of middle range sound pressure levels, between 55 and 85 dB. It is not commonly used as it does not give good correlation to subjective tests of hearing perception.

C scale

This scale gives most sensitivity in low frequencies and is, therefore, of limited use. As with the B scale, it does not give good correlation to subjective tests.

D scale

This scale is generally limited to the measurement of aircraft noise and has little or no application in the measurement of industrial noise.

Noise control

In any strategy to reduce noise two factors must be considered: first, the source of the noise, and second, the actual pathway taken by the noise to the recipient. Personal protective equipment, e.g. ear-plugs, ear defenders or acoustic wool, may go some way towards preventing people from going deaf at work, but such a strategy should be regarded as secondary since it relies too heavily upon the exposed person wearing potentially uncomfortable and inconvenient protection for the correct amount of time. The better and primary way of preventing noise and, therefore, the risk of persons sustaining noise-induced hearing loss is, if practicable, to tackle the potential problem at the design stage, rather than endeavouring to control noise once the machinery or noise-emitting item is installed and has become operational.

Different methods of noise control are suitable for dealing with different sources and for the different possible stages in the pathway to the recipient. These may be summarised as in Table 21.3. The sequence does not necessarily apply in all cases, and is reversible or interchangeable. Control of the main or primary noise pathway is the most important factor in noise control. For instance, the noise pathway for a vibration-induced noise has three distinct stages:

(a) structure-borne noise emission;

(b) radiation of the noise from the structure into the air;

(c) the actual airborne noise pathway.

Table 21.3 ◆ Methods of noise control

Sources and pathways	*Control measures*
Vibration produced through machinery operation	Reduction at source, e.g. substitution of nylon components for metal, tapered tools on presses
Structure-borne noise (vibration)	Vibration isolation, e.g. resilient mounts and connections, anti-vibration mounts
Radiation of structural vibration	Vibration damping to prevent resonance
Turbulence created by air or gas flow	Reduction at source or use of silencers
Airborne noise pathway	Noise insulation – reflection; heavy barriers Noise absorption – no reflection; porous lightweight barriers

E

In this case, the way to reduce structure-borne noise is to isolate the noise-generating parts of the machine from the radiating surface with vibration isolators (resilient mounts). Isolators consist of steel springs or flexible materials such as rubber or cork. The amount of noise radiated from machine surfaces will be increased substantially if the surfaces resonate or oscillate. In this case it would be appropriate to apply a damping treatment using a spray-on, stick-on or magnetic damping compound. Panel-damping materials can be extremely effective in a wide range of applications, and the application of damping treatments can result in noise reductions of as much as 10 dBA if the material is initially poorly damped.

Once the noise has been radiated into the air, two completely different methods of control are possible. The recipient can be insulated from the noise with a heavy limp airtight screen, which simply acts as an impervious barrier to prevent noise transmission. Alternatively, noise can be absorbed in porous materials mounted on walls or suspended from a ceiling. Absorption of noise will only be helpful if most of the noise which reaches the subject has been reflected off walls and ceilings.

The Noise at Work Regulations 1989

These Regulations impose requirements on employers with respect to:

(a) the making and review of noise assessments (Reg 4);

(b) the keeping of records of noise assessments and reviews thereof (Reg 5);

(c) the reduction of risk of damage to the hearing of their employees from exposure to noise (Reg 6);

(d) the reduction of exposure to noise of their employees (Reg 7);

(e) the provision to their employees of personal ear protectors (Reg 8);

(f) the marking of, and entry of their employees into, ear protection zones (Reg 9);

(g) the use and maintenance of equipment provided by employers pursuant to the provisions of the Regulations (Reg 10);

NOTE. Similar requirements relating to use and maintenance of equipment apply to employees also in this case.

(h) the provision of information, instruction and training to such of their employees as are likely to be exposed to specified noise levels (Reg 11);

Duties under sec 6 of HSWA on the part of manufacturers, designers, etc, are modified to include a duty to provide certain information relating to noise generation (Reg 12).

Assessment of exposure

The Regulations bring in the concepts of 'daily personal noise exposure' and 'action levels' when undertaking assessment of exposure. *Daily personal noise exposure* means the level of daily personal noise exposure of an employee ascertained in accordance with Part I of the Schedule to the Regulations, but taking no account of the effect of any personal ear protector used. The formulae used for assessing both daily personal noise exposure, and the weekly average of daily personal noise exposure of employees, specified in Parts I and II to the Schedule are shown in Fig. 21.4. A number of *action levels* are also specified in the Regulations, as follows:

(a) the first action level – means a daily personal noise exposure of 85 dB(A);

(b) the peak action level – means a level of peak sound pressure of 200 pascals;

(c) the second action level – means a daily personal noise exposure of 90 dB(A)

Reg 4 requires that every employer shall, when any of his employees is likely to be exposed to the first action level or above, or to the peak action level or above, ensure that a competent person makes a noise assessment which is adequate for the purposes:

Fig. 21.4 ◆ Formulae for assessing personal noise exposure

THE SCHEDULE Regulations 2(1) and 13(1)

PART I

DAILY PERSONAL NOISE EXPOSURE OF EMPLOYEES

The daily personal noise exposure of an employee ($L_{EP,d}$) is expressed in dB(A) and is ascertained using the formula:

$$L_{EP,d} = 10 \log_{10} \left\{ \frac{1}{T_0} \int_0^{T_e} \left[\frac{P_A(t)}{p_0} \right]^2 dt \right\}$$

where–

T_e = the duration of the person's personal exposure to sound;

T_0 = 8 hours = 28,800 seconds;

p_0 = 20 μPa; and

P_A (t) = the time-varying value of A–weighted instantaneous sound pressure in pascals in the undisturbed field in air at atmospheric pressure to which the person is exposed (in the locations occupied during the day), or the pressure of the disturbed field adjacent to the person's head adjusted to provide a notional equivalent undisturbed field pressure.

PART II

WEEKLY AVERAGE OF DAILY PERSONAL NOISE EXPOSURE OF EMPLOYEES

The weekly average of an employee's daily personal noise exposure ($L_{EP,w}$) is expressed in dB(A) and is ascertained using the formula:

$$L_{EP,w} = 10 \log_{10} \left[\frac{1}{5} \sum_{k=l}^{k=m} 10^{0.1(L_{EP,d})k} \right]$$

where–

$(L_{EP,d})_k$ = the values of $L_{EP,d}$ for each of the m working days in the week being considered.

(a) of identifying which of his employees are so exposed;

(b) of providing him with such information with regard to the noise to which those employees may be exposed as will facilitate compliance with his duties under Regs 7, 8, 9 and 11.

The noise assessment must be reviewed when:

(a) there is reason to suspect that the assessment is no longer valid; or

(b) there has been a significant change in the work to which the assessment relates;

and, where as a result of the review changes in the assessment are required, those changes shall be made.

Under Reg 4, the employer must ensure that an adequate record of that assessment, and of any review thereof, is kept until a further noise assessment is made for reasons shown above. (*See* Fig. 21.5.)

Fig. 21.5 ◆ Record of noise exposure

Name and address of premises, department etc.____________________

____

____

____

Date of survey__________ Survey made by____________________

Workplace Number of persons exposed	Noise level (Leq(s) or sound level)	Daily exposure period	$L_{EP,d}$ dB(A)	Peak pressure (where appropriate)	Comments/ remarks

General comments ____________________

Instruments used ____________________

Date of last calibration________________ Signature________________

Date________________

(Noise at Work: Noise Guide No. 3: *Noise assessment, information and control*, HMSO, London)

Noise control programmes

In considering how an organisation should approach compliance with the Noise at Work Regulations, it is appropriate to consider the elements of occupational hygiene practice, i.e. identification/recognition of the health risk, measurement, evaluation against current standards (legal and otherwise), and control.

1. Identification/recognition

This may be achieved through routine observation in a working area. Generally, if it is necessary to shout to make one's self understood, it is fairly certain that sound pressure levels are around or above 90 dB(A). Alternatively, there may have been complaints from safety representatives, shop stewards or employees, claims against the company for occupational deafness, or action by the local environmental health officer in the event of noise nuisance to local residents. A preliminary survey, using a sound pressure level meter, will indicate variations in sound pressure level from one part of the premises to another.

2. Measurement

The second stage of the operation is the carrying out of a full-scale noise survey of the premises using a precision grade sound level meter with facilities for octave band analysis. Such a survey will produce an indication of those items of plant and machinery producing unacceptable noise emissions, the frequency ranges involved and the risk of occupational deafness to operators.

3. Evaluation

Reference to action levels specified in the Regulations, the actual results of the noise survey, the number of people exposed generally and in specific locations, and to the transmission pathways of the noise, will make it possible to decide on the relative urgency of action necessary. Short-term, medium-term and long-term measures for eliminating or reducing such exposure can then be made.

4. Control strategies

Many options are available based on the degree of risk, cost and sheer practicability of implementation. The ultimate objective must be control through recognised engineering methods as opposed to the provision of ear protection. The options are:

(a) Control –
   (i) reduction at source;
   (ii) installation of soundproof enclosures and close shields; or
   (iii) installation of noise refuges/havens (*see* Table 21.3, p. 411);

(b) Prevention –
   (i) new plant specification, including liaison with designers, manufacturers and suppliers of plant and machinery;
   (ii) preventive maintenance schedules which incorporate attention to existing and future potential noise emission;
   (iii) designation of demarcated ear protection zones;
   (iv) provision and use of ear protection by all persons who may be exposed;
   (v) staff training to recognise the hazards from noise exposure;
   (vi) propaganda aimed at informing operators as to the causes and effects of exposure to noise;
   (vii) health surveillance through audiometric testing of new employees and of existing employees on an annual basis, e.g. as part of an annual health examination.

It is further recommended that an organisation should produce and publish a Statement of Policy on Noise which states the intention of the company to take the above measures with a view to preventing employees going deaf at work. Such a Statement should incorporate a statement of intent, organisation and arrangements for implementing the policy and individual responsibilities of all concerned from chief executive to shop floor employees.

A typical structure for a company noise control programme is shown in Fig. 21.6.

Fig. 21.6 ◈ Typical structure of a noise control programme

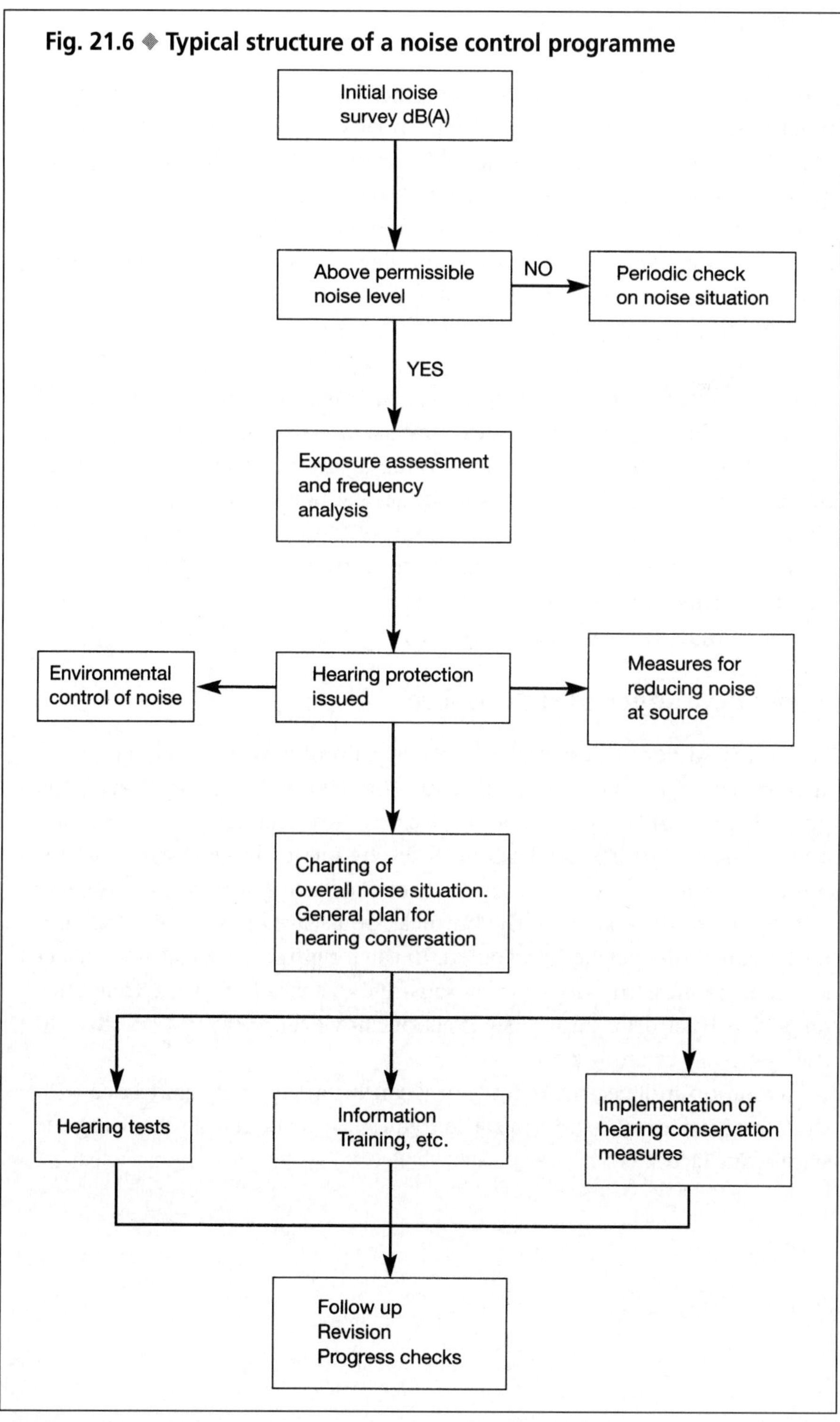

Vibration

A body is said to vibrate when it describes an oscillating motion about a fixed position. As with sound, the number of times a complete motion cycle takes place during the period of one second is referred to as the 'frequency', which is measured in hertz (Hz). The motion can consist of a single component occurring at a single frequency, as with a tuning fork, or of several components occurring at different frequencies simultaneously, e.g. with the piston motion of an internal combustion engine.

Vibration signals in practice usually consist of very many frequencies occurring simultaneously, so that it is not possible to see immediately, just by examination of the amplitude-time pattern, how many components there are and at what frequencies they occur. These components can be revealed by plotting vibration amplitude against frequency, the process being known, as in the case of sound, as 'frequency analysis'. The graph showing the vibration level as a function of frequency is known as a 'frequency spectrogram', and the vibration amplitude is the characteristic which describes the severity of the vibration. Typical frequency analyses and spectrograms are shown in Fig. 21.7.

Frequency ranges of significance

The principal hazards associated with vibration are whole body vibration and the condition known as vibration-induced white finger (VWF), the physiological aspects of which are discussed in Chapter 19. The human body is most sensitive to vibration in the frequency range 1–80 Hz and is principally subjected to vibration in three supporting surfaces, viz. the feet of a person while standing, the buttocks of a seated person and the supporting areas of a person lying down. In the longitudinal direction, i.e. feet to head, the human body is most sensitive to vibration in the frequency range 4–8 Hz. In the transverse direction, however, it is most sensitive to the frequency range 1–2 Hz.

Vibration-induced white finger is a condition generally associated with the use of vibratory hand tools, the frequency of the hand tool being the significant factor.

Fig. 21.7 ◆ Vibratory motion

(a) Tuning fork – a single component vibrating at a single frequency.
(b) Internal combustion engine – several components vibrating at different frequencies simultaneously.
(c) Factory machinery – a large number of components vibrating at different frequencies simultaneously.

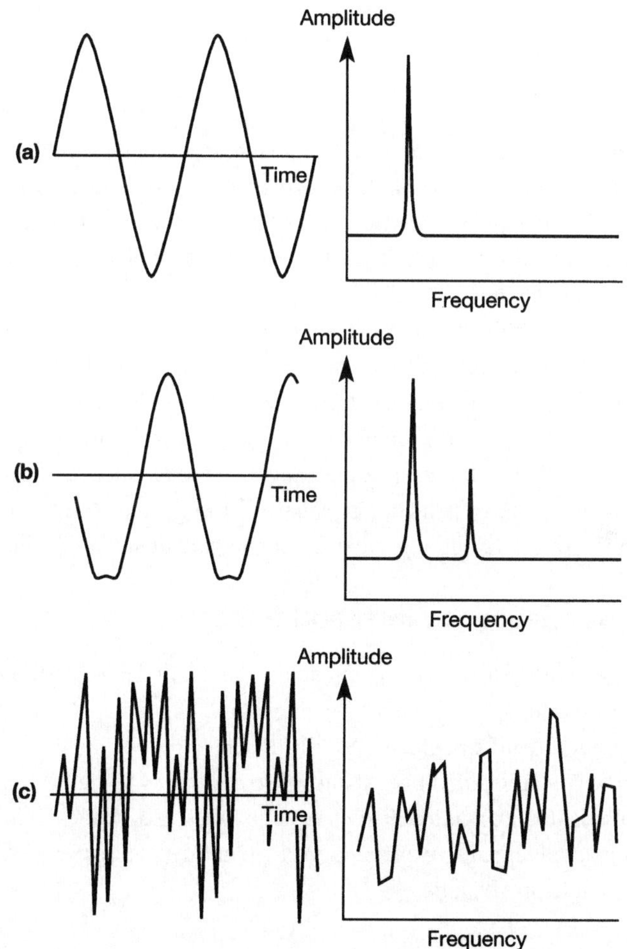

E

22 Dust and fumes

Dust

Dust is defined by the ILO as 'an aerosol composed of solid inanimate particles'. The term aerosol implies that airborne particles are carried in or contained in air, which may be inhaled. An aerosol can embrace liquid droplets as well as solid particles.

Some dusts are fibrogenic, i.e. they cause fibrotic changes to lung tissue, or toxic, in that they eventually poison the body systems. Examples of fibrogenic dusts are silica, cement dust and certain metals, whereas toxic dusts may include arsenic, mercury, beryllium, phosphorus and lead. Some toxic dusts, such as arsenic, have an acute effect. Others, such as mercury, may have a chronic effect. A number of dusts, although not harmful to health, can have a nuisance effect, e.g. dust from the combustion of solid fuels.

Definitions relevant to dust and fumes

Particulate – a collection of solid particles, each of which is an aggregation of many molecules.
Mist – airborne liquid droplets, e.g. oil mist.
Fumes – airborne fine solid particulates formed from the gaseous state usually by vaporisation or oxidation of metals, e.g. lead fume.
Vapour – airborne liquid droplets given off from the surface of a volatile liquid, e.g. trichlorethylene.

The behaviour of dusts

All dusts are potential aerosols and the behaviour of particles is influenced by:

(a) the rate of air movement;
(b) Brownian motion, that is the 'joggling' movement or effect imparted to submicron particles by molecular bombardment; and

(c) the size, density and shape of the particle.

The unit of particle size is the micron, which equals one thousandth of a millimeter, designated μm.

When a particle falls in air it does not accelerate indefinitely. Eventually it reaches a speed at which air resistance equals its weight, and thereafter it falls at constant speed, its 'terminal velocity'. This depends to a great extent upon its size and density.

Physiology of the human lung

The lungs are enclosed in the thoracic cavity and have a sponge-like elastic texture. There are expanded or compressed by movements of the thorax in such a way that air is repeatedly taken in and expelled. They communicate with the atmosphere through the trachea or windpipe, which opens into the pharynx. In the lungs, gaseous exchange takes place. Some of the atmospheric oxygen is absorbed and carbon dioxide from the blood is released into the lung cavities. The trachea divides into two bronchi which enter the lungs and divide into smaller branches. These divide further into bronchioles which terminate in a mass of minute thin-walled, pouch-like air sacs or alveoli (*see* Figs 22.1 and 22.2).

Fig. 22.1 ◆ Diagram of lungs

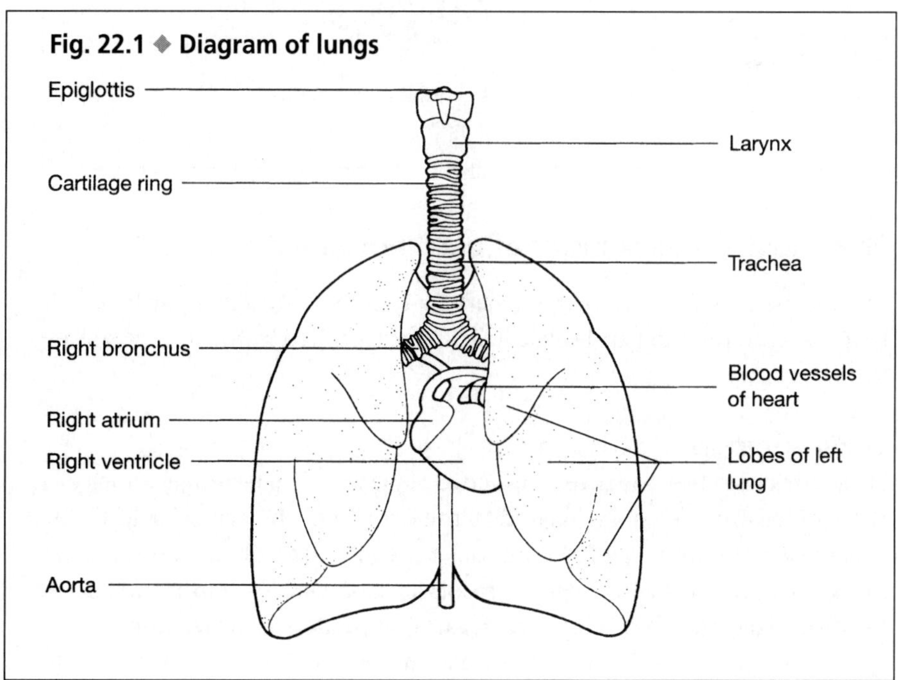

E

Fig. 22.2 ◆ The respiratory tract

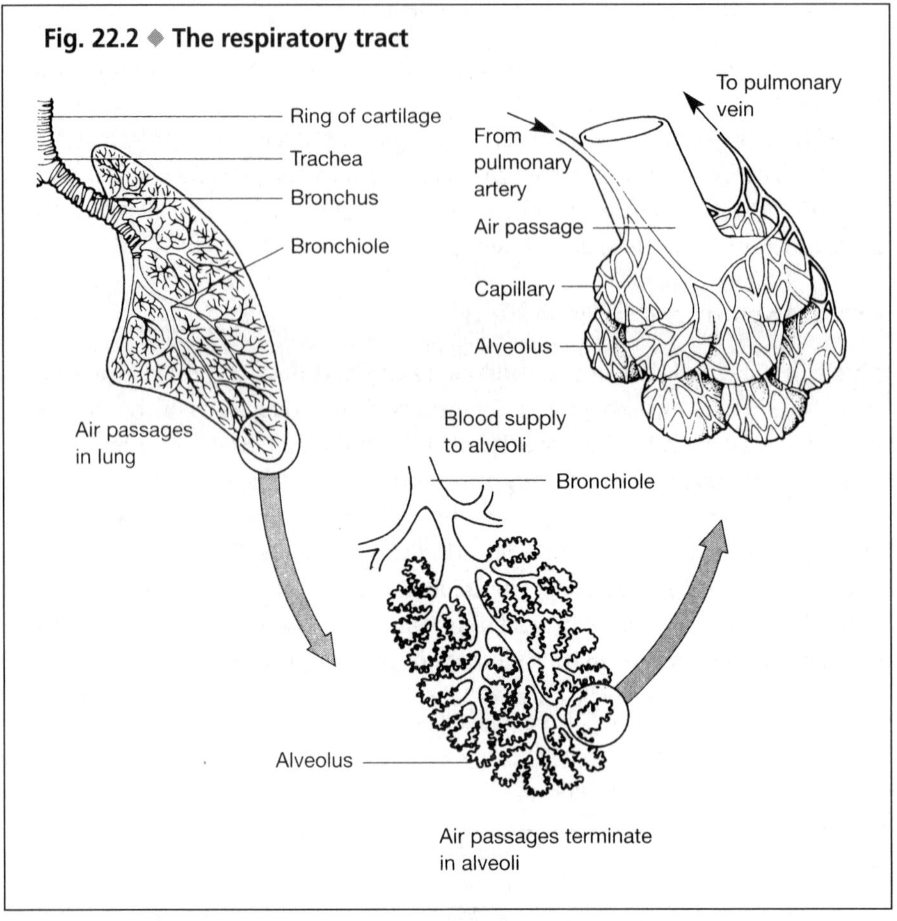

Physiological mechanisms of dust movement

The mechanisms which induce a particle to move through a particular air pathway into the lung and subsequently be deposited in the lung tissue are the following.

Sedimentation

Dust particles settle under the influence of gravity. The terminal velocity of the sedimenting aerosol is related to the density of the aerosol and to the square of the diameter of the aerosol for those aerosols in the diameter range 1–20μm. Many industrial aerosols are not spheres of uniform shape, however. They may be clumps (aggregates) of particles. The terminal velocity of aggregated aerosols cannot be determined by the above relationship; instead, the aerodynamic diameter must be considered. This is the diameter of a uniform sphere which has the same terminal velocity as the aggregate of other irregular particles. Where the uniform sphere has unit density, the aerodynamic diameter is expressed as the diameter of an equivalent unit density sphere, but where the uniform sphere has the same density as the irregular sphere the diameter is expressed as the Stokes diameter. Both are expressed in microns.

Interception

This is the process whereby irregular particles such as asbestos fibres become caught on the walls of small airways. The length and size of the fibres in relation to the dimensions of the airway are important.

Impaction

Impaction takes place through curving in airstream. Suspended aerosols continue under momentum and collide with the wall of the airway. Impaction is related to the velocity of aerosol movement and angular change of direction.

Diffusion

This is the process whereby small aerosols behave like molecules and move freely throughout an air space. It is brought about by the random bombardment of the aerosols by the molecules of the gas in which they are suspended.

The body's protective mechanisms

There are a number of mechanisms by which the body endeavours to prevent dust entering the lungs. The operation of a particular mechanism depends upon the shape and size of particles. The principal protective mechanisms are the following.

The nose

The very coarse hairs lining the nostrils have a filtering effect and trap the larger particles. The cyclonic effect caused by the sudden changes in direction of the nasal passages also causes dust to impinge on the mucous membrane of the nose. In many cases the particle may be expelled by sneezing or blowing the nose.

Ciliary escalator

The surface of the respiratory tract (trachea and bronchi) is lined with special cells, each of which has a cilium growing from its head. The mucous membrane contains mucous glands which secrete a tacky fluid. This forms a sticky film bound up with the cilia. The cilia exhibit a wave-like motion and a particle falling on to the cilia is carried by this motion back to the pharynx, after which it may be swallowed or expectorated. This mechanism is assisted by mucus which causes particles to adhere to the cilia.

Dust deposition locations in the respiratory tract (*see* Fig. 22.2) can be broadly classified as follows:

Above 7 microns – mouth and throat only

4.7 to 7.0 microns – pharynx

3.3 to 4.7 microns – trachea and bronchi

2.1 to 3.3 microns – bronchioles

1.1 to 2.1 microns – terminal bronchi

0.43 to 1.1 microns – alveoli

Particles less than 0.43 microns tend to remain airborne and are exhaled.

'Respirable range' particles are, therefore, those particles in the size range 0.43 to 7.0 microns which enter the various parts of the respiratory tract, those entering the respiratory bronchioles, terminal bronchi and alveoli being the most significant. Consequently, control over particles in this size range is important.

Fibres have different deposition characteristics from uniform density spheres. Broadly, asbestos and man-made mineral fibres having a diameter less than 3.5 microns may be regarded as aerodynamically respirable.

Macrophages (phagocytes)

These are wandering scavenger cells with a large nucleus and irregular outline. They move freely through tissue, engulfing bacteria and dust particles in the process. They secrete hydrolytic enzymes which attack the foreign body, neutralising its activity to some extent. They are found in the alveoli where, after carrying out this scavenging action, they migrate back along the respiratory pathway. At the terminal bronchioles they meet the lowest reaches of the ciliary escalator on which they are carried, ultimately to be swallowed or expectorated. In this way, macrophage action supplements respiratory filtration processes.

Lymphatic system

The lymphatic system acts as a form of drainage system throughout the body for the removal of foreign bodies. Lymphatic glands or nodes at specific points in the lymphatic system act as selective filters preventing infection from entering the bloodstream. In many cases a localised inflammation occurs in the node.

Sources of dust

There are many industrial sources of dust, and they may be classified thus:

(a) dust produced in the cleaning and preliminary treatment of raw materials, e.g. dust resulting from sand-blasting operations in foundries, abrasive treatments for the removal of rust;

(b) dust produced in processes such as refining, grinding, milling and other size reduction processes;

(c) manufactured dusts for specific treatments or dressings, e.g. in the dressing of seed corn with powdered mercury-based fungicides;

(d) environmental or background dusts, such as those produced by routine sweeping of factory floors, combustion of fuels, the use of packaging materials or road dust.

Dust control measures

The following aspects are important in the selection of dust control measures:

(a) the type of dust in terms of particle size, weight, density, air velocity and toxicity;

(b) the source of dust in a particular process;

(c) the number of personnel exposed, the duration of exposure (continuous or intermittent) per day, and the number of days per week this emission takes place;

(d) methods of monitoring emissions, e.g. static sampling, personal dosemeters, and the results of past monitoring activities;

(e) the efficiency of cleaning procedures; manual methods should be replaced by the use of industrial vacuum cleaners;

(f) the efficiency of dust arrestment plant, including the system for the maintenance of, and testing, the efficiency of such plant.

Emphasis should always be placed on control at source by means of dust arrestment plant, in preference to the provision and use of respiratory and other protection.

Control strategies

Replacement or substitution

Replacement or substitution of the hazardous dust-producing process or material by a suitable alternative should always be considered first. For instance, the use of a liquid mercury-based dressing for seed corn instead of the powder form, or the replacement of toxic dust-producing materials by non-toxic materials, is an effective control strategy.

Suppression

In many cases, the use of a wet process, as opposed to a dry process, will be sufficient to reduce the dust hazard. A typical example is in the pottery industry where flint is ground under water due to the danger of fibrogenic dust emission in a dry process.

Isolation

Isolation entails enclosure of the complete process or the actual point of dust production, for instance the total enclosure of large grinding processes or of tipping points for certain dust-producing materials, such as coal, to the total exclusion of the workforce. Tipping points should be provided with efficient dust arrestment plant to prevent dust nuisance to people living in the immediate vicinity.

Local exhaust ventilation (LEV)

Local exhaust ventilation points linked to collection and filtration plant must be considered. In most cases it is necessary to install a system of total or partial enclosure in conjunction with cyclone arrestors, dry deduster units, wet arrestors or electrostatic precipitators. It is vital that factors such as particle size, weight and density, together with efflux velocity, are evaluated prior to the selection of a particular form of dust arrestment. (*See* further Chapter 26 'Prevention and control strategies in occupational hygiene'.)

Cleaning and housekeeping

High standards of cleaning and housekeeping should be maintained wherever workers are exposed to a dusty process. Failure to do so can lead to an action for breach of statutory duty and/or common law duty. Whilst dust suppression plant, depending upon its efficiency, will remove the majority of dust from the working environment, small quantities may escape as a result of handling, plant defects or plant malfunction. Hand sweeping, using brushes or brooms, should be replaced by mechanical vacuum-cleaning equipment. Operators should be trained in the correct use of the equipment, which should be serviced and maintained on a regular basis. Such activities should form part of a general cleaning schedule for the area. In recent years in situ vacuum systems (ring mains) have been introduced to facilitate the removal of dust from process and storage areas. With this system dust is removed to a central collection point through fixed pipework connected via hosing to hand-held suction devices.

Personal protection

This aspect subdivides into the following areas:

(a) medical supervision of exposed personnel for early detection of respiratory conditions, supported by annual health screening by occupational health nurses, with referral to the occupational physician where appropriate;

(b) the supply, maintenance and use of personal protective equipment, which implies the provision of the correct type of respiratory protection according to the dust hazard involved, and which the operator should use all the time that he may be exposed to dust; also he should wear a one-piece boiler suit, cap and gloves;

(c) the provision of a high standard of welfare amenities, in particular showering and separate workwear and personal clothing storage facilities;

(d) the frequent training of management and operators in these procedures.

Dust explosions

Many solid particulates, particularly organic materials, in the right combination with air, will form an explosive mixture. Some particulates are relatively harmless in their traditional form, but when reduced to dust by grinding, sanding or other forms of size reduction or refining they can become highly explosive. In fact, some of the most serious dust explosions have been associated with dusts created during the processing of tea, sugar, starch and potato as well as metals such as zinc and aluminium. Other materials such as coal, wood, cork, grain and many plastics can form explosive dust clouds.

Although an intimate mixture of flammable dust and air may burn with explosive violence, not all mixtures will do so. There is a range of concentrations of the dust and air within which the mixture can explode, but above or below this range an explosion will not take place. The lowest concentration of dust capable of exploding is referred to as the lower explosive limit and the concentration above which an explosion will not take place as the upper explosive limit. Furthermore, the range of the explosive concentrations of a dust cloud is not solely a function of the chemical composition of the dust. The limits vary, inter alia, with the size and shape of the particles in the dust cloud.

For an explosion to take place there must be some form of ignition source available. This can be a hot surface, electrical spark, frictional spark or direct flame. The ignition temperature for sugar is 350°C, coal 610°C, wood 430°C, zinc 600°C, polystyrene 490°C and magnesium 520°C. The lower explosive concentration for sugar is 350 mg/m^3, coal 550 mg/m^3, wood 400 mg/m^3, zinc 4,800 mg/m^3, polystyrene 150 mg/m^3 and magnesium 200 mg/m^3.

There are several clearly defined stages of a typical factory dust explosion. The preliminary stage, similar to the situation where fine coal dust is thrown on to an open fire, is the typical 'flare-up', where there is a sudden release of flame for an instant. This can, however, be sufficient to raise locally deposited dust into suspension in air and cause a localised explosion. This primary explosion stage may not result in a great degree of damage but is sufficient to send pressure waves in all directions, causing further liberation into the air of deposited dust. The secondary explosion stage, which is much more devastating than the primary stage, follows quickly, resulting in extensive damage and often loss of life. Depending upon the layout of the premises and the presence of walls, which may act as temporary baffles, the secondary stage may take place as one great explosion or a series of lesser explosions in different parts of the premises.

Most dust explosions take place, however, in specific items of plant such as spray driers, cyclones, settling chambers, powder silos, pneumatic conveying equipment, grinding plant, disintegrators, milling plant and dust collection systems.

Precautions against dust explosions

The frequent removal of deposited dust by industrial vacuum cleaners is one of the most important strategies in preventing dust explosions. Moreover, dust-producing plant should be checked frequently for leakages. Items of plant such as evaporator driers, storage silos and bins, grain elevators, fluid beds and cyclones should be fitted with explosion reliefs, which minimise the devastation by relieving the explosive pressure to a safe area or to atmosphere. Explosion reliefs (vents) may take the form of lightweight panels installed at the top of evaporator driers, elevators and silos. The size of the explosion relief is related to the volume of the installation and its mechanical strength. There are several methods for calculating the size of explosion relief according to the type of installation and particulate under consideration.

E

In general, any explosion of a flammable mixture, whether dust or gaseous, which, when ignited in a confined space, reaches its maximum pressure in not less than 40 milliseconds, can be brought under control by methods which include suppression, venting, advance inheriting, isolation and automatic plant shutdown.

As a dust explosion is not an instantaneous occurrence but requires a definite time for the development of maximum pressure, it is possible, by the introduction of a suppressant, to arrest the rise of pressure before it reaches dangerous levels. The explosion suppression system in its simplest form consists of a detector, an electrical power unit and a number of suppressors (*see* Fig. 22.3).

Fig. 22.3 ◆ Explosion suppression using an explosion detector, electrical power unit and a hemispherical suppressor

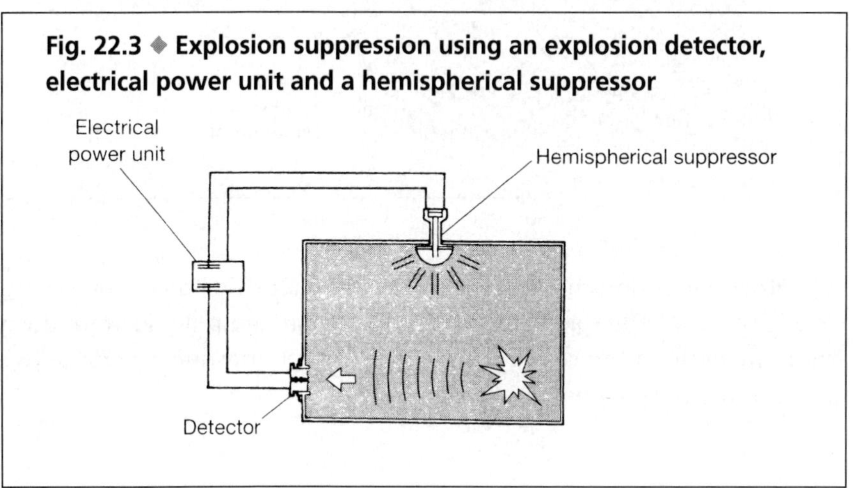

An explosion detector and the associated electrical equipment may also be used to open detonator-operated bursting discs, to close high-speed isolation valves, to inert automatically parts of the plant remote from the seat of the explosion and to shut down the plant immediately an explosion occurs. These methods may be used individually but more often are used in combination, depending upon the type and construction of plant and its operating conditions.

Although the fitting of explosion reliefs may prevent devastation of plant by an explosion this may not be sufficient to stop flame or smouldering material from spreading elsewhere through rotary valves, worms, conveyors or other inlets or outlets for the plant. The use of an explosion detector to initiate inerting and isolating arrangements coupled with automatic plant shutdown, therefore, offers an important additional degree of safety which it is often difficult, if not impossible, to achieve in any other way (*see* Fig. 22.4).

Fig. 22.4 ◆ Explosion venting using a detonator operated bursting disc

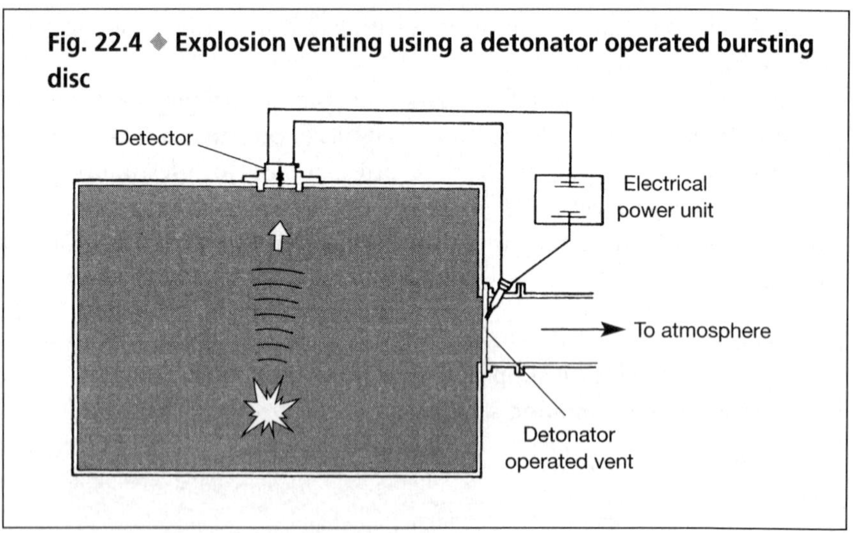

Other precautions include the installation of baffle walls in processing areas to prevent the spread of explosion, regular damping down of dusty areas, enclosure of processes and the use of dust arrestment plant appropriate to the type of dust produced.

Fumes

Fume is formed by the vaporisation or oxidation of metals. Typical metallic fumes encountered in industry are lead fume and welding fume, each of which creates ill-effects following inhalation.

Lead fume

Environmental control of fume should include damping of process and raw materials, the control molten lead well below 500°C – the temperature at which fume is produced – and the use of dust and fume control equipment. Any lead process which emits dust and fume should be enclosed and maintained under negative pressure by an enclosing hood, or a hood fitted as close as possible to the source of emission with a capture velocity not less than 1.0 m/s. Fume must be treated before discharge to atmosphere. Dust and fume arrestment plant should incorporate cyclone dust arrestors for the removal of coarse particles, and fabric filters or high-efficiency wet scrubbers for fine dust and fume produced.

Control measures should be supported by meticulous levels of cleaning and housekeeping, environmental and biological monitoring, strict control over personal hygiene and welfare amenities, and personal protection measures to prevent the contamination of the body and clothing worn by process workers.

The Control of Lead at Work Regulations 1998, together with the ACOP, aim to protect people at work exposed to lead by controlling such exposure. These Regulations apply to any work which exposes people to lead. 'Lead' includes alloys, any compounds of lead and lead as a constituent of any substance or material, which is liable to be inhaled, ingested or otherwise absorbed. Duties on the employer include assessing exposure to lead and the provision of adequate controls, including adequate washing arrangements, through to the provision of respiratory protective equipment and medical surveillance. Control measures, respiratory protection and protective clothing provided must be maintained in an efficient state, in efficient working order and good repair. Employers must ensure that employees use the measures provided properly, report defects immediately and provide information, instruction and training in the use of such controls.

Control limits, which are equivalent to maximum exposure limits (MELs) under the COSHH Regulations, for exposure to lead in air are set out in Appendix 1 of the ACOP. These are:

Tetraethyl lead (as Pb) (8-hour time-weighted average)	0.10 mg m^{-3}
Lead (and all compounds other than tetraethyl lead) (8-hour time-weighted average)	0.15 mg m^{-3}

The ACOP draws attention to the fact that lead absorption will depend not only on the lead-in-air value. Factors such as composition, solubility and particle size are also relevant. Some departure from the lead-in-air standard is therefore permitted, subject to there being evidence that it is justified.

Under the Regulations adequate records must be maintained of assessments, maintenance of controls, air monitoring, medical surveillance undertaken and biological tests.

Welding fume

During welding a wide range of airborne particulates are produced according to the base metals and electrodes used. Inhalation of these fumes, gases and dusts may lead to the condition known as 'welder's lung'. Metallic fumes in the form of oxides are the main constituent of welding fume, together with dust and fumes from flux coatings and metals being welded. The action of heat and ultraviolet light during the process can lead to the evolution of ozone, carbon monoxide and oxides of nitrogen. These gases are harmful. Heavier particulate matter is also produced as smoke and metal spatter. Most spasmodic welding operations are relatively safe because the fumes are readily diluted by fresh air in the workshop. A serious situation can develop, however, where welding is carried out in confined or unventilated areas.

The following control measures are recommended wherever welding is to be undertaken regularly. Welding workshops should be provided with mechanical ventilation capable of achieving six to ten air changes per hour. Local exhaust ventilation should be provided at the point of fume production to supplement general ventilation. Portable extraction and filtration units should be used when welding is undertaken *in situ* on production plant. Where welding is carried out in confined spaces, a permit to work system should be operated, together with a system for environmental monitoring. Welders should know the composition of different welding materials in use and any new materials introduced, together with fumes, dust and gases which could be evolved during the welding process. They must also understand the need for different forms of respiratory protection.

23 Radiation and radiological protection

E

The structure of matter

All matter is composed of elements such as hydrogen, lithium, carbon, iron, lead and oxygen. Elements consist of characteristic atoms, which contain a relatively small nucleus and a number of electrons. The nucleus contains protons, which carry positive electric charges, and neutrons, which carry no charge. The electrons are negatively charged and may be imagined as encircling the nucleus, most frequently within shells with indefinite boundaries. Atoms are relatively empty structures and symbolic diagrams, such as Fig. 23.1, cannot fully convey this.

Fig. 23.1 ◆ Diagram of a lithium atom

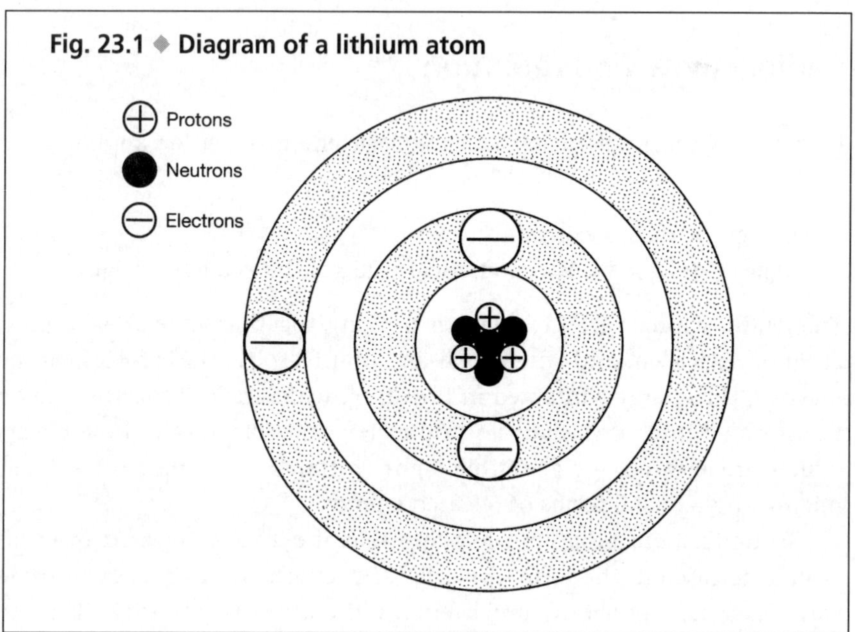

An atom contains equal numbers of protons and electrons and is electrically neutral. It is the number of protons, called the atomic number, that characterises an element. The atomic number of lithium, for instance, is three, and of oxygen, eight. Atoms of the same or different elements combine to form uncharged entities called molecules. Thus one atom of oxygen combines with two atoms of hydrogen to form a molecule of water. In the wave way, two atoms of oxygen combine to form one molecule of oxygen.

The mass of an atom is concentrated in the nucleus and the number of protons plus neutrons is called the mass number. Most species of atom can be characterised by the atomic number and mass number, or simply by the name of the element and the mass number. Thus characterised, they are called nuclides. Lithium-7, as Fig. 23.1 shows, is a nuclide with three protons, the unique atomic number of the element, lithium, plus four neutrons. Carbon-12 is a nuclide with six protons plus six neutrons. Lead-208 is a nuclide with 82 protons plus 126 neutrons.

Nuclides of an element that have different numbers of neutrons are called isotopes of that element. Hydrogen, for instance, has three isotopes – hydrogen-1, hydrogen-2, called deuterium, and hydrogen-3, called tritium. Iron has ten isotopes from iron-52 to iron-61.

Radioactivity and radiation

Radiation is a form of energy, and the basic energy equation applies:

$$\begin{aligned} \text{E} &= mc^2 \\ \text{Energy} &= \text{mass} \times \text{constant}^2 \\ \text{Joules} &= \text{kg} \times 3 \times 10^8 \text{ (numerical value of the velocity of light)} \end{aligned}$$

Tremendous forces are involved in holding together even the simplest atom of hydrogen, and when such an atom is 'split', these forces can be released, the energy contained in the atom, which held it together, being transformed into heat, light and other forms of radiation. These effects which are created – e.g. visible light, infrared radiation, ultraviolet, microwaves – are all forms of released energy.

To understand radiation, imagine that the energy released from an atom is released in the form of waves. The length and frequency of these waves depend upon how much energy the atom is releasing. It is the length and frequency of the waves that control the form of energy and the effect it can have on the human body, and it is possible to list these energy types in a table indicating increasing wavelength. This is called the electromagnetic spectrum (*see* Table 23.1).

Table 23.1 ◆ The electromagnetic spectrum

Radiation	*Frequency (Hz)*	*Wavelength*	*Energy*	*Radiation sources*
Gamma	10^{21}	Short	High	Cosmic sources
X-ray	10^{18}	↑	↓	Atomic struck by high-energy particles
Ultraviolet light				Excited gases
Visible light	10^{15}			Hot bodies
Infrared	10^{12}–10^{14}			Hot bodies
Microwaves generator	10^{9}			Microwave
Radio waves	-10^{6}	Long	Low	Radio transmitter

The most important division in the spectrum is between ionising and non-ionising radiation. (An ion is defined as a charged atom or group of atoms.) Ionising radiation can produce chemical changes as a result of ionising molecules upon which it is incident. Non-ionising radiation, however, does not have this effect and is usually absorbed by the molecules on which it is incident, with the result that the material will heat up, as in the case of microwaves.

To assess the hazard associated with radiation, it is necessary to consider the type of radiation, the energy, the extent of penetration of tissue, and the duration of exposure. This is particularly so with particulate radiation which involves the emission of streams of particles, such as protons, neutrons and electrons. These kinds of particulate radiation are often involved in spontaneous emissions from radioactive materials.

Forms of radiation

Alpha particles

An alpha particle may be said to consist of two protons and two neutrons bound together. It is therefore heavy and double charged. These are helium nuclei, i.e. helium atoms that have lost their two orbiting electrons and are, therefore, positively charged. Though they are ejected from the nucleus of the radioactive substance with considerable energy, they are relatively large particles and are easily absorbed by matter.

A nucleus that ejects an alpha particle will form a new nucleus with a mass number reduced by four and an atomic number reduced by two. For example, radium ejects an alpha particle and becomes an inert gas, radon, as follows:

$$^{226}_{88}\mathrm{Ra} \rightarrow {}^{222}_{86}\mathrm{Rn} + {}^{4}_{2}\mathrm{He}$$

(In this equation the subscript denotes the atomic number of the nucleus while the superscript denotes the mass number.)

Beta particles

These are electrons, not from the orbiting electrons of the atom but from within the nucleus. They are ejected with great speed and have a range of up to 15 cm in air. A nucleus that ejects a beta particle will form a new particle with the same mass number but with one added to its atomic number. For example, thorium ejects a beta particle and becomes proactinum, thus:

$$^{234}_{90}\mathrm{Th} \rightarrow {}^{234}_{91}\mathrm{Th} + {}^{0}_{-1}\mathrm{e}$$

Gamma rays

These are very high-energy electromagnetic waves that are emitted at the same time as the alpha and beta particles. They are similar in nature to light but of very much shorter wavelength. They are very highly penetrating, being capable of passing through several centimetres of lead.

X-rays

Similar to gamma rays, X-rays are emitted from metals when bombarded with high-energy electrons. They are produced by changes in the energy state of planetary electrons (e^-). As with gamma rays, X-rays are a discrete quantity of energy, without mass or charge, that are propagated as a wave.

Neutrons

A neutron is an elementary particle with unit atomic mass and no electric charge. The most powerful source of neutrons is the nuclear reactor.

Bremsstrahlung

These are weak X-rays, produced by negative beta particles impinging on heavy materials, i.e. those with a high atomic number.

Cosmic rays

These are fundamentally high-energy ionising radiations from outer space. They have a complex composition at the surface of the earth.

Radiation energy

It is customary to express the energy with which some radiations are produced in units of electron volt (symbol eV). This is equivalent to the energy gained by an electron passing through a potential difference of one volt. Multiples of this unit are frequently employed, especially one million or 10^6 electron volts (symbol MeV).

Sources of radiation

These sources can be split into non-ionising and ionising radiation sources.

Sources of non-ionising radiation

New processes and developments in industry continue to increase the uses of non-ionising radiation. Lasers are being used in industry and medicine more and more. (The word 'laser' broadly implies light amplification by stimulated emission of radiation.) These very high-energy beams of light are used widely in the construction industry for reference lines, since the beam is straight and does not spread like a torch beam. Lasers are also used for welding and cutting, and in surgery for the sealing of fine blood vessels. The organ most vulnerable to all sources of non-ionising radiation is the eye. A laser beam can destroy the retina, causing blindness. Use of the appropriate goggles can, in most cases, protect the operator from this danger, however.

Ultraviolet radiation is produced in arc welding and exposure may cause the condition known as 'arc eye'. As with laser, protection may be obtained by the use of goggles. Ultraviolet radiation also burns the skin and may cause cataracts and inflammation of the cornea. All workers exposed to ultraviolet radiation must wear the correct type of protective glasses.

Infrared radiation is emitted by all hot bodies, particularly radiant fires. Long-term exposure to even low doses of infrared radiation may damage the eyes, burning the lens and causing heat cataracts. Shields which reflect the heat may be installed as a form of protection. Personnel in frequent contact should wear the correct type of protective glasses.

Microwaves are emitted at extremely high radio frequencies. Typical examples of use are radar in military installations, certain drying processes and medical diathermy. It is relatively simple to shield such sources since microwaves cannot penetrate even thin metal. Microwave ovens have caused public concern. It is vital to ensure that the door fits perfectly and that the appliance is regularly maintained.

Sources of ionising radiation

Radioactive sources producing ionising radiation vary considerably. They include radioactive gases, certain rocks and dust, and X-ray machines. Small doses of radiation are received by wearers of luminised watches, though the watch face acts as a filter. Radioactive materials and X-rays are used mainly in medicine. X-rays are used increasingly in industry, for example in thickness gauges and in checking welded joints. The treatment of various types of cancer frequently involves radium therapy which may expose hospital staff. Workers in the nuclear power industry may be exposed to radiation, especially during the mining of uranium ore and in the reprocessing of spent fuel.

Radioactive decay

The quantity of a radionuclide is described by its activity, the rate at which spontaneous decay occurs in it, expressed in becquerels (Bq). One becquerel corresponds to the decay of one radionuclide per second. (Activity was formerly expressed in curies, i.e. the radioactivity of one gram of radium, namely 3.7×10^{10} disintegrations or nuclear transformations per second.) Thus 1 curie = 3.7×10^{10} becquerels.

The time taken for the radioactivity of a radionuclide to lose half its value by decay is called the half-life (symbol $t_{1/2}$). Each radionuclide has a unique and unalterable half-life. For carbon-14 it is 5,730 years, for barium-140, 12.8 days. Values for radionuclides range from fractions of a second to millions of years. In successive half-lives, the activity of a radionuclide is reduced by decay to ½, ¼, ⅛, etc., so that it is possible to predict the activity remaining at any future time.

Units of measurement

Rontgen (R) is the unit of measurement of a material's absorption of gamma or X-radiation.

Rad is 'radiation absorbed dose', and is the quantity of ionising radiation energy absorbed per unit mass. The unit for absorbed dose has undergone various changes in definition and numerical value over the years. Whilst the rontgen and rad are actually different in concept and magnitude, they may be considered to be equivalent to 100 ergs/gram of water. The modern unit of absorbed dose is the gray, which corresponds to 1 J/kg, abbreviated Gy.

Rem means 'radiation equivalent man' and is a unit of biological damage to living tissue per unit weight. It is generally taken to mean the degree of damage due to one rad of alpha, beta or X-radiation. Each type of radioactive particle has its own 'relative biological effectiveness' (rbe) to destroy tissue by ionisation. The rem is equal to the number of rads multiplied by the relative biological effectiveness. The use of the rem for this purpose is being overtaken by that of the sievert (Sv). One rem is one rad multiplied by one for gamma or beta rays, by ten for alpha rays or neutrons, and by other factors for some other radiations, e.g. 'heavy recoil nuclei'. The sievert is similar, but is based on the gray rather than the rad.

1 Sv = 100 rems; 1 rem = 10 mSv

Definitions associated with radiation

Dose is a general term to indicate the quantity of radiation in relation to the duration of exposure.

Dose equivalent is the quantity obtained by multiplying the absorbed dose by a factor to allow for the different effectiveness of the various ionising radiations in causing harm to tissue, as discussed under 'rem' above. The unit is the sievert. For gamma rays, X-rays and beta particles, the factor is one, and the gray (Gy) and sievert are numerically equal. For alpha particles, the factor is 20, so that one Gy of alpha radiation corresponds to a dose equivalent of 20 Sv. The dose equivalent thus provides an index of risk of harm from exposure of a particular tissue to various radiations.

Effective dose equivalent (EDE) is the quantity obtained by multiplying the dose equivalents to various tissues and organs by a risk weighting factor appropriate to each and adding the various fractions. The sum of the weighted dose equivalents is the dose equivalent to the whole body that would yield the same overall risk. Typical risk weighting factors are testes and ovaries – 0.25, thyroid – 0.03, breast – 0.15 and red bone marrow – 0.12. For instance, if a radionuclide causes irradiation of the liver and lungs the following calculation would apply. If the dose equivalents to the liver and lungs are respectively 70 mSv and 100 mSv, the effective dose equivalent is calculated thus:

$$(70 \times 0.06) + (100 \times 0.12) = 4.2 + 12.0 = 16.2 \text{ mSv}$$

The effective dose equivalent is, therefore, the dose equivalent weighted for the susceptibility to harm of different tissues.

Collective dose equivalent is the quantity obtained by multiplying the average effective dose equivalent by the number of persons exposed to a given source of radiation. It is expressed in man-sieverts (symbol man-Sv)

Maximum permissible dose is the strict limit on the effective dose equivalent that a person may receive. For a radiation worker this is 50 mSv, and for a member of the public, 5 mSv. These values must be observed without regard to cost, and are designed to control the incidence of effects, such as cancer, that involve an element of probability.

The type, energy, radioactive half-life, relative biological effectiveness and any concentration in particular tissues or organs are all combined to give maximum permissible body burdens, which also take into account the differing radiosensitivities of the different body tissues.

Radiological protection

In considering radiological protection strategies it is necessary to distinguish between sealed and unsealed sources of radiation.

Sealed sources

As the term suggests, the source is contained in such a way that the radioactive material cannot be released, e.g. X-ray machines. The source of radiation can be a piece of radioactive metal, such as cobalt, which is sealed in a container or held in another material which is not radioactive. It is usually solid and the container and any bonding material are regarded as the source.

Unsealed sources

Unsealed sources may take a variety of forms – gases, liquids and particulates. Because they are unsealed, entry into the body is comparatively easy.

Criteria for radiological protection

The basic criteria for radiological protection rest on three specific considerations – *time, distance* and *shielding*. The principle is to ensure that no one receives a harmful dose of radiation:

(a) Radiation workers may be protected on a time basis by limiting the duration of exposure to certain predetermined limits.

(b) Alternatively, they may be protected by ensuring that they do not come within certain distances of radiation sources. This may be achieved by the use of restricted areas, barriers and similar controls. The Inverse Square Law applies in this case.

(c) They may be shielded by the use of absorbing material, such as lead or concrete, between themselves and the source to reduce the level of radiation to below the maximum dose level. The quality and quantity (thickness) of shielding varies for the radiation type and energy level and varies from no shielding through lightweight shielding (e.g. 1 cm thick Perspex) to heavy shielding (e.g. centimetres of lead or metres of concrete).

'Half thickness' is the thickness of shielding required to reduce incident ionising radiation to half intensity.

Ionising Radiations Regulations 1999

These extensive and specialised regulations deal with the general principles and procedures involving work with ionising radiation (Part II), arrangements for the management of radiation protection (Part III), designated areas (Part IV), classification and monitoring of persons (Part V), arrangements for the control of radioactive substances, articles and equipment (Part VI), together with the duties of employees and certain miscellaneous provisions (Part VII).

Important definitions

Accelerator means an apparatus or installation in which particles are accelerated and which emits ionising radiation with an energy higher than 1 MeV.

Classified person means:

(a) a person designated as such pursuant to regulation 20(1);

(b) in the case of an outside worker employed by an undertaking in Northern Ireland or in another Member State, a person who has been designated as a Category A exposed worker within the meaning of Article 21 of the Directive.

Contamination means the contamination by any radioactive substance of any surface (including any surface of the body of clothing) or any part of absorbent objects or materials or the contamination of liquids or gases by any radioactive substance.

Controlled area means:

(a) in the case of an area situated in Great Britain, an area which has been so designated in accordance with Reg 16(1);

(b) in the case of an area situated in Northern Ireland or in another Member State, an area subject to special rules for the purposes of protection against ionising radiation and to which access is controlled as specified in Article 19 of the Directive.

Dose means, in relation to ionising radiation, any dose quantity or sum of dose quantities mentioned in Schedule 4.

Dose assessment means the dose assessment made and recorded by an approved dosimetry service in accordance with Reg 21.

Dose constraint means a restriction on the prospective doses to individuals which may result from a defined source.

Dose limit means, in relation to persons of a specified class, the limit on effective dose or equivalent dose specified in Schedule 4 in relation to a person of that class.

Dose rate means, in relation to a place, the rate at which a person or part of a person would receive a dose of ionising radiation from external radiation if he were at that place being a dose rate at that place averaged over one minute.

Dose record means, in relation to a person, the record of the doses received by that person as a result of his exposure to ionising radiation, being the record made and maintained on behalf of the employer by the approved dosimetry service in accordance with Reg 21.

Ionising radiation means, the transfer of energy in the form of particles or electromagnetic waves of a wavelength of 100 nanometres or less or a frequency 15 of 3 × 10 hertz or more capable of producing ions directly or indirectly.

Overexposure means any exposure of a person to ionising radiation to the extent that the dose received by that person causes a dose limit relevant to that person to be exceeded or, in relation to Reg 26(2), causes a proportion of a dose limit relevant to any employee to be exceeded.

Practice means work involving:

(a) the production, processing, handling, use, holding, storage, transport or disposal of radioactive substances; or

(b) the operation of any electrical equipment emitting ionising radiation and containing components operating at a potential difference of more than 5kV, which can increase the exposure of individuals to radiation from an artificial source, or from a radioactive substance containing naturally occurring radionuclides which are processed for their radioactive, fissile or fertile properties.

Radiation accident means an accident where immediate action would be required to prevent or reduce the exposure to ionising radiation of employees or any other persons.

Radiation employer means an employer who in the course of a trade, business or other undertaking carries out work with ionising radiation and, for the purposes of Regs 5, 6 and 7, includes an employer who intends to carry out such work.

Sealed source means a source containing any radioactive substance whose structure is such as to prevent, under normal conditions of use, any dispersion of radioactive substances into the environment, but it does not include any radioactive substance inside a nuclear reactor or any nuclear fuel element.

Supervised area means an area which has been so designated by the employer in accordance with Reg 16(3).

Part II: General principles and procedures

Reg 5 – Authorisation of specified procedures

A radiation employer shall not, except in accordance with a prior authorisation granted by the HSE in writing, carry out the following practices:

(a) the use of electrical equipment intended to produce X-rays for the purpose of:
   (i) industrial radiography;
   (ii) the processing of products;
   (iii) research; or
   (iv) the exposure of persons for medical treatment:

(b) the use of accelerators, except electron microscopes.

An authorisation granted may be subject to conditions.

Reg 6 – Notification of specified work

A radiation employer shall not for the first time carry out work with ionising radiation unless at least 28 days before commencing that work he has notified the HSE of his intention to do so and has provided particulars specified in Schedule 2.

Reg 7 – Prior risk assessment, etc.

Before a radiation employer commences a new activity, he shall make a suitable and sufficient assessment of the risk to any employee or other person for the purposes of identifying the measures he needs to take to restrict exposure of that employee or other person to ionising radiation.

Reg 8 – Restriction of exposure

Every radiation employer shall, in relation to any work with ionising radiation that he undertakes, take all necessary steps to restrict so far as is reasonably practicable the extent to which his employees and other persons are exposed to ionising radiation. Restriction of exposure shall, so far as is reasonably practicable, be by:

(a) engineering controls and design features, together with safety features and warning devices;

(b) the provision of safe systems of work to restrict exposure;

(c) the provision of adequate and suitable personal protective equipment (including respiratory equipment).

Reg 9 – Personal protective equipment

Any personal protective equipment shall comply with any provision in the Personal Protective Equipment (EC Directive) Regulations 1992.

Reg 10 – Maintenance and examination of engineering controls etc. and personal protective equipment

A radiation employer shall ensure:

(a) that any such control feature or device is properly maintained;

(b) where appropriate, thorough examinations and tests of such controls, features or devices are carried out at suitable intervals.

Every radiation employer shall.ensure that all personal protective equipment is, where appropriate, thoroughly examined at suitable intervals and is properly maintained and that, in the case of respiratory protective equipment, a suitable record of that examination is made and kept for at least two years.

Reg 11 – Dose limitation

Every employer shall ensure that his employees are not exposed to ionising radiation to an extent that any dose limit is exceeded in one calendar year.

Reg 12 – Contingency plans

Where an assessment shows that a radiation accident is reasonably foreseeable, the radiation employer shall prepare a contingency plan designed to secure, so far as is reasonably practicable, the restriction of exposure to ionising radiation and the health and safety of persons who may be affected by such accident.

Part II: Arrangements for the management of radiation protection

Reg 13 – Radiation protection adviser

Every radiation employer shall consult such suitable radiation protection advisers as are necessary for the purpose of advising the radiation employer as to the observance of these Regulations.

Reg 14 – Information, instruction and training

Every employer shall ensure that:

(a) those of his employees who are engaged in work with ionising radiation are given appropriate training in the field of radiation protection and receive such information and instruction as is suitable and sufficient for them to know –

  (i) the risks to health created by exposure to radiation;

  (ii) the precautions which should be taken;

  (iii) the importance of complying with the medical, technical and administrative requirements of these Regulations;

(b) adequate information given to other persons who are directly concerned with the work with ionising radiation carried on by the employer to ensure their health and safety so far as is reasonably practicable;

(c) those female employees of that employer who are engaged in work with ionising radiation are informed of the possible risks arising from radiation to the foetus and to a nursing infant and of the importance of those employees informing the employer in writing as soon as possible –

  (i) after becoming aware of their pregnancy; or

  (ii) if they are breastfeeding.

Reg 15 – Co-operation between employers

Where work with ionising radiation is likely to give rise to the exposure to ionising radiation of the employee of another employer, the employers concerned shall co-operate by the exchange of information.

Part IV: Designated areas

Reg 16 – Designation of controlled or supervised areas

Every employer shall designate as a controlled area any area under his control which has been identified by an assessment as an area in which:

(a) it is necessary for any person to follow special procedures to restrict significant exposure or prevent or limit the probability and magnitude of radiation accidents or their effects; or

(b) any person working in the area is likely to receive an effective dose greater than 6 mSv a year or an equivalent dose greater than three-tenths of any relevant dose limit in respect of an employee aged 18 years and above.

An employer shall designate as a supervised area under his control:

(a) where it necessary to kept the conditions of the area under review to determine whether the area should be designated as a controlled area; or

(b) in which any person is likely to receive an effective dose greater than 1 mSv a year or an equivalent dose greater than one-tenth of any relevant dose limit in respect of an employee aged 18 years and over.

Reg 17 – Local rules and radiation protection supervisors

Every radiation employer shall, in respect of any controlled area and supervised area, make and set down in writing such local rules as are appropriate to the radiation risk and the nature of the operations undertaken in that area.

The radiation employer shall take all reasonable steps to ensure local rules are observed and brought to the attention of appropriate employees and other persons.

The radiation employer shall appoint one or more radiation protection supervisors to ensure compliance with these Regulations in respect of any area made subject to local rules.

Reg 18 – Additional requirements for designated areas

Every employer shall ensure that any designated area is adequately described in local rules and that:

(a) in the case of any controlled area –
   (i) the area is physically demarcated or, where this is not reasonably practicable, delineated by some other suitable means;
   (ii) suitable and sufficient signs are displayed in suitable positions indicating that the area is a controlled area, the nature of the radiation sources and the risks arising; and

(b) in the case of any supervised area, suitable and sufficient signs are displayed indicating the nature of the sources and the risks arising.

Only persons designated as classified persons shall enter and remain in a controlled area.

No person shall enter a controlled area unless he can demonstrate, by personal dose monitoring or other suitable measurements, that the doses are restricted.

Employers who undertake monitoring or measurements shall keep the results for a period of two years from the date they were recorded.

Where there is a significant risk of the spread of radioactive contamination from a controlled area, the employer shall make adequate arrangements to restrict, so far as is reasonably practicable, the spread of such contamination. The arrangements shall, where appropriate include:

(a) the provision of suitable and sufficient washing facilities;
(b) the proper maintenance of such facilities;
(c) the prohibition of eating, drinking or smoking or similar activity;
(d) the means for monitoring for contamination of any person, article or goods leaving a controlled area.

Reg 19 – Monitoring of designated areas

The levels of radiation in controlled and supervised areas shall be adequately monitored and the working conditions kept under review.

Monitoring equipment shall be properly maintained and adequately tested and examined, and its accuracy established before initial use under the supervision of a qualified person.

The employer shall make and maintain suitable records of monitoring and testing required above.

Part V: Classification and monitoring of persons

Reg 20 – Designation of classified persons

The employer shall designate as classified persons those of his employees who are likely to receive an effective dose in excess of 6 mSV per year or an equivalent dose which exceeds three-tenths of any relevant dose limit and shall forthwith inform those employees that they have been so designated.

A classified person must be 18 years and over and certified as fit for work with ionising radiation by an appointed doctor of employment medical adviser.

Reg 21 – Dose assessment and recording

Every employer shall ensure, in the case of classified persons, that an assessment of all doses received by such employees, is made and recorded.

The arrangements that the employer makes with the approved dosimetry service shall include requirements, for instance, that records must be made and maintained until the person has or would have attained the age of 75 years but in any event for at least 50 years from when they were made.

Regs 22–26

These regulations cover procedures in respect of:

(a) estimated doses and special entries in dose records;

(b) dosimetry for accidents, etc.;

(c) medical surveillance arrangements;

(d) investigation and notification of overexposure;

(e) dose limitation for overexposed employees.

Part VI: Arrangements for the control of radioactive substances, articles and equipment

Reg 27 – Sealed sources and articles containing or embodying radioactive substances

Whenever reasonably practicable, a radioactive substance shall be in the form of a sealed source. The design, construction and maintenance of any article containing or embodying a radioactive substance, including its bonding, immediate container or other mechanical protection, must be such as to prevent leakage:

(a) in the case of a sealed source, so far as is practicable; or

(b) in the case of any other article, so far as is reasonably practicable.

Suitable tests shall be carried out at suitable intervals to detect leakage of radioactive substances from any article. Records to be made and maintained for at least two years.

Regs 28 – 33

These Regulations deal with:

(a) procedures for accounting for radioactive substances;

(b) precautions for the moving, transporting or disposing of radioactive substances;

(c) notification of certain occurrences, e.g. releases into the atmosphere, spillages;

(c) losses, and evidence of stealing, of radioactive substances;

(d) investigation of occurrences;

(e) duties of manufacturers, etc. of articles for use in work with ionising radiation;

(f) equipment used for medical exposure.

E

Reg 33 – Misuse of or interference with sources of ionising radiation

No person shall intentionally or recklessly misuse or without reasonable excuse interfere with any radioactive substance or any electrical equipment to which these Regulations apply.

Part VII: Duties of employees and miscellaneous

Reg 34 – Duties of employees

An employee who is engaged in work with ionising radiation shall:

(a) not knowingly expose himself or any other person to ionising radiation to an extent greater than is reasonably necessary for the purposes of the work;

(b) exercise reasonable care while carrying out such work;

(c) make full and proper use of any personal protective equipment;

(d) forthwith report any defect in his personal protective equipment;

(e) take all reasonable steps to return after use the personal protective equipment to the accommodation provided;

(f) comply with any reasonable requirement imposed on him by his employer for the purposes of making the measurements and assessments required under the Regulations;

(g) present himself during working hours for such medical examination and tests as may be required;

(h) forthwith notify his employer if he has reasonable cause to believe that –

  (i) he or some other person has received an overexposure;

  (ii) an occurrence has occurred; or

  (iii) an incident has occurred.

Schedule 4

Limits on effective dose

Limits on effective dose (dose to the whole body) are:

(a) 20 mSv a year for employees aged over 18 years – in special cases, employers may apply a dose limit of 100 mSv in five years with no more than 50 mSv in a single year, subject to strict conditions;

(b) 6 mSv a year for trainees;

(c) 1 mSv for any other person, including members of the public.

An ACOP, together with HSE Guidance accompanies the regulations.

Radiological emergencies

With any process involving unsealed sources of radiation, a suitable emergency or contingency plan is needed (*see* Chapter 9). There is a special scheme called NAIR (National Arrangements for Incidents Involving Radioactivity) which can be implemented via the police if there is a risk to the public following an incident, such as the loss of a source. The principal radiological emergency is a major release of radioactivity from a nuclear reactor. Following such accidental release, the radioactivity would be dispersed from the power station in a spreading plume or cloud, and in a manner determined by the characteristics of the release coupled with the prevailing weather conditions.

Emergency plans are required by law and exist for each station in the

UK. They depend upon co-operation from the licensee, local authorities and emergency services. The administration of plans is facilitated by local liaison committees. Essentially the plans require:

(a) an early assessment of the magnitude and nature of the release;

(b) its dispersion in the environment;

(c) the doses that might arise.

Measurement and detection of radioactivity

Geiger-Muller counter

E

The operation of the instrument is based on the fact that ionising radiations produce electrical charges which can be detected. One 'click' of a Geiger counter corresponds to one atom of radioactivity being detected. The Geiger counter (Fig. 23.2) is an extremely sensitive instrument and can be used for monitoring background radiation to the extent that a dose of one-fiftieth (background) of the maximum permissible dose can be detected.

Fig. 23.2 ◆ Geiger counter

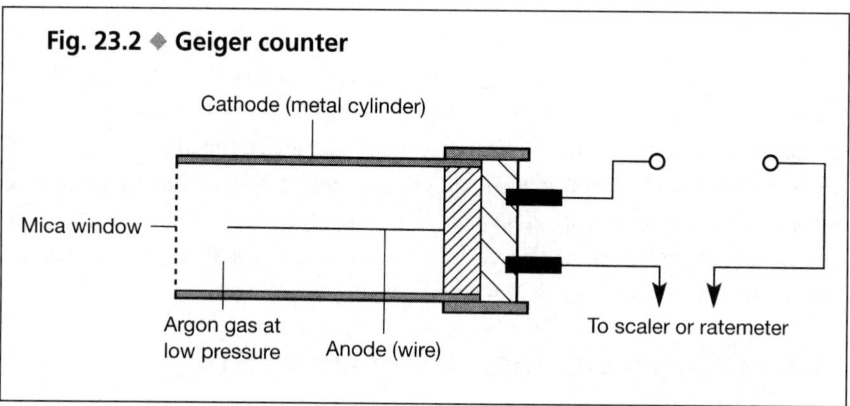

When radiation enters the tube, either through a thin window made of mica, or, if it is very penetrating, through the wall, it creates argon ions and electrons. These are accelerated towards the electrodes and cause more ionisation by colliding with other argon atoms. On reaching the electrodes, the ions produce a current pulse which is amplified and fed either to a scaler or a ratemeter. A scaler counts the pulses and shows the total received in a certain time. A ratemeter has a meter marked in counts per second (or minute) from which the average pulse rate can be read. It usually incorporates a loudspeaker which gives the characteristic click for each pulse. This instrument cannot deal with count rates in excess of 1,000 to 2,000 counts per second. It is very useful, however, for measuring low levels of radiation. It will not generally detect the presence of beta particles due to the relative thickness of the window.

Scintillation counter

This instrument measures radiation intensity and uses a screen of material which emits flashes of light when bombarded with alpha, beta, gamma and/or slow neutron radiation. These light flashes are converted to an electric current which increases as the bombardment increases. The current is amplified and indicated in the same way as with a Geiger-Muller counter.

Airborne sampler

As with airborne dust, certain radiations can be sampled on a filter paper using a high-volume air sampler. A known volume of air is drawn through a filter which is removed at the end of the sampling period and scanned for its radioactivity, using a counter.

Film badges

A film sensitive to radiation is housed in a specially designed plastic casing containing windows of various materials which shield certain kinds of radiation, but which allow others to pass through. This allows assessment of the various types of radiation. The device is worn during periods of exposure, one badge lasting a week. The film is developed and analysed for the accumulated dose of the various types of radiation, and a permanent record of the worker's personal exposure is produced.

Thermoluminescent personal dosemeter (TLD)

Some materials, such as lithium fluoride, can convert to an 'excited' state when bombarded with ionising radiation. This state is reversed only on the application of heat when the crystals return to normal, but with a measurable emission of light. Thus, a small badge containing these crystals can be used as a dosemeter, since the degree of irradiation can be related to the amount of light produced on heating. An advantage with this type of dosemeter is that it is small and its analysis can be quickly and automatically performed.

Quartz fibre detector (packet electrometer)

This detector consists of a metal cylinder with a loop of quartz fibre in the middle which is charged up to approximately 200 volts, electrical attraction or repulsion taking place between the fibre and the outside case. Ionising radiation will allow the charge to leak away slowly, permitting the loop to move over a scale back to the centre of the cylinder. This instrument will measure down to five millirems of dose and is ideal for use in short-term exposure situations, e.g. one to two weeks.

Assessment of information

E

Wherever radioactive sources or radio isotopes are under consideration for future use, as part of the assessment to ensure maximum safety, the following information/data must be known:

(a) The quantity of radiation, i.e. the activity in becquerels.

(b) The concentration of radioactivity in becquerels per gram.

(c) The types of radiation emitted – alpha, beta or gamma.

(d) The energy of each type of radiation in MeV.

(e) The penetrating power of the radiations, i.e. half-thicknesses or range in air, living tissue and the appropriate shielding materials (concrete, lead, etc.), though these depend upon (*c*) and (*d*) above.

(f) The radioactive half-life.

(g) The biological half-life, i.e. the time taken for the original level of radioactivity in the body to fall to exactly half.

(h) The relative biological effectiveness, which depends on (*c*) and (*d*) above.

(i) Any concentrations of the substance in particular tissues or organs after uptake in the body.

(j) The physical nature of the radioactive substance, i.e. solid, liquid or gas. (If solid, the degree of solubility is important, and if in liquid form, the vapour pressure.) Volatile liquids, gases and particulate solids are especially dangerous because of the inhalation risk. Where the substance can be absorbed through the skin, this fact should be known.

(k) The most appropriate monitoring devices for detecting the radiation, which will depend upon the types of radiation and energies involved.

(l) Maximum permissible body burdens and maximum permissible concentrations should be clear. Maximum permissible concentrations are listed, for instance, by the International Commission on Radiological Protection (ICRP) for air and water – i.e. air for breathing and drinking water – which are also derived from maximum permissible body burdens.

(m) The exact location, quantity and nature of all radioactive materials to be stored and used.

(n) The procedure for training personnel and the maintenance of dose records.

(o) The complete procedures in the event of accidental release together with safety drills necessary. These should include decontamination measures, warning systems and evacuation procedures.

The effects of exposure to radiation

The effects of a dose of ionising radiation vary according to the type of exposure, for instance, whether the dose was local, affecting only a part of the body surface, or general, affecting the whole body. Furthermore, the actual duration or length of time of exposure determines the severity of the outcome of such exposure.

Local exposure, which is the most common form of exposure, may result in reddening of the skin with ulceration in serious cases. Where exposure is local and the dose small, but of long duration, loss of hair, atrophy and fibrosis of the skin are known to occur.

The effects of acute general exposure range from mild nausea to severe illness, with vomiting, diarrhoea, collapse and eventual death. General exposure to small doses may result in chronic anaemia and leukaemia. The ovaries and testes are particularly vulnerable and there is evidence that exposure to radiation reduces fertility and causes sterility.

Apart from the danger of increased susceptibility to cancer, radiation can damage the genetic structure of reproductive cells, causing increases in the number of stillbirths and malformations. It is a maxim of radiological protection that all exposures must be the absolute minimum allowable. Dose equivalent limits for workers are:

(a) 0.5 Sv (50 rem) in a year for all tissues except the lens of the eye;

(b) 0.15 Sv (15 rem) in a year for the lens of the eye;

(c) 0.05 Sv (5 rem) in a year for uniform irradiation of the whole body or its equivalent if the body is not uniformly irradiated.

The limits of 0.5 and 0.15 Sv in a year are to prevent those effects which only occur above relatively high doses, e.g. erythema (reddening/inflammation of the skin) together with increased cancer and genetic damage risks.

Dose equivalent limits for members of the public are:

(a) 0.05 Sv (5 rem) in a year for all tissues;

(b) 0.005 Sv (0.5 rem) in a year for uniform irradiation of the whole body or its equivalent if the body is not uniformly irradiated.

Personal protective equipment

One-piece overalls or chemical grade suits are necessary when dealing with unsealed sources, together with respiratory protection in the form of breathing apparatus. However, thick overclothing will generally protect the individual from alpha and beta radiation provided dust respirators are worn and proper decontamination is carried out afterwards.

Disposal of radioactive wastes

Under the Radioactive Substances Act 1960, no person may dispose of, or accumulate for disposal, radioactive wastes except in accordance with an authorisation issued by the Department of the Environment. Where such disposal can safely be carried out, the conventional methods of waste disposal should be followed, i.e. discharge into sewers or disposal on refuse tips. In the case of refuse tips, the waste should be taken direct to the tip without passing through any preliminary processing or handling, and immediately buried under at least 1.5 metres of refuse or refuse ash.

Radiological protection agencies in the UK

The International Commission on Radiological Protection (ICRP)

The ICRP was established in 1928. Today its activities, which are mainly concerned with setting dosage standards, cover all aspects from individual exposure to population exposure.

Health Safety Commission and Health and Safety Executive

The HSC has overall responsibility for matters of public health and safety arising from industrial activities, and this includes radiation. The HSE enforces current legal requirements relating to ionising radiations in the workplace.

Nuclear Installations Inspectorate

Specific responsibility for nuclear installations rests with the Nuclear Installations Inspectorate, whose approval of any nuclear installation is a prerequisite for the granting of an operating licence. The Advisory Committee on Nuclear Installations advises the HSC on the siting, design, operation, etc., of nuclear installations.

National Radiological Protection Board (NRPB)

This board was set up in 1970 with the following functions:

(a) by means of research and otherwise, to advance the acquisition of knowledge about the protection of mankind from radiation hazards;

(b) to provide information and advice to persons, including Government Departments, with responsibilities in the UK in relation to the protection from radiation hazards either of the community as a whole or particular sections of the community.

The board maintains a register of all radiation workers, except those employed by the Ministry of Defence. In fulfilment of its statutory functions, the board produces reports from time to time reviewing the radiation exposure of the population.

24 Personal protection

E

Reference was made earlier, in Chapter 12, to the twin strategies of 'safe place' and 'safe person'. Safe place strategies characterise much legislation, particularly provisions relating to access and egress, machinery and plant, environmental factors and the structural safety of premises. Wherever possible, a safe place strategy is preferred to a safe person strategy.

Personal protection, implying the provision and use of various items of personal protective equipment – e.g. safety boots, ear defenders – should be considered as a last resort when all other measures have failed, or purely as an interim form of protection until the hazard can be eliminated at source or by a form of safe place strategy. As a form of protection it relies too heavily on, first, the individual wearing the equipment, and second, wearing it correctly and for the full length of time that he may be exposed to the hazard, such as dust or noise. Above all, many forms of personal protection are uncomfortable and/or inconvenient to wear, particularly over long periods. It is, therefore, an imperfect solution for preventing occupational ill-health or personal injury. However, total safety can rarely be achieved. The technology for ensuring total protection may not have been developed, the cost of such protection may be out of proportion to the risks involved or particularly in situations where workers frequently move from one location to another, as with welding operations, total protection may be impracticable. There is, therefore, a need for personal protective equipment with certain tasks, which is recognised in statutes and specific regulations.

Personal Protective Equipment at Work Regulations 1992 (PPEWR)

These Regulations:

(a) amend certain Regulations made under the HSWA which deal with personal protective equipment, so that they fully implement the European Directive in circumstances where they apply;

(b) cover all aspects of the provision, maintenance and use of personal protective equipment at work in other circumstances;

(c) revoke and replace almost all pre-HSWA and some post-HSWA legislation which deals with personal protective equipment.

Specific requirements of current Regulations dealing with personal protective equipment, namely the Control of Lead at Work Regulations 1980, the Ionising Radiations Regulations 1985, the Control of Asbestos at Work Regulations 1987, the Control of Substances Hazardous to Health (COSHH) Regulations 1999, the Noise at Work Regulations 1989 and the Construction (Head Protection) Regulations 1989 take precedence over the more general requirements of the Personal Protective Equipment at Work Regulations.

Reg 2 – Interpretation

Personal protective equipment means all equipment (including clothing affording protection against the weather) which is intended to be worn or held by a person at work and which protects him against one or more risks to his health and safety, and any addition or accessory designed to meet this objective.

Reg 3 – Disapplication of these Regulations

1. These Regulations shall not apply to or in relation to the master or crew of a sea-going ship or to the employer of such persons in respect of normal shipboard activities of a ship's crew under the direction of a master.

2. Regs 4 to 12 shall not apply in respect of personal protective equipment which is:

(a) ordinary working clothes and uniforms which do not specifically protect the health and safety of the wearer;

(b) an offensive weapon within the meaning of sec 1(4) of the

Prevention of Crime Act 1953 used as self-defence or deterrent equipment;

(c) portable devices for detecting and signalling risks and nuisances;

(d) personal protective equipment used for protection while travelling on a road;

(e) equipment used during the playing of competitive sports.

Reg 4 – Provision of personal protective equipment

1. Every employer shall ensure that suitable personal protective equipment is provided to his employees who may be exposed to a risk to their health and safety while at work except where and to the extent that such risk has been adequately controlled by other means which are equally or more effective.

2. Similar provisions as above apply in the case of self-employed persons.

3. Personal protective equipment shall not be suitable unless:

(a) it is appropriate for the risk or risks involved and the conditions at the place where exposure to the risk may occur;

(b) it takes account of ergonomic requirements and the state of health of the person or persons who may wear it;

(c) it is capable of fitting the wearer correctly, if necessary after adjustments within the range for which it is designed;

(d) *so far as is practicable*, it is effective to prevent or adequately control the risk or risks involved without increasing overall risk.

Reg 5 – Compatibility of personal protective equipment

1. Every employer shall ensure that where the presence of more than one risk to health or safety makes it necessary for his employees to wear or use more than one item of personal protective equipment, such equipment is compatible and continues to be effective against the risk or risks in question.

2. Similar provisions as above apply in the case of self-employed persons.

Reg 6 – Assessment of personal protective equipment

1. Before choosing any personal protective equipment which he is required to provide, an employer or self-employed person shall ensure an assessment is made to determine whether the personal protective equipment he intends to provide is suitable.

2. The assessment shall include:
   (a) an assessment of any risk or risks which have not been avoided by other means;
   (b) the definition of the characteristics which personal protective equipment must have in order to be effective against the risks referred to above, taking into account any risks which the equipment itself may create;
   (c) comparison of the characteristics of the personal protective equipment available with the characteristics referred to in (b) above.
3. The assessment shall be reviewed if:
   (a) there is reason to suspect that it is no longer valid; or
   (b) there has been a significant change in the matters to which it relates.

   Where, as a result of any such review, changes in the assessment are required, the relevant employer or self-employed person shall ensure that they are made.

Reg 7 – Maintenance and replacement of personal protective equipment

1. Every employer shall ensure that any personal protective equipment provided to his employees is maintained (including replaced or cleaned as appropriate) in an efficient state, in efficient working order and in good repair.
2. Similar provisions as above apply in the case of self-employed persons.

Reg 8 – Accommodation for personal protective equipment

Every employer and every self-employed person shall ensure that appropriate accommodation is provided for personal protective equipment provided when it is not being used.

Reg 9 – Information, instruction and training

1. Where an employer is required to ensure personal protective equipment is provided to an employee, the employer shall also ensure that the employee is provided with such information, instruction and training as is adequate and appropriate to enable the employee to know:
   (a) the risk or risks which the personal protective equipment will avoid or limit;

(b) the purpose for which and the manner in which the personal protective equipment is to be used;

(c) any action to be taken by the employee to ensure that the personal protective equipment remains in an efficient state, in efficient working order and in good repair.

2. The information and instruction provided shall not be adequate and appropriate unless it is comprehensible to the persons to whom it is provided.

Reg 10 – Use of personal protective equipment

1. Every employer shall take all reasonable steps to ensure that any personal protective equipment provided to his employees is properly used.

2. Every employee shall use any personal protective equipment in accordance with both any training in the use of the personal protective equipment concerned which has been received by him and the instructions respecting that use which have been provided to him.

3. Every self-employed person shall make full and proper use of any personal protective equipment.

4. Every employee and self-employed person shall take all reasonable steps to ensure that it is returned to the accommodation provided for after use.

Reg 11 – Reporting loss or defect

Every employee who has been provided with personal protective equipment shall forthwith report to his employer any loss of or obvious defect in that personal protective equipment.

HSE guidance

These Regulations are accompanied by extensive guidance on the selection, use and maintenance of personal protective equipment, including a 'Specimen risk survey table for the use of personal protective equipment' (*See* Fig.24.1.)

Fig. 24.1 ◆ Specimen risk survey table for the use of personal protective equipment

			Risks																				
			The PPE at Work Regulations 1992 apply except where the Construction (Head) Protection Regulations 1989 apply										The CLW, IRR, CAW, COSHH and NAW Regulations[(1)] will each apply to the appropriate hazard										
			Mechanical					Thermal															
			Falls from a height	Blows, cuts, impact, crushing	Stabs, cuts, grazes	Vibration	Slipping, falling over	Scalds, heat, fire	Cold	Immersion	Non-ionising radiation	Electrical	Noise	Ionising radiation	Dust fibre	Fume	Vapours	Splashes, spurts	Gases, vapours	Harmful bacteria	Harmful viruses	Fungi	Non-micro biological antigens
P	Head	Cranium																					
A		Ears																					
R		Eyes																					
T		Respiratory tract																					
S		Face																					
		Whole head																					
of	Upper limbs	Hands																					
the		Arms (parts)																					
	Lower limbs	Feet																					
B		Legs (parts)																					
O	Various	Skin																					
D		Trunk/abdomen																					
Y		Whole body																					

(1) The Control of Lead at Work Regulations 1980, The Ionising Radiations Regulations 1985, The Control of Asbestos at Work Regulations 1987, The Control of Substances Hazardous to Health Regulations 1988, The Noise at Work Regulations 1989.

The choice and use of personal protective equipment

A systematic approach is needed to ensure that workers at risk are properly protected. The main components of this approach are choice, introduction and use, maintenance, and the system for monitoring the effectiveness of personal protective equipment.

Choice

When considering the type and form of equipment to be provided, the following factors are relevant:

(a) the needs of the user in terms of comfort, ease of movement, convenience in putting on, use and removal, and individual suitability;

(b) the number of personnel exposed to a particular hazard, e.g. noise;

(c) the type of hazard, e.g. fume, dust, molten metal splashes, etc.;

(d) the scale of the hazard;

(e) standards representing recognised 'safe limits' for the hazard, e.g. British Standards, HSE guidance notes or codes of practice, in-house codes of practice (here, it should be understood, a standard, code of practice, or guidance note is not a legal instrument and cannot be enforced by the enforcing authorities. Their recommendations can, however, have legal consequences, particularly in criminal cases. *See* Chapter 4);

(f) specific Regulations currently in force;

(g) specific job restrictions or requirements, e.g. work in confined spaces, roof work;

(h) the presence of environmental stressors such as extremes of temperature, inadequate lighting and ventilation, background noise;

(i) ease of cleaning, sanitisation, maintenance and replacement of equipment and/or its component parts.

E

Introduction and use

Users must be educated as to the reasons for personal protection, the nature of the protection provided and the correct way to wear the equipment. Moreover, where hazards are assessed on the basis of exposure over a period of time, e.g. noise and airborne particulates, the level of protection afforded will reduce dramatically with only partial use. This is particularly true in the case of exposure to noise levels in excess of 90 dBA, and workers must appreciate this fact. Other ways of promoting and reinforcing positive attitudes to the wearing of personal protective equipment are:

(a) involving workers in the choice and selection of equipment;
(b) requiring management and visitors to wear/use the personal protection in a hazardous environment or situation (it should be remembered that the General Duties of the Health and Safety at Work etc., Act 1974 apply to all levels of management and, therefore, failure to observe them, e.g. not wearing ear protection in a designated ear protection area, can result in a criminal offence);
(c) the use of safety propaganda such as safety posters, displays and demonstrations of the equipment to be used.

Maintenance

An effective maintenance system should ensure the routine repair and maintenance of personal protective equipment where this may be required, e.g. breathing apparatus. It may also be necessary to provide a sanitisation facility for some equipment, e.g. safety spectacles, ear muffs. (Provision and maintenance of personal protective equipment is both a common law and a statutory requirement, giving rise to both civil and criminal liability.)

Monitoring

There should be a permanent system for assessing the effectiveness, use and maintenance of equipment. Users should be encouraged to report any difficulties or problems.

Types of personal protective equipment

The following are the main categories of personal protective equipment.

Head, face and neck protection

Helmets: These protect the head from falling objects or overhead hazards. They should meet the British Standard BS 5240: *Specification for General Purpose Industrial Safety Helmets*. Helmets should be strong but light. If they are too heavy, they will cause discomfort, particularly if worn for long periods. Individual head protection should be identified by means of a helmet numbering system or by the use of name tabs on helmets.

The Construction (Head Protection) Regulations 1989 impose requirements for the provision of suitable head protection for, and the wearing of suitable head protection by, persons at work on building operations or works of engineering construction within the meaning of the FA. Reg 3 requires an employer to provide suitable head protection of his employees, and the self-employed must provide their own. Head protection must be maintained and replaced by the employer whenever necessary, and employees must report to their employers losses of or defects in their head protection.

Reg 4 imposes requirements on employers and others having control over workers, e.g. contractors and sub-contractors, to ensure that head protection is worn and Reg 5 allows rules to be made or directions given as to the wearing of head protection in specified circumstances.

'Suitable head protection' means head protection which:

(a) is designed to provide protection, so far as is reasonably practicable, against foreseeable risks of injury to the head to which the wearer may be exposed;
(b) after any necessary adjustment, fits the wearer;
(c) is suitable having regard to the work or activity in which the wearer may be engaged.

Employees and self-employed persons engaged in construction work are required to wear suitable head protection, 'unless there is no foreseeable risk of injury to his head other than by his falling'.

It should be noted that prosecutions of individuals by HSE inspectors were taken immediately following the introduction of the Regulations.

By virtue of sec 11 of the Employment Act 1989, the Regulations (and any other legal requirements for the provision and use of head protection on construction sites) do not require a Sikh who is wearing a turban to wear head protection.

Caps and hair nets: These prevent the hair coming into contact with moving machinery, e.g. the terrible 'scalping' accidents when long hair becomes entangled in the bits and chucks of vertical drilling machines. Long-haired workers, both male and female, should, where circumstances demand, be provided with a form of snood attachment, which completely encloses the hair. To refuse to wear such protection is a dismissible offence (*Marsh* v. *Judge International Housewares Ltd* [1976] COIT No. 511/57, Case No. 23119/76).

Face shields: These protect the area from the forehead to the neck against flying particles and splashes, i.e. metal dusts and chips, molten metal, glass splinters and chemicals. They are also used in spot welding. Manufactured in clear or tinted plastic or polycarbonate, they can be provided with chin guards to provide protection from upward splashing of molten metals or chemicals.

Eye protection

Eye protection takes a number of forms:

Safety spectacles: These may have toughened glass or plastic lenses with plastic or metal frames. Lenses should not be removable as they could fall out. Spectacles can be supplied with side shields and should be fitted for the individual concerned, particularly if he has prescription lenses.

Safety goggles: These are generally cheaper and more versatile than spectacles. Users can experience discomfort when they are worn for long periods, however. Some designs can be worn over prescription lenses but this, too, may cause discomfort. Cup-type goggles protect against flying particles, welding glare or radiation, whereas wide vision goggles (specific to purpose) give protection against flying particles, welding glare, radiation, dust, fumes and splashes.

Shields: Face shields, which can be hand-held, fixed to the helmet or strapped to the head, protect the face and eyes. The shield can also be fixed between the operator's head and his job.

Hearing protection

Under the Noise at Work Regulations 1989 employers must provide hearing protection to employees at their request, where they are likely to be exposed to the first action level (85 dBA) or above. Where employees are exposed to the second action level (90 dBA) or above, or to the peak action level or above, they must ensure that they are provided with hearing protection which, when properly worn, can reasonably be expected to keep the risk of damage to those employees' hearing to below that arising from exposure to the second action level or, as the case may be, the peak action level. When selecting it, it is important to ensure that the form of hearing protection – acoustic wool, ear plugs, ear muffs – will produce the necessary attenuation (sound pressure reduction) at the operator's ear (*see* Chapter 21).

The need for workers to wear hearing protection for long periods can be reduced by job rotation and reorganisation of tasks, so that part of the work may be done without exposure to noise, and short breaks during which exposed workers can relax in a quiet environment, perhaps in a sound haven. The principle should always be, however, to reduce noise at source, and the wearing of hearing protection regarded as an unavoidable last resort.

Problems associated with hearing protection

The wearing of hearing protection incurs the risk that accidents could occur if audible warning signals or human speech are not heard, or if audible malfunction of machinery or equipment is not noticed. This situation can easily arise, particularly with intermittent noise and, unless this hazard can be eliminated, the use of hearing protection may introduce dangers far worse than those of impaired hearing. The problem of disorientation, through wearing hearing protection for long periods, is a further risk which management must consider.

Forms of hearing protection

Hearing protection takes three principal forms:

Ear plugs: There are fitted into the auditory canal. They tend to move out of place with jaw movements. Ear plugs are manufactured in plastic, rubber, glass down or combinations of these materials.

Ear defenders, muffs and pads: These cover the whole ear and can reduce exposure by up to 50 dBA at certain frequencies. They can be uncomfortable in hot conditions and difficult to wear with safety spectacles or goggles. Many safety helmets incorporate a fitting for attaching ear defenders.

Ear valves: These are inserted into the auditory canal and, theoretically, allow ordinary conversation to continue while preventing harmful noise reaching the ear.

Respiratory protection

Use of respiratory protection (*see* Fig. 24.2) is essential wherever workers are exposed to dangerous concentrations of toxic or fibrogenic dusts, or fumes – e.g. from paint spraying – or where they may work in unventilated spaces. The correct selection and use of respiratory protection is absolutely

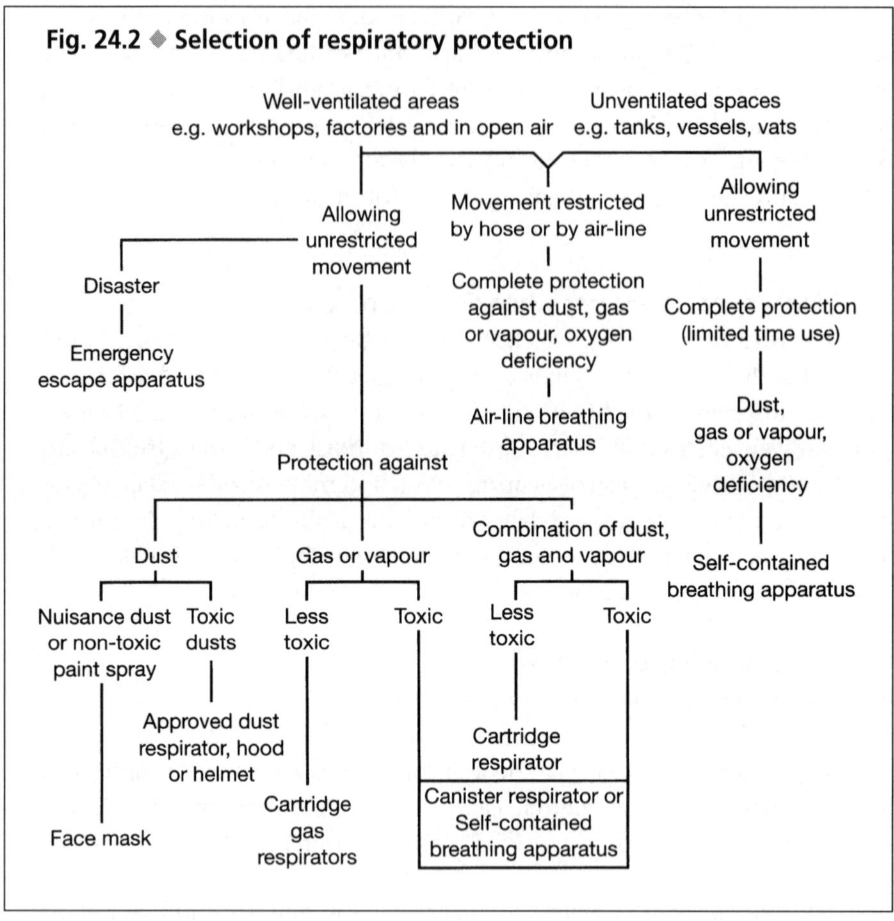

Fig. 24.2 ◆ Selection of respiratory protection

vital and a guide to the use of different forms of protection is shown above and in Fig. 24.3. Management should decide on the respiratory protection required in individual circumstances. The use of face masks (*see* page 435) is definitely not recommended as a means of protection against anything other than low concentrations of nuisance particulates and atomised liquids. Reference should be made to BS 4275: *Recommendations for the Selection, Use and Maintenance of Respiratory Protective Equipment* and to the quoted nominal protection factor for the equipment.

Nominal protection factor

BS 4275 refers to the selection of respiratory equipment and lists nominal protection factors for different forms of equipment. The nominal protection factor (NPF) measures the theoretical capability of respiratory protection and is calculated thus:

$$\text{NPF} = \frac{\text{Concentration of contaminant in the atmosphere}}{\text{Concentration of contaminant in the face-piece}}$$

Forms of respiratory protection

Face masks: These are a simple device for holding filtering media against the nose and mouth to remove coarse nuisance dust particles or non-toxic paint sprays. They should not be used for protection against dangerous or toxic substances.

General purpose dust respirators: These take the form of an ori-nasal face mask and a particulate filter to trap finely divided solids or liquid particles. BA 2091 *Specification for Respirators for Protection against Harmful Dusts, Gases and Scheduled Agricultural Chemicals* provides a classification and specification for such forms of general purpose respiratory protection. In addition, the HSE specify 'approved' types for certain industries. For industries involved in work with substances subject to specific legislation, e.g. asbestos, the HSE has instituted an approval scheme for respirators suitable for compliance with the hygiene provisions of that legislation. There are three groups, which are listed on forms available from HMSO, namely:

> F2486: 'Dust Respirators of All Sorts Suitable for Work with Asbestos, Lead and Other Dangerous Dusts.'
>
> F2500: 'Blasting Helmets (for Operations Involving Shot Blasting and Allied Operations).'
>
> F2506: 'Chemical Works Respirators.'

An approval scheme is dependent on a respirator manufacturer applying to the HSE and paying for the testing.

Positive pressure powered dust respirators: These comprise an ori-nasal face mask fitted to a power-driven pack carried on the individual and connected by a flexible hose. They are more efficient than the simple form of dust respirator, as they utilise a much more efficient filtering medium and operate with positive pressure in the face-piece. Their construction is specified in BS 4558: *Specification for Positive Pressure Powered Dust Respirators*.

Helmet-contained positive pressure respirators: This device provides head, eye, face and lung protection together with a high degree of comfort (*see* Fig. 24.3). It incorporates a helmet and visor with a high-efficiency axial fan mounted at the rear of the helmet, which draws the dust-laden air through

Fig. 24.3 ◆ Helmet-contained positive pressure respirator

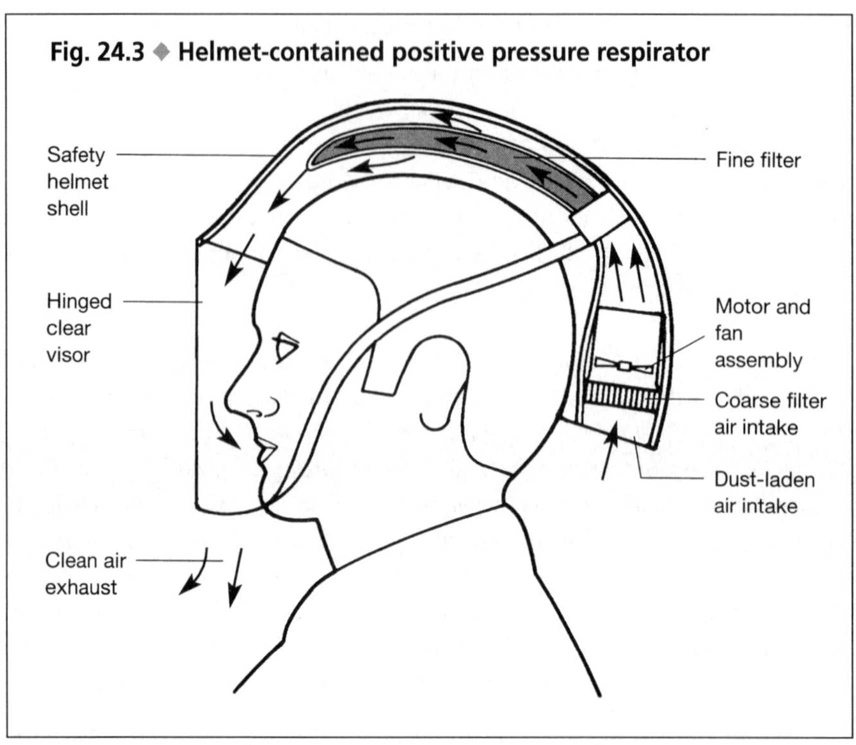

a coarse filter. The partially filtered air is then passed through a fine filter bag. The filtered air provides a cool pleasant airstream over the entire facial area, and is finally exhausted at the bottom of the visor at a flow rate sufficient to prevent dust entering the mouth or nose. The low voltage electrical power is supplied by a lightweight rechargeable battery pack, connected to the helmet by means of a flexible cable. This portable battery pack may be clipped to a belt or carried in an overall pocket.

Gas respirators: This respirator takes two forms, cartridge and canister. The cartridge respirator is similar to the dust respirator. It uses a chemical cartridge filter and is effective against low concentrations of relatively non-toxic gases or vapours which have an acceptable level of concentration exceeding 100 ppm. Canister respirators, on the other hand, are normally of the full facepiece type with exhalation valves, incorporating goggles and visor. They are connected to a chemical canister filter for protection against low concentrations of designated toxic gases or vapours. The manufacturer's instructions on the avoidance of cartridge/canister saturation, maximum periods of use in relation to concentrations of gas in air, shelf-life, etc., must in all cases be carefully followed. They are effective against toxic gases and vapours in limited concentrations. A particulate filter can be incorporated to remove dust particles.

Emergency escape respirators: These are specially designed respirators using a chemical filter which will enable persons to escape from dangerous atmospheres in an emergency. They are intended for very short-term use and should never be used for normal industrial protection.

Air-line breathing apparatus: This apparatus consists of a full face mask or half mask connected by flexible hose either to a source of uncontaminated air (short distance) or to a compressed air-line via a filter and demand valve. The apparatus is usually safe for use in any contaminated atmosphere (see manufacturer's stated nominal protection factor) but is limited by the length of the air-line, which also places some restriction on movement. When using a fresh-air hose, a pump is necessary for lengths over 10 metres. Reference should be made to BS 4667 *Breathing Apparatus*.

Self-contained breathing apparatus: This appliance can be of the open or closed circuit type. The open circuit type supplies air by a lung-governed demand valve or pressure reducer connected to a full face-piece via a hose supply. The hose is connected to its own compressed air or oxygen supply which is carried by the wearer in a harness. The closed circuit type incorporates a purifier to absorb exhaled carbon dioxide. The purified air is fed back to the respirator after mixing with pure oxygen. Both types of apparatus may be used in dangerous atmospheres or where there is a deficiency of oxygen, or for rescue purposes from confined spaces. British Standard 4667 is relevant in the case of self-contained breathing apparatus:

NOTE. The UK is a multiracial society, yet many respirators are designed for the European face. Negroid and Asiatic faces are not normally catered for, and this can leave gaps in the fit between face and mask. Beards, moustaches and sideburns can also create gaps in the fit or reduce the efficiency of fit. Thus, in certain cases, it may be necessary to provide certain people with positive pressure powered dust respirators as opposed to general purpose dust respirators.

Skin protection

Where industrial dermatitis develops it is often due to failure by the individual to wash off immediately any chemicals from the skin. Personal cleanliness is of utmost significance in many processes. For instance, cotton waste or rag, if used to wipe off oils from the skin, should not be kept in the trouser pocket owing to the risk of scrotal cancer. In *Stokes* v. *GKN Sankey Ltd* [1968] 1 WLR 1776, the defendant employer was held vicariously liable for the failure of the company doctor to warn the plaintiff employee of the danger arising from rubbing oil on his overalls, i.e. scrotal cancer. The plaintiff later contracted cancer and brought a successful suit. (For 'vicarious liability' *see* Chapter 2.)

Chemical substances causing dermatitis include strong acids and alkalis, chromates and bichromates, formaldehyde, organic solvents, resins, certain adhesives, suds, degreasing compounds and lubricants. Paraffin and trichlorethylene remove the natural fats from the skin and render it liable to damage by other substances. Where dermatitis is identified, medical aid should be sought. The principal cause of dermatitis is poor personal hygiene. The presence, however, of dermatitis is often a first indication of exposure to dangerous substances, in particular primary irritants and secondary cutaneous sensitisers (*see* Chapter 19). As an additional precaution, therefore, the use of barrier creams is recommended.

A range of barrier creams is available to meet varying work conditions. They provide skin protection in wet conditions, and for workers handling acids, alkalis and other potentially dangerous substances. The barrier cream must be applied to the hands and forearms before work commences. Failure on the part of employees to use a barrier cream may well mean that, if they contract dermatitis, they are not entitled to damages against their employer (*Clifford* v. *Challen & Sons Ltd* [1951] 1 AER 72).

Barrier creams should be used only as an *aid* to protection and should not be regarded as a protection against powerful skin damage agents. The greatest care should be taken to avoid contact with corrosive chemicals. The use of barrier creams before work generally dispenses with the need to use skin cleansers other than soap. Where skin cleansers are used, these should be as weak as possible, consistent with effective cleansing, and should be completely removed from the skin by washing with soap and water. The use of perfumed barrier creams in catering and food preparation can result in taint in the foods under preparation.

Body protection

This classification includes one-piece and two-piece overalls, donkey jackets, aprons, warehouse coats, etc., which can be of the washable, disposable or semi-disposable (short life) type. In selecting the correct body protection, the following factors are relevant:

(a) the degree of personal contamination from the task or process, e.g. dust, oil, general soiling;
(b) the level of hygiene control necessary for the product, e.g. food manufacture;
(c) whether a wet or dry process is involved;
(d) the ease and cost of washing or dry cleaning;
(e) individual preferences shown for specific types of protection (*see above* and Chapter 5);
(f) the degree of exposure to temperature and humidity variations;
(g) possible discomfort produced by moisture/sweating when wearing impervious garments;
(h) ease of storage.

It is now common practice for employers to provide overalls, donkey jackets, foul weather clothing, Wellington boots, etc., for the use of employees whilst at work. Company rules should prohibit employees returning home in such equipment due to the risk of contamination from processes, dangerous substances and other agents being taken into public transport and the home. Moreover, this would reduce the losses of protective garments and increase the life of such garments.

Hand and arm protection

A wide variety of hand and arm protection is available, from general purpose fibre gloves to PVC fabric gauntlets and sleeves, depending upon the hazards. Persons exposed to risk of injury through handling operations, or exposure to dangerous substances, or through the use of hand tools, such as knives, chisels, hammers and strapping equipment, should be provided with protection. A recent development has been a range of chain-mail protective wear for personnel using knives, particularly in abattoirs, meat-packing establishments and butchers' shops.

A chart outlining the principal hazards to hands, the typical operations involved, suitable protective materials and suggested glove types is incorporated in *The Hand Book* published by RoSPA (Hamilton, 1983).

Leg and foot protection

Many injuries are caused as a result of poor standards of leg and foot protection; others are caused because management fails to provide protection or workers fail to wear it. Foot protection should be waterproof, resistant to acids, alkalis, oils and other substances, and incorporate steel toecaps where workers handle items such as drums, crates or other heavy containers. Safety footwear should incorporate a safety toecap to either Grade 1 or Grade 2 standard under BS 1870 *Safety Footwear*. Where safety boots are provided, e.g. for drivers and maintenance fitters, the toecap should be to Grade 1 standard. With wet processes, Wellington boots provided should incorporate protective fins for ankle, instep and leg protection, a deep tread patterned outsole and internal steel toecap. Moreover, the provision of gaiters for workers exposed to flying sparks and molten metal during casting operations is well established in law and practice in the UK. Where, on construction sites, workers are likely to tread on sharp objects, such as nails projecting through timber, consideration should be given to safety footwear incorporating steel sole inserts in addition to safety toecaps.

Safety belts, harnesses and lanyards

All three items must conform to BS 1397 *Industrial Safety Belts, Harnesses and Safety Lanyards*. A safety harness must incorporate simply fitted quick action buckles and fixed rings positioned in a manner which avoids body contact. Harnesses should be adjustable over the range of body sizes, and should incorporate quickly adjustable waist and shoulder straps, fixed leg straps, a safety line and two canvas pockets for carrying tools and other items. This requirement is particularly relevant in the case of self-employed or contract window cleaners working on high-rise buildings, e.g. office blocks. The design, construction and materials used should minimise adverse effects on the body in the event of a fall. The safety line should not exceed 1.83 metres in length and should be of double thickness tubular-woven high-elasticity nylon cord with a minimum breaking strength of 35 kg/cm^2, attached to a belt by a 'D' ring and with a steel Karabinier safety hook at the anchorage end. Safety harnesses are recommended in preference to safety belts.

Safety belts should only be considered as a safeguard when it is impracticable to wear the full harness. Belts should carry a record card indicating the date of the last inspection by a responsible employee.

Safety lanyards are used when workers may be moving within a fixed area above ground level, e.g. on a mobile access tower or loading stage. They incorporate a two-metre length of webbing contained in a spring-loaded drum. One end of the webbing is hooked to a fixed anchorage point, the other end to the worker's belt. The webbing extends from and retracts to the fixed drum as the worker moves. In the event of a fall, the belt locks, the nylon webbing stretching slightly to cushion the operator as the fall is arrested.

Eye and face washes, emergency showers

Wherever workers are exposed to contact with strong acids and alkalis, an emergency eye and face wash facility is necessary. This can be an emergency eye and face irrigation bottle, a permanently installed eye and face wash point, particularly in laboratories, or a larger combined emergency shower unit and eye wash station adjacent to a bulk chemical storage area. A safety shower can be stationed in an internal or external location. It delivers a large quantity of water quickly in order to prevent chemical burns.

25 Occupational hygiene practice

Occupational hygiene

Hygiene

One definition of hygiene is 'the science of the preservation of health' (*see* Chapter 17 on this subject). People need to protect their health in terms of the food they eat (food hygiene) and the air they breathe (atmospheric hygiene). Occupational hygiene, therefore, is concerned with the preservation of health whilst at work.

Occupational hygiene

Occupational hygiene has been defined as 'the identification, measurement and control of contaminants and other phenomena, such as noise and radiation, which would otherwise have unacceptable adverse effects on the health of people exposed to them' (Annual Report of HM Chief Inspector of Factories, 1973). It is concerned with the monitoring and control of the working environment to ensure that contaminants are kept to as low a level as is reasonably practicable and, in all cases, to a level that is above the appropriate hygiene standard, i.e. below the exposure limit. The provision of a safe working environment forms part of the legal duty of all employers (HSWA, sec 2). (*See* Chapter 3.)

Principles and practice of occupational hygiene

Wherever environmental contamination is present, or suspected to be present, the following sequence of operations is necessary.

Recognition and identification

The recognition and subsequent identification of the specific contaminant – e.g. dust, fume, gas, vapour, mist, virus, sound pressure level – is the first stage in the sequence. A number of spot check devices are used such as detector stain tubes for gases, coated slides for dust or, in the case of noise, a sound pressure level meter.

Measurement

Once the contaminant has been identified, it is necessary to measure the extent of the contamination. A wide range of equipment is available for the accurate measurement of environmental contamination.

Evaluation

Evaluation is an important part of the procedure. Measured levels of contamination must be compared with existing hygiene standards (always assuming there is such a standard applicable to the material in question), such as occupational exposure limits (maximum exposure limits and occupational exposure standards). (*See* Chapter 18.) In addition, the duration and frequency of exposure to the contaminant must be taken into account. Following a comprehensive evaluation, a decision must be made as to the actual degree of risk to workers involved. This degree of risk will determine the control strategy to be applied (*see* Chapter 26).

Control

Prevention of exposure or the implementation of a specific control strategy is the last of the four stages of occupational hygiene practice. In extreme cases it may be necessary to implement a total prohibition strategy; in less severe situations the risk to the health of workers may be eliminated by the installation of a system of exhaust ventilation.

Measurement techniques

Stress in the working environment can be created by the presence of dust, gases, fumes and vapours, extremes of lighting, temperature, ventilation and humidity or by the presence of high sound pressure levels, radiation or vibration. Whatever the cause of the stress, there must be a system of control to ensure adequate protection for the worker. Whilst systems for the measurement of noise and radiation are dealt with in Chapters 21 and 23, this chapter is principally concerned with the measurement of airborne particulates, a process which involves the taking of air samples for measurement purposes (*see* Fig. 25.3).

Air sampling

Air sampling can be undertaken on either a short-term or a long-term basis.

Short-term sampling techniques (grab and snap sampling)

Grab sampling implies taking an immediate sample of air and, in most cases, passing it through a particular chemical reagent which responds to the contaminant being monitored. The simplest device for the measurement of concentrations of gases and vapours is the hand pump and bellows device (*see* Fig. 25.1) which incorporates a specific detector tube.

Fig. 25.1 ◆ Multi-gas detector

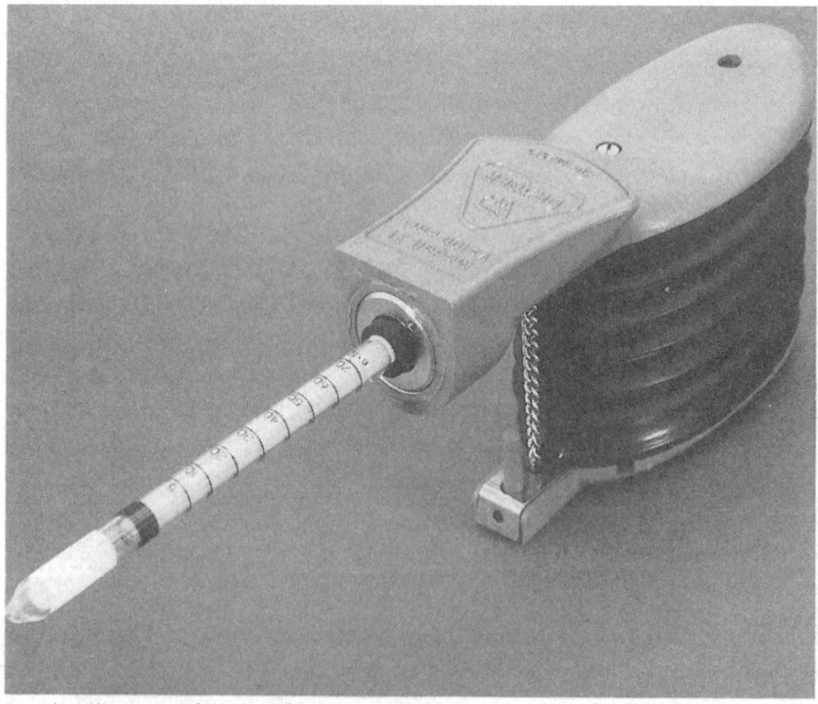

Reproduced by courtesy of Draeger Safety

The detector tube is a glass tube, sealed at both ends, and filled with porous granules of an inert material, such as silica gel. The granules are impregnated with a chemical reagent which changes colour in the presence of the contaminant gas. In order to detect and measure this gas, the ends are broken off the tube and the tube inserted into the tube holder of the sampling pump. The correct volume of air is drawn through the tube (*see* Fig. 25.2), according to the number of times the bellows are depressed, and the resultant colour stain indicates the presence of the gas. The actual concentration can be determined either by the length of the stain, which increases proportionally with the concentration, or by comparing the intensity of the colour with a prepared standard. The sampling pump is small, light and hand-operated. The combination of pump and detector tube provides a convenient method of on-the-spot evaluation of atmospheric contamination.

Fig. 25.2 ◆ Gas and vapour detector tubes

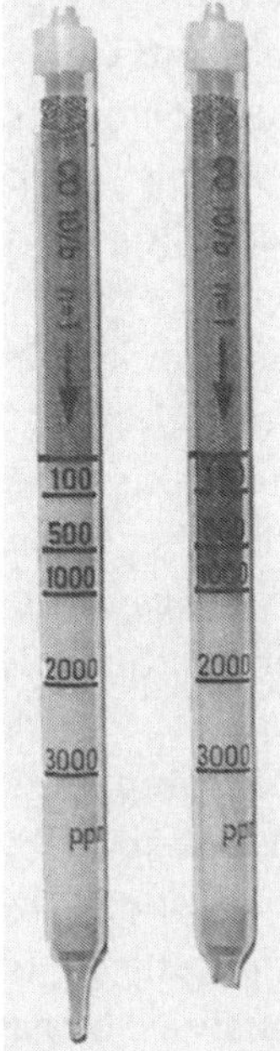

Reproduced by courtesy of Draeger Safety

The hand pump and bellows device is a useful instrument for the early detection of atmospheric contamination. Users do not require extensive training and it can be used relatively cheaply in random sampling exercises or routine day-to-day monitoring. However, its accuracy must always be suspect owing to the problems of cross-sensitivity of detector tubes, tubes being used that have gone past the expiry date or even the operator failing to give the appropriate number of pump strokes when operating the bellows. Therefore, when frequent monitoring must be undertaken, an automatic system should be provided. Furthermore, normal stain tubes will only give 'point in time' results and, dependent on circumstances, may not give an adequate picture of major fluctuations in concentration across a working shift.

Long-term sampling techniques

Instruments which carry out long-term sampling are, broadly, of two types, personal samplers and static samplers.

Personal sampling instruments: There are several kinds of personal sampling instruments, e.g. gas monitoring badges, filtration devices and impingers. With gas monitoring badges, the worker wears a badge containing a solid sorbent or a chemically impregnated carrier and the air sample comes into contact with the badge by diffusion. The results can be read directly by a colour change or determined by analytical instruments.

Filtration devices comprise a low-flow or constant-flow sample pump, which is motor-operated from a rechargeable battery, and a sampling head, which incorporates a specific filter, attached close to the operator's breathing zone. Many dusts, mists, etc., can be collected by passing a known volume of air through a filter, the pore diameter of which is selected to remove the chemical hazard completely. The quantity of dangerous material collected may be determined gravimetrically (weighing the filter before and after collection) or by solvent extraction and analysis by gas chromatography, atomic absorption, etc. Filtration methods are particularly useful in the sampling of large particles or aggregates of particles.

The impinger method is used for collecting such chemical compounds as acetic anhydride, hydrogen chloride, etc. The impinger is a specially designed glass bubble tube. A known volume of air is bubbled through the impinger containing a liquid medium chosen to react chemically to, or physically to dissolve, the contaminant. The liquid is then analysed by gas chromatography, spectrophotometry, etc. An impinger operates in conjunction with a constant-flow sample pump. It may be mounted on the side of the sample pump, which is worn on the operator's belt, or in a holster near the breathing zone.

Another method of personal sampling is with the aid of a sorbent tube, which is a method for collecting a large percentage of the hazardous chemical vapours in a work environment. A glass sample tube is used, normally filled with two layers of a solid adsorbent capable of completely removing chemicals from the air. The tube has breakable end tips. To collect a sample, the end tips are broken and known volume of air is drawn through the tube. Airborne chemicals are trapped by the first adsorbent layer, with the back-up layer assuring removal of all the chemicals from the air. The tube is then sealed with a push-on cap prior to analysis. The tube may be inserted into a tube holder, located in the breathing zone of the operator, which is connected to a constant-flow pump, or direct into the pump by means of a short extension piece.

Static samplers: These are devices stationed in the working area. They sample continuously over the length of a shift, or longer period if necessary. Mains or battery-operated pumps are used. They can sample contaminants which may be present in the general atmosphere in very small quantities but which nevertheless may be dangerous. Such pumps can handle large or small quantities of air per minute and pass it through a variety of sampling devices. For obtaining samples of harmful dusts such as asbestos or silica, filters of various ranges of porosity are used. Where the particle size of the dust is significant, size-selective filters are used. Moreover, when the substance to be sampled is volatile, e.g. a solvent vapour, or must be analysed in solution, the contaminant must be trapped on to a suitable medium, such as activated charcoal, or in an absorbing liquid, for subsequent laboratory analysis.

Long-term stain detector tubes are now available for this purpose. The device is connected to a constant-flow pump and air is drawn through the tube at a steady rate. Examination of the detector tube at the termination of the sampling period will indicate the amount of contaminant absorbed, which is directly related to the average level of contamination present over the period.

A number of direct monitoring devices are also available for the detection and measurement of gases and vapours. They operate on several different principles, e.g. infrared absorption, and give an instant read-out on a chart, meter or display. In many cases they can be linked to an alarm device which sounds once a particular concentration of gas or vapour reaches a predetermined level.

Monitoring strategies

The COSHH Regulations 1999 require the employer to undertake air monitoring in certain cases. Fig. 25.3 'Monitoring strategies flow diagram' indicates a step-by-step approach to monitoring recommended by the HSE.

Fig. 25.3 ◆ Monitoring strategies flow diagram

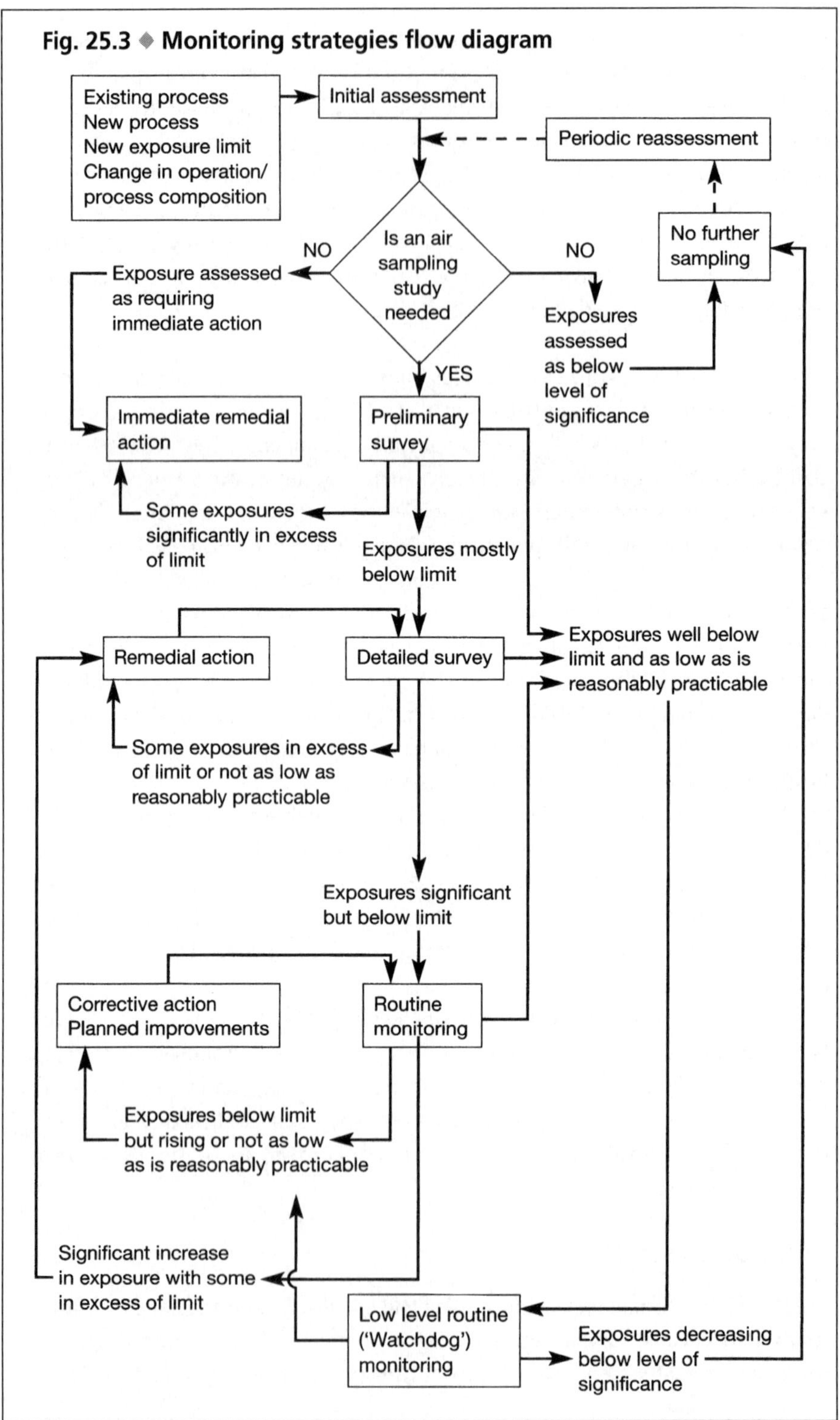

Conclusion

The control of occupational health hazards is a specific duty laid on the employer under HSWA and the COSHH Regulations. As indicated in Chapter 12 'Principles of accident prevention', a 'safe place' strategy should always take precedence over a 'safe person' strategy. Fig. 25.4 summarises an organisation's approach to dealing with health hazards.

Fig. 25.4 ◆ Standard approach to occupational hygiene

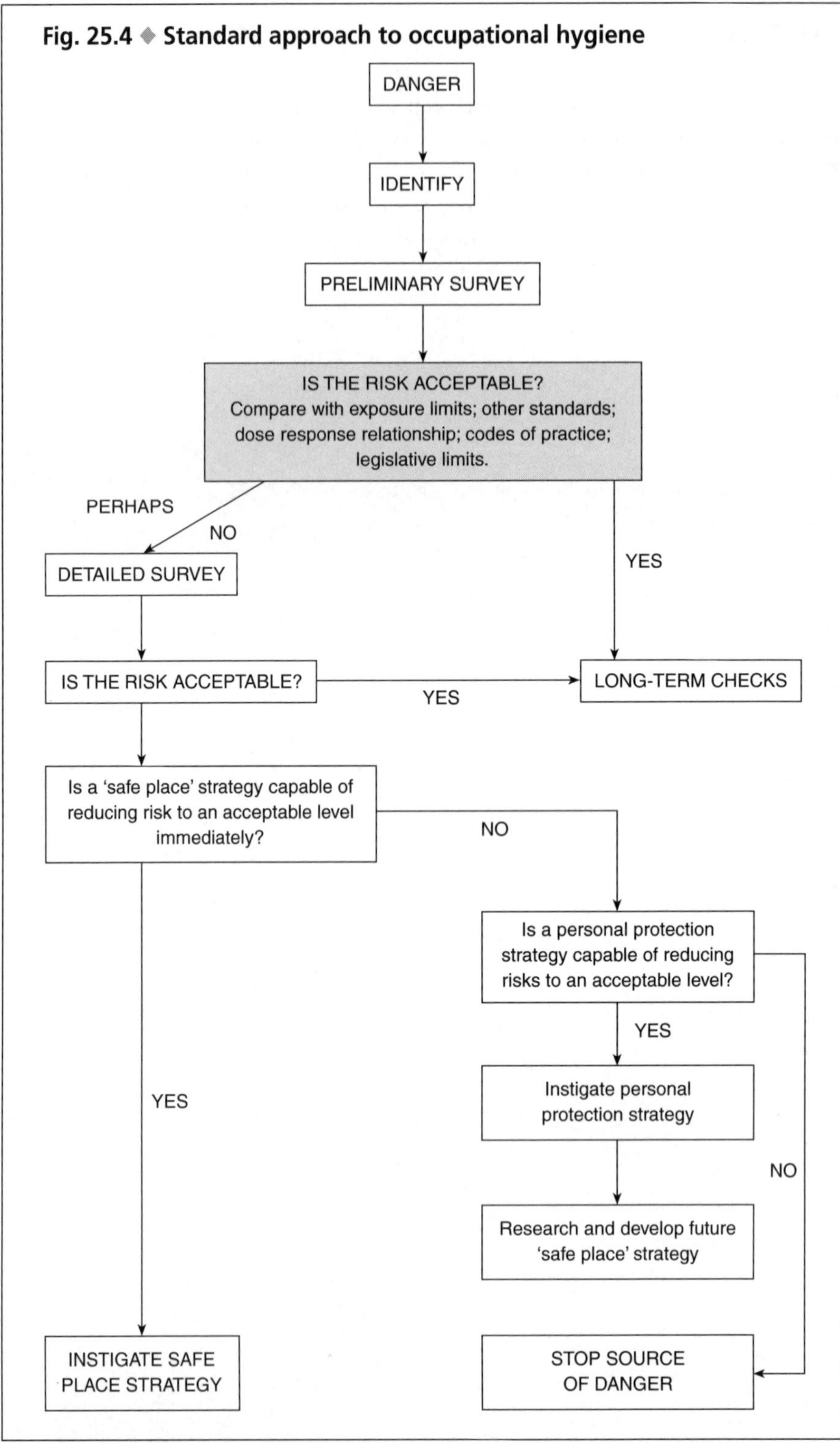

26 Prevention and control strategies in occupational hygiene

Occupational health practice and occupational hygiene are closely related, their basic objective being the prevention or reduction of stress in the working environment, the outcome of which can be occupational ill-health and disease.

E

Principles of control

HSWA clearly defines the duty of the employer to provide a safe working environment (HSWA, sec 2(2)(*e*); see Chapter 3). The COSHH Regulations greatly reinforce this duty under Reg 7. Reg 7(1) requires every employer to 'ensure that the exposure of his employees to substances hazardous to health is either prevented or, where this is not reasonably practicable, adequately controlled'. Moreover, under Reg 7(2) 'so far as is reasonably practicable, the prevention or adequate control of exposure of employees to a substance hazardous to health shall be secured by measures other than the provision of personal protective equipment'. In other words, personal protective equipment must never be considered as the sole means of controlling exposure to hazardous substances. Under Reg 7(3) it should be used, in certain cases, in addition to control measures, particularly where such measures do not prevent or provide adequate control of exposure to substances hazardous to health.

In the prevention and control of health risks, many approaches are available, depending upon the severity and nature of the offending agent. The principle prevention and control strategies, together with support strategies, are outlined below.

Prevention strategies

Prohibition

This is the most extreme form of prevention strategy applicable and can form the basis for a prohibition notice (see Chapter 4). As a strategy, prohibition characterises much of the legislation relating to known carcinogens, as well as other hazards where there is no form of protection for the operator. Prohibition, therefore, implies a total ban on a particular system, substance or the operation of a practice where the danger level is unacceptable.

Elimination

Reviews of the needs of specific processes often reveal potentially hazardous substances whose use is no longer necessary. Such substances should be eliminated thereby negating the need for a form of control.

Substitution

The substitution of a less toxic material in place of a more highly toxic one is a frequently used strategy. Typical examples are the substitution of toluene for benzene and trichloroethane 1,1,1 in place of carbon tetrachloride. Wherever possible, substances with an assigned Maximum Exposure Limit (MEL) detailed in HSE Guidance Note EH 40 should be substituted by substances not quoted in the list of MELs. Health risk assessments undertaken to comply with the COSHH Regulations will, in many cases, identify safer substances which can be substituted for the more dangerous ones.

Control strategies

Enclosure/containment

This strategy is based on the containment of an offending agent or environmental stressor to prevent its liberation into the working environment. Total enclosure or containment of a process may be possible by the use of bulk tanks and pipework to deliver a liquid directly into a closed production vessel. Complete enclosure is practicable if the materials are in liquid form, used in large quantities, and where the range of materials is very limited. Enclosure may take a number of forms, e.g. acoustic enclosures for noisy machinery, dust enclosures, paint spray booths, laboratory fume cupboards (*see* Chapters 15 and 22).

Isolation/separation

The isolation of a process using potentially dangerous substances may simply mean relocating it in a controlled area, thereby separating the majority of the workforce from the risk. Alternatively, it could involve the construction of a chemical manufacturing plant in a remote geographical area. In the first case, well-established procedures for limiting access to only trained and authorised operators will be necessary.

NOTE. Both isolation and enclosure will present maintenance requirements and the operation of a permit to work system will be necessary to safeguard those who undertake this work.

Ventilation

Infiltration of air into buildings through openings in their fabric and even planned natural ventilation give no continuing protection wherever toxic gases, fumes, vapours, etc. are emitted from a process. Local exhaust ventilation (LEV) systems must therefore be operated. In certain cases (*see below*), dilution ventilation may be appropriate.

E

LEV systems

LEV systems (Fig. 26.1) are designed to intercept the contaminant as soon as it is generated and direct it into a system of ducting connected to an

Fig. 26.1 ◆ A typical LEV system

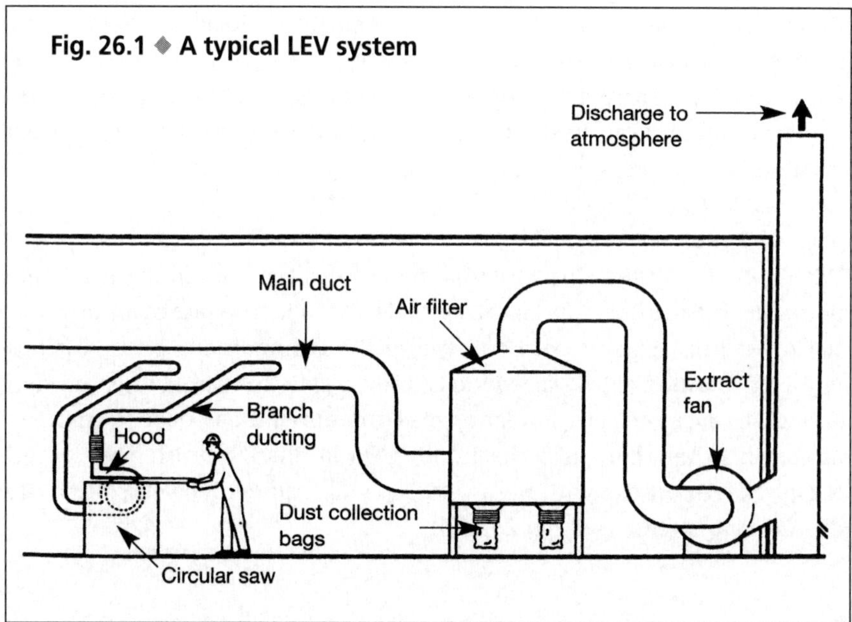

extract fan. They ensure that the contaminant is removed from the workplace before it can be inhaled. LEV systems incorporate a number of principal features, namely:

(a) a hood, enclosure or other inlet to collect and contain the offending agent close to the source of its generation;
(b) ductwork to convey the contaminant away from the source;
(c) a filter or other air-cleaning device to remove the contaminant from the extracted airstream; (NOTE. The filter should normally be located between the hood and the fan.)
(d) a fan or other air-moving device to provide the necessary airflow;
(e) further ductwork to discharge the cleaned air to the outside atmosphere at a suitable point.

HSE booklet HS(G)37 *An Introduction to Local Exhaust Ventilation* provides excellent guidance in the design and specification of LEV systems.

LEV systems may take a number of forms, as follows.

Receptor systems

In receptor systems the contaminant enters the system without inducement. The fan in the system is used to provide air flow to transport the contaminant from the hood through ducting to a collection system. The hood may form almost a total enclosure around the source, as with highly toxic contaminants, such as beryllium, or with radioactive sources; or it may form a partial enclosure, e.g. a spray booth in which all spraying takes place, or a laboratory fume cupboard. Generally, hoods which receive the contaminant air as it flows from its origin under the influence of thermal currents are receptors (*see* Fig. 26.2(*a*)).

Captor systems

With a captor system the moving air captures the contaminant at some point outside the hood and induces its flow into it. The rate of air flow into the hood must be sufficient to capture the contaminant at the furthest point of origin, and the air velocity induced at this point must be high enough to overcome any tendency the contaminant may have to go in any direction other than into the hood. Contaminants emitted with high energy (large particles with high velocity) will require high velocities in the capturing airstream (*see* Fig. 26.2(*b*)).

Fig. 26.2 ◆ (a) Receptor systems; (b) Captor systems; (c) Low-volume high-velocity system

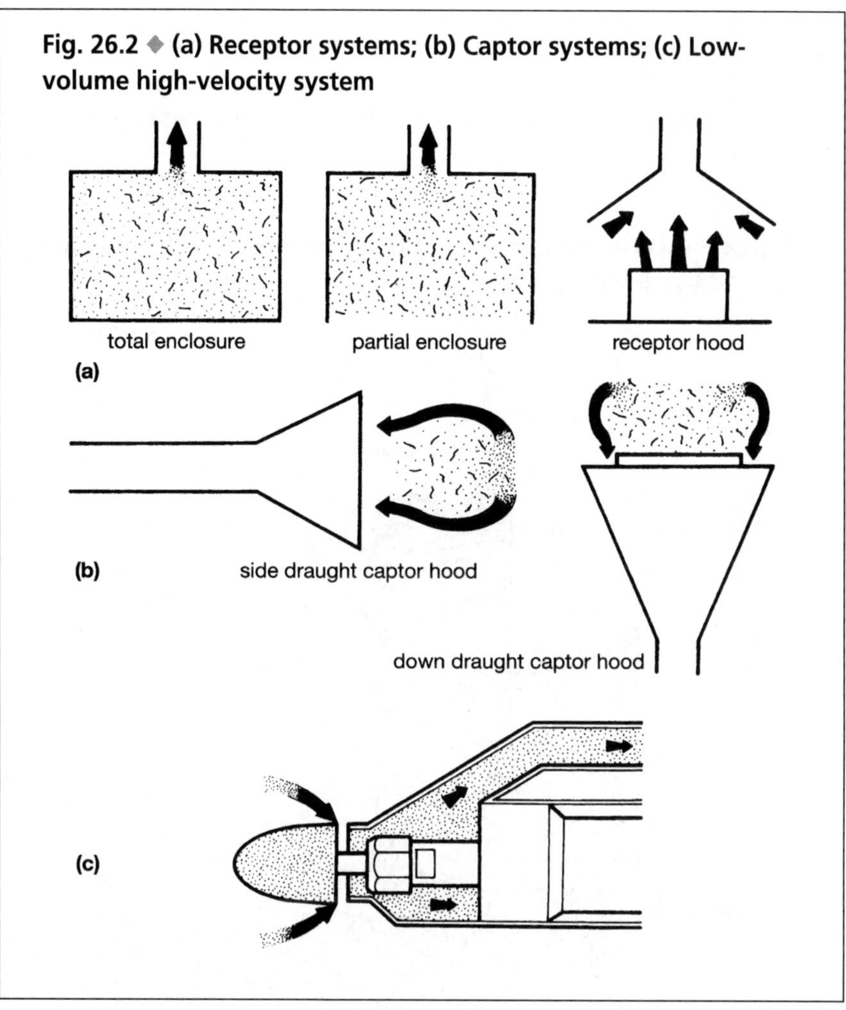

Source: American Conference of Government Industrial Hygienists, 1980

E

Low-volume high-velocity (LVHV) systems

Dust particles emitted by high-speed grinding machines or pneumatic chipping tools require very high capture velocities. One method of achieving very high velocities at the source is to extract from small apertures very close to the source of the contaminant (*see* Fig. 26.2(*c*)). The high velocities can be achieved with quite low air-flow rates. The following factors are essential in LVHV design:

(a) appropriate ergonomic design of the cowl and hoses;

(b) correct cowl adjustment;

(c) regular thorough maintenance;

(d) training of operators.

Further examples of local exhaust ventilation systems are shown in Figs 26.3 to 26.6.

Fig. 26.3 ◆ Local exhaust ventilation by means of a proximity hood applied to a paddle mixer

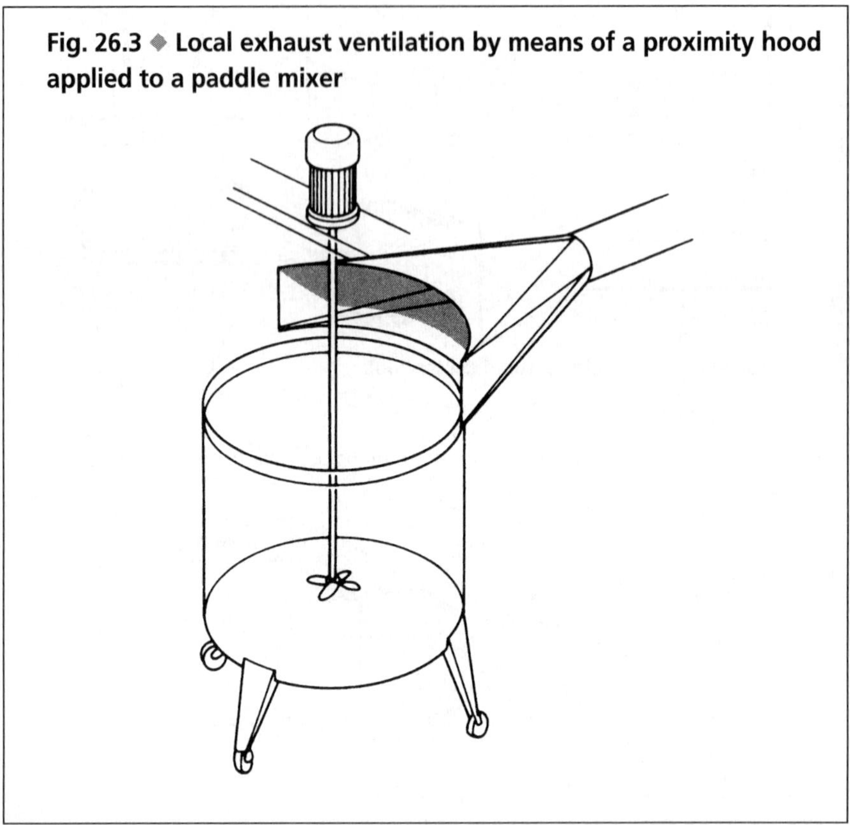

Reproduced by courtesy of DCE Limited

Dilution ventilation

Occasionally it is not possible to extract a contaminant close to the point of origin. If the quantity of contaminant is small, uniformly evolved and of low toxicity, it may be possible to dilute the contaminant by inducing large volumes of air to flow through the contaminated region. Dilution ventilation is most successfully used to control vapours, e.g. organic vapours from low-toxicity solvents, but is seldom successfully applied to dust and fumes, as it will not prevent inhalation. In cold weather this method has self-evident implications for cost and thermal discomfort.

Fig. 26.4 ◆ Application of combination of local exhaust ventilation (proximity hoods) and enclosure to a weighing station

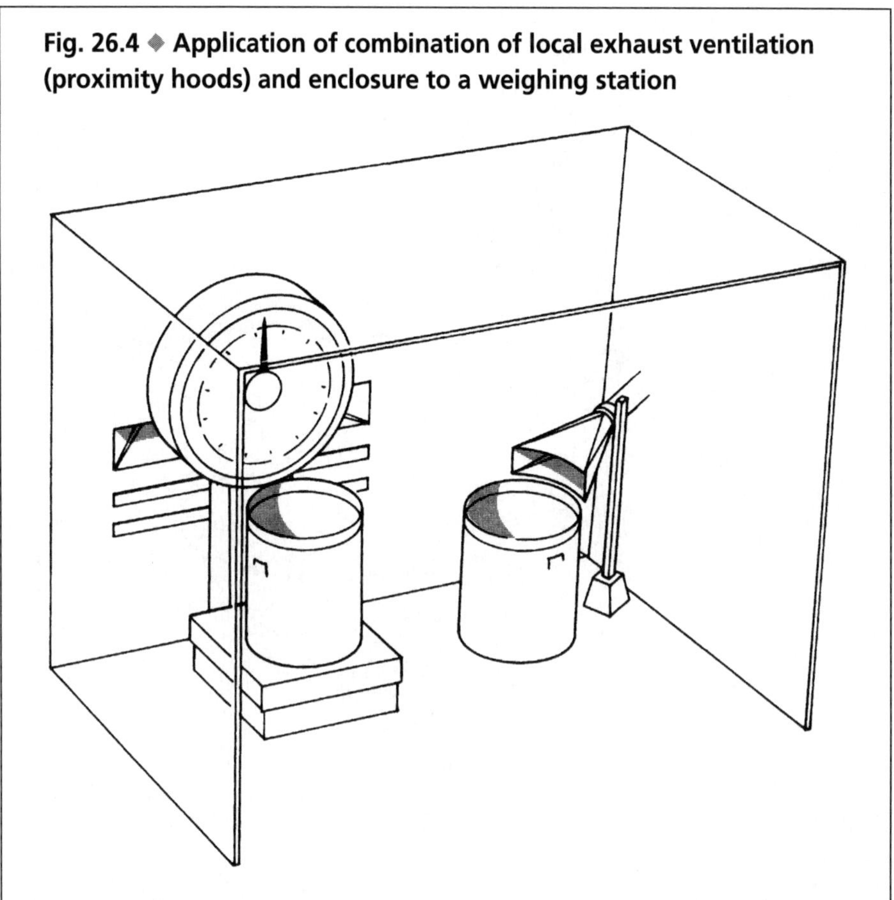

Reproduced by courtesy of DCE Limited

E

Segregation

Segregation is a method of controlling the risks from toxic materials or physical hazards such as noise and radiation. It can take a number of forms.

Segregation by distance (separation)

This is the relatively simple process where a person separates himself from the source of the danger. This is appropriate in the case of noise where, as the distance from the noise source increases, the risk of occupational deafness reduces. Similar principles apply to radiation. Segregation by distance protects those at secondary risk, if those at primary risk are protected by other forms of control.

Fig. 26.5 ◆ Enclosure applied to a ribbon mixer

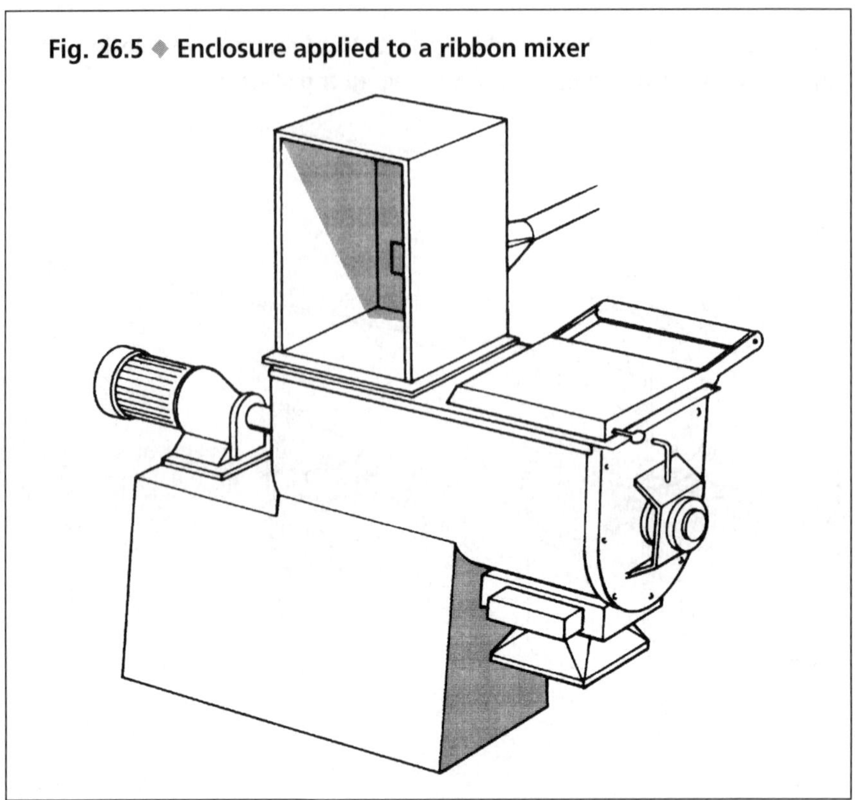

Reproduced by courtesy of DCE Limited

Segregation by age

The need for protection of young workers has reduced over the last 50 years but, where the risk is marginal, it may be necessary to exclude young persons, particularly females, from an activity. An example of such segregation occurs in the Control of Lead at Work Regulations 1980, which exclude the employment of young persons in lead processes.

Segregation by time

This refers to the restriction of certain hazardous operations to periods when the number of workers present is small, for instance at night or during weekends, and when the only workers at risk are those involved in the operation. An example of such an operation is examination by radiation of very large castings.

Segregation by sex

There is always the possibility of sex-linked vulnerability to certain toxic materials, particularly in the case of pregnant women, where there can be damage to the foetus, e.g. in certain processes involving lead.

Change of process

Improved design or process engineering can bring about changes to provide better operator protection. This is appropriate in the case of machinery noise or dusty processes. A typical example is the dressing of seed corn with mercury-based fungicides. Traditionally, this was carried out in a seed-dressing machine in which the fungicidal powder was brushed on to the seed corn as it passed along a conveyor inside a dressing chamber. The fungicide escaped through apertures in the construction of the chamber, contaminating the atmosphere and surrounding structural items. A change to liquid seed dressing using a sealed spray chamber has eliminated this toxic dust hazard.

Controlled operation

Controlled operation is closely related to the duty under HSWA, sec 2(2)(*c*), to provide a safe system of work. It is particularly appropriate where there is a high degree of foreseeable risk. It implies the need for high standards of supervision and control, and may take the following forms:

(a) isolation of processes in which dangerous substances are used or where there may be a risk of heat stroke;

(b) the use of mechanical or remote control handling systems, e.g. with radioactive substances;

(c) the use of permit to work systems, e.g. entry into confined spaces such as closed vessels, tanks and silos, or fumigation processes using dangerous substances such as methyl bromide;

(d) restriction of certain activities to highly trained and supervised staff, e.g. competent persons working in high-voltage switchrooms.

Enclosure

This strategy is based on the containment of an offending agent or environmental stressor to prevent its liberation into the working environment. Enclosure may take a number of forms, e.g. acoustic enclosures for noisy machinery, dust enclosures, paint spray booths, laboratory fume cupboards (*see* Chapters 15 and 22).

Reduced time exposure (limitation)

Risks to health from dangerous substances or physical phenomena such as noise can be reduced by limiting the exposure of workers to certain predetermined maxima. This strategy forms the basis for the establishment of occupational exposure limits, i.e. long-term exposure limits (eight-hour time-weighted average value) and short-term exposure limits (ten-minute time-weighted average value). (*See* Chapter 18.) This strategy is encompassed in the Noise at Work Regulations 1989.

Dilution

There is always some danger in handling chemical compounds in concentrated form. Handling and transport in dilute form reduce the risk. This strategy is appropriate where it is necessary to feed strong chemicals into processing plant regularly or carry quantities of dangerous substances for short distances in open containers. Generally, such practices should be discouraged and, wherever possible, eliminated by process change. Dilution is a poor form of control strategy.

Neutralisation

This is the process of adding a neutralising compound to another strong chemical compound, e.g. acid or alkali, thereby reducing the immediate danger. This strategy is practised commonly in the transportation of strong liquid waste chemical substances, e.g. acid-based wastes, where a neutralising compound is added prior to transportation, and in the treatment of factory effluents prior to their passing to a public sewer.

Support strategies

Reference was made in Chapter 13 to the twin concepts of 'safe place' and 'safe person'. The above control strategies are essentially 'safe place' strategies, where there is a duty upon the employer to provide a safe working environment to protect the health of workers. In order to comply fully with these duties, the following support strategies should also be noted.

Cleaning, housekeeping and preventive maintenance

Cleaning and housekeeping is an important support strategy in the prevention of accidents and occupational disease. Emphasis should be placed, where appropriate, on the use of portable mechanical cleaning equipment rather than on manual methods of cleaning. Moreover, housekeeping inspections should feature in any general safety inspection system. Within the field of preventive maintenance, the potential for leakage of dangerous substances from pipework, ducting and processing plant is considerable. Maintenance schedules, therefore, should include parts of the premises and plant where there is a high degree of risk, e.g. pressure vessels, bulk chemical stores (*see also* Chapter 17).

Welfare amenity provisions and personal hygiene

Good standards of sanitation, hand cleansing, showers and clothing storage facilities are a prerequisite to personal hygiene. Welfare facilities should cater for the direct needs of workers and the elimination of the health hazards to which they may be exposed, e.g. the risk of dermatitis (*see* Chapter 17). Personal hygiene control measures include the following:

(a) strict control over decontamination procedures, particularly before eating, drinking, smoking or leaving the premises at termination of work;

(b) a prohibition on eating, drinking and smoking wherever there may be a risk of hand-to-mouth contamination, e.g. handling dangerous substances;

(c) use of barrier creams and other forms of skin protection directly related to the risks;

(d) procedures for sanitisation of eye, face and hearing protection, or the use of disposable forms of protection;

(e) total prohibition of workers returning home whilst wearing contaminated protective clothing;

(f) training at induction and on a regular basis in the principles of personal hygiene and its relationship to existing health risks;

(g) linking personal hygiene requirements with current industrial relations policy;

(h) formulation and formalisation of a company hygiene policy accompanied by strict enforcement.

Personal protection

Although HSWA, sec 2, generally requires the provision of personal protective equipment by employers and the wearing of it by employees, this strategy must be considered a last resort or a short-term strategy until control can be achieved by more permanent means. Clearly, it is also wrong to expect employees to wear potentially uncomfortable and inconvenient protective equipment if alternative control strategies can be employed. (*See* Chapter 24; also Chapter 5 for the legal position concerning failure to make use of protective measures.)

Health surveillance and first aid

There is now increasing emphasis on occupational health, including health surveillance, as a support strategy, together with first aid and other services (*see* Chapters 20 and 28).

Training, propaganda and joint consultation

The training of staff to identify health risks and measures to avert them should feature in every company training programme. Operators must be made aware of the potential hazards, understand management actions and directives, and appreciate that their co-operation is needed to make the environment a safe and healthy one for themselves and fellow workers. Co-operation by employees is a legal requirement under HSWA, sec 7, and is also an implied condition of an employment contract at common law.

The use of various forms of health and safety propaganda, with a view to drawing the attention of workers to health risks, is a continuing process, and a wide range of posters, films and other forms of propaganda is available.

Joint consultation between employers and employees, through safety representatives, by the work of a health and safety committee or by the use of in-company working parties is an important tool in the prevention of health risks. The hazards must be identified and systems established to reduce, or preferably eliminate, the risks.

27 Manual handling

E

More than a third of all industrial injuries result from manual handling activities. Statistics from the last 60 years indicate that in almost every year the number of people injured in this way has increased. As a result, more than 70,000 workers are off work for variable periods of time. This amounts to nearly 30 per cent of all reportable accidents. Even more surprising is the fact that, during this period, there have been greater advances in the technology and engineering aspects of mechanical handling than ever before. Certainly workers today perform fewer manual handling tasks than their grandfathers, and it would be reasonable, therefore, to expect a comparable reduction in injuries associated with manual handling.

Not only manual workers contribute to the handling injuries statistics, however. Those in sedentary occupations are similarly at risk, e.g. offfice workers, library staff, catering staff, hospital workers and people working in shops.

The Manual Handling Operations Regulations (MHOR) 1992 came into operation on 1st January 1993 . The Introduction to the HSE Guidance on the Regulations *Manual Handling – Guidance on Regulations* (HMSO) indicates that more than a quarter of the accidents reported each year to the enforcing authorities are associated with manual handling – the transporting or supporting of loads by hand or by bodily force. The vast majority of reported manual handling accidents result in over-three-day injury, most commonly a sprain or strain, often of the back.

The Guidance classifies the types of injury caused by handling accidents thus: sprains and strains 65 per cent; superficial injuries – 9 per cent; contusions 7 per cent; lacerations 7 per cent; fractures 5 per cent; and other forms of injury 5 per cent. Sites of bodily injury caused by handling are: backs 45 per cent; fingers/thumbs 16 per cent; arms 13 per cent; lower limbs 9 per cent; rest of torso 8 per cent; hands 6 per cent; other sites 3 per cent.

What comes out of the statistical information on manual handling injuries is the fact that four out of five people will suffer some form of back condition at some time in their lives, the majority of these conditions being associated with work activities.

Manual handling injuries and conditions

Typical injuries and conditions associated with manual handling can be both external and internal. External injuries include cuts, bruises, crush injuries and lacerations to fingers, hands, forearms, ankles and feet. Generally, such injuries are not as serious as the internal forms of injury which include muscle and ligamental tears, hernias (ruptures), prolapsed intervertebral discs and damage to knee, ankle, shoulder and elbow joints. One of the most significant injuries, and the one which results in frequent incapacity and even permanent crippling, is the prolapsed intervertebral disc. The various features of internal handling injuries are discussed below, together with certain conditions resulting from manual handling.

Muscle and ligamental strain

Muscle is the most abundant tissue in the body, and accounts for some two-fifths of the body weight. The specialised component is the muscle fibre, a long slender cell or agglomeration of cells which becomes shorter and thicker in response to a stimulus. These fibres are supported and bound by ordinary connective tissue, and are well supplied with blood vessels and nerves. When muscles are utilised for manual handling purposes, they are subjected to varying degrees of stress. Carrying generally imposes a pronounced static strain on many groups of muscles, especially those of the arms and trunk. This is a particularly unsuitable form of labour for human beings because the blood vessels in the contracted muscles are compressed and the flow of blood, and with it the oxygen and sugar supply, is thereby impeded. As a result, fatigue very soon sets in, with pains in the back muscles, which perform static work only, occurring sooner than in the arm muscles, which perform essentially dynamic work.

Ligaments are fibrous bands occurring between two bones at a joint. They are flexible but inelastic, come into play only at the extremes of movement, and cannot be stretched when they are taut. Ligaments set the limits beyond which no movement is possible in a joint. A joint can be forced beyond its normal range only by tearing a ligament: this is a sprain. Fibrous tissue heals reluctantly, and a severe sprain can be as incapacitating as a fracture. There are many causes of torn ligaments, in particular jerky handling movements which place stress on the joint, uncoordinated team lifting, and dropping a load half-way through a lift, often caused by failing to assess the load prior to lifting.

Hernia

A hernia is a protrusion of an organ from one compartment of the body into another, e.g. of a loop of intestine into the groin or through the frontal abdorninal wall. Both these forms of hernia can result from incorrect handling techniques and particularly from the adoption of bent back stances, which produce compression of the abdomen and lower intestines.

The most common form of hernia or 'rupture' associated with manual handling is the inguinal hernia. The weak point is the small gap in the abdominal muscles where the testis descends to the scrotum. Its vessels pass through the gap, which therefore cannot be sealed. Excessive straining, and even coughing, may cause a bulge at the gap and loop of intestine or other abdominal structure easily slips into it. An inguinal hernia sometimes causes little trouble, but it can, without warning, become strangulated, whereby the loop of intestine is pinched at the entrance to the hernia. Its contents are obstructed and fresh blood no longer reaches the area. Prompt attention is needed to preserve the patient's health, and even his life will be at risk if the condition does not receive swift attention. The defect, in most cases, must be repaired surgically.

Prolapsed disc

The spine consists of a number of small interlocking bones or vertebrae (*see* Fig. 27.1). There are seven neck or cervical vertebrae, twelve thoracic vertebrae, five lumbar vertebrae, five sacral vertebrae and four caudal vertebrae. The sacral vertebrae are united, as are the caudal vertebrae, the others being capable of independent but co-ordinating articulating movement. Each vertebra is separated from the next by a pad of gristle-like material (intervertebral disc). These discs act as shock absorbers and help to protect the spine. A prolapsed or 'slipped' disc occurs when one of these intervertebral discs is displaced from its normal position and is no longer performing its function properly. In other cases, there may be squashing or compression of a disc. This results in a painful condition, sometimes leading to partial paralysis, which may be caused when the back is bent while lifting, as a result of falling awkwardly, getting up out of a low chair or even through over-energetic dancing.

Fig. 27.1 ◆ The human spine

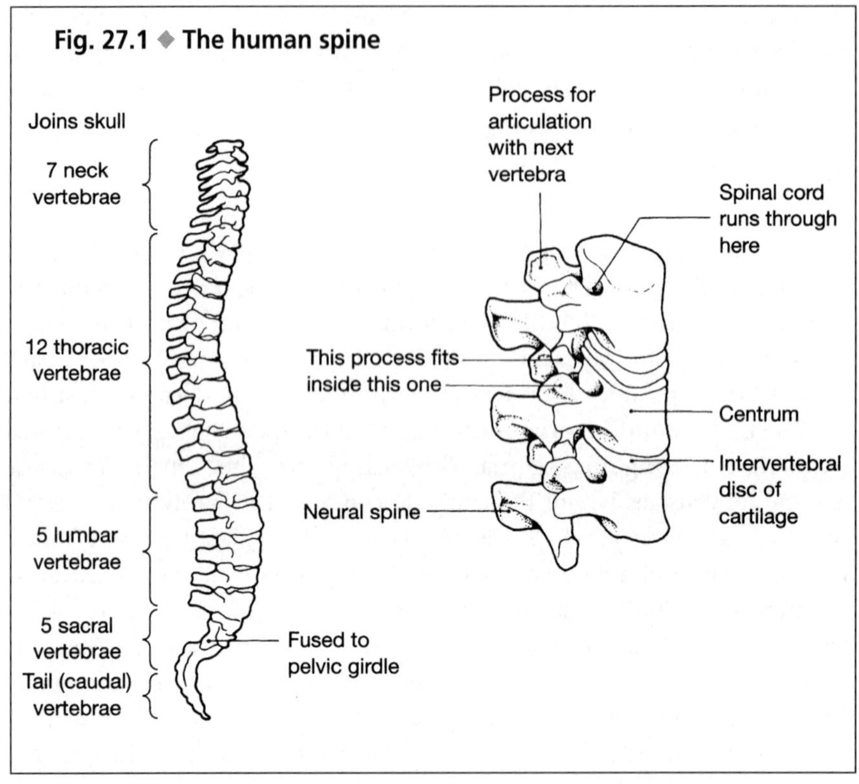

Rheumatism

Rheumatism is a painful disorder of joints or muscles not directly due to infection or injury. This rather ill-defined group includes rheumatic fever, rheumatoid arthritis, osteoarthritis, gout and 'fibrositis', itself an ill-defined group of disorders in which muscular and/or joint pain are common factors. There is much evidence to support the fact that stress on the spine, muscles, joints and ligaments during manual handling activities in early life results in rheumatic disorders as people get older.

In industrial handling situations, therefore, there is an urgent need to instruct operators in correct techniques of movement and posture. Above all, they must be made aware of the importance of the correct assessment and planning of a task such as moving a particular load prior to tackling it.

Assessment prior to handling

Before examining the various problems of handling, the first important rule must be recognised, namely: 'If you think you can't manage to move the load, get help.' This may be help from another person or the use of a mechanical handling aid. The decision must ultimately be left to the person who is doing the job. There will rarely be a single deciding factor, however. The size, shape, weight, degree of rigidity, outside surface of the load, conditions such as the height of the load, state of the ground surface, headroom and temperature and, most important of all, the physical and temperamental characteristics of the individual concerned, must be considered. These factors are discussed below.

Size

Sizes of loads vary in terms of their actual volume. For instance, a sack of cement weighing 50 kg needs a totally different lifting technique from a box of feathers, measuring 1 m × 1 m × 1 m but of the same weight. The centre of gravity of the load should be as near to the body as possible (as with any lifting machine). Moreover, the wider the arms are stretched and the further the hands are in front of the body, the greater the tension on the shoulders, chest and back muscles. Often the use of straps, hooks and other handling aids will assist when moving large loads. There is also a considerable difference between handling indoors and outdoors, where wind velocity can affect individual lifting ability, especially where the handling of sheet materials or large lightweight loads is concerned.

Shape

It is essential to carry a load at the point of balance. Many loads are carried off balance and, in the case of a moving load in a container – e.g. liquids, loose items – the point of balance constantly changes.

Where more than one person is moving a large object, it is essential that the weight be evenly distributed among the people concerned, and that the contents of any container, box, drum, packing case or sack are known.

Weight

Many organisations and individuals have attempted to identify a maximum weight that people of different sexes and ages could handle safely. Indeed, this concept was embodied in several sets of Regulations already mentioned in this chapter. However, such philosophy is not practicable for a number of reasons:

(a) all people are different in terms of physical strength, height, degree of physical fitness and body weight;
(b) if statute stipulated a maximum weight, the implication would be that the handling of a weight in excess of that figure was illegal and vice versa; this philosophy is suspect, since many strain injuries occur when moving the lightest of loads;
(c) the weight of an object is only one factor in determining whether or not to move a load. For instance, reducing the overall weight of a load might be considered salutary, but a person is likely to take greater care when moving a large heavy load than a light one, and he can abuse his body more easily with small light loads than with large heavy ones.

Lack of rigidity

A load likely to change its overall shape during handling can create difficulties in terms of grip and hold. Many materials are packed in sacks or bags. When these are pressure-filled, the 'floppy' nature of the load can pull a person off balance. A typical example here is in the nursing profession. Lifting patients, who may be totally or partially incapable of assisting, is the principal reason why this profession has such a bad record of muscular strain injuries. The fact that the load is live aggravates the problem, coupled with the need to provide careful support not only to the torso but to individual limbs. The difficulty in obtaining a satisfactory grip without harming the patient, who may have been injured, compounds the problem.

Outside surfaces

The material in which a load is packed can directly affect the ease with which it is handled. In the days of hessian sacks the handler used hooks, such as the docker's hook, and the roughness of the material enabled him to acquire a better grip and hold on the load. The increasing use of plastics has reduced this ability for obtaining a good hold. Moreover, the practice of shrink wrapping of goods has created its own handling problems in terms of gripping the load properly. The use of purpose-designed gloves can help in overcoming many of these problems. Such gloves are useful for moving smooth-surfaced loads, whether plastic-sacked goods, domestic appliances or even glass sheets.

Height

The location of a load and the positioning of the person's hands will affect the ease with which the load is moved. Hands can only perform a task efficiently when they are placed directly in front of the body, and close to it, in an area between the chest and thigh levels. Handling loads below the feet or above the head is inadvisable, but there are times when this is unavoidable. The use of hooks and other aids can assist when the load is below foot level. A typical example is keys used to lift manhole tops. Handling loads above head level is made the more hazardous because the handler cannot see the top of the load. This exposes him to risk of other items falling on his head. The weight of a load should always be known when it is to be taken from a shelf, so that the person is not 'taken off guard' when he initially receives it. Here the use of staging or steps to permit the load to be at waist level is advantageous.

Ground or floor surface

Balance depends upon the stability of the base. If a person stands or moves on an unstable base, muscles will automatically be tensed to safeguard balance. Icy or wet surfaces also create this stiffening. When a person walks over a loose surface, this creates tension in the legs and lower parts of the body, e.g. walking over sand dunes. Suitable footwear should, therefore, be worn. When selecting footwear there is a tendency to consider the protection afforded to the toes only. However, on certain surfaces, the sole of the boot or shoe can stabilise the body. The use of safety shoes, with high-grip soles, generally improves body stability.

Headroom

Whenever a person needs to lower his head, he tends to adopt a top-heavy bending action. Many tasks are performed with restricted headroom, e.g. mining, and loading and unloading vans and other forms of transport. The removal of unconscious passengers from window seats of many airliners creates almost insurmountable problems for airline staff. Acquiring the ability to perform instinctively good 'base movements' as distinct from 'top-heavy bending' will reduce the risk of strain in such cases.

Temperature and humidity

Temperature and humidity affect the way and speed with which a person moves. If it is too cold, muscles tend to stiffen; a 'warming up' period should be allowed – not, however, using movements that cause 'stiffening up' but rather using gentle movements to stretch and shorten muscles by relaxing and tightening. An advantage of working in high temperatures is that the worker tends to move more slowly, reducing the discomfort of perspiring, and the tendency to employ sudden or snatching movements. Injury statistics for tropical climates indicate fewer strain injuries. Moreover, use of the correct clothing is important. Clothing should not be tight yet should be sufficiently close fitting to give freedom of movement. Gaps, especially around the waist, should be avoided when working outdoors.

Physical and psychological characteristics

It is not only the physical shape and size of a person that is important but posture and muscular condition. There are quite heavy jobs which can be undertaken successfully by small people. What they lack in body bulk, they may compensate for in dexterity, suppleness and timing. Age is also relevant, particularly the inevitable stiffening which accompanies the ageing process. So, too, is the temperament of an individual and the degree of mental stress which can affect muscular tension. The more placid the individual, the more likely he is to be a relaxed mover, making more fluid and 'segmental' movements than the jerky staccato-type movements associated with a tense person. Equally, a person's own physical performance will often vary from day to day, depending upon factors such as general state of health, food intake, amount of sleep, tasks performed the day before and general mental state.

All the above factors must be considered when assessing handling procedures. No single factor should be considered in isolation. It is incorrect to adopt a black-or-white approach to every situation, and frequently a compromise must be reached. Often, the really awkward job does not create difficulties when extra care and thought are given, but the simple everyday job, apparently with few risks, in some cases leads to severe injuries being sustained.

Handling techniques

General rules for safe lifting

(a) No one should ever attempt to lift anything beyond their capacity. If in doubt, get help.

(b) Where mechanical lifting aids are provided, they should be used.

(c) Extra care should be taken when lifting awkwardly shaped objects.

(d) In the analysis of individual posture and movement considerable observation of the operator is necessary. The following should occur at the moment the force is exerted on the load –

  (i) Position the feet correctly. The feet should be placed hip-width apart to provide a large base. One foot should be put forward and to the side of the object which gives better balance.

  (ii) Bend or 'unlock' the knees and crouch to the load. The weight will then be safely taken down the spine and the strong leg muscles will do the work.

  (iii) Get a firm grip. The load should be gripped by the roots of the fingers and the palm of the hand. This keeps the load under control and permits it to be distributed more evenly up the arms. Use of the finger tips only can produce excessive tension in the forearms and possible loss of grip.

  (iv) Extend the neck upwards by tucking in the chin. This will automatically straighten the back as the load is taken. This does not mean in a vertical position, but inclined at an angle of approximately 15°. This prevents pressure on the abdomen, reduces the risk of hernia and ensures an even pressure on the intervertebral discs.

  (v) Keep the arms close to the body. This reduces muscle fatigue in the arms and shoulders and the effort required by the arms. It ensures that the load moves with the body and becomes, in effect, part of the body.

  (vi) Use the leg muscles. Lifting should utilise the strong thigh muscles. Lifting should proceed by straightening the legs lifting in one smooth and progressive movement from floor to carrying position. Push off with the rear foot.

NOTE. The above points should be considered as features of an overall smooth lifting movement and not a sole means of instruction.

(e) Hand protection, and arm protection where appropriate, should always be used, particularly when lifting rough loads, or loads with sharp edges or projections.

Specific handling activities

Some of these are illustrated in Fig. 27.5, p. 510.

Human kinetics

This section examines the concepts of kinetics, first by observing the human being as a handling machine and, second, by relating these observations to the problems involved in handling which were discussed earlier, e.g. size, shape, weight, lack of rigidity. Correct and incorrect methods of movement are also considered.

Human kinetics follows from the detailed study of body reactions, and may be defined as 'the study of mechanical, nervous and psychological factors which influence the functions and structure of the human body as a means of producing higher standards of skill and reducing cumulative strain'. A simpler definition might be 'a technique of moving in a more relaxed and effficient way'.

Physiology of kinetics

Many injuries and disabilities arise from a common cause, namely cumulative strain, or stiffening of the body structures. When body structures are unduly tense their adaptability is reduced, and they are more easily injured and tired. Body structures are elastic, given to stiffening and relaxing, shortening and lengthening. If they are maintained in either state over a period of time, they will lose their elasticity. Indeed, one of the inevitable effects of growing old is the loss of suppleness accompanied by general stiffening throughout the body.

Human tissues undergo structural changes as a result of reasonable activity regularly performed. Moreover, structural adjustment of tissue can be beneficial. It enables people to perform a particular task more easily, thereby developing a skill or 'knack' in the job, especially if the working method is related to sound kinetic principles. The kinetic method of movement depends upon good balance adjustment and tissue elasticity. In this case structural changes produced by regular, but not excessive, use of the tissues tend to be healthy and beneficial. In contrast, where the actions are carried out by sheer muscular effort and brute force, this increases the risk of strain and expedites development of other harmful conditions, such as rheumatism and fibrositis.

Rhythmic muscle contraction and relaxation has a salutary effect on body tissues whether movements are carried out performing a task or a specific exercise. Sustained tension in the muscle inhibits their nourishment, restricts absorption of muscle waste products and causes fatigue, thereby undermining efficiency of effort. Where tension is sustained in a muscle, a 'muscle-bound' condition develops. The connective tissue which binds the muscle fibres becomes contracted so that it tends to resist expansion of muscle fibres and makes proper relaxation of the muscles impossible. Conversely. when the muscles are relaxed, the connective tissue will become sufficiently loose to allow the blood to circulate freely in the small blood vessels which carry nourishment to and waste materials from the muscle fibres. If the connective tissue is contracted for too long, the muscle fibres begin to lack vitality, become slower in action and more easily tired.

When a muscle contracts or tightens, it does not lose bulk. What it loses in length, it gains in breadth (*see* Fig. 27.2). Connective tissue is important to the economy of body movement, and sustained tension, which causes connective tissue to shorten, undermines the health and efficiency of the body. A good muscle is one that reacts quickly and accurately to the demands imposed by retaining or reacquiring elasticity in the connective tissue.

A large percentage of strains occur in experienced operators who have been doing a particular job for years prior to injury. It is sometimes cynically suggested that this problem arises as a result of carelessness brought about by overfamiliarity. Such suggestions are fallacious, however. The

Fig. 27.2 ◆ Connection of biceps muscle

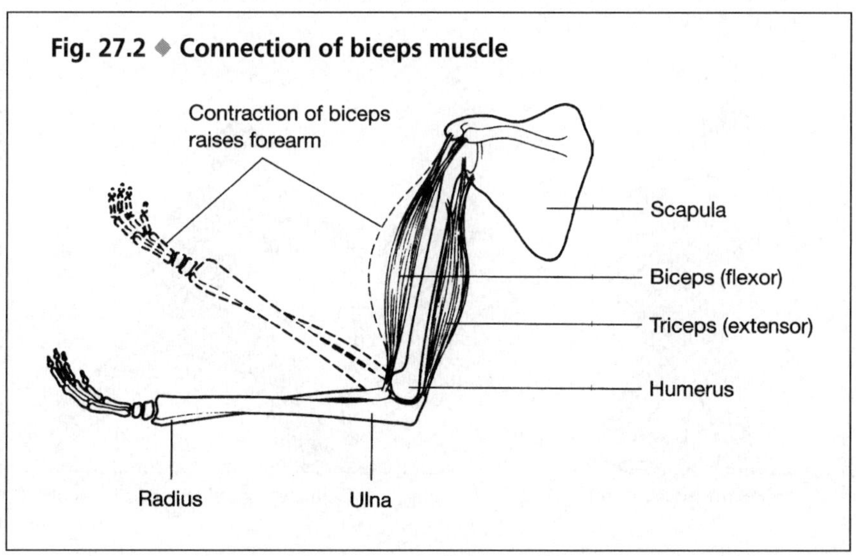

most common cause of injury in experienced workers is cumulative strain, usually arising from incorrect methods of working. Most strains and injuries arise as a result of repetitive incorrect body usage, contributed to by poor workplace design, systems of work or environmental factors. Additionally, however, body conditions dictate the way the body is used. Muscles are powerful in relation to their size, and the natural tendency to be proud of physical strength leads to muscle abuse. To use the body intelligently, individuals must learn to employ skill instead of brute force, thereby cultivating sensitivity rather than muscle bulk.

This can be illustrated by the two methods of pulling on a rope *(see* Fig. 27.3). If a 'doubled-up' position is adopted, brute force and sheer muscular effort are used, thereby explaining why there are so many strains and ruptures throughout industry. In a 'top-heavy' movement, the initial head-bending action causes concentration of pressure on the toes, simultaneously stimulating a chain of stiffening reactions throughout the body. Similarly, the doubled-up position puts a needless load on the abdominal and lower back muscles, causing congestion in the lower abdomen and pelvis, and leading to rupture or hernia. However, in the correct 'skilful' position the feet are so placed that the maximum amount of body weight is deployed and arm muscles, which are strongly contracted in the 'brute

Fig. 27.3 ◆ Conflicting ideas of movement
Real strength depends upon skilful use of body weight, good balance adjustment and elasticity of body structures

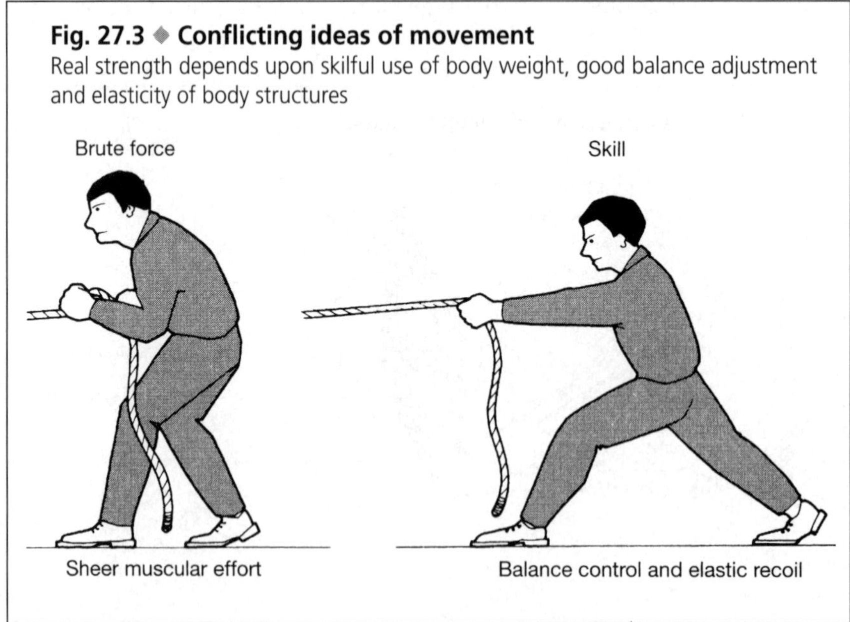

force' method, are under less strain when the arms are kept straight *(see* Fig. 27.4).

Fig. 27.4 ◆ Healthful movement
Progressive relaxation results from unlocking both knees as hands are lowered. Raising the head as the hand takes the load automatically

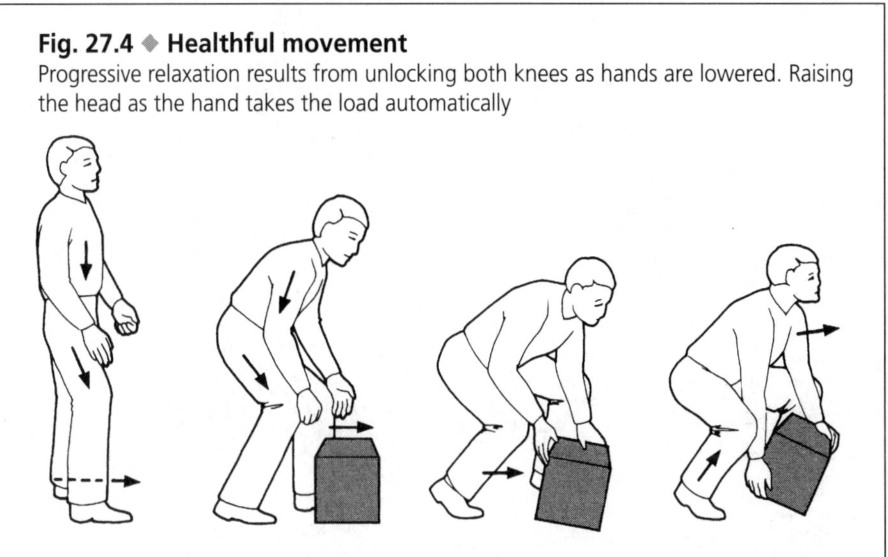

E

The character and effects of a movement are determined by how the movement begins. For example, if the position in a pushing action has been assumed by leaning forward in a top-heavy manner, the operator will increase resistance by pushing the load into the ground. There would also be excessive tension throughout the body. On the other hand, if the action had first started by unlocking the knees, the feet would adjust and the operator could push the load over the ground, thereby reducing resistance and requiring less muscular effort with correspondingly more effective use of body weight.

The human being is like a puppet on a string. Any task that requires lowering of the head or hands starts by unlocking the knees and the feet and works progressively upwards. Any upward movement starts by raising the head and works progressively downwards. This principle applies irrespective of the movement to be performed, e.g. lifting, pulling, pushing, thrusting, down pulling or just sitting or kneeling down. All movements start by relaxing or unlocking the knees to allow the feet to adjust to safeguard balance (*see* Figs 27.5–27.9).

Fig. 27.5 ◆ Bag and sack handling

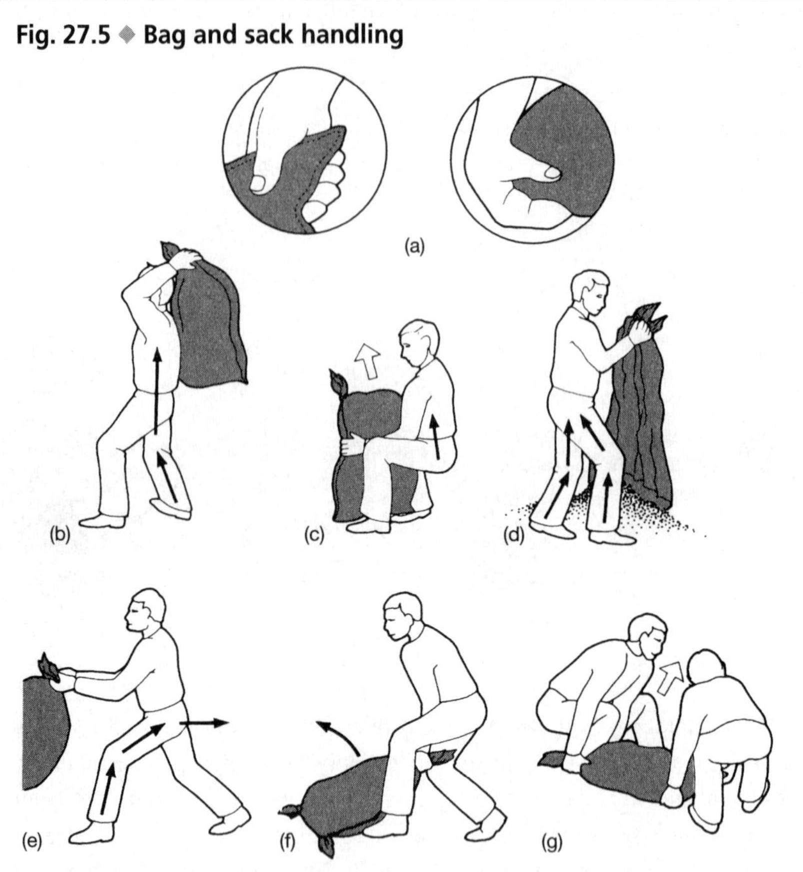

(a) **Gripping** Take hold from below. Grip with the palms and the roots of the fingers
(b) **Carrying** Relax the knees, arms in, and forward foot pointing in the direction of travel. Straighten knees to lift the load clear. Thrust strongly with rear foot to pivot and move off
(c) **Lifting** Raise the head and straighten the back. Correct feet positioning enables the handler to swing the load forward then upwards. The leg muscles then do the work
(d) **Emptying** Relax the knees. Keep the elbows tucked into the body
(e) **Laying a sack down** Use front foot straight forward to thrust, with knee bent to let body move back as far as possible. Stretch rear foot well out to improve balance. Tuck the chin in to keep the shoulders stable. Keep the arms straight
(f) **Standing a sack up** Lower the hands by relaxing the knees. Take a proper hold and position the feet correctly, one between the lugs, and one forward alongside the sack. Move forward and upwards in one rhythmic movement
(g) **Team lifting** The same principles apply as for single lifting. The lift must be co-ordinated by the team leader.

Source: Creber, 1967

The correct and the incorrect way of moving

Good movement

Good movement is that which fulfils its function efficiently with the minimum of effort and cumulative strain. The key to this is coordination of muscular action, i.e. reciprocation between:

(a) muscles which contract to produce movement and those which elongate to allow it to take place;

(b) muscles concerned with maintaining and readjusting balance throughout the movement;

(c) muscles which stabilise the spinal joints and the bases of the limbs during movement.

The most 'expensive' form of muscle work, in terms of energy expended and cumulative strain, is that involving sustained contraction of muscles. Housewives find that the most tiring jobs are working at the sink and ironing, which involve standing up with the feet side by side and bending the upper truck, resulting in gradual build up of excessive tension and stiffening of body tissues, i.e. cumulative strain.

Body balance

Good balance is a prerequisite of most physical activities, from maintaining stationary positions to balance readjustment whilst performing movements. The slightest limb movement can change the centre of gravity, requiring balance readjustment. There are two ways of maintaining balance:

(a) by simply stiffening the legs; or

(b) by relaxing or unlocking the legs so that the feet can readjust to safeguard balance.

All movements performed in an upright position are either top-heavy or base movements. Top-heavy movements are those which begin by bending the head, upper trunk and arms, so that the legs and back stiffen to prevent the body falling. This leads to a staccato-type movement which concentrates stresses in the shoulders. neck and lower back. Sustained tension in the legs leads to circulatory deficiency, loss of resilience and predisposition to injury in the legs and back.

A good movement (*see* Fig. 27.8) always begins as a base movement and is segmental. In contrast to the initial stiffening of the legs in a top heavy movement, it begins by relaxation of the legs so that one foot can

Fig. 27.6 ◆ Pulling and pushing

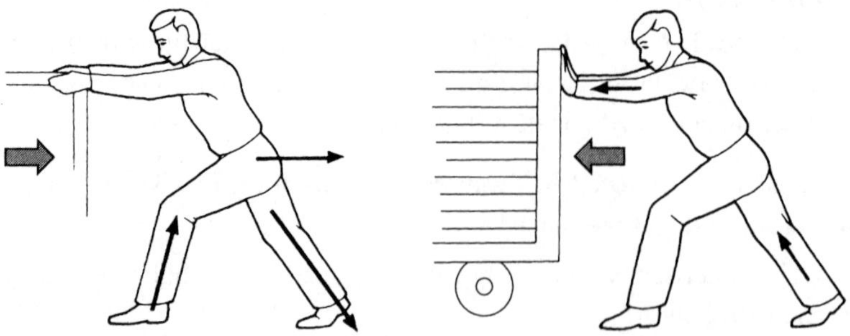

Tuck the chin in, keeping the back and arms straight. In pushing, the front foot should balance the body, while the rear foot, pointing forward, gives the thrust. In pulling, the back foot safeguards balance while the front leg, with knee bent to allow the body to move back, does the thrusting

Source: Creber, 1967

move automatically to safeguard balance. Simply bending the knees, however, is not necessarily relaxing to the legs. People tend to bend their knees excessively, leading to awkwardness in movement. In a good movement, action in the knees is rather a 'giving' or unlocking of the joints.

Fig. 27.7 ◆ Stowing and stacking

When stacking to high level, make the legs do the work. Relax both knees when approaching the stack and thrust upwards with a swinging movement, one foot following through

Source: Creber, 1967

Fig. 27.8 ◆ Box handling

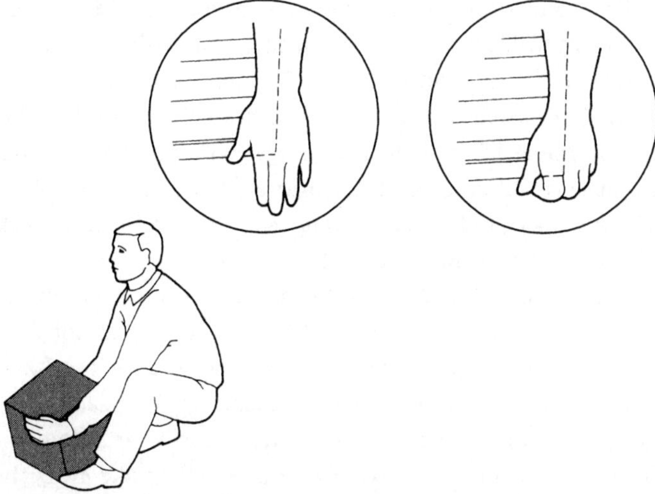

Grip with the palms to reduce finger strain. Position the feet as shown to maintain balance and give a strong thrust forward and upwards off the back foot. The arms should be close to the sides and the hands placed diagonally. The legs will then do the lifting

Source: Creber, 1967

Fig. 27.9 ◆ Spool handling

The head is raised, the chin tucked in and the back straight. To lift or lower, relax the knees, keep the arms into the body, stabilising one forearm inside the thigh, feet apart, with one foot forward, pointing in direction of travel. Grip with the palms and the roots of the fingers

Source: Creber, 1967

Conclusion

Human kinetics is not simply a new type of drill routine to be employed in performing certain tasks, but a method of thinking which involves fundamental changes in attitude and physical habits. First, the problem itself must be clarified before deciding on the methods to be employed in dealing with that problem. Much confusion arises when body movements are discussed because of different interpretations of the terms employed. For instance, if twelve people in a class were asked to 'arch' their backs, half would bend forwards and the other half backwards.

Whilst some people have a natural sense of good movement, in most instances individuals must be encouraged to change long-established physical habits and acquire a new way of thinking regarding the mechanical use of their bodies. Instructors themselves must learn to feel reactions in their own bodies and acquire skill in getting others to do the same. Teaching should concentrate on practical experiments which make clear the basic principles of good movements, followed by explanations of how and why they influence the body structures. Practical demonstration speaks louder than words!

Even when trainees are fully convinced that kinetic methods of moving are more beneficial, there is a likelihood that old habits will reassert themselves when the individual ceases to think about his movements. It is for this reason that the main emphasis in teaching should be on cultivating good base movements, so that when movements begin correctly the correct movements will follow.

No matter how efficiently the primary instruction is carried out, long-established habits cannot be changed overnight. There should always be some form of follow-up. The role of immediate supervision is essential. The man who supervises the job has the most influence on how the individual carries out the job. Because of this, all supervisors should have an understanding of the basic aspects of good movement.

Human kinetics is a way of life and should be taught at a very early age. Waiting until people reach adulthood, before introducing them to the subject, may be too late.

Manual Handling Operations Regulations 1992

These Regulations came into force on 1 January 1993 and implement the European Directive on the manual handling of loads. They supplement the general duties placed upon employers and others by the Health and Safety at Work Act 1974 and the broad requirements of the MHSWR. They are supported by Guidance issued by the HSE.

Reg 2

A number of important terms are defined in Reg 2 as follows.

Injury

Does not include injury caused by any toxic or corrosive substance which:

(a) has leaked or spilled from a load;

(b) is present on the surface of a load but has not leaked or spilled from it; or

(c) is a constituent part of a load.

Load

Includes any person and any animal.

Manual handling operations

Means any transporting or supporting of a load (including the lifting, putting down, pushing, pulling, carrying or moving thereof) by hand or by bodily force.

Reg 4 – Duties of employers

1. Each employer shall:
   (a) so far as is reasonably practicable, avoid the need for his employees to undertake any manual handling operations at work which involve a risk of their being injured;
   (b) where it is not reasonably practicable to avoid the need for his employees to undertake any manual handling operations at work which involve a risk of their being injured –
      (i) make a suitable and sufficient assessment of all such manual handling operations to be undertaken by them, having regard to the factors which are specified in column 1 of Schedule 1 to these Regulations and considering the questions which are specified opposite thereto in column 2 of that Schedule;
      (ii) take appropriate steps to reduce the risk of injury to those employees arising out of their undertaking any such manual handling operations to the lowest level reasonably practicable;

(iii) take appropriate steps to provide any of those employees who are undertaking such manual handling operations with general indications and, where it is reasonably practicable to do so, precise information on –
   (aa) the weight of each load;
   (bb) the heaviest side of any load whose centre of gravity is not positioned centrally.

2. Any assessment such as referred to in para 1(*b*)(i) of this regulation shall be reviewed by the employer who made it if:
   (a) there is reason to suspect it is no longer valid; or
   (b) there has been a significant change in the manual handling operations to which it relates: and where as a result of any such review changes to an assessment are required, the relevant employer shall make them.

Reg 5 – Duties of employees

Each employee while at work shall make full and proper use of any system of work provided for his use by his employer in compliance with Reg 4(1)(*b*)(ii) of these Regulations.

HSE Guidance on the Regulations

The hierarchy of measures

Implementation of, and compliance with, the Regulations involves a specific hierarchy of measures, namely:

(a) the avoidance of hazardous manual handling operations so far as is reasonably practicable;
(b) assessment of any manual handling operations that cannot be avoided;
(c) reduction of the risk of injury so far as is reasonably practicable.

Once the measures to reduce risk of injury in manual handling practices have been introduced, the employer must monitor handling operations to ensure these measures are being followed.

Work away from employers' premises

Many handling injuries occur whilst employees are working on other people's premises, for example to contractors' employees and those delivering goods. In such cases, the respective employers must liaise to ensure visiting employees are not exposed to risk of injury.

Avoidance of manual handling

The principal emphasis of the Regulations is that of avoiding manual handling wherever possible and replacing manual handling operations by the use of mechanised equipment, e.g. hand trucks, wheelbarrows, lift trucks, conveyors, elevators, etc. (These factors should be taken into account in the general risk assessment carried out under the MHSWR.) Particular emphasis should be placed on assessing the risk of injury in terms of the nature of such injury and the likelihood of same. Here reference should be made to 'Guidelines for lifting and lowering' shown below (*see* Fig. 27.10).

Fig. 27.10 ◆ Lifting or lowering

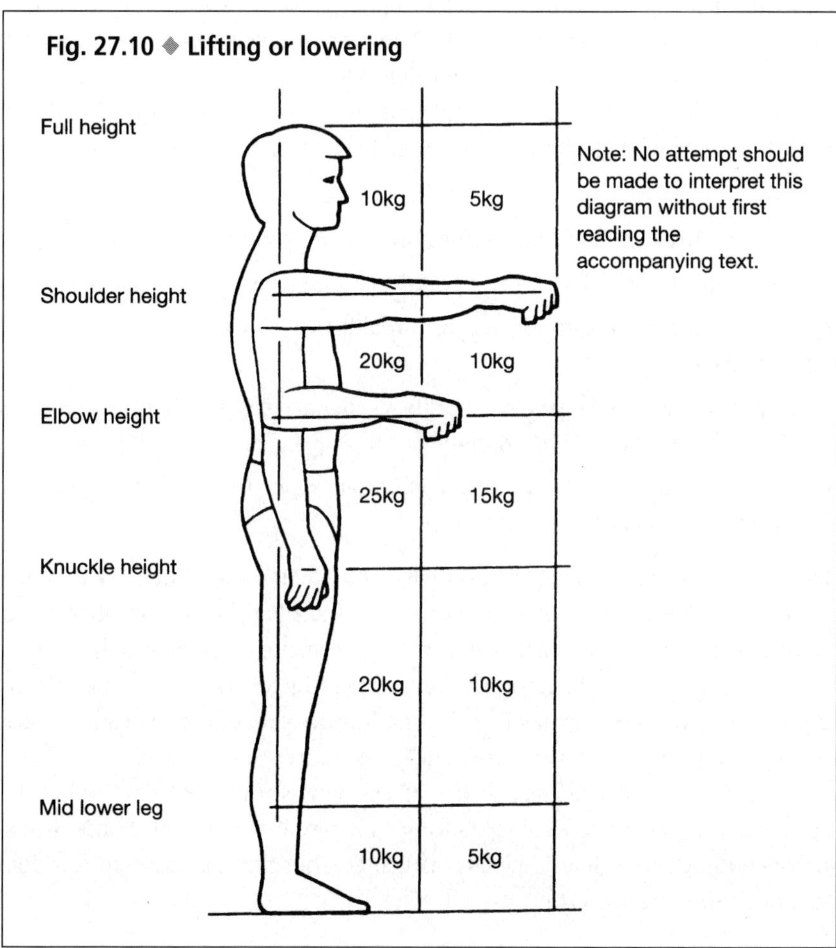

The guidelines' figures take into consideration the vertical and horizontal position of the hands as they move the load during the handling operation, as well as the height and reach of the individual handler. It will be apparent that the capability to lift or lower is reduced significantly if, for example, the load is held at arm's length or the hands pass above shoulder height.

If the hands enter one of the box zones during the operation the smallest figure should be used. The transition from one box zone to another is not abrupt; an intermediate figure may be chosen where the hands are close to the boundary. Where lifting or lowering with the hands beyond the box zones is unavoidable a more detailed assessment should be made.

Assessment of risk

Schedule 1 to the Regulations specifies factors which should be taken into account in the assessment of risk. These include the *task*, the *load*, the *working environment* and *individual capability*.

It is recommended that a small risk assessment team combining different areas of knowledge and expertise be established. Such a team could comprise:

(a) the requirements of the Regulations – *manager, safety professional*;

(b) the nature of the handling operations – *supervisor, industrial engineer*;

(c) a basic understanding of human capabilities – *occupational health nurse, safety professional*;

(d) identification of high-risk activities – *manager, supervisor, occupational health nurse, safety professional*;

(e) practical steps to reduce risk – *manager, supervisor, industrial engineer, safety professional*.

In some cases, it may be appropriate to seek outside assistance, particularly in terms of training the assessors or where handling risks may be unusual or difficult to assess. In particular, the views of the employees who will be undertaking the handling operations should be sought, both specifically and as part of the normal joint consultation procedures through trade union safety representatives and safety committees.

The significant findings of the assessment should be recorded and a record kept, readily accessible, as long as it remains relevant. Some assessments will be more detailed than others. Furthermore, an assessment does not need to be recorded if:

(a) it could very easily be repeated and explained at any time because it is simple and obvious; or

(b) the manual handling operations are quite straightforward, of low risk, are going to last only a very short time, and the time taken to record them would be disproportionate.

Manual Handling Operations Regulations 1992

Schedule 1

Factors to which the employer must have regard and questions he must consider when making an assessment of manual handling operations.

1. The tasks

Do they involve:

- holding or manipulating loads at distance from trunk?
- unsatisfactory bodily movement or posture, especially:
  - twisting the trunk?
  - stooping?
  - reaching upwards?
- excessive movement of loads, especially:
  - excessive lifting or lowering distances?
  - excessive carrying distances?
- excessive pushing or pulling of loads?
- risk of sudden movement of loads?
- frequent or prolonged physical effort?
- insufficient rest or recovery periods?
- a rate of work imposed by a process?

2. The loads

Are they:

- heavy?
- bulky or unwieldy?
- difficult to grasp?
- unstable, or with contents likely to shift?
- sharp, hot or otherwise potentially damaging?

3. The working environment

Are there:

space constraints preventing good posture?
uneven, slippery or unstable floors?
variations in level of floors or work surfaces?
extremes of temperature or humidity?
conditions causing ventilation problems or gusts of wind?
poor lighting conditions?

4. Individual capability

Does the job:

require unusual strength, height, etc.?
create a hazard to those who might reasonably be considered to be pregnant or have a health problem?
require special information or training for its safe performance?

5. Other factors

Is movement or posture hindered by personal protective equipment or clothing?

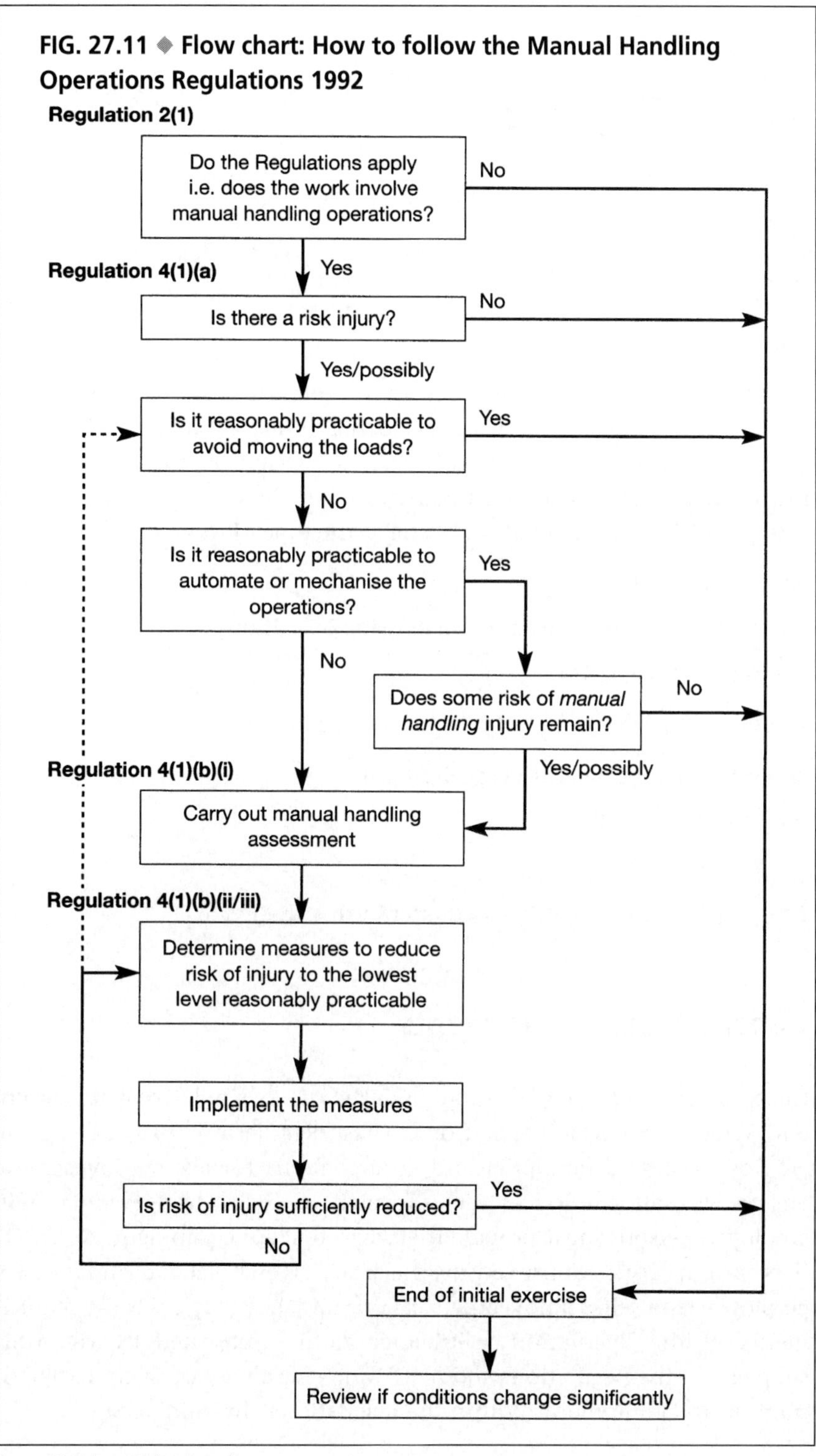

Source: Health and Safety Executive

28 First aid

First aid is a post-accident strategy vital to prevent loss of life and future deterioration in health following accidental injury or sudden illness. First aid is defined as 'the skilled application of accepted principles of treatment on the occurrence of an accident or in the case of sudden illness, using facilities and materials available at the time'. It is given:

(a) to sustain life;

(b) to prevent deterioration in an existing condition;

(c) to promote recovery.

The most important areas of first aid treatment are:

(a) restoration of breathing (resuscitation);

(b) control of bleeding;

(c) prevention of collapse.

These various areas are discussed later in this chapter.

General legal requirements

Generally, first aid activities throughout the UK are regulated by the Health and Safety (First Aid) Regulations. These Regulations provide that all employers should make first aid arrangements for all employees, and ensure that all employees are informed of such arrangements. Self-employed persons must provide their own first aid equipment. An ACOP 'First aid at work', approved by the HSC, gives practical guidance for employers and self-employed persons on how they may meet the requirements of the Regulations. Guidance notes, published by the HSE, supplement the Regulations and ACOP and give advice on such matters as training and equipment. Within the Regulations, 'first aid' means:

(a) in cases where a person will need help from a medical practitioner or nurse, treatment for the purpose of preserving life and minimising the consequences of injury or illness until such help is obtained;

(b) treatment of minor injuries which would otherwise receive no treatment or which do not need treatment by a medical practitioner or nurse.

First aid equipment and first aid rooms

The criteria for deciding on the scale of first aid provision, together with equipment and facilities, depend upon the range of work activities undertaken and the hazards to which workers are exposed. Certain activities, such as office work, involve relatively low levels of risk compared with manufacturing activities, such as operations in chemical works and factories. However, in all establishments first aid provision should be readily available at all times, and the equipment and facilities to be provided will vary from a small travelling first aid kit to a first aid room. All establishments must provide at least one first aid box, and every box should be placed in a clearly identified and readily accessible location. Every first aider should have access to first aid equipment and, where appropriate, facilities.

Soap and water and disposable drying materials should be provided for first aid purposes. However, where soap and water are not available, individually wrapped moist cleaning wipes which are not impregnated with alcohol may be used. The use of antiseptics is not necessary for the first aid treatment of wounds.

First aid equipment

First aid boxes and travelling first aid kits should contain a sufficient quantity of suitable first aid materials and *NOTHING ELSE*. Contents of boxes and kits should be replenished as soon as possible after use, and a system should be established to ensure first aid equipment is regularly checked. First aid boxes should be proofed against dust and damp and clearly identified with a white cross on a green background in accordance with the Safety Signs Regulations 1980. Where first aid boxes form part of an establishment's first aid provision, they should contain only those items which a first aider has been trained to use. Sufficient quantities of each item should always be available in every first aid box or container. In most cases, these will be:

(a) one guidance card (Fig. 28.1);

(b) twenty individually wrapped sterile adhesive dressings (assorted sizes) appropriate to the work environment (which may be detectable for the food and catering industries);

(c) two sterile eye pads, with attachment;

(d) six individually wrapped triangular bandages;

(e) six safety pins;

(f) six medium sized individually wrapped sterile unmedicated wound dressings (approx. 10 cm × 8 cm);

(g) two large sterile individually wrapped unmedicated wound dressings (approx. 13 cm × 9 cm);

(h) three extra large sterile individually wrapped unmedicated wound dressings (approx. 28 cm × 17.5 cm).

Where mains tap water is not available for eye irrigation, sterile water or sterile normal saline (0.9 per cent) in sealed disposable containers should be provided. Each container should hold at least 300 ml and should not be re-used once the sterile seal is broken. At least 900 ml should be provided. Eye baths/eye cups/refillable containers should not be used for eye irrigation.

Where an employee has received additional training in the treatment of specific hazards which require the use of special antidotes or special equipment, these may be stored near the hazard area or may be kept in the first aid box.

Travelling first aid kits

In the case of employees who regularly work away from their employer's establishment in isolated locations or where they are involved in travelling long distances in remote areas from which access to accident and emergency facilities may be difficult, it may be necessary for first aid equipment to be carried by, or made available to, employees where potentially dangerous tools or machinery are used. The equipment should be suitable for the numbers involved and the potential hazards to which employees are exposed.

The contents of travelling first aid kits should be appropriate for the circumstances in which they are used. At least the following should be included:

Fig. 28.1 ◆ RoSPA emergency card

EMERGENCY AID

Name ....

Address ....

....

Telephone number ....

In case of accident or illness, please inform next of kin

Name ....

Relationship..

Address ....

....

Telephone number ....

For emergency medical information, see below.

Height...................... Weight.................. Date of birth ....

Blood group............. Religion ....

Allergic to ....

....

Date of last tetanus injection..

Medical condition (tick ☑ where applicable)

☐ Heart condition	☐ Pacemaker	☐ Asthma
☐ Contact lenses	☐ Deaf	☐ Epilepsy
☐ Diabetic	☐ Hypertension	☐ Haemophilia
☐	☐	☐

(enter other conditions in spaces above)

RESCUE BREATHING AND HEART COMPRESSION INSTRUCTIONS INSIDE

Cannon House, The Priory Queensway, Birmingham B4 6BS IS222

(a) card giving the general first aid guidance;
(b) six individually wrapped sterile adhesive dressings;
(c) one large sterile unmedicated dressing;
(d) two triangular bandages;
(e) two safety pins;
(f) individually wrapped moist cleaning wipes.

First aid rooms

Where an establishment presents a high risk from hazards the employer should provide a suitably equipped and staffed first aid room. In situations where access to accident and emergency facilities are difficult, or where there is dispersed working, a first aid room may also be necessary. (The ACOP stresses the fact that the need for a first aid room is no longer solely dependent upon the number of persons employed in the undertaking.)

Where a first aid room is considered appropriate for an establishment, the following conditions should be met:

(a) A 'suitable person' should be responsible for the room and its contents.
(b) A suitable person should be available at all times when employees are at work.
(c) The room should be readily available at all times when employees are at work and should not be used for any purposes other than the rendering of first aid or health screening.
(d) The room should be positioned as near as possible to a point of access for transport to hospital, taking into account the location and layout of the establishment.
(e) The room should be large enough to hold a couch, with space for people to work around it, and a chair.
(f) The room's entrances should be wide enough to accommodate a stretcher, wheelchair or carrying gear.
(g) The room should contain suitable facilities and equipment, have impervious floor covering and should be effectively ventilated, heated, lighted and maintained; all surfaces should be easy to clean; the room should be cleaned each working day and suitable arrangements for refuse disposal should be provided.
(h) Suitable facilities, e.g. one or more chairs, should be provided close to the first aid room if employees have to wait for treatment.

(j) The room should be clearly identified as a first aid room by means of a sign complying with the Safety Signs Regulations 1980.
(k) A notice should be attached to the door of the first aid room clearly showing names and locations of the nearest first aiders/appointed persons.

When siting a new first aid room the necessity to have toilets nearby and for the room to be on the ground floor should be considered. Corridors, lifts, doors, etc., which lead to the first aid room should allow access for a stretcher, wheelchair or carrying chair. Emergency lighting should be provided.

The following facilities and equipment should be provided in first aid rooms:

(a) sink with running hot and cold water always available;
(b) drinking water when not available on tap and disposable cups;
(c) soap and paper towels;
(d) smooth-topped working surfaces;
(e) a suitable store for first aid materials;
(f) first aid equipment equivalent in range and standard and quantities to those listed on pp. 524–6;
(g) suitable refuse containers lined with a disposable plastic bag;
(h) a couch with waterproof surface, together with a frequently-cleaned pillow and blankets;
(j) clean protective garments for use by first aiders;
(k) a chair;
(l) an appropriate book for recording first aid treatments;
(m) a bowl.

'Suitable persons'

A suitable person, for the purposes of the Health and Safety (First Aid) Regulations is:

(a) a first aider who holds a current first aid certificate issued by an organisation whose training and qualifications were, at the time of issue of the certificate, approved by the HSE for the purposes of the Regulations; in certain circumstances a first aider will need additional or specific training to be a 'suitable person';

(b) any other person who has undergone training and obtained qualifications approved by the HSE for the purposes of the Regulations.

Practising registered medical practitioners, and practising nurses whose names are entered on Parts 1, 2 or 7 of the Single Professional Register maintained by the United Kingdom Central Council for Nursing, Midwifery and Health Visiting may be regarded as first aiders for the purposes of the ACOP.

Occupational health services

Where a full-time occupational health service, which is in the charge of a registered medical practitioner or qualified occupational health nurse, is provided by an employer, the first aid arrangements for the establishment should be made by those persons. Such arrangements may differ from those set out in the ACOP provided that they are of at least equivalent standard. The occupational health service need not be staffed continuously by a full-time registered medical practitioner or qualified occupational health nurse, provided that arrangements are made by him/her for suitable coverage for all employees during working hours.

First aid training and qualifications

Any organisation, or individual employer, may seek approval to train and examine first aiders and award certificates of qualification in first aid. The criteria to which the HSE will have regard in deciding whether to approve the training and qualifications given include:

(a) that the qualifications and training of trainers conform to guidance issued by the HSE;
(b) that the proposed syllabus includes both theoretical and practical work and conforms to guidance issued by the HSE;
(c) that the equipment listed for inclusion in first aid boxes is used for training and examination purposes;
(d) that suitable arrangements are made for conducting examinations which should be carried out by independent examiners who have not been involved in the training of the candidates they examine;
(e) that training organisations only accept for first aid at work courses individuals whose intention is to practise first aid in the workplace during the validity of the first aid certificate;
(f) that suitable premises are available for training and examination purposes.

Courses will be monitored by HSE staff to ensure compliance with the above criteria. Training courses, including examinations, should be of at least four full days' duration (six contact hours per day) or the equivalent, allowing the course to run over a longer period. Each session should be of not less than two hours' duration and the whole course (including the examination) must be completed within 13 weeks from the date of its commencement. The following subjects should be included in the syllabus:

(a) resuscitation;

(b) treatment and control of bleeding;

(c) treatment of shock;

(d) management of the unconscious casualty;

(e) contents of first aid boxes and their use;

(f) purchasing first aid supplies;

(g) transport of casualties;

(h) recognition of illness;

(i) treatment of injuries to bones, muscles and joints;

(j) treatment of minor injuries;

(k) treatment of burns and scalds;

(l) eye irrigation;

(m) poisons;

(n) simple record keeping;

(o) personal hygiene in treating wounds; reference to Hepatitis B and Human Immunodeficiency Virus with regard to first aiders;

(p) communications and delegation in an emergency.

Certificates of Qualification in First Aid are valid for such a period of time as the HSE directs, currently three years. A refresher course, followed by examination, is required before recertification. In cases where there is:

(a) a danger of poisoning by certain cyanides or related compounds;

(b) a danger of burns from hydrofluoric acid; or

(c) a need for oxygen as an adjunct to resuscitation,

training should be carried out by organisations approved by the HSE for training in these specific hazards.

Emergency first aid training

Training in emergency first aid may be given at the workplace by occupational health staff or in short courses run by organisations whose training and qualifications for first aiders are approved by the HSE. Emergency first aid training should be considered for appointed persons and employees working in small groups away from their employers' establishments or where a specific hazard exists. Short courses, of at least four contact hours' duration, should include the following items:

(a) resuscitation;

(b) control of bleeding;

(c) treatment of the unconscious casualty;

(d) communication, contents of first aid boxes and, where appropriate, treatment of the effects of specific hazards existing in the workplace.

This training should be repeated, as a minimum, every three years.

First aid procedures

As many employees as possible should be trained in emergency first aid procedures, namely resuscitation, control of bleeding and treatment of the unconscious patient, so as to ensure that casualties receive prompt attention. It may also be appropriate to train employees in other aspects of first aid, such as treatment for burns, scalds, broken bones and electric shock and in the event of gassing.

The HSE Guidance Leaflet INDG347 '*Basic advice on first aid at work*' (Figures 28.2 and 28.3) makes the following recommendations on what to do in an emergency. Procedures for dealing with broken bones and spinal injuries, burns and eye injuries are also incorporated in the leaflet.

Specific advice covering severe bleeding, broken bones and spinal injuries, burns and eye injuries, together with record keeping requirements, is also incorporated in the leaflet, thus:

Severe bleeding

- apply direct pressure to the wound;
- raise and support the injured part (unless broken);
- apply a dressing and bandage firmly in place.

Broken bones and spinal injuries

If a broken bone or spinal injury is suspected, *obtain expert help. Do not move casualties* unless they are in immediate danger.

Burns

E

Burns can be serious so if in doubt, *seek medical help*. Cool the part of the body affected with cold water until pain is relieved. Thorough cooling may take 10 minutes or more, but this must not delay taking the casualty to hospital.

Certain chemicals may seriously irritate or damage the skin. Avoid contaminating yourself with the chemical. Treat in the same way as for other burns but flood the affected area with water for 20 minutes. Continue treatment even on the way to hospital, if necessary. Remove any contaminated clothing which is not stuck to the skin.

Eye injuries

All eye injuries are potentially serious. If there is something in the eye, wash out the eye with clean water or sterile fluid from a sealed container, to remove loose material. *Do not attempt to remove anything that is embedded in the eye.*

If chemicals are involved, flush the eye with water or sterile fluid for at least 10 minutes, while gently holding the eyelids open. Ask the casualty to hold a pad over the injured eye and send them to hospital.

Record keeping

It is good practice to record in a book any incidents involving injuries or illness which have been attended. Include the following information in your entry:

- date, time and place of incident;
- name and job of injured or ill person;

Fig. 28.2 ◆ What to do in an emergency

Priorities

- assess the situation – do not put yourself in danger;
- make the area safe;
- assess all casualties and attend first to any *unconscious* casualties;
- *send for help – do not delay*,
- follow the advice given below.

Check for consciousness

If there is no response to gentle shaking of the shoulders and shouting, the casualty may be *unconscious*. The priority is then to check the **A**irway, **B**reathing and **C**irculation. This is the **ABC** of resuscitation.

A Airway

To open the airway:

- place one hand on the casualty's forehead and gently tilt the head back;
- remove any obvious obstruction from the casualty's mouth;
- lift the chin with two fingertips.

B Breathing

Look along the chest, listen and feel at the mouth, for signs of normal breathing, for no more than 10 seconds.

If the casualty is breathing:

- place in the recovery position and ensure the airway remains open;
- send for help and monitor the casualty until help arrives.

If the casualty is not breathing:

- send for help;
- keep the airway open by maintaining the head tilt and chin lift;
- pinch the casualty's nose closed and allow the mouth to open;
- take a full breath and place your mouth around the casualty's mouth, making a good seal;
- blow slowly into the mouth until the chest rises;
- remove your mouth from the casualty and let the chest fall fully;
- give a second slow breath, then look for signs of a circulation (see opposite);
- **if signs of a circulation are present**, continue breathing for the casualty and recheck for signs of a circulation about every 10 breaths;
- if the casualty starts to breathe but remains unconscious, put them in the recovery position, ensure the airway remains open and monitor until help arrives.

C Circulation

Look, listen and feel for normal breathing, coughing or movement by the casualty, for no more than 10 seconds.

If there are no signs of a circulation, or you are at all unsure, immediately start chest compressions:

- lean over the casualty and with straight arms, press vertically down 4–5 cm on the breastbone, then release the pressure;
- give 15 rapid chest compressions (a rate of about 100 per minute) followed by two breaths;
- continue alternating 15 chest compressions with two breaths until help arrives or the casualty shows signs of recovery.

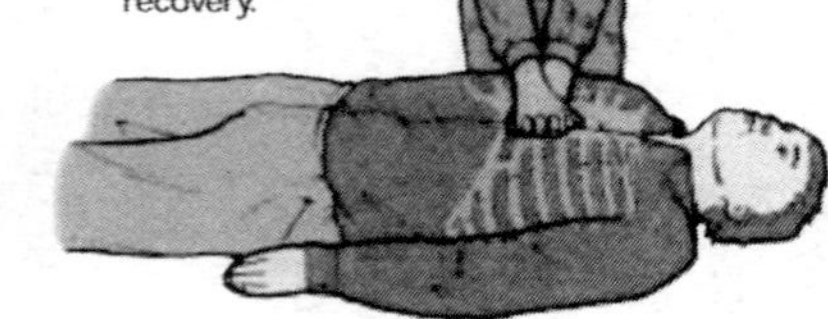

Source: HSE

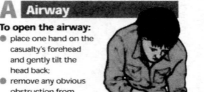

- details of injury/illness and any first aid given;
- what happened to the casualty immediately afterwards (for example went back to work, went home, went to hospital);
- name and signature of the person dealing with the incident.

This information can help identify accident trends and possible areas for improvement in the control of health and safety risks.

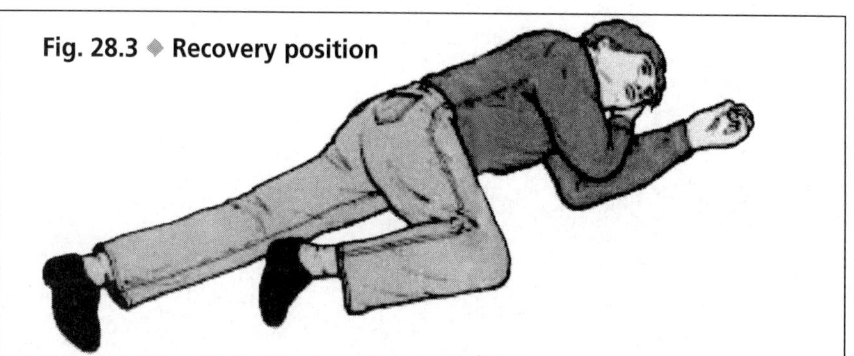

Fig. 28.3 ◆ Recovery position

E

SECTION F

Human factors and safety at work

29 Human factors and safety

Human factors

The term human factors is used to cover a range of issues including:

(a) the perceptual, physical and mental capabilities of people and the interaction of individuals with their job and working environments;

(b) the influence of equipment and system design on human performance;

(c) the organisational characteristics which influence safety-related behaviour.

It is now widely accepted that the majority of accidents are in some measure attributable to human as well as technical factors in the sense that actions by people initiated or contributed to the accidents, or people might have acted better to avert them.

The maintenance of a 'safe place' strategy is readily identified in current health and safety legislation, and the role of the enforcement authorities has always been geared to producing improvements in the physical conditions in the workplace. This has been brought about by a continuing enforcement programme in matters relating to machinery safety, the control of hazardous substances, fire prevention and protection and the maintenance of a sound structural environment.

A system which relies heavily on the enforcement of legal standards to bring about these improvements in the physical conditions, however, has not always been successful in reducing accidents or ill-health, due to the failure to examine the 'human factors' aspect of safety and accident prevention.

The role of management

Although most UK health and safety legislation places the duty of compliance firmly on the body corporate, i.e. the organisation, this duty can only be discharged by the effective actions of its managers. Studies by the HSE's Accident Prevention Advisory Unit have shown that the vast majority of fatal accidents, and those causing major injury, could have been prevented by management action. For example:

1. During the period 1981–5, 79 people were killed in the construction industry; 90 per cent of these deaths could have been prevented. In 70 per cent of cases, positive action by management could have saved lives.
2. A study of 326 fatal accidents during maintenance activities occurring between 1980 and 1982 showed that in 70 per cent of cases positive management action could have saved lives.
3. A study of maintenance accidents in the chemical industry between 1982 and 1985 demonstrated that 75 per cent were the result of management failing to take reasonable precautions.

These studies tend to emphasise the crucial role of the organisation in the management of job and personal factors, principally aimed at preventing human error.

Areas of influence on people at work

The three areas of influence on people at work are:

(a) the organisation;

(b) the job;

(c) personal factors.

These areas are directly affected by:

(a) the system for communication within the organisation;

(b) the training systems and procedures in operation,

all of which are directed at preventing human error.

Features of organisations

Organisations operate on the basis of a hierarchy, which is based on power. Orders pass down and information passes back up. The flow is one-way in both cases. Promotion within the hierarchy is allegedly based on merit and hard work. Conflict in the selection and effectiveness of line managers can arise, and supervisors and foremen frequently find themselves in conflict.

The formal organisation of industry

1. It is deliberately impersonal.
2. It is based on ideal relationships.
3. It is based on the 'Rabble Hypothesis' of the nature of people, i.e. 'Each man for himself'.

It comprises as a rule:

1. *The functional/line organisation,* which is based on the type of work being done.
2. *The staff organisation,* which is in direct contrast to the line organisation and represented by those people who have advisory (e.g. health and safety specialists), service or control functions.

Potential sources of conflict

Conflict may arise between these two parts of the organisation for a number of reasons, such as:

(a) differing motivations;

(b) misunderstanding of individual roles;

(c) differing cultures and objectives;

(d) differing priorities and levels of commitment.

Weaknesses of formal organisations

1. Communications failures.
2. They ignore emotional factors in human behaviour.
3. They are frequently seen as uncaring and lacking in interest or commitment to ensuring appropriate levels of safety, health and welfare within the organisation.

The job

Successful management of human factors and the control of risk involves the development of systems designed to take proper account of human capabilities and fallibilities. Tasks should be designed in accordance with *ergonomic principles* so as to take into account limitations in human performance. Matching the man to the job will ensure that he is not overloaded, and that he makes the most effective contribution to the enterprise.

Physical match includes the design of the whole workplace and working environment. *Mental match* involves the individuals' information and decision making requirements, as well as their perception of tasks. Mismatches between job requirements and workers' capabilities provide potential for human error.

Factors which influence compliance with health and safety practices

These factors include:

(a) the interaction of individuals with their job and the working environment;

(b) the influence of equipment and system design on human performance.

The design of the job and working environment should be based on *task analysis* of the activities required of the operator. This provides the information for evaluating the suitability of tools and equipment, procedures, work patterns and the worker's physical and social surroundings.

Major considerations in job design

These include:

(a) identification and comprehensive analysis of the critical tasks expected of individuals and appraisal of likely errors;

(b) evaluation of required operator decision making and the optimum balance between human and automatic contributions to safety actions;

(c) application of ergonomic principles to the design of man-machine interfaces, including displays of plant process information, control devices and panel layouts;

(d) design and presentation of procedures and operating instructions;

(e) organisation and control of the working environment, including the extent of the workspace, access for maintenance work, and the effects of noise, lighting and thermal conditions;

(f) provision of the correct tools and equipment;

(g) scheduling of work patterns, including shift organisation, control of fatigue and stress, and arrangements for emergency operations/ situations;

(h) efficient communications, both immediate and over periods of time.

The potential for human error

Limitations in human capacity to perceive, attend to, remember, process and act on information are all relevant in the context of human error. Typical human errors are associated with:

(a) lapses of attention;

(b) mistaken actions;

(c) misperceptions;

(d) mistaken priorities;

(e) in some cases, wilfulness.

Perceptual capabilities

How people perceive risk is associated with their attitude, personality, their ability to process information, memory, extent of training received, the level of arousal and individual skills available.

The principal sensory inputs are sight and hearing, whereas touch, taste and smell are more secondary sensory inputs. However, no two people necessarily perceive danger in the same way. Perception of risk is affected by past experience, the context in which the information or stimulus is presented and the extent of training received.

Mental and physical capabilities

These capabilities vary considerably from person to person, and clearly need consideration in job selection, job design, in the design of safe systems of work and in the assessment of hazards.

Both the physical and mental requirements of tasks should be assessed as part of task analysis and job safety analysis.

In certain cases, persons with mental or physical incapacities may need to be restricted in their activities, particularly where continuing forms of danger may be present.

Human error

The HSE publication *Human Factors and Industrial Safety* (HS(G)48, 1989) lists a number of factors that can contribute to human error which can be a significant causative feature of accidents at work.

1. Inadequate information

People do not make errors merely because they are careless or inattentive. Often they have understandable (albeit incorrect) reasons for acting in the way they did. One common reason is ignorance of the production processes in which they are involved and of the potential consequences of their actions.

2. Lack of understanding

This often arises as a result of a failure to communicate accurately and fully the stages of a process that an item has been through. As a result, people make presumptions that certain actions have been taken when this is not the case.

3. Inadequate design

Designers of plant, processes or systems of work must always take into account human fallibility and never presume that those who operate or maintain plant or systems have a full and continuous appreciation of their essential features. Indeed, failure to consider such matters is, itself, an aspect of human error.

Where it cannot be eliminated, error must be made evident or difficult. Compliance with safety precautions must be made easy. Adequate information as to hazards must be provided. Systems should 'fail safe', that is, refuse to produce unsafe modes of operation.

4. *Lapses of attention*

The individual's intentions and objectives are correct and the proper course of action is selected, but a slip occurs in performing it. This may be due to competing demands for (limited) attention. Paradoxically, highly skilled performers, because they depend upon finely tuned allocation of their attention, to avoid having to think carefully about every minor detail, may be more likely to make a slip.

5. *Mistaken actions*

This is the classic situation of doing the wrong thing under the impression that it is right. For example, the individual knows what needs to be done, but chooses an inappropriate method to achieve it.

6. *Misperceptions*

Misperceptions tend to occur when an individual's limited capacity to give attention to competing information under stress produces tunnel vision or when a preconceived diagnosis blocks out sources of inconsistent information. There is a strong tendency to assume that an established pattern holds good so long as most of the indications are to that effect, even if there is an unexpected indication to the contrary.

One potent source of error in such situations is an inability to analyse and reconcile conflicting evidence deriving from an imperfect understanding of the process itself or the meaning conveyed by the instruments. Full analysis of the preventative measures required involves the need for people to understand the process as well as technical and ergonomic considerations concerned with the instrumentation.

The official report on the accident in 1979 at the Three Mile Island nuclear power station in the USA cited human factors as the main causes. Misleading and badly presented operating procedures, poor control room design, inadequate training and poorly designed display systems all, in one way or another, gave the operators misleading or incomplete information. In the event, the radioactive exposure off the site was very small indeed. However, the official enquiry emphasised how failures in human factors design, inadequate training and procedures and inadequate management organisation led to a series of relatively minor technical faults being magnified into a near disaster with significant economic and possible human consequences.

7. *Mistaken priorities*

An organisation's objectives, particularly the relative priorities of different goals, may not be clearly conveyed to, or understood by, individuals. A crucial area of potential conflict is between safety and other objectives, such as output or the saving of cost or time. Misperceptions may then be partly intentional as certain events are ignored in the pursuit of competing objectives. When top management's goals are not clear, individuals at any level in the organisation may superimpose their own.

8. *Wilfulness*

Wilfully disregarding safety rules is rarely a primary cause of accidents. Sometimes, however, there is only a fine dividing line between mistaken priorities and wilfulness. Managers need to be alert to the influences that, in combination, persuade staff to take (and condone others taking) short cuts through the safety rules and procedures because, mistakenly, the perceived benefits outweigh the risks, and they have perhaps got away with it in the past.

9. *Elimination of human error*

For the potential for human error to be eliminated or substantially reduced, all the above factors need consideration in the design and implementation of safe systems of work, processing operations, work routines and activities. Training and supervision routines should take account of these factors and the various features of human reliability.

Personal factors

The factors affecting the way people behave in a potential accident situation are largely associated with individual perception of risk (*see* Fig. 29.1). People perceive risk in many different ways, all of which are determined by psychological factors such at attitude, personality, perception, memory, motivation, training and, in many cases, the skills available to the individual. Stress in individuals can be an immediate cause of accidents (*see* Chapter 31 'Stress').

In order to appreciate the relevance of psychological factors in the assessment and control of hazards, the various components of human behaviour are outlined below.

Fig. 29.1 ◆ Performance-shaping factors

MAN-MACHINE INTERFACE CHARACTERISTICS (DISPLAYS AND CONTROLS)
Compatibility
Ease of operation
Reliability
Meaning
Feedback
Sufficiency
Location
Readability
Distinguishability
Identification

TASK DEMANDS
Perceptual
Physical
Memory
Attention
Vigilance

TASK CHARACTERISTICS
Frequency
Repetitiveness
Workload
Criticality
Continuity
Duration
Interaction with other tasks

INDIVIDUAL FACTORS
Capacities
Training
Experience
Skills
Knowledge
Personality
Physical condition
Attitudes
Motivation

INSTRUCTIONS AND PROCEDURES
Accuracy
Sufficiency
Clarity
Meaning
Readability
Ease of use
Applicability
Format
Level of detail
Selection and location
Revision

SOCIO-TECHNICAL FACTORS
Manning
Work hours/breaks
Resource availability
Actions of others
Social pressures
Organisation structure
Team structure
Communication
Authority
Responsibility
Group practices
Rewards and benefits

ENVIRONMENT
Temperature
Humidity
Noise
Vibration
Lighting
Workspace

STRESSES
Time pressure
Workload
High risk environment
Monotony
Fatigue/pain/discomfort
Conflicts
Isolation
Distraction
Vibration
Noise
Lighting
Temperature
Movement constriction
Shiftwork
Incentives

Attitude

Attitude can be defined as a 'predetermined set of responses, built up as a result of experience of similar situations', 'a shorthand way of responding to a situation', or 'a tendency to respond positively (favourably) or negatively (unfavourably) to certain persons, objects or situations'. It is, to some extent, connected with a person's self-image. Everyone has some form of self-image by way of attitudes which are particularly strong and well integrated in that person. It is extremely difficult to change a person's self-image, his resistance to such change often being a defence mechanism against realities, and forced changes of self-image can sometimes have traumatic results. Exactly when an attitude will result in certain behaviour, and when people will behave differently to their normal attitude, are very important. Opinion polls, for instance, endeavour to link attitudes and behaviour more strongly. On the other hand, attitudes can bear no relationship to facts. This is particularly so in the case of the superstitions held by people, for instance the frequently-encountered view that all accidents are 'acts of God' over which they have no control.

Social groups also have a strong influence in the formation of attitudes. Social groups of all types, such as students' groups, clubs, political parties, trade unions, sports teams, professional bodies and skilled workers' groups, all set 'group norms' or standards, whereby membership of the group entails sharing the attitudes of the group and conforming with the norms set. It is generally difficult to change the central attitude of a group and people will often, after training aimed at changing their attitude to, for instance, the use of safe systems of work, revert to the group attitude after completing such training. Hence the importance of training people in their individual work groups if any significant change in attitude, particularly relating to safe working, is to be achieved.

To be effective, the changing of a person's attitude must take place in a series of stages or steps. The first stage must be that of attracting the attention of the person to the fact that a change of attitude is needed, and secondly of convincing him that his attitude is incorrect or wrong. In the safety field it may be a question of convincing someone that he is acting in an unsafe manner, which could result in personal injury or injury to fellow workers.

A further barrier to attitude-change is the problem of 'cognitive dissonance', the conflict situation where a person holds an attitude which is not compatible with the information being presented, for instance in the case of eye injuries through failure to wear eye protection. To remedy this dissonant situation, training and discussion with the individual should be designed to direct his thoughts through a number of stages, thus:

1. I do not wear eye protection.
2. People could be blinded through using this machine.
3. I could be blinded through using this machine.
4. If I wear the eye protection, I should avoid being blinded.
5. Therefore I shall wear the eye protection in future.

This is the ideal chain of events, but people seek other ways of getting out of the dissonant situation, e.g.

> 'You can't see what you're doing when you use the eye protectors!'

or

> 'I've been doing this job for 30 years and haven't been hit in the eye yet!'

These statements all insulate the person against the facts by refusing to recognise them. What is important is that when views are in conflict, there is motivation to remedy the situation. Training, supervision and propaganda aimed at improving attitudes should, therefore, take into account why people tend to rationalise on these issues.

The following factors can affect the changing of attitudes by individuals or groups:

(a) *The individual* – built-in opinion, degree of conservatism, past experience, level of intelligence and extent of education, motivating factors and the extent that he is prepared to follow blindly the example of others.

(b) *Attitude currently held* – the problem of cognitive dissonance, self-image, group norms, masculine versus feminine behaviour (sex identification), financial gain, both short-term and long-term, the opinions of other people, skills. These are all strong motivators which can bring about attitude change. Other factors include the 'daredevil' attitude, i.e. risk for its own sake, and certain past events which may have helped form a particular attitude.

(c) *Situation* – group situations, group norms, the influence of change agents, e.g. HSE inspectors, health and safety specialists and insurance company liability surveyors, possible sanctions that could be imposed by management or trade unions, prestige, the climate for change at a particular time.

(d) *Management example* – perhaps the strongest of all strategies for bringing about a change in people's attitudes at work, particularly in terms of compliance with procedures, wearing of personal protective equipment and acceptance of responsibilities for health and safety.

(e) *The company culture* – where health and safety is ranked of equal importance to production, sales, communications, quality assurance, etc.

(f) *Publicity* – the use of various forms of propaganda aimed at raising awareness of health and safety issues, including safety posters, films, displays, competitions and demonstrations. Some forms of planned motivation scheme, such as safety incentive schemes, may bring about short-term changes in attitude, but there may be difficulties in maintaining this attitude change achieved once the incentive is removed.

Personality

Defined as 'the dynamic organisation within the individual of the psychophysical systems that determine his characteristic behaviour and thought', personality relates to people's behaviour as perceived by others, e.g. rigid, honest, bumptious, sincere, etc. The term 'dynamic' implies that personality is composed of interacting parts, and this interaction produces flexibility in response to situations, i.e. the ability to adapt to change. The degree that an individual is able to accept change is important, particularly in the selection of people who may be exposed to continuing forms of danger.

Perception

How people perceive danger is significant. This generally takes place through the two principal sensory inputs, sight and hearing, although the senses of touch, taste and smell play an important part in many situations.

Perception relies to a great extent on learning and past experience. It is heavily influenced by motivational factors, the context in which a stimulus is produced, e.g. the context in which people hear something, the level of arousal of the individual, ergonomic factors, such as the layout and design of controls and displays to machinery, and the level of training of operators.

Generally, no two people perceive danger in exactly the same way.

Memory

We have both short-term and long-term memory. Memory is essentially associated with how people learn things. Short-term memory, the amount of material one can take in and retain at any one time is, in many cases, a limiting factor on individual ability and safety consciousness. There is a direct connection between short-term memory and human errors or omissions, many of which result in accidents.

Long-term memory, on the other hand, operates largely through the repetition of facts or information, such as children learning tables at school and codifying them to produce a meaning. A mnemonic is an example of such codifying systems. There is a characteristic drop in memory over a period of time associated with ageing, and interference with memory can be caused by:

(a) events of close similarity which tend to confuse; or

(b) memory overload situations where what has been committed to memory earlier interferes with material subsequently memorised, thereby creating confusion.

Motivation

The factors which motivate people to behave in particular ways are important in accident prevention. In the last 50 years a number of theories of motivation, particularly those of Herzberg, have been applied to the safety situation. Much of this thinking hinged around the concept of 'The Quality of Working Life' which was concerned with promoting industrial harmony and increasing productivity during a period of history fraught with the problems of inflation (*see* Fig. 29.2). The ideas of Herzberg have had a significant impact in this field and should form an important consideration in the executive decision-making process, particularly in the design and implementation of health and safety programmes.

Herzberg's original hypothesis followed the notion that the opposite of dissatisfaction is not satisfaction, but simply no dissatisfaction, and that

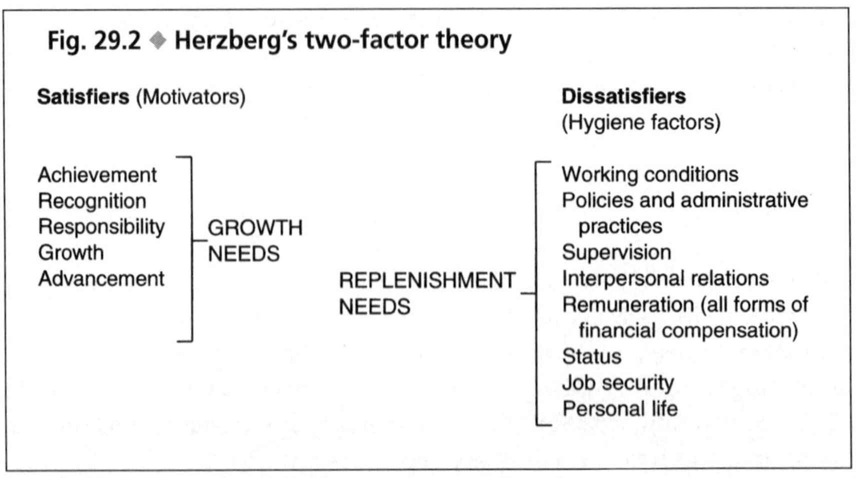

Fig. 29.2 ◆ Herzberg's two-factor theory

the absence of satisfaction is not dissatisfaction but no satisfaction. His postulation of these factors as being different in character caused the theory to be called the '*Herzberg two-factor theory of job satisfaction*'. The following is a list of the basic factors which he referred to as 'satisfiers' and 'dissatisfiers'. It illustrates the basic difference between the two.

Further analysis suggests that the satisfiers are all integral to the performance of the job, and are referred to as 'job content factors'. Herzberg further called the satisfiers 'motivators' and the dissatisfiers 'hygiene factors'. He referred to the dissatisfiers as 'replenishment needs' since provision must always be made for them, but their importance is realised only when they are inadequate or absent. In other words, people expect their replenishment needs to be met at all times. The motivators, on the other hand, he called 'growth needs' since they are the work elements that provide real motivation in this theory of job satisfaction.

Abraham Maslow in 1943 developed the concept of 'self-actualising man', in which he identified a 'hierarchy of needs' as follows:

(a) *Physiological needs* – the needs for the basic necessities of life, i.e. food, water, air to breathe.

(b) *Safety needs* – the needs for physical and psychological safety and security, for shelter, freedom from attack both physically and mentally.

(c) *Social needs* – the need to relate to other people, the need for friendship and affection, and for belonging to a group, both at home and at work.

(d) *Esteem needs* – the need for self-respect and the respect of others, for competence, independence, self-confidence and prestige.

(e) *The need for self-actualisation* – 'What a man can be; he must be', the need to become what one is capable of becoming; self-fulfillment, self-expression and creativity.

The principal lesson to be learnt from Maslow is that all needs must be considered in an attempt to motivate people at work. This must follow a specific pattern starting with people's physiological needs progressing upwards through the other forms of need. Only when one need has been completely satisfied should the next need directly above be considered. Moreover, any particular need will be more dominant at any one time than another as far as people are concerned. Any attempt to stimulate motivation by means of incentives relevant to needs at a lower level or one higher than that which is dominant is likely to fail. Individuals and groups differ with regard to the needs that are dominant and the same incentives cannot be expected to motivate everyone at the same time. Attempts must be made, therefore, to identify needs of a group which are dominant at a particular time if members of the group are to be successfully motivated.

What comes out of a study of Maslow's work is that if companies and organisations are to succeed in the marketplace and become highly profitable, they must look to the physiological and safety needs of the workforce first. By providing better standards of health and safety in the workplace, the motivation of the workforce will be increased. It is regrettable that many managements simply fail to recognise this fact, taking the line of 'Comply with the law, but no more!'

Information processing

The relative speed with which people process information is a precondition of many accidents, or the errors or omissions which result in accidents. Here we need to consider Single Channel Theory which reveals that the mechanism for processing information has only a finite capacity. Thus, once a stimulus has been received by the brain, second and subsequent stimuli must wait until the first stimulus has been dealt with. Each stimulus produces (a reaction) and the response can be divided into two elements:

(a) specific reaction time, i.e. the actual time it takes to perceive and process the response;

(b) movement time, i.e. the time taken to move in order to execute the response.

If a second stimulus arrives during the movement time of the first stimulus, it has to wait. The actual movement in response is being monitored, the single channel ensuring the execution of the original response was accurate. The particularly crucial parts of the movement are the beginning and the end, the middle part generally being partially neglected. The significant point about information processing is that people cannot do more than one thing at a time, the speed and sequence of response varying from person to person.

With well-known tasks, such as driving a car, the monitoring action of the brain is frequently reduced, much depending upon the speed with which one can respond to stimuli and not monitor specific movements. Results are achieved by continual practice or the speed-accuracy tradeoff, where the monitoring is voluntarily removed.

The relationship between skills and accidents

Basic skills have a direct relationship to the events leading to an accident. Factors affecting individual skills in relation to accidents include:

(a) *Reaction time* – simple speed of reaction bears little relationship to accident causation. Choice reaction, on the other hand, where several stimuli are presented together with several responses available to choose from can have a direct effect, and is frequently the cause of what are classed as 'human error' type accidents.

(b) *Co-ordination* – manipulation and dexterity tests, e.g. the loop on the electric wire, which require hand and eye co-ordination, can be used to predict accident potential in certain cases, e.g. driving and flying activities.

(c) *Attention* – where attention is divided between several stimuli, and the operator must respond to all at the same time, a considerable degree of skill is needed. This is an important causative factor in many accidents.

Conclusion

Human factors are emerging as a significant factor in safety and accident prevention. Evidence now shows that too much emphasis in the past has been given to compliance with legal requirements and the promotion of 'safe place' strategies, with very little or no attention being paid to the way people perform within the work system and environment and the reasons for, on occasions, unsafe behaviour.

There is no doubt that human factors can be managed successfully, however. By adopting a systematic approach, by examining the various features of human factors and concentrating on the practical application of sound safety management techniques, most failings associated with people could be prevented.

30 Ergonomics

The scope of ergonomics

Ergonomics is the scientific study of the interrelationships between people and their work. Ergonomics, or 'human factors engineering' or 'the scientific study of work', seeks to create working environments in which people receive prime consideration. It is a multidisciplinary study which incorporates the expertise of engineers, occupational physicians, occupational health nurses, occupational hygienists, health and safety specialists, organisation and method study people and research scientists in a team approach to the examination of numerous aspects of the working environment.

Ergonomics is also sometimes narrowly referred to as the study of the 'man-machine interface'. This interface is significant in the design of working layouts and safe systems of work and in setting work rates. Ergonomics covers four main areas.

The human system

This is the study of the principal characteristics of people, in particular the physical elements of body dimensions, strength and stamina, coupled with psychological elements of learning, perception, and reaction to given situations.

Environmental factors

This area examines the effects on the human system of the working environment, in particular the effects of temperature, light, ventilation, humidity, sound and vibration (*see* Chapters 15 and 21).

The man-machine interface

This is the study of displays, controls and design features of machinery, automation and communication systems, with a view to reducing operator error.

Total working system

Here consideration is given to fatigue, stress, work rate, productivity, systems effectiveness and the related aspects of health and safety. This area would cover, for instance, the aspect of working posture, in particular the need to ensure a correct and comfortable stance, in order to avoid long-term bad posture effects. These areas of study may be summarised as in Table 30.1

Table 30.1 ◆ The total working system – areas of study

Human characteristics	*Environmental factors*
Body dimensions Strength Physical and mental limitations Stamina Learning Perception Reaction	Temperature Humidity Light Ventilation Noise Vibration
Man-machine interface	*Total working system*
Displays Controls Communications Automation	Fatigue Work rate Posture Stress Productivity Accidents Safety

Ergonomic design

Important features in the design of the interface include the following.

Layout

Layout of working areas and operating positions should allow for free movement, safe access and egress, and unhindered visual and oral communication. Congested, badly planned layouts result in operator fatigue and increased potential for accidents.

Vision

The operator should be able to set and read easily control switches, dials and displays. This reduces fatigue and accidents arising from inadequate or faulty perception.

Posture

The more abnormal the working posture, the greater the potential for fatigue and long-term injury. Work processes and systems should be designed to permit a comfortable posture which reduces excessive job movements. This must be considered in the siting of controls and displays on machinery, plant and vehicles, and in the organisation of working systems, e.g. assembly work.

Work rate

Rates should be set to suit the operator. They need constant reassessment and revision. Movements which are too fast or too slow cause fatigue.

Comfort

The comfort of the operator, whether driving a vehicle or operating machinery, is essential for physical and mental well-being. Environmental factors directly affecting comfort – e.g. lighting, temperature, ventilation and humidity levels – should be given priority.

Anthropometry

A key feature of ergonomic design is matching people to their equipment. Anthropometry is the study and measurement of body dimensions, the orderly treatment of resulting data and the application of those data in the design of workspace layouts and equipment. Few workstations are 'made to measure' owing to the wide range of human dimensions and the sheer cost of designing individual workstations and machines to conform with individual body measurements. The fact that this is not done creates many problems, best demonstrated by research undertaken at the Cranfield Institute of Technology, who created 'Cranfield Man' (*see* Fig. 30.1). Using a horizontal lathe, researchers examined the positions of controls and compared the locations of these controls with the physical dimensions of the average operator. Table 30.2 shows the wide differences between the two. The ideal operator would be 1.35m tall with a 2.44m arm span. Clearly, no effort had been made to design the machine even remotely within the bodily limitations of the average operator. More recently, however, greater attention has been paid to this aspect, particularly in the design of machinery. This can be seen in the improved design of controls

Fig. 30.1 ◆ 'Cranfield Man' – 1.35m tall with a 2.44m arm span

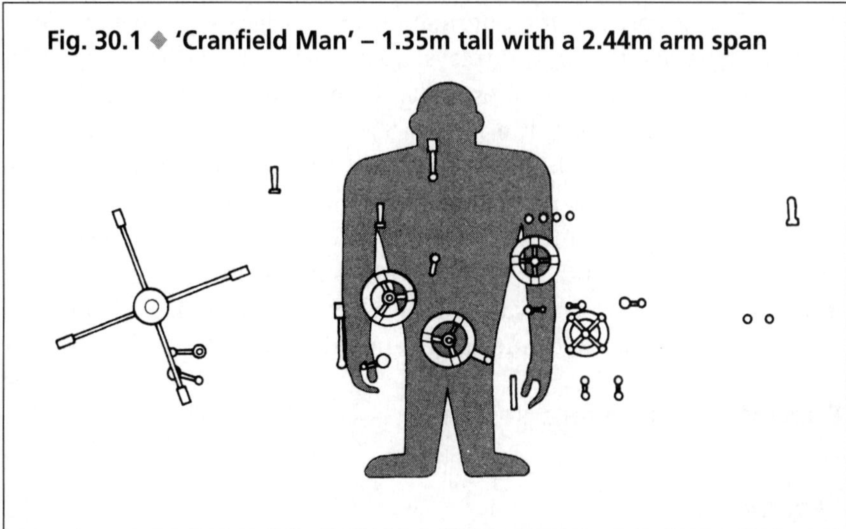

Table 30.2 ◆ Physical dimensions of an average operator compared with those of 'Cranfield Man'

Average operator	*Dimension*	*Operator who would suit these controls*
1.75m	Height	1.35m
0.48m	Shoulder width	0.61m
1.83m	Arm span	2.44m
1.07m	Elbow height	0.76m

and displays on motor vehicles, cranes, aircraft, computer terminals and processing machinery, the objective being to reduce operator fatigue and error, which lower productivity and contribute to accidents.

Interface design

The principal objective of interface design is to minimise the potential for error in the operation of plant and equipment. Equipment can include anything from a horizontal lathe to a jet aircraft. Design is concerned with two aspects, namely controls and displays. Controls take many forms, e.g. the steering wheel of a fork lift truck, the lever on a crane, an electrically operated stop-start button on a vertical drill.

Displays supply specific information to the operator. They are either static or dynamic. Dynamic displays, such as gauges, give an instant indication of variations, e.g. in pressure and temperature. (The most obvious form of dynamic display is a clock.) Static displays include instruction labels on plant and machinery, wiring diagrams and specific data relating to dangerous substances or emergency fire procedures.

In the design of controls and displays, it is necessary to consider the potential for error. This has been most apparent in the motor vehicle industry with its standardisation of gear-change routines, signalling and the layout of speedometer, fuel gauge and other gauges.

Principles of interface design

Separation

Physical controls should be separated from displays. The safest routine is achieved where there is no relationship between them.

Comfort

If separation cannot be achieved, control and display elements should be mixed to produce a system which can be operated with ease.

Order of use

Controls and displays can be set in the order in which they are used, e.g. left to right for start-up and the reverse direction for closedown of plant.

Priority

Where there is no competition for space, controls most frequently used should be sited in key positions. Controls such as emergency stop buttons should be sited in the most easily seen and reached position.

Function

With large consoles, controls can be divided according to functions. This division is commonly found in power stations. The layout relies heavily on the ability of the operator and speed of reaction. A well-trained operator, however, benefits from such functional division and the potential for error is reduced.

Fatigue

Convenient siting of controls is paramount. In designing a layout, the hand movements and body positions of the operator should be observed and studied.

Working conditions, ergonomics and health

The quality of the working environment, the health of workers and the adoption of ergonomic principles are increasingly seen by many employers as essential components of the 'human factors' approach to health and safety at work. These relationships are summarised in Table 30.3.

Table 30.3 ◆ Relationships between industrial conditions, ergonomics and health

ANATOMY AND PHYSIOLOGY		
Industrial condition	*Human factors*	*Health hazards*
Dimension of seats, benches	Anthropometry – body dimensions	Bad posture Discomfort General fatigue
Motion study – workplace layout – workrate	Anthropometry – body dimensions – strengths of muscle groups Structures of joints Functions of muscles	Bad posture Discomfort General fatigue Local muscular fatigue (incl. tenosynovitis)
Design of hand tools	Anthropometry – body dimensions – strengths of muscle groups Structures of joints Functions of muscles	Local muscle strain Local muscular fatigue (incl. tenosynovitis)
Design of controls – levers – hand-wheels – knobs – buttons, etc	Anthropometry – body dimensions – strengths of muscle groups Structures of joints Functions of muscles	Local muscle strain Local muscular fatigue (incl. tenosynovitis)
Manual handling	Kinetic methods: based on – structures of joints – strengths of muscle groups	Injuries – muscle strains – hernias – skeletal damage – slipped discs
Heavy manual work	Physical fitness Physiological cost of work – oxygen consumption – heart rate – body temperature	General fatigue
Control of – air temperature – radiant heat – humidity – air movement	Physiological cost of work – heart rate – body temperature Temperature regulating mechanisms	Heat stress and disorders Cold stress and disorders
Vibration Cold		Raynaud's phenomenon
Flying Diving Caisson work	Ear anatomy Gases in blood	Ear damage Bends Anoxia Oxygen poisoning
Noise	Hearing	Deafness Auditory discomfort

Table 30.3 ◆ continued

PSYCHOLOGY		
Industrial condition	*Human factors*	*Health hazards*
Control of – air temperature – radiant heat – humidity – air movement	Subjective feelings	Thermal discomfort
Design of indicators – dials – warning lights, etc. Design of controls – levers – buttons, etc.	Sensory and perceptual abilities – especially visual 'Natural' directions	Accidents Stress
Design of – labels – notices – posters	Visual abilities – sensory – perceptual Mental abilities – learning – thinking, etc.	Accidents Stress
Lighting – quantity – distribution – glare Colour – environment – colour coding	Visual abilities – sensory – perceptual Visual abilities – sensory – perceptual Mental abilities – learning – thinking, etc.	Accidents Stress Visual fatigue Visual discomfort Depression Accidents Stress
Inspection Fine assembly (Arrangements of – lighting – contrasts – colours – movement, etc.)	Visual abilities – sensory – perceptual Vigilance	Boredom Stress Visual fatigue
Job design – duties	Sensory and perceptual abilities – especially visual Mental abilities – learning – thinking, etc. Motivation	Boredom Stress Accidents
Human relations	Personality	Stress Neuroses

31 Stress

The meaning of stress

Stress is a word which is rarely clearly understood. It means different things to different people. Indeed, almost anything one can think of, pleasant or unpleasant, has been described as a source of stress, e.g. getting married, being made redundant, getting older, getting a job, too much or too little work, solitary confinement or excessive noise. A number of definitions of 'stress' are shown below.

Definitions

(a) Any influence that disturbs the natural equilibrium of the living body.

(b) The common response to attack (Hans Selye, 1936).

(c) Some taxation of the body's resources in order to respond to some environmental circumstance.

(d) The common response to environmental change.

(e) A psychological response which follows failure to cope with problems.

(f) A feeling of sustained anxiety which, over a period of a time, leads to disease.

(g) The non-specific response of the body to any demands made upon it.

Stress could be defined simply as the rate of wear and tear of the body systems caused by life. The acknowledged father of stress research, Dr Hans Selye, a Vienna-born endocrinologist of the University of Montreal, comments thus: 'It is important that people understand what they are talking about when they speak about stress. Whenever anyone experiences something unpleasant, for lack of a better word they say they are under stress.' In his classic book *The Stress of Life*, Selye (1936) corrects several notions relating to stress, in particular:

(a) Stress is not nervous tension.

(b) Stress is not the discharge of hormones from the adrenal glands: the common association of adrenalin with stress is not totally false, but the two are only indirectly associated.

(c) Stress is not simply the influence of some negative occurrence: stress can be caused by quite ordinary and even positive events, such as a passionate kiss.

(d) Stress is not an entirely bad event; we all need a certain amount of stimulation in life and most people can thrive on some forms of stress.

(e) Stress does not cause the body's alarm reaction, which is the most common misuse of the expression; what causes a stress reaction is a stressor.

Whilst (a) to (e) above eliminate virtually everything that people associate with stress, a number of common factors emerge from these definitions and comments. Stress is a state manifested by a specific syndrome of biological events. Specific changes occur in the biological system, but they are caused by such a variety of agents that stress is, of necessity, non-specifically induced. The key to understanding the nature of this overall biological impact lies in the fact that some stress response, however slight, will result from any stimulus.

Quite simply, a 'stressor' produces stress. First, the stressors may be extremes of temperature, lighting or ventilation, or the emission of noise, dust or fume from a process (environmental stressors). Second, stress may be induced by isolation, rejection or the feeling that one has been badly treated (social stress). Third, stress can be viewed as a general overloading of the body systems (distress). Stress has a direct association with the autonomic system which controls an individual's physiological and psychological responses . This is the 'flight or fight' system, characterised by two sets of nerves, the sympathetic and parasympathetic, which are responsible for the automatic and unconscious regulation of body function. The sympathetic system is concerned with answering the body's call to fight, i.e. increased heart rate, more blood to organs, stimulation of sweat glands and the tiny muscles at the roots of the hairs, dilation of the pupils, suppression of the digestive organs, accompanied by the release of adrenalin and noradrenalin. The parasympathetic system is responsible for emotions and protection of the body, which have their physical expression in reflexes, such as widening of the pupils, sweating, quickened pulse, blushing, blanching, digestive disturbance, etc.

The general adaptation syndrome

Stress response is a mobilisation of the body's defences, an ancient biochemical survival mechanism perfected during the evolutionary process, allowing human beings to adapt to threatening circumstances. In 1936, in *The Stress of Life*, Selye defined the 'general adaptation syndrome', recognised as a major advance in biological research. It comprises three stages:

(a) *The alarm reaction stage*. This is typified by receiving a shock, at which time the body's defences are down, followed by a counter shock when the defences are raised. In physiological terms, once a stressor is recognised, the brain sends forth a biochemical 'messenger' to the pituitary gland which secretes adrenocortitrophic hormone (ACTH). ACTH causes the adrenal glands to secrete corticoids, such as adrenalin. The result is a general 'call to arms' of the body's systems.

(b) *The resistance stage*. This stage is concerned with two responses. The body will either resist the stressor or adapt to the effects of the stressor. It is the opposite of the alarm reaction stage, whose characteristic physiology fades and disperses as the organism adapts to the derangement caused by the stressor.

(c) *The exhaustion stage*. If the stressor continues to act on the body, however, this acquired adaptation is eventually lost and a state of overloading is reached. The symptoms of the initial alarm reaction stage return and, if the stress is unduly prolonged, the wear and tear will result in damage to a local area or death of the organism as a whole (*see* Fig. 31.1).

The causes of stress

The causes of stress are diverse. They are classified thus.

Environmental stressors

Stress in the working environment is caused by extremes of temperature, inadequate lighting and ventilation, presence of dust, fumes, vapours and gases, noise and vibration and poor amenity provision. Noise, with its attendant risk of occupational deafness, is one of the greatest environmental stressors.

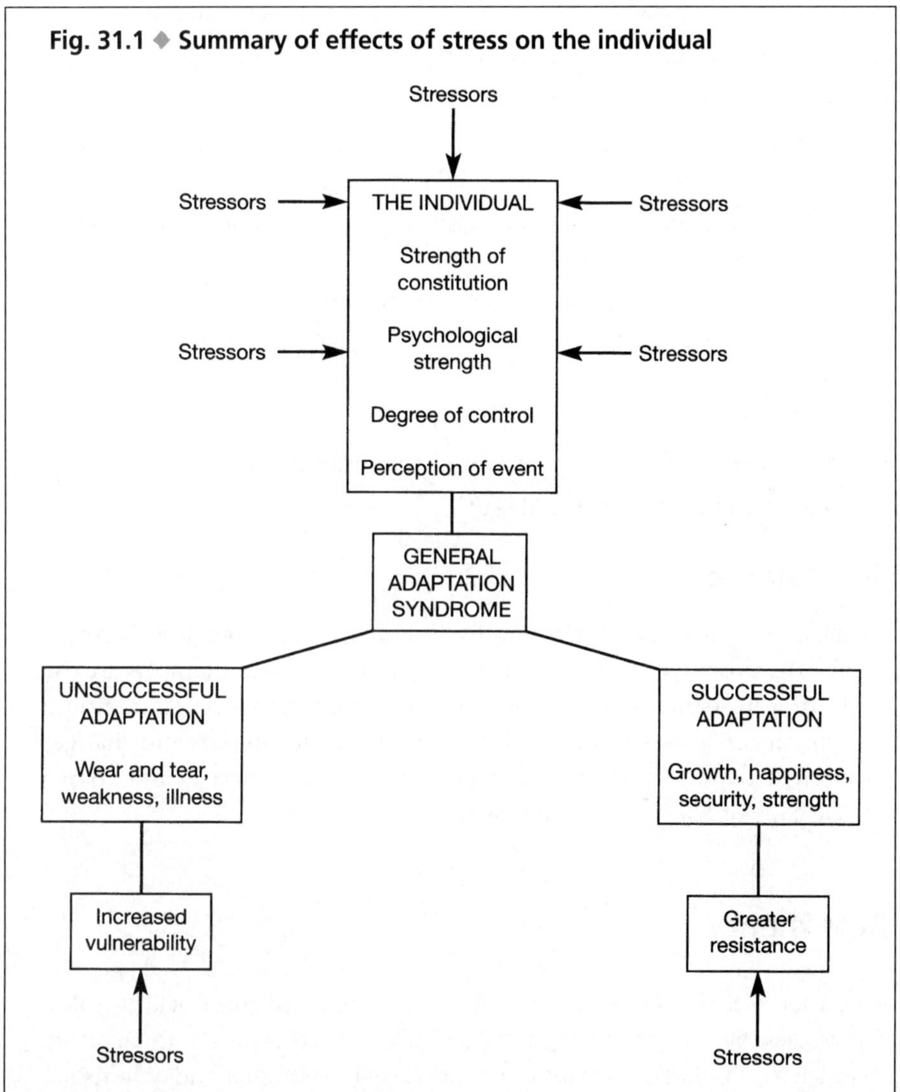

Fig. 31.1 ◆ Summary of effects of stress on the individual

Occupational stressors

Stressful conditions at work can lead to accidents or health deterioration. Typical conditions include:

(a) too heavy or too light a workload;

(b) a job which is too difficult or too easy;

(c) working excessive hours, e.g. 60 or more hours per week;

(d) conflicting job demands – the 'servant of two masters' situation;
(e) too much or too little responsibility;
(f) poor human relationships;
(g) incompetent superiors, in terms of their ability to make decisions, their level of performance and their job knowledge;
(h) lack of participation in decision making and other activities where a joint approach would be beneficial;
(i) middle-age vulnerability associated with reduced career prospects or the need to change career, the threat of redundancy or premature retirement;
(j) overpromotion or underpromotion;
(k) interaction between work and family commitments;
(l) deficiencies in interpersonal skills.

Social stressors

Social stress is associated with family life, marital relationships, bereavement – the everyday problems of coping with life. Typical examples are the death of a husband or wife, parent or child, moving house, maintaining repayments on a large mortgage, trouble with the neighbours and changes in social activities. These situations result in worry and anxiety and no two people cope or adapt in the same way.

Role theory

Role theory views large organisations as systems of interlocking roles. These roles relate to what people do, and what others expect of them rather than their individual identities. An individual's thoughts and actions are influenced by identification with a role and with the duties and rights associated with that role. Everyone in a role has contact with people – superiors, subordinates, external contacts or contractors who communicate their expectations of the role holder, trying to influence his behaviour and subjecting him to feedback. The individual, therefore, has certain expectations about how people should behave according to their status, age, function and responsibility. These expectations form the basis for a standard by which individual behaviour is evaluated, as well as a guide for reward. Stress arises in this framework due to role ambiguity, role conflict and role overload.

Role ambiguity

This is the situation where the role holder has insufficient information for adequate performance of his role, or where the information is open to more than one interpretation. Potentially ambiguous situations are in jobs where there is time lag between action taken and visible results, or where the role holder is unable to see the results of his actions.

Role conflict

Role conflict arises where members of the organisation, who exchange information with the role holder, have different expectations of his role. Each may exert pressure on the role holder. Satisfying one expectation could make compliance with other expectations difficult.

Role overload

This results generally from a combination of role ambiguity and role conflict. The role holder works harder to clarify normal expectations or to satisfy conflicting priorities which are impossible to achieve within the time limits specified.

Research has shown that when experience of role conflict, ambiguity and overload is high, then job satisfaction is low. This may well be coupled with worry and anxiety. These factors may add to the onset of stress-related diseases such as peptic ulcers, coronary heart disease and nervous breakdowns.

Evaluation of stress

Major changes in people's lives – such as marital separation, changes in responsibility at work, job loss and even getting married – can be stressful. Research by Dr Holmes and Dr Rahe of the School of Medicine, University of Washington, USA, into the clinical effects of major life changes has identified the concept of the 'life change unit' (LCU), a unit of individual stress measurement in terms of the impact of stress on health (Table 31.1). Over a period of 20 years Holmes and Rahe were able to assign a numerical value to a range of 'life events', such as a son or daughter leaving home, change in residence or the death of a family member, and rank them according to their magnitude and importance. Then they compared the LCU scores of some 5,000 individuals with their respective medical

Table 31.1 ◆ Holmes-Rahe scale of life change events

Event	*LCUs*	*Event*	*LCUs*
Death of a spouse	100	Change in work responsibilities	30
Marital separation	65	Son or daughter leaving home	29
Death of a close family member	63	Trouble with in-laws	29
Personal injury or illness	53	Outstanding personal achievement	29
Marriage	50	Wife beginning or stopping work	29
Loss of job	47	Revision of personal habits	24
Marital reconciliation	45	Trouble with business superior	23
Retirement	45	Change in work hours or conditions	20
Change in health of a family member	44	Change in residence	20
Wife's pregnancy	40	Change in schools	20
Sex difficulties	39	Change in recreation	19
Gain of new family member	39	Change in social activities	18
Change in financial status	38	Taking out a small mortgage on your home	17
Death of a close friend	37	Change in sleeping habits	16
Change to a different kind of work	36	Change in number of family get-togethers	15
Increase or decrease in arguments with spouse	35	Change in eating habits	15
Taking out a bigger mortgage on home	31	Vacation	13
Foreclosure of mortgage or loan	30	Minor violations of law	11

histories. They concluded that those with a high rating on the life change index were more likely to contract illness.

The 'schedule of recent events', based on the total number of LCUs received in a year, has since been applied to many groups, confirming the notion that the higher the degree of life change within a period of time, the greater the risk of subsequent illness, regardless of whether the change is perceived as desirable or undesirable. According to Holmes and Rahe, if an individual's LCUs total 150–99, he stands a mild chance of illness in the following year. If the total is in the range 200–99, then he stands a moderate risk. Over 300 LCUs puts him in the group very likely to suffer serious physical or emotional illness.

The lesson to be learnt from this theory is that people should try to regulate the changes in their lives, most of which are under their control, and endeavour to stagger their incidence and intensity. Table 31.1 shows life change events and LCU ratings. The values assigned to the Holmes-Rahe scale are, of course, averages.

Effects of stress

The signs and symptoms of stress disorders vary. They can include headaches, inability to sleep, fatigue, over-eating, constipation, lower back pain, allergies, nervousness, nightmares, high blood pressure, alcohol abuse, indigestion, dermatitis, menstrual distress, nausea, irritability, loss of appetite, asthma attacks, depression, arthritis, minor accidents, peptic ulcers, heart palpitations, sexual problems, feelings of anger and many others. Typical signs of this 'flight or fight' response include rapid pulse, increased perspiration, pounding heart, tightened stomach, tensing of limb muscles, shortness of breath, gritting the teeth, clenching the jaw, an inability to sit still, racing thoughts and gripping emotions. No two people react in the same way to a stressful occurrence; one may become withdrawn and depressed, whilst another is hyperactive, compulsive or abnormally gregarious. One loses his appetite, while another becomes gluttonous; one sleeps incessantly, another gets insomnia.

The common psychological effects of stress are anxiety and depression. Anxiety is a state of tension coupled with apprehension, worry, guilt, insecurity and a constant need for reassurance. It is accompanied by a number of psychosomatic symptoms, such as profuse perspiration, difficulty in breathing, gastric disturbances, rapid heartbeat, frequent urination, muscle tension or high blood pressure. Insomnia is a reliable indicator of a state of anxiety. On the other hand, depression is much more a mood, characterised by feelings of dejection and gloom, and other permutations such as feelings of hopelessness, futility and guilt. The well-known American psychiatrist, David Viscott, described depression as 'a sadness which has lost its relationship to the logical progression of events'. It may be mild or severe. Its milder form may be a direct result of a crisis in work relationships. Severe forms may exhibit biochemical disturbances, and the extreme form may lead to suicide.

Stress-related diseases and conditions

Research indicates that many diseases and conditions incorporate a stress element. Whilst stress is not the sole cause, it can play a key role in the promotion of the disease or condition. Some of the more common stress-related diseases are cardiovascular disease, arteriosclerosis including stroke, angina pectoris and heart attack, hypertension, duodenal and stomach ulcers, diabetes, migraine and disorders of the immune system – e.g. allergies, infections (viruses, influenza, colds) – rheumatoid arthritis and ulcerative colitis. There is a correlation with the incidence of certain forms of cancer, e.g. breast cancer.

Smoking, alcohol and drugs

Stress can result in increased addiction to smoking and alcohol consumption and, occasionally, drug addiction (*see* Chapter 20).

Drugs and drug addiction

The expression 'drug' is popularly identified with a narcotic or habit-forming substance. Technically, however, a drug is a substance taken medicinally to assist recovery from sickness or relieve symptoms, or to modify any natural body process. Consequently, many people see drug taking as the panacea for stress, relying on tranquillisers to reduce anxiety and amphetamines (pep pills) to counter fatigue. Such drug taking represents a major health menace, particularly if the individual consumes alcohol. The prescription of drugs should be seen, in most cases, as a stopgap measure, not a permanent one.

Stress at work

The civil implications

A ticket collector was awarded £375,000 damages against his employer, London Underground Ltd., for the condition known as post-traumatic stress disorder suffered as a result of the King's Cross fire. Whilst he did not suffer any physical injury in the fire in 1987, he did, however, suffer depression and other psychological effects which, he alleged, ruined his life.

In *Walker* v. *Northumberland County Council* [1995] IRLR 35 (judgment delivered by Colman J on 16th November 1994), the High Court held an employer liable for the psychiatric damage suffered by an employee after it failed to take reasonable steps to avert a second nervous breakdown. Traditionally, the requirement for employers to take reasonable care for the employee's health and safety was limited to physical injury, or the risk of physical injury, to an employee at the workplace. This decision has, however, extended this duty of care to embrace mental injury suffered by an employee.

Breach of contract

The *Walker* decision raises a number of practical considerations. It makes it clear that an employer may be in breach of an implied term in failing to take reasonable care of the employee's health and safety at the workplace, which includes mental injury. This may be considered to be a fundamental breach of contract which goes to the root of the contract of employment.

The employee may be entitled to treat the contract as having been repudiated by the employer because of the employer's unreasonable conduct in taking steps to minimise stress in the workplace, and also by exposing the employee to high levels of stress through a heavy workload. The employee could claim constructive dismissal and claim for damages. Thus some consideration should be given to identifying factors that could lead to stress at the workplace and to establish ways of minimising the stress levels.

The criminal implications

The HSWA and various regulations, such as the Management of Health and Safety at Work Regulations and the Health and Safety (Display Screen Equipment) Regulations, are concerned with protecting the health, in addition to, the safety of employees at work.

One definition of the term 'health' is 'a state of physical and mental wellbeing' (International Labour Organisation). Much has been done over the last century to improve the physical health of people at work, but what of their mental health, particularly where mental ill-health may be stress-induced?

It has taken many years for industry and, indeed, the enforcement authorities, to recognise the existence of stress at work. Indeed many organisations now run stress management courses to assist employees to cope with stress, both at work and in their domestic lives, and have formally documented policies on the subject.

Whilst there would appear to be a total lack of criminal case law associated with stress at work, a number of factors need consideration, particularly by employers on this matter.

1. Do risk assessments consider the potential for stress and stress-related ill-health?
2. Do the management employment systems and procedures consider this matter?
3. What provision is made for the health surveillance of staff identified from the risk assessment who may be subject to stress in their day-to-day activities, such as dealing with members of the public, use of display screen equipment, coping with violence or, in the case of social workers, dealing with child abuse cases?
4. What is the quality and extent of information given to employees on the potentially stressful aspects of their jobs?
5. From a human capability viewpoint, what criteria are used to decide whether a particular employee has 'got what it takes' to cope with the stresses associated with various jobs?
6. What factors should an employer consider at the selection stage and subsequently where an employee may be complaining of stress and seeking alternative work?
7. Bearing in mind the absolute duty to provide training for employees under the MHSWR, what measures are taken from a training viewpoint to alert employees to potentially stressful situations that can arise in their work and measures for coping with same?

And what of the role of the enforcement authorities in this matter? Is it conceivable that an HSE inspector, on consideration of a complaint from a worker or group of workers of stress-induced injury, would be prepared to serve an Improvement Notice on the employer requiring him to take appropriate stress-reduction measures?

HSE Guidance on stress in the workplace

HSE Guidance Note HS(G)116 *Stress at Work* indicates that a great deal can be done to alleviate occupational stress through the development and maintenance of 'sensible work organisations, two-way communication and plain good management'.

Factors recognised as contributing to occupational stress are:

(a) excessive periods of repetitive or monotonous work;

(b) uncertainty;

(c) lack of clear objectives;

(d) interpersonal conflict;

(e) inflexible schedules;

(f) over-demanding schedules.

The Guidance Note stresses that it is not concerned about normal day-to-day pressures which are part of any job, but about 'excessive and unreasonable pressures which pose a risk to health'.

It does not prescribe any action for employers to take but offers a broad framework for action which employers can adapt to their needs (*see* Fig. 31.2).

Fig. 31.2 ◆ Signs that stress may be a problem in your workplace

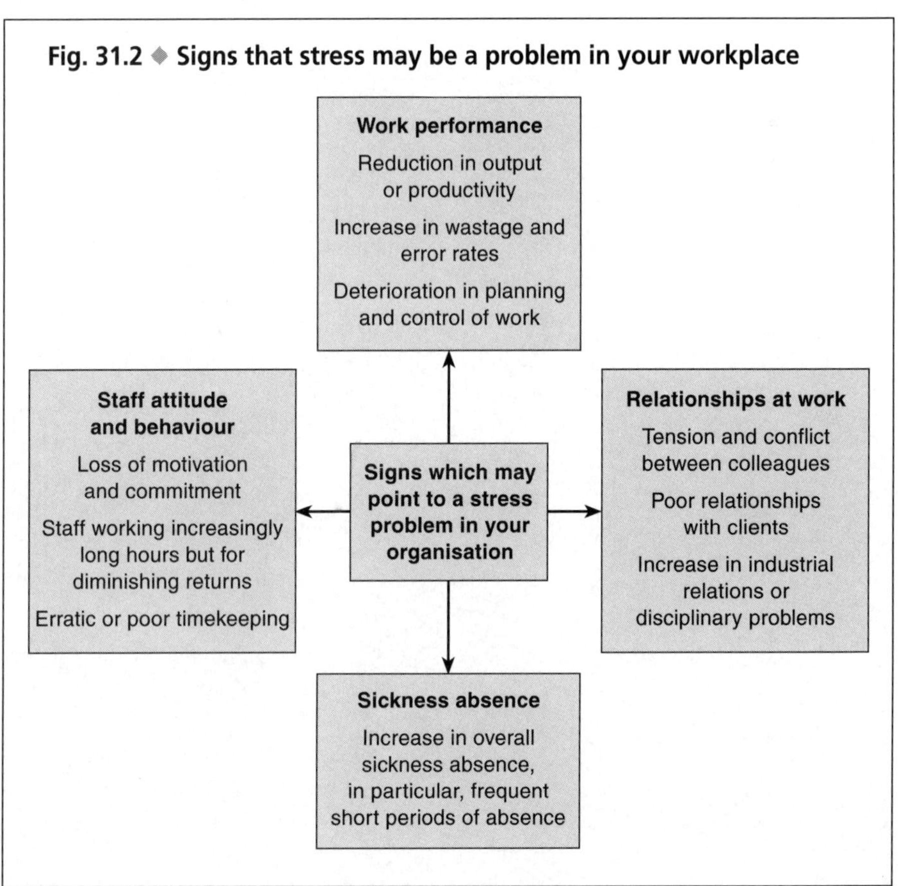

SECTION G

Safety technology

32 Work equipment

Machinery – general aspects

Any analysis of machinery safety involves an examination of the first principles of design. By definition, a machine is an 'apparatus for applying power, having fixed and moving parts, each with definite functions' (British Standard Code of Practice No. 5304 (BSEN 292): 1988 *Safeguarding of machinery*). Machines have:

G

(a) operational parts, which perform the primary output function of the machine, namely the manufacture of a product, e.g. the chuck and drill bit of a vertical drill;
(b) non-operational or functional parts, which convey power or motion to the operational parts, e.g. drives to motors.

The functional parts comprise the prime mover and transmission machinery, which are defined in the Factories Act 1961, sec 176(1), as follows:

> *Prime Mover* – means any engine, motor or other appliance which provides mechanical energy derived from steam, water, wind, electricity, the combustion of fuel or other source.
>
> *Transmission Machinery* – means every shaft, wheel, pulley, drum, system of fast and loose pulleys, coupling, clutch, driving belt or other device by which the motion of a prime mover is transmitted to or received by any machine or appliance.

A motor car is a typical machine. The engine is the prime mover: it provides the power to drive the transmission machinery, in this case the transmission shaft(s), which results in the wheels turning. All machines operate on this principle.

The principal legal requirements relating to machinery are dealt with in the Provision and Use of Work Equipment Regulations 1998 (PUWER).

Provision and Use of Work Equipment Regulations (PUWER) 1998

These comprehensive Regulations are intended:

(a) to implement the Machinery Safety Directive;

(b) to simplify and clarify existing laws on the provision and use of work equipment by the reform of older legislation;

(c) to form a coherent single set of key health and safety requirements concerning the provision and use of work equipment.

The Regulations are supported by guidance prepared by the HSE and HSC containing practical advice on implementation of the Regulations. It must be recognised that, as with most modern health and safety legislation, these Regulations do not stand on their own, but must be read in conjunction with the more general duties of employers under the Management of Health and Safety at Work Regulations 1992. Because of the importance of these Regulations, and the extensive nature of same, the main requirements are dealt with below.

The majority of the requirements are of an absolute nature implying no form of defence is available to an employer, self-employed person or manufacturer, designer, supplier, importer or installer of work equipment when charged with an offence under the Regulations. The Regulations incorporate a number of important definitions.

Inspection in relation to an inspection under paragraph 1 or 2 of Reg 6:

(a) means such visual or more rigorous inspection by a competent person as is appropriate for the purpose described in that paragraph;

(b) where it is appropriate to carry out testing for the purpose, includes testing the nature and extent of which are appropriate for the purpose.

Thorough inspection in relation to a thorough examination under paragraph 1, 2, 3 or 4 of Reg 32:

(a) means a thorough examination by a competent person;

(b) includes testing the nature and extent of which are appropriate for the purpose described in the paragraph.

Work equipment means any machinery, appliance, apparatus or tool or installation for use at work (whether exclusively or not).

Use in relation to work equipment means any activity involving work equipment and includes starting, stopping, programming, setting, transporting, repairing, modifying, maintaining, servicing and cleaning.

The requirements imposed by these Regulations on an employer also apply:

(a) to a self-employed person in respect of work equipment he uses at work;

(b) subject to paragraph 5, to a person who has control to any extent, of –

  (i) work equipment;

  (ii) a person at work who supervises or manages the use of work equipment; or

  (iii) the way in which work equipment is used at work,

  and to the extent of his control.

Any reference above to a person having control of any premises or matter is a reference to the person having control of the premises or matter in connection with the carrying on by him of a trade, business or other undertaking (whether for profit or not).

G

The requirements imposed by these regulations shall not apply to a person in respect of work equipment supplied by him by way of sale, agreement for sale or hire-purchase agreement.

General requirements for work equipment

Part I of PUWER deals with the more general requirements for work equipment.

Suitability of work equipment

1. Every employer shall ensure that work equipment is so constructed or adapted as to be suitable for the purpose for which it is used or provided.
2. In selecting work equipment, every employer shall have regard to the working conditions and to the risks to health and safety of persons which exist in the premises or undertaking in which that work equipment is to be used and any additional risk posed by the use of that work equipment.
3. Every employer shall ensure that work equipment is used only for operations for which, and under conditions for which, it is suitable.

4. In this regulation 'suitable' means suitable in any respect which it is reasonably foreseeable will affect the health or safety of any person (Reg 4).

Maintenance of work equipment

1. Every employer shall ensure that work equipment is maintained in an efficient state, in efficient working order and in good repair.
2. Every employer shall ensure that where any machinery has a maintenance log, the log is kept up to date (Reg 5).

Inspection of work equipment

1. Every employer shall ensure that, where the safety of work equipment depends on the installation conditions, it is inspected:

   (a) after installation and before being put into service for the first time; or

   (b) after assembly at a new site or in a new location,

   to ensure that it has been installed correctly and is safe to operate.
2. Every employer shall ensure that work equipment exposed to conditions causing deterioration which is liable to result in dangerous situations is inspected:

   (a) at suitable intervals;

   (b) each time that exceptional circumstances which are liable to jeopardise the safety of the work equipment have occurred,

   to ensure that health and safety conditions are maintained and that any deterioration can be detected and remedied in good time.
3. Every employer shall ensure that the result of an inspection made under this regulation is recorded and kept until the next inspection under this regulation is recorded.
4. Every employer shall ensure that no work equipment:

   (a) leaves his undertaking; or

   (b) is obtained from the undertaking of another person; or

   (c) is used in his undertaking,

   unless it is accompanied by physical evidence that the last inspection required to be carried out under this regulation has been carried out.
5. This regulation does not apply to:

   (a) a power press to which Regs 32 to 35 apply;

(b) a guard or protection device for the tools of such power press;

(c) work equipment for lifting loads including persons;

(d) winding apparatus to which the Mines (Shafts and Winding) Regulations 1993 apply;

(e) work equipment required to be inspected by Reg 29 of the Construction (Health, Safety and Welfare) Regs 1996 (Reg 6).

Specific risks

1. Where the use of work equipment is likely to involve a specific risk to health or safety, every employer shall ensure that:

   (a) the use of that work equipment is restricted to those persons given the task of using it;

   (b) repairs, modifications, maintenance or servicing of that work equipment is restricted to those persons who have been specifically designated to perform operations of that description (whether or not also authorised to perform other operations).

2. The employer shall ensure that the persons designated for the purposes of paragraph l(b) above have received adequate training related to any operations in respect of which they have been so designated (Reg 7).

G

Information and instructions

1. Every employer shall ensure that all persons who use work equipment have available to them adequate health and safety information and, where appropriate, written instructions pertaining to the use of that work equipment.

2. Every employer shall ensure that any of his employees who supervises or manages the use of work equipment has available to him adequate health and safety information and, where appropriate, written instructions pertaining to the use of that work equipment.

   Without prejudice to the generality of paragraphs 1 and 2, the information and instructions required by either of these paragraphs shall include information and, where appropriate, written instructions on:

   (a) the conditions in which and the methods by which the work equipment may be used;

   (b) foreseeable abnormal situations and the action to be taken if such a situation were to occur;

(c) any conclusions to be drawn from experience in using the work equipment.

4. Information and instruction required by this regulation shall be readily comprehensible to those concerned (Reg 8).

Training of users of work equipment

1. Every employer shall ensure that all persons who use work equipment have received adequate training for the purposes of health and safety, including training in the methods which may be adopted when using the work equipment, any risks which such use may entail and the precautions to be taken.

2. Every employer shall ensure that any of his employees who supervises or manages the use of work equipment has received adequate training for purposes of health and safety, including training in the methods which may be adopted when using the work equipment, any risks which such use may entail and precautions to be taken (Reg 9).

Conformity with Community requirements

1. Every employer shall ensure that any item of work equipment provided for use in the premises or undertaking of the employer complies with any enactment (whether in an Act or instrument) which implements in Great Britain any of the relevant Community Directives listed in Schedule 1 which is applicable to that item of work equipment.

2. Where it is shown that an item of work equipment complies with an enactment (whether an Act or instrument) to which it is subject by virtue of paragraph 1, the requirements of Regs 11 to 24 shall apply in respect of that item of work equipment only to the extent that the relevant Community Directive implemented by that enactment is not applicable to that item of work equipment (Reg 10).

Dangerous parts of machinery

1. Every employer shall ensure that measures are taken in accordance with paragraph 2 which are effective:

(a) to prevent access to any dangerous part of machinery to any rotating stock-bar; or

(b) to stop the movement of any dangerous part of machinery or rotating stock-bar before any part of a person enters a danger zone.

2. The measures required by paragraph 1 shall consist of:

(a) the provision of fixed guards enclosing every dangerous part or rotating stock-bar where and to the extent that it is not; then

(b) the provision of other guards or protection devices where and to the extent that it is practicable to do so, but where or to the extent that it is not; then

(c) the provision of jigs, holders, push-sticks or similar protection appliances used in conjunction with the machinery where and to the extent that it is practicable to do so, but where or to the extent that it is not; then

(d) the provision of information, instruction, training and supervision.

3. All guards and protection devices provided under sub-paras (a) or (b) of paragraph 2 shall:

(a) be suitable for the purpose for which they are provided;

(b) be of good construction, sound material and adequate strength;

(c) be maintained in an efficient state, in efficient working order and in good repair;

(d) not give rise to any increased risk to health or safety;

(e) not be easily bypassed or disabled;

(f) be situated at sufficient distance from the danger zone;

(g) not unduly restrict the view of the operating cycle of the machinery, where such a view is necessary;

(h) be so constructed or adapted that they allow operations necessary to fit or replace parts and for maintenance work, restricting access so that it is allowed only to the area where the work is to be carried out and, if possible, without having to dismantle the guard or protection device.

4. All protection appliances provided under sub-para (c) of paragraph 2 shall comply with sub-paragraphs (a) to (d) and (g) of paragraph 3.

5. In this regulation:

danger zone – means any zone in or around machinery in which a person is exposed to a risk to health or safety from contact with a dangerous part of machinery or a rotating stock-bar;

stock-bar – means any part of a stock-bar which projects beyond the head-stock of a lathe (Reg 11).

Protection against specified hazards

1. Every employer shall take measures to ensure that the exposure of a person using work equipment to any risk to his health or safety from any hazard specified in paragraph 3 is either prevented, or, where that is not reasonably practicable, adequately controlled.
2. The measures required by paragraph 1 shall:
   (a) be measures other than the provision of personal protective equipment or of information, instruction, training and supervision, so far as is reasonably practicable;
   (b) include, where appropriate, measures to minimise the effects of the hazard as well as to reduce the likelihood of the hazard occurring.
3. The hazards referred to in paragraph 1 are:
   (a) any article or substance falling or being ejected from work equipment;
   (b) rupture or disintegration of parts of work equipment;
   (c) work equipment catching fire or overheating;
   (d) the unintended or premature discharge of any article or of any gas, dust, liquid, vapour or other substance which, in each case, is produced, used or stored in the work equipment;
   (e) the unintended or premature explosion of the work equipment or any article or substance produced, used or stored in it.
4. For the purposes of this regulation, adequate means adequate having regard only to the nature of the hazard and the nature and degree of exposure to the risk, and adequately shall be construed accordingly (Reg 12).

High or very low temperature

Every employer shall ensure that work equipment, parts of work equipment and any article or substance produced, used or stored in work equipment which, in each case, is at a high or very low temperature shall have protection where appropriate so as to prevent injury to any person by burn, scald or sear (Reg 13).

Controls for starting or making a significant change in operating conditions

1. Every employer shall ensure that, where appropriate, work equipment is provided with one or more controls for the purposes of:

(a) starting the work equipment (including re-starting after a stoppage for any reason); or

(b) controlling any change in the speed, pressure or other operating conditions of the work equipment where such conditions after the change result in risk to health and safety which is greater than or of a different nature from such risks before the change.

2. Subject to paragraph 3, every employer shall ensure that where a control is required by paragraph 1, it shall not be possible to perform any operation mentioned in sub-paragraph (a) or (b) of that paragraph except by a deliberate action on such control.

3. Paragraph 1 shall not apply to re-starting or changing operating conditions as a result of the normal operating cycle of an automatic device (Reg 14).

Stop controls

1. Every employer shall ensure that, where appropriate, work equipment is provided with one or more readily accessible controls, the operation of which will bring the work equipment to a safe condition in a safe manner.

2. Any control required by paragraph 1 shall bring the work equipment to a complete stop where necessary for reasons of health and safety.

3. Any control required by paragraph 1 shall, if necessary for reasons of health and safety, switch off all sources of energy after stopping the functioning of the work equipment.

4. Any control required by paragraph 1 shall operate in priority to any control which starts or changes the operating conditions of the work equipment (Reg 15).

Emergency stop controls

1. Every employer shall ensure that, where appropriate, work equipment is provided with one or more emergency stop controls unless it is not necessary by reason of the nature of the hazards and the time taken for the work equipment to come to a complete stop as a result of the action of any control provided by virtue of Reg 15(1).

2. Any control required by paragraph 1 shall operate in priority to any control required by regulation 15(1) (Reg 16).

Controls

1. Every employer shall ensure that all controls for work equipment shall be clearly visible and identifiable, including by appropriate marking where necessary.

2. Except where necessary, the employer shall ensure that no control for work equipment is in a position where any person operating the control is exposed to risk to his health or safety.

3. Every employer shall ensure where appropriate:

(a) that, so far as is reasonably practicable, the operator of any control is able to ensure from the position of that control that no person is in a place where he would be exposed to any risk to his health or safety as a result of the operation of that control, but where or to the extent that it is not reasonably practicable;

(b) that, so far as is reasonably practicable, systems of work are effective to ensure that, when work equipment is about to start, no person is in a place where he would be exposed to a risk to his health or safety as a result of the work equipment starting, but where neither of these is reasonably practicable;

(c) that an audible, visible or other suitable warning is given by virtue of Reg 24 whenever work equipment is about to start (Reg 17).

Control systems

1. Every employer shall ensure, so far as is reasonably practicable, that all control systems of work equipment are safe.

2. Without prejudice to the generality of paragraph 1, a control system shall not be safe unless:

(a) its operation does not create an increased risk to health or safety;

(b) it ensures, so far as is reasonably practicable, that any fault in or damage to any part of the control system or the loss of supply of any source of energy used by the work equipment cannot result in additional or increased risk to health or safety;

(c) it does not impede the operation of any control required by Reg 15 or 16 (Reg 18).

Isolation from sources of energy

1. Every employer shall ensure that where appropriate work equipment is provided with suitable means to isolate it from all its sources of energy.
2. Without prejudice to the generality of paragraph 1, the means mentioned in that paragraph shall not be suitable unless they are clearly identifiable and readily accessible.
3. Every employer shall take appropriate measures to ensure that reconnection of any energy source to work equipment does not expose any person using the work equipment to any risk to his health or safety (Reg 19).

Stability

Every employer shall ensure that work equipment or any part of work equipment is stabilised by clamping or otherwise where necessary for purposes of health or safety (Reg 20).

Lighting

Every employer shall ensure that suitable and sufficient lighting, which takes account of the operations to be carried out, is provided at any place where a person uses work equipment (Reg 21).

Maintenance operations

Every employer shall take appropriate measures to ensure that work equipment is so constructed or adapted that, so far as is reasonably practicable, maintenance operations which involve a risk to health or safety can be carried out while the work equipment is shut down or, in other cases:

(a) maintenance operations can be carried out without exposing the person carrying them out to a risk to his health or safety; or

(b) appropriate measures can be taken for the protection of any person carrying out maintenance operations which involve a risk to his health or safety (Reg 22).

Markings

Every employer shall ensure that work equipment is marked in a clearly visible manner with any marking appropriate for reasons of health and safety (Reg 23).

Warnings

1. Every employer shall ensure that work equipment incorporates any warnings or warning devices which are appropriate for the reasons of health and safety.
2. Without prejudice to the generality of paragraph 1, warnings given by warning devices on work equipment shall not be appropriate unless they are unambiguous, easily perceived and easily understood (Reg 24).

Mobile work equipment

Because of the high incidence of both fatal and major injury accidents arising from the use of mobile work equipment, including fork lift trucks, specific provisions for this type of equipment were written into PUWER 1998. Part II deals with these provisions.

Employees carried on mobile work equipment

Every employer shall ensure that no employee is carried by mobile work equipment unless:

(a) it is suitable for carrying persons;

(b) it incorporates features for reducing to as low as is reasonably practicable risks to their safety, including risks from wheels and tracks (Reg 25).

Rolling over of mobile work equipment

1. Every employer shall ensure that, where there is a risk to an employee riding on mobile work equipment from its rolling over, it is minimised by:

    (a) stabilising the work equipment;

    (b) a structure which ensures that the work equipment does no more than fall on its side;

   (c) a structure giving sufficient clearance to anyone being carried if it overturns further than that; or
   (d) a device giving comparable protection.
2. Where there is a risk of anyone being carried by mobile work equipment being crushed by its rolling over, the employer shall ensure that it has a suitable restraining system for him.
3. This regulation shall not apply to a fork-lift truck having a structure described in sub-paragraph (b) or (c) of paragraph 1.
4. Compliance with this regulation is not required where:
   (a) it would increase the overall risk to safety;
   (b) it would not be reasonably practicable to operate the mobile work equipment in consequence; or
   (c) in relation to an item of work equipment provided for use in the undertaking or establishment before 5 December 1998 it would not be reasonably practicable (Reg 26).

Overturning of fork-lift trucks

Every employer shall ensure that a fork-lift truck to which Reg 26(3) refers and which carries an employee is adapted or equipped to reduce to as low as is reasonably practicable the risk to safety from its overturning (Reg 27).

Self-propelled work equipment

Every employer shall ensure that, where self-propelled work equipment may, while in motion, involve risk to the safety of persons:

(a) it has facilities for preventing it being started by an unauthorised person;
(b) it has appropriate facilities for minimising the consequences of a collision where there is more than one item of rail-mounted work equipment in motion at the same time;
(c) it has a device for braking and stopping;
(d) where safety constraints so require, emergency facilities operated by readily accessible controls or automatic systems are available for braking and stopping the work equipment in the event of failure of the main facility;
(e) where the driver's direct field of vision is inadequate to ensure safety, there are adequate devices for improving his vision so far as is reasonably practicable;

(f) if provided for use at night or in dark places –
   (i) it is equipped with lighting appropriate to the work being carried out;
   (ii) it is otherwise sufficiently safe for such use;

(g) if it, or anything carried or towed by it, constitutes a fire hazard and is liable to endanger employees, it carries appropriate fire-fighting equipment unless such equipment is kept sufficiently close to it (Reg 28).

Remote-controlled self-propelled work equipment

Every employer shall ensure that where remote-controlled self-propelled work equipment involves a risk to safety while in motion:
(a) it stops automatically once it leaves its control range; and
(b) where the risk is of crushing or impact it incorporates features to guard against such risk unless other appropriate devices are able to do so (Reg 29).

Drive shafts

1. Where the seizure of the drive shaft between mobile work equipment and its accessories or anything towed is likely to involve a risk to safety every employer shall:
   (a) ensure that the work equipment has a means of preventing such seizure; or
   (b) where such seizure cannot be avoided, take every possible measure to avoid an adverse effect on the safety of an employee.

2. Every employer shall ensure that:
   (a) where mobile work equipment has a shaft for the transmission of energy between it and other mobile work equipment;
   (b) the shaft could become soiled or damaged by contact with the ground while uncoupled,

   the work equipment has a system for safeguarding the shaft (Reg 30).

Power presses

Specific provisions for power presses were incorporated in Part IV to PUWER 1998.

Power presses to which Part IV does not apply

Regs 32 to 35 shall not apply to a power press which is described in Schedule 2 (Reg 31).

Thorough examination of power presses, guards and protection devices

1. Every employer shall ensure that a power press is not put into service for the first time after installation, or after assembly at a new site or in a new location unless:

(a) it has been thoroughly examined to ensure that it –

(i) has been installed correctly;

(ii) would be safe to operate;

(b) any defect has been remedied.

2. Every employer shall ensure that a guard, other than one to which paragraph 3 relates, or protection device is not put into service for the first time on a power press unless:

(a) it has been thoroughly examined when in position on that power press to ensure that it is effective for its purpose;

(b) any defect has been remedied.

3. Every employer shall ensure that a part of a closed tool which acts as a fixed guard is not used on a power press unless:

(a) it has been thoroughly examined when in position on any power press in the premises to ensure that it is effective for its purpose;

(b) any defect has been remedied.

4. For the purpose of ensuring that health and safety conditions are maintained, and that any deterioration can be detected and remedied in good time, every employer shall ensure that:

(a) every power press is thoroughly examined, and its guards and protection devices are thoroughly examined when in position on that power press –

(i) at least every 12 months, where it has fixed guards only; or

(ii) at least every six months, in other cases;

(iii) each time that exceptional circumstances have occurred which are liable to jeopardise the safety of the power press or its guards or protection devices; a

(b) any defect is remedied before the power press is used again.

5. Where a power press, guard or protection device was before the coming into force of these regulations required to be thoroughly examined by Reg 5(2) of the Power Presses Regs 1965, the first thorough examination under paragraph 4 shall be made before the date by which a thorough examination would have been required by Reg 5(2) had it remained in force.
6. Paragraph 4 shall not apply to that part of a closed tool which acts as a fixed guard.
7. In this regulation, 'defect' means a defect notified under Reg 34 other than a defect which has not yet become a danger to persons (Reg 32).

Inspection of guards and protection devices

1. Every employer shall ensure that a power press is not used after the setting, re-setting or adjustments of its tools, save in trying out its tools or save in die proving, unless:
   (a) its every guard and protection device has been inspected and tested while in position on the power press by a person appointed in writing by the employer who is –
      (i) competent; or
      (ii) undergoing training for that purpose and acting under the immediate supervision of a competent person; and who has signed a certificate which complies with paragraph 3; or
   (b) the guards and protection devices have not been altered or disturbed in the course of the adjustment of its tools.
2. Every employer shall ensure that a power press is not used after the expiration of the fourth hour of a working period unless its every guard and protection device has been inspected and tested while in position on the power press by a person appointed in writing by the employer who is:
   (a) competent; or
   (b) undergoing training for that purpose and acting under the immediate supervision of a competent person; and who has signed a certificate that complies with paragraph 3.
3. A certificate referred to in this regulation shall:
   (a) contain sufficient particulars to identify every guard and protection device inspected and tested and the power press on which it was positioned at the time of the inspection and test;
   (b) state the date and time of the inspection and test;

   (c) state that every guard and protection device on the power press is in position and effective for its purpose.

4. In this regulation 'working period', in relation to a power press, means:
   (a) the period in which the day's or night's work is done; or
   (b) in premises where a shift system is in operation, a shift (Reg 33).

Reports

1. A person making a thorough examination for an employer under Reg 32 shall:
   (a) notify the employer forthwith of any defect in a power press or its guard or protection device which in his opinion is or could become a danger to persons;
   (b) as soon as is practicable, make a report of the thorough examination to the employer in writing authenticated by him or on his behalf by signature or equally secure means and containing the information specified in Schedule 3;
   (c) where there is in his opinion a defect in a power press or its guard or protection device which is or could become a danger to persons, send a copy of the report as soon as is practicable to the enforcing authority for the premises in which the power press is situated.
2. A person making an inspection and test for an employer under Reg 33 shall forthwith notify the employer of any defect in a guard or protection device which in his opinion is or could become a danger to persons and the reason for his opinion (Reg 34).

Keeping of information

1. Every employer shall ensure that the information in every report made pursuant to Reg 34(1) is kept available for inspection for two years after it is made.
2. Every employer shall ensure that a certificate under Reg 33(1)(a)(ii) or 2(b) is kept available for inspection:
   (a) at or near the power press to which it relates until superseded by a later certificate; and
   (b) after that, until six months have passed since it was signed (Reg 35).

Transitional provision

The requirements in Regs 25 to 30 shall not apply to work equipment provided for use in the undertaking or establishment before 5 December 1998 until 5 December 2002.

Schedule 2

Power presses to which Regs 32 to 35 do not apply

1. A power press for the working of hot metal.
2. A power press not capable of a stroke greater than six millimetres.
3. A guillotine.
4. A combination punching and shearing machine, turret punch press or similar machine for punching, shearing or cropping.
5. A machine, other than a press brake, for bending steel sections.
6. A straightening machine.
7. An upsetting machine.
8. A heading machine.
9. A riveting machine.
10. An eyeletting machine.
11. A press-stud attaching machine.
12. A zip fastener bottom stop attaching machine.
13. A stapling machine.
14. A wire stitching machine.
15. A power press for the compacting of metal powders.

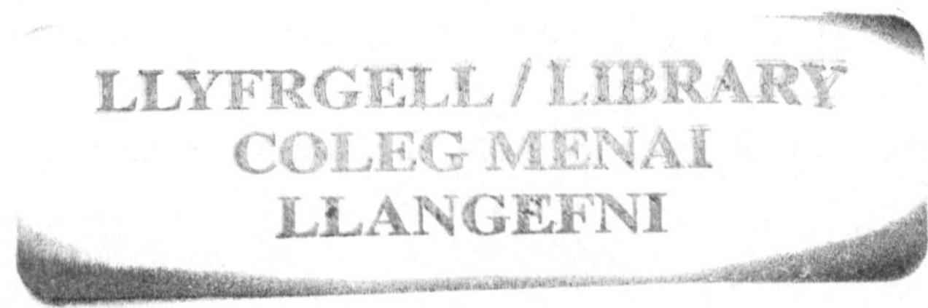

Schedule 3

Information to be contained in a report of a thorough examination of a power press, guard or protection device

1. The name of the employer for whom the thorough examination was made.
2. The address of the premises at which the thorough examination was made.
3. In relation to each item examined:
   (a) that it is a power press, interlocking guard, fixed guard or other type of guard or protection device;
   (b) where known its make, type and year of manufacture;
   (c) the identifying mark of –
      (i) the manufacture;
      (ii) the employer.
4. In relation to the first thorough examination of a power press after installation or after assembly at a new site or in a new location:
   (a) that it is such thorough examination;
   (b) either that it has been installed correctly or would be safe to operate or the respects in which it has not been installed correctly or would not be safe to operate;
   (c) identification of any part found to have a defect, and a description of the defect.
5. In relation to a thorough examination of a power press other than one to which paragraph 4 relates:
   (a) that it is such other thorough examination;
   (b) either that the power press would be safe to operate or the respects in which it would not be safe to operate;
   (c) identification of any part found to have a defect which is or could become a danger to persons, and a description of the defect.
6. In relation to a thorough examination of a guard or protection device:
   (a) either that it is effective for its purpose or the respects in which it is not effective for its purpose;
   (b) identification of any part found to have a defect which is or could become a danger to persons, and a description of the defect.

7. Any repair, renewal or alteration required to remedy a defect found to be a danger to persons.

8. In the case of a defect which is not yet but could become a danger to persons:

   (a) the time by which it could become such danger;

   (b) any repair, renewal or alteration required to remedy it.

9. Any other defect which requires remedy.

10. Any repair, renewal or alteration referred to in paragraph 7 which has already been effected.

11. The date on which any defect referred to in paragraph 8 was notified the employer under Reg 34(1)(a).

12. The qualification and address of the person making the report; that he is self-employed or if employed, the name and address of his employer.

13. The date of the thorough examination.

14. The date of the report.

15. The name of the person making the report and where different the name of the person signing or otherwise authenticating it.

Implications for employers

These Regulations incorporate important provisions of both a general and specific nature. It must be appreciated that the term 'work equipment' covers just about every form of machine, appliance and hand tool used by people at work. The regulations impose specific provisions for mobile work equipment and power presses.

The implications for employers are systems related, namely:

(a) the assessment of risks at the selection stage of new work equipment in terms of –

   (i) the actual construction of the equipment;

   (ii) the intended use of the equipment;

   (iii) its suitability for use in the workplace;

   (iv) the conditions under which it is used;

(b) ongoing safety assessment of existing work equipment;

(c) the implementation of formally documented planned maintenance systems;

(d) designation of certain trained persons to undertake identified high-risk activities, such as maintenance or fault-finding without guards in position;

(e) the provision of information, instruction and training for staff using any form of work equipment;

(f) the development, documentation and implementation of management procedures aimed at ensuring safe use of work equipment in all work situations, including requirements for the inspection of work equipment.

Second-hand, hired and leased work equipment

Second-hand equipment

In situations where existing work equipment is sold by one company to another and brought into use by the second company, it becomes 'new equipment' and must meet the requirements for such equipment, even though it is second-hand. This means that the purchasing company will need to ascertain that the equipment meets the specific hardware provisions of Regs 11–24 before putting it into use.

G

Hired and leased equipment

Such equipment is treated in the same way as second-hand equipment, namely that it is classed as 'new equipment' at the hire/lease stage. On this basis, organisations hiring or leasing an item of work equipment will need to check that it meets the requirements of Regs 11 to 24 before putting it into use.

Machinery hazards

A person may be injured at machinery through:

(a) coming into contact with it, or being trapped between the machinery and any material in or at the machinery or any fixed structure;

(b) being struck by, or becoming entangled in motion in the machinery;

(c) being struck by parts of the machinery ejected from it;

(d) being struck by material ejected from the machinery (BS EN 292).

(For the effect of claims litigation on statutory fencing requirements, etc., *see* the end of this chapter.)

Assessment of danger from machinery

The objective of machine safety strategy is the prevention of injury to operators and other persons. Assessment of new and existing machinery should, therefore, consider design features, the circumstances involving operators and other persons in the use of the machine, and specific circumstances or events which can lead to injury.

Design features

Many machines, including new machines, incorporate hazards in their basic design. BS EN 292 classifies these hazards as follows.

Traps

There are three principal forms of trap:

(a) Reciprocating trap – featured in the vertical or horizontal motion of machinery such as presses. At the point where the injury occurs, the limb is stationary (Fig. 32.1).

Fig. 32.1 ◆ Vertical reciprocating motion of a power press

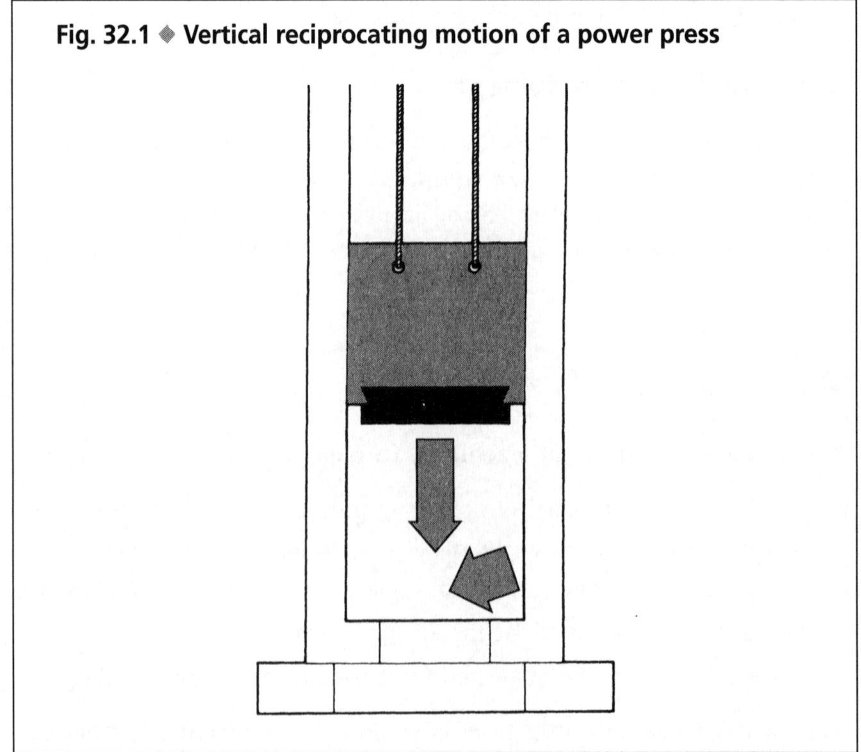

Source: BS 5304 *Safeguarding of Machinery*

(b) Shearing trap – this is the guillotine effect produced by a moving part traversing a fixed part or by two moving parts traversing each other, similar to the operation of garden shears (Fig. 32.2).

(c) In-running nips – these are traps created where a moving belt or chain meets a roller or toothed wheel, or at the point where two revolving drums, rollers or toothed wheels (gears) meet (Fig. 32.3).

Fig. 32.2 ◆ Guillotine

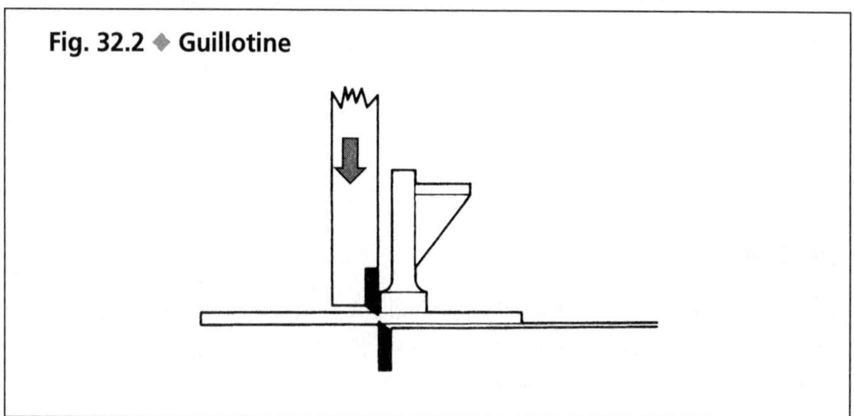

Fig. 32.3 ◆ In-running nips

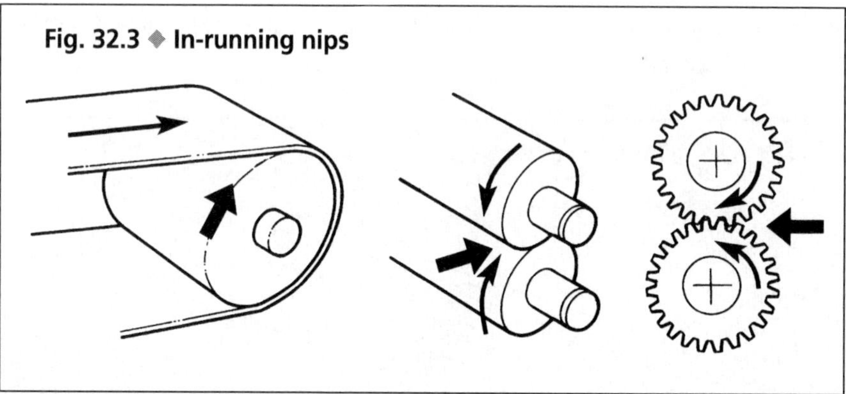

Source: BS 5304

G

Entanglement

Wherever there are unguarded revolving shafts, drills or chucks to drills, there will always be a risk of entanglement of limbs, hair or clothing (Fig. 32.4).

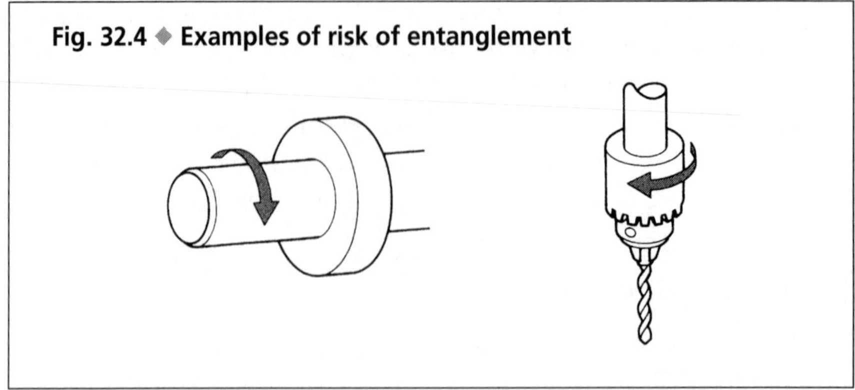

Fig. 32.4 ◆ Examples of risk of entanglement

Source: BS 5304

Ejection

Many machines can eject particles of metal or actual parts of the machine. Grinding machines can emit particles of the metal being ground or chips and parts of the grinding wheel.

Contact

Contact with the machine may cause injury, for instance burns from hot exposed surfaces or lacerations from the metal fastenings of a belt to a belt conveyor (Fig. 32.5).

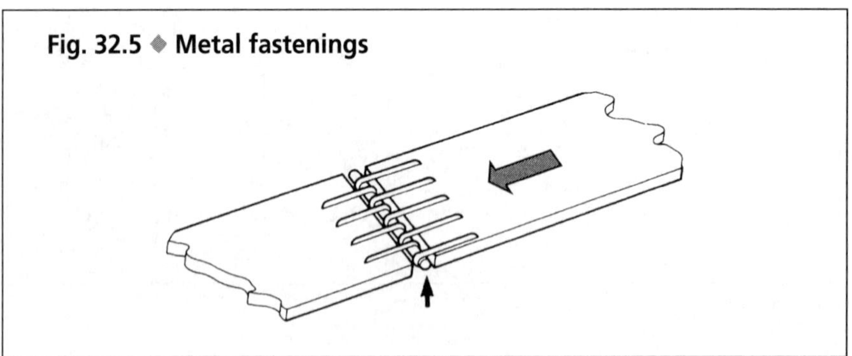

Fig. 32.5 ◆ Metal fastenings

Source: BS 5304

Circumstances involving operators and other persons

Specific activities associated with machinery can result in injuries to operators and others in the immediate vicinity. These include job loading and removal, tool changing, the removal of waste and scrap items, actual operation of the machine, particularly if located in a congested area or one frequented by other personnel, routine maintenance and adjustment, gauging, breakdown situations, trying out after adjustment or setting, and unauthorised presence, e.g. persons taking short cuts through the machining area. (The fact that a person's presence is unauthorised does not prevent his obtaining damages for injury, though damages are likely to be reduced owing to his contributory negligence. The case of *Uddin* v. *Associated Portland Cement Manufacturers Ltd* [1965] 2 AER 213 concerned a worker who trespassed in order to retrieve a pigeon from machinery and was injured, but still obtained partial damages.)

Specific circumstances or events which can lead to injury

A number of circumstances or events need particular consideration, e.g. the likelihood of unexpected start-up or movement, uncovenanted stroke of the machine, mechanical failure, and operators reaching into the feeding device or danger area of the machine. Access to and egress from the machine area must also be considered, in particular the provision of sound floor surfaces, freedom from slippery substances and machinery waste products which could cause falls on to machinery, and sufficient free workspace around the machine.

The 17 dangerous parts of machinery

Certain parts, or combinations of parts, of machinery are classified by the HSE as dangerous should workers operate unsafely or should an unsafe action develop in respect of their motion. Such parts must be securely fenced. The 17 dangerous parts are listed below, with examples:

(a) Revolving shafts, spindles, mandrels and bars, e.g. line and counter shafts, machine shafts: drill spindles; chucks and drills, etc.; boring bars; stock-bars; traverse shafts.

(b) In-running nips between pairs of rotating parts, e.g. gear wheels; friction wheels; calendar bowls; mangle rolls; metal manufacturing rolls; rubber washing, breaking and mixing rolls; dough brakes; printing machines; paper-making machines.

(c) In-running nips of the belt and pulley type, e.g. belts and pulleys, plain, flanged (i.e. V-belts) or grooved; chain and sprocket gears; conveyor belts and pulleys; metal coiling and the like.

(d) Projections on revolving parts, e.g. key heads; set screws; cotter pins; coupling bolts.

(e) Discontinuous rotating parts, e.g. open arm pulleys; fan blades; spoked gear wheels and spoked flywheels.

(f) Revolving beaters, spiked cylinders and revolving drums, e.g. scutchers; rag flock teasers; cotton openers; carding engines; laundry washing machines.

(g) Revolving mixer arms in casings, e.g. dough mixers; rubber solution mixers.

(h) Revolving worms and spirals in casings, e.g. meat mincers; rubber extruders; spiral conveyors.

(i) Revolving high-speed cages in casings, e.g. hydro-extractors; centrifuges.

(j) Abrasive wheels, e.g. manufactured wheels; natural sandstone.

(k) Revolving cutting tools, e.g. circular saws; milling cutters; circular shears; wood slicers; routers; chaff cutters; woodworking machines such as spindle moulders, planing machines and tenoning machines.

(l) Reciprocating tools and dies, e.g. power presses, drop stamps; relief stamps; hydraulic and pneumatic presses; blending presses; hand presses; revolution presses.

(m) Reciprocating knives and saws, e.g. guillotines for metal, rubber and paper; trimmers; corner cutters; perforators.

(n) Closing nips between platen motions, e.g. letterpress platen printing machines; paper and cardboard platen machine cutters; some power presses; foundry moulding machines.

(o) Projecting belt fasteners and fast-running belts, e.g. bolt and nut fasteners; wire pin fasteners and the like; woodworking machinery belts; centrifuge belts; textile machinery side belting.

(p) Nips between connecting rods or links, and rotating wheels. cranks or discs, e.g. side motion of certain nat-bed printing machines; jacquard motions on looms.

(q) Traps arising from the traversing carriages of self-acting machines, e.g. metal-planing machines.

Safeguarding machinery

The authoritative guide to machinery guarding is BS EN 292 *Safeguarding of Machinery*. Here a *safeguard is* defined as 'a guard or device to protect persons from *danger*'. The British Standard defines *danger* as follows: 'When applied to machinery in motion, it is a situation in which there is a reasonably foreseeable risk of injury from the mechanical hazards referred to in clause 6.' (These mechanical hazards are itemised under 'Machinery Hazards' earlier in this chapter.) To ensure maximum safety with machinery, therefore, the appropriate guard or safety device – or, in some cases, both – must be used. The forms of guards and safety devices are outlined below, using the definitions contained in BS EN 292. It should be noted that the law defines machinery as *dangerous* when 'it is a possible cause of injury to anybody acting in a way in which a human being may be reasonably expected to act in circumstances which may reasonably be expected to occur' (*Walker* v. *Bletchley-Flettons Ltd* [1937] 1 AER 170).

Machinery guards

Fixed guard

This is 'a guard which has no moving parts associated with it, or dependent upon the mechanism of any machinery, and which, when in position, prevents access to a danger point or area' (BS EN 292). (Case law supports the view that a fixed guard should not be readily removable, i.e. its removal should require recourse to a hand tool. The use of wing nuts for securing fixed guards, for instance, is not acceptable.) An example of a fixed guard constructed of wire mesh and angle section to prevent access to transmission machinery is shown in Fig. 32.6.

Adjustable guard

This is 'a guard incorporating an adjustable element which, once adjusted, remains in that position during a particular operation' (BS EN 292). An adjustable guard for use on a band-saw is shown in Fig. 32.7. The height at which the guard is set can be adjusted according to the thickness of the material being cut.

Distance guard

This is 'a guard which does not completely enclose a danger point or area but which places it out of normal reach' (BS EN 292). A typical example of a distance guard is a tunnel guard. Figure 32.8 shows a tunnel guard appropriate for use on a metal-cutting machine. The strip metal is fed through

Fig. 32.6 ◆ Fixed guard to transmission machinery

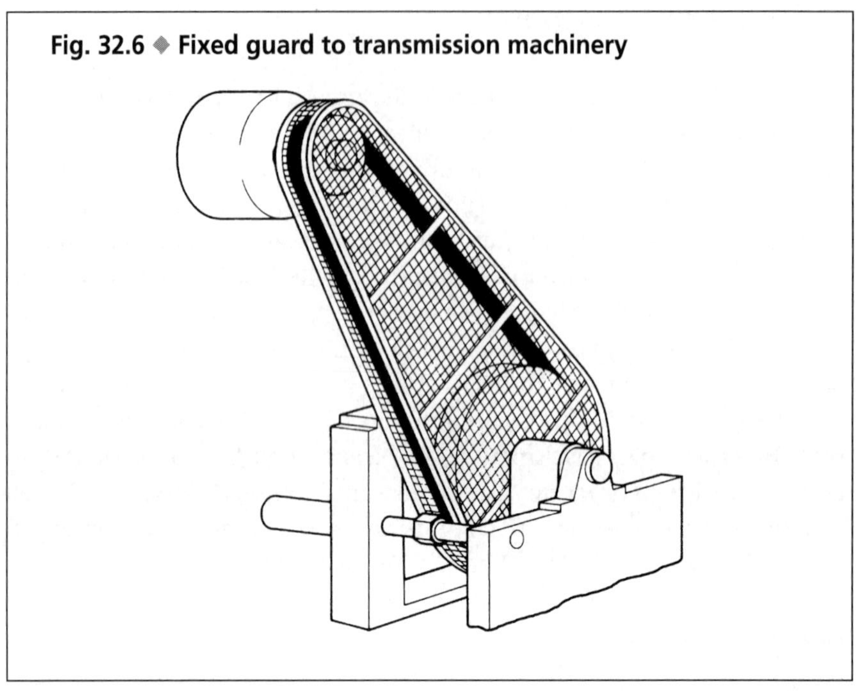

Fig. 32.7 ◆ Adjustable guard to a band-saw blade

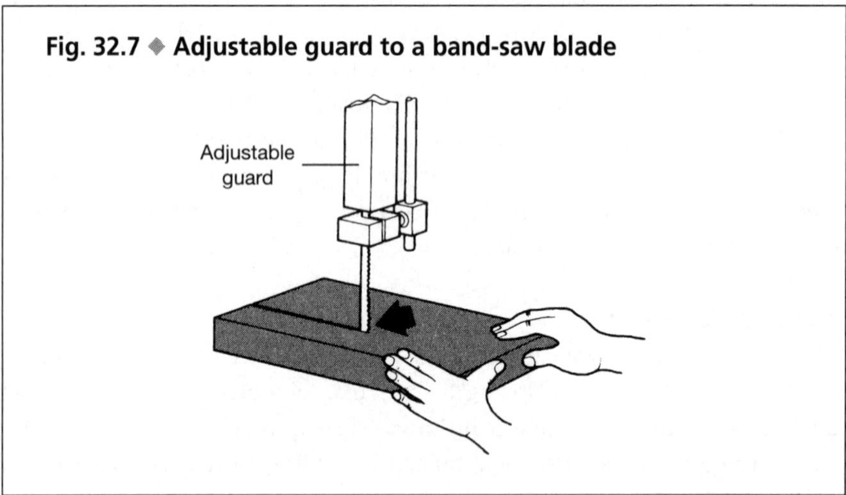

the tunnel guard to the cutters. In this case an interlocking device is fitted so that if the guard is lifted the power supply to the machine is cut off.

Fig. 32.8 ◆ Tunnel guard for a metal-cutting machine

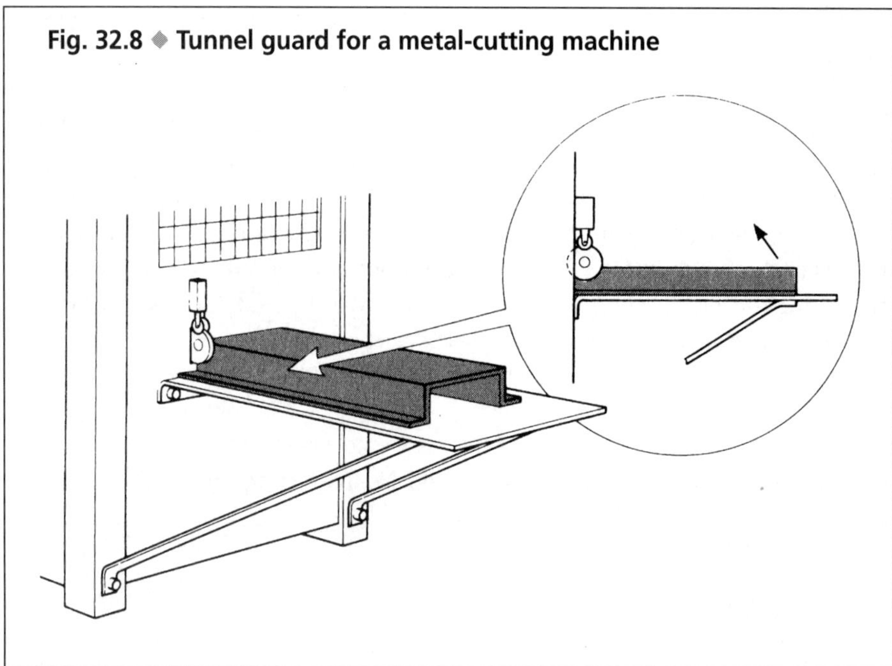

G

Interlocking guard

This is a guard which has a movable part so connected with the machinery controls that:

(a) the part(s) of the machinery causing danger cannot be set in motion until the guard is closed;

(b) the power is switched off and the motion braked before the guard can be opened sufficiently to allow access to the dangerous parts;

(c) access to the danger point or area is denied while the danger exists (BS EN 292).

In order to achieve the same level of safety as that attained with fixed guards, reliability and maintenance of interlocking guards are important.

Interlocking systems may take a number of forms – mechanical, electrical, hydraulic, pneumatic or a combination of these. Whatever the form, the system should fail to safety (fail-safe).

NOTE. The term 'failure to safety (fail-safe)' is commonly used in machinery guarding. It implies that 'any failure in, or interruption of, power supply will result in the prompt stopping or, where appropriate, stopping and reversal of the movement of the dangerous parts before injury can occur, or the safe guard remaining in position to prevent access to the danger point or area' (BS EN 292).

Mechanical interlocking: Mechanical interlocking incorporates two specific elements: first, the operation or actuation of a device, which may be a hydraulic or pneumatic valve, and second, as a result of the operation of the device, the movement of a particular component, generally a guard. In the example shown in Fig. 32.9 the guard slides horizontally to close. When the guard is open the control level is held down, i.e. 'safe'. Only after the guard has been closed can the control lever be raised so as to initiate the machine sequence. The lever then holds the guard in the closed position.

Fig. 32.9 ◆ Horizontal sliding guard fitted to upstroking hydraulic press

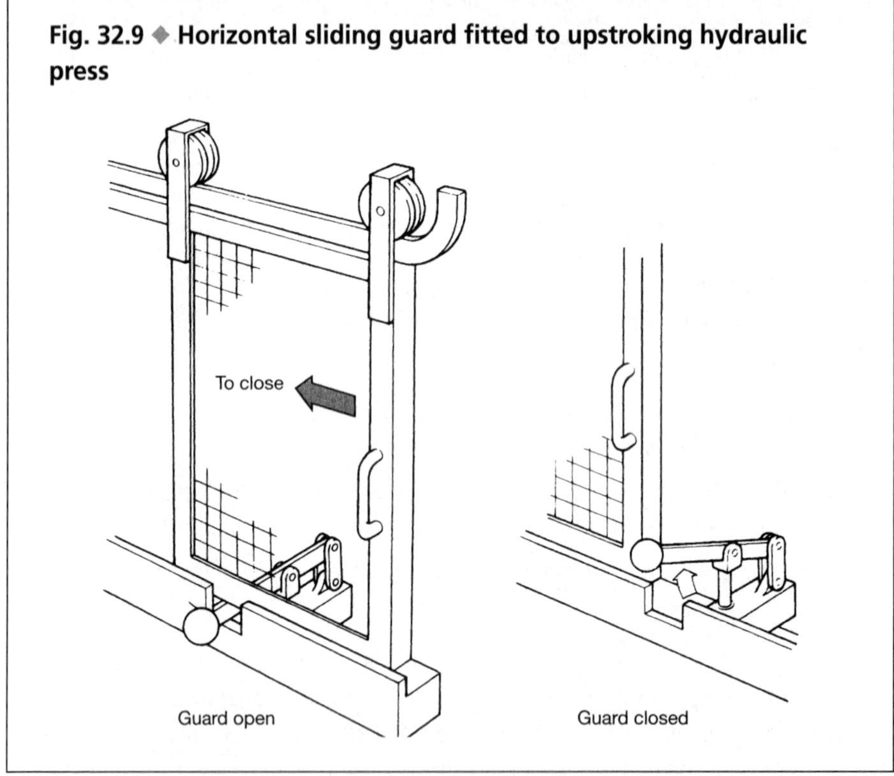

Electrical interlocking: BS EN 292 identifies four methods of electrical interlocking of guards, as follows:

(a) Control interlocking

This incorporates an actuating switch operated by the guard, interposed electromechanical relays and/or solid state switching devices, if any, the electromagnetic contactor (or solid state equivalent e.g. thyristor), and/or a pneumatic or hydraulic solenoid valve controlling power to the drive. Failures of any of these elements or of the wiring interconnecting them can all be failures to danger. All elements of the system should therefore be designed to give the maximum degree of reliability.

The range of switching methods used for associating guard movement with an on/off electrical control signal includes –

(i) cam or track operated limit switches,

(ii) captive key switches,

(iii) trapped key control of electrical switches,

(iv) magnetic switches, and

(v) diode links [cl. 44].

Specific aspects of these switching methods are shown in BS EN 292.

(b) Power interlocking

Power interlocking is achieved by direct mechanical control of a switch in series with the main power supply to the drive of the machinery. The direct mechanical control may be by links etc., by captive key or by trapped key.

Power interlocking is inherently superior to control interlocking, and thus acceptable for high-risk situations, because the mechanical link between the guard and the switch ensures that the guard cannot be opened if, for any reason, the switch contacts stick in the 'on' position. However, because direct power interlocking involves the stopping of the drive motor(s) it should only be applied to machinery where the requirement to open the guard is infrequent, or the motor is of low power (*see further* BS EN 292).

(c) Control interlocking with back-up

In high-risk situations where frequent access to the danger area is required control interlocking is acceptable if combined with a backup power drive interlock incorporated in the cyclic control element, for example direct pneumatic or hydraulic actuation of the ram etc. or air-operated or electromagnetic clutch. The basic requirement of back-up power drive interlocking is that the air or hydraulic pressure to the drive cylinder or clutch has to be exhausted or dumped automatically in the event of control interlock failure (*see further* BS EN 292).

(d) Dual circuit interlocking

Where direct mechanical linkage between the guard gate and the back-up exhaust or dump valve is not practicable electrical actuation of the back-up is acceptable. It is essential that the normal control interlock and back-up circuits are kept wholly separate from each other, except for connection to the supply, to minimise the possibility of common faults. The control and back-up limit switches should be arranged in opposite modes, the control being negative and the back-up positive (*see further* BS EN 292).

Automatic guard

This is a 'guard which is associated with, and dependent upon, the mechanism of the machinery and operates so as to remove physically, from the danger area any part of a person exposed to the danger' (BS EN 292). Automatic guards are frequently used on power presses and paper-cutting guillotines. In the first case, the guard is connected to the moving part of the press. The guard closes automatically when the machine cycle commences. Trip devices are generally fitted on power press guards where trapping points occur as the two parts of the guard meet.

Self-adjusting automatic guard: 'This is a guard which prevents accidental access of a person to a danger point or area but allows the access of a workpiece which itself acts as part of the guard, the guard automatically returning to its closed position when the operation is completed.' This form of guarding is particularly appropriate to portable electric circular saws, as shown in Fig. 32.10.

Fig 32.10 ◆ Self-adjusting guard to a portable circular saw

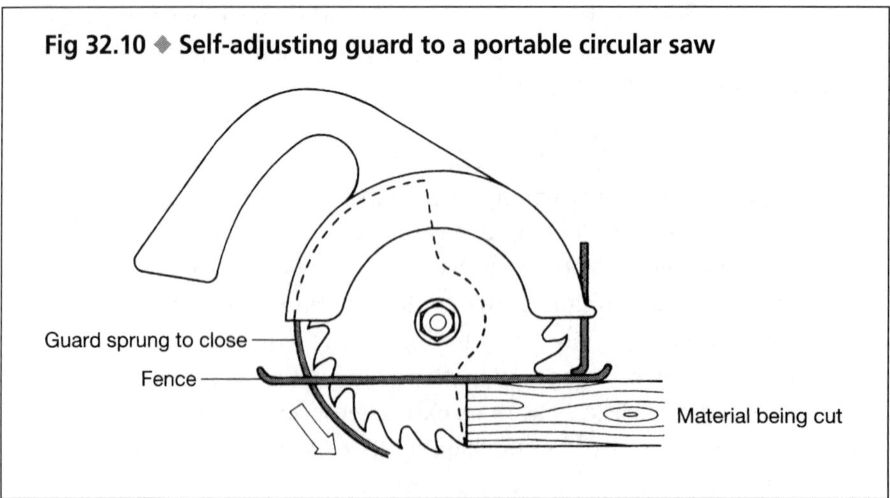

Safety devices

A safety device is a 'protective appliance, other than a guard, which eliminates or reduces danger before access to a danger point or area can be achieved'. There are several types of safety device.

Trip device

This is a means whereby any approach by a person beyond the safe limit of working machinery causes the device to actuate and stop the machinery or reverse its motion, thus preventing or minimising injury at the danger point (BS EN 292). Trip devices take several forms.

Mechanical trip device: This device incorporates a barrier which is contacted by part of the body as it approaches the danger area. Contact with the barrier operates the device which brings the machine to rest. Fig. 32.11 shows a safety trip bar for horizontal two-roll mills used in the rubber industry.

Movement of the trip bar A towards the front roll switches off the drive to the rolls by means of limit switch B and applies a brake. The position of the trip bar is important. Its height above the floor and its horizontal distance from the in-running nip should be such that the

Fig. 32.11 ◆ Safety trip bar on a horizontal two-roll mill

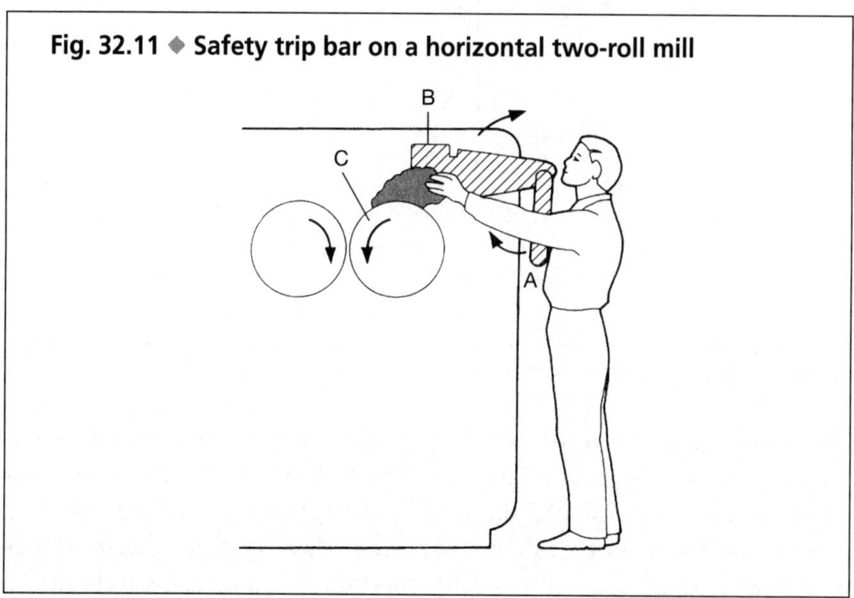

Source: BS EN 292 *Safeguarding of Machinery*

operator cannot reach beyond the safety limit C which is dependent upon the efficiency of the brake, making allowance for brake wear. After the trip bar A has been tripped the brake should arrest the motion of the rolls before a hand can be drawn into the nip.

Photo-electric trip device: A photo-electric trip device provides a curtain of light which can be arranged in either a horizontal (A) or vertical (B) configuration (as shown in Fig. 32.12). Interruption of the curtain while the dangerous parts of the machine are moving results in a signal being given for the machine to stop. The speed of stopping should be such as to ensure that the dangerous parts have come to rest before they can be reached by the operator. Access to the danger area from any direction not protected by the device should be prevented by effective fixed or interlocking guards.

Photo-electric trip devices are particularly suitable for guarding power presses, hydraulic presses and guillotines and trimmers.

Fig. 32.12 ◆ Photo-electric trip device

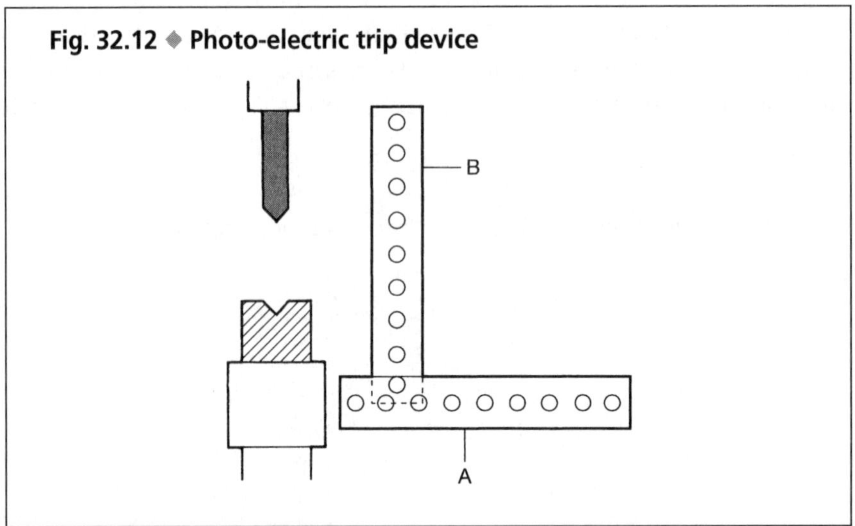

Source: BS EN 292 *Safeguarding of Machinery*

Pressure sensitive mat: This device operates by means of a number of suitably spaced electrical or fluid switches/valves contained within a mat connected to a control unit and covering the approaches to the danger area. Pressure on the mat operates one or more of these switches. Electrical pressure-sensitive mats are connected into machine control circuits and their use should therefore be restricted to normal risk situations. A pressure sensitive mat may be appropriate in circumstances where the use of a fixed guard is impracticable, and is particularly suitable for use as an emergency stopping device, as a means of protecting a person who may be inside machinery, or as a secondary safety device to augment a conventional guard.

Ultrasonic device: With these devices, inaudible high-frequency sound senses the presence of an object or person in the danger area. Ultrasonics are not affected by strong light or dirt, but sound attenuates over a distance so the width of protection is limited. This variation in sensitivity with distance can create difficulties in ensuring the effectiveness of the device, however, and BS EN 292 recommends that such devices should be restricted to the guarding of machinery at which normal interlocking is appropriate.

Two-hand control device

This is a device which requires both hands to operate the machinery controls, thus affording a measure of protection from danger only to the machinery operator and not other persons. The provision of two-hand controls at the clicking press shown in Fig. 32.13, used in the manufacture of footwear, ensures that the operator has both hands in a safe position while the press head descends. To protect against accidental operation, the buttons should be shrouded (*see* inset).

A two-hand control device should be designed in accordance with the following requirements outlined in BS EN 292:

(a) The hand controls should be so placed, separated and protected as to prevent spanning with one hand only, being operated with one hand and another part of the body, or being bridged by a tool.

(b) It should not be possible to start the machinery unless the controls are operated within approximately one second of each other. This prevents the operator from locking one control in the start position so allowing him to operate the machinery by means of the other control leaving one hand free.

(c) Movement of the dangerous parts should be arrested immediately or, where appropriate, arrested and reversed if one or both controls are released while there is yet danger from the movement of these parts. This should ensure that both hands of the operator are clear of the danger area during the whole of the dangerous movement.

(d) It should not be possible to initiate a subsequent cycle until both controls have first been returned to their original position. This prevents the possibility of one control being locked in the start position.

Fig. 32.13 ◆ Two-hand control on a clicking press

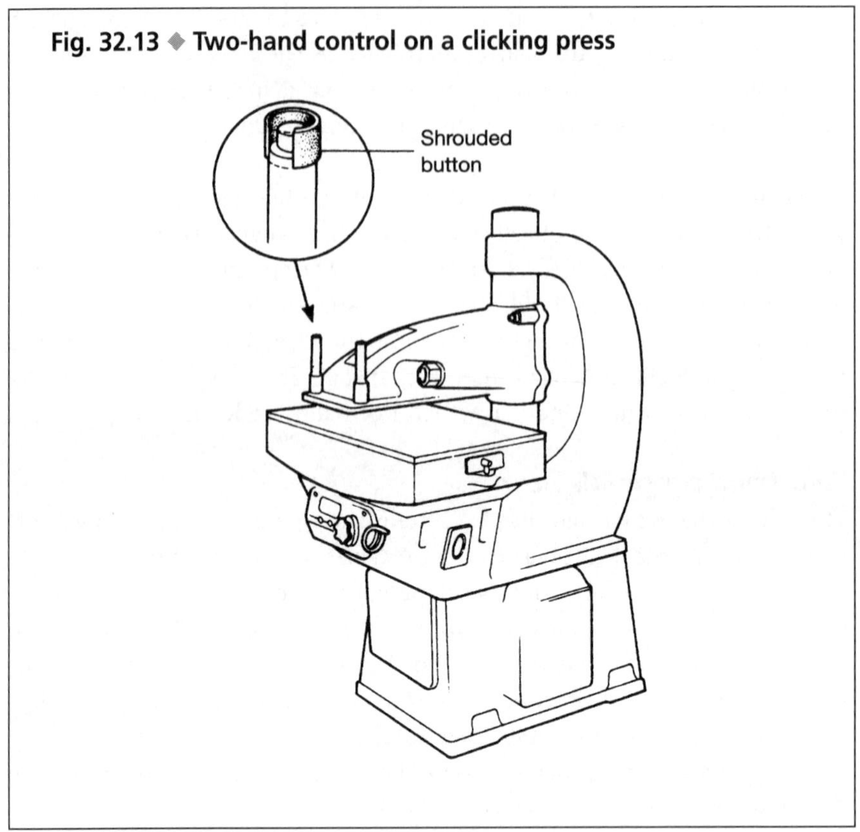

Source: BS EN 292 *Safeguarding of Machinery*

Overrun device

'This is a device which, used in conjunction with a guard, is designed to prevent access to machinery parts which are moving by their own inertia after the power supply has been interrupted, so as to prevent danger' (BS EN 292). Where a machine is liable to overrun after the power supply has been switched off it is necessary to ensure that the guard cannot be opened until the motion has ceased. This can be achieved by one of three means:

(a) A 'rotation sensing device – which ensures that after the power has been cut off the guard remains locked closed until the device has sensed that rotation of the dangerous parts has ceased'. These are used with highspeed mixers and centrifuges.

(b) A 'timing device – which ensures that after the power has been cut off the guard remains locked closed until the dangerous parts have come to rest'.

(c) A 'brake – which is interlocked with the guard and the machine controls so that the act of cutting off the power to the dangerous parts or opening the guard applies the brake'. A brake is virtually instantaneous in action.

Mechanical restraint device

'This is a device which applies mechanical restraint to a dangerous part of machinery which has been set in motion owing to failure of the machinery controls or other parts of the machinery, so as to prevent danger' (BS EN 292). Mechanical restraint devices are commonly found on pressure die-casting machines and plastics injection-moulding machines. With these hydraulically or pneumatically powered machines a trap is created between a fixed and a moving platen to which access is required usually once in every cycle. With horizontally moving platens, a simple method of applying mechanical restraint is to provide a scotch in the form of a strut which falls into place between the platens as soon as they are fully open. A device of this kind gives adequate protection provided the guard remains locked closed until the platens are fully open.

Criteria for assessment – machinery guards and safety devices

Design considerations

(a) Whenever practicable, dangerous parts should be eliminated or effectively enclosed in the initial design of the machinery. If they cannot be eliminated, then suitable safeguards should be incorporated as part of the design. If this is impossible, provision should be made for safeguards to be easily incorporated at a later stage.

(b) Provision should be made to facilitate the fitting of alternative types of safeguards on machinery where it is known that this will be necessary because the work to be done on it will vary.

(c) Where a movable guard, cover, etc., is used as a safeguard, it should be interlocked with the drive, of whatever kind, to the parts being safeguarded; maintenance operations may require complete isolation of the machinery from the power supply.

(d) The guard must be securely attached to the machine in such a way that it can only be removed by a tool.

(e) Lubrication and routine maintenance facilities should be incorporated remote from the danger area.

(f) Suitable supplementary lighting should be provided at operating points: a light fitting which is portable or which relies on manual action for directional adjustment should preferably be supplied at extra-low voltage, i.e. normally not exceeding 50 V between conductors and not exceeding 30 V a.c. or 50 V d.c. between any conductor and earth.

(g) Every mechanism and control forming part of a safeguard should, so far as is practicable, be of fail-safe design.

Construction of safeguards

(a) All safeguards should be of sound design and adequate strength.

(b) Guards may be made of metal, timber, laminated or toughened glass, suitable plastics or a combination of these, as may be appropriate to the conditions; the use of shatter-resistant materials may be an advantage.

(c) The size of openings between the interstices of wire mesh guards should be properly considered in relation to finger and hand access.

(d) Whatever safeguard is selected, it should not itself present a hazard such as trapping or shear points, splinters, rough or sharp edges, or other sources likely to cause injury. In the case of food processing machinery, the safeguard should not constitute a source of contamination of the product.

(e) Where an opening in a fixed guard is necessary for the purpose of feeding material by hand, it should not allow the operator access to the dangerous parts. Where it is necessary to provide such an opening, it should be at a sufficient distance from a danger point. Fig. 32.14 provides a guide to show the relationship between the guard opening and the distance of the guard from the danger point which, if followed in the design, should prevent unsafe access.

(f) The guard must be securely attached to the machine and must require use of a tool to remove it.

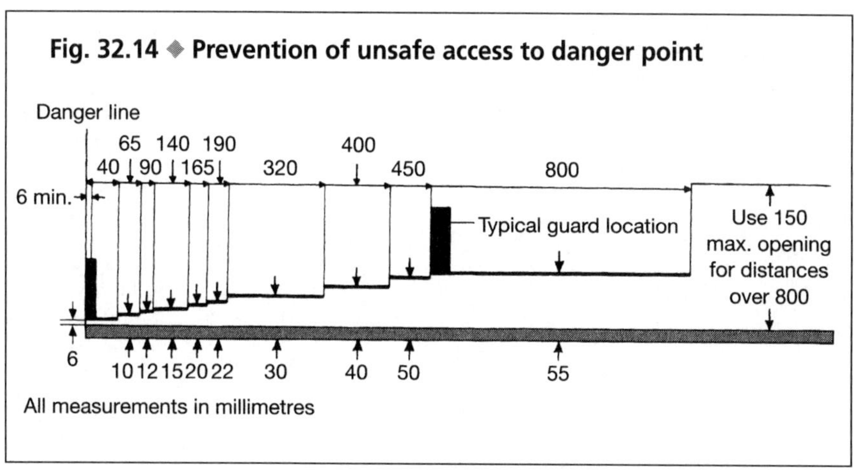

Fig. 32.14 ◆ Prevention of unsafe access to danger point

Source: BS EN 292 *Safeguarding of Machinery*

The more common machinery hazards

Drilling machines

Contact with revolving spindles, chucks and drills

Spindles and chucks should be guarded by means of telescopic or spring-loaded guards which completely enclose the danger point. Alternatively, the 'Quickstop'or 'Deadstop' type of trip device (*see* Fig. 32.15) can be fitted. This consists of a telescopic vertical trip bar suspended from the drilling head approximately 75 mm from the spindle. At the top of the trip bar is an electrical switch, and when the bar is pivoted a few degrees from the vertical in any direction the switch is operated and a brake is applied.

Injury from broken or spintered drills

Only properly sharpened drills should be used and these should be checked to ensure that they are running true. The drill should be run at the correct speed and never be forced.

Injury from the workpiece

The work should always be clamped securely to the table. No attempt should be made to hold the work while drilling. All burrs in drilled holes should be filed.

Fig. 32.15 ◆ Telescopic trip device for a drilling machine

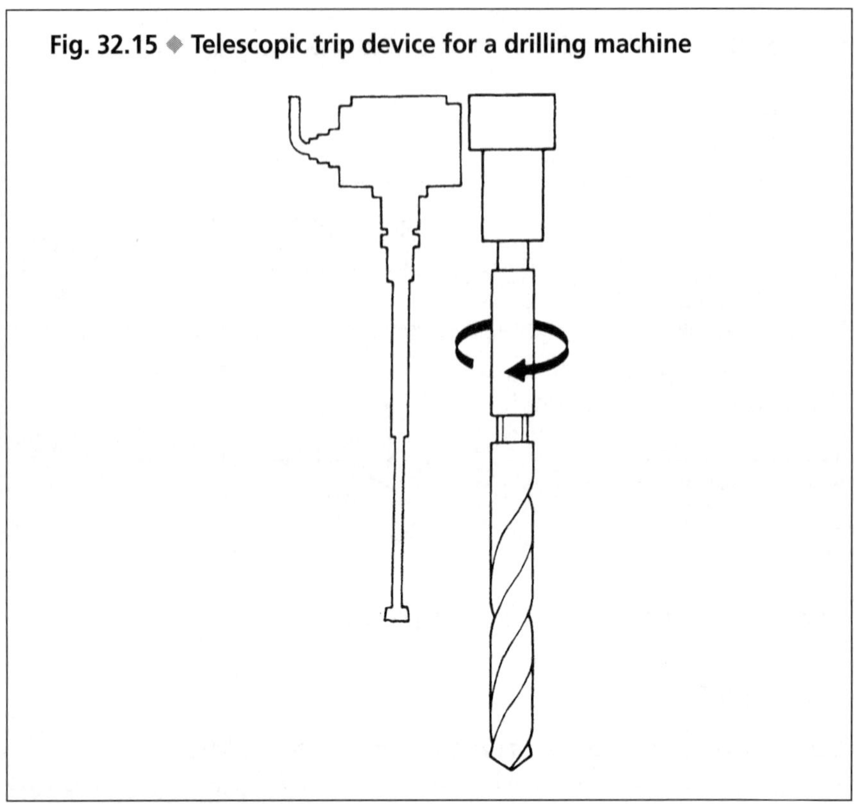

Source: BS EN 292 *Safeguarding of Machinery*

Pedestal and bench-mounted grinding machines

Shattering of artificially bonded wheels

Before mounting, all wheels should be closely inspected and 'rung' to ensure they have not been mishandled in transit or storage. A sound wheel will have a clear ring when tapped gently with a small hammer, whereas a cracked wheel will not ring. 'Ringing' should be undertaken by a trained operator, whose appointment is recorded in the Factories Act register, otherwise tapping can weaken the wheel and cause it to burst when mounted. The wheel should be mounted only by this trained and appointed operator, at the correct tension and who should ensure that the safe working speed marked on the wheel is correct for the speed of the driving motor or shaft. An unmarked wheel should never be used. The face of the wheel should be flat and ungrooved, and dressing of the wheel (if necessary) should be carried out with the proper tool. (This may consist of star wheels separated by washers on a spindle attached to a handle, or a diamond in a proper holder.) The tool rest should be adjusted as close as possible to the face of the wheel, the clearance not exceeding 3 mm. Too much clearance may allow the workpiece to jam and burst the wheel.

Contact with the wheel; trapping between the wheel and the machine casing

Correct setting of the tool rest is vital. Operators should not set the tool rest while the machine is in motion. The work should be held firmly to prevent slipping. When grinding small jobs, a clamp or holding device should be used. The wheel should never be left running when not in use, nor left unattended during the run-down period after switching off. Guarding for a bench-mounted grinder is illustrated in Fig. 32.16.

Fig. 32.16 ◆ Guarding for a bench-mounted grinder

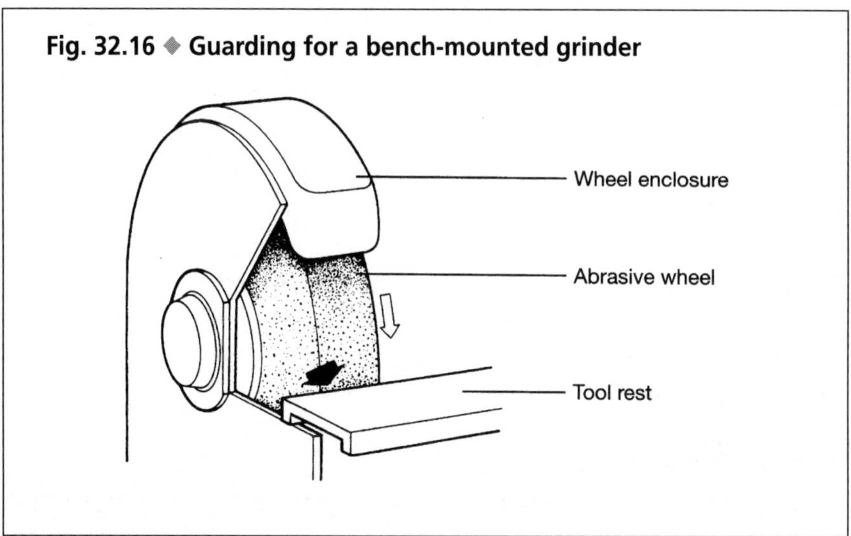

Eye injuries

Eye protection should always be used, either the fixed transparent guard on the machine (*see* Fig. 32.17) or goggles or safety spectacles. These precautions are dictated by both current good practice and the requirements of the PPEWR 1992.

Power presses

Trap between the tool and the die

Secure fencing must be provided, and there must be no access to the danger point or area from any direction. Guards should always be in position when the press is under power, whether for production or for test after setting up. Young persons must not operate presses unless they have

Fig. 32.17 ◆ Fixed transparent guard to a grinder

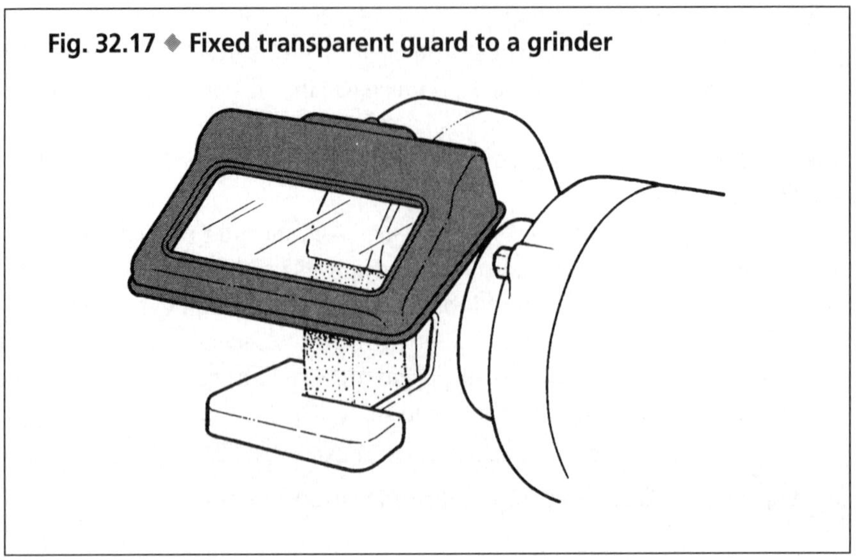

been thoroughly instructed and are either fully trained or under supervision.

Traps between the ram and parts of the guard

Guards should be so constructed and set that no trap exists between the ram or any projection on it and the guard itself. If this occurs the guard must be readjusted.

Uncovenanted strokes

An uncovenanted stroke occurs when a press makes a second stroke while the guard is open. On positive clutch presses, uncovenanted strokes can occur through the extractor failing to return to the fully disengaging position when the control is released, breakage of the clutch key, failure of the extractor mounting and excessive wear on the rubbing faces of the extractor and/or key tail. On positive clutch or friction clutch presses, there may also be seizure or collapse of the flywheel bearing, or other clutch bearing, and on friction clutch machines, clutch drag, and the unintended entry, or the retention, of air in the clutch operating cylinder due to electrical or mechanical faults. Good design and planned maintenance should prevent most of these occurrences. It should be appreciated that some presses may be inherently unsuited for use with certain types of guard. The fitting of single stroke devices can eliminate this problem. Such a device is provided for the purpose of disengaging the clutch withdrawal device, e.g. extractor, from the influence of the operating control so that it will disengage the clutch before a second stroke can occur. An arrestor brake is also used to arrest the crankshaft, flywheel and ram of a press within specified limits in the event of an uncovenanted stroke.

Overrun

Overrun occurs when the crankshaft does not come to rest, after disengagement of the clutch, in the correct position at the end of a cycle or, in some cases, at part stroke. On some types of press the crankshaft can 'run away' from the flywheel. The fitting of a press brake, correctly adjusted and regularly maintained, including the replacement of brake linings as necessary, correct lubrication and correct braking by the operator should considerably reduce this problem.

Horizontal milling machines

Contact with the cutter; splintering of the tool and injury through flying particles

Operators should ensure that the cutter is sharp and in good condition. The guard should also be maintained in good condition and correctly adjusted. When setting up, the traverse table should be taken to the maximum out-run and not in close proximity to the cutter. The arbour nuts should not be adjusted when the machine is running.

Eye injuries

Eye protection should always be worn when machining, and chips of metal and swarf removed by brush, never by hand.

Guillotines

Trapping between the blades or under the hold-down pads

The blades of a guillotine should always be enclosed by fixed guards, front and rear, which should effectively prevent all access of fingers to the danger zone (*see* Figs 32.18 and 32.19). Guards should be in sound condition and properly adjusted.

Fig. 32.18 ◆ Guard fitted to a hand-operated paper guillotine

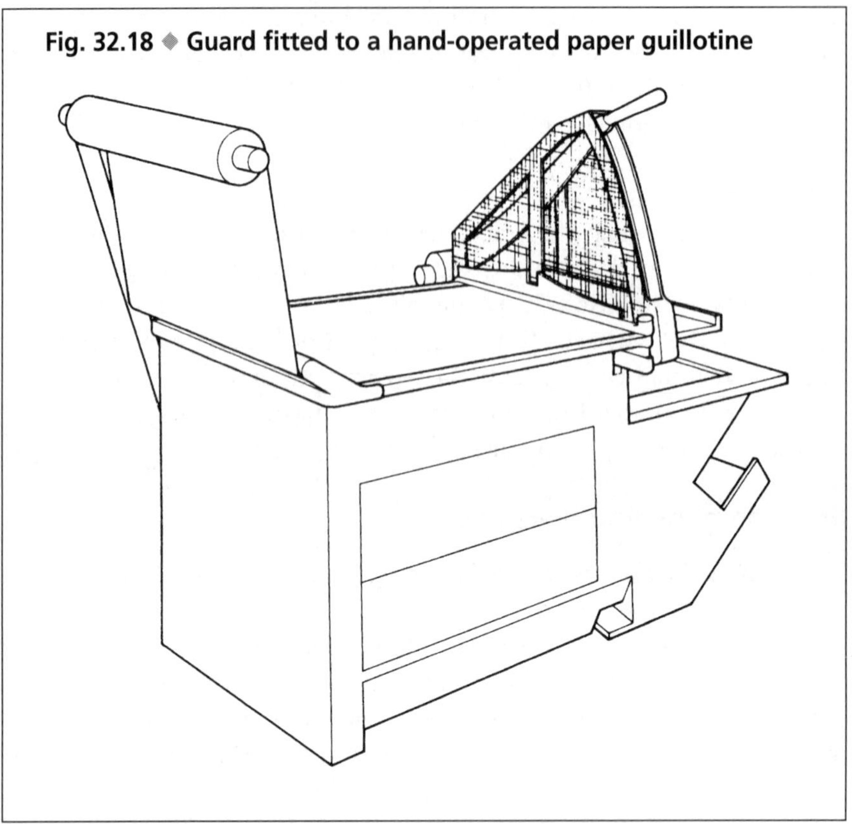

Lathes

Contact with revolving parts; contact with work; trapping between work and tool; contact with transmission machinery

The chuck or face plate should be removed or replaced only when the machine is stationary. The work, tool-holder and tail-stock should be securely clamped before the machine is switched on, and no adjustment to the tool or measurements of the work should be undertaken whilst the machine is operating. Belts should be shifted and gears changed only when the machinery is stationary. (These items must be fully guarded when the machine is in motion.) A proper pushing stick should always be used.

Sudden ejection of work or tool from the machine

Any wrench or other tool should always be removed from the chuck before switching on the machine. Only light cuts on long thin work should be taken due to the risk of the job flying from the machine.

Fig. 32.19 ◆ Guarding to a power-operated guillotine

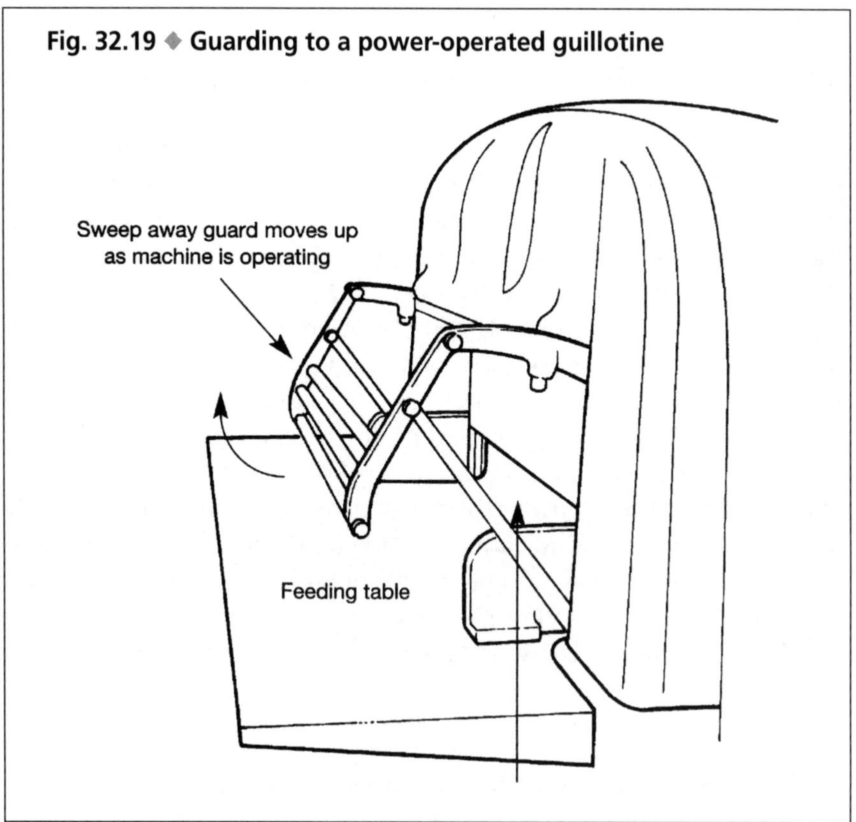

G

Lacerations from swarf

A rake or similar tool should always be used for the removal of swarf. Operators should wear heavy-duty gloves.

Contact with stock-bar

All stock-bars must be adequately guarded for the whole of their length, including the section nearest the machine.

Vertical spindle bowl mixers

Trapping and entanglement through contact with the rotating spindle

The bowl and rotary arm should be completely enclosed using two semi-circular guards, hinged at the rear and interlocked at the front, with a top mesh cover incorporated each side. A funnel inlet in the top or side can be incorporated for the addition of ingredients and liquids during mixing.

Circular saws

Contact with the moving saw blade; disintegration of the saw blade

The guarding arrangements for a floor-mounted circular saw are shown in Fig. 32.20. Such a saw should be provided with guards for three portions of the blade. i.e. the part below the bench table, the up-running part above the bench at the rear of the blade, and the crown and front cutting part of the blade. Guarding for the part below the bench table incorporates two plates of metal or other suitable material (A), one on each side of the saw set not more than 153 mm apart and extending outward from the axis of the saw to a distance not less than 51 mm beyond the teeth of the saw. The metal plates must normally be of a thickness at least equal to 14 gauge, but if the edges are beaded they must be of a thickness not less than 20 gauge. The up-running part is guarded by a riving knife (B) set directly behind the saw and extending upwards from the table to within 25 mm of the top of the saw. The riving knife must be capable of horizontal adjustment so that its front edge at table level can be kept within 12 mm of the teeth of any size of saw which may be used in the bench. The crown and front cutting

Fig. 32.20 ◆ Guards for a floor-mounted circular saw

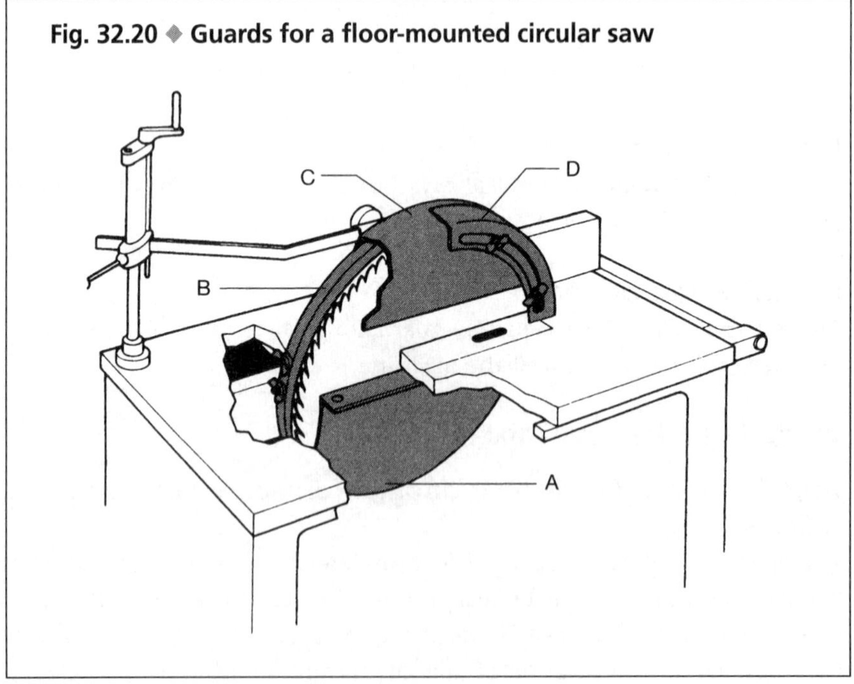

part of the blade incorporates a U-section or L-section cover (C) for the top of the saw supplemented by an L-section extension piece (D). The vertical flanges on (C) and (D) should be of sufficient width to extend below the roots of the teeth on the side of the saw remote from the fence. The guard should be strong and capable of easy adjustment, to allow for variations in the thickness of wood being cut.

Push sticks must be provided and used with a circular saw. Where wood which has been cut is removed from the saw table by a second operator during operation of the saw, he must stand at the delivery end of the machine and the machine table must extend at least 1,200 mm from the up-running part of the saw blade.

Circular saws frequently overrun or continue in motion after the power has been switched off. The installation of a braking system is, therefore, recommended. Provision for lowering the saw below the bench top when not in use is also an effective safeguard.

Band-saws

Contact with the moving blade; breaking of the blade

These saws should be provided with a guard or guards (A), as shown in Fig. 32.21, which totally enclose the pulleys around which the saw band runs and all parts of the band except the part between the bench table and the top pulley. The part of the band between the friction disc or rollers and the top pulley should be provided with a U-section guard (B) with the sides of the guard extending behind the saw band.

Small printing machines

Traps between gear wheels and chain and sprocket drives

These parts should normally be completely enclosed by a guard. However, on small hand-turned machines an interlocked guard should be provided.

In-running nips between the plate and blanket cylinders, between inking rollers, or between the cylinder and inking roller

When two cylinders or rollers revolve, a finger trap is created at the in-running nip. A common method of guarding is to fit a nip bar which prevents access to the nip (*see* Fig. 32.22). On small machines, where this is impracticable, a hinged interlocked guard should be fitted so as to enclose completely the rollers or cylinders.

Fig. 32.21 ◆ Guards to a band-saw

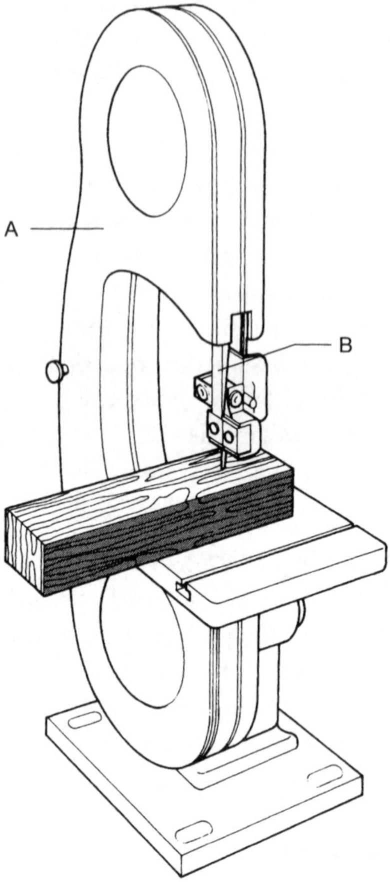

A

B

Fig. 32.22 ◆ Nip bar fitted to revolving rollers on a small printing machine

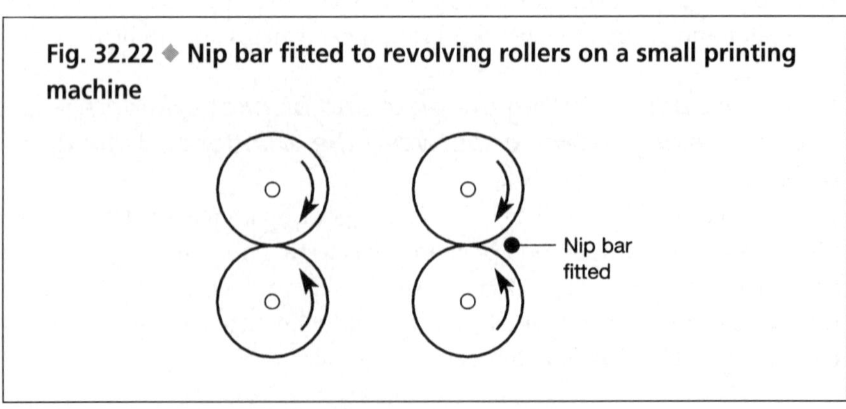

Traps between the gap in the plate cylinder and fixed parts

Offset printing machines often incorporate a gap in one of the two revolving cylinders. As the gap passes the nip bar a trap is created between the rear edge of the gap and the bar. Traps will also occur between the rear edge of the gap and any fixed member of the machine in the vicinity of its travel. There are two methods of guarding, either:

(a) to fit a gap plate (*see* Fig. 32.23) which is normally only practicable on large machines; or

(b) to fit a hinged interlocked guard without a nip bar, which can usually only be adopted for smaller machines.

Fig. 32.23 ◆ Positioning of a gap plate

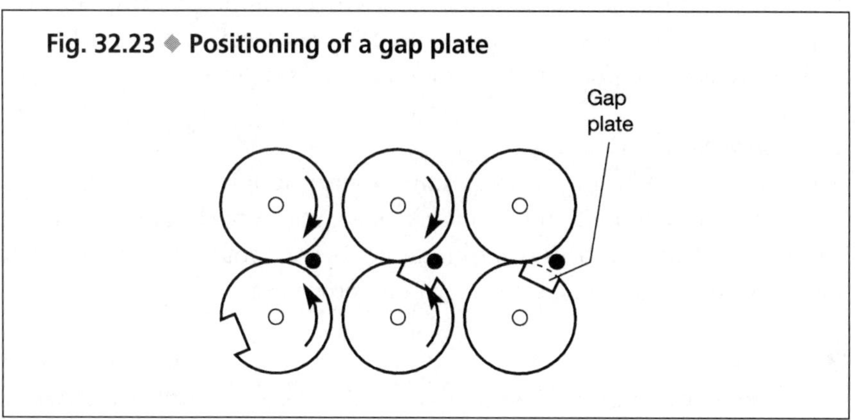

G

Summary

Figure 32.24 (p. 627) presents in diagrammatic form a check-list for ensuring the safety of machinery.

Hand tools

The definition of 'work equipment' under PUWER includes hand tools. The abuse and misuse of hand tools frequently result in injuries, many of which are of a serious nature, e.g. amputations of fingers, blinding, severing of arteries as a result of deep cuts, and account for approximately 10 per cent of all lost-time injuries.

Reg 6 of PUWER places a strict duty on employers to ensure that work equipment is 'maintained in an efficient state, in efficient working order and in good repair'. This duty is particularly appropriate in the case of hand tools, and implies the need for frequent inspection of hand tools to ensure this duty is complied with. Furthermore, the correct use of hand tools should be ensured through training and regular supervision of users.

Hand tool inspections

A number of points should be considered when examining hand tools.

1. Chisels

'Mushroomed' chisel heads are a frequent cause of blinding and eye injuries, and any mushrooming should be removed through grinding. Chisel heads should be kept free from dirt, oil and grease.

2. Hammers

The shaft should be in sound condition and soundly fixed to the head. Where it is split, loose to the head or broken, the shaft should be replaced. Chipped, rounded or badly worn hammer heads should not be used, and hammer heads should be kept free of oil and grease.

3. Files

A file should never be used without a handle, and the handle should be in sound condition. Evidence of chips and other signs of damage indicate the file could be dangerous.

4. Spanners

Open-end spanners which are splayed or box spanners with splits should be disposed of. Adjustable spanners and monkey-wrenches should be examined regularly for evidence of free play and splaying of the jaws.

5. Screwdrivers

Handles and tips should be in sound condition and worn-ended screwdrivers should never be used. A screwdriver should never be used as a chisel and, when using a screwdriver, the work should be clamped or secured, never held in the hand. Employees must be trained to use the correct size screwdriver at all times.

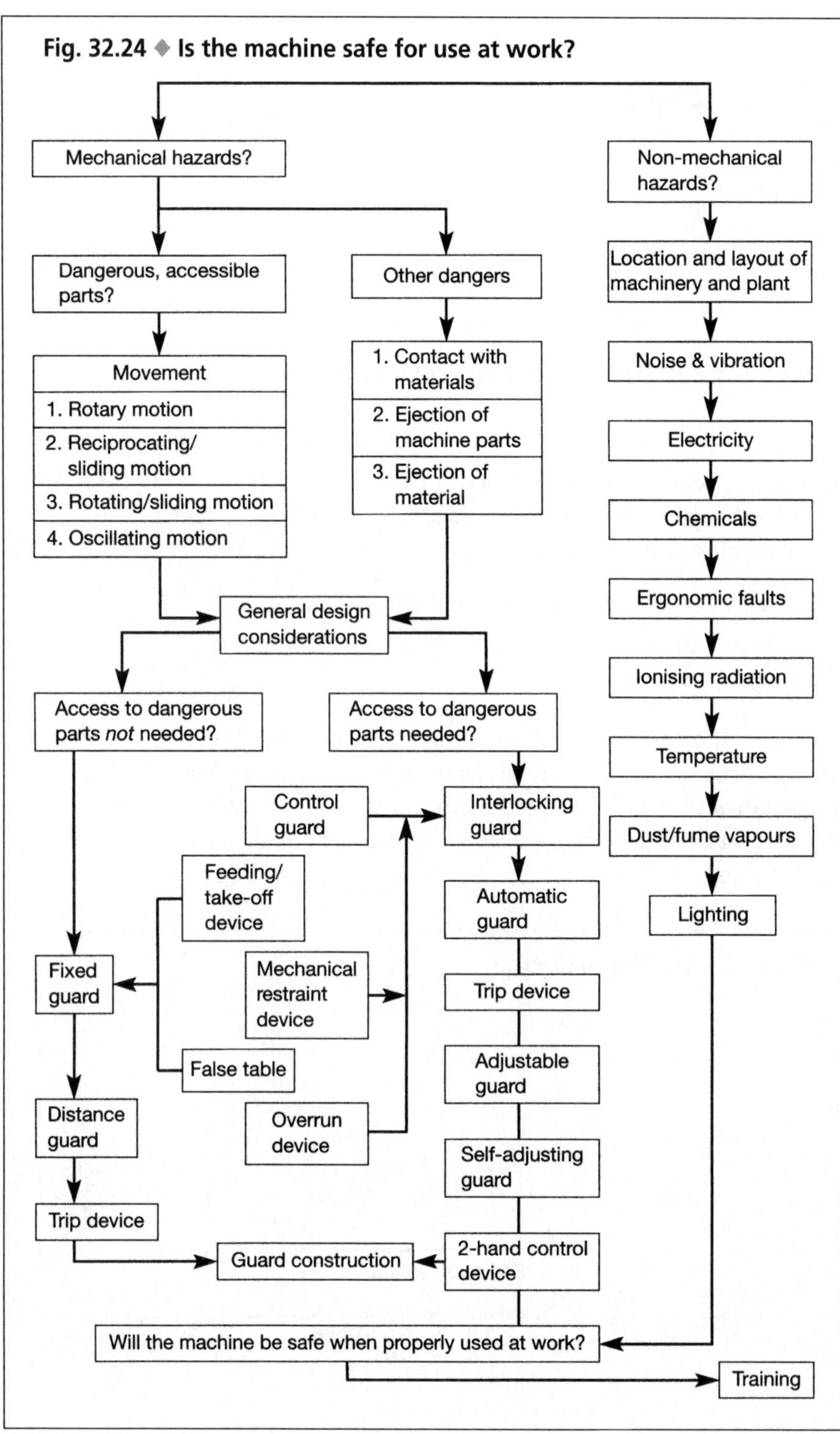
Fig. 32.24 ◆ Is the machine safe for use at work?
Mechanical hazards?
Non-mechanical hazards?
Dangerous, accessible parts?
Other dangers
Location and layout of machinery and plant
Movement
1. Rotary motion
2. Reciprocating/ sliding motion
3. Rotating/sliding motion
4. Oscillating motion
1. Contact with materials
2. Ejection of machine parts
3. Ejection of material
Noise & vibration
Electricity
Chemicals
Ergonomic faults
Ionising radiation
Temperature
Dust/fume vapours
Lighting
General design considerations
Access to dangerous parts not needed?
Access to dangerous parts needed?
Control guard
Interlocking guard
Feeding/ take-off device
Automatic guard
Fixed guard
Mechanical restraint device
Trip device
False table
Adjustable guard
Distance guard
Overrun device
Self-adjusting guard
Trip device
Guard construction
2-hand control device
Will the machine be safe when properly used at work?
Training

G

33 Fire

Occurrence through combustion

Fire is a spectacular example of a fast chemical reaction between a combustible substance and oxygen accompanied by the evolution of heat. An explosion is an even more spectacular example of the same reaction. The three requirements for fire are:

(a) oxygen, except in very special circumstances;

(b) a fuel or combustible substance;

(c) a source of energy.

These three requirements are easily remembered in the form of the fire triangle (Fig. 33.1). The triangle, however, says nothing about the relative importance of each of the requirements or their states of matter. These come from a full knowledge of the chemistry of combustion.

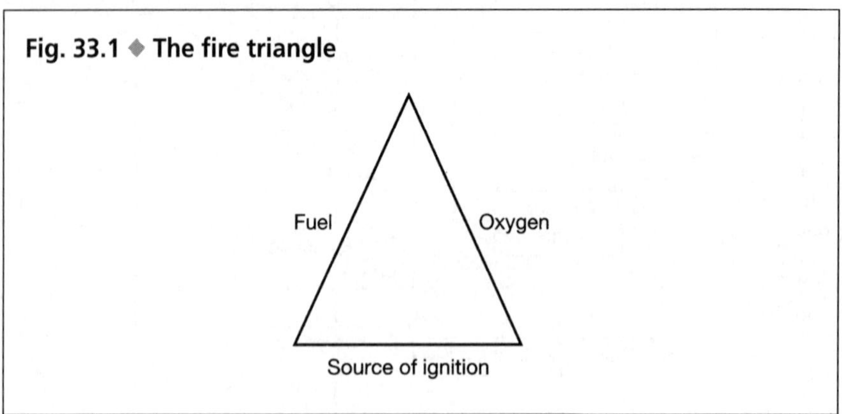

Fig. 33.1 ◆ The fire triangle

Heat transmission

Heat may be transmitted or transferred by the processes of convection, conduction and radiation.

Convection

When matter is heated the range of vibration of its molecules increases and the matter expands. In a solid this has little significance, but in a fluid, where the particles are free to move throughout the substance, it gives rise to a *convection current* in the body of the fluid.

These currents are produced due to the fact that the unheated particles of a fluid are more dense than the newly heated particles and thus gravitate to the lowest point of the body of fluid. In doing so they upwardly displace the hot particles. These hot particles, whether of liquid or gas, have no specific ability to rise. They rise simply because they are displaced from beneath, and the difference in weight between hot and cold particles is the sole force operating. The greater the difference in temperature between the hottest and coldest particles, the more vigorous will be the convection current produced. If the source of heat is maintained, the cooler particles, already in motion as a result of their greater weight, become heated in turn, whilst the previously heated particles, having been displaced from the heat source, lose some of their heat to the surroundings and become cooler. A convection current is thus continuous and is described as a *circulation* within the body of the fluid.

Gases are also fluids and act in the same way as fluids. For instance, systems of natural ventilation in buildings operate on the basis of convection currents being established and maintained. Hot air rises and cold air sinks.

Conduction

With solids the increased molecular vibration due to an increase in temperature of one part is imparted by contact to adjoining molecules. Thus if a length of metal wire is heated at one end it would, by the process of conduction, ultimately reach uniform temperature through its entire length, but for the fact that some of the heat is lost to the air from its exposed surface. This method of transmission is called *conduction* and is confined largely to solids because of their rigid structure.

Solids are generally classed as good or bad conductors according to their ability or otherwise to conduct heat. Metals are good conductors, whereas wood is a bad conductor.

Conduction also takes place in liquids and gases. In any normal body of liquid or gas the particles are displaced by convection before any appreciable transfer of heat by conduction can take place. Generally, liquids are poor conductors and gases even worse.

Radiation

In *convection* heat is transferred from one point to another by the relative movement of particles. In *conduction* transference is between adjoining particles which do not move relative to each other. In *radiation* the molecular heat motion of a body causes to be emitted or radiated into space rays or waves travelling in straight lines, in all directions from the source, and at a speed of about 186,400 miles per second.

All hot bodies emit waves, the larger and hotter the body the more intense the waves. These are waves of energy which, when directed on to a body, produce sensible heat in the body. Heat radiation travels best through empty space but can also travel through a variety of media.

Heat radiation is subject to the *Inverse Square Law*. This law states that the intensity of heat produced by radiation is inversely proportional to the square of the distance from the source.

$$\text{Intensity} = \frac{1}{\text{distance}^2}$$

Thus if the distance from the source is doubled, the same amount of radiation affects four times the area and the intensity is thus one quarter. The same law applies in the case of light emission.

Elements of fire

Oxygen

A fire always requires oxygen for it to occur or, having started, to continue. The chief source of oxygen is air, which is a mixture of gases comprising nitrogen (78 per cent) and oxygen (21 per cent). The remaining one per cent is made up of water vapour, carbon dioxide, argon and other gases. A number of substances can be a source of oxygen in a fire, e.g. oxidising agents. These are substances which contain oxygen that is readily available under fire conditions and include sodium chlorate (Na_2ClO_3), hydrogen peroxide (H_2O_2), nitric acid (HNO_3) and organic peroxides ($R—O—O—R^1$). A third source of oxygen is the combustible substance itself, e.g. ammonium nitrate (NH_4NO_3).

Combustible substance

This is the second requirement for fire and includes a large group of organic substances, i.e. those with carbon in the molecule, e.g. natural gas (methane) (CH_4), butane (C_4H_{10}), petrol, plastics, natural and artificial fibres, wood, paper, coal and living matter. Inorganic substances, i.e. those not containing carbon in the molecule, are also combustible, e.g. hydrogen (H_2), sulphur (S), sodium (Na), phosphorus (P), magnesium (Mg) and ammonium nitrate (NH_4NO_3).

Ignition source

This is the energy that has to be applied to the oxygen/fuel mixture to start the fire. Usually this energy is in the form of heat, but not necessarily. The heat can be simply that contained in the combustible substance. This is often the source of ignition energy when hot fuel leaks from a pipe and fires, but it can be heat generated by friction such as rubbing a match against sandpaper or a hot bearing in a machine. Electrical energy in the lightning of a thunderstorm or when an electrical contact, such as a switch, is made or broken, would also qualify.

Chemistry of combustion

Combustion chemistry is an example of a larger group of chemical reactions known as oxidation reactions. Other examples are the rusting of iron and the process of breathing. Chemically the process can be written as:

Fuel plus oxygen gives products of combustion and heat

If the fuel is natural gas (i.e. methane), the reaction is written as:

$$\underset{\text{methane}}{CH_4} + \underset{\text{oxygen}}{2O_2} \rightarrow \underset{\text{carbon dioxide}}{CO_2} + \underset{\text{water}}{2H_2O} + \text{Heat}$$

If the fuel were hydrogen, the equation would be:

$$\underset{\text{hydrogen}}{2H_2} + \underset{\text{oxygen}}{O_2} \rightarrow \underset{\text{water}}{2H_2O} + \text{Heat}$$

Or if it were sulphur, it would be:

$$\underset{\text{sulphur}}{S} + \underset{\text{oxygen}}{O_2} \rightarrow \underset{\text{sulphur dioxide}}{SO_2} + \text{Heat}$$

However, for ammonium nitrate, which has its own source of oxygen, the equation is:

$$\underset{\text{ammonium nitrate}}{NH_4NO_3} \rightarrow \underset{\text{nitrous oxide}}{N_2O} + \underset{\text{water}}{2H_2O} + \text{Heat}$$

Generally, the carbon in the fuel is oxidised to carbon dioxide and hydrogen to water. The other elements will be oxidised to a variety of substances.

Initiation energy – explanation of ignition

When the chemical reaction

$$2H_2 + O_2 \rightarrow 2H_2O + \text{Heat}$$

is taking place, it can be thought of as molecules hitting one another and sometimes bouncing off like balls or otherwise mutually breaking up into different molecules. In the latter case a chemical reaction has occurred. Characteristic of the 'ball' theory (more properly known as the Kinetic Theory of Gases) is that kinetic energy possessed by the balls, i.e. their speed, is a function of the temperature of the gas. In particular, the higher the temperature the higher the kinetic energy of the molecules. The reaction may be described as one where the energy of the collision has to be greater than some value for the molecules to hit and break up into different molecules. If the collision energy is below this value the molecules will just bounce off one another and no reaction will take place. Collision energy is the physical representation of the initiation energy required for a fire to start. It is usually heat, i.e. increasing the temperature of the combustion mixture, but it can be electrical energy from a spark.

The final feature of all combustion reactions is that they emit energy. These reactions are known as exothermic reactions.

Table 33.1 shows the heat of combustion in kilocalories per molecular weight of the substance when combustion takes place at atmospheric pressure and 20°C. Table 33.2 gives the minimum ignition energy of substances in millijoules.

The fire process – a summary

A fire may be described as a mixture in gaseous form of a combustible substance and oxygen, with sufficient energy being put into the mixture to start the fire. Once started, the energy output from the fire provides a continuous source of energy for it to be sustained and excess is given off as sensible heat (*see* Fig. 33.2). For all practical purposes, fire takes place in the gaseous state and division of fires into those of solids (e.g. wood), liquid

Table 33.1 ◆ Standardised values of heat output

Fuel	*kcal*
Acetone $(CH_3)_2CO$	126.8
Benzene C_6H_6	782.3
Carbon disulphide CS_2	246.6
Cyclohexane C_6H_{12}	937.8
Dimethyl ether $(CH_3)_2O$	347.6
Ethane C_2H_6	368.8
Ethyl alcohol C_2H_5OH	327.6
n-Hexane C_6H_{14}	989.8
Hydrogen H_2	58.3
Methane CH_4	210.8
Methyl alcohol CH_3OH	170.9
Propane C_3H_8	526.3

Table 33.2 ◆ Standardised values of ignition energy

Minimum ignition energy	*mJ*
Benzene	0.22
Carbon disulphide	0.01–0.02
Ethane	0.24
n-Hexane	0.25
Hydrogen	0.019
Methane	0.39
Propane	0.25

(e.g. petrol) and gas (e.g. a gas flame) is convenient but not accurate. This can be seen with burning wood and liquid where close observation reveals that the flame burns at a small distance from the wood or liquid.

Fig. 33.2 ◆ The fire process

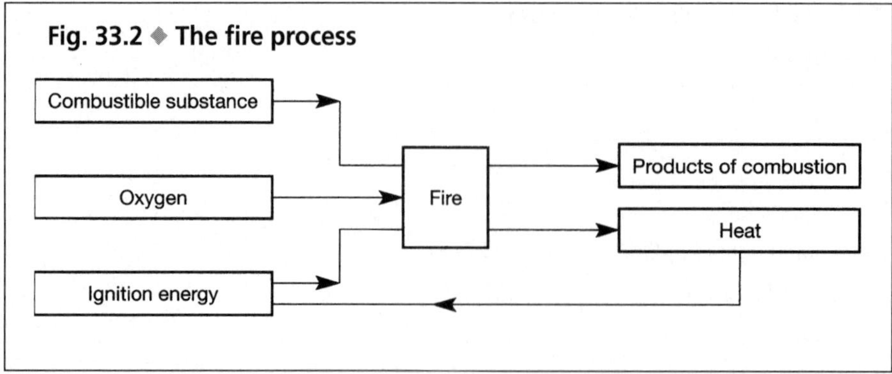

Ignition energy – liquids and gases

Ignition energy of liquids and gases can be measured, and is expressed in different ways. The three measures are:

(a) *Flash point.* The flash point is the minimum liquid temperature at which sufficient vapour is given off to form a mixture with air capable of ignition under prescribed test conditions. There are two sets of prescribed conditions, the Abel and the Pensky-Martin. The former usually yields slightly lower results. It is necessary to cite the method used when quoting flash point values.

(b) *Fire point.* This is the lowest temperature at which the heat from combustion of a burning vapour is capable of producing sufficient vapour to sustain combustion.

(c) *Spontaneous ignition temperature*, or ignition temperature. This is the lowest temperature at which the substance will ignite spontaneously.

Usually, for a given combustible substance the flash point is less than the fire point, which is less than the ignition temperature. Since energy is measured by temperature and two effects are being studied, i.e. the evaporation of the liquid and combustion of evaporated vapour, the ignition temperature must be the highest value because the whole of the required energy comes from the heat, i.e. temperature, of the liquid. The flash point must be the lowest because the source of ignition is external to the evaporation of the liquid.

Hitherto it has been assumed that a mixture will only react if sufficient energy is put into it. This is not always the case. Consider an oxygen/methane mixture. If there are very few methane molecules in the mixture these will react very quickly and the fire stops. As the concentration of methane increases, a point will be reached when the heat output from the fire will eventually be sufficient to provide the source of ignition for the rest of the methane and the fire is sustained. Similar conditions apply if the oxygen/methane mixture is rich in methane and weak in oxygen. Here it is the oxygen, however, that is instantaneously used up until only sufficient oxygen is present for the heat output to sustain combustion. These two concentration values which exist for all gaseous combustible substances and oxygen are known as the lower and upper flammability limits.

Lower flammable limit

The lower flammable limit (lower explosive limit) is the smallest concentration of flammable gas or vapour which, when mixed with air, is capable of ignition and subsequent flame propagation under prescribed test conditions.

Upper flammable limit

G

The upper flammable limit (upper explosive limit) is the highest concentration of flammable gas or vapour which, when mixed with air, is capable of ignition and subsequent flame propagation under prescribed test conditions. Table 33.3 gives values of flammable limits for some common substances.

Spontaneous combustion

The final combustion characteristic is spontaneous combustion. Sometimes material bursts into flames without apparent means of ignition. Examples are haystacks, rags soaked in linseed oil and oil-soaked lagging. This is called spontaneous ignition. It is clear from the mechanism of fire that there must be a source of ignition but the source itself is not obvious. Some organic substances, when exposed to oxygen, undergo slow oxidation, a process similar to fire, releasing little sensible heat. Slow oxidation does not result in carbon dioxide and water but other substances chemically smaller and more easily combustible than the original substance. In effect, the original substance is being made more combustible by oxidation and, in consequence, little heat is generated. When reduction in combustibility and heat output match, a fire occurs. The source of ignition is the heat of oxidation.

Table 33.3 ◆ Properties of some flammable substances

Substance	*Flash point (°C)*	*Ignition temperature (°C)*	*Flammable range (% v/v in air)*
Acetic acid	40	485	4–17
Acetone	–18	535	2.1–13
Acetylene	–18	305	1.5–80
Ammonia	Gas	630	1.5–2.7
Benzene	–17	560	1.2–8
n-Butane	–60	365	1.5–8.5
Carbon disulphide	–30	100	1–60
Carbon monoxide	Gas	605	12.5–74.2
Cyclohexane	–20	259	1.2–8.3
Ether	–45	170	1.9–48
Ethanol	12	425	3.3–19
Ethylene	Gas	425	2.7–34
Hydrogen	Gas	560	4.1–74
Methane	Gas	538	5–15
Toluene	4	508	1.2–7
Vinyl chloride	–78	472	3.6–33

The main causes of fire and fire spread

Various studies by the Fire Protection Association of a range of industrial fires have indicated the following as the principal sources of fire in production and storage areas.

Production areas

1. Heat-producing plant and equipment
2. Frictional heat and sparks
3. Refrigeration plant
4. Electrical equipment

– setting fire to:

1. Materials being processed

2. Dust
3. Waste and packing materials

Storage areas

1. Intruders, including children
2. Cigarettes and matches
3. Refuse burning
4. Electrical equipment

– setting fire to:

1. Stored goods
2. Packing materials

Ignition temperature

G

Generally, fire is spread by a range or combination of factors, namely through heat transmission, direct contact and/or through the release of flammable gases or vapours. In all cases, some form of ignition source must be present which is sufficient to create the energy necessary to raise a volume of combustible material to its ignition temperature. This is the temperature at which a small amount of combustible material (fuel) will spontaneously ignite in a given atmosphere and continue to burn without any further input of heat. There must be specific conditions present for ignition to take place, as seen with the Fire Triangle, namely the appropriate temperature, the right source of ignition and an appropriate mixture of combustible material and oxygen.

Extinction

Extinction means putting out a fire. It can be achieved by one or more of the following:

(a) *starvation* – a reduction in concentration of the fuel;

(b) *smothering* – a reduction in concentration of oxygen: and

(c) *cooling* – a reduction in the rate of energy input to the fire.

It is a reduction that is required and not necessarily the removal of any one of them, since a fire occurs when the fuel/oxygen mixture is within two concentrations and it receives a minimum amount of energy.

Starvation

There are three ways that this can be achieved:

(a) take the fuel away from the fire;

(b) take the fire away from the fuel;

(c) reduce the quantity or bulk of the fuel.

The first is achieved every day on the gas stove when the tap is turned off. For large chemical plants this means isolating the feed at the remote isolation valve. Examples of taking the fire away from the fuel include breaking down stacks and dragging away the burning debris. Breaking down a fire into smaller units is an example of reducing the quantity or bulk of the fuel.

Smothering

There are two ways of achieving this:

(a) allow the fire to consume the oxygen while preventing the inward flow of more oxygen;

(b) add an inert gas to the burning mixture.

Wrapping a burning person in a blanket is an example of smothering. Other examples include pouring foam on top of a burning pool of oil or putting sand on a small fire. A danger inherent in extinguishing fires by smothering occurs when the fire is out but everything is still hot. Any inrush of oxygen, caused by disturbing the foam layer or opening the door to a room, could result in reignition as there may still be suffficient energy in the form of sensible heat present.

If an inert gas is to be used as an extinguisher, carbon dioxide or halogenated hydrocarbons such as BCF are suitable. Alternatively, nitrogen can be used, and is in fact more common for petrochemical plant fires. If a flammable gas pipeline leaks and the escaping gas fires, nitrogen blanketing can be achieved by injecting nitrogen into the gas stream downstream of the release. Smothering is only effective when the source of oxygen is air. It is totally ineffective when the burning substance contains oxygen, such as ammonium nitrate.

Cooling

This is the most common means of fighting a fire, water being the cheapest and most effective medium. For a fire to be sustained, some of the heat output from the combustion is returned to the fuel, providing a continuous source of ignition energy. When water is added to a fire, the heat output serves to heat and vaporise the water, i.e. the water provides an alternative heat sink. Ultimately, insufficient heat is added to the fuel and continuous ignition ceases. In order to assist rapid absorption of heat, water is applied to the fire as a spray rather than a jet, the spray droplets being more efficient in absorbing heat than the stream of water in a jet. Another example of heat absorption is provided by dry chemical extinguishers. These are very fine powders which readily absorb heat and in one case, MONEX, break up into small even particles. Dry powders can also be used for the smothering method.

Classification of fires

Fires are commonly classified into four categories according to the fuel type and means of extinction.

G

Class A

Fires involving solid materlals, normally of an organic nature, in which the combustion occurs with the formation of glowing embers, e.g. wood, paper, coal and natural fibres. Water applied as a jet or spray is the most effective way of achieving extinction.

Class B

Fire involving (i) liquids (ii) liquefiable solids. Liquids fall into two groups:

(i) miscible with water – methanol, acetone, acetic acid;

(ii) immiscible with water- petrol, benzene, fats and waxes.

Foam, light water, vaporising liquids, carbon dioxide and dry powder can be used on both B(i) and B(ii) type fires. Water spray can be used on type B(i) but not on type B(ii). There may also be some restriction on the type of foam which can be used because some foams break down on contact with alcohols. In all cases extinction is mainly achieved by smothering. However, water on a B(i) fire also acts by cooling, and by removal of the fuel in that the fuel dissolves in the water.

Class C

Fires involving gases or liquefied gases, e.g. methane, propane and butane. Both foam and dry chemicals can be used on small liquefied gas spillage fires, particularly when backed up by water to cool the leaking container or spillage collector. A fire from a gas leak can be extinguished either by isolating the fuel remotely or by injecting an inert gas into the gas stream. Direct flame extinguishment is difficult and may be counterproductive in that if the leak continues there may be reignition, often in the form of an explosion. Extinguishers used on liquid gas spillage fires work by smothering.

Class D

Fires involving metals, e.g. magnesium or aluminium. They can only be extinguished by use of dry powders which include talc, soda ash, limestone and dry sand. All the extinguishers work by smothering.

'Electrical' fires

This is now an obsolete classification. Fires which involve electrical equipment must always be tackled first by isolating the electricity and then by use of carbon dioxide, vaporising liquid or dry powder. The use of these agents minimises damage to equipment.

Portable fire extinguishers

These are appliances designed to be carried and operated by hand. They contain an extinguishing medium which can be expelled by action of internal pressure and directed on to a fire. The pressure may be stored, or obtained by chemical reaction or by release of gas from a cartridge. The maximum mass of a portable extinguisher in working order is 23 kg. Portable extinguishers must be coloured red and incorporate a colour-coded label as shown in Table 33.4.

Water-containing extinguishers

There are three kinds, i.e. soda acid, gas cartridge and stored pressure.

Table 33.4 ◆ Grouping and coding of portable fire extinguishers

Extinguishers	*Colour code*
Water	Red
Foam	Cream
Carbon dioxide	Black
Dry chemical powder	Blue
Vaporising liquid	Green

Soda acid

This is the original form of water-containing extinguisher, and is gradually being replaced by the gas cartridge and stored pressure types.

Two common types are shown in Fig. 33.3. Gas is generated in the cylinder when the acid phial is broken, and this expels all the water through the discharge tube.

G

Fig. 33.3 ◆ Sections through a soda acid extinguisher (a) with plunger at the top (b) with hammer at side

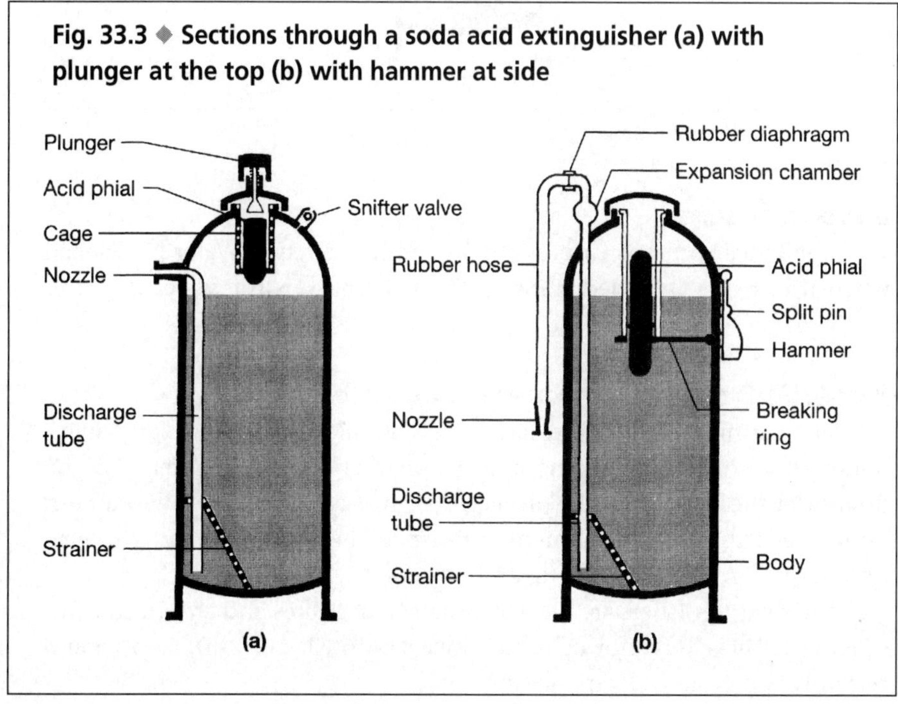

Source: Home Office, *Manual of Firemanship*

Gas cartridge

With this type, carbon dioxide is held in a small pressure cylinder, the seal being broken by a plunger. The gas so released expels the water out of the nozzle (*see* Fig. 33.4).

Fig. 33.4 ◆ Section through a water (gas cartridge) type of water extinguisher

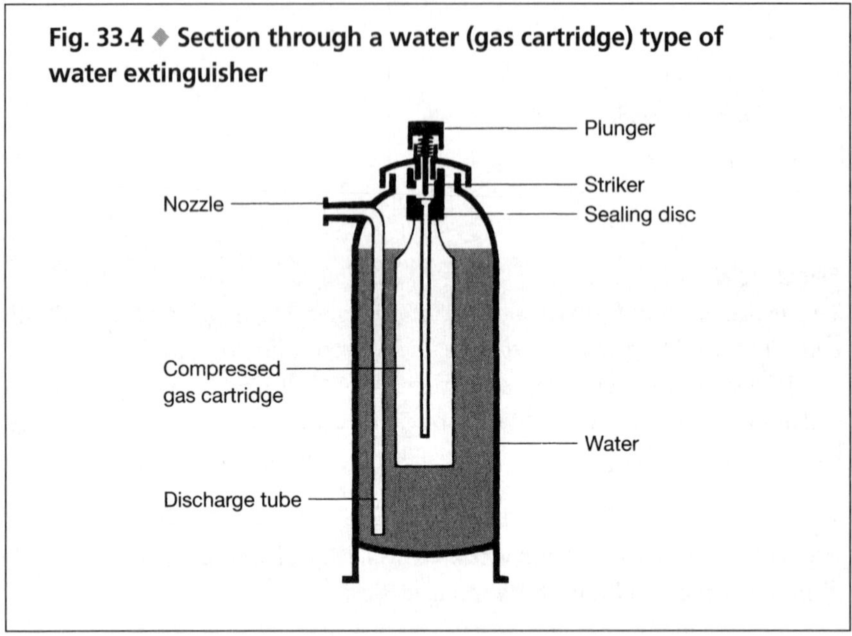

Source: Home Office, *Manual of Firemanship*

Stored pressure

This appliance contains carbon dioxide under pressure. Water is expelled when the trigger is pulled. Closing the trigger stops the water flow (*see* Fig. 33.5).

Precautions relating to water extinguishers

All water extinguishers are pressure vessels, so they must be regularly maintained in accordance with the maker's instruction (*see* Chapter 35). The interior of the vessel must be protected against corrosion and the water in it may need anti-freeze treatment. It must be operated in the upright position in order to discharge water and not gas.

Water extinguishers are ideal for small Class A fires and are effective for Class B(i). They must not be used, however, on Classes B(ii), C or D, and certainly not on live electric wiring.

Fig. 33.5 ◆ Section through a stored pressure type extinguisher

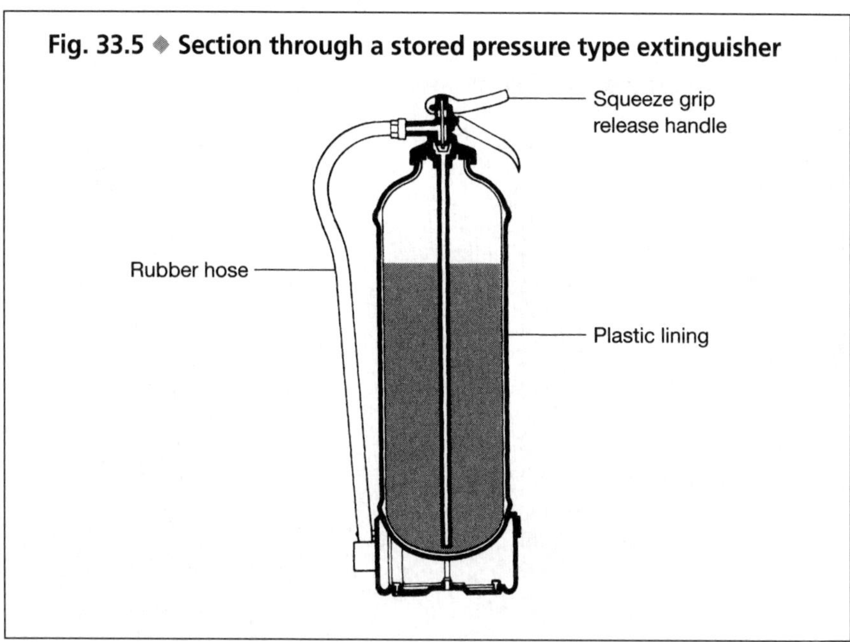

Source: Home Office, *Manual of Firemanship*

Foam extinguishers

Of the four types of foam extinguisher (*see* Figs 33.6, 33.7, 33.8 and 33.9), three contain foam solution which is expelled by carbon dioxide. With the fourth (gas cartridge type), foam is generated at the exit by carbon dioxide.

Chemical foam

The cylinder contains two solutions which are mixed on inversion, viz. aluminium sulphate and sodium bicarbonate. The cylinder itself is filled with sodium bicarbonate solution containing about three per cent of a stabiliser, such as saponin, liquorice or turkey red oil. The inner compartment contains 13 per cent of aluminium sulphate. The foaming mixture is expelled in the inverted position by carbon dioxide generated in the chemical reaction. In some chemical foam extinguishers, the inner cylinder is sealed and has to be broken with a plunger before inversion (Fig. 33.6).

Stored pressure

The alternative to the chemical foam extinguisher is the self-aspirating type. Here the cylinder contains foam concentrate which is expelled either by stored pressure or by gas cartridge. There are four types of foam

Fig. 33.6 ◆ Sections through foam extinguishers (a) Turn-over type without seal; (b) Bayonet seal turn-over type

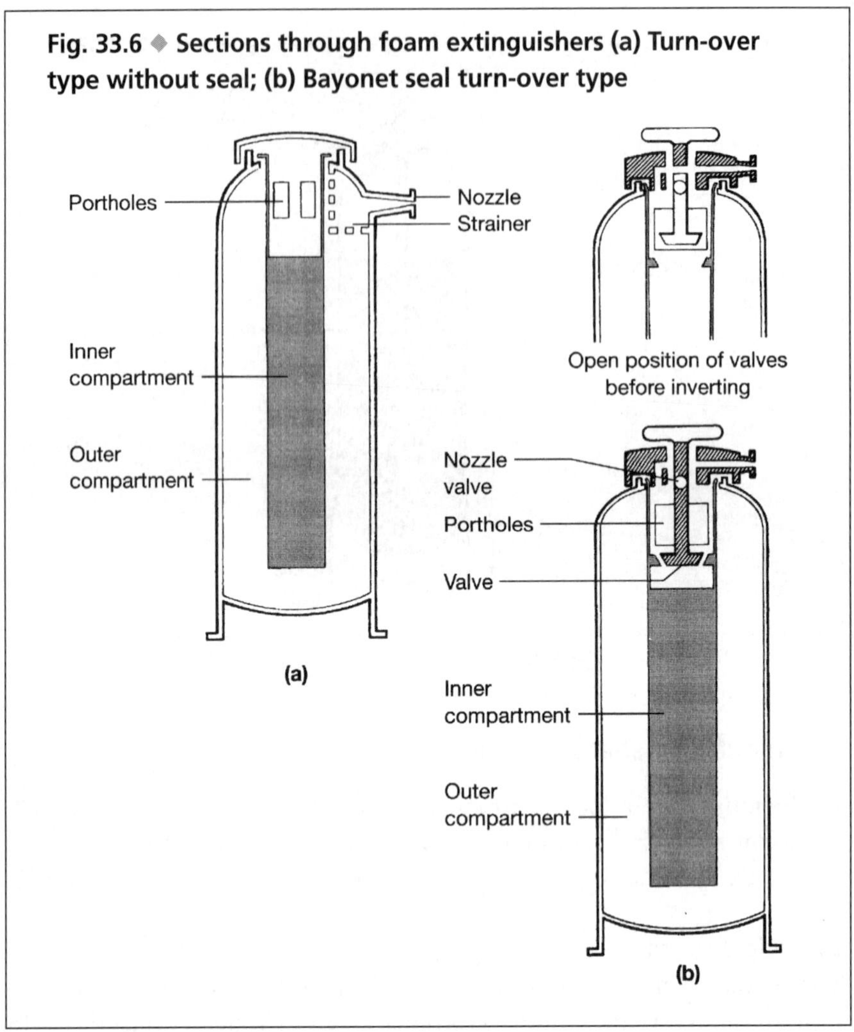

Source: Home Office, *Manual of Firemanship*

concentrate protein, fluor-protein, fluor-chemical and synthetic. In a stored pressure foam extinguisher the cylinder is filled with foam concentrate and pressurised to 10 bar with air or nitrogen. Operation of the trigger valve allows the pressure to expel the foam concentrate through the exit pipe where foam is generated at the end of the hose (Fig. 33.7).

In the gas cartridge extinguisher the foam concentrate is expelled by breaking a seal on a carbon dioxide cartridge (Fig. 33.8). Alternatively, the cylinder can be filled with water, the foam concentrate being contained in a plastic bag which surrounds the gas cartridge (Fig. 33.9). Breaking the seal

Fig. 33.7 ◆ Foam extinguisher (stored pressure) type

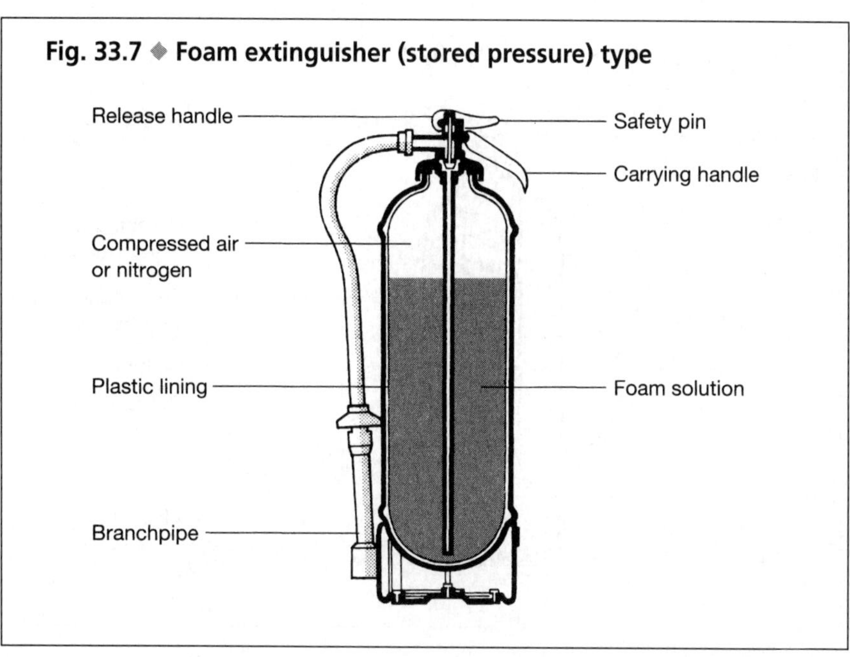

Source: Home Office, *Manual of Firemanship*

Fig. 33.8 ◆ Gas cartridge foam solution type of foam extinguisher

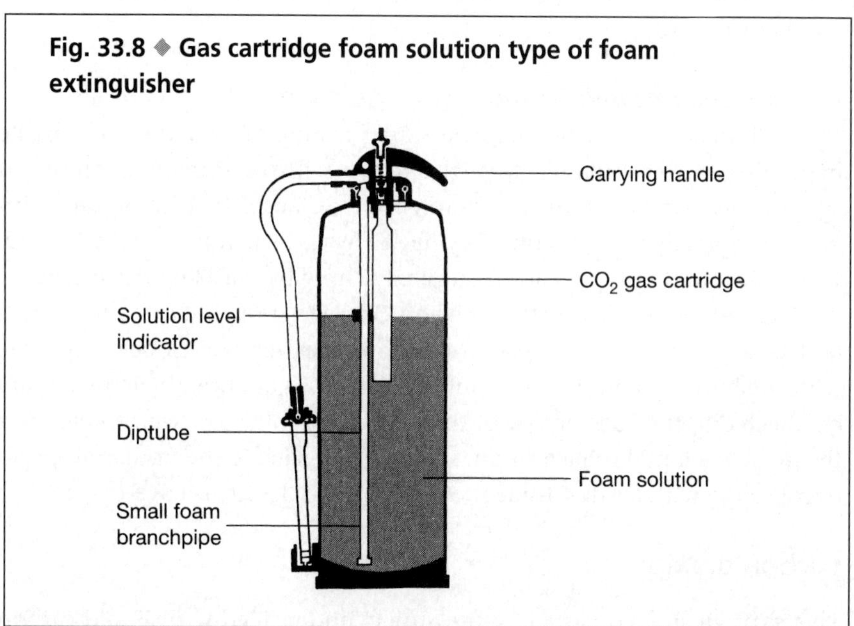

Source: Home Office, *Manual of Firemanship*

Fig. 33.9 ◆ Foam extinguisher (gas cartridge) type

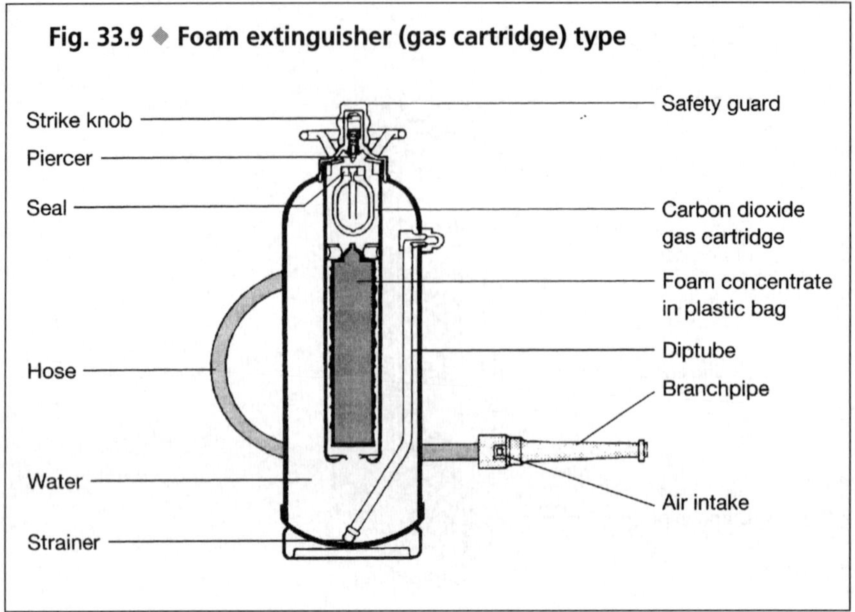

Source: Home Office, *Manual of Firemanship*

on the cartridge releases the carbon dioxide which then bursts the plastic bag. Pressure expels the foam concentrate through the exit hose, releasing foam at the exit.

Precautions relating to foam extinguishers

The cylinders of foam extinguishers are pressure vessels and so must be maintained in accordance with the maker's instructions and protected against internal corrosion (*see* Chapter 35). Because chemical foam extinguishers operate by inversion, they must always be completely discharged. The self-aspirating types can be stopped simply by releasing the trigger.

Foam extinguishers are best used on Class B(ii) fires, but can also be used on Classes A and B(i). In the latter case, however, foam must be compatible with the burning liquid. In use on a contained liquid fire, the foam should be dlrected to the back or side of the containers and allowed to spread over the fire. On a liquid spillage, foam should be directed to the front and spread over the fire with a side-to-side movement until the fire is covered.

Carbon dioxide

This extinguisher consists of a pressure cylinder filled with liquid carbon dioxide. A trigger allows the liquid to be discharged through a horn under its own pressure. On discharge the liquid is converted into carbon dioxide snow in the nozzle which is converted to gas in the fire. It is noisy on discharge and the horn can become very cold (Fig. 33.10).

Carbon dioxide can be used on both Classes A and B fires and on fires involving electrical equipment. On Class A fires it is less efficient than water because of limited cooling, and on Class B it is less efficient than vaporising liquid. A second extinguisher should be available because of the limited cooling effect. Carbon dioxide is an asphyxiant and is heavier than air. It can collect in pits and hollows where it may be a risk to people entering the premises.

Fig. 33.10 ◆ A carbon dioxide extinguisher showing the piercing mechanism, control valve and discharge horn

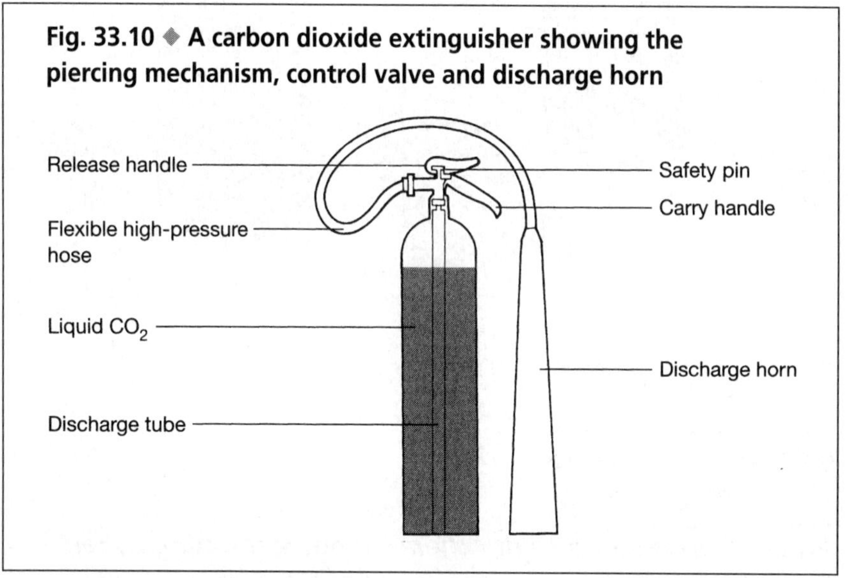

Source: Home Office, *Manual of Firemanship*

Dry chemical powder

These are of two types – stored pressure and gas cartridge. In both cases the cylinder is fed with dry powder, but in the stored pressure type the cylinder is pressurised to 10 bar with dry air or nitrogen. Operating the trigger allows the pressure to expel the dry powder through the hose. With the gas cartridge type the seal on the cartridge is broken by a plunger and the flow of powder is controlled by a trigger valve on the hose (Fig. 33.11).

Fig. 33.11 ◆ (a) One type of dry powder (gas cartridge) extinguisher; (b) Another type with a squeeze grip control

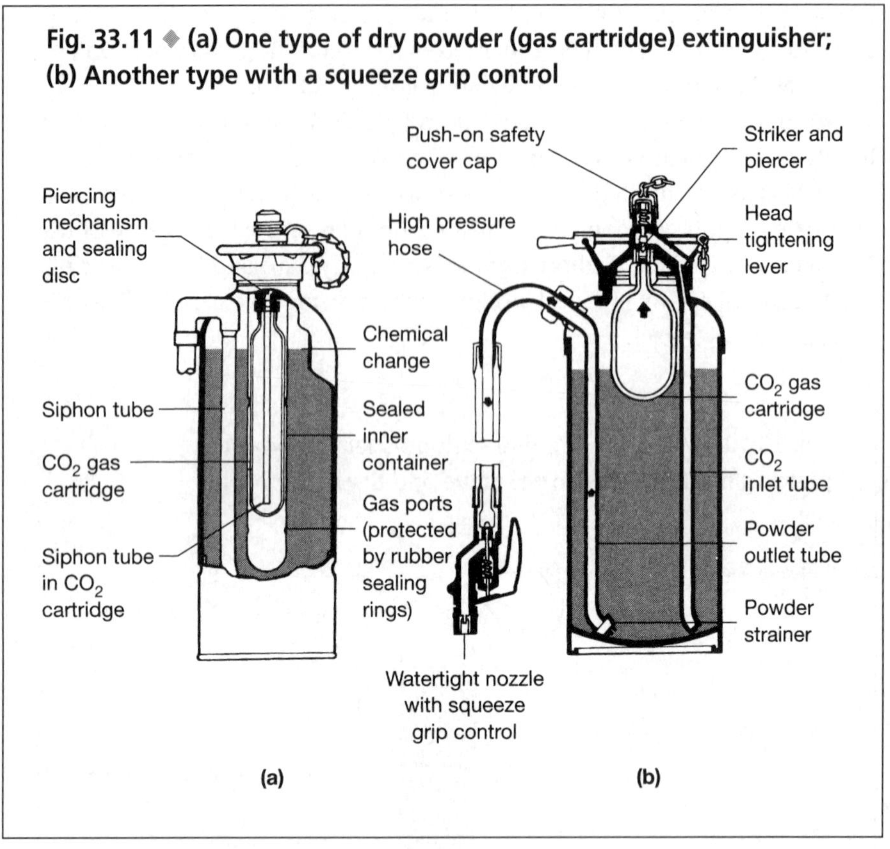

Source: Home Office, *Manual of Firemanship*

Precautions relating to dry chemical powder extinguishers

The cylinder is a pressure vessel so it needs to be maintained in accordance with the maker's instructions (*see* Chapter 35). Dry powder has a tendency to cake during prolonged storage in the cylinder. Moreover, after discharge, the trigger end of the hose can become blocked, necessitating thorough cleaning. These extinguishers can be used on Class A and Class B fires provided that they contain general purpose powder.

Vaporising liquid

All vaporising liquid extinguishers consist of a cylinder containing the liquid which is pressurised to 10 bar with dry carbon dioxide or nitrogen. Striking a knob allows the pressure to expel the liquid (Fig. 33.12).

These extinguishers are useful for Class A and Class B fires and small or incipient fires involving burning liquids. They are less efficient than foam

Fig. 33.12 ◆ Section through vaporising liquid (stored pressure) extinguisher

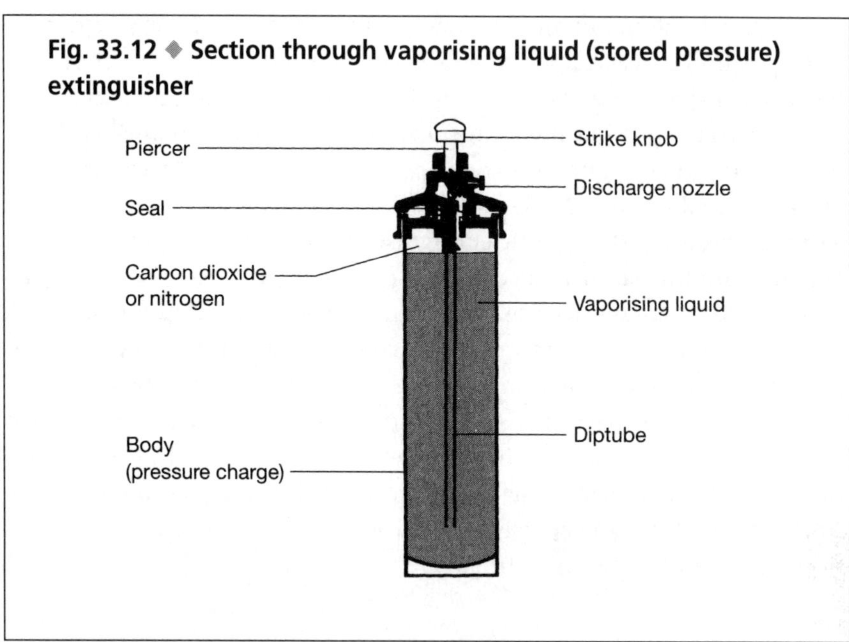

Source: Home Office, *Manual of Firemanship*

on large liquid fires because the heat dissipates the vaporising liquid, and less efficient than water on Class A fires as they have a lower cooling effect. On fires involving electrical equipment they are particularly effective because they are non-conducting and do not cause damage to the equipment. Vaporising liquids should not be used in a confined space as the liquids and their combustion products are toxic. Discharge should be started at one edge of the fire, then swept across the surface of the burning material to concentrate on the heart of the fire.

Fire-fighting equipment

Portable fire extinguishers

Portable extinguishers provide first aid fire fighting for Class A and Class B fires, but not in general for Class C and Class D fires. They are only useful, however, if sufficient are provided of the right type, in the right place, if they are properly maintained and if people are available who have been trained in their use. They should be located in conspicuous positions – identified with an approved sign – usually on exit routes by doors in corridors or landings. In multi-storey buildings they should be stored in the same position on each floor, to be available for use at all times. They must be easily accessible at all times and free from obstruction. The travel distance from a possible fire to the nearest appropriate extinguisher should not be more than 30 metres. For a special fire risk, appropriate fire extinguishers should be provided (*see* Table 33.5). (No one should have to make a choice regarding the specific extinguisher necessary for a particular fire.) Where a fire can occur in a confined space, extinguishers should be kept outside and be suitable for use in a confined space. All extinguishers should be protected against excessive heat or cold; storage in easily opened containers will provide some protection against external corrosion.

A monthly inspection routine should be undertaken for all extinguishers and the maker's instructions followed for all routine and non-routine maintenance. Maintenance and internal replacements should be carried out by trained personnel and accurate records kept of all inspections and maintenance of extinguishers.

Table 33.5 ◆ Classification of fires which can be controlled by portable extinguishers

Class of fire	*Description*	*Appropriate extinguisher*
A	Solid materials, usually organic, with glowing embers	Water, foam, dry powder, vaporising liquid, CO_2
B	Liquids and liquefiable solids:	
(i)	miscible with water, e.g. methanol, acetone	Water, foam (but must be stable on miscible solvents), CO_2, dry powder
(ii)	Immiscible with water, e.g. petrol, benzene, fats, waxes	Foam, dry powder, vaporising liquid, CO_2

Hose reels

Hose reels are a form of fixed fire-fighting installation and consist of a coil of 25 mm ID flexible hose directly connected to a rising main. They should be located in a recess so as not to obtrude into an access way, each container being clearly marked with the standard notice. The complete fixed installation, with the hose reel as the terminal point, consists of either a wet or a dry rising main and a landing valve or fire hydrant. The main consists of a heavy quality wrought steel pipe of not less than 100 mm ID. A wet rising main should be full of water and connected directly into the fire main with the water at fire main pressure, whereas with a dry rising main the pipe has to be charged with water prior to use. This can be done either by opening out the main to the fire main or by charging the main via a water pump. Wet rising mains are subject to frost damage if not suitably protected. The inlet to a dry rising main must be in a convenient position for the fire brigade to gain access to it, and it must be provided with hard standing for pumps. The inlet must be suitably identified and kept free for access at all times. In addition to hose reels, rising mains are fitted with landing valves which allow the connection of a standard fire hose. Both landing valves and hose reels must be kept free from obstruction at all times and should be sited not more than 30 metres from a possible fire location.

Sprinkler systems

Sprinkler systems provide an automatic means of detecting and extinguishing or controlling a fire in its early stages. The system consists of an overhead pipe installation on which sprinkler heads are fitted at suitable intervals. The installation is supplied with water from a head tank and/or water main. Each sprinkler head acts as a valve which is preset to open at a given temperature and release water onto the fire. For the system to be effective, the water supply must be automatic, reliable and not subject to freezing or drought. As each head is temperature operated only, those heads nearest the fire open and are not subject to smoke or fume. Being overhead, they do not block or restrict access routes and can be arranged to operate an alarm on releasing water (Fig. 33.13).

Fire detection

Fire can be detected in one of three main ways:

(a) by sensing heat – actual temperature or the rate of rise of temperature;

(b) by detecting the presence of smoke; or

(c) by detecting flame.

Heat detectors

Heat detectors are of two kinds: fusion and expansion.

Fig. 33.13 ◆ Typical sprinkler heads (a) A fusible solder type sprinkler head; (b) Bulb type sprinkler head

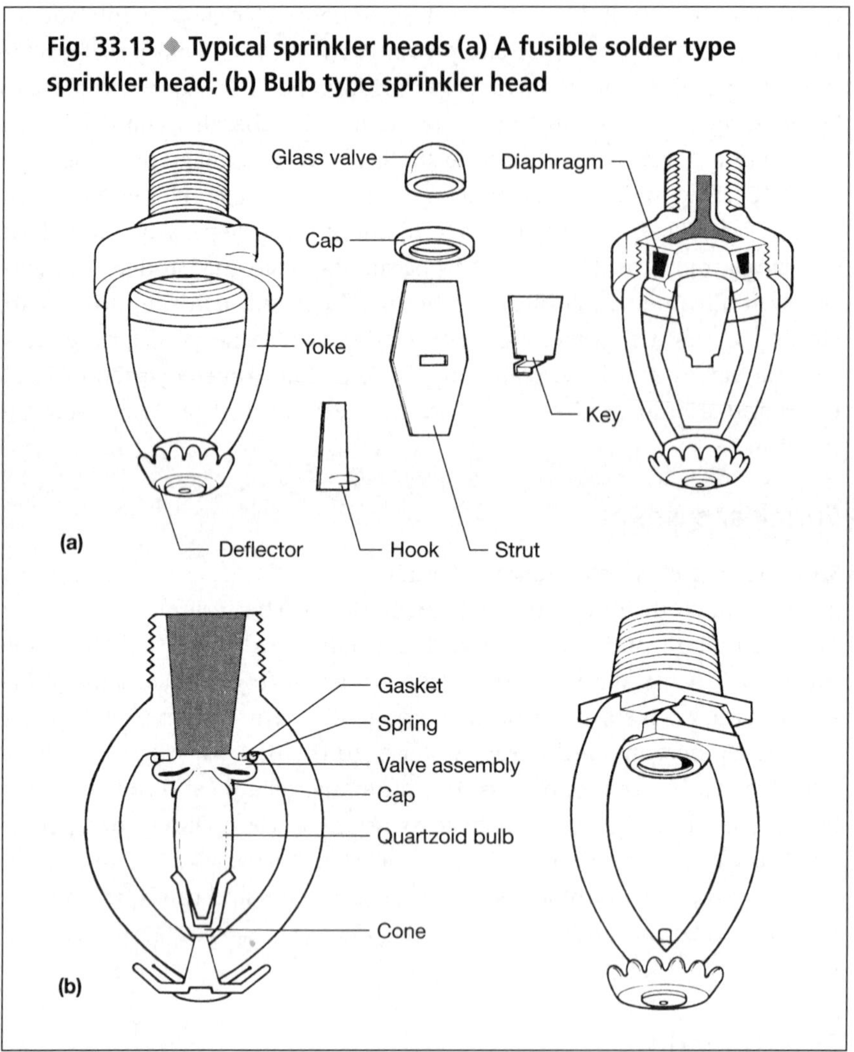

Source: Home Office, *Manual of Firemanship*

Fusion heat detectors

Here a metal melts, completing an electrical circuit and releasing water. Alternatively, the expansion of a solid, liquid or gas activates some other device. The simplest form of fusion detector consists of an electrical circuit containing a switch held in either the open or the closed position by a piece of low melting alloy. In the heat of the fire, this alloy melts, the switch is released and the circuit conditions change (Fig. 33.14).

Fig. 33.14 ◆ Chubb fixed temperature detector

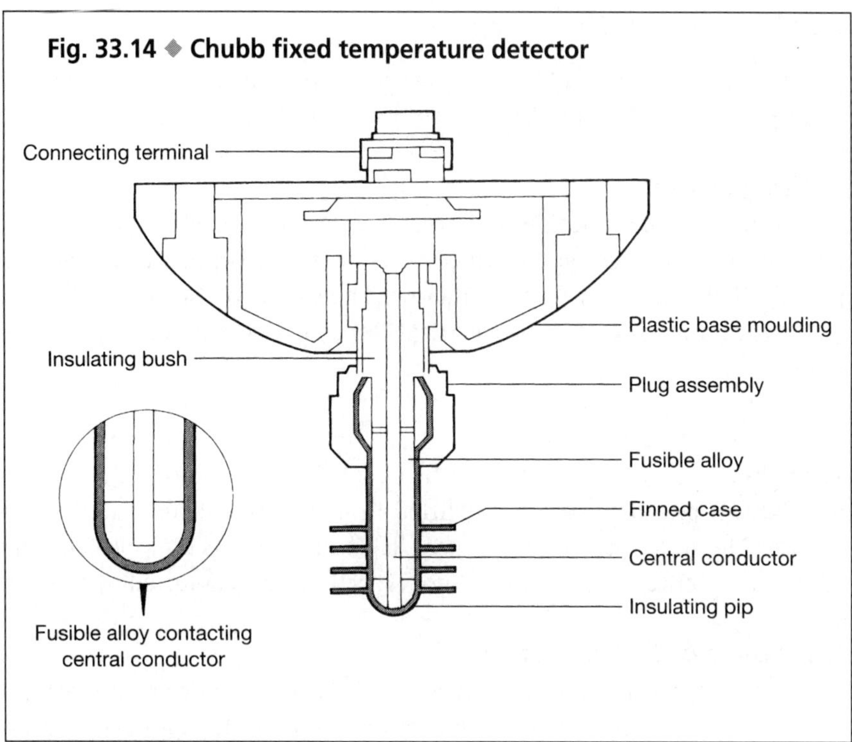

Source: Home Office, *Manual of Firemanship*

Expansion heat detector

The simplest form of a thermal expansion heat detector consists of a bimetallic strip which expands in a circular mode under the influence of heat, closing off an electrical circuit. Gases and liquids have greater coefficients of heat expansion than metals and as such are potentially more sensitive than metals as heat detectors. The most common use of liquid expansion occurs in the quartzite bulb used in sprinkler systems. The final 'valve' consists of a quartzite bulb filled with liquid and on contact with fire the liquid expands. At a predetermined temperature the bulb bursts and water is released.

The flame of a fire emits not only visible light but ultraviolet and infrared radiation. Each one of these can be used to detect fire. Although visible radiation is obscured by smoke, ultraviolet and infrared radiation remain unaffected. Flame detectors operate on the principle of detecting either or both of these. The disadvantages of these detectors is that they react to any source of ultraviolet or infrared radiation, such as the sun or the moon. Hence they can be subject to spurious alarms as well as circuit failure.

Smoke detectors

Smoke detectors are of different types operating on the basis of ionising radiation, light scatter or obscuration.

Ionising detectors

This form of detector utilises a small radioactive source which maintains a level of ionisation in two chambers, one of which is open to the atmosphere. When smoke enters this chamber, the smoke particles absorb some of the ionisation, causing an electrical imbalance detected by the instrument.

Light scatter detectors

The light scatter detector works on the easily observed fact that smoke scatters light. A photo-electric cell is fitted in a chamber at right angles to a source of light. In a fire-free condition the cell would receive no light, but when smoke enters the chamber, light is scattered and detected by the cell.

Light obscuration detectors

With this detector the photo-electric cell is mounted opposite a light source in a chamber so when smoke enters some of the light is obscured. The cell detects decreases in light intensity.

Flammable gas detectors

Technically, a flammable gas detector is not a fire detector because it works on the pre-fire condition, measuring the concentration of flammable gas in the atmosphere and raising the alarm when the concentration reaches a predetermined fraction of the lower explosive limit. The gas mixture is drawn over a catalytic surface on which the flammable gas is oxidised. In turn, the heat of oxidation raises the temperature of the catalyst surface and the device responds to the rise in temperature dependent on the concentration of the air/flammable gas mixture passing over it. These detectors have to be calibrated to take into account the different gases to which they will be exposed; for instance, different calibrations are required for propane and petrol.

Fire alarms

The most effective fire alarm system is the human voice, but this can be very expensive when 24-hour coverage is needed. Routine patrols of premises, carried out diligently and in a disciplined way, will detect most fires before they become a serious risk. Alarm should be raised by telephoning a local centre or the local fire brigade.

A fire can also be detected by any of the detectors outlined above. Each device can be made to operate an alarm system which alerts employees or an in-house fire brigade. Some detection devices will automatically start to fight the fire – e.g. sprinkler systems, where the flow of water sets off the alarm.

The major disadvantage of all automatic alarms is the frequency of false alarms. All systems suffer from this problem and should be tuned to deal with the local situation.

Fire prevention

It is impossible to prevent the occurrence of a fuel/oxygen mixture in many situations. Most fire prevention activities, therefore, consist of controlling sources of ignition. The most common are lighting, heating, ventilation equipment and machinery, tools and people. Moreover, electric sparks can cause fires (not, however, electricity itself – a popular myth).

G

Lighting

Here it is not the light which is the source of ignition but the heat generated by the light. Incandescent lights present a greater risk than discharge lights because they give off more heat. The heat from a light will transfer to any object by a combination of conduction, convection and radiation. For conduction, the object must be in contact with the light: obviously, any flammable object in actual contact with a light will heat up. With convection the heat transfer medium is air which travels upwards when heated. Any object, therefore, which is above a light will heat up by convection, and the nearer the object is to the top of the light source the hotter it will get. Radiation decreases rapidly with distance, the rate of decrease being dependent on the temperature of the radiating obiect.

Lights should be so placed that the heat rising from them will not be transferred by convection to objects which are placed above them. A good practice is to mount each lamp in a cage so that objects cannot be inadvertently placed in direct contact with the lamp. An unusual form of lighting is burning gas or oil. Both the hot wick and the mantle are sources of ignition, and so is the heat from the flame.

Heating

Heating is a source of ignition, emanating from stoves, boilers, open fires, electric strip and bar heaters, as well as lagged and unlagged steam heating and hot water pipes. Heat transfer occurs through conduction and convection. In order to prevent conduction, objects should never be placed in direct contact with sources of heat. Convection takes place in an upward direction, being more effective the nearer the object is to the source of heat. If heat transfer by convection is to be avoided, therefore, objects should not be placed above a source of heat.

Stoves and portable heating appliances should always stand on a non-combustible surface and, if possible, be surrounded by a low kerb and guard rail. The area itself should be kept free from all rubbish and debris. Any portable stoves should be secured to prevent overturning. When they are in use the handling of flammable fluids should be forbidden. Moreover, permanent heating systems require installation, maintenance, operation and repair in accordance with the manufacturer's recommendations. Strict control of access to boiler rooms is necessary and they should be kept clean and tidy.

Ventilation equipment, machinery, etc.

Dangers arising from ventilation equipment and machinery result from electrical connections, overheated bearings and hot surface. With machinery these are often detected by the operators. Ventilation equipment is often installed in inaccessible places and dangers remain undetected for long periods. Such equipment is designed to operate in a good air draught, thereby assisting the spread of fire.

Tools can be a source of ignition. Burning gear, blow lamps and soldering irons present obvious dangers as well as uncooled cutting edges, electrically powered tools and steel scrapers. Control methods include a high standard of housekeeping, particularly in the case of burning and welding gear, when the surrounding area must be free from all combustible material. Ideally, soldering irons should never be put down except on to an insulated rest. These and other 'hot' tools should never be used close to flammable materials.

Construction and layout

All buildings contain an extensive variety of materials without taking into account working materials and people. When a fire starts each of these materials will react in a differing way and it is virtually impossible to predict what will happen. Fire normally spreads upwards as the hot gaseous products of combustion rise. It can also spread horizontally due to conduction and radiation, but it very rarely spreads downwards.

The layout of a room or factory floor affects fire spread. Where possible, all high fire risk processes should be grouped together to permit appropriate fire prevention of protection. Stores containing flammable materials must be separate from other stores. In general storage areas, combustible goods should be segregated from non-combustible ones. If it is not possible to segregate work processes or stores according to flammability, then the highest risk must be assumed to apply to the whole area.

Waste paper, scrap products and by-products, wrapping materials and redundant items form the fuel for fires, the spread of which can be rapid. The most effective fire prevention activity is good housekeeping. Refuse should be removed at least daily and stored in a fireproof location, such as a metal skip kept in a non-smoking area. Spillages, particularly of oil and other flammable substances, should be removed. Flammable liquids must always be handled in special containers. Housekeeping inspections should be frequently undertaken, not only static situations being examined but the housekeeping standards of operators too (*see* Chapter 14).

Means of escape in case of fire

A *means of escape* in case of fire is a continuous route by way of a space, room, corridor, staircase, doorway or other means of passage, along or through which persons can travel from wherever they are in a building to the safety of the open air at ground level by their own unaided efforts. An *alternative means of escape* is a second route, usually in the opposite direction, but which may join the first means of escape. The alternative means is not the second best route, but rather another main means, and should be planned and maintained as such.

Considerations to be taken into account in planning and maintaining a means of escape are travel distance, doors, signs, lighting and protection of the route.

Travel

There are three travel stages:

(a) travel within rooms;

(b) travel from rooms to a stairway or final exit;

(c) travel within stairways and to a final exit.

The following general rules apply to means of escape:

(a) The total travel distance between any point in a building and the nearest final exit or protected stairway should not be more than –
   (i) 18 m if there is only one exit; or
   (ii) 45 m if more than one exit.
(b) Two or more exits are necessary –
   (i) from a room in which more than 60 people work; or
   (ii) if any point in the room is more than 12 m from the nearest exit.
(c) Minimum width of exit should be 750 mm.
(d) (i) Corridors should not be less than 1 m in width;
   (ii) in the case of offices, where corridors are longer than 45 m, they should be subdivided by fire-resisting doors.
(e) Stairways should be at least 800 mm in width, and fire resistant, along with doors connecting them.
(f) A single stairway is sufficient in a building of up to four storeys only.
(g) The following are not acceptable as means of escape –
   (i) spiral staircases;
   (ii) escalators;
   (iii) lifts;
   (iv) lowering lines;
   (v) portable or throw-out ladders.
(h) Fire doors must open outwards only.
(i) Doors providing means of escape should never be locked. (If they have to be kept locked for security purposes, panic bolts should be fitted or keys maintained in designated key boxes close to the exit.) A notice should indicate that the doors can be opened in the case of fire.
(j) A fire exit notice should be fitted to or above fire exit doors.
(k) Appropriate notices should be affixed along fire escape routes, which should be provided with emergency lighting.
(l) Corridors and stairways forming a means of escape should have half-hour fire resistance, i.e. no fire should be able to break through within 30 minutes. This means that a corridor or stairway should be built from non-combustible materials, i.e. brick or concrete. The surface finish should also be non-combustible.

(m) Fire alarm warnings must be audible throughout the building. In larger buildings this will require the provision of electrically operated alarms, whereas in smaller buildings a manually operated gong or bell may suffice.

(n) Normally no person should have to travel more than 30 m to the nearest alarm point.

Fire instructions

A fire instruction is a notice informing people of the action they should take on either:

(a) hearing the alarm; or

(b) discovering a fire.

A typical notice is shown in Fig. 33.15.

However, fire instructions do not end at issuing a notice to the employees and displaying a copy in a prominent position. People need training as well, i.e. regular fire drills and/or simulated fire exercises. These can disrupt working, but are the only way employees can learn what they must do in a real situation. The alarm should be sounded weekly at the same time so that employees become familiar with its sound. An evacuation exercise

Fig. 33.15 ◆ Fire instruction

WHEN THE FIRE ALARM SOUNDS

1. Close the windows, switch off electrical equipment and leave the room, closing the door behind you.
2. Walk quickly along the escape route to the open air.
3. Report to the fire warden at your assembly point.
4. Do not attempt to re-enter the building.

WHEN YOU FIND A FIRE

1. Raise the alarm by . . . (If the telephone is to be used, the notice must include a reference to name and location.)
2. Leave the room, closing the door behind you.
3. Leave the building by the escape route.
4. Report to the fire warden at the assembly point.
5. Do not attempt to re-enter the building.

should be held annually. In all workplaces, trained employees should be designated as fire wardens, i.e. to carry out a head count on evacuation. They may also act as 'last one out' in large buildings and as helpers to the public. Training of selected employees in the correct use of fire extinguishers may well be necessary, so that they can act as first aid firefighters.

Liaison with the fire authority

Apart from visits made by officers of the fire authority in connection with fire certificates, the authority is a valuable source of practical information on fire protection, fire precautions and training. Before any change of use is made in premises, consultation should take place, so that the views of the fire authority can be accommodated in the design, thereby saving expense later. In many cases, the fire authority will give assistance in training. In large establishments, it is advantageous for a fire officer to visit regularly to familiarise himself with the layout, the people and any special fire-fighting needs. These visits can save valuable minutes in the event of fire.

Fire certificates

The Fire Precautions Act 1971 (FPA) required that a fire certificate be issued for certain classes of factory and commercial premises, based on the concept of 'designated use'. The Fire Safety and Safety of Places of Sport Act 1987 (FSSPSA) in effect 'deregulated' many premises which formerly required certification under the 1971 Act. These include:

(a) factories, offices and shops where –
   (i) more than 20 persons are employed at any one time;
   (ii) more than ten persons are employed at any one time elsewhere than on the ground floor;
   (iii) buildings containing two or more factory and/or office premises, where the aggregate of persons employed in all of them at any one time is more than 20;
   (iv) buildings containing two or more factory and/or office premises, where the aggregate of persons employed at any one time in all of them, elsewhere than on the ground floor, is more than ten;
   (v) factories where explosive or highly flammable materials are stored, or used in or under the premises, unless, in the opinion of the fire authority, there was no serious risk to employees from fire;

(b) hotels and boarding houses where sleeping accommodation –

(i) for six or more persons; or

(ii) at basement level; or

(iii) above the first floor;

was provided for either guests or staff.

The effect of this deregulation was to exempt certain low-risk premises from the certification requirements. (It should be appreciated that the FSSPSA amended the FPA. It did not revoke or repeal that Act.) Thus, the FSSPSA empowers local fire authorities to grant exemptions in respect of certain medium and low-risk premises which were previously 'designated use' premises.

The exemption certificate must, however, specify the maximum number of persons who can safely be in or on the premises at any one time. Furthermore, the exemption can be withdrawn by the fire authority without prior inspection and by service of notice of withdrawal, where the degree of risk associated with the premises increases.

However, the FSSPSA still attaches great importance to the provision of means of escape in the event of fire and the provision of fire fighting appliances in such premises. Under the Act, 'escape' is defined as 'escape from the premises to some place of safety beyond the building, which constitutes or comprises the premises, and any area enclosed by it or within it; accordingly, conditions or requirements can be imposed as respects any place or thing by means of which a person escapes from premises to a place of safety' (FSSPSA, sec 5(5)).

A fire certificate specifies:

(a) the use or uses of the premises it covers;

(b) the means of escape in the case of fire indicated on a plan of the building;

(c) the means for ensuring the safety and effectiveness of the means of escape, such as fire and smoke stop doors, emergency lighting and direction signs;

(d) the means of fighting fire for the use of persons on the premises;

(e) the means of raising the alarms;

(f) particulars of explosive or highly flammable liquids stored and used on the premises.

The fire certificate may also impose requirements relating to:

(a) the maintenance of the means of escape and keeping it free from obstruction;

(b) the maintenance of other fire precautions;

(c) the training of people and the keeping of records;

(d) limitations on numbers of persons in the premises;

(e) any other relevant fire precautions.

When a fire certificate has been issued the fire precautions specified must be kept in accordance with the specification and all other requirements must be observed. The fire certificate must be kept on the premises to which it refers. When changes to conditions or alterations to premises are being considered, the fire authority must be notified in advance of any proposal such as:

(a) to make a material extension of, or material structural alteration to, the premises; or

(b) to make a material alteration in the internal arrangements of the premises or in the furniture or equipment with which the premises are provided.

By 'material' is meant any alteration which would render the means of escape and related fire precautions inadequate in relation to the normal conditions of use of the premises. If changes are proposed they may be considered as though a new application has been made for a certificate.

Enforcement provisions under the FSSPSA

Where a fire authority is of the opinion that an occupier has not fulfilled his duty with regard to the provision of:

(a) means of escape in case of fire; and/or

(b) means for fighting fire.

the authority may serve on the occupier an improvement notice detailing the steps that should be taken by way of improvements, alterations and other measures to remedy this breach of the Act. The occupier must undertake remedial work within 21 days unless, of course, he appeals against the notice. Such an appeal must be lodged within 21 days from the date of service of the improvement notice, and has the effect of suspending operation of the notice. Where such an appeal fails, the occupier must undertake the remedial work specified in the improvement notice, and failure to do so can result in:

(a) a maximum fine of £2,000 on summary conviction; and

(b) on conviction on indictment, an indefinite fine or imprisonment for up to two years, or both.

Where, in the opinion of the fire authority, there is a serious risk of personal injury to persons from fire, the authority may serve on the occupier a prohibition notice, requiring the occupier to undertake remedial work or alternatively, cease use of the premises, i.e. close down. Prohibition notices (FSSPSA, sec 10B(3)) can be served on the following types of premises:

(a) providing sleeping accommodation;

(b) providing treatment/care;

(c) for the purposes of entertainment, recreation or instruction, or for a club, society or association;

(d) for teaching, training or research;

(e) providing access to members of the public, whether for payment or otherwise;

(f) places of work.

A prohibition notice is most likely to be served where means of escape are inadequate or non-existent, or where there is a need to improve them. As with prohibition notices served under HSWA, they can be either of immediate effect or deferred. An appeal does not suspend operation of a prohibition notice.

Places of sport

The 1985 Bradford City Football Stadium disaster was instrumental in identifying a need for much greater standards of fire safety in such establishments, resulting in specific provisions being incorporated in the FSSPSA 1987. This legislation extends the existing Safety of Sports Grounds Act 1975, which only applied to sports stadia, i.e. sports grounds where the accommodation for spectators wholly or substantially surrounds the activity taking place, to all forms of sports ground.

The legal situation relating to sports grounds under the FSSPSA is as follows:

(a) general safety certificates are required in respect of any sports ground;

(b) the Safety of Sports Grounds Act 1975 is extended to any sports ground which the Secretary of State considers appropriate;

(c) validity of safety certificates no longer requires the provision at sports grounds of a police presence, unless consent has been given by a chief constable or chief police officer;

(d) there is provision for the service of prohibition notices in the case of serious risk of injury to spectators, prohibiting or restricting the admission of spectators in general or on specified occasions.

In the last case, the notice may specify steps which must be taken to reduce risk, particularly from fire, to a reasonable level, including structural alterations (irrespective of whether this may contravene the terms of a safety certificate for the ground issued by the local authority, or for any stand at the ground). (*See below* 'Safety certificates'.) Where a prohibition notice requires provision of a police force, such requirements cannot be specified without the consent of the chief constable or chief police officer. A prohibition notice (FSSPSA, sec 23) may be served on any of the following persons:

(a) the holder of a general safety certificate;

(b) the holder of a specific safety certificate, i.e. a safety certificate for a specific sporting activity or occasion;

(c) where no safety certificate is in operation, the management of the sports ground;

(d) in the case of a specific sporting activity for which no safety certificate is in operation, the organisers of the activity;

(e) where a general safety certificate is in operation for a stand at a ground, the holder of same;

(f) where a specific safety certificate is in operation for a stand, the holder of the certificate.

Under sec 25 of the FSSPSA sports grounds must be inspected at least once per annum.

Safety certificates

Where a sports ground provides covered accommodation in stands for spectators, a safety certificate, issued by the local authority, is required for each stand providing covered accommodation for 500 or more spectators, i.e. a regulated stand. In certain cases, safety certificates may well be required for stands accommodating smaller numbers.

Sec 27 of the Act gives the local authority power to require the keeping of the following records in the case of stands at sports grounds:

(a) number of spectators in covered accommodation; and
(b) procedures relating to the maintenance of safety in the stand.

Sports grounds with regulated stands must be inspected periodically.

Offences and penalties under the FSSPSA – regulated stands

The principal criminal offences under the Act occur in respect of regulated stands. Where:

(a) spectators are admitted to a regular stand at a sports ground on an occasion when a safety certificate should be, but is not, in operation; or
(b) any term or condition of a safety certificate for a regulated stand at a sports ground is contravened,

both the management of the sports ground and, in the second case, the holder of the certificate, are guilty of an offence (FSSPSA, sec 36).

A person or organisation convicted of the above offences is liable to:

(a) on summary conviction, to a maximum fine of £20,000; or
(b) on conviction on indictment, to an indefinite fine or a maximum of two years' imprisonment or both.

Under sec 36 of the Act, a number of defences are available, namely:

(a) that either –
  (i) the spectators were admitted with no safety certificate in operation; or
  (ii) the contravention of the safety certificate occurred without his consent; *and*
(b) that he took all reasonable precautions and exercised all due care to avoid the commission of the offence himself or by other persons under his control.

Where a person or corporate body is charged with an offence under sec 36 (1) above, i.e. no safety certificate in respect of a regulated stand, the defendant may plead that he did not know that the stand had been determined to be a regulated stand.

G

Fire precautions in workplaces

The Fire Precautions (Workplace) Regulations 1997 apply to all workplaces (see definition of 'workplace' under the WHSWR) other than 'excepted workplaces'.

Moreover, these regulations do not stand on their own, extending duties under existing health and safety and fire safety legislation. As such, they must be read in conjunction with a range of duties on employers and controllers of workplaces under the HSWA, FPA, FSSPSA, WHSWR and, in particular, the MHSWR. The majority of the duties are of an absolute nature.

Part II (Regs 3 to 6) deals with the general application of the regulations with regard to fire precautions in the workplace.

Application of Part II (Reg 3)

Reg 3 places a general duty on employers to comply with this Part of the regulations in respect of any workplace, other than an excepted workplace, which is to any extent under his control, so far as the requirements relate to matters within his control.

Every person who has, to any extent, control of a workplace, other than an excepted workplace, shall ensure that, so far as the requirements relate to matters within his control, the workplace complies with any applicable requirement of this Part.

Similar provisions apply in respect of persons who, by virtue of any contract or tenancy, have an obligation in respect of the maintenance or repair, or safety, of a workplace.

Any reference to a person having control of any workplace is a reference to a person having control in connection with the carrying on by him of a trade, business or undertaking (whether for profit or not).

Excepted workplaces

Reg 5 defines an excepted workplace thus:

(a) any workplace which comprises premises for which a fire certificate is in force or for which an application is pending under the FPA 1971;

(b) any workplace which comprises premises in respect of which there is a safety certificate under the Safety of Sports Grounds Act 1975 or under Part III of the FSSPSA 1987, and which are in use for the activities specified;

(c) any workplace which comprises premises to which the Fire Precautions (Sub-surface Railway Stations) Regulations 1989 apply;

(d) any workplace which is or is on a construction site to which the Construction (Health, Safety and Welfare) Regulations 1996 apply;

(e) any workplace which is in or on a ship within the meaning of the Docks Regulations 1988, including any such ship which is in the course of construction or repair;

(f) any workplace which comprises premises to which the Fire Certificates (Special Premises) Regulations 1976 apply;

(g) any workplace which is deemed to form part of a mine for the purposes of the Mines and Quarries Act 1954;

(h) any workplace which is or is in an offshore installation within the meaning of the Offshore Installations and Pipelines Works (Management and Administration) Regulations 1995;

(i) any workplace which is or is in or on an aircraft, locomotive or rolling stock, trailer or semi-trailer used as a means of transport or a vehicle for which a licence is in force under the Vehicle Excise and Registration Act 1994 or a vehicle exempted from duty under that Act;

(j) any workplace which is in fields, woods or other land forming part of an agricultural or forestry undertaking but which is not inside a building and is situated away from the undertaking's main buildings.

Fire-fighting and fire detection (Reg 4)

Where necessary (due to features of a workplace, activities undertaken, hazards present or other relevant circumstances) in order to safeguard the safety of employees in case of fire, workplaces must be equipped with appropriate fire-fighting equipment and with fire detectors and alarms. Any non-automatic fire-fighting equipment must be easily accessible, simple to use and indicated by signs. What is appropriate must be determined by the dimensions and use of the building housing the workplace, the equipment it contains, the physical and chemical properties of substances likely to be present and the maximum number of people present at any one time.

An employer must take measures for fire-fighting which are adapted to the nature of the activities undertaken and the size of his undertaking and of the workplace concerned, taking into account persons other than employees who may be present.

He must nominate employees to implement these measures and ensure that the number of such employees, their training and the equipment available to them are adequate, taking into account the size of, and specific hazards involved in, the workplace concerned, and arrange any necessary contacts with external emergency services.

Emergency routes and exits (Reg 5)

Emergency routes and exits must be kept clear at all times. Where necessary, the following requirements must be complied with:

(a) emergency routes and exits shall lead directly as possible to a place of safety;

(b) in the event of danger, it must be possible for employees to evacuate the workplace quickly and as safely as possible;

(c) the number, distribution and dimensions of emergency routes and exits shall be adequate having regard to the use, equipment and dimensions of the workplace and the maximum number of persons that may be present there at any one time;

(d) emergency doors shall open in the direction of escape;

(e) sliding and revolving doors shall not be used for exits specifically intended as emergency exits;

(f) emergency doors shall not be so locked or fastened that they cannot be easily and immediately opened by any person who may require to use them in an emergency;

(g) emergency routes and exits must be indicated by signs;

(h) emergency routes and exits requiring illumination shall be provided with emergency lighting of adequate intensity in the case of failure of the normal lighting.

Maintenance (Reg 6)

The workplace and any equipment and devices provided under Regs 4 and 5 shall be subject to a suitable system of maintenance and be maintained in an efficient state, in efficient working order and in good repair.

Amendment of the MHSR

Part III covers amendments to the general and specific provisions of the MHSWR.

Reg 7 states that the general provisions of the MHSWR must be taken into account when interpreting these regulations. Reg 8 requires that specific provisions of Part II be inserted into the following provisions of the MHSWR, namely:

(a) definition of 'the preventive and protective measures';

(b) risk assessment;

(c) health and safety assistance;

(d) co-operation and co-ordination (shared workplaces);

(e) persons working in host employers' undertakings.

Reference to these regulations must be made in Reg 8 (information to employees) of the MHSWR

The requirements of Part II of these regulations must be considered in the interpretation of Reg 9 (co-operation and co-ordination) of the MHSWR.

The workplace fire precautions legislation (Reg 9)

Specific provisions with regard to enforcement, disclosure of information, provisions as to offences, service of notices and civil liability under the HSWA do not apply, and these regulations do not form part of the relevant statutory provisions. The workplace fire precautions legislation means:

(a) Part II of these regulations;

(b) Regs 1 to 4, 6 to 10 and 11(2) and (3) of the MHSWR as amended by Part III of these regulations insofar as those regulations –

    (i) impose requirements concerning general fire precautions to be taken or observed by an employer;

    (ii) have effect in relation to a workplace in Great Britain other than an excepted workplace,

    and for this purpose general fire precautions means measures which are to be taken or observed in relation to the risk to the safety of employees in the case of fire in a workplace, other than any special precautions in connection with the carrying on of any manufacturing process.

Part IV – Enforcement and offences

Part IV provides fire authorities and their inspectors with a range of enforcement procedures according to the severity of the situation. This may be by way of:

(a) a written opinion from the authority, stating the breach of the fire precautions legislation and the action that could be taken to remedy it (Reg 10);

(b) prosecution, where –

  (i) a person fails to comply with the fire precautions legislation;

  (ii) that failure places employees at serious risk; and

  (iii) that failure is intentional or is due to the person being reckless as to whether he complies or not (Reg 11);

(c) service of a prohibition notice (Reg 12);

(d) (except where an enforcement notice cannot be delayed), following a written notice of intent, service of an enforcement notice where there is a failure to comply with any provision of the workplace fire precautions legislation and that failure places employees at serious risk in the case of fire (Reg 13);

(e) application to a court for an enforcement order where a person has failed to comply with any requirement imposed upon him by the workplace fire precautions legislation (Reg 14).

A person guilty of an offence under Reg 11 shall be liable:

(a) on summary conviction to a fine not exceeding the statutory maximum; or

(b) on conviction on indictment, to a fine, or to imprisonment for a term not exceeding two years, or both.

Enforcement notices: rights of appeal (Reg 14)

A person on whom an enforcement notice is served may, within 21 days from the date of the notice, appeal to the court. Such an appeal may either cancel or affirm the enforcement notice, with or without modifications.

Enforcement notices: offences (Reg 15)

It is an offence for any person to contravene any requirement imposed by an enforcement notice. A person guilty of such an offence shall be liable:

(a) on summary conviction to a fine not exceeding the statutory maximum; or

(b) on conviction on indictment, to a fine, or to imprisonment for a term not exceeding two years, or both.

In any proceedings for an offence under this regulation it shall be a defence for the person charged to prove that he took all reasonable precautions and exercised all due diligence to avoid the commission of the offence.

34 Lifting machinery and equipment

Definitions

Lifting machinery

The statutory definition of lifting machinery is 'a crane, crab, winch, teagle, pulley block, gin wheel, transporter or runway' (FA, sec 27(9)). Also included are hoists, cranes, elevators and lifts, whether used for carrying people, goods or both.

Lifting tackle

The statutory definition of lifting tackle or lifting equipment is 'chain slings, rope slings, rings, hooks, shackles and swivels' (FA, sec 26(3)). Hence the general meaning of lifting tackle is taken to include various types of ropes, chains, hooks, eyebolts, 'D' rings and other items.

Lifting machinery – legal aspects

Lifting machinery may be classified thus:

(a) hoists;

(b) cranes;

(c) lifts.

The Lifting Operations and Lifting Equipment Regulations (LOLER) 1998

These Regulations replace most sector-specific legislation, e.g. that relating to the construction industry, on lifting equipment, creating a single set of regulations that apply to all sectors. LOLER applies over and above the general requirements of the Provision and Use of Work Equipment Regulations 1998 in dealing with specific hazards and risks associated with lifting equipment and lifting operations. The regulations are supported by an ACOP and HSE Guidance *Safe use of lifting equipment.*

Under LOLER there is an absolute duty on employers and others to undertake lifting operations safely.

Important definitions

The 1992 Regulations means the Supply of Machinery (Safety) Regulations 1992.

Accessory for lifting means work equipment for attaching loads to machinery for lifting.

EC declaration of conformity means a declaration which complies with:

(a) Reg 22 of the 1992 Regulations;
(b) Article 12.1 of the Council Directive 89/686/EEC on the approximation of the laws of the Member States relating to personal protective equipment; or
(c) Reg 8(2)(d) of the Lifts Regulations 1997.

Examination scheme means a suitable scheme drawn up by a competent person for such thorough examinations of lifting equipment at such intervals as may be appropriate for the purpose described in Reg 9(3).

Lifting equipment means work equipment for lifting or lowering loads and includes its attachments used for anchoring, fixing or supporting it.

Lifting operation means an operation concerned with the lifting or lowering of a load (Reg 8).

Load includes a person.

Thorough examination in relation to a thorough examination under paragraph 1, 2 or 3 of Reg 9:

(a) means a thorough examination by a competent person;
(b) where it is appropriate to carry out testing for the purpose described in the paragraph, includes such testing by a competent person as is appropriate for the purpose.

Work equipment means any machinery, appliance, apparatus, tool or installation for use at work (whether exclusively or not).

The more important requirements of LOLER are outlined below.

Strength and stability

Every employer shall ensure that:

(a) lifting equipment is of adequate strength and stability for each load, having regard in particular to the stress induced at its mounting or fixed point;

(b) every part of a load and anything attached to it and used in lifting it is of adequate strength (Reg 4).

Lifting equipment for lifting persons

1. Every employer shall ensure that lifting equipment for lifting persons:
   (a) subject to sub-paragraph (b), is such as to prevent a person using it being crushed, trapped or struck or falling from the carrier;
   (b) is such as to prevent so far as is reasonably practicable a person using it, while carrying out activities from the carrier, being crushed, trapped or struck or falling from the carrier;
   (c) subject to paragraph 2, has suitable devices to prevent the risk of a carrier falling;
   (d) is such that a person trapped in any carrier is not thereby exposed to danger and can be freed.
2. Every employer shall ensure that if the risk described in paragraph 1(c) cannot be prevented for reasons inherent in the site and height differences:
   (a) the carrier has an enhanced safety coefficient suspension rope or chain;
   (b) the rope or chain is inspected by a competent person every working day (Reg 5).

Positioning and installation

1. Every employer shall ensure that lifting equipment is positioned or installed in such a way as to reduce to as low as is reasonably practicable the risk:
   (a) of the lifting equipment or a load striking a person; or
   (b) from a load –

(i) drifting;
(ii) falling freely; or
(iii) being released unintentionally,

and it is otherwise safe.

2. Every employer shall ensure that there are suitable devices to prevent a person from falling down a shaft or hoistway (Reg 6).

Marking of lifting equipment

Every employer shall ensure that:

(a) subject to sub-paragraph (b), the machinery and accessories for lifting loads are clearly marked to indicate their safe working loads (SWLs);

(b) where the SWL of machinery for lifting depends upon its configuration –
   (i) the machinery is clearly marked to indicate its SWL for each configuration; or
   (ii) information which clearly indicates its SWL for each configuration is kept with the machinery;

(c) accessories for lifting are clearly marked in such a way that it is possible to identify the characteristics necessary for their safe use;

(d) lifting equipment which is designed for lifting persons is appropriately and clearly marked to this effect;

(e) lifting equipment which is not designed for lifting persons but which might be so used in error is appropriately and clearly marked to the effect that it is not designed for lifting persons (Reg 7).

Organisation of lifting operations

1. Every employer shall ensure that every lifting operation involving lifting equipment is:
   (a) properly planned by a competent person;
   (b) appropriately supervised;
   (c) carried out in a safe manner (Reg 8).

Thorough examination and inspection

1. Every employer shall ensure that before lifting equipment is put into service for the first time by him it is thoroughly examined by him for any defect unless either:

(a) the lifting equipment has not been used before;

(b) in the case of lifting equipment for which an EC declaration of conformity could or (in the case of a declaration under the Lifts Regulations 1997) should have been drawn up, the employer has received such declaration made not more than 12 months before the lifting equipment is put into service;

or, if obtained from the undertaking of another person, it is accompanied by physical evidence referred to in paragraph 4.

2. Every employer shall ensure that, where the safety of lifting equipment depends upon the installation conditions, it is thoroughly examined:

(a) after installation and before being put into service for the first time;

(b) after assembly and before being put into service at a new site or in a new location,

to ensure that it has been installed correctly and is safe to operate.

3. Subject to paragraph 6, every employer shall ensure that lifting equipment which is exposed to conditions causing deterioration which is liable to result in dangerous situations is:

(a) thoroughly examined –

(i) in case of lifting equipment for lifting persons or an accessory for lifting, at least every six months;

(ii) in the case of other lifting equipment, at least every 12 months; or

(iii) in either case, in accordance with an examination scheme; and

(iv) each time that exceptional circumstances which are liable to jeopardise the safety of the lifting equipment have occurred;

(b) if appropriate for the purpose, is inspected by a competent person at suitable intervals between thorough examinations.

4. Every employer shall ensure that no lifting equipment:

(a) leaves his undertaking; or

(b) if obtained from the undertaking of another person, is used in his undertaking, unless it is accompanied by physical evidence that the last thorough examination required to be carried out under this regulation has been carried out (Reg 9).

Reports and defects

1. A person making a thorough examination for an employer under Reg 9 shall:

(a) notify the employer forthwith of any defect in the lifting equipment which in his opinion is or could become a danger to persons;

(b) as soon as is practicable make a report of the thorough examination in writing authenticated by him or on his behalf by signature or equally secure means and containing the information specified in Schedule 1 to –

(i) the employer;

(ii) any person from whom the lifting equipment has been hired or leased;

(c) where there is in his opinion a defect in the lifting equipment involving an existing or imminent risk of serious personal injury send a copy of the report as soon as is practicable to the relevant enforcing authority.

G

2. A person making an inspection for an employer under Reg 9 shall:

(a) notify the employer forthwith of any defect in the lifting equipment which in his opinion is or could become a danger to persons;

(b) as soon as is practicable make a record of his inspection in writing.

3. Every employer who has been notified under paragraph 1 shall ensure that the lifting equipment is not used:

(a) before the defect is rectified; or

(b) in a case to which sub-paragraph (c) of paragraph 8 of Schedule 1 applies, after a time specified under that sub-paragraph and before the defect is rectified.

4. In this regulation relevant enforcing authority means:

(a) where the defective lifting equipment has been hired or leased by the employer, the HSE;

(b) otherwise, the enforcing authority for the premises in which the defective lifting equipment was thoroughly examined (Reg 10).

Keeping of information

1. Where an employer obtaining lifting equipment to which these regulations apply receives an EC declaration of conformity relating to it, he shall keep the declaration for so long as he operates the lifting equipment.
2. The employer shall ensure that the information contained in:
   (a) every report made to him under Reg 10(1)(b) is kept available for inspection –
      (i) in the case of a thorough examination under paragraph 1 of Reg 9 of lifting equipment other than an accessory for lifting, until he ceases to use the lifting equipment;
      (ii) in the case of a thorough examination under paragraph 1 of Reg 9 of an accessory for lifting, for two years after the report is made;
      (iii) in the case of a thorough examination under paragraph 2 of Reg 9, until he ceases to use the lifting equipment at the place it was installed or assembled;
      (iv) in the case of a thorough examination under paragraph 3 of Reg 9, until the next report is made under that paragraph or the expiration of two years, whichever is later;
   (b) every record made under Reg 10(2) is kept available until the next such record is made (Reg 11).

Schedule 1

Information to be contained in a report of a thorough examination

1. The name and address of the employer for whom the thorough examination was made.
2. The address of the premises at which the thorough examination was made.
3. Particulars sufficient to identify the lifting equipment including where known its date of manufacture.
4. The date of the last thorough examination.
5. The SWL of the lifting equipment or (where its SWL depends upon the configuration of the lifting equipment) its SWL for the last configuration in which it was thoroughly examined.

6. In relation to the first thorough examination of lifting equipment after installation or after assembly at a new site or in a new location:

(a) that it is such thorough examination;

(b) (if such be the case) that it has been installed correctly and would be safe to operate.

7. In relation to a thorough examination of lifting equipment other than a thorough examination to which paragraph 6 relates:

(a) whether it is a thorough examination –

(i) within an interval of six months under Reg 9(3)(a)(i);
(ii) within an interval of 12 months under Reg 9(3)(a)(ii);
(iii) in accordance with an examination scheme under Reg 9(3)(a)(iii); or
(iv) after the occurrence of exceptional circumstances under Reg 9(3)(a)(iv);

(b) (if such be the case) that the lifting equipment would be safe to operate.

G

8. In relation to every thorough examination of lifting equipment:

(a) identification of any part found to have a defect which is or could become a danger to persons, and a description of the defect;

(b) particulars of any repair, renewal or alteration required to remedy a defect found to be a danger to persons;

(c) in the case of a defect which is not yet but could become a danger to persons –

(i) the time by which it could become such danger;
(ii) particulars of any repair, renewal or alteration required to remedy it;

(d) the latest date by which the next thorough examination must be carried out;

(e) where the thorough examination included testing, particulars of any test;

(f) the date of the thorough examination.

9. The name, address and qualifications of the person making the report; that he is self-employed or, if employed, the name and address of the employer.

10. The name and address of the person signing or authenticating the report on behalf of its author.

11. The date of the report.

Implications of the regulations

1. The regulations impose health and safety requirements with respect to lifting equipment.
2. They make provision with respect to:
   (a) the strength and stability of lifting equipment;
   (b) the safety of lifting equipment for lifting persons;
   (c) the way lifting equipment is positioned and installed;
   (d) the marking of machinery and accessories for lifting, and lifting equipment which is designed for lifting persons or which might be so used in error;
   (e) the organisation of lifting operations;
   (f) the thorough examination and inspection of lifting equipment in specified circumstances;
   (g) the evidence of examination to accompany it outside the undertaking;
   (h) the exception of winding apparatus at mines from Reg 9;
   (i) the making of reports of thorough examinations and records of inspections.

Teagle openings

A teagle opening is an opening in the fabric of a building above ground level through which goods can be hoisted into the building. Such openings are common in flour and agricultural mills in particular. They also feature in older types of factories and in warehouses. The FA, sec 24, requires that every teagle opening or similar doorway used for hoisting or lowering goods or materials, whether by mechanical power or otherwise, must be securely fenced and provided with a secure handhold at each side. The fencing must be properly maintained and kept in position, except when hoisting or lowering goods or materials.

Work at teagle openings and platform edges

What constitutes 'secure fencing' within the requirements of sec 24 is a matter for legal interpretation, but it is quite common to see openings fenced solely by a metal bar or wooden beam set horizontally approximately one metre from the floor. This arrangement relies heavily on operators replacing the bar after loading or unloading operations, and there is always the risk of operators falling over or under the bar. Furthermore, the sole provision of handholds at each side of the opening is of limited value in preventing falls from teagle openings and platform edges.

In recent years a number of safety barriers for high-level openings have been developed, the Ajax safety barrier being the most common. It is designed to give continuous protection to anyone working at a high-level loading point or teagle opening. The barrier operates on the basis that when a load has to be placed on a platform or through a teagle opening, the barrier is pivoted laterally through 180 degrees and comes to rest in a similar position some distance back from the edge of the platform or opening. Once the materials have been placed on the platform, the barrier is pivoted upwards and over the load until it rests in its original position at the platform edge so that the load can be removed. The moving part of the barrier is spring-loaded for ease of operation and a simple automatic catch locks the barrier in the closed position.

While the standard application is at loading points on platforms or mezzanine floors, the barrier can also be made to suit difficult installations where it is not practicable to make structural alterations, such as teagle openings, where doors may be involved or the pallet load is too large for standard type of barrier.

Safety aspects of lifting machinery

Cranes

Cranes have numerous applications in industrial activities, construction, docks and shipbuilding, and on railways. The principal hazard associated with any crane operation is the risk of collapse or overturning of the crane which can be caused by a variety of factors such as overloading, incorrect slewing or even incorrect construction of the crane. One of the principal causes of crane overturning is associated with the crane operator exceeding the 'maximum permitted moment' which is the product of the load and the radius of operation of the crane. (The radius of operation is the horizontal distance between the crane's centre of rotation and a vertical line drawn through the crane hook.) If the maximum permitted moment is exceeded, the crane is in danger of overturning or collapse.

There are many types of crane in use. Several of the more common forms of crane are discussed below.

Some defects in cranes which could result in accidents are illustrated in Figs. 34.1–34.4.

Fixed cranes

This type of crane is permanently fixed in one location, such as wharf, loading bay, dock or rail siding. It may incorporate a fixed angle or adjustable angle jib, and may rotate through 360 degrees. Accidents involving fixed cranes with resultant crane collapse, fall of the load and/or injury to operatives can occur in many ways. One principal cause of accidents is the failure to lift vertically. This may arise through the physical impossibility of getting the load directly below the lifting point, or the use of a fixed crane in a deliberate attempt to drag a load sideways, a very dangerous practice. Loads treated in this way can overstress the crane and cause collapse, or the load may swing violently, crushing people and damaging property. Alternatively, loads being raised or lowered may catch in a fixed structure causing damage. The 'snatching' of loads, instead of operating a slow and steady lifting action, can cause crane failure. Moreover, attempting to pull an object from under other material can impose loads of up to one hundred times that anticipated, often with disastrous results. Cranes with adjustable angle jibs have collapsed through the operator's failure to observe the reduction in the safe working load as the jib moves towards the horizontal.

Fig. 34.1 ◆ Defects in cranes

Unsatisfactory welds between braces to the jib of a mobile crane. This could seriously reduce the strength of the crane and lead to failure

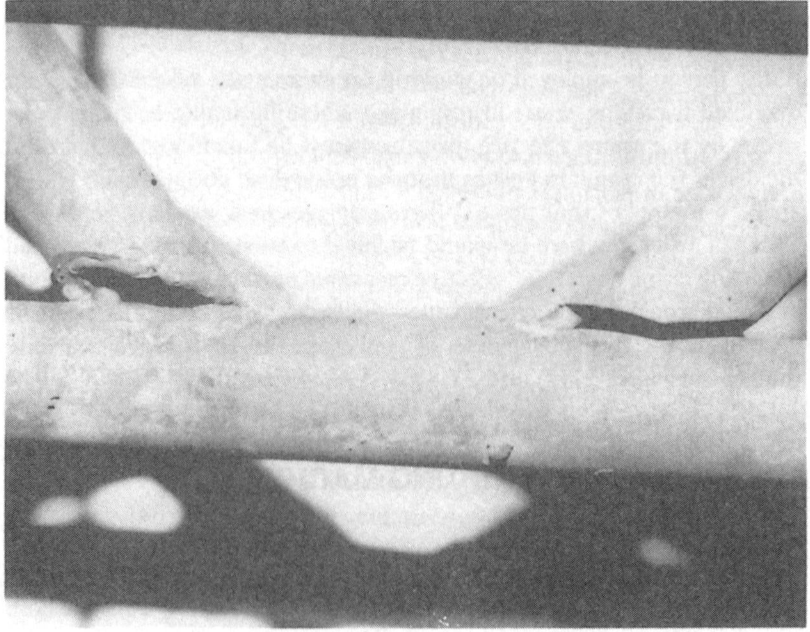

Fig. 34.2 ◆ A close-up of cracking originating from a weld in a 57 mm diameter tubular section of a part of a crane jib head frabrication
Further development of this defect, aided by internal corrosion of the member, could lead to jib failure

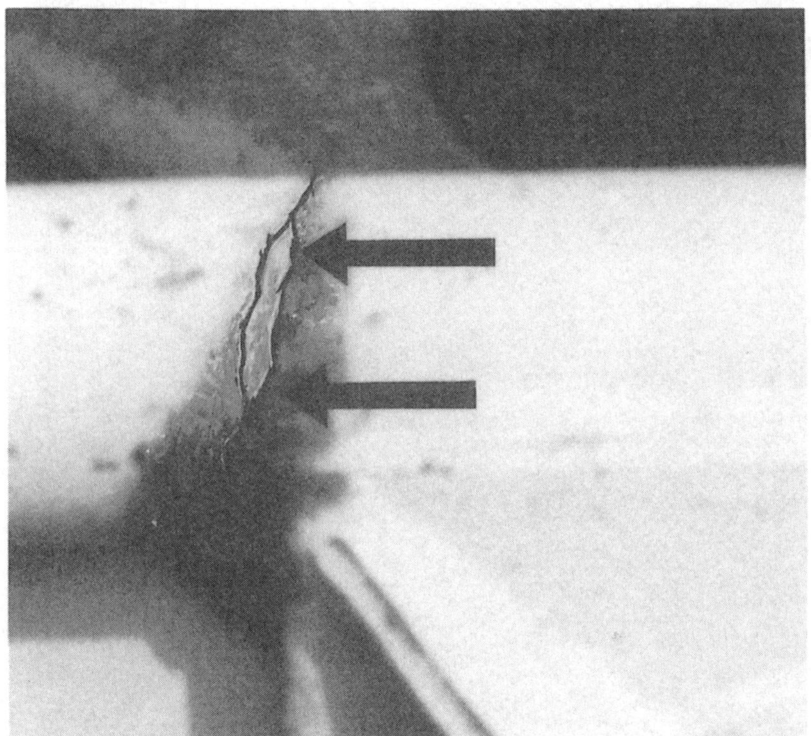

Similar observations apply in the case of rotational cranes. Accidents are caused by incorrect lifting and slewing, failure of the rotating gear when slewing and, more commonly, slewing too fast. Variations in wind speed, particularly while slewing, have a direct effect on the strength and stability of this type of crane.

Tower cranes

These cranes are often covered by the Lifting Operations and Lifing Equipment Regulations 1998, and are highly complex items of plant. Accidents are caused through incorrect assembly of the crane, and insufficient access to the jib, mast and driver's cab. The need for the driver to reach the cab safely is well recognised, but safe access to other parts of the crane is necessary during inspection, maintenance and repair, and in the course of erection or dismantling. Modifications to tower cranes may affect

Fig. 34.3 ◆ A dangerous situation with a manual hydraulic portable jib crane
The hook-retaining nut is only engaging one or two threads. Provision was made for locking the nut with a split pin which is missing

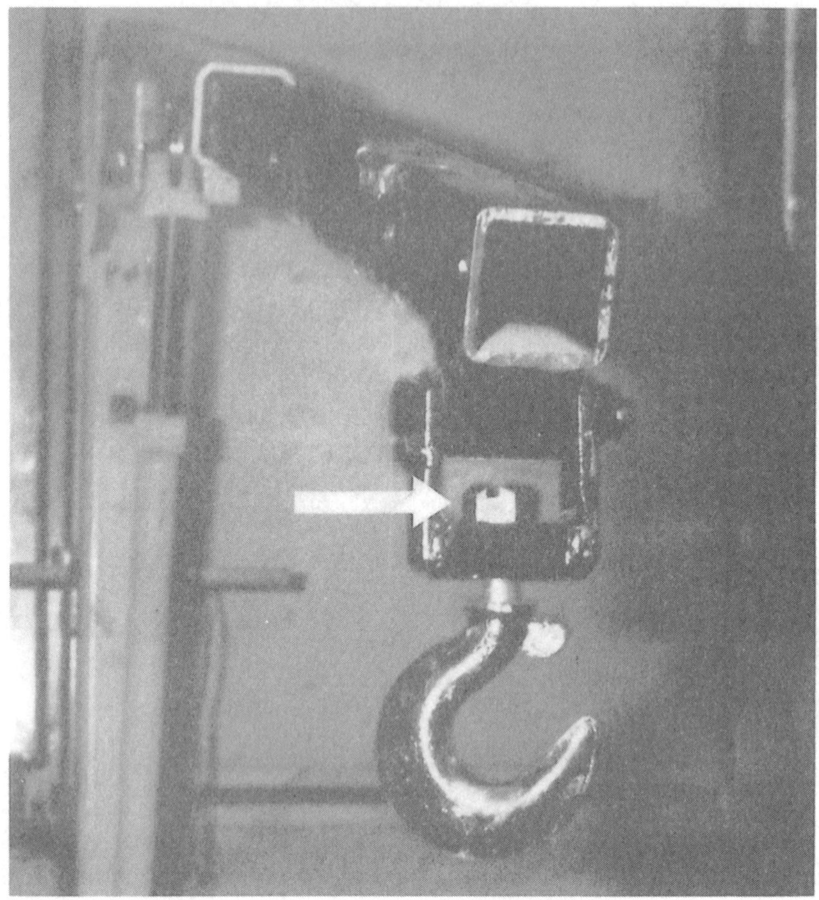

their strength and stability, and the manufacturer's advice should be sought prior to any modification. Rail-mounting arrangements and the system for maintenance of such cranes must be considered in any assessment of safe working operations.

Mobile cranes

Mobile cranes are used increasingly for lifting heavy items into specific locations. Some incorporate a telescopic or articulated boom and rotate through 360 degrees on the chassis of a purpose-built road vehicle. In

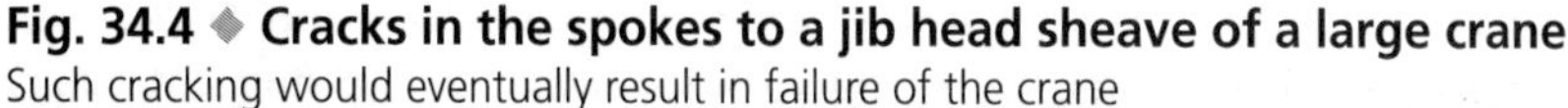

Fig. 34.4 ◆ Cracks in the spokes to a jib head sheave of a large crane

Such cracking would eventually result in failure of the crane

Reproduced by courtesy of National Vulcan Engineenng Insurance Group Ltd

addition to the precautions outlined previously for fixed cranes, it is imperative that any lift takes place on solid level ground, using the vehicle's outriggers fully extended to spread the load through the vehicle to the ground. The principal cause of accidents is the use of cranes on uneven sloping ground, where the centre of gravity of the load combined with that of the crane has fallen outside the wheel base of the vehicle, resulting in overturning.

Overhead travelling cranes

The most common application of this crane is in heavy fabrication shops and foundries where the crane runs along a fixed traverse. The crane may be fixed to operate in one position or to rotate through 360 degrees. The main hazards are derailment due to overloading, obstructions on the traverse or rail track, and the absence of adequate stops at each end of the traverse or rails. With rail-mounted cranes, either the crane must be fitted with effective brakes for the travelling motion, or sprags, scotches or chocks must be provided and used.

Many accidents are attributed to overhead travelling cranes crushing or striking operators working in the vicinity of the track. Previous reference has been made to the requirements of the FA, sec 27(7), whereby effective measures must be taken to prevent a crane approaching within about six metres of any place where a person may be working on or near the wheel track of such a crane. The only reliable 'effective measures' are the complete isolation and locking off of the electrical supply to the crane, coupled with the issue of a permit to work indicating that the isolation procedure has been carried out and that it is safe for work to proceed in the vicinity of the crane tract (*see also* Chapter 13). The mere switching off of the electrical supply at a control box, even with the display of a cautionary notice on the switchbox itself, is not considered sufficient. The starter switch should be physically locked in the OFF position, or where this is not possible, the fuses removed from the operating circuit.

Powered working platforms

These platforms are now commonly used where quick and safe access to overhead machinery and plant, electrical installations, lighting equipment and stored goods is needed, and for lifting people and equipment into position where high-level maintenance of buildings, ships, aircraft and public service vehicles is undertaken. Where height, reach and mobility are required, these aerial working platforms have substantial advantages over other systems such as fixed or rolling scaffolding, access towers, bosun's chairs and working platforms fitted to fork-lift trucks. They are completely mobile, operating from either a self-propelled electric trolley or a light trailer. Their operations take three specific forms:

(a) self-propelled hydraulic boom operation;

(b) semi-mechanised articulated boom operation;

(c) self-propelled scissor lift operation.

Safety procedures for powered working platforms

This equipment is classed as 'mobile work equipment' under the Provision and Use of Work Equipment Regulations (PUWER) 1998 and the provisions of Part III of PUWER apply. Specific requirements for mobile work equipment include:

(a) precautions necessary where employees are carried on mobile work equipment (Reg 25);
(b) minimising the risk of rolling over of mobile work equipment (Reg 26).

Above all, the user and operator must be aware, from a study of the manufacturer's specification, of the scope and capabilities of the machine, and the machine must not be used beyond its recommended design capabilities. The following points must be considered in their operation and use.

Siting

Platforms should be sited on firm level working surfaces, and attention paid to the risk of aerial collision with nearby platforms and cranes, and possible contact with electrical conductors and power lines. Ample room must be allowed for passing vehicles; traffic cones and barriers, to warn approaching vehicles, should always be used when these platforms are in use.

Transportation

Care must be taken when loading and unloading platforms from road transport trailers, and ramps should always be used. Platforms should be driven by trained personnel only. When towing platforms on sites, it must be appreciated that the majority of self-propelled platforms do not have automatically applied brakes when under tow. Platforms should be lowered and correctly stowed before transporting and wheels chocked. Care should be taken in lifting any powered platform by crane or fork-lift truck. The manufacturer should always be consulted prior to any lifting operation of this type.

Overturning

Hazards from overturning may be created by overloading of the platform, wind loading, impact or shock loading, and improper use. The maximum lifting capacity must be marked on the platform and shown in the manufacturer's specification. In no circumstances should this be exceeded. If any tools or equipment are carried, their extra weight must be considered. When assessing the weight of person to be carried in terms of overall safe working load, it is good practice to work on the basis of the first person weighing 100 kg and each subsequent person 75 kg.

For normal applications, it is considered impracticable to operate a working platform in wind speeds above Force 4 (Beaufort Scale), i.e. fair breeze, which is equal to 16 mph or 7.3 m/sec or 14 knots. Care must be exercised when using a platform on high-rise structures, e.g. bridges, elevated roadways.

Shock loading through dropping heavy materials or articles on to the platform should be avoided, and particular care exercised when using a platform or folding extension for overreach purposes. Platforms should never be used for jacking or as lifting appliances unless specifically designed for this purpose, and only when written approval, together with a test certificate, has been granted by the manufacturer. Moreover, the platform should not be moved in an elevated condition on uneven surfaces. Outriggers must only be operated with a platform in the lowered position. They must always be used where recommended by the manufacturer and for greater stability. The maximum gradient for safe use must not be exceeded, and on soft ground or unsurfaced areas supporting plates of timber mats should be used. (All platforms are rated for use on firm level ground, even when supplied with rough-terrain wheels.) Care must be taken when using the machine close to excavation or trenches, or on embankments. (Maximum gradient = 1:40.)

Falls and trapping

Personnel working on a platform should wear safety harness with lanyards attached to the platform and not to an adjacent structure. Loose tools and articles should be properly stowed on the platform. No attempts to extend the working height of the platform with boxes, steps, planks, trestles, etc., should be permitted. Operators should not use excessive force against the hand rails.

Trapping can occur between the chassis and a wall or by the working platform against roof trusses, overhead cranes, pipelines and bridges. Guards fitted around a scissor mechanism and baseframe must be adequately maintained and kept in position.

Electric cables and bare conductors present a major hazard. No platform must come within six metres, with the platform at maximum elevation, of a power cable. Inside this limit, a permit to work system should be used, in conjunction with the power supply authority. (*See also* HSE Guidance Note GS6, *Avoidance of Danger from Overhead Electric Lines*.)

Maintenance

Regular maintenance is crucial to platform safety. The manufacturer's instructions should be followed and fitters trained in specific maintenance procedures. When undertaking maintenance on, or inspecting, the working platform mechanism, a 'scotch' or mechanical locking bar should be used to sustain the platform in a raised position should the hydraulics, etc., fail. (Numerous fatal and severe accidents are on record where this vital advice has not been followed.)

Operators

No one should be allowed to drive a power-operated platform unless he has been selected, trained and authorised. Operators should be reasonably fit and intelligent. Persons suffering from disorders such as epilepsy, poor hearing and/or poor eye-sight should be health screened before a decision as to their fitness to operate is made. Operator training should be taken in three stages by the company operating the platform, i.e. basic operating skills and knowledge of the platform operation; specific job training for the particular needs of the site employer; and familiarisation training at the work area, e.g. emergency procedures.

Safety aspects of chains, ropes and lifting tackle

Ropes

Natural fibre ropes

This is a rope made from material of vegetable origin, generally comprising manila hemp, sisal, coir and cotton. Manufacturers do not normally provide a certificate stating the SWL. However, the purchaser should obtain the guaranteed breaking strength from the manufacturer in order to assess the SWL. For new ropes used on a direct lift, the factor of safety (i.e. ratio of ultimate stress to the maximum design stress) should not be less than 6 under favourable conditions. For ropes used for slings, the factor of safety should be increased to at least 8. Fibre ropes are not legally required to have a test certificate before being taken into service. They do, however, require examination every six months (and hence be capable of identification), and when deterioration takes place to such an extent that the SWL cannot be guaranteed, the rope must be withdrawn and destroyed. Great care must be taken with fibre ropes. When they have become damp or wet, they should be dried naturally, as direct heat will cause brittleness. They should be kept in a well-ventilated store, hung on wooden or galvanised steel pegs, within a temperature range of 13–19°C.

When lifting loads with sharp edges, the rope should be protected with packing. When reeved through blocks, the sheaves must be of adequate diameter, with the grooves of the sheaves of adequate diameter and in sound condition to allow the rope to seat correctly. Fibre ropes should be inspected before use by opening the strands slightly and checking for serviceability. If discoloured, weak or rot (mildew) has occurred, the rope should be destroyed. A reduction in circumference will generally indicate that the rope has been overloaded. The rope should also be examined for chemical action, particularly by acids. (Nylon or terylene ropes are recommended where there is any possibility of acid attack.)

Figs 34.5 and 34.6 illustrates ropes which are dangerously worn.

Wire ropes

There are many types of wire rope specified for cranes, lifts, hoists, elevators, construction site equipment, such as excavators, and those for use in general engineering. Wire rope is designated by diameter, except when

Fig. 34.5 ◆ Dangerous ropes
This is the condition of a closing rope found during examination of a 1.13 cubic metre grab. The frequent examination of ropes is crucial in these situations

Fig. 34.6 ◆ A badly worn rope on a trolley crane
Failure of the rope would result in the trolley travelling towards the mast with possible impact and consequent risk to people below and to the main structure of the crane itself

Reproduced by courtesy of National Vulcan Engineenng Group Ltd

G

used in shipping, and the tensile strength of the wire used ranges within 1,550–700 N/mm^2 (100–10 tons/in^2) and 1,700–850 N/mm^2 (110–20 tons/in^2). The diameter is normally measured by rope calipers and the average of three measurements, taken at intervals of about 127 mm, is construed as the rope diameter. A wire rope is made up of strands and the number of wires per strand is termed the 'construction' of the rope. Thus a rope of 6 × 19 construction has six strands each having 19 individual wires. Ropes of ordinary or regular 'lay' have the direction of the lay of the strands opposite to the direction of helix of the individual wires which balance the rope against twist. Where flexibility is needed, then a rope with a greater number of wires should be selected. Thus a 6 × 24 rope is more flexible than a 6 × 19 rope, and a 6 × 37 rope even more flexible than a 6 × 24 rope for the same approximate strength. If, however, hard wear is significant, a 6 × 19 rope would be more suitable because of the larger size of wires.

In a wire rope, broken wires must be regarded as a warning sign, but the actual position of the breaks is significant. If the breaks occur over a short distance, or occur in one or two strands, then the rope should be removed from service and the cause investigated. If breaks occur as a result of normal service over a reasonable period, this may well indicate that the rope is reaching the end of its life and should be replaced. In the interests of safety, however, there must be a limit to the number of broken strands. When broken wires appear, more frequent examination of the rope is needed. The ends of the wires should be manipulated until they break off inside the rope and should never be trimmed with pliers.

During manufacture, wire ropes are thoroughly impregnated with a lubricant to reduce wear, exclude moisture and delay corrosion. Frequent relubrication is necessary depending upon the nature of service and the degree of use. A number of proprietary lubricants are available.

Wire ropes should never be knotted as a means of joining ropes. They may be joined by the use of sockets, swaged ferrules, bulldog clips or by splicing. Proof testing may be required in any event as a splice is never as strong as the parent rope and strength may vary from 85 to 95 per cent according to rope size. Where a mechanical splice is used, proof loading to twice the safe working load is required before placing the wire rope into service.

Wire ropes should be stored in a clean dry place. Wire rope slings should be cleaned after use, inspected and hung on pegs to prevent corrosion and kinking.

Man-made fibre ropes

Ropes of this type, generally manufactured in nylon or terylene, have the following advantages over natural fibre ropes:

(a) higher tensile strength;

(b) greater capacity for absorbing shock loading;

(c) freedom from rotting, mildew formation, etc.;

(d) they can be stored away whilst still wet;

(e) their performance is the same whether wet or dry;

(f) some degree of immunity from degradation due to contact with oil, petrol and many solvents;

(g) resistance to acids and other corroding agents.

Procedures for the care of man-made fibre ropes are similar to those for other types of rope, viz. clean dry storage, frequent inspection for cuts, abrasions, signs of overloading and evidence of chemical attack and exposure to extreme heat. Exposure to strong sunlight may weaken the fibres and unnecessary exposure should be avoided.

Chains

Despite the increased use of wire rope, chain is still one of the principal components of lifting gear. For the same safe working load, it is five to six times heavier than rope but it has a longer life, stands up to rough usage, is almost 100 per cent flexible and can be stored externally for long periods without deterioration. Chains do not kink or curl; they grip the load better and possess superior shock-absorbing properties. Moreover, chain is available in several grades – mild steel, high tensile steel and alloy steel – and it is common practice to designate these grades by their minimum ultimate strength in terms of the diameter of the bar from which the links are made i.e. breaking strength = $30d^2$ – measured in tons or tonnes.

Classification of chain strengths

Wrought iron: Each link is hand-welded by a scarf weld at the end or crown of the link. The weld is distinctive and serves to identify the chain. The safe working load of the chain under normal working conditions is $6d^2$ and minimum breaking strength is $27d^2$. The skin of wrought iron chain is liable to embrittlement due to impact in service causing fine hairline cracks. These act as stress leading to failure without stretching. Under the FA, sec 26, wrought iron chains must be annealed every 14 months or more frequently when used in connection with molten metal. Annealing involves the chain being heated uniformly until the whole of the metal has attained a temperature of between 600 and 650°C, after which it is withdrawn from the furnace and allowed to cool in still air. Wrought iron chain has largely been superseded by alloy steel chain.

Mild steel: Mild steel chain incorporates a machine-made butt or flash resistance welded joint in the middle length of the link. The safe working load is assessed at $6d^2$ and the minimum breaking strain is $30d^2$, identified by the figure '3' on the links. Mild steel is liable to embrittlement and should be subject to 'normalising', a form of heat treatment, at frequent intervals. The use of mild steel chain is rapidly declining.

High-tensile steel grade 40: This grade of steel has mechanical properties which are superior to those of mild steel, the safe working load being 30 per cent higher. The safe working load is assessed at $8d^2$ and the minimum breaking strain is $40d^2$, identified by the figures '4' or '04' on the links.

Alloy steel grade 60: This type is 50 per cent stronger than high-tensile steel grade 40 chain and twice as strong as mild steel chain. It is suitable for most lifting purposes. The safe working load is assessed at $12d^2$ and the minimum breaking strain is $60d^2$. Such chain is subjected to hardening and tempering during manufacture and is marked '06' on each twentieth link.

Alloy steel grade 80: Grade 80 chain is designed for specific purposes, for instance as load chain for pulley blocks and similar applications which require an accurately calibrated chain of great strength and wear resistance. Its proof load is only one and a half times the safe working load due to the fact that the standard proof load of twice the safe working load for other chains listed above can disturb the pitch of the links. The safe working load is rated at $24d^2$, and the minimum breaking strain is $80d^2$. This type of chain is not suitable for slings. It requires considerable technical and manufacturing resources, as does the actual servicing of it. The length of the chain is regarded as a complete unit; individual links are not marked but the chain is identified by a disc attached.

Chain breakages

Generally, chains break for one of three reasons:

(a) from a defect in one of the links;

(b) through the application of a static load in excess of the chain's breaking load; or

(c) through the sudden application of a load which, but for the shock, the chain would have been capable of withstanding.

In no circumstances should nuts and bolts be used to replace broken chain links, nor must chains ever be knotted.

A comparison of the strengths of the various types of chain is shown in Table 34.1.

Table 34.1 ◆ Comparison of chain strengths (12.7mm (½″) diameter)

Type	*SWL (tonnes)*	*Proof load (tonnes)*	*Minimum breaking (tonnes)*	*Marking*
Wrought iron	1.5	3.0	6.75	–
Mild steel grade 30	1.5	3.0	7.15	3
HT steel grade 40	2.0	4.0	10.00	4 or 04
Alloy steel grade 60	3.0	6.0	15.00	06
Alloy steel grade 80	3.5	8.0	20.00	08

G

Lifting tackle

The same safety principles as those for ropes and chains apply to items of lifting tackle such as slings, hooks, grips and eyebolts. In the majority of cases, the complete set of tackle in a lifting situation must be viewed as one specific unit. For instance, it is unwise to use a particular sling for a particular lifting job if the rope or chain being used for the lift is of inadequate strength.

Slings

Slings should be made of chains, wire ropes or fibre ropes (whether man-made or natural) of adequate strength. Rings, hooks, swivels and end links of hoisting chains should be fabricated in the same metal as the chain. Tables showing the maximum safe working loads for slings at various angles should be conspicuously displayed (*see* Fig. 34.7 and Tables 34.2 and 34.3), and workers using slings should be trained in the use of these tables, particularly the fact that sling leg tension increases rapidly with increase in leg angle. For instance, for a 1 tonne load, the tension in the leg increases

Fig. 34.7 ◆ Maximum safe working loads for slings at various angles

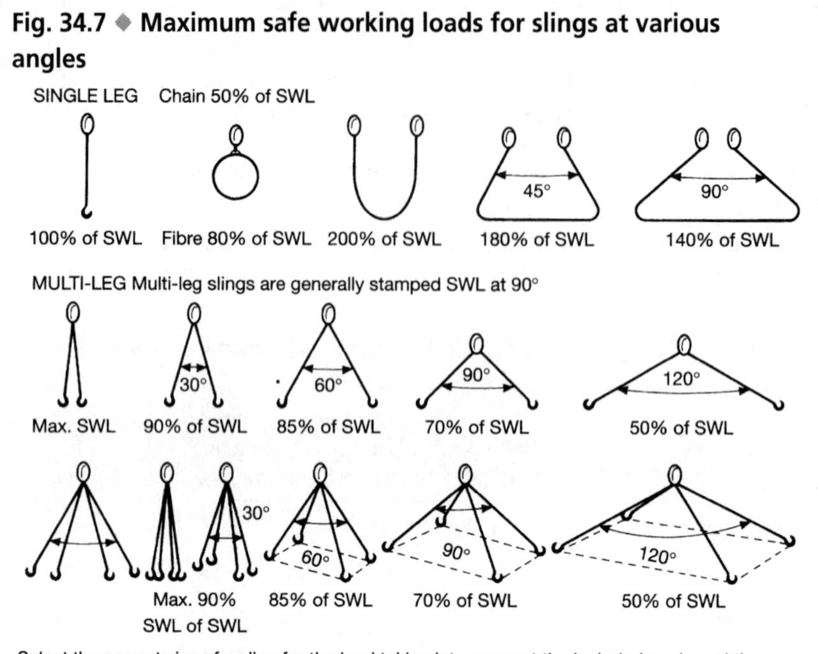

Select the correct size of a sling for the load taking into account the included angle and the possiblity of unequal loading in the case of multi-leg slings

Table 34.2 ◆ Estimation of sling leg angles

Sling angle	*Distance apart of legs*
30°	½ leg length
60°	1 leg length
90°	1⅓ leg length
120°	1⅔ leg length

Table 34.3 ◆ Relationship between sling leg angle and tension in leg

Sling leg angle	*Tension in leg (tonnes)*
90°	0.7
120°	1.0
151°	2.0
171°	6.0

as shown in Table 34.3. Reference to this table indicates how important it is for anyone dealing with multi-leg slings to appreciate fully how changes in the angles of the legs affect the safe working load, particularly for angles in excess of 120 degrees.

Slings that show evidence of cuts, excessive wear, distortion (*see* Fig 34.8) or other dangerous defects should be withdrawn. Wire rope slings should be well lubricated. To prevent sharp bends in slings, corners of loads should be packed. When multiple slings are used, the load should be distributed evenly among the ropes, and where double or multiple slings are used for hoisting loads, the upper ends of the sling should be connected by means of a ring or shackle and not put separately into a lifting hook. When bulky objects are being raised, the correct number of slings should be selected to ensure stability and support the weight of the load (*see* Fig. 34.9).

Hooks

Hooks for lifting appliances should be manufactured from forged steel or an equivalent material, and fitted with a safety catch shaped so as to prevent the load from slipping off (*see* Figs 34.10 and 34.11). In certain situations a hook must be provided with an efficient device to prevent displacement of a sling, or be of such a shape as to reduce, as far as possible, the risk of displacement.

In potentially dangerous lifting operations, hooks should be provided with a hand rope (tag line) long enough to enable workers engaged in loading to keep clear. Parts of hooks likely to come into contact with ropes and chains should have no sharp edges.

Pulley blocks

Pulley blocks should be manufactured from shock-resistant metal, e.g. mild steel. Axles of pulleys should be made of metal of suitable quality and of adequate dimensions. The diameter of the pulley should be at least twenty times the diameter of the rope to be used. The axle in the blocks should be capable of lubrication and a suitable lubricating device provided. (Regular and adequate lubrication of pulley blocks is essential.) The sheaves and housing of blocks should be so constructed that the rope cannot become caught between the sheaf and the side of the block, and the grooves in the sheaves should be such that the rope cannot be damaged in the sheaf. (Badly worn blocks should be taken out of use.) Blocks designed for use with fibre rope should not be used with wire rope. A pulley within reach of

Fig. 34.8 ◆ Badly distorted chain links in a sling which could lead to failure of the sling

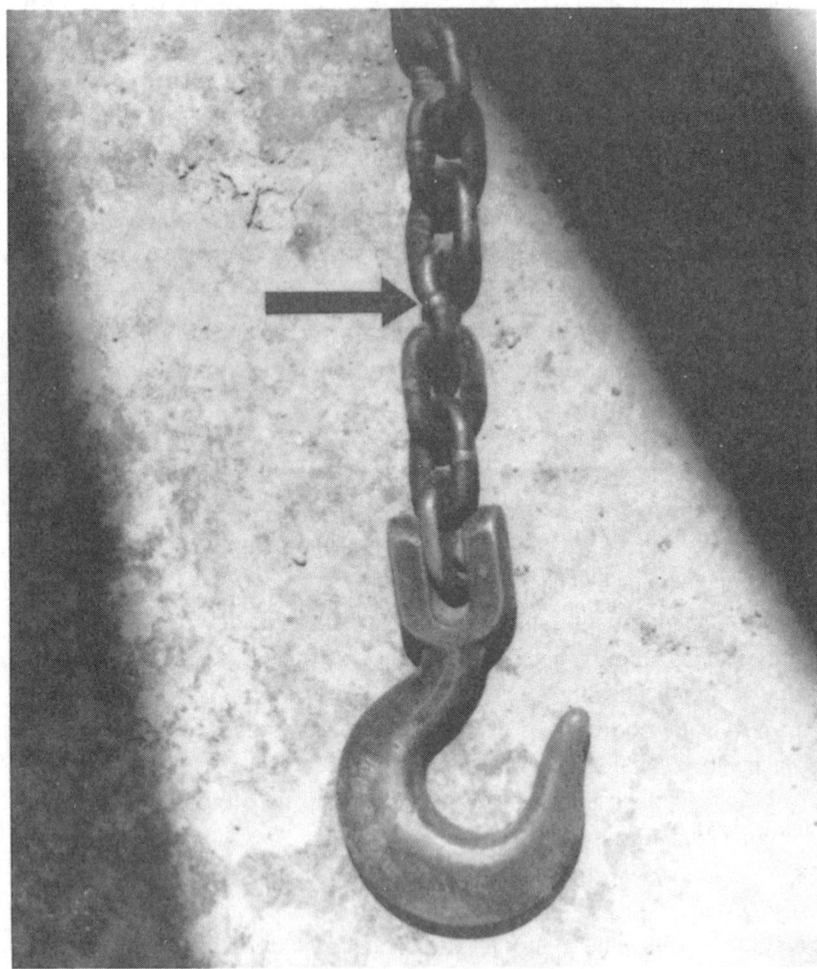

Reproduced by courtesy of National Vulcan Engineenng Insurance Group Ltd

Fig. 34.9 ◆ When lifting a bulky object, use a lifting ring and pack the edges of the object

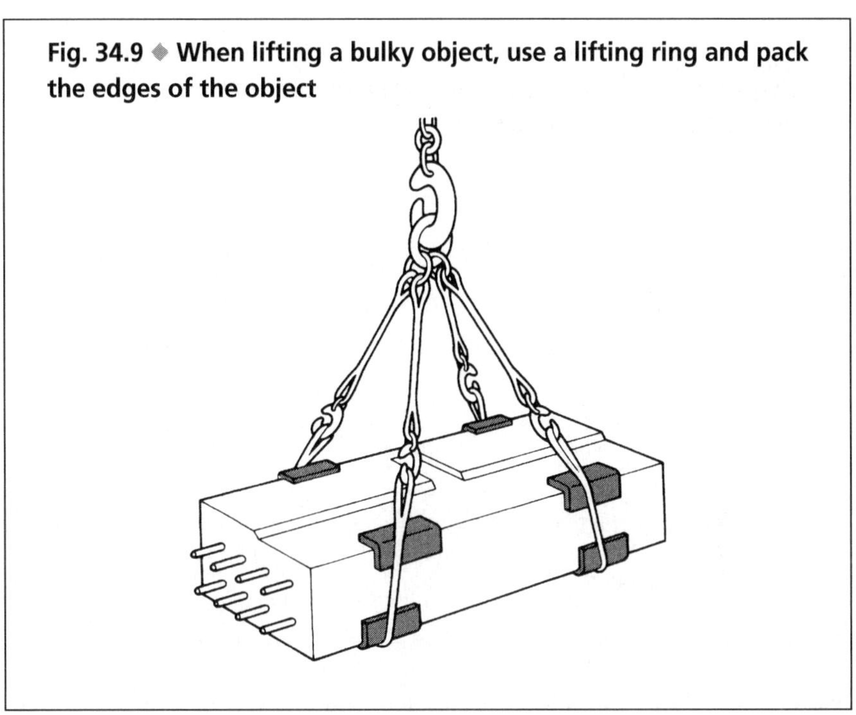

G

Fig. 34.10 ◆ Defects in lifting tackle

A comparison between a new hook (left) and one found at examination. This hook has been opened out by about 15 per cent and the safety latch is missing. A hook in this state will increase the possibility of a sling slipping off, and the distortion will get progressively worse

Fig. 34.11 ◆ Hooks for lifting appliances (a) Hook specially shaped to prevent the load from slipping off; (b) Hook incorporating a safety catch

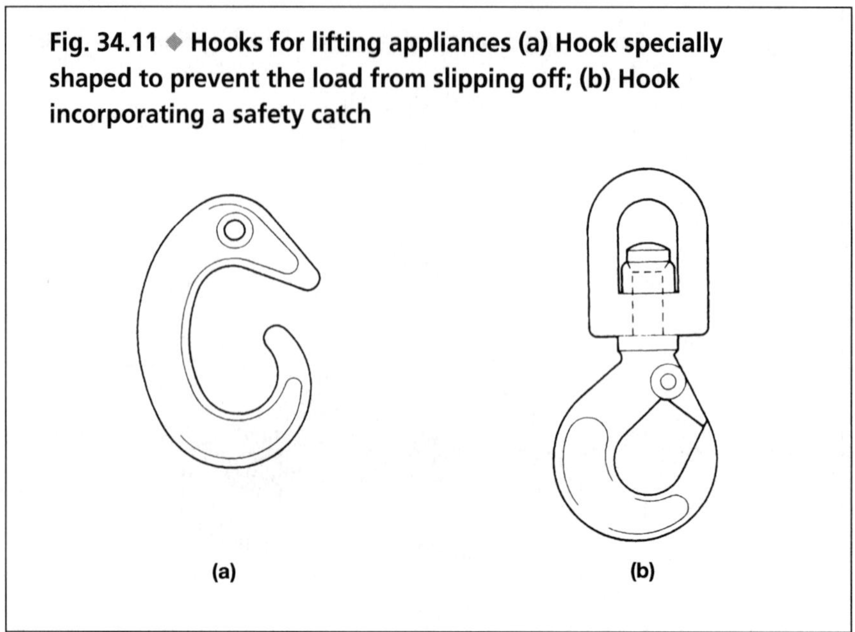

workers should be provided with a guard that effectively prevents a hand being drawn in.

Shackles

Shackles used for joining lines should have a breaking strength of at least 1.5 times that of the lines joined. In the case of shackles used for hanging blocks, the breaking strength should be at least twice that of the pulling lines, and the pins should be secured by locked nuts or other suitable means. Shackle pins should be secured by keys or wire, unless bolts are employed.

Eyebolts

Eyebolts are used for lifting loads which may be heavy and concentrated. (Three common types of eyebolts are shown in Fig. 34.12.) The use of the wrong type of eyebolt is a common contributory cause of accidents. A typical example of this is when a dynamo eyebolt is used for other than a vertical lift.

Dynamo eyebolts are large enough to receive a hook of a comparable safe working load, but should only be used for a vertical lift because the eye is so large that it is likely to bend, should an inclined load be placed upon it. Furthermore, a load which is only slightly out of the vertical plane places an undue stress on the screw threads of the eyebolt shank. Where

Fig. 34.12◆ Eyebolts (a) Dynamo eyebolt; (b) Collar eyebolt; (c) Eyebolt with link

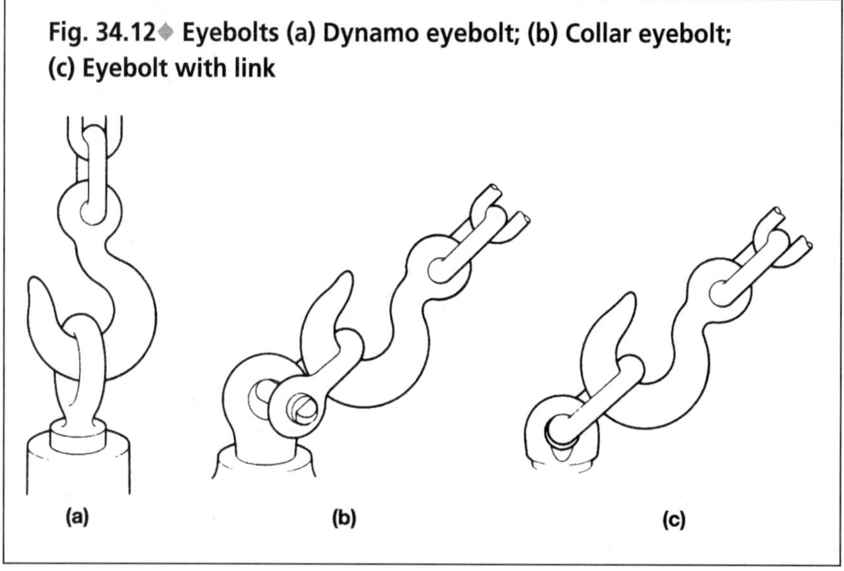

dynamo eyebolts are fitted in pairs, a spreader bar should form an integral part of the lifting gear used to move the load. The dynamo eyebolt is designed to receive the hook directly, and in all cases the hook should be able to operate freely. It is extremely dangerous to use a hook which jams in an eyebolt because, under load, serious weakening of both hook and eyebolt could occur leading to failure at some later date. If the hook available for the eyebolt is too large, a shackle of adequate size should be fitted to the eyebolt to accommodate the hook.

Where inclined loads are encountered, e.g. when a multi-leg sling is in use, collar eyebolts or eyebolts with links must be used. The collar eyebolt, or service eyebolt, has a squat eye that is too small to accommodate a hook, so a shackle is always necessary. Collar eyebolts are intended for permanent attachment to heavy pieces of equipment and are usually fitted in pairs for use with shackles or two-leg slings. When two pairs of eyebolts are fitted to a single load, then two-leg slings and a spreader bar (lifting beam) should be used in lifting (*see* Fig. 34.13).

The third type of eyebolt, that incorporating a link, is intended for general lifting. Although its rated load decreases as the angle of the load to the axis of the screw thread increases, by virtue of its special construction these rated loadings are greater than those of a collar eyebolt of equivalent vertical safe working load.

Fig. 34.13 ◆ Typical lifting beam (spreader bar)

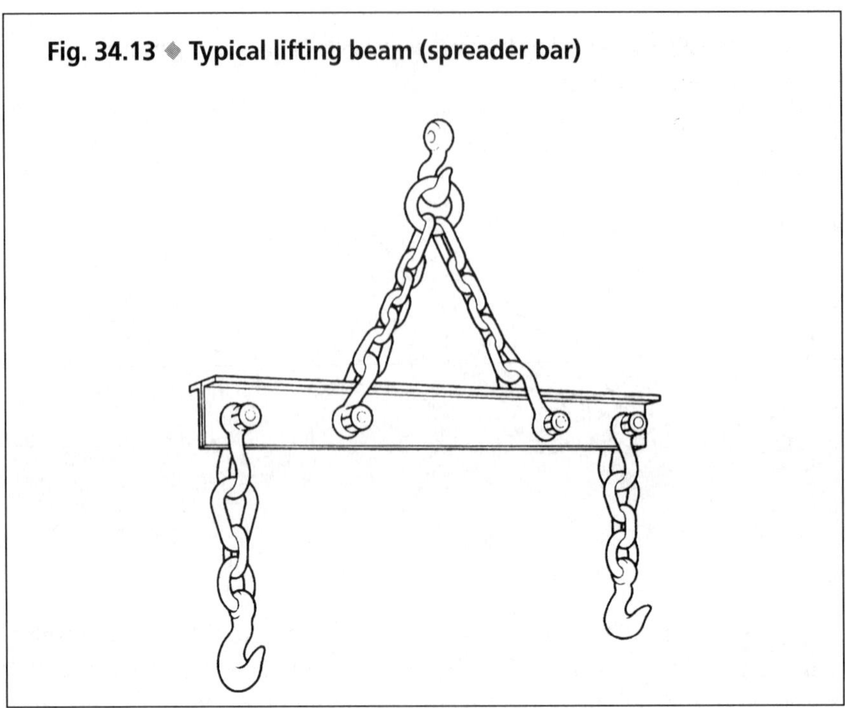

Lifting beams (spreader bars)

A lifting beam is a special purpose device which enables a particular load to be lifted in a particular way, often to prevent horizontal stressing of eyebolts. Lifting beams are commonly used in foundries for transporting vats of molten metal prior to casting or for the transport of engines in vehicle assembly. This is far safer than the use of multiple slings.

Lifting beams incorporate three basic parts:

(a) the beam, which is made from rolled steel section or plate;

(b) the means of attaching the beam to the lifting machine, such as a ring, shackle, eyebolt or hole in the main structure of the beam; and

(c) the means of fixing the beam to the actual load to be lifted, such as chain or wire rope slings which are fixed to the beam with fittings at the ends for securing the load, such as shackles or locking pins. (*see* Fig. 34.13).

Lifting beams are designed using standard steel rolled section or plate, angles, tees, universal beams or hollow sections. To ensure rigidity and resistance to accidental damage, the thickness of metal should not be less than 6 mm. This thickness can be reduced to 4.5 mm with hollow sections provided that they are adequately sealed against ingress of water or other damaging material. Many beams will be left outside and exposed to all weather conditions, so care must be taken in their design to avoid places where water might lodge; where this is not practicable, suitable drainage holes should be provided. If corrosion is the agent most likely to affect the beam, an allowance of up to 2 mm should be added to the minimum dimensions determined from all other strength calculations.

35 Pressure systems

Pressure Systems Safety Regulations 2000

These Regulations impose safety requirements with respect to pressure systems which are used or intended to be used at work. They further impose safety requirements with a view to preventing certain pressure vessels from becoming pressurised. The majority of duties are of an absolute nature.

Important definitions

Competent person means a competent individual person (other than an employee) or a competent body of persons corporate or unincorporate; and accordingly any reference in the Regulations to a competent person performing a function includes a reference to his performing it through his employees.

Danger in relation to a pressure system means reasonably foreseeable danger to persons from a system failure, but (except in the case of steam) it does not mean danger from the hazardous characteristics of the relevant fluid other than from its pressure.

Examination means a careful and critical scrutiny of a pressure system or part of a pressure system, in or out of service as appropriate, to assess:

(a) its actual condition;

(b) whether, for the period up to the next examination, it will not cause danger when properly used if normal maintenance is carried out, and for this purpose *normal maintenance* means such maintenance as is reasonable to expect the user (in the case of an installed system) or owner (in the case of a mobile system) to ensure is carried out independently of any advice from the competent person making the examination.

Installed system means a pressure system other than a mobile system.

Mobile system means a pressure system which can be readily moved between and used in different locations, but it does not include a pressure system of a locomotive.

Owner in relation to a *pressure system* means the employer of self-employed person who owns the pressure system or, if he does not have a place of business in Great Britain, his agent in Great Britain or, if there is no such agent, the user.

Pipeline means a pipe or system of pipes used for the conveyance of relevant fluid across the boundaries of premises, together with any apparatus for inducing or facilitating the flow of relevant fluid through, or through a part of, the pipe or system, and any valves, valve chambers, pumps, compressors or similar works which are annexed to, or incorporated in the course of, the pipe or system.

Pipework means a pipe or system of pipes together with associated valves, pumps, compressors and other pressure containing components and includes a hose or bellows, but does not include a pipeline or any protective device.

G

Relevant fluid means:

(a) steam;

(b) any fluid or mixture of fluids which is at a pressure greater than 0.5 bar above atmospheric pressure, and which fluid or mixture of fluids, is –

  (i) a gas; or

  (ii) a liquid which would have a vapour pressure greater than 0.5 bar above atmospheric pressure when in equilibrium with its vapour at either the actual temperature of the liquid or 17.5 degrees Celsius; or

(c) a gas dissolved under pressure in a solvent contained in a porous substance at ambient temperature and which could be released from the solvent without the application of heat.

Safe operating limits means the operating limits (incorporating a suitable margin of safety) beyond which failure is liable to occur.

Scheme of examination means the written scheme referred to in Reg 8.

System failure means the unintentional release of stored energy (other than from a pressure relief system) from a pressure system.

User in relation to a pressure system or a pressure vessel means the employer or self-employed person who has control of the operation of the pressure system or pressure vessel.

Part II of the Regulations applies generally.

Reg 4 – Design, construction, repair and modification

Any person who designs, manufactures, imports or supplies any pressure system or any article which is intended to be a component part of any pressure system shall ensure the following requirements are complied with. The pressure system or article, as the case may be:

(a) shall be properly designed and properly constructed from suitable material, so as to prevent danger;

(b) shall be so designed and constructed that all necessary examinations for preventing danger can be carried out.

Where the pressure system has any means of access to its interior, it shall be so designed and constructed as to ensure, so far as is practicable, that access can be gained without danger.

The pressure system shall be provided with such protective devices as may be necessary for preventing danger; and any such device designed to release contents shall do so safely, so far as is practicable.

Reg 5 – Provision of information and marking

Reg 5 requires designers and suppliers of pressure systems to provide sufficient information concerning the design, construction, examination, operation and maintenance of same as may reasonably foreseeably be needed to enable the provisions of the Regulations to be complied with. Similar requirements apply in the case of employers of persons who modify or repair such systems.

Manufacturers must mark pressure systems with specified information (see Schedule 3), and no person shall remove such a mark from or falsify any mark on a pressure system, or on a plate attached to it, relating to its design, construction, test or operation.

Reg 6 – Installation

The employer of a person who installs a pressure system at work shall ensure that nothing about the way in which it is installed gives rise to danger or otherwise impairs the operation of any protective device or inspection facility.

Reg 7 – Safe operating limits

This Regulation places specific duties on the users of installed systems and owners of mobile systems. They must not operate the system or allow it to be operated unless they have established the safe operating limits of that system. Furthermore, the owner of a mobile system shall, if he is not also the user of it:

(a) supply the user with a written statement specifying the safe operating limits of that system; or

(b) ensure that the system is legibly and durably marked with such safe operating limits and that the mark is clearly visible.

Reg 8 – Written scheme of examination

This regulation is, perhaps, the most important requirement. The user of an installed system and owner of a mobile system shall not operate the system or allow it to be operated unless he has a written scheme for the periodic examination, by a competent person, of the following parts of the system, namely:

(a) all protective devices;

(b) every pressure vessel and every pipeline in which (in either case) a defect may give rise to danger;

(c) those parts of the pipework in which a defect may give rise to danger.

Such parts of the system shall be identified in the scheme.

The user or owner shall:

(a) ensure that the scheme has been drawn up, or certified as being suitable, by a competent person;

(b) ensure that –

  (i) the content of the scheme is reviewed at appropriate intervals by a competent person for the purpose of determining whether it is suitable in current conditions of use of the system;

  (ii) the content of the scheme is modified in accordance with any recommendations made by that competent person arising out of that review.

No person shall certify or draw up a scheme of examination as above unless the scheme is suitable and:

(a) specifies the nature and frequency of examination;

(b) specifies any measures necessary to prepare the pressure system for safe examination other than those it would be reasonable to expect the user or owner respectively to take without specialist advice;

(c) where appropriate provides for an examination to be carried out before the pressure system is used for the first time.

References above to the suitability of the scheme are references to its suitability for the purposes of preventing danger from those parts of the pressure system included in the scheme.

Reg 9 – Examination in accordance with the written scheme

This Regulation goes into considerable depth on the procedures to be followed in the examination of pressure systems. The user of an installed system and owner of a mobile system shall:

(a) ensure that those parts of the system included in the scheme are examined by a competent person within the intervals specified and, where appropriate, before the system is used for the first time;

(b) before each examination take all appropriate safety measures to prepare the system for examination, including any such measures as are specified in the scheme of examination.

Where a competent person undertakes an examination, he shall carry out that examination properly and in accordance with the scheme of examination.

Where a competent person undertakes an examination, he shall make a written report of examination, sign or add his name to it, date it and send it to the user or owner respectively; and the said report shall be sent as soon as is practicable after completing the examination, and in any event to arrive:

(a) within 28 days of the completion of the examination; or

(b) before the date specified in the report outlined below;

whichever is the sooner.

Where the competent person is the user or owner respectively, the requirement to send the report to the user or owner shall not apply, but he shall make the report by the time it would have been required to have been sent to him if he had not been the competent person. The report specified above shall:

(a) state which parts of the pressure system have been examined, the condition of those parts and the results of the examination;

(b) specify any repairs or modifications to, or changes in the established safe operating limits of, the parts examined which, in the opinion of the competent person, are necessary to prevent danger or to ensure the continued effective working of the protective devices, and specify the date by which any such repairs or modifications must be completed or any such changes to the safe operating limits must be made;

(c) specify the date within the limits set by the scheme of examination after which the pressure system may not be operated without a further examination under the scheme of examination;

(d) state whether in the opinion of the competent person the scheme of examination is suitable (for the purpose of preventing danger from those parts of the pressure system included in it) or should be modified, and if the latter, state the reasons.

The user of an installed system and the owner of a mobile system a system which has been examined under this Regulation shall ensure that the system is not operated, and no person shall supply such a mobile system for operation, after (in each case):

(a) the date specified under paragraph (5)(b), unless the repairs or modifications specified under that paragraph have been completed, and the changes in the safe operating limits so specified have been made;

(b) the date specified under paragraph (5)(c) (or if that date has been postponed under paragraph (7), the postponed date) unless a further examination has been carried out under the scheme of examination.

The date specified in a report under paragraph (5)(c) may be postponed to a later date by agreement in writing between the competent person who made the report and the user or owner respectively if:

(a) such postponement does not give rise to danger;

(b) only one such postponement is made for any one examination;

(c) such postponement is notified by the user or owner in writing to the enforcing authority for the premises in which the pressure system is situated, before the date specified in the report under paragraph (5)(c).

Where the competent person above is the user or owner respectively the reference to an agreement in writing shall not apply, but there shall be included in the notification under sub-paragraph (c) a declaration that the postponement will not give rise to danger.

The owner of a mobile system shall ensure that the date specified under paragraph (5)(c) is legibly and durably marked on the mobile system and that the mark is clearly visible.

Reg 10 – Action in case of imminent danger

Where a competent person is of the opinion that the pressure system or part of the pressure system will give rise to imminent danger unless certain repairs or modifications have been carried out or unless suitable changes to the operating conditions have been made, then without prejudice to the requirements of the above Regulation, he shall forthwith make a written report to that effect identifying the system and specifying the repairs, modifications or changes concerned and give it:

(a) in the case of an installed system, to the user; or

(b) in the case of a mobile system, to the owner and to the user, if any.

The competent person shall within 14 days of the completion of the examination send a written report containing the same particulars to the enforcing authority for the premises at which the pressure system is situated.

Where a report is given in accordance with the above, the user or owner respectively shall ensure that the system is not operated until the repairs, modifications or changes have been carried out or made.

Reg 11 – Operation

The user or owner respectively shall provide for any person operating the system adequate and suitable instructions for:

(a) the safe operation of the system;

(b) the action to be taken in the event of an emergency.

The user of a pressure system shall ensure that it is not operated except in accordance with the instructions provided in respect of that system.

Reg 12 – Maintenance

This Regulation places a duty on users and owners respectively to ensure that systems are properly maintained in good repair, so as to prevent danger.

Reg 13 – Modifications and repair

The employer of a person who modifies or repairs a pressure system at work shall ensure that nothing about the way in which it is modified or repaired gives rise to danger or otherwise impairs the operation of any protective device or inspection facility.

Reg 14 – Keeping of records, etc.

This Regulations details the record-keeping requirements in respect of installed and mobile systems. Such records must include:

(a) the last report by the competent person;

(b) previous reports if they contain information which will materially assist in assessing whether –

  (i) the system is safe to operate; or

  (ii) any repairs or modifications to the system can be carried out safely;

(c) any documents provided pursuant to Reg 5 which relate to those parts of the pressure system included in the written scheme of examination; and

(d) any agreement made pursuant to Reg 9(7) and, in a case to which Reg 9(8) applies, a copy of the notification referred to in Reg 9(7)(c), until a further examination has been carried out since that agreement or notification under the scheme of examination.

G

Reg 15 – Precautions to prevent pressurisation

In the case of a vessel:

(a) which is constructed with a permanent outlet to the atmosphere or to a space where the pressure does not exceed atmospheric pressure;

(b) which could become a pressure vessel if that outlet were obstructed,

the user of a vessel shall ensure that the outlet referred to is at all times kept open and free from obstructions when the vessel is in use.

Reg 16 – Design standards, approval and certification

No person shall:

(a) supply for the first time;

(b) import; or

(c) manufacture and use,

a TGC unless the conditions specified below have been met. These conditions are as follows:

(a) the container has been verified (either by certificate in writing or by means of stamping the container) as conforming to a design standard or design specification approved by the HSE –

- (i) by a person or body of persons corporate or unincorporate approved by the HSE for the purposes of this paragraph; or
- (ii) in accordance with a quality assurance scheme approved by the HSE; or

(b) the container is an EEC-type cylinder, that is –

- (i) there is an EEC Verification Certificate in force in respect of it issued by an inspection body which, under the law of any Member State, is authorised to grant such a Certificate for the purposes of the framework directive and the separate directive relating to that type of cylinder, or, in the case of a cylinder not subject to EEC verification under any of the separate directives, it conforms to the requirements of the framework directive and the separate directive relating to that type of cylinder;
- (ii) it bears all the marks and inscriptions required by the framework directive and the separate directive relating to that type of cylinder.

Any approval under this regulation shall be by a certificate in writing, may be made subject to conditions and may be revoked by a certificate in writing at any time.

Reg 17 – Filling of containers

The employer of a person who is to fill a TGC with a relevant fluid at work shall ensure that before it is filled that person:

(a) checks from the marks on the cylinder that –

- (i) it appears to have undergone proper examination at appropriate intervals by a competent person (unless the manufacturer's mark reveals that such an examination is not yet due);
- (ii) it is suitable for containing that fluid; and

(b) makes all other appropriate safety checks.

The employer of a person who fills a TGC with a relevant fluid at work shall ensure that that person:

(a) checks that after filling it is within its safe operating limits;
(b) checks that it is not overfilled;
(c) removes any excess fluid in a safe manner in the event of overfilling.

An employer shall ensure that no person employed by him refills at work a non-refillable container with a relevant fluid.

Reg 18 – Examination of containers

The owner of a TGC shall, for the purpose of determining whether it is safe, ensure that the container is examined at appropriate intervals by a competent person. Where a competent person undertakes such an examination, he shall carry out that examination properly, and if on completing the examination he is satisfied that the container is safe, he shall ensure that there is affixed to the container a mark showing the date of the examination. No person other than the competent person or person authorised by him shall affix to a TGC the above mark or a mark liable to be confused with it.

Reg 19 – Modification of containers

G

1. Subject to paragraph 2:

(a) an employer shall ensure that no person employed by him modifies at work the body of a TGC –

(i) of seamless construction; or

(ii) which has contained acetylene;

(b) an employer shall ensure that no person employed by him modifies at work the body of another type of TGC if that modification would put the TGC outside the scope of the design standard or design specification to which it was originally constructed;

(c) a person shall not supply any modified TGC for use unless following such work a person or body of persons approved by the HSE has marked or certified it as being fit for use or, in the case of an EEC-type cylinder, an inspection body has so marked or certified it.

2. Paragraph 1 shall not apply to the remaking of a thread if this is done in accordance with a standard approved by the HSE.

Reg 20 – Repair work

An employer shall ensure that no person employed by him carries out at work any major repair on the body of a TGC:

(a) of seamless construction; or

(b) which has contained acetylene.

An employer shall ensure that no person employed by him carries out at work any major repair on the body of any other type of TGC unless he is competent to do so.

No person shall supply a TGC which has undergone a major repair unless following such work a person or body of persons approved by the HSE has marked or certified it as being fit for use or, in the case of an EEC-type cylinder, an inspection body has so marked or certified it.

In this regulation *major repair* means any repair involving hot work or welding on the body of the TGC but, except in the case of a TGC which has contained acetylene, it does not mean heat treatment applied for the purpose of restoring the metallurgical properties of the container.

Reg 21 – Re-rating

This regulation applies to the re-rating of a TGC, that is, the reassessment of its capability to contain compressed gas safely with a view to improving its capacity by means of an increase in the charging pressure (or in the case of liquefied gas, the filling ratio) from that originally assessed and marked on the container at the time of manufacture.

An employer shall ensure that no person employed by him re-rates a TGC at work unless he is competent to do so and does it in accordance with suitable written procedures drawn up by the owner of the container.

No person shall supply a TGC which has been re-rated unless following the re-rating a person or body of persons approved by the HSE has certified it as being safe for use. In this regulation *filling ratio* means the ratio of the volume of liquefied gas in the container to the total volume of the container.

Reg 22 – Records

The manufacturer, or his agent, or the importer of a TGC:

(a) made to an approved design specification, shall keep a copy of the said specification together with any certificate of conformity issued in accordance with Reg 16(2)(a);

(b) made to an approved design standard, shall keep a copy of any certificate of conformity issued in accordance with Reg 16(2)(a);

(c) which is an EEC-type cylinder, shall keep the EEC Verification Certificate referred to in Reg 16(2)(b)(i) where one has been issued.

The owner of a hired out TGC:

(a) made to an approved design specification shall keep a copy of the said specification together with a copy of any certificate of conformity issued in accordance with Reg 16(2)(a);

(b) made to an approved design standard, shall keep a copy of any certificate of conformity issued in accordance with Reg 16(2)(a);

(c) which is an EEC-type cylinder, shall keep a copy of the EEC Verification Certificate referred to in Reg 16(2)(b)(i) where one has been issued;

(d) which –

  (i) is a refillable container;

  (ii) is used solely for containing liquefied petroleum gas;

  (iii) has a water capacity up to and including 6.5 litres,

shall keep a copy of the design specification for the container.

The owner of a TGC for acetylene shall keep records of the tare weight of the container, including the porous substance and acetone or other solvent, the nature of the solvent and the maximum pressure allowed in the container. Miscellaneous provisions of the Regulations are covered in Part VI.

G

Reg 23 – Defence

1. In any proceedings for an offence for a contravention of the provisions of these Regulations, it shall, subject to paragraphs 2 and 3, be a defence for the person charged to prove:

  (a) that the commission of the offence was due to the act or default of another person not being one of his employees (hereinafter called 'the other person');

  (b) that he took all reasonable precautions and exercised all due diligence to avoid the commission of the offence.

2. The person charged shall not, without leave of the court, be entitled to rely on the above defence unless, within a period ending seven clear days before the hearing, he has served on the prosecutor a notice in writing giving such information of the other person as was then in his possession.

3. For the purpose of enabling the other person to be charged with and convicted of the offence by virtue of sec 36 of HSWA, a person who establishes a defence under this regulation shall nevertheless be treated for the purposes of that section as having committed the offence.

Schedule 4 – Marking of pressure vessels

The information referred to in Reg 5(4) is as follows:

1. The manufacturer's name.

2. The serial number to identify the vessel.

3. The date of manufacture of the vessel.

4. The standard to which the vessel was built.

5. The maximum design pressure of the vessel.

6. The maximum design pressure of the vessel where it is other than atmospheric.

7. The design temperature.

The Simple Pressure Vessels (Safety) Regulations 1991

These Regulations apply to:

(a) simple pressure vessels, i.e. welded vessels made of certain types of steel or aluminium, intended to contain air or nitrogen under pressure and manufactured in series; and

(b) relevant assemblies, i.e. any assembly incorporating a pressure vessel.

The Regulations incorporate a number of definitions, in particular, the following:

Safe

Means that when a vessel is properly installed and maintained and used for the purpose for which it is intended, there is no risk (apart from one reduced to a minimum) of its being the cause of or occasion of death, injury or damage to property (including domestic animals).

Manufacturer's instructions

Instructions issued by or on behalf of the manufacturer, and including the following information:

(a) manufacturer's name or mark;

(b) the vessel type batch identification or other particulars identifying the vessel to which the instructions relate;

(c) particulars of maximum working pressure in bar, maximum and minimum working temperature in °C and capacity in litres;
(d) intended use of the vessel;
(e) maintenance and installation requirements for vessel safety;

and written in the official language of the Member State accordingly.

Series manufacture

Where more than one vessel of the same type is manufactured during a given period by the same continuous manufacturing process, in accordance with a common design.

Vessel

This means a simple pressure vessel being a welded vessel intended to contain air or nitrogen at a gauge pressure greater than 0.5 bar, not intended for exposure to flame, and having the following characteristics:

(a) the components and assemblies contributing to the strength of the vessel under pressure are made either of non-alloy quality steel, or of non-alloy aluminium, or of non-age hardening aluminium alloy;
(b) the vessel consists either –
    (i) of a cylindrical component with a circular cross-section, closed at each end, each end being outwardly dished or flat and being also co-axial with the cylindrical component; or
    (ii) of two co-axial outwardly dished ends;
(c) the maximum working pressure (PS) is not more than 30 bar, and the PS.V not more than 10,000 bar litres;
(d) the minimum working temperature is not lower than –50°C and the maximum working temperature is not higher than 300°C in the case of steel vessels; and 100°C in the case of aluminium or aluminium alloy vessels.

The Regulations apply only to vessels manufactured in series. They do not apply to:

(a) vessels designed specifically for nuclear use, where vessel failure might or would result in an emission of radioactivity;
(b) vessels specifically intended for installation in, or for use as part of the propulsive system of, a ship or aircraft; or
(c) fire extinguishers.

Principal requirements of the Regulations

1. Vessels with a stored energy of *over 50 bar* litres when supplied in the UK must:
   (a) meet the essential safety requirements, i.e. with regard to materials used in construction, vessel design, manufacturing processes and placing in service of vessels;
   (b) have safety clearance, i.e. checks by an approved body;
   (c) bear the EC mark and other specified inscriptions;
   (d) be accompanied by manufacturer's instructions;
   (e) be safe (as defined).
2. Vessels with a stored energy up *to 50 bar* litres, when supplied in the UK must:
   (a) be manufactured in accordance with engineering practice recognised as sound in the Community country;
   (b) bear specific inscriptions (but not the EC mark);
   (c) be safe.
3. Similar requirements as above apply to such vessels when taken into service in the UK by a manufacturer or importer.
4. The Regulations do not apply to exports to countries outside the Community, or, for a transitional period, to the supply and taking into service in the UK of vessels that comply with existing UK safety requirements.
5. Failure to comply with these requirements:
   (a) means that the vessels cannot be sold legally;
   (b) could result in penalties of a fine of up to £2,000 or, in some cases, of imprisonment for up to three months, or both.

Categories of vessels

Different provisions are made for different categories of vessels depending upon their stored energy expressed in terms of the product of the maximum working pressure in bar and its capacity in litres (PS.V).

Category A vessels

These are graded according to PS.V range thus:
A.1 – 3,000 to 10.000 bar litres
A.2 – 200 to 3,000 bar litres
A.3 – 50 to 200 bar litres

Category B vessels

These are vessels with a PS.V of 50 bar litres or less.

The safety requirements in each case are according to items 1 and 2 of the principal requirements (*see above*).

Safety clearance

A vessel in Category A has safety clearance once an approved body has issued an EC Verification Certificate or an EC Certificate of Conformity in respect of that vessel.

G

Approved bodies

These are bodies designated by Member States, in the case of the UK, by the Secretary for Trade and Industry.

EC mark and other specified inscriptions

1. Where an approved body has issued an EC verification certificate, that approved body has responsibility for the application of the EC mark to every vessel covered by the certificate.
2. Where a manufacturer has obtained an EC certificate of conformity, he may apply the CE mark to any vessels covered by the certificate where he executes an EC declaration of conformity that they conform with a relevant national standard or the relevant prototype.
3. The EC mark must consist of the appropriate symbol, the last two digits of the year in which the mark is applied and, where appropriate, the distinguishing number assigned by the EC to the approved body responsible for EC verification or EC surveillance.

Other specified inscriptions to be applied to Category A and B vessels are:

(a) maximum working pressure in bar;
(b) maximum working temperature in °C;
(c) minimum working temperature in °C;
(d) capacity of the vessel in litres;
(e) name or mark of the manufacturer;
(f) type and serial or batch identification of the vessel.

EC surveillance

This implies surveillance by the approved body which issued a certificate. The approved body has the following powers with respect to surveillance:

(a) powers of entry –
  (i) to take samples;
  (ii) to acquire information;
  (iii) to require additional information;
(b) to compile reports on surveillance operations;
(c) to report to the Secretary of State cases of wrongful application, and failures by manufacturers to –
  (i) carry out their undertakings;
  (ii) authorise access;
  (iii) provide other facilities.

Steam boilers

The purpose of a steam boiler or, more specifically, a steam generator, is to produce steam under pressure from the raw materials, fuel, air and water. The potential heat of the fuel is made available through combustion, and this is transmitted to and stored by water vapour in the form of sensible and latent heat. There are two principal types of steam boiler: the vertical boiler and the horizontal boiler.

The vertical boiler

In its simplest form the vertical boiler would be a metal cylinder containing water, with a firebox at the bottom and a flue passing up the centre to carry hot gases away. However, such a boiler would be inefficient, as there would be little opportunity for hot gases to give up their heat through the sides of the flue to the water. In order to ensure that more heat is transferred, a greater part of the metal surface must be exposed to hot gases. One way of achieving this is by means of water tubes located across the central flue, as in the vertical cross-tube boiler shown in Fig. 35.1. In addition, extra exposed surface can be provided by the use of smoke tubes or fire tubes to carry the flue gases through water space, as in the fire-tube boiler shown in Fig. 35.1(*b*). This latter is more efficient mainly because the firebox is shaped to expose a larger area to heat from fire. Vertical fire-tube boilers are used mainly in small factories where steam requirements are not excessive. They are moderately cheap and do not occupy a great deal of room.

Fig. 35.1 ◆ Vertical boilers (a) Vertical cross-tube boiler; (b) Vertical fire-tube boiler with horizontal tubes

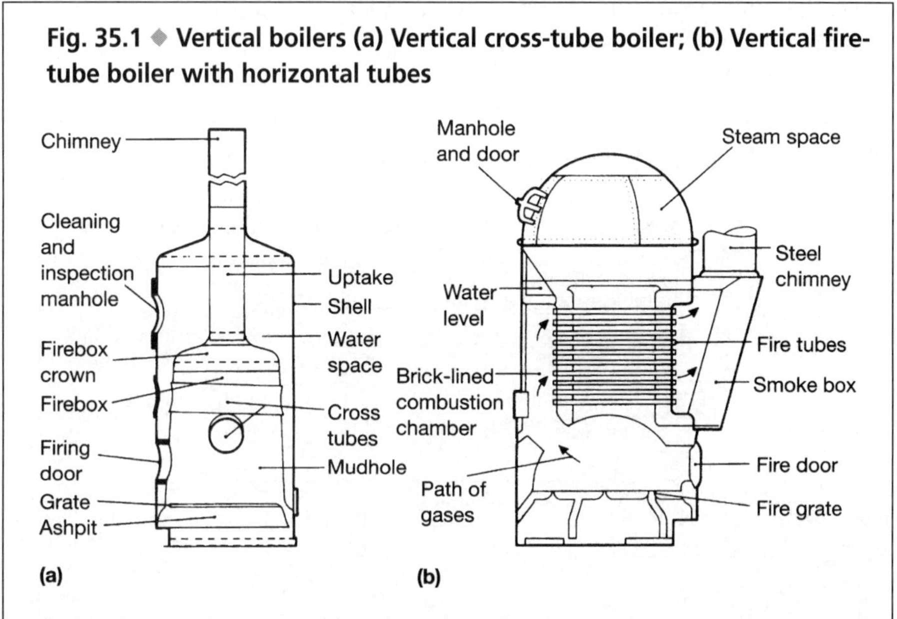

The horizontal boiler

This comprises a horizontal cylinder three-quarters full of water, with one or more furnace tubes passing through the water space. The fire is located at the front of the furnace tubes and hot gases travel through these tubes, heating the surrounding water prior to reaching the flue. The most common form of horizontal boiler is the Economic boiler, shown in Fig. 35.2. This boiler consists of a cylindrical shell with two flat end plates. One or more flue tubes are disposed between the end plates below the centre of the boiler, the grate or other fuel burning equipment, e.g. pressure jet oil burners, being arranged at the front end of these flues. The hot gases traverse the flue tubes to the back of the boiler where they enter a brick-lined combustion chamber in the case of a 'dry back' boiler or a water-cooled combustion chamber in the case of a 'wet back' boiler. Here the gas path is reversed and the gases travel to the front of the boiler through a bank of smoke tubes located above the flues. These tubes are normally about 8 cm in diameter and, by their use, the gas stream is broken up into a number of small elements, materially increasing the rate and efficiency and heat transfer. After passing through the smoke tubes, the gases are collected in a smoke box, whence they are led to atmosphere. This is the normal arrangement of a 'double pass' Economic boiler. A 'treble pass' Economic boiler has a second bank of tubes superimposed through which gases traverse the boiler from front to back, being finally collected in a smoke box, whence they are led to atmosphere.

Fig. 35.2 ◆ Economic boiler (solid fuel fired)

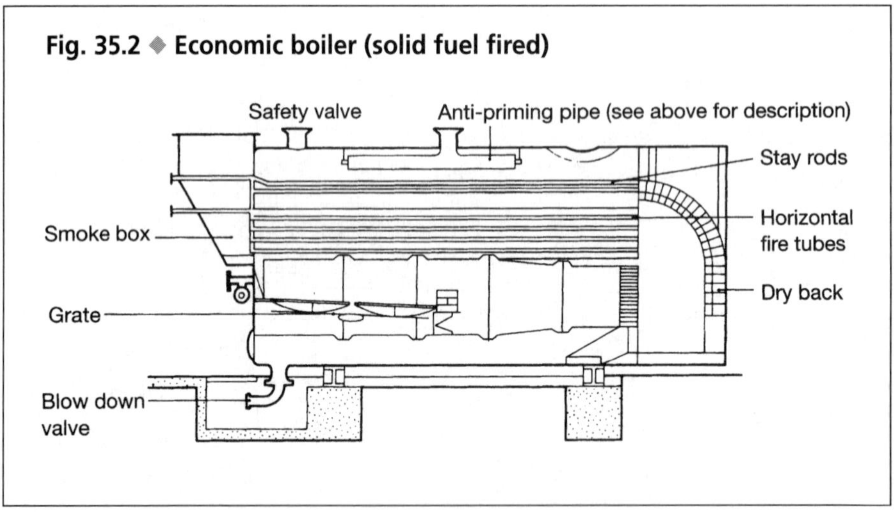

The testing of pressure vessels

A pressure vessel consists mainly of a series of sheets of metal suitably shaped and welded together. Most pressure vessel failures are associated with the blowing of a welded joint and, in order to ensure that a welded joint has properties comparable to those of the original materials and is metallurgically satisfactory, it is necessary for the actual physical properties of the weld to be investigated by destructive as well as non-destructive tests. Of course, the former cannot be carried out on the actual seams and it is undertaken using test plates. The normal system is to approve the manufacturer's procedures for the type of welding and the materials used in the vessel manufacture, as well as individual welders for certain classes of work, by the mechanical testing of separate test plates prepared by them before the actual production welding is commenced. Depending upon the requirements of the pressure vessel code used, the thickness of the materials, the difficulty in welding the materials selected, the future use of the vessel and the safety criteria used in the design code, non-destructive testing will be called for on the materials of construction and on the welding undertaken.

Although radiography has been used for non-destructive testing of pressure vessels for many years and produces a permanent record (radiograph), its effectiveness for identifying defects is dependent on the techniques used. Therefore, the radiograph should never be regarded as evidence of quality unless full details of the procedures are also known. Ultrasonic techniques are now also widely used, and in the construction of nuclear reactor vessels they are, on balance, more effective than radiography. There is, however, only limited scope for a permanent record of the results of the inspection.

As for inspection of boilers and other pressure vessels by non-destructive methods, it is essential to decide what defects require identification and to choose the particular method to suit. As the most serious defect likely to be encountered in a pressure vessel is a crack, it is important to recognise that radiography only discloses a fine crack if it is parallel to the direction of the radiation, i.e approximately perpendicular to the X-ray film. Ultrasonic methods do not identify a crack in the direction of the ultrasonic beam, and dye penetrants only show a crack which breaks the surface. These limitations should be appreciated prior to testing.

Installations and fittings – safety requirements

To ensure maximum safety of operations, the following installations and fittings are recommended for all pressure vessels, in particular steam boilers:

(a) two water gauges;

(b) two safety valves;

(c) a pressure gauge;

(d) a fusible plug or high and low water alarm;

(e) a blow down valve;

(f) a stop valve (steam);

(g) a feed check valve;

(h) an anti-priming pipe.

The functions of these installations and fittings are outlined below.

Water gauges

All boilers with an evaporative capacity exceeding 136 kg of steam per hour should be fitted with two water gauges. For boiler pressures up to 4.400 kg/cm^2 it is usual to fit tubular gauges which must be protected, protection usually consisting of a shield of specially toughened glass. Water gauges should be so situated that the water level can be easily seen by the operator and so arranged that the lowest visible section of the glass is higher than the minimum working level. It is good practice to test water gauges and cocks (a form of hand-operated tap or valve) once per shift by opening and closing the cocks in the prescribed manner.

Safety valves

Every boiler should have at least two safety valves, each capable of discharging the total peak evaporation of the boiler. The valves can be arranged on a single chest. This recommendation for two valves on each boiler is made on the assumption that if one valve fails to act, the other will function. However, this is not by itself a sufficient safeguard, and the safety valves should receive frequent attention and maintenance.

There are three types of safety valve – deadweight, lever arm or steelyard, and spring-loaded (*see* Fig. 35.3 for the latter two).

A safety valve is fitted to open at a set pressure, higher than the normal working pressure but below the maximum working pressure established by the insurance company. Steam will be discharged until the pressure drops sufficiently for the valve to close again. Each type is manufactured in a variety of patterns (the more modern having means for locking and adjustment). Each valve is so arranged that no unauthorised person can tamper with its setting. Valves should be tested periodically to ensure their reliable operation, in accordance with boiler insurance requirements. Failure to comply with insurance requirements, although not in itself a breach of law, may wholly or partially invalidate cover, as it is an implied term of insurance contracts that the insured will take steps to mitigate loss, e.g. will take practical safety measures.

Fig. 35.3 ◆ Safety valves (a) Spring-loaded safety valve; (b) Lever arm safety valve

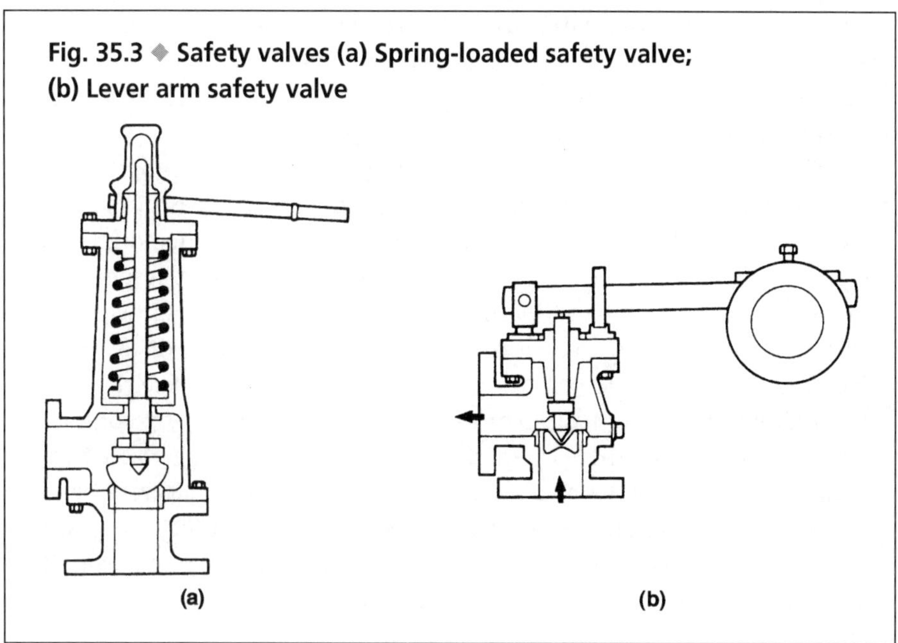

G

Pressure gauge

This must be connected to the steam space and indicate the pressure of steam in the boiler. The maximum permissible working pressure should be clearly marked on the gauge or the gauge glass.

Fusible plug, and high and low water alarm

A steam boiler must be equipped with either a fusible plug (if of the shell type) or a high and low water alarm, which makes a sound that can easily be recognised by the operator. A fusible plug is a plug of metal with a low melting point and is set into the boiler shell at low level. If the water level falls, and the boiler overheats, the plug melts, allowing the boiler water to escape and douse the fire.

There are two types of high and low water alarm, internal and external. The internal type of alarm consists of two floats, one at each end of a long arm suspended from the crown of the boiler shell. The arm is attached in such a manner as to pivot about an axis so that one float acts at a dangerously high level, the other at a dangerously low level. At correct levels of the water in the boiler the low-water float lies on the surface of the water, the high-water float being suspended clear of the surface, the system being maintained in equilibrium, When the water level falls, the low-level float drops and actuates a lever which operates a steam whistle. When the water level rises to the level of the top float, the upward movement of the bottom float being restricted, buoyancy occurs and causes the arm to tilt in the same direction as before so that the whistle again sounds.

The internal type of alarm is now being superseded by the external types: float and thermostatic. The first system, i.e. float, consists of chambers mounted at the normal working level of the water in the boiler and they are connected to the steam and water spaces. The floats respond to changes in the level of the water in the boiler and, at predetermined high and low positions, actuate a steam whistle, or two whistles of different notes, one for high-water conditions and the other for low. The thermostatic type, on the other hand, consists of rods that expand and contract according to whether they are in steam or water. When the water is at normal working level the upper rod is immersed in steam and lower one in water. If, through a rise in the water level, the upper rod becomes immersed in water, it contracts, and if the lower rod becomes immersed in steam, it expands. Either movement actuates an electric circuit which causes a bell to ring.

Blow down valve

A blow down valve has three important functions:

(a) deconcentration of the boiler water to prevent the solids content rising above prescribed limits;

(b) ejection of sludge and solids precipitated from the boiler water which settle at the bottom of the boiler;

(c) for emptying the boiler prior to inspection or for other purposes.

Blow downs may be continuously or intermittently operated. The former method is more effective in deconcentrating the boiler to prevent priming or foaming. In many cases the blow down is located near the surface of the water in the boiler shell. The intermittent blow down valve is used for the systematic ejection of unwanted solid matter deposited from the boiler water, as well as for emptying the boiler. It is an essential fitting and must be of first-class structure. It is necessarily situated at the lowest part of the boiler. Consequently it may be in a position where it is not continuously under the eye of the attendant. Discharge from the valve should be piped to a place where it can easily be inspected, as undetected leakage can cause serious wastage of fuel, together with corrosion.

When two or more boilers discharge their blow downs into the same pipe, each valve should be operable by only one key that remains locked in position when the valve is open and is removable only when it is completely closed. In this way only one blow down in a bank of boilers can be operated at any one time.

Stop valve (steam)

This valve is located between the boiler and the steam pipe or outlet, and is used to control the flow of steam from the boiler.

Feed check valve

This valve is situated on the boiler shell or steam drum, usually just below the low water level. It is essentially a non-return valve to prevent water escaping from the boiler should the pressure in the feed line be less than that of the boiler. A stop valve must be inserted between this non-return valve and the boiler. This may be incorporated in the feed check valve but the arrangement should be such that when the stop valve is closed it will be possible to remove the non-return valve for inspection, adjustment or minor repair while the boiler continues in operation. When more than one boiler is being fed by a single pump, the stop valve can be manipulated to control the rate of feed to the boilers.

Anti-priming pipe

Priming is the phenomenon whereby water is carried over from one part of the boiler into another part, such as a superheater. The action is mainly siphonic in nature. To prevent the loss of water which would result, an anti-priming pipe is fitted to break this siphonic action.

Hazards associated with boiler operation

The two principal hazards are overheating caused by low water level, which is the most frequent cause of boiler explosions and other damage, and the long-term effects of corrosion, which often result in explosion but, more frequently boiler failure.

Overheating in boilers

The main causes of overheating incidents are:

(a) lack of testing and maintenance of controls and alarms, leading to malfunction;

(b) occasional inadequate standards of control;

(c) (less frequently nowadays) isolation of control chambers.

Causes (a) and (b) above are associated with poor standards of boiler operation. Isolation of the control chambers (caused by the attendant closing and leaving closed either the water or the steam isolating valve, or both, after closing the drain valve) has in the past resulted in cases of explosion and damage from overheating of the boiler brought about by the resulting low water level. This hazard has largely been eliminated through improved boiler design.

Boiler corrosion

The long-term effects of corrosion in boilers can be both explosions and boiler failure. The principal sites and causes of corrosion are as follows.

Vertical boilers (see Fig. 35.1)

(a) External shell crown 'wastage' (loss of metal thickness, and therefore strength, due to corrosion) and internal and external uptake wastage. due to the effects of damp lagging and gases.

(b) Internal wastage of the upper firebox due to heat intensity, the effects of gases and ingress of water.

(c) Cross-tube wastage (gas side) due to the effects of gases and moisture.

(d) Lower firebox wastage caused by the excessive accumulation of ash in the ashpit which can be further accelerated by moisture.

(e) Wastage around the attachments of boiler mountings to the shell, in most cases due to leakage.

(f) Wastage at leaking mudhole doors, in many cases due to faulty or unsuitable jointing material; the lagging retains moisture and the wastage occurs unseen.

(g) 'Grooving', a form of mechanical corrosion, due to expansion and contraction, and accelerated by a build-up of solids, occurring at the junction of the firebox and shell.

(h) Grooving at the junction of the firebox and uptake due to pressure and

Fig. 35.4a ◆ Corrosion in pressure vessels
Corrosion of a safety valve to a calorifier making it totally inoperable. The seat is completely corroded and the spring actually corroded away. In addition, this valve was not fitted with a test lever so that, without dismantling it at examination, no evaluation of the operational performance of the valve could be made

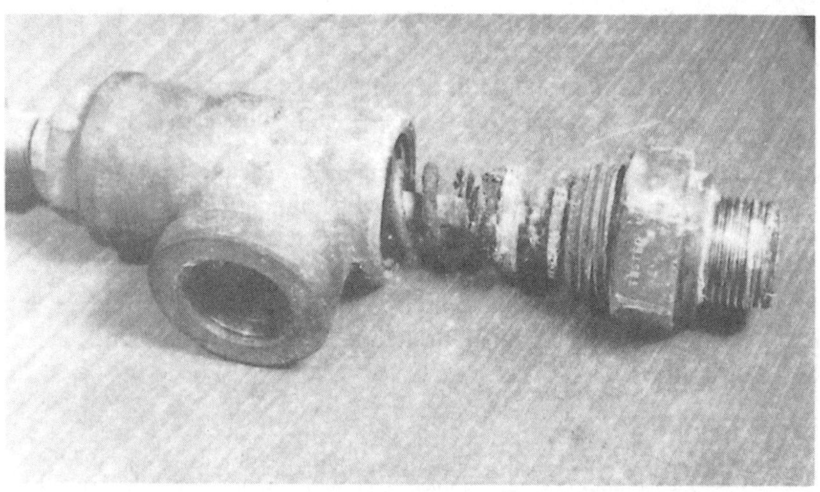

G

Fig. 35.4b ◆ Corrosion in pressure vessels
The flues of a cast iron water-heating boiler showing evidence of extensive corrosion

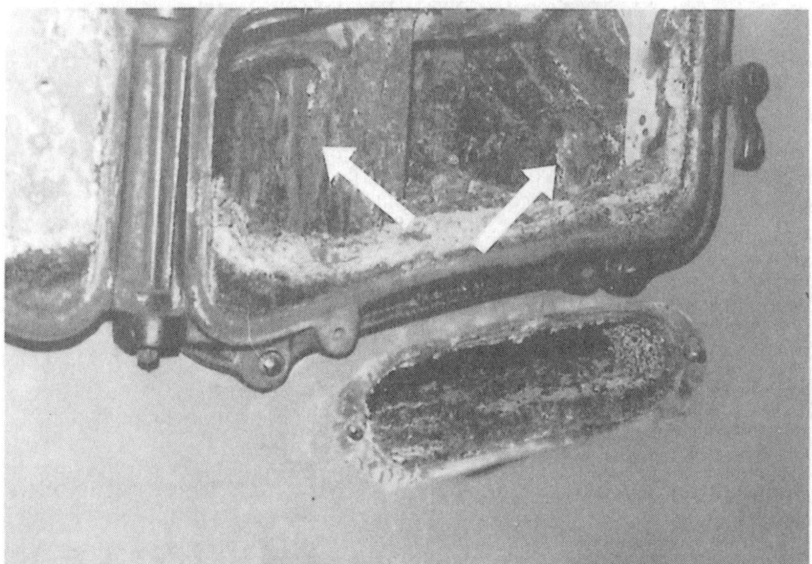

Fig 35.4c ◆ Corrosion in pressure vessels

General corrosion in a cast-iron water-heating boiler can occur extremely rapidly. In this case the surveyor's hammer has gone through the shell where the cast iron had been corroded to paper thinness

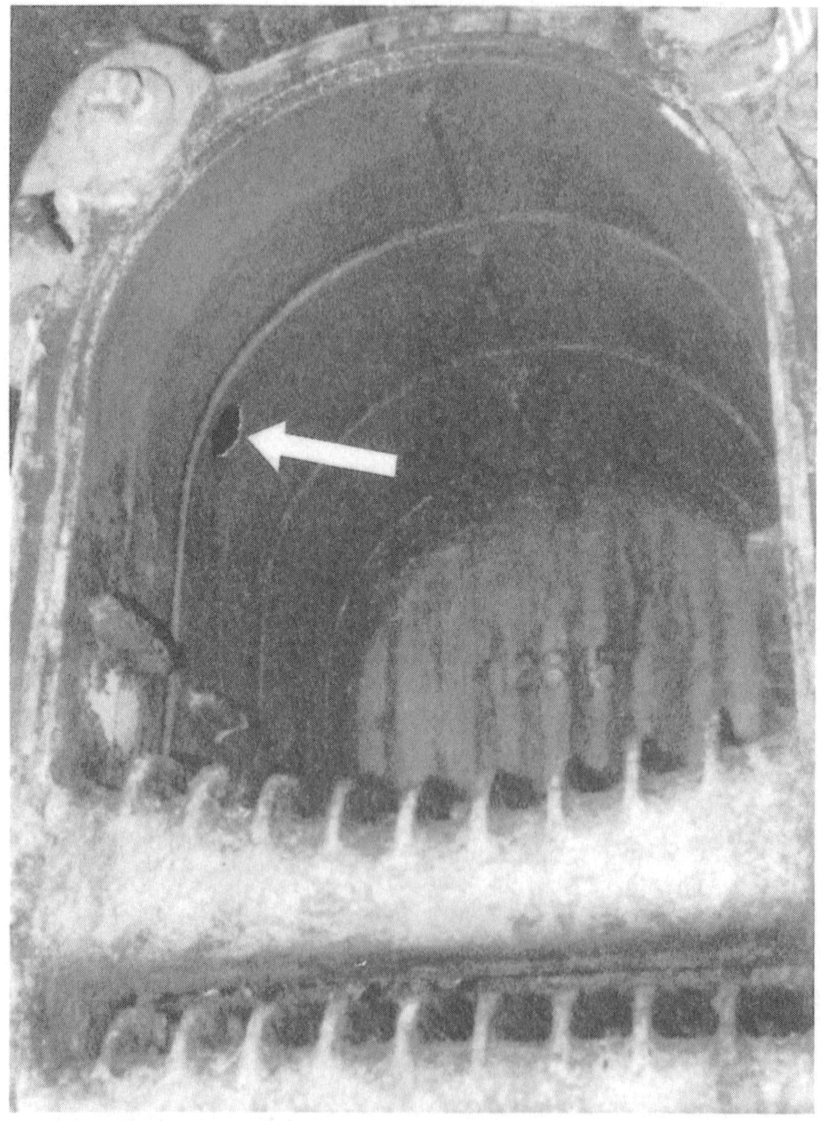

temperature fluctuations and, possibly, continued forced firing of the boiler.

(i) Wastage on the water side of the uptake between high and low water levels due to ebullience at the water surface and oxygen release on the uptake surfaces.

Fig. 35.4d ◆ Corrosion in pressure vessels
Primary to secondary leakage in heat exchangers results in loss of efficiency or in overpressure failure of the lower pressure side. The photograph shows the effect of a hydraulic test on the tube nest of a calorifier which indicated failure of one tube and prevented a more serious failure later

Reproduced by courtesy of National Vulcan Engineenng Insurance Group Ltd

(j) Wastage around the mudhole door and compensating ring caused by leakage or a badly fitting door to door joint.

(k) Grooving at the vertical lap joint of boiler shell plates due to steam pressure tending to improve the circular shell form; this is largely mechanical action which is localised at seam edges.

(l) General wastage of the internal surfaces of the ashpit plating and wastage of rivet heads, largely caused by sulphur in ash deposits and moisture.

Horizontal boilers (see Fig. 35.2)

(a) Distorted chamber tube plates, a situation where the tubes have been allowed to develop excessive scale and push the plate inwards.

(b) 'Necked' stays – wastage of the stays or stay tubes at plate entry due to leakage at the thread; this is accelerated through the tensile forces acting.

(c) Shell wastage at the mountings due to leakage at the joints, especially under the lagging.

(d) Combustion chamber crown distortion resulting from overheating.

(e) Girder stay wastage caused by leakage at the entry of the stay into the plate.

(f) Stay tube leakage as a result of incorrect welding giving inadequate support.

(g) Fractured stays caused by forcing the boiler.

(h) Radial grooving around the stays, a mechanical action caused by varying expansion of the heating surfaces; this weakens and breaks down the material stucture.

(i) Cracks from rivet holes to the plated edge due to overheating and subsequent corrosion.

(j) Front end plate grooving caused by cyclic variations in pressure; this accelerates with the age of the boiler.

(k) Front tube plate wastage due to leakage past defective tube ends.

(l) Sulphur attack from the slaking of ashes; this attacks most metal surfaces and will waste rivets away.

Automatic controls on steam and hot water boilers

The majority of steam boilers are now automatically controlled, and the most common water level and firing controls are float-operated controls situated outside the boiler. The floats are housed in chambers which are connected to the steam and water spaces in the boiler so that the water level in the chambers will be approximately the same as that in the boiler. Water level controls, low-water alarms and firing controls may be incorporated in the same chamber, but an additional chamber, with an independent electrical control circuit and independently connected to the boiler, is required for overriding low-water alarm and fuel cut-off in the case of fully automatically controlled steam boilers.

Standards for automatic controls

Automatic water level controls are of two basis standards:

(a) controls to assist the boiler attendant who constantly surpervises the boiler;

(b) controls intended to replace continuous supervision with occasional supervision.

The minimum recommended requirements for automatic controls for boilers not continuously supervised are as follows.

Automatic water level controls

These should be so arranged that they positively control the boiler or the feed pumps, or regulate the water supply to the boiler and effectively maintain the level of water in the boiler between certain predetermined limits.

Automatic firing controls

These should be so arranged that they effectively control the supply of fuel to the burners of oil- or gas-fired boilers, and shut off the supply in the event of any one or more of the following circumstances:

(a) flame/pilot flame failure on oil- or gas-fired boilers; the control should be of the lock-out type requiring manual resetting;
(b) failure to ignite the fuel on oil- or gas-fired boilers within a predetermined time; the control should be of the lock-out type requiring manual resetting;
(c) when a predetermined high pressure at or below the safety valve set pressure is reached;
(d) when the water level falls to a predetermined point below the normal operating level; this control should also cause an audible alarm to sound;
(e) failure of forced or induced draught fans, or any automatic flue damper, when these are provided.

Independent overriding control

This control should cut off the fuel supply to oil- or gas-fired boilers or air to the mechanical stokers of solid fuel fired boilers and cause an alarm to sound when the water level in the boiler falls to a predetermined low water level. The control or its electrical circuit should be so arranged that it has to be reset manually before the boiler can be brought back into operation.

Electrical failure to safety

All electrical equipment for water level and firing controls should be so designed that faults in the circuits cause the fuel and air supply to the boiler to be automatically shut off. Positive means, requiring manual resetting, should be provided to cut off the fuel and air supplies to the boiler, should there be a failure of electrical supply to water level and firing control equipment. All electrical conductors and equipment in connection with water levels and firing controls should be of adequate size, properly insulated and protected to prevent danger and including, where necessary, adequate protection against the ingress of moisture or the effects of high temperature.

36 Electrical safety

Electricity is perfectly safe if treated with respect. If misused, like many other things, if can have harmful effects on people and plant. The purpose of this chapter is to explain certain facts about electricity and how to use it safely. We all use electricity in our daily lives at work, in the home and in a host of other pursuits. Most of the time, we take it for granted. Occasionally, things go wrong, usually for simple reasons; however, if the advice given in this chapter is heeded, death, injuries and damage should be prevented.

Causes of electric shock and effects on the body

Electric shock is a possible outcome of electric current flowing through the human body, which causes disturbance in the normal functions of the body's organs and nervous system. Death occurs if the rhythm of the heart is upset for long enough to stop the flow of blood to the brain. It is crucial to act quickly in such emergencies, i.e. with first aid and resuscitation treatment, particularly simple oral resuscitation treatment. (*See* the standard St John's Ambulance Association *First Aid Manual* or RoSPA's factory Electric Shock placard; also Chapter 28.) Fortunately, death and serious injury from electric shock are relatively rare. Most electrical injuries, in fact, arise from burns received at the point of contact with the body. However, some of these burns can be deep-seated and immediate careful treatment is required.

To understand the electric shock phenomenon, the relationship between voltage (the 'driving force' of electrical energy), the current (the actual 'flow' of electricity) and the resistance (the characteristic of the circuit or path through which electricity flows and which offers resistance to the current) must be understood. In alternating current circuits, because resistive components are added vectorially, the term used is 'impedance'.

Ohm's Law

Ohm's Law is a simple mathematical relationship and it may be expressed as:

$$\text{Current} = \frac{\text{Voltage}}{\text{Resistance}}$$

or, using the actual measurement terms for these values, as:

$$\text{Amps} = \frac{\text{Volts}}{\text{Ohms}}$$

from which, by simple transposition of the formula, may be derived the following:

$$\text{Volts} = \text{Amps} \times \text{Ohms and Ohms} = \frac{\text{Volts}}{\text{Amps}}$$

It is usual to derive the simple power equation from Ohm's Law. Hence power (measured in watts) may be found from the following formula, although in strict alternating current theory it is somewhat more complicated mathematically:

$$\begin{aligned}\text{Watts} &= \text{Volts} \times \text{Amps} \\ &= \text{Amps}^2 \times \text{Ohms} \\ &= \frac{\text{Volts}^2}{\text{Ohms}}\end{aligned}$$

Effect of current

If a person is in contact with a live conductor (a conductor being a material that readily conducts electricity) and another part of his body is touching a conducting path, such as an earthed metal pipe, then the voltage to earth of that conductor will cause current to flow, through the body's resistance, to earth. The amount of current flowing will depend upon the voltage (which is usually the standard 240 volts supply) and upon the resistance of the body and other parts of the conducting path for the current to earth.

For a given current flow through a body, the severity of electric shock depends upon the length of time that the current flows, but it must be realised that only a small current flowing for a moment in time can be dangerous. Thus a current of 0.05 amp (50 milliamps) flowing for up to four to five seconds will probably not cause harm, but a current ten times the value but still only 0.5 amp (500 milliamps), flowing for 0.05 seconds (50 milliseconds), could be fatal.

The effect of electric shock varies with age, sex, medical and physical condition and the body's resistance to current flow, but a normal mains voltage of 240 volts, with an average body resistance of 1,000 ohms, would result in a current flow of 0.24 amps (240 milliamps) (*see* Ohm's Law above). This is a dangerously high value.

Fortunately, the shock current path contains more than the human body. Additional resistance to current flow may be found in the circuit, the contact with earth, any footwear worn, or the surface (a wooden floor, for example) on which the person may be standing. *It follows that contact with live electrical parts must be avoided. This is the first step to the prevention of electric shock.*

Legal requirements

The Electricity at Work Regulations 1989 impose standard health and safety requirements in respect of electricity at work. They apply to all places of work and cover all electrical equipment ranging from battery-operated equipment to high voltage equipment. A Memorandum of Guidance, issued by the HSE, should be read in conjunction with the Regulations.

Certain definitions in the Regulations are significant. *Electrical equipment* is defined as 'anything used or intended to be used and installed, to generate, provide, transmit, transform, rectify, convert, conduct, distribute, control, store, measure or use electrical energy'. *System* is defined as 'an electrical system in which all electrical equipment is or may be electrically connected to a common source of electrical energy and includes such source and electrical equipment'.

Reg 3 places a general duty on employers, self-employed persons and the manager of a mine or quarry to comply with the Regulations in so far as they relate to matters which are within their control. Similarly, an employee must, while at work, co-operate with his employer so far as is necessary to enable any duty placed on that employer to be complied with, and comply with the provisions of the Regulations in so far as they relate to matters which are within his control. All four groups, employers, employees, self-employed persons and mine/quarry managers are classified as 'duty holders' under the Regulations.

Under Reg 5 electrical equipment must not be put into use where its strength and capability may be exceeded so as to give rise to danger. This means that engineering management must take into account potential fault conditions, certain electrochemical, electromagnetic and thermal conditions, and ensure use of the equipment within the manufacturer's rating.

Systems, work activities and protective equipment are covered under Reg 4. All systems shall, so far as is reasonably practicable, be constructed and maintained so as to prevent danger. *Constructed* implies the need for testing, commissioning, maintenance and safe operation, together with the correct loading, rating and use of protective devices. The employer must take into account both fault conditions and environmental conditions. *Maintained,* on the other hand, relates to the quality and frequency of maintenance undertaken, and the need for proper record keeping of maintenance activities. Work activities must be carried out in such a manner as not to give rise to danger. This requires written procedures for safe operation, use and maintenance, for instance the isolation of equipment when carrying out repairs. In the case of protective equipment, this must be provided, maintained and properly used.

Particular attention is paid to operations in 'adverse or hazardous environments'. In this case, Reg 6 requires that electrical equipment which may reasonably be foreseen to be exposed to mechanical damage, the effects of weather, natural hazards, temperature or pressure, the effects of wet, dirty, dusty or corrosive conditions, or to any flammable or explosive substance, including dusts, vapours or gases, shall be of such construction or as necessary protected to prevent, so far as is reasonably practicable, danger. (BS5490, which deals with flameproof equipment, is significant in this case.)

Specific requirements are laid down with regard to the insulation of conductors (Reg 7), earthing or 'other suitable means' to prevent danger, e.g. double insulation, safe voltages, earth-free non-conducting environments and current limitation, and the relative suitability of electrical connections. For instance, all connections must be both mechanically and electrically suitable for use. This requirement applies to plugs and sockets, and to temporary and permanent connections, taking into account the conditions of use (Reg 10). Reg 11 deals with means for protecting from excess current. Here efficient means, suitably located, must be provided for protecting from excess of current every part of a system as may be necessary. The use of fuses and circuit breakers must be considered, taking into account possible fault conditions. Means for cutting off the supply, and for isolation where necessary, to prevent danger must be provided, together with precautions for work on equipment. In the latter case, adequate precautions, such as locking off, the operation of defined safe systems of work, electrical permits to work and other precautions, must be taken to prevent electrical equipment, which has been made dead in order to prevent danger while work is carried out on or near the equipment, from being electrically charged during the work.

Where work is being carried out on or near live conductors no person shall be engaged in any work activity on or near any live conductor, unless it is unreasonable in all the circumstances for it to be dead, and it is reasonable in all the circumstances for the person concerned to be at work on or near it while it is live, and suitable precautions are taken to prevent injury. This requirement is particularly appropriate in the testing of live equipment and in excavation work. Here the precautions necessary would include the use of personal protective equipment, earth-free areas, isolating transformers, insulated tools and cable detection equipment.

Persons must be competent to prevent danger and injury. This means that electrical engineering personnel must possess such 'knowledge and experience' or be under the appropriate supervision having regard to the nature of the work. Technical knowledge and experience would include adequate knowledge of electricity, and experience of electrical work, adequate understanding of the system to be worked on and practical experience of that class of system, understanding of the hazards which may arise during the work and of the precautions which need to be taken, and an ability to recognise at all times whether it is safe for work to continue.

IEE regulations

The Institution of Electrical Engineers publishes the *Institution of Electrical Engineers Regulations for Electrical Installations* on a regular basis. Known as the 'Wiring Regulations', these Regulations should be read in conjunction with the Electricity at Work Regulations and the Memorandum of Guidance. The Wiring Regulations are obtainable from the IEE, PO Box 26, Hitchin, Herts SG5 1SA.

British Standards

A wide range of British Standards covering many areas of electrical safety is currently available from the British Standards Institution, Sales Department, Linford Wood, Milton Keynes MK14 6LE.

Principles of electrical safety

The prime objective of electrical safety is to protect people from electrical shock, and also from fire and burns, arising from contact with electricity. There are two basic preventive measures against electric shock, namely:

(a) protection against direct contact, e.g. by providing proper insulation for parts of equipment liable to be charged with electricity; and
(b) protection against indirect contact, e.g. by providing effective earthing for metallic enclosures which are liable to be charged with electricity if the basic insulation fails for any reason.

When it is not possible to provide adequate insulation as protection against direct contact, a range of measures is available, including protection by barriers or enclosures, and protection by position, i.e. placing live parts out of reach.

Earthing

The provision of effective earthing, to give protection against indirect contact, can be achieved in a number of ways, including connecting the extraneous conductive parts of premises (water pipe, taps, radiators) to the main earthing terminal of the electrical installation. This would create an 'equipotential' zone and eliminate the risk of shock that could occur if a person touched two different parts of the metalwork liable to be charged, under earth fault conditions, at different voltages. It is crucial to ensure that in the event of earth fault, such as when a live part touches an enclosed conductive part (usually metalwork), that the electricity supply is automatically disconnected. Such disconnection is achieved by the use of overcurrent devices (correctly rated fuses or circuit breakers) or by correctly placed and rated residual current devices. (More details of these and other circuit components will be given later.) Maintenance of earth continuity is also vital.

Reduced low voltage

Another protective measure against electric shock is the use of reduced low voltage systems, the most commonly used being the 110 volt centre point earthed system, i.e. the secondary winding of the transformer providing the 110 volt supply is centre tapped to earth, thus ensuring that at no part of the 110 volt circuit can the voltage to earth exceed 55 volts. (*See* BS 4363:1968 and BS Code of Practice CP 1017.) Safe extra-low voltage systems are also available. These operate at up to 50 volts a.c. and obviously have limited though safer application.

Electric generators

Simple generation

A simple basic generator comprises two main parts, the rotor, which is the rotating part, and the stator, the fixed part. The rotor contains insulated conductors wound around a metal cylinder and, as it rotates under the action of an external mechanical power source, these conductors cut across a magnetic field from the electromagnet housed in the stator. The laws of electromagnetism, which apply to the interrelationship between electricity and magnetism, are such that this action causes voltage to be induced in the rotor conductors. By suitably connecting these conductors to an external circuit, the electric current thus produced can be tapped and used.

Alternatively, the conductors and electromagnets can be housed in the stator and rotor respectively, without affecting the generation of electricity in the conductors. It is the motion of the conductors in relation to the magnetic field that matters. The value of the current, voltage, etc., depends upon the characteristics of the generator. In modern power stations, for example, the generators (usually called 'alternators') may each have output in excess of 500 MW which is 500 million watts!

Small portable and mobile generators

These are used for a variety of purposes. On construction sites, for example, they are used to provide an emergency and temporary supply of electricity or to serve particular items of plant, such as electric arc welding sets. Most generators on construction sites are capable of supplying at 240 volts and 120 volts, which enables them to supply a 55 volt system via special transformers available for this purpose. The generator's frame, metallic parts of the transformer and centre point of the secondary winding of the transformer should be bonded electrically and suitably earthed.

It is important that the manufacturer's instructions on the installation and use of generators are known and understood by operators and other staff involved. (For further information *see* BS Code of Practice CP 1017.)

Alternating and direct currents

The natural current generated is alternating in kind (a.c.) which, in a normal single cycle of 1/50 second, alternates in value from zero to maximum in one direction, back to zero and to maximum in the opposite direction, and back to zero again, thus producing the standard 50 cycles per second (50 hertz) system used in the UK.

To change an a.c. generator to one that produces direct current (d.c.) it is necessary to rectify the positive and negative surges of the a.c. current so that it flows only in one direction at a constant value. One way of doing this is to use a commutator.

So far as safety is concerned, a.c., because of its effects on the heart, is rather more dangerous than d.c., but it is best to assume that the same preventive measures apply to both and to act accordingly.

Single and three phase systems

A single phase a.c. system, normally at 240 volts, is one in which the circuit comprises two conductors (called 'phase' and 'neutral') usually with an additional conductor for 'earth'. A three phase a.c. system, normally at 415/240 volts, has four conductors. Three of them are live and the fourth is neutral. The three phase supply is for larger equipment and the maximum voltage of the system (415 volts) is utilised. But 240 volts may be obtained between one of the live phase conductors and neutral.

Electrical installation

An electrical installation comprises such items as cables, conduit or other mechanical protection, main and local switches, distribution boards, fuses, socket outlets, etc.

Circuits

For electric current to flow and provide a source of energy, it must be contained within a circuit comprising conductors. These conductors contain a suitable inner metal core (e.g. tinned copper or aluminium) which conducts the electricity, together with an outer sheath of insulating material (e.g. rubber or similar man-made substance) which normally safely 'contains' the electricity within the circuit.

Fuses and circuit breakers

In the event of a fault or overload, it is necessary for a circuit to be disconnected or excessive current could flow and the resulting overheating could cause fire. Automatic disconnection is provided by the fuse or circuit breaker.

A fuse is a protected strip of thin metal which melts at a value well below an excessive value of current and cuts off supply. It will be readily appreciated that a fuse should be of the type and rating appropriate to the circuit and appliance it protects. Clearly, it would be folly to provide a 30 amp fuse, or a makeshift one of the same value, when only a 13 amp fuse is required. It should also be remembered that fuses will protect against overheated circuits and fire, but *not* against receipt of a potentially lethal or dangerous electric shock. (The fault current required to blow a fuse will invariably be far higher than the minimal current which would present danger to human beings.) Fuses protect the equipment but not necessarily the user.

A circuit breaker is a device that looks like an enclosed switch. It has a mechanism that trips the switch from 'on' to 'off' position if an excess current flows in the circuit. As with the fuse, the circuit breaker would be of the type and rating for the circuit and appliance it protects.

It must be stressed that whereas fuses and circuit breakers, of themselves, provide protection from excess current flow, they may not, unaided, provide complete protection against electric shock. A special type of circuit breaker, known as a 'residual current circuit breaker', helps here in that it provides a good standard of protection (i.e. at very low fault currents) against earth leakage faults, particularly at locations where effective earthing cannot otherwise be achieved.

Socket outlets and plugs

Socket outlets are the means of 'tapping' into the circuit to allow electricity to be used to supply an appliance. A plug, attached to a flexible cable supplying the apparatus, is used to 'plug in' to the socket outlet and thus form another circuit.

Fixed apparatus, such as large electric motors which drive equipment, do not require socket outlets and are supplied from their own circuit, controlled by a separate motor control which not only incorporates 'stop' and 'start' facilities, but also protection against overload.

Nature and purpose of electrical use

Equipment

Electricity is an efficient and convenient form of energy used for lighting, heating and power applications in occupational life. Consider, for example, just one group of uses, electric heating processes, which are finding ever-increasing applications in industry. Typical applications are electric furnaces and ovens, resistance and induction heating, electric arc melting, microwave heating, lasers and electron beams, and infrared and ultraviolet processes. Such equipment operates at frequencies from the standard 50 Hz for resistance heating to THz (THz = 10^{12} Hz) for lasers. Power ranges of typical heating processes are from 1 kW to 100 MW.

Use

The flexibility and ease of use of equipment brings its own dangers, and the design, installation, operation, control, use and safety requirements of the type of equipment just described are highly specialised. However, certain basic requirements apply for all electrical equipment:

(a) The equipment should be of adequate design, construction and performance for the intended use. Reference to appropriate British Standards and codes should ensure a satisfactory product.

(b) The equipment should be properly installed with a comparable standard for circuitry and control equipment to ensure that location and environmental conditions do not invalidate the specification.

(c) Adequate information should be supplied to the user so that when the equipment is properly used it will be safe and without risk to health.

(d) The equipment should be properly used for the purpose for which it is intended. Any unreasonable departure from the maximum loading, rating, temperature and other characteristics, or any misuse, is liable to lead to breakdown, plant damage or even shock, burns or fire.

(e) Operators should receive proper training, instruction and supervision so that skills match operational requirements.

(f) There should be regular on-site inspection of equipment and associated circuitry to detect any obvious deterioration or other defect.

(g) There should be an adequate maintenance system undertaken at the required periods by competent staff. Records of all maintenance, testing and repairs should be kept.

(h) Where work on high-voltage electrical apparatus is to be undertaken, a permit to work system should be operated (*see* Chapter 13). The permit to work should indicate that, prior to work commencing, the apparatus to be worked on is dead, is properly isolated from all live conductors, has been discharged of electricity and is efficiently connected to earth. The operation should be supervised and controlled by an authorised person, namely a person appointed in writing by management to issue and cancel permits to work, and who is competent, through training and experience, to undertake specific operations and/or work on high-voltage systems and apparatus.

(i) Where work of a general nature needs to be undertaken in high-voltage switchrooms by maintenance staff, general building workers or electrical engineers who may be competent persons, but not authorised persons, they must always be accompanied by and under the supervision of, an authorised person or senior authorised person.

G

Portable tools

Portable electrical apparatus is a common cause of electric shock and burns. Whereas some of these accidents occur as a result of defects in the apparatus itself, such as missing covers or lack of adequate earthing for exposed metallic parts, others result from defects in the flexible cable supplying the apparatus. Defects in cables arise through damage from vehicles and other objects passing over them as they lie strewn over the floor, or from excessive wear and tear from arduous construction site use. Plugs are sometimes wrongly connected or repairs are attempted with the equipment still switched on.

As with other apparatus, portable apparatus requires protection against overload and short circuit. Earth fault protection must be provided and the integrity of the earth core of flexible cables is crucial to such protection.

Various proprietary additional protective systems are available for portable apparatus, including reduced voltage systems (110 volts), circulating current earth monitoring, and residual current devices. The risk of electric shock can also be reduced by using all-insulated or double-insulated tools.

Flameproofing

The use of electrical equipment in flammable atmospheres is a highly specialised topic, detailed treatment of which is beyond the scope of this book.

Clearly, electrical equipment which can give rise to heat or sparking will provide a source of ignition for a flammable atmosphere of the right mixture, and it is therefore necessary to take adequate precautions. In industry, potentially flammable hazardous areas are classified according to a graded probability of an explosive gas or vapour concentration occurring. Three classifications, or zones, are usually referred to as follows:

(a) Zone 0 – which is a zone in which a flammable atmosphere is known to be continuously present, or present for long periods.

(b) Zone 1 – which is a zone in which a flammable atmosphere is likely to occur, at least during normal working.

(c) Zone 2 – which is a zone in which a flammable atmosphere is unlikely to occur save under abnormal conditions, such occurrence being of only short duration.

The particular zone is predeterrnined at the design stage and it is in everyone's interests to 'design out' the more hazardous zones so far as possible.

Types of protection

The type of protection required for electrical equipment which is to be placed at a particular location depends upon the zone. First, the possibility of excluding electrical equipment from the zone should be considered. If this is not a practicable proposition, consideration should be given to segregation of the electrical equipment within the zone, say by suitable barriers. However, the installation of electrical equipment in hazardous areas may be unavoidable, in which case special types of protection are available and designated as in the following examples.

Type N equipment

This equipment is designed for use in zone 2 conditions and is so constructed that, properly used, it will not ignite flammable atmospheres under normal conditions.

Type E equipment

This includes equipment such as transformers and squirrel cage motors. They do not normally produce sparks and are suitable for zone 2 conditions.

Pressurising or purging methods

These are methods of using pressurised or inert gases to prevent ingress of flammable gases, and are suitable for all zone conditions.

Intrinsically safe equipment

This is used for instrumentation and low-energy equipment. Intrinsically safe equipment is categorised 'ia' or 'ib' if it can be used in zones 1 and 2. Exceptionally, category ia may be used in zone 0 if sparking contacts are not part of the equipment.

Flameproof equipment

This is probably the most familiar term and is a well-tried British approach to the problem. The principle of 'flameproofing' is that the apparatus so described is constructed to withstand any explosion, within the apparatus, arising from ignition of flammable gas that may enter through the casing or other enclosure. All flanges and other joints of the casing or enclosure of the apparatus are well designed and constructed so as to prevent any internal ignition of gas from moving out of the enclosure and igniting a surrounding flammable atmosphere. By its nature, flameproof equipment is substantial, heavy and rather costly. It is intended for application in zones 1 or 2 but is not suitable for use in zone 0.

Marking

Each item of electrical equipment for use in hazardous atmospheres should be marked with appropriate group number for the range of gases it was designed for, and also with a temperature classification indicating the maximum allowable surface temperature of the equipment used.

Circuitry

Electrical apparatus for use in hazardous atmospheres has to be supplied with electricity via a cable circuit and control gear. As much as possible of this should be located outside the hazardous zone. Within the zone, however, the suitability of the metallic cable sheathing, armouring or conduit must be ensured, with provision of special cable sealing glands and fittings. Earthing requires particular attention.

Principles of welding – dangers and precautions

Electric arc welding may be either by alternating current (a.c.) or by direct current (d.c.). Given similar conditions, d.c. is safer but is probably less convenient.

An a.c. supply for welding sets is obtained from a transformer, but for d.c. a transformer must be supplemented by a rectifier. Alternatively, a d.c. generator could be employed.

Output voltages for welding sets are up to 100 volts for a.c. and 80 volts for d.c.

Particular care is required when electric arc welding, if electric shock or burns from contact with parts of the electrode holder or the live electrode are to be avoided. 'All insulated' electrode holders are recommended.

Principal components

The principal components of a welding circuit are:

(a) the welding lead, which carries the current from the welding set to the electrode and to the workpiece;
(b) the welding return lead, which returns the current from the workpiece to the welding set;
(c) the welding earth, which must be connected to the workpiece.

The welding return lead is essential to prevent current taking random paths say, via structural steelwork, metal pipes, railway lines, etc. A particular danger is that if a random path includes a loosely bolted connection, a high resistance is set up which will cause heat to generate as the current flows through it, possibly resulting in fire. The welding return lead should be firmly clamped to the workpiece.

The welding earth is necessary to maintain the workpiece at earth potential and thus prevent shock by safeguarding against the possibility of the workpiece becoming energised at mains voltage, or energised because of lack of continuity in the welding return.

A.c. welding sets can be provided with a special device that prevents the voltage of the electrode holder from rising above 50 volts during the periods when no arc is being struck. This is a recommended precaution.

Personal protection

Apart from electric shock and fire, there is also a danger to the eyes and skin from the effects of ultraviolet light from the arc. This can cause painful maladies. One is known as 'arc eye', a form of conjunctivitis, and the other is a nasty 'sunburn' effect on the exposed skin.

As an essential precaution, the welder's eyes (and those of his mate) must be protected by a suitable electric arc welding filter lens in the welding helmet or in a hand-held face shield. Additional protection must be provided for the skin in the form of suitable welder's gloves and sensible footwear and clothing.

For regular welding jobs in a workshop, a properly constructed welding booth is necessary to protect the eyes of people working or passing in the vicinity. When working at heights, screens and fire blankets are necessary to prevent, so far as possible, arc eye and/or sparks from the process falling on to people or on to flammable materials below.

Training of electrical personnel

The need for training

A key to safe working on electrical installations and apparatus is that all those engaged on such work shall have been adequately trained. The training, backed up by relevant experience in a working environment, should be aimed to ensure that the trainee acquires the skills, related knowledge and attitudes necessary for safe and efficient working. Both off-the-job and on-the-job training must be properly supervised, and at each stage the trainee must be made aware of the extent and limitations of the job in hand, the hazards that may be present, and the precautions that have to be taken to ensure safe working.

Precautions

Before any trainee commences work, the electrical circuit or apparatus to be worked on must be properly isolated and locked off from all sources of electricity supply, and proved dead by means of a suitable instrument. Where necessary, appropriate measures should be taken to ensure that the isolated parts cannot be re-energised during the course of the work. At an early stage of training it is necessary that trainees should be familiar with emergency procedures and with the treatment for electric shock.

Training programme

The training programme itself will normally follow the recommendations of the particular training board, company scheme or apprenticeship applicable, and will be incorporated in the academic syllabus of any associated training establishment, possibly leading to the attainment of a qualification. Particular topics that should be covered, at a fundamental level, to ensure a sound knowledge of electrical safety include:

(a) rudiments of electrical circuits, including the terms used, the function of earthing, and circuit protection;

(b) the physiological effects of electricity, electric shock and burns, including measures both to avoid shock and to treat its consequences;

(c) basic principles of generation, accumulation and dissipation of static electricity;

(d) safety measures for both dead and live work, and the application, limitation and use of voltage indicators and test instruments;

(e) use of insulated tools, rubber mats and rubber gloves;

(f) use of temporary screens, shrouds and covers to prevent inadvertent contact with live conductors and earthed metalwork;

(g) special hazards of particular locations such as hazardous atmospheres and work on switchboards, in confined or restricted spaces, or in the vicinity of crane trolley wires and the like;

(h) special hazards of particular plant or equipment, including portable tools;

(i) scope and application of company safety rules, including standardised safety procedures, permits to work and safety documentation;

(j) the law on electrical safety, relevant HSE Guidance Notes and BSI standards and codes of practice.

Updating training

Training in electrical safety is not a once and for all activity, but must be updated from time to time so as to take account of such matters as changes in the law or in methods of work, or following the introduction of new equipment, changed workshop procedures and new techniques. Persons so trained will also require adequate instructions for the job in hand. It must never be presumed that the trained person will always be familiar with likely hazards in a particular job and the means of overcoming them. This is where proper planning and control of the job comes in and reinforces the need for adequate information and instruction.

Principles and practice of electrical testing

The need for testing

G

Electrical testing is necessary to ensure that the design, construction and performance specifications of the items being tested are maintained at an adequate standard for the anticipated continued use. Electrical testing also enables faults to be detected so that remedial measures can be taken before the fault develops and damage or personal injury arises.

Types of testing

Routine testing of production lines

Equipment such as electric motors and various types of industrial, commercial and domestic apparatus requires individual or batch testing. Procedures include resistance and insulation tests for correct polarity, connection and operation. It is important that those undertaking the testing have adequate skills and proper equipment. The layout of the work area should be such as not to expose anyone to live and other dangerous parts. 'Anyone' means not only the tester but other persons present.

Testing of electrical installations

A prerequisite to such testing is visual inspection to ensure that all circuits and equipment comply with an accepted standard and are properly installed, and that circuit protection and earthing arrangements appear in order.

The method adopted for testing should not, of course, be liable to endanger people or plant. Thus the testing operation must be under proper control and the testing equipment suitable for the required use.

Ordinary test lamps, and leads with excessively exposed test prods or even bare ends, have caused numerous flashovers, and the dangerous practice of using metal lampholders is, unfortunately, still encountered. Properly designed, protected and approved test equipment, employing current limiting resistors, well-shrouded test prods, and properly insulated handles are available, and should always be used.

Apart from tests to determine correct polarity, the measurement of earth fault loop impedance is required to check that the earthing arrangements will ensure a safe and automatic disconnection of electricity supply in case of a fault.

It is a requirement of the IEE Wiring Regulations that on final circuits the circuit protective device (e.g. fuse or miniature circuit breaker) will, in the event of a fault to earth, operate within 0.4 seconds for socket outlets in outdoor circuits and within five seconds for circuits supplying fixed equipment. The time taken for protective devices to operate and clear a fault depends upon the device's time/current characteristics and the earth fault impedance, the maximum values of which should not be exceeded. Maximum values to meet a range of conditions are laid down in the IEE Wiring Regulations. For example, with a 0.4 second disconnection and a 50 amp rated fuse, the value should not exceed 0.6 ohms.

Testing of electronic telecommunications and similar equipment

The dangers of working on and testing components inside television sets, without first having disconnected the set from the power supply, are obvious. Repair and testing bays in workshops should be in earth-free areas, set apart, with special arrangements such as the provision of a single isolating transformer and proper test benches and equipment.

The testing of electronic equipment with earthed metal casings requires special precautions to remove the risk of electric shock through the casing. This may be achieved by provision of current-limiting devices for the testing circuit. Testing is a specialised operation and for further information and advice the latest edition of the Institution of Electrical Engineers' wiring regulations and HSE's (1980) booklet should be consulted.

Testing of portable electrical applicances

Approximately 25 per cent of accidents involving electricity are associated with portable electrical appliances. To ensure compliance with the general provisions of the Electricity at Work Regulations 1989, there is an implied duty on employers, in particular, to undertake some form of testing of electrical equipment. Further guidance and information on portable appliance testing is incorporated in the Memorandum of Guidance which accompanies the Regulations and HSE Guidance Note PM 32: *The safe use of portable appliances*.

Electrical equipment is very broadly defined in the Regulations as including anything used, intended to be used or installed for use, to generate, provide, transmit, transform, rectify, convert, conduct, distribute, control, store, measure or use electrical energy. *Portable appliances* include such items as electric drills, kettles, floor polishers and lamps, in fact any item that will connect into a 13 amp socket. 110 volt industrial portable electrical equipment should also be considered as portable appliances.

Safety of appliances

The operator or user of an electrical appliance is protected from the risk of electric shock by insulation and earthing of the appliance, which prevent the individual from coming into contact with a live electrical part. For insulation to be effective it must offer a high resistance at high voltages. In the case of earthing, it must offer a low impedance to any potentially high fault current that may arise.

A principal of electrical safety is that there should be two levels of protection for the operator or user and this results in two classes of appliance.

Class 1 appliances incorporate both earthing and insulation (earthed appliances), whereas *Class 2* appliances are doubly insulated. The testing procedures for Class 1 and Class 2 appliances differ according to the type of protection provided.

Appliance testing programmes

Testing should be undertaken on a regular basis and should incorporate the following:

(a) inspection for any visible signs of damage to or deterioration of the casing, plug terminals and cable sheath;

(b) an earth continuity test with a substantial current capable of revealing a partially severed conductor;

(c) high voltage insulation tests.

The test results should be recorded, thus enabling future comparisons to determine any deterioration or degradation of the appliance.

Control system

The control system should include:

(a) clear identification of the specific responsibility for appliance testing;

(b) maintenance of a log listing portable appliances, date of test and a record of test results;

(c) a procedure for labelling appliances when tested with the date for the next inspection and test.

Any appliance that fails the above tests should be removed from use.

Frequency of testing

An estimation of the frequency of testing must take into account the type of equipment, its usage in terms of frequency of use and risk of damage, and any recommendations made by the manufacturer/supplier.

The use of portable appliance testing equipment

Electrical tests of appliances should confirm the integrity or otherwise of earthing and insulation. To simplify this task a competent person may use a proprietary portable appliance testing (PAT) device. In this case, the unit under test is plugged into the socket of the testing device. Some tests are carried out through the plug, others through both the plug and an auxiliary probe to the casing of the appliance.

The tests

Two basic tests are offered by PAT device.

Earth bond test

This applies a substantial test current, typically around 25 amps, down the earth pin of the plug to an earth test probe which should be connected by the user to any exposed metalwork on the casing of the unit under test. From this the resistance of the earth bond is determined by the PAT device.

Insulation test

This applies a test voltage, typically 500 volts DC, between the live and neutral terminals bonded together and earth, from which the insulation resistance is calculated by the PAT device.

Other tests

Flash test

This tests the insulation at a higher voltage, typically 1.5 kV for Class 1 appliances and 3 kV for Class 2 appliances. From this test the PAT device derives a leakage current indication. This is a more stringent test of the insulation that can provide an early warning of insulation defects developing in the appliance. It is recommended that this test should not be undertaken at a greater frequency than every three months to avoid overstressing the insulation.

Load test

This test measures the load resistance between live and neutral terminals to ensure that it is not too low for safe operation.

Operation test

This is a further level of safety testing which proves the above tests were valid.

Earth leakage test

This is undertaken during the operation test as a further test of the insulation under its true working conditions. It should also ensure that appliances are not responsible for nuisance tripping of residual current devices (RCDs).

Fuse test

This will indicate the integrity of the fuse and that the appliance is switched on prior to other tests.

Earthed Class 1 appliances

The following tests are undertaken:

(a) earth bond test;

(b) insulation test;

(c) in certain cases, flash test.

Double-insulated Class 2 appliances

The following tests are undertaken:

(a) insulation test;

(b) flash test.

Conclusion

Electricity is a technical subject, a comprehensive study of which requires complex theory and a treatment far beyond the scope of this chapter. Even an attempt at a basic approach has its limitations, particularly if explanation has necessarily to be oversimplified. It is hoped that readers will, nevertheless, understand the problem of safety and will be encouraged to take their studies of electrical matters much further. Some excellent textbooks and courses are available for this purpose.

The safe use of electricity requires competence, concern, skill, related knowledge and care not only on the part of those who may be exposed to danger or who supervise and manage. Such 'care and competence' qualities are also required by those who design, install and maintain electrical equipment and – dare we say it – by those who practise safety and health in areas where electricity is generated, distributed and used.

37 Construction safety

Construction safety is a very broad area covering work above and below ground, demolition, the use of machinery, plant and vehicles on site, fire protection procedures, the use of explosives and specialised operations, such as tunnelling. In the past its safety record was poor; a fact which led to the need for specific regulation by the enforcement authorities. Construction operations are specifically covered by:

(a) Construction (Design and Management) Regulations 1994 (which must be interpreted in conjunction with duties under the Management of Health and Safety at Work Regulations 1992).

(b) Construction (Health, Safety and Welfare) Regulations 1996 together with –

  (i) Lifting Operations and Lifting Equipment Regulations 1998.

  (ii) Constuction (Head Protection) Regulations 1989.

G

Accidents in the construction industry

The most common fatal type of accident in construction remains falling from a height, and the most common non-fatal type of accident is that of manual handling. A more complete categorisation of accidents is as follows.

Ladder accidents

These may be due to the ladder slipping outwards at the base, to the use of defective or rotten wooden ladders, or as a result of overreaching when working at the top of a ladder, causing the ladder to fall sideways. Even if a ladder is of sound construction and in good condition, it is still dangerous if improperly used. Placing it too close to, or too far away from, a wall can have tragic consequences and anyone using a ladder should be aware of the '1 out, 4 up' rule. This rule is that the vertical height from ground to point of rest of the ladder should, wherever practicable, be four times the distance between the base of the vertical dimension and the foot of the ladder. Failure to adopt this rule can result in the ladder becoming unstable (*see* Fig. 37.1).

A standing ladder must be securely fixed near to its upper resting place (or its upper end if vertical). Where such fixing is impracticable, the ladder must be fixed at or near its lower end. If this is also impracticable, a person must be stationed at the foot of the ladder to prevent it from slipping. Except where there is an adequate handhold, ladders must rise to a height of at least 1,070 mm above the landing place or above the highest rung reached by the feet of persons using the ladder. When this is impracticable, the ladder must rise to the greatest practicable height.

Fig. 37.1 ◆ The '1 out, 4 up' rule for ladders.

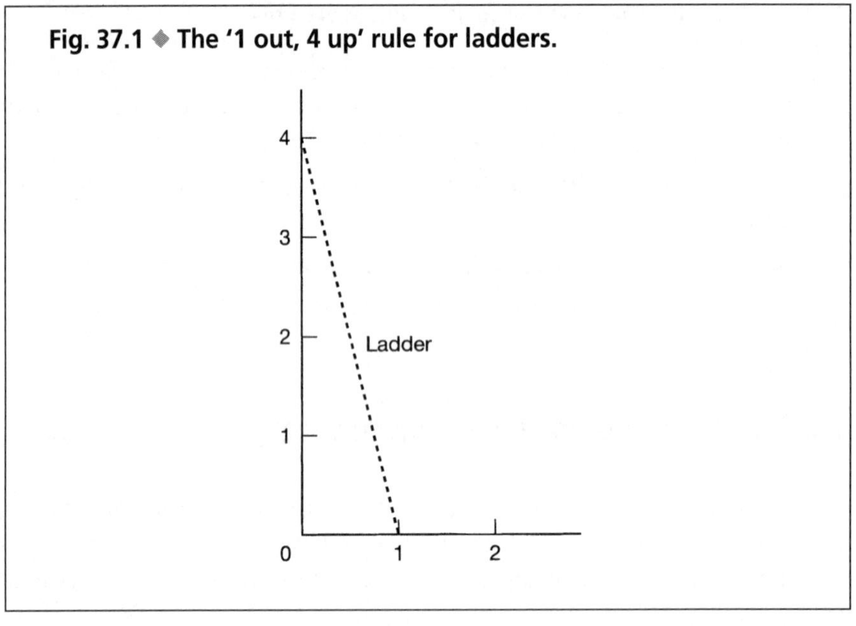

Falls from working platforms

Working platforms include fixed scaffolds and lightweight, readily assembled mobile access equipment commonly used in factories and commercial premises for maintenance work, painting, cleaning and installation work. Causes of accidents include the use of unfenced and inadequately fenced platforms, insufficient or inadequate boarding, or even boarding which is rotten and defective. The need for structural stability of lightweight mobile platforms is important, particularly where they may be subject to wind loading. Mobile platforms should be stationed on a firm level base and, where practicable, tied to the structure to prevent sideways movement. They should incorporate a wheel-locking device. For routine work the following height: base ratios should be taken into account: work indoors – 3.5:1; work outdoors – 3:1.

Falls of materials

Small objects such as hammers, rooftiles, bricks and even six-inch nails, dropped from a height, can have serious consequences if they hit someone below. The main causes of falls of materials are poor housekeeping by people working above, absence of toe boards and edge barriers on working platforms, incorrect assembly of gin wheels for raising materials, incorrect hooking and slinging prior to raising and, in many cases, failure to install catchment platforms or 'fans' which are designed to catch small objects which may fall during construction. The provision of fans is particularly appropriate where construction work may take place above or adjacent to a public thoroughfare.

Falls from pitched roofs or through fragile roofs

Workers sliding down a slippery pitched roof, particularly if it is wet after rain or covered in moss growth, and falling to the ground below is a common construction accident. There is often a need, therefore, for eaves and roof edge protection, coupled with the use of crawl boards or firmly fixed ladders. The practice of stacking building materials on roofs should be prohibited, as should working on roofs without crawl boards.

Falls through openings in flat roofs and floors

The principal cause of this type of accident is failure to cover openings or to replace covers after use. In other cases, covers have not been properly marked 'Hole Below', with the result that workers have walked on them, causing the cover to collapse. Edge protection, in the form of rails and toe boards, is preferable to covers as the worker is given a clear indication of the hazard involved. (Numerous accidents are on record where two employees have been involved in removing a cover to a floor hole, such as a large sheet of plywood, without actually realising that the hole existed under the cover. The employee at the leading edge of the horizontally held sheet has walked off in a certain direction unharmed, but his colleague at the trailing edge has followed him and fallen through the hole which he could not see.)

Collapse of excavations

Absence of timbering to trenches, or insufficient timbering in relation to depth and width of the trench and nature of the surrounding ground, are the main causes of excavation collapse, resulting in workers being buried alive. Careful examination of the subsoil, particularly where water and/or shifting sand may be present, is essential prior to excavation. A further cause of trench collapse is the stacking of excavated ground and building materials such as pipes, blocks and bricks too close to the edge of the excavation, causing excessive loading on supporting timbers.

Site transport

Many accidents result from site transport. They include men falling off machines not designed to carry passengers, such as dumper trucks, being run down or crushed by reversing lorries, or crushed when vehicles overturn. Great care is needed in the maintenance of all site vehicles, particularly their braking and reversing systems. Furthermore, only competent and trained personnel should drive site vehicles. Site road maintenance is very important, as poor standards of maintenance and housekeeping can cause skidding and overturning. Site transport can be subject to misuse. Where, for instance, an employee gives another employee, whether from the same workforce or not, a lift on site in a dumper truck, as a result of which the latter is injured or killed, the former employee's employer is almost certainly liable at civil law for the injury, even though the driver was acting contrary to orders. An act committed during the course of, and within the scope of, employment will attract the principle of vicarious liability. (*See* the case of *Rose* v. *Plenty* [1976] 1 AER 97.)

Machinery and powered hand tools

All moving parts of machinery and plant with which operatives may come into contact should be securely guarded. This includes power take-offs, engines, cooling fans and belt drives, together with various items of woodworking machinery such as circular saws and planing machines.

Electric hand tools should comply with BS 2769:1964 and, unless 'all insulated' or 'double insulated', must be effectively earthed. Reduced voltage distribution for portable tools and temporary lighting, using 110 volt mains isolation transformers with the secondary winding centre tapped to earth, is always recommended.

Housekeeping

Poor standards of housekeeping are frequently a contributory cause of accidents resulting in trips and falls over debris and building materials, and puncture wounds to the feet from nails projecting from pieces of timber left lying around: the careless use and storage of chemical-based compounds and other materials is another hazard. Inadequate facilities for the storage of flammable refuse is frequently a contributory cause of fires on sites.

Construction (Health, Safety and Welfare) Regulations 1996

G

These regulations impose requirements with respect to the health, safety and welfare of persons at work carrying out 'construction work' as defined and of others who may be affected by that work. Specified regulations apply in respect of 'construction work carried out on a construction site' as defined. Where a workplace on a construction site is set aside for purposes other than construction, the regulations do not apply.

Subject to specific exceptions, the regulations impose requirements on duty holders, i.e. employers, the self-employed and others who control the way in which construction work is carried out. Employees have duties in respect of their own actions. Every person at work has duties as regards co-operation with others and the reporting of danger.

The principal duties of employers, the self-employed and controllers

Safe places of work

A general duty to ensure a safe place of work and safe means of access to and from that place of work.

Specific provisions include the following.

Precautions against falls

(a) Prevention of falls from heights by physical precautions or, where this is not possible, provision equipment that will arrest falls.

(b) Provision and maintenance of physical precautions to prevent falls through fragile materials.

(c) Erection of scaffolding, access equipment, harnesses and nets under the supervision of a competent person.

(d) Specific criteria for using ladders.

Falling objects

(a) Where necessary to protect people at work and others, taking steps to prevent materials or objects from falling.

(b) Where it is not reasonably practicable to prevent falling materials, taking precautions to prevent people from being struck, e.g. covered walkways.

(c) Prohibition of throwing any materials or objects down from a height if they could strike someone.

(d) Storage materials and equipment safely.

Work on structures

(a) Prevention of accidental collapse of new or existing structures or those under construction.

(b) Ensuring any dismantling or demolition of any structure is planned and carried out in a safe manner under the supervision of a competent person.

(c) Only firing explosive charges after steps have been taken to ensure that no one is exposed to risk or injury from the explosion.

Excavations, cofferdams and caissons

(a) Prevention of the collapse of ground both in and above excavations.

(b) Identification and prevention of risk from underground cables and other services.

(c) Ensuring cofferdams and caissons are properly designed, constructed and maintained.

Prevention or avoidance of drowning

(a) Taking steps to prevent people falling into water or other liquid so far as is reasonably practicable.

(b) Ensuring that personal protective and rescue equipment is immediately available for use and maintained, in the event of a fall.

(c) Ensuring sure transport by water is under the control of a competent person.

Traffic routes, vehicles, doors and gates

(a) Ensuring construction sites are organised so that pedestrians and vehicles can move both safely and without risks to health.

(b) Ensuring routes are suitable and sufficient for the people or vehicles using them.

(c) Prevention or control of the unintended movement of any vehicle.

(d) Ensuring arrangements for giving a warning of any possible dangerous movement, e.g. reversing vehicles.

(e) Ensuring safe operation of vehicles including prohibition of riding or remaining in unsafe positions.

(f) Ensuring doors and gates which could prevent danger, e.g. trapping risk of powered doors and gates, are provided with suitable safeguards.

Prevention and control of emergencies

(a) Prevention of risk from fire, explosion, flooding and asphyxiation.

(b) Provision of emergency routes and exits.

(c) Provision of arrangements for dealing with emergencies, including procedures for evacuating the site.

(d) Where necessary, provision of fire-fighting equipment, fire detectors and alarm systems.

Welfare facilities

(a) Provision of sanitary and washing facilities and an adequate supply of drinking water.

(b) Provision of rest facilities, facilities to change and store clothing.

Site-wide issues

(a) Ensuring sufficient fresh or purified air is available at every workplace, and that associated plant is capable of giving visible or audible warning of failure.

(b) Ensuring a reasonable working temperature is maintained at indoor workplaces during working hours.

(c) Provision of facilities for protection against adverse weather conditions.

(d) Ensuring suitable and sufficient emergency lighting is available.

(e) Ensuring suitable and sufficient lighting is available, including the provision of secondary lighting where there would be a risk to health or safety if the primary or artificial lighting failed.

(f) Maintaining construction sites in good order and in a reasonable state of cleanliness.

(g) Ensuring the perimeter of a construction site to which people, other than those working on the site could gain access, is marked by suitable signs so that its extent can be easily identified.

(h) Ensuring all plant and equipment used for construction work is safe, of sound construction and used and maintained so that it remains safe and without risks to health.

Training, inspection and reports

(a) Ensuring construction activities where training, technical knowledge or experience is necessary to reduce risks of injury are only carried out by people who meet these requirements or, if not, are supervised by those with appropriate training, knowledge or experience.

(b) Before work at height, on excavations, cofferdams or caissons begins, ensuring the place of work is inspected, (and at subsequent specified periods) by a competent person, who must be satisfied that the work can be done safely.

(c) Following inspection, ensuring written reports are made by the competent person.

Work on scaffolds

The basic requirements for safe working on scaffolds are:

(a) correct erection of the scaffold;

(b) a suitable system for the inspection and maintenance of the scaffold, taking into account the frequent likelihood of alterations to the scaffold as the work progresses.

Specific safety requirements relating to work on scaffolds are incorporated in Schedules 1 and 2 of the Construction (Health, Safety and Welfare) Regulations 1996.

Schedule 1 (Regs 6(2), 6(3)(a) and 8(2))

Requirements for guard rails etc.

1. A guard rail, toe board, barrier or other similar means of protection shall:

(a) be suitable and of sufficient strength and rigidity for the purpose or purposes for which it is being used;

(b) be so placed, secured and used as to ensure, so far as is reasonably practicable, that it does not become accidentally displaced.

2. Any structure or any part of a structure which supports a guard rail, toe board, barrier or other similar means of protection or to which a guard rail, toe board, barrier or other similar means of protection is attached shall be of sufficient strength and suitable for the purpose of such support or attachment.

3. The main guard rail or other similar means of protection shall be at least 910 mm above the edge from which any person is liable to fall.

4. There shall not be an unprotected gap exceeding 470 mm between any guard rail, toe board, barrier or other similar means of protection.

5. Toe boards or other similar means of protection shall be at not less than 150 mm high.

6. Guard rails, toe boards, barriers or other similar means of protection shall be so placed as to prevent, so far as is reasonably practicable, the fall of any person, or any material or object, from any place of work.

Schedule 2 (Regs 6(2), 6(3)(b) and 8(2))

Requirements for working platforms

Interpretation

1. In this Schedule, *supporting structure* means any structure used for the purpose of supporting a working platform and includes any plant and equipment used for that purpose.

Condition of surfaces

2. Any surface upon which any supporting structure rests shall be stable, of sufficient strength and of suitable composition safely to support the supporting structure, the working platform and any load intended to be placed on the working platform.

Stability of supporting structure

3. Any supporting structure shall:

(a) be suitable and of sufficient strength and rigidity for the purpose or purposes for which it is being used;

(b) be so erected and, where necessary, securely attached to another structure as to ensure that it is stable;

(c) when altered or modified, be so altered or modified as to ensure that it remains stable.

Stability of working platform

4. A working platform shall:

(a) be suitable and of sufficient strength and rigidity for the purpose or purposes for which it is being used;

(b) be so erected and used as to ensure, so far as is reasonably practicable, that it does not become accidentally displaced so as to endanger any person;

(c) when altered or modified, be so altered or modified as to ensure that it remains stable;

(d) be dismantled in such a way as to prevent accidental displacement.

Safety on working platforms

5. A working platform shall:

(a) be of sufficient dimensions to permit the free passage of persons and the safe use of any equipment or materials required to be used and to provide, so far as is reasonably practicable, a safe working area having regard to the work there being carried out;

(b) without prejudice to paragraph (a), be not less than 600 mm wide;

(c) be so constructed that the surface of the working platform has no gap giving rise to the risk of injury to any person or, where there is a risk to any person below the platform being struck, through which any material or object could fall;

(d) be so erected and used, and maintained in such condition, as to prevent, so far as is reasonably practicable –

(i) the risk of slipping or tripping; or

(ii) any person being caught between the working platform and any adjacent structure;

(e) be provided with such handholds and footholds as are necessary to prevent, so far as is reasonably practicable, any person slipping from or falling from the working platform.

Loading

6. A working platform and any supporting structure shall not be loaded so as to give rise to a danger of collapse or to any deformation which could affect its safe use.

Roof work and work at high level

Accidents associated with people falling from or through roofs, or from high-level working positions, are common in the construction industry and maintenance activities. Many of these accidents could be avoided by the operation of a safe system of work. Such systems include careful planning prior to the operation, use of trained and experienced operators, effective supervision and control, and used of permit to work systems where there is a high degree of foreseeable risk coupled with complex precautions. (See HSE Guidance Note GS10, *Roofwork: Prevention of Falls*.)

Two specific aspects merit closer attention, namely:

(a) the means of access to the roof or high-level working position;

(b) the system of work adopted once the working position has been reached.

Access by ladder

The most common means of access is by portable ladder. There are many types: they can be manufactured from aluminium, fibreglass or timber. Most accidents, however, arise in connection with timber ladders. The following points are relevant to the safe use of ladders:

(a) The ladder should be strong enough for the work to be undertaken.

(b) Only sound ladders should be used. Ladders with split uprights, broken feet or loose rungs, or which have become distorted, should not be used but rather destroyed. Ladders should not be painted since this can hide any defect which may have developed. However, this prohibition does not extend to ladders being varnished or treated with wood preservative.

(c) A ladder should be long enough for the work to be undertaken.

(d) The ladder should be placed at the correct angle. (*See* earlier in this chapter, 'Ladder accidents'.)
(e) Support for the ladder should be strong enough to withstand the thrust imposed by the ladder.
(f) 'Home-made' ladders, the rungs of which depend for support solely on nails, must not be used.
(g) Subject to size, the ladder should be securely fixed near to its upper resting place or, if impracticable, at its lower end, to prevent it slipping sideways. Someone should foot the ladder until secured.
(h) All ladders should have clearly marked identification numbers for company records and inspection purposes. It is recommended that they be inspected at regular intervals of not more than six months and details of this inspection entered on a record card.
(i) Ladders should not be stored on wet ground or left exposed to weather. They should be stored at normal ambient temperature under cover to prevent warping, and never be hung from brackets, as this tends to separate the rungs from the stiles.

The above points are important in preventing typical ladder accidents. However, not all work at high level is undertaken through the use of ladders. In many cases this work is undertaken using personal suspension equipment and other equipment designed to arrest falls. Schedules 3, 4 and 5 of the Construction (Health and Safety) Welfare Regulations (CHSWR) lay down requirements for work at high level.

Schedule 3 (Reg 6(3)(c))

Requirements for personal suspension equipment

1. Personal suspension equipment shall be of suitable and sufficient strength for the purpose or purposes for which it is being used having regard to the work being carried out and the load, including any person, it is intended to bear.

2. Personal suspension equipment shall be securely attached to a structure or to plant and the structure and the means of attachment thereto shall be suitable and of sufficient strength and stability for the purpose of supporting that equipment and the load, including any person, it is intended to bear.

3. Suitable and sufficient steps shall be taken to prevent any person falling or slipping from personal suspension equipment.

4. Personal suspension equipment shall be installed or attached in such a way as to prevent uncontrolled movement of that equipment.

Schedule 4 (Reg 6(3)(d))

Requirements for means of arresting falls

1. In this Schedule, 'equipment' means any equipment provided for the purpose of arresting the fall of any person at work and includes any net or harness provided for that purpose.

2. The equipment shall be suitable and of sufficient strength safely to arrest the fall of any person who is liable to fall.

3. The equipment shall be securely attached to a structure or to plant and the structure or plant and the means of attachment thereto shall be suitable and of sufficient strength and stability for the purpose of safely supporting the equipment and any person who is liable to fall.

4. Suitable and sufficient steps shall be taken to ensure, so far as is practicable, that in the event of a fall by any person the equipment does not itself cause injury to that person.

Schedule 5 (Reg 6(6))

Requirements for ladders

1. Any surface upon which a ladder rests shall be stable, level and firm, of sufficient strength and of suitable composition safely to support the ladder and any load intended to be placed on it.

2. A ladder shall:

(a) be suitable and of sufficient strength for the purpose or purposes for which it is being used;

(b) be so erected as to ensure that it does not become displaced;

(c) where it is of a length when used of 3 metres or more, be secured to the extent that it is practicable to do so and where it is not practicable to secure the ladder a person shall be positioned at the food of the ladder to prevent it slipping at all times when it is being used.

3. All ladders used as a means of access between places of work shall be sufficiently secured so as to prevent the ladder slipping or falling.

4. The top of any ladder used as a means of access to another level shall, unless a suitable alternative handhold is provided, extend to a sufficient height above the level to which it gives access so as to provide a safe handhold.

5. Where a ladder or run of ladders rises a vertical distance of nine metres or more above its base, there shall, where practicable, be provided at suitable intervals sufficient safe landing areas or rest platforms.

Working on roofs

Flat roofs and roofs up to 10 degrees pitch

The main hazard is falling from the edge of the roof, especially where work is being carried out, or through openings in the roof. Guard rails and toe boards should be provided in these circumstances. One method of guarding the edge of a flat roof is with freestanding frames (*see* Fig. 37.2). A series of triangulated tubular steel frames are anchored to the roof by means of precast concrete counterweights attached to the inner ends of the frames. The frames are placed at approximately 2.4 metre centres. Guard rails and toe boards are attached to the frames, the toe boards being held in position by purpose-made brackets. The frames can be moved to allow roofing felt to be laid. The frames can form part of a roofer's equipment

Fig. 37.2 ◆ Use of freestanding frames

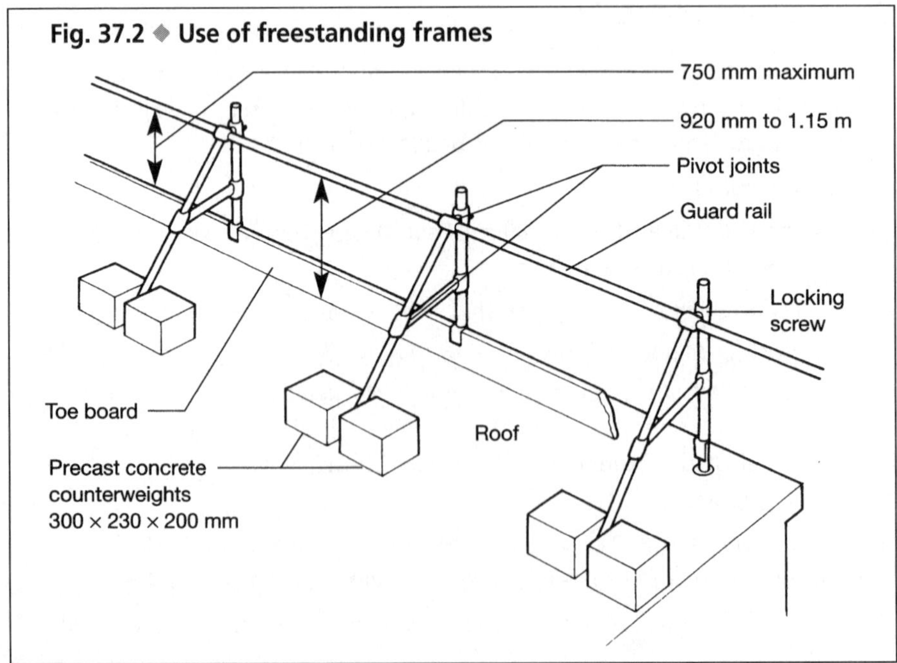

which can be carried from site to site. As the concrete counterweights each weigh about 30 kg, care must be exercised when lifting or moving them manually.

Sloping roofs over 10 degrees pitch

Three hazardous situations exist, namely working close to the edge, at the edge, or at some part where a person is likely to slide down the roof and fall from the edge. In the latter case, the degree of danger depends upon the pitch of the roof, the possibility of the surface being made slippery by ice, rain, moss or moisture, and the type of footwear worn by the individual. It can also be hazardous working on the roofs in extreme weather conditions such as high winds. In such cases, interlocking crawl boards or roof ladders should be provided and used. (*Provided* means 'made available in a reasonably accessible place' by the employer; *Ginty* v. *Belmont Building Supplies Ltd* [1959] 1 AER 414.) This rule may well have been superseded by HSWA, sec 2(2)(*c*), which requires employers to inform, train, instruct and supervise their employees in the use of safety equipment, as well as merely 'providing' the requisite equipment.

Fragile roofs

Many accidents are associated with workers falling through fragile roofs. In many cases these are fatal accidents caused by a false sense of security created through long experience of working on such roofs. Asbestos cement roof covering, for instance, may be capable of carrying some distributed load and give the impression of a surface which is solid enough to bear a man's weight. It will not, however, carry a concentrated load such as that applied by the heel of a man walking, or the shock load imposed by a man stumbling and falling. Many types of single thickness asbestos cement sheeting simply shatter without warning, and experience shows that walking along the lines of the sheeting bolts, i.e. on the purlins, does not always offer sufficient security. Asbestos cement becomes brittle with age.

Before any work is undertaken on a roof, or where a roof may be used for access, it is essential to identify the parts covered by fragile materials and to decide on precautions to be taken. For instance, on roofs covered with fragile material, at least two crawl boards or roof ladders should be used so that the worker will always have one ladder at his side to stand on when moving the other ladder to a new working position. Where a valley or parapet gutter is used as means of access and the adjacent roof is of fragile material, suitable precautions, such as the provision of covers, should be taken to prevent a person falling through the fragile material. Valley gutters which are overhung by roof sheets, to such an extent that there is inadequate clearance for a man's feet, should not be used for access along the roof. Prominent warning notices (e.g. 'CAUTION – FRAGILE ROOF – USE CRAWL BOARDS') should be fixed at the approaches to each fragile roof. With large buildings, such notices should be displayed at 50 metre intervals along all faces of the building and at all normal access points, e.g. valleys and junctions with vertical structural members.

Practical precautions when working at high levels

Work at high levels covers a wide range of activities from the simple replacement of roof tiles to the complete stripping and recovering of existing roofs. With each job, the requisite practical precautions must be considered before anyone is permitted to start work.

Consideration of weather conditions

Hazards resulting from adverse weather conditions must be anticipated. In windy weather, work involving the fixing of roof sheets, in particular, should not be carried out, as a man can easily be blown off balance whilst carrying a sheet up to or on the roof. The presence of moisture, ice or snow can turn an apparently safe foothold into a hazard.

Safety harnesses and belts

Where harnesses and belts are to be used because safe means of access cannot be arranged, suitable anchorage points must be provided which are capable of sustaining an anticipated shock load. Excessive shock loads on the anchorage, the equipment and the wearer should be avoided. 'Free all' distance should not be more than two metres in the case of a safety harness or 0.6 metres where a safety belt is used. Use of inertia-controlled reels or slides designed to allow greater freedom of movement without excessive slackness in the rope, and to act as an arrestor if a sudden pull is exerted as the result of a fall, should be considered. (Guidance on standards for the design and construction of safety harnesses and belts is given in BS 1397 *Industrial Safety Belts, Harnesses and Safety Lanyards*.) The use of safety harnesses and safety anchorage points is particularly relevant to cleaning outer surfaces of windows of high-rise office and industrial properties. (For a more detailed examination of this subject, *see* later in this chapter, 'Window cleaning'.)

Safety nets

Where safety nets are installed to catch men and materials falling from working positions, the nets should be rested as close to the working level as possible.

Personal protective equipment

Like all construction workers, workers engaged in high-level work should be provided with safety helmets and other personal protective equipment appropriate to the task, e.g. gloves, one-piece overalls and non-slip safety shoes or boots. Safety helmets should be fitted with chin straps. Failure to wear such personal protective clothing can result in:

(a) prosecution (*see* Chapter 4);

(b) job dismissal, including summary dismissal (*see* Chapter 5);

(c) loss, either partial or total, of compensation for injury (*see* Chapter 2).

Work below ground level

This covers activities such as tunnelling, excavating and the digging of shafts, all of which have been responsible for many deaths. Often it involves working in confined spaces, which may require the operation of permit to work systems (*see* Chapter 13). The following precautions are necessary in any type of work below ground, irrespective of the depth of the excavation:

(a) Except where an excavation is relatively shallow, or where the slopes of the sides make it impossible for slides to occur, supports – e.g. poling boards, walings, struts, sheet piling – whether constructed in timber or other materials, must be available and in position as from the start of work. Supports must be of sound construction and inspected daily by a competent person.

(b) Construction of the support system and any work involving alterations to, additions to or dismantling of the system, must be carried out under the supervision of a competent person.

(c) In the event of flooding, there must be adequate means of egress from the excavation to a position of safety.

(d) Excavation work must not affect the security or stability of a structure.

(e) Excavations more than two metres deep must be adequately fenced by barriers to prevent people falling into them (To ensure maximum safety, barriers should be provided for lesser depths.)

(f) Excavations and approaches to them should be well lighted.

(g) No materials (or excavated ground), plant or equipment should be located near the edge of an excavation owing to the risk of the sides collapsing.

(h) Cables and underground services should be located before any form of excavation is undertaken. Careful consultation with the supply authorities, use of cable detection equipment and hand digging of trial holes may well be necessary.

(i) The possibility of ingress of toxic or asphyxiant gas must be considered. Many fatalities have occurred, for example, as a result of carbon dioxide accumulation in pits, chambers, ducts, etc. This gas arises from chalk or limestone in the ground reacting with acidic rain. Where gas ingress is suspected, the air purity must be tested by a competent person before entry of workers is permitted, and a permit to work system should be operated.

Demolition

Demolition is one of the most dangerous activities undertaken in the construction industry. Yet there will always be a need for it as buildings deteriorate with age to the point of instability and where existing buildings must be replaced in the cause of redevelopment. For maximum safety, the following features of demolition work need consideration.

Predemolition survey

Prior to demolition taking place, a safe system of work must be established. The system will be determined by a predemolition survey – including perusal of the original building plans, if they are still in existence – undertaken by a competent person. The use of a demolition hazard check-list is recommended (*see* Fig. 37.3). This survey should identify:

(a) the nature and method of construction of the building;

(b) the arrangement of buildings adjacent to that for demolition and the condition of this adjoining property;

(c) the location of underground services, e.g. water mains, electricity cables, gas pipes, drains, sewers, telephone cables, etc.;

Fig. 37.3 ◆ Demolition hazard check-list

Services

☐ Gas ☐ Electricity ☐ Water
☐ Telephone ☐ Sewers ☐ Others

Glass

☐ Doors/windows ☐ Partitions ☐ Skylights

Roof

☐ Fragile ☐ Weak

Basement

☐ Sumps/wells ☐ Extensions under pavements/other buildings

Structural timber

☐ Decayed ☐ Damaged

Adjacent buildings

☐ Bracing or shoring required ☐ Weather-proofing required

Overhead hazards

☐ Materials ☐ Cables

Industrial plant

☐ Carboniferous dust deposits ☐ Gas tests required

After-effects of fire, flooding, blasting

☐ Bracing or shoring required

(d) the previous use of premises, e.g. for the storage of inflammable substances;

(e) the presence of dangerous substances, e.g. asbestos lagging;

(f) the method of bonding of the main load-bearing walls;

(g) the system of shoring or other supports necessary during demolition;

(h) the presence of cantilevered structures, their form of construction and the nature of the danger;

(i) the presence of basements, cellars, vaults or other spaces affecting the structure of adjoining properties;

(j) the potentially dangerous effects of removing superstructure stabilising loads from an old basement or vault retaining walls;

(k) the presence of storage tanks below and above ground, and the nature of their contents;

(l) the actual sequence of operations, which should generally take place in the reverse order of building erection.

Demolition work should be undertaken in accordance with the BS 6187 Code of Practice for Demolition by registered demolition contractors listed in the *Demolition and Dismantling Industry.* A 'Guide to typical methods of demolition' (Table 1 in BS 6187) is shown as Table 37.1.

Action prior to demolition

Prior to actual demolition, a number of actions must be taken. Form 10 (Notice of Commencement of Building Operations or Works of Engineering Construction) must be submitted to the local offices of the HSE. Local authorities, statutory undertakings (Gas Board, Water Authority, etc.) and the owners of adjoining property must be notified and consulted. Services, such as water, gas and electricity, must be isolated. A competent supervisor, with experience of this type of work, must be appointed to take charge of the operation, and all persons employed on the site adequately briefed. All dangerous areas, particularly those affecting members of the public, should be fenced off or barricaded and appropriate warning notices displayed.

NOTE. The effect of such notices, in so far as they might attempt to 'contract out' of liability in the event of death and/or personal injury, either to employees or to an indirect labour force on site, or even to members of the public, is severely inhibited by the Unfair Contract Terms Act 1977, sec 2(1). This Act makes it illegal to contract out of negligence, whether by means of a notice or otherwise.

'Fans' or catching platforms should be installed not more than six metres below the working level, when there is risk to the public. Personal protection equipment, including safety helmets with chin straps, goggles, heavy duty gloves and safety boots with steel insoles, must be provided and worn during the total period when demolition is in progress. Respiratory protection, together with the use of safety belts or harnesses, may also be necessary.

Action during demolition

Where possible, demolition should be carried out in the reverse order of building erection. No isolated freestanding wall should be left unless judged to be secure by the competent person in charge. Scaffold working platforms should be used, all refuse and debris being removed from these

Table 37.1 ◆ BS618.7: 1982 A guide to typical methods of demolition (see Note 1)

Type of structure	*Type of construction*	*Method of demolition*			
		Detached building isolated site	*Detached building confined site*	*Attached building isolated site*	*Attached building confined site*
Small and medium two-storey buildings	Loadbearing walls	ABCDM	ABDM	ABDM	ADM
Large buildings three storeys and over	Loadbearing walls	ABDM	ABDM	ABDM	AD
	Loadbearing walls with wrought iron and cast iron members	ABDM	AM	AM	AM
Framed structures	Structural steel	ACM	AM	AM	AM
	In situ reinforced concrete	ADM	ADM	ADM	AM
	Precast reinforced concrete	ADM	ADM	ADM	AM
	Prestressed reinforced concrete	ADM See 7.3.4(BS6187: 1982)	ADM	ADM	AM
	Composite (structural steel and reinforced concrete)	ADM	ADM	ADM	AM
	Timber	ABCDM	ABDM	ABDM	ABDM
Independent cantilevers (canopies, balconies and staircases)		ADM	ADM	ADM	ADM
Bridges		ABCDM	ABCDM	AM	AM
Masonry arches		ACDM	ACDM	ACDM	ACDM
Chimneys	Brick or masonry	ACD	A	ACD	A
	Steel	AC	A	A	A
	In situ and precast reinforced concrete	AD	A	AD	A
	Reinforced plastics	AC	A	A	A
Spires		ACD	A	A	A
Pylons and masts		AC	A		

Table 37.1 ◆ continued

Type of structure	*Type of construction*	*Method of demolition*			
		Detached building isolated site	*Detached building confined site*	*Attached building isolated site*	*Attached building confined site*
Petroleum tanks (underground)					
Above ground storage tanks					
Chemical works and similar establishments					
Basements					
Special structures					

Note 1. This table is a general guide to the methods of demolition usually adopted in particular circumstances. In addition, subject to local restraints, explosives may be used by experienced personnel in many of the circumstances listed. This table should be read in conjunction with the main text. The indication of a particular method does not necessarily preclude the use of another method, or the use of several methods in combination.

Note 2. *Legend*
A denotes hand demolition
B denotes mechanical demolition by pusher arm
C denotes mechanical demolition by deliberate collapse
D denotes mechanical demolition by demolition ball
M denotes demolition by other mechanical means excluding wire pulling

temporary structures on a regular basis to avoid overloading. Debris which has accumulated behind walls should also be removed. Independently supported working platforms over any reinforced concrete slabs should be demolished. Support for members of framed structures must be provided before gutting, along with temporary props, bracing or guys to restrain remaining parts of the building. On no account must operators work from the noor being demolished, and site control must ensure that all personnel are kept at a safe distance from the scene of operations, when pulling arrangements, demolition balls, pusher arms and/or explosives are being used.

Above all, an ongoing system of inspection must be maintained during demolition to detect further hazards which may result from the demolition process, e.g. loosened materials, overloaded floors.

Competent persons in construction operations

Safety of construction operations has always relied heavily on the use of designated competent persons to undertake a range of inspections and examinations of, and supervise operations in the case of, potentially high risk situations, for instance, in the inspection of scaffolds and excavations, the use of explosives and where lifting operations are undertaken.

Schedule 7 of the CHSWR indicates places of work requiring inspection by a competent person (*see* Table 37.2).

Construction (Health, Safety and Welfare) Regulations 1996

Schedule 7

See Table 37.2.

Regulating the work of the contractor

From the viewpoint of legal liability and practical safety, the relationships between the occupier of the premises and the main contractor, subcontractors, etc., are of key importance. In the case of legal liability, building owner(s), main contractor(s) and subcontractors, nominated or domestic, may often be said to be in joint occupation or control. (The main contractor has the greater liability under the construction safety legislation as he has the greatest control.) This has practical consequences in relation to the HSWA, sec 4, as far as criminal liability is concerned. In the case of civil liability, all three parties can be construed as being in occupation for the purposes of the OLA, and can be proceeded against accordingly including, if necessary, any architect, civil engineer or geotechnical consultant who has been negligent.

The legal position notwithstanding, organisations should endeavour to operate some form of practical regulation of contractors on their premises. Contracting operations can cover a wide range of situations, from activities such as window cleaning or maintenance of plant and equipment, to large-scale construction works, such as extensions to premises or the building of new premises on a 'green field' site. Consultation with the contractors prior to the commencement of the work and/or the contract being signed, and during the course of the work, is of the utmost significance if a safe site is to be maintained. The various facets of this relationship between the contractor and the occupier of the premises are as follows.

Table 37.2 ◆ Reg 29(1): Places of work requiring inspection by a competent person

Column 1: place of work	*Column 2: time of inspection*
1. Any working platform or part thereof or any personal suspension equipment provided pursuant to paragraph (3)(b) or (c) of Reg 6	**1.** (i) Before being taken into use for the first time; and (ii) after any substantial addition, dismantling or other alteration; and (iii) after any event likely to have affected its strength or stability; and (iv) at regular intervals not exceeding seven days since the last inspection.
2. Any excavation which is supported pursuant to paragraphs (1), (2) or (3) of Reg 12	**2.** (i) Before any person carries out work at the start of every shift; and (ii) after any event likely to have affected the strength or stability of the excavation or any part thereof; and (iii) after any accidental fall of rock or earth or other material.
3. Cofferdams and caissons	**3.** (i) Before any person carries out work at the start of every shift; and (ii) after any event likely to have affected the strength or stability of the cofferdam or caisson or any part thereof.

Consultation prior to commencement of contract work

Before any contract work is commenced, a responsible person representing the main contractor must discuss with the occupier the safety precautions necessary as far as his own workforce is concerned and any other parties on site. This requirement is graphically underlined by the decision in *R* v. *Swan Hunter Shipbuilders Ltd* (reported in *The Times*, July 1981) where the main contractor and the subcontractor, Telemeter Installations Ltd, were fined £15,000 and £3,000 respectively for failing to inform incoming sub-contract labour of the dangers produced by oxygen-enriched atmospheres in confined spaces. As a result eight workmen were burnt to death whilst working below decks on HMS *Glasgow*. Hence, except on clearly defined 'green field' sites, where the main contractor has full contractual responsibility for safety, both occupier and contractor should ensure that:

(a) the site of operations is clearly defined, if necessary on a factory plan, including those areas which contractors' staff are not permitted to enter (for the purposes of the FA, sec 175(6), there will be two separate factories here);

(b) agreement is reached as to whether any amenities are to be made available by the occupier for the contractors' employees. e.g. catering, washing, sanitation, first aid and clothing storage;

(c) where applicable, the contractor has full information concerning the occupier's processes or activities which may affect or involve contract work, e.g. parking limitations, specific hazards, hygiene requirements;

(d) the contractor has, or will shortly obtain, adequate insurance cover to indemnify the occupier in respect of any negligence resulting in personal injury and/or death, or damage to property and plant, arising out of or in connection with contract work.

Responsibilities of contractors, subcontractors and other persons

As part of the consultation process prior to the commencement of work, the responsibilities of all parties, i.e. main contractor, subcontractor and other persons coming on the construction site, must be clearly defined in writing and agreed with the main contractor. In most cases, this should be incorporated in a Statement of Health and Safety Policy for the site, particularly in the case of long-term projects.

Where work with a foreseeably high hazard content is to be undertaken, e.g. where a fire risk may exist or work involving lifting operations, the main contractor should be required to produce a Method Statement indicating the safe systems of work to be operated during that particular exercise. The use and contents of Method Statements are covered later in this chapter.

Use of owner's equipment

Contractors should generally be expected to provide all their own tools, plant, equipment and materials necessary for the satisfactory performance of work. On no account should use be made of the owner's electricity, gas or air mains without appropriate authority. Where such permission is granted, the method of connection should be approved by the appropriate manager.

Reporting, recording and investigation of accidents and dangerous occurrences

The contractor should be aware of the requirements of the RIDDOR, and of details of any internal system for the reporting, recording and investigation of accidents. Except on 'green field' sites, all accidents, including traffic accidents, and all dangerous occurrences involving contractors should be reported to the occupier.

Permit to work systems and other procedures

The contractor should be informed by the occupier of the need to follow the latter's safe systems of work, including permit to work and other systems, particularly where contract work may involve the occupier's staff and/or other subcontractors, or where failure to do so might endanger the plant, etc., involved.

Plant and machinery

Only in specifically controlled circumstances should a contractor remove the guard, fencing or other safety equipment from machinery and plant, and any guards or safety devices should be reinstated and operational before the machinery or plant is handed back for use. Plant and machinery belonging to the contractor should be adequately guarded before being operated on the premises. Electrically operated hand tools should be of the low-voltage type and connected to a 110 volt mains-isolated circuit where applicable. Alternatively, contractors should provide their own step-down transformer (250/110-volts, 50 cycles) with the mid-point of the secondary winding efficiently earthed. In all cases, the metalwork on portable equipment and any flexible metallic covering of conductors should be efficiently earthed and in all other respects constructed and maintained in compliance with the Electricity at Work Regulations 1989. Cables supplying portable apparatus should be of the correct size and properly connected to standard plugs and sockets. Makeshift and/or unsafe connections are dangerous and must not be permitted. In the case of a.c. welding equipment, contractors should limit d.c. to 40 volts, and ensure the correct use of earth leads at all times, including the earthing down wherever possible of the article being welded.

Noise

Where noisy equipment is used in close proximity to a working area on site, operatives should be provided with a recognised form of hearing protection (e.g. ear muffs), use of which must be encouraged. Moreover, noise from equipment should be minimised in working areas, e.g. by the use of bag mufflers on pneumatic drills. Contractors should also take all reasonable steps to prevent the inhabitants of the neighbourhood being troubled by noise.

Fire protection

Every year fires of varying degrees of severity cause loss of life and devastation on building sites. The Fire Triangle in Chapter 33 shows the three requisites for fire to take place, i.e. an ignition source, fuel and oxygen. Ignition sources are many and varied – welding and cutting activities, people smoking whilst working on site or in site offices, cooking on open flames in site huts, the use of blow lamps and gas or liquid fuel fired heating appliances. The fuel, again, can take numerous forms – timber, bitumen impregnated paper, plastics, flammable substances such as paraffin, and combustible refuse produced during the erection of buildings and other site work. Oxygen, which is naturally present in air, will, on a windy day, cause fire to spread at a rapid rate. The need, therefore, for effective fire protection procedures on construction sites cannot be overemphasised.

Fire protection procedures will vary to some extent according to the type of site. For example, they will be different on a 'green field' site, where no construction work has previously taken place, from those on an existing site, where modification or extension to existing buildings may be taking place.

'Green field' sites

Planning of fire protection procedures should take place well before site work commences. Here consideration should be given to the following:

(a) provision of access for fire brigade appliances;

(b) location of buildings and the separation of high-risk buildings from low-risk buildings;
(c) the provision of adequate space between buildings;
(d) the construction of buildings, e.g. site huts, canteens, offices, flammable stores, equipment stores, etc.;
(e) the establishment of areas where smoking and the use of naked lights are forbidden;
(f) the system for the storage and disposal of combustible and flammable refuse;
(g) the provision of a separate flammable materials store;
(h) the availability of a water supply for fire brigade appliances and onsite fire fighting;
(i) the provision of adequate fire-fighting appliances, located according to the fire risk involved, and an effective fire alarm system;
(j) evacuation procedures in the event of fire, including the training of site operators in such procedures;
(k) the appointment of fire wardens and the use of fire patrols, particularly at night and weekends;
(l) the system for liaison with the local fire authority, including frequent inspections by the fire protection officer, not only to ensure compliance with the terms of the fire certificate but to see that any new risks which may have arisen are quickly and effectively controlled.

Existing sites

All the contractor's employees should understand the fire warning systems currently in operation throughout the premises. Instructions on action to be taken in the event of fire should be made available to the contractor and workforce by being clearly displayed on site. Fire-fighting equipment installed by the occupier should also be made known to contractors; alternatively, fire-fighting equipment should be provided by the contractors at the consultation stage. Where any work involves interference with or removal of fire-fighting appliances, alarms or systems, prior notification should be given to the contractor. Moreover, use of petroleum, petroleum mixtures, liquefied petroleum gas, celluloses and other highly flammable and/or explosive substances should comply with the requirements of the Petroleum Consolidation Acts and the Highly Flammable Liquids and Liquefied Petroleum Gases Regulations (HFL&LPGR) 1972. Where any form of protection material or covering is used, whether against dust or climatic conditions, it should be fire-resistant or treated with fire-resistant solution.

Further guidance on fire precautions and procedures is given in Chapter 33.

Welding operations on site

Operations involving the use of oxy-acetylene welding and cutting equipment, electric arc welding, blow lamps or other flame-producing equipment should not be commenced until authority to do so (written or verbal) has been received from the occupier. In certain cases this may require issue of a permit to work (hot work permit). Gas cylinders, particularly acetylene, should be stored in a manner required by the HFL&LPGR. The contractor should provide and ensure use of welding screens to give protection to all persons in respect of any arc flash caused by his workforce. Combustible material such as paper, timber, rags, should be placed in suitable refuse containers. Non-combustible blankets, i.e. fibreglass, should be used to afford protection against 'welding batter'. Burning refuse on site should be prohibited.

Dangerous substances and wastes

Two aspects must be considered here:

(a) the legal requirements to protect the contractor's workforce and workers employed by any subcontractors on site from exposure to dangerous substances (HSWA, secs 2 and 3, and in particular, the control of Substances Hazardous to Health Regulations 1994;

(b) requirements of environmental health legislation, covering air, water and ground pollution, and waste disposal procedures.

Dangerous substances on site

A wide range of toxic and other dangerous substances can be encountered on construction sites. These include asbestos, lead and flammable substances. Site management must be aware of the hazards associated with such substances, ensuring correct storage, use and disposal (*see* Chapter 39).

Personal protective equipment

Personal protective equipment should be provided wherever hazards to the contractor's workforce exist. This includes safety helmets, eye protection, respirators and safety boots. Contractors should comply with the Construction (Head Protection) Regulations 1989. (*See* further Chapter 24 'Personal protection'.)

Contractors' vehicles

The movement of vehicles within premises on which construction work is taking place should be consistent with safety, and any speed limit specified on notices displayed there should not be exceeded. Drivers should comply with any traffic direction systems and signs in use on site. Vehicles used by contractors or their employees should be parked only in locations specified for that purpose. All vehicles on premises should be prohibited from moving during times when access roads are crowded with people arriving at or returning home from work, as the risk of accidents is greater at these times. This prohibition extends to cranes, mobile cranes, dumpers, concrete supply vehicles and vehicles. All mechanically propelled vehicles and trailers used on sites must be in efficient working order and in good repair, and must not be used in an improper manner.

Site clearance

On completion of work, contractors should be required to remove all unused materials and leave the site clean and tidy. This may include reinstatement of the perimeter fencing, removal of mud and debris from roads, removal of waste building materials and site refuse, and the levelling of disturbed ground. On no account should items such as empty gas cylinders, oil drums or paint cans be buried. Excavations and trenches should be filled in and levelled.

Window cleaning and external painting

Cleaning the outer surfaces of high-rise office blocks and industrial premises, as well as the external painting of such buildings, has spotlighted many hazards.

Window cleaning

There is a general legal requirement to keep windows and skylights, used for the natural lighting of workrooms and offices, clean. Much of this work is now undertaken by window-cleaning contractors.

In order to ensure safe working when cleaning windows or painting, the following should be considered.

(a) The contractor should ensure that any ladder, safety harness or other appliance used, or intended for use, by his employee, be of sound construction, adequate strength, sufficient length and properly maintained.

(b) Where it is not practicable to clean windows or paint external surfaces from a ladder, and the contractor's employee has to work at a height of more than 1.98 metres, or otherwise in conditions where any specific danger or risk might be involved, the contractor, or his authorised representative – e.g. manager, supervisor – should inspect the place before work is commenced . The contractor should take all precautions to prevent an accident and instruct his employees in the precautions to be taken. (Failure to do so on the part of an employer/contractor can result in prosecution for breach of HSWA, sec 2(2)). Moreover, in the event of an employee being injured as a result of lack of instruction, a costly action for damages at common law may well follow. (*See* Chapters 2 and 3.)

(c) The contractor should satisfy himself that any structural handhold and/or foothold likely to be used by his employees is secure. Where the reliability of any handhold or foothold is in doubt, he should warn his employees and instruct them that it is not to be used. Failure to do so would result in breach of HSWA, sec 2 (*see* Chapter 3), and the common law duty to take reasonable care owed by all employers to their employees (*see* Chapter 2).

Further information is given in HSE Guidance Note GS10 *Roof Work Prevention of Falls*.

The use of Method Statements

A Method Statement is a formally written safe system of work, agreed between occupier and contractor or main contractor and subcontractor, where work with a foreseeably high hazard content has to be undertaken. It should specify the operations to be undertaken on a stage-by-stage basis and indicate the precautions necessary to protect site operators, staff occupying the premises where the work is undertaken and members of the public, including local residents, who may be affected by the work. It may incorporate information and specific requirements stipulated by health and safety advisers, enforcement officers, the police, site surveyor and the manufacturers and suppliers of plant and equipment, or substances, used at work. In certain cases it may identify training needs or the use of specifically trained operators.

Method Statements may be necessary to ensure safe systems of work in activities involving:

(a) the use of substances hazardous to health;
(b) the use of explosives;
(c) lifting operations;
(d) potential fire risk situations;
(e) electrical hazards;
(f) the use of sealed sources of radiation;
(g) the risk of dust explosions or inhalation of toxic dusts;
(h) certain types of excavation adjacent to existing buildings;
(i) demolition;
(j) the removal of asbestos from existing buildings.

Contents of a Method Statement

The following features should be incorporated in a Method Statement:

(a) the technique to be used;
(b) access provisions;
(c) safeguarding of existing work locations;
(d) structural stability requirements;
(e) safety of others, including members of the public and local residents;
(f) health precautions, including the use of personal protective equipment;
(g) the plant and equipment to be used;
(h) procedures for prevention of area pollution;
(i) segregation of certain areas;
(j) procedures for disposal of toxic wastes;
(k) procedures to ensure compliance with specific legislation, e.g. the COSHH Regulations, Control of Asbestos at Work Regulations.

Asbestos

In addition to the above requirements, the following features should be incorporated in a Method Statement where work involves the removal, stripping and disposal of asbestos from a building:

(a) the specific safe system of work to be followed;

(b) procedures for segregation of the asbestos stripping area;
(c) personal protective equipment requirements;
(d) welfare amenity provisions, e.g. hand washing and showers, sanitation arrangements, separation of protective clothing area from personal clothing area, catering facilities, drinking water provision;
(e) ventilation requirements for the working area;
(f) personal hygiene requirements for operators;
(g) supervision arrangements;
(h) air monitoring procedures, including action to be taken following the receipt of unsatisfactory results;
(i) notification requirements under the Control of Asbestos at Work Regulations 1987.

The need for Method Statements

The need for contractors to produce Method Statements prior to high risk operations should be raised in any pre-contract discussions between occupiers and main contractor. In certain cases, e.g. asbestos removal, standard forms of Method Statement are used and signed by the main contractor as an indication of his intent to follow the particular safe system of work agreed between occupier and himself.

Construction (Design and Management) Regulations 1994

These Regulations specify the relationships which must exist between a client, principal contractor and other contractors from a health and safety viewpoint.

Under these regulations a number of people have both general and specific duties. Thus a *client*, namely a person for whom a project is carried out, must:

(a) appoint a planning supervisor and a principal contractor in respect of each project;
(b) ensure that the planning supervisor has been provided with information about the state or condition of specified premises;
(c) ensure that information in a *health and safety file* is available for the inspection of specified persons.

A *planning supervisor* must ensure that specified pariculars of a notifiable project are notified to the HSE. (Particulars to be notified to the HSE are detailed in Schedule 1 to the regulations.) A notifiable project is one where the construction phase:

(a) will be longer than 30 days; or

(b) will involve more than 500 person days of construction work.

Planning supervisors have specific duties in respect of:

(a) the design of any structure comprised in a project;

(b) the co-operation between designers;

(c) the giving of adequate advice to specified persons;

(d) the preparation, review and necessary amendment of a health and safety file;

(e) the delivery of the health and safety file to the client.

The *principal contractor* has a number of duties in respect of:

(a) co-operation between contractors;

(b) compliance with the *health and safety plan*;

(c) the exclusion of unauthorised persons;

(d) the display of notices;

(e) the provision of information to the planning supervisor;

(f) the provision of certain health and safety information to contractors and the provision of specified information and training to the employees of those contractors;

(g) ensuring that the views and advice of persons at work on the project or their representatives concerning matters relating to their health and safety are received, discussed and co-ordinated;

The principal contractor is empowered, for certain purposes, to give directions to contractors and to include rules in the health and safety plan. In the case of a *designer*, he:

(a) is prohibited from preparing a design unless the client for the project is aware of his duties under the regulations and of the requirements of any practical guidance issued by the HSC;

(b) must ensure that the design he prepares, and which is to be used for the purposes of construction work or cleaning work, takes into account among design considerations certain specified matters.

A *contractor* (other than the principal contractor) must co-operate with the principal contractor in order to enable him to comply with his duties as indicated above.

A *person who appoints a planning supervisor, or who arranges for a designer to prepare a design, or a contractor to carry out or manage construction work*, is prohibited from doing so unless he is reasonably satisfied:

(a) as to the competence of those so appointed or arranged;

(b) as to the adequacy of the resource so allocated or to be allocated for the purposes of performing their respective functions by those so appointed or arranged.

Both the planning supervisor and principal contractor must comply with the requirements relating to the health and safety plan, and the commencement of the construction phase of a project is prohibited unless a health and safety plan has been prepared in respect of the project.

Competent persons

No person shall be appointed by a client as his agent unless the client is reasonably satisfied as to that person's competence to perform the duties imposed on the client under the regulations.

Similarly, planning supervisors, designers and contractors must be competent in terms of complying with any requirements and conducting his undertaking without contravening any prohibitions imposed by or under health and safety law.

Health and safety files

The following information must be incorporated in a health and safety file:

(a) information included with the design which will enable the planning supervisor and designers to comply with legal requirements;

(b) any other information relating to the project which it is reasonably foreseeable will be necessary to ensure the health and safety of all persons involved in the project.

Health and safety plans

An important requirement is that the planning supervisor shall ensure that a health and safety plan for the project has been prepared in sufficient time for it to be provided to a contractor before arrangements are made to carry out or manage the construction work (*see* Figure 37.4).

Under Reg 15, the following information must be incorporated in a health and safety plan:

(a) a general description of the construction work comprised in the project;
(b) details of the time in which it is intended that the project, and any intermediate stages, will be completed;
(c) details of the risks to the health or safety of any person carrying out the construction work so far as such risks are known to the planning supervisor or are reasonably foreseeable;
(d) any other information concerning the competence of the planning supervisor and the availability of resources to enable him to undertake his specified duties;
(e) information held by the planning supervisor and needed by the principal contractor to enable him to comply with his specified duties;
(f) information held by the planning supervisor and which any contractor should know in order for him to comply with statutory provisions in respect of welfare.

The principal contractor must then take suitable measures to ensure that the health and safety plan incorporates appropriate features for ensuring the health and safety of all persons involved. The regulations are accompanied by:

(a) an Approved Code of Practice *Managing construction for health and safety* (HSC);
(b) *A guide to managing health and safety in construction* (HSC);
(c) *Designing for health and safety in construction* (HSC).

Items (b) and (c) above were prepared, in consultation with the HSE, by the Construction Industry Advisory Committee (CONIAC) which was appointed by the HSC as part of its formal advisory structures.

Fig 37.4 ◆ Construction health and safety check-list

1. Health and welfare	
First Aid boxes	Responsible person
Ambulance arrangements	Stretcher
Foul weather shelter	Mess room
Clothing storage/changing facilities	Food heating facilities
Sanitation – urinals, water closets	Washing facilities
Drinking water	Facilities for rest
Emergency procedure	
2. Environmental aspects	
Access to site and all parts/safe egress	Housekeeping, order and cleanliness
Site lighting and emergency lighting	Segregation from non-construction activities
Dust and fume control	Effective ventilation
Waste storage and disposal	Temperature control – indoor workplaces
Ventilation arrangements	Perimeter signs
Adverse lighting	
Secondary lighting	
3. Fire protection	
Access for fire brigade appliance	System for summoning fire brigade
Siting of huts	Space between huts
Flammable refuse storage	Fireproof hut construction
Prohibited areas – notices displayed	Specific fire risks
Flame-producing plant and equipment	Heaters and heating in huts
Vehicle parking arrangements	Fire appliances
Storage of flammable substances	Fire detectors and alarms
4. Storage of materials	
Siting	Stacking
Storage huts	Separation of flammable materials
Compressed gases	Hazardous substances
Segregation of compressed gases	Explosives

G

Fig 37.4 ◆ continued

5. Plant, machinery and hand tools	
Lifting appliances	Woodworking machinery
Electrical equipment	Abrasive wheels
Welding equipment	Hand tools
Maintenance, examination and testing	Construction, strength and suitability
Guarding and fencing arrangements	
6. Access equipment and working places	
Scaffolding	Overhead cables
Ladders	Working platforms
Trenches/excavations	Movable access equipment
Fragile material	Falling objects
Means for arresting falls	Personal suspension equipment
Competent persons	Unstable structures
7. Cofferdams and caissons	
Design and construction	Materials
Strength and capacity	Maintenance
Competent person	
8. Prevention of drowning	
Drowning risks	Rescue equipment
Transport over water	Vessels
Flooding risks	Work over water
9. Radiography and radioactive materials	
Competent person	Classified workers
Medical supervision arrangements	Personal dose monitoring
Control of sealed sources	Records
Dose records	Personal dose meters/film badges

Fig 37.4 ◆ continued

10. Site transport	
Cautionary signs and notices	Directional signs
Site layout	Authorised drivers
Lift trucks	Mobile access equipment
Dangerous vehicles	Unsafe driving
Controlled access/egress to site	Separate parking areas
Segregation of traffic routes	Suitability of traffic routes
Clear traffic routes	Towing procedures
Passenger-carrying vehicles	Safe loading of vehicles
Prevention of overrunning	Emergency routes and exits
11. Personal protective equipment	
Safety helmets	Eye/face protection
Gloves/gauntlets	Respiratory protection
Foul weather clothing	Hearing protection
Personal suspension equipment	Donkey jackets
Safety footwear	
12. Demolition	
Pre-demolition survey	Competent person
Method Statement	Asbestos
13. Personnel	
General safety training	Competent person training
First Aider training	Health surveillance
14. Inspection and reports	
Guard rails, etc.	Working platforms
Excavations	Personal suspension equipment
Means for arresting falls	Ladders
Welfare facilities	Cofferdams and caissons
Reporting arrangements	

38 Mechanical handling

A host of equipment is available to ease the task of handling goods and, whenever possible, mechanical handling systems should be used in preference to manual handling. This chapter examines four principal forms of mechanical handling, namely conveyorised systems, elevators, internal factory transport and the use of goods vehicles. The choice of mechanical handling systems will depend on several criteria such as the weight, shape, size, form, distance and frequency of movement of loads, together with space restrictions, storage systems and the nature of the material to be handled.

Conveyors

The following are the most commonly used conveyors.

Belt conveyors

These may be flat or troughed and are commonly used for transporting materials over long distances. The flat type is largely used to convey bulky packages or boxed goods, whilst the trough type is employed to carry loose materials, such as coal and aggregates. These materials are prevented from falling over the sides of the belt because it forms a cross-sectional 'trough' which prevents side spillage unless the belt has been overloaded.

Roller conveyors

Roller conveyors, which can be of the gravity type or the powered type, are used for the movement of unit loads. The powered type is used where level or slightly rising runs are installed, where manual pushing of loads is impracticable, or where the incline necessary for gravity movement is not possible.

Chain conveyors

These are often of the 'scraper' type, used for pushing or pulling materials along a fixed trough. Overhead chain types employ 'hangers' attached to the chaln from which are suspended the objects requiring transfer. 'Trolley' types comprise specially designed trolleys mounted on a guide system, and are used, for instance, for the transfer of vehicle bodies during vehicle assembly.

Screw conveyors

This type of conveyor is used mainly for the transfer of loose or freeflowing solid materials, generally over short distances, e.g. solid fuel from bunker to boiler furnace or grain from storage silo to processing plant.

Slat conveyors

These conveyors comprise a series of spaced wooden or metal slats moving on side chains. They are commonly used for the transfer of boxed or sacked goods, and can operate on inclined levels for the transfer of goods between floors.

Hazards associated with conveyors

Whilst the more general aspects of conveyor safety are dealt with in Chapter 32, the main hazards associated with conveyors of different types are:

(a) traps or 'nips' between moving parts of a conveyor, e.g. between a conveyor chain and chain wheels, or between a moving belt and rollers, particularly drive and 'end' rollers and also belt tensioning rollers;

(b) traps between moving and fixed parts of a conveyor, e.g. between the screw of a screw conveyor and the edge of the feed opening in the transfer tube;

(c) hazards associated with sharp edges, e.g. on worn conveyor chains and belts, which may be exposed;

(d) traps and nips created by the drive mechanism, e.g. V-belts and pulleys, chains and sprockets;

(e) traps created at transfer points between two conveyors, e.g. between a belt conveyor and roller conveyor.

Specific aspects of conveyor guarding

Whilst all conveyors present similar trapping and contact hazards, the various forms of conveyor need guarding in different ways. The hazards and guarding requirements are outlined below.

Belt conveyors

Traps formed between belt and rollers: Traps formed between the belt and drive, driven and tension rollers (*see* Fig. 38.1) should be covered with fixed guards extending to 850 mm from the trapping point. Side guards should be provided along the whole length of the conveyor and extend to 25 mm below the return belt. Where there is pedestrian access, and the underside of the belt is carried by return roller, this section of the conveyor should be enclosed.

Fig. 38.1 ◆ A reversible conveyor showing trapping points

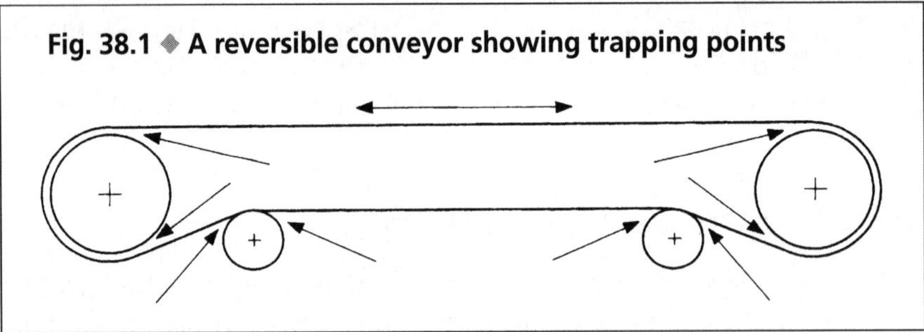

Traps between belts and end plates: In addition to providing guards as outlined above, a horizontal guard plate should be fitted (*see* Fig. 38.2). Clearance between plate and belt should not exceed 4 mm.

Fig. 38.2 ◆ Guarding of conveyor end using a horizontal guard plate

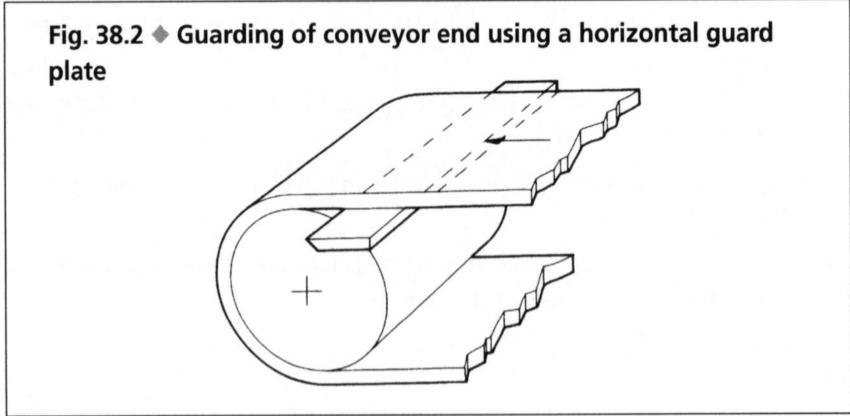

Traps at belt conveyor transfer points: Where static transfer plates (dead plates) are fitted at the junction of two belt conveyors, the gap between the top surface and the belt should not exceed 4 mm (*see* Fig. 38.3).

Fig. 38.3 ◆ Use of a dead plate where two conveyors meet

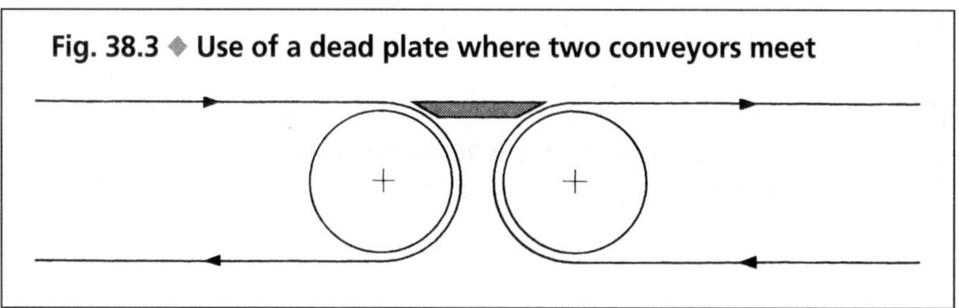

Traps between items conveyed and fixed structures: The risk associated with traps formed between a heavy item conveyed, e.g. a heavy crate, and fixed structures should be minimised by ensuring that there is a clearance of at least 50 mm between the item and any fixed structure. Support members should be free from sharp edges.

G

Hazards from worn and defective belts: Worn and defective belts and belt joints can create a hazard. Belts should be examined regularly and belt joints should be designed to be as smooth as possible. Worn belts or belts with loose fasteners should be replaced.

Roller conveyors

Traps between rollers and fixed structures across the conveyor: The gap between the rollers and fixed structures should be either 4 mm or less, or greater than 50 mm, to minimise trapping.

Hazards from rotating ends of exposed roller shafts: Rotating ends of roller shafts should be contained within the supporting frame for the rollers or flush with the bearing housing. They may, however, be exposed without guards outside the bearing housing up to a distance of 25 mm, provided they are smooth. If they extend more than 25 mm beyond the bearing housing, or if they have protrusions or irregularities, they should be fitted with fixed caps which completely enclose them.

Hazards from missing or jammed rollers: Missing, worn or jammed rollers may cause instability of the conveyed load and increased risk of trapping. These rollers should be replaced.

Hazards at transfer points: Where a roller conveyor is fed from another conveyor and the gap between the conveying surface is less than 50 mm, the first roller should not be power-driven, but so arranged that it would be displaced from its position should a hand be trapped between conveyors (*see* Fig. 38.4). Such a roller is termed a jump-out roller.

Fig. 38.4 ◆ Junction of a roller conveyor and belt conveyor showing position of jump-out roller

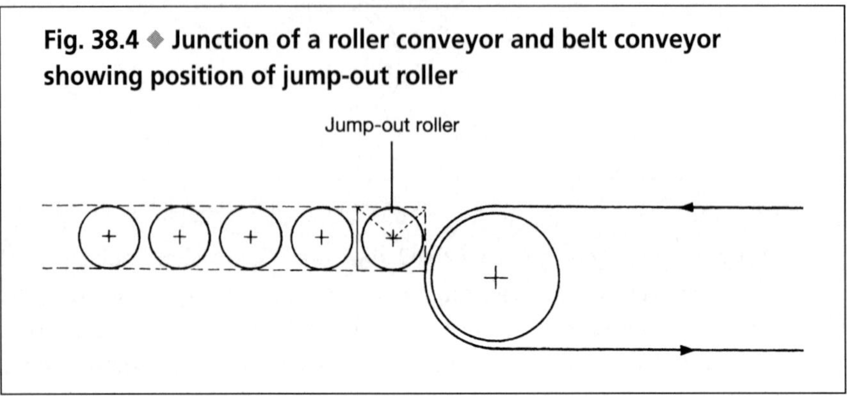

Chain conveyors

Traps between the conveyor chain and chain wheels: Fixed side and end guards (*see* Fig. 38.5) should be fitted at both the powered and free chain wheel ends of the conveyor, except where the conveyor is fitted in to the floor.

Fig. 38.5 ◆ Guarding at the end of a chain conveyor by enclosure of the sprocket wheels and nip points

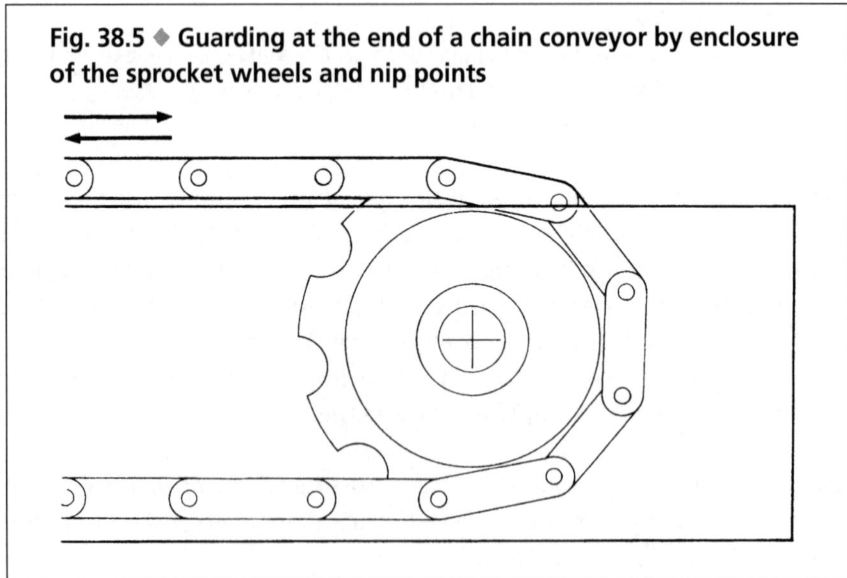

Clearance between the lower edge of the guards and the floor, so as to facilitate cleaning, is allowable up to 180 mm, provided that the return chains are carried in guides and the guard extends to not less than 50 mm below the guide. On the top run of the convevor, fixed in-fill plates should be fitted in the gaps between the chain guides and the side and end guards or the adjacent floor. All guards should be extended to 850 mm from the trapping points or otherwise prevent access. Clearance between guard and conveyor chain, where it runs over the chain wheel, should be 6 mm or less. Where a chain passes through a guard, clearance should not be less than 50 mm.

Traps at overlapping sections between chains and chain wheels: At overlapping sections, side guards or in-fill plates should be fitted. Alternatively, only side guards need to be fitted, provided that the reach distance to the trapping points over the side of the guards corresponds with Table 2 of Appendix A of BS 5667: Part 19 *Specification for Continuous Mechanical Handling Equipment Safety Requirements: Belt Conveyors – Examples for Guarding of Nip Points.*

Traps between return chains and guide shoes, guide wheels and adjacent supporting structures: The return and carrying chain, where practicable, should be carried in continuous guides. Where this is not practicable and short guides, guide wheels or guide shoes are used, the chain should be free to lift 100 mm above the guide so as to reduce the risk of trapping. Guide wheels which have spokes or similar apertures should be fitted with discs to prevent hand and finger access.

Traps between carrying and return chains: Where carrying and return links of a chain run close together in guides, the return links should be positioned below guide level so as to reduce risk of trapping.

Hazards on bends caused by the chain slipping from its guide: The tendency for a chain to slip out of its guide on a bend can be reduced by:

(a) inclining the guide away from the centre of the bend;
(b) use of wedge-shaped strips and plastic wearing strips so as to deflect the chain downwards in the guides;
(c) fitting of specially coated metal blocks in place of guides.

On the return chain, a plate can be fitted on top of the guide to prevent the chain from slipping out.

Screw conveyors

Traps between the rotating auger (screw) and the fixed parts: A fixed or interlocked guard should be fitted to prevent access to the screw. Where mesh or bar guards are fitted to allow free passage of materials, there must be sufficient distance between the guard members and the screw flight to prevent contact with the screw.

Slat conveyors

Traps between the conveyor chain and chain wheels: Fixed side guards should be fitted below conveyor track level along the whole length of the conveyor. The clearance between the guard and the chain should be 4 mm or less, and the guards should extend below the return chain so that the distance between the chain and the bottom of the guard is not less than 25 mm. Guards to enclose the chain underneath are not generally necessary, except where there is pedestrian access beneath the conveyor.

Traps between conveyor chain wheels and fixed structure: Safeguarding should be provided as above.

Traps between conveyor chain and end plates, dead plates, rails and fixed structures: Safeguarding should be provided as above. In addition, fixed guards should be fitted at the driven end of the conveyor where the chain passes close to the end plate. The clearance between the guard or a dead plate and the conveyor chain should be 4 mm or less.

Traps between the slats of biplanar chain conveyors where the slats open and close on bends and are inadequately supported: Biplanar chains should, at bends where the slats open and close, be supported underneath by solid fixed plates extending to the edge of the slats, and guarded by protective hoods, as shown in Fig. 38.6, over the return point to a distance of not less than 850 mm.

Traps between rotating corner plates and fixed structures or conveyor chains: Fixed nip guards should be provided at circular corner plates where the plate is in-running with the conveyor chains or runs up to a fixed structure.

Hazards from cornerplates with exposed sharp edges: Before the edge of the corner plate becomes sufficiently sharp, it should be replaced or protected by a guard around the entire periphery.

Fig. 38.6 ◆ Protective hood over the return point of a conveyor

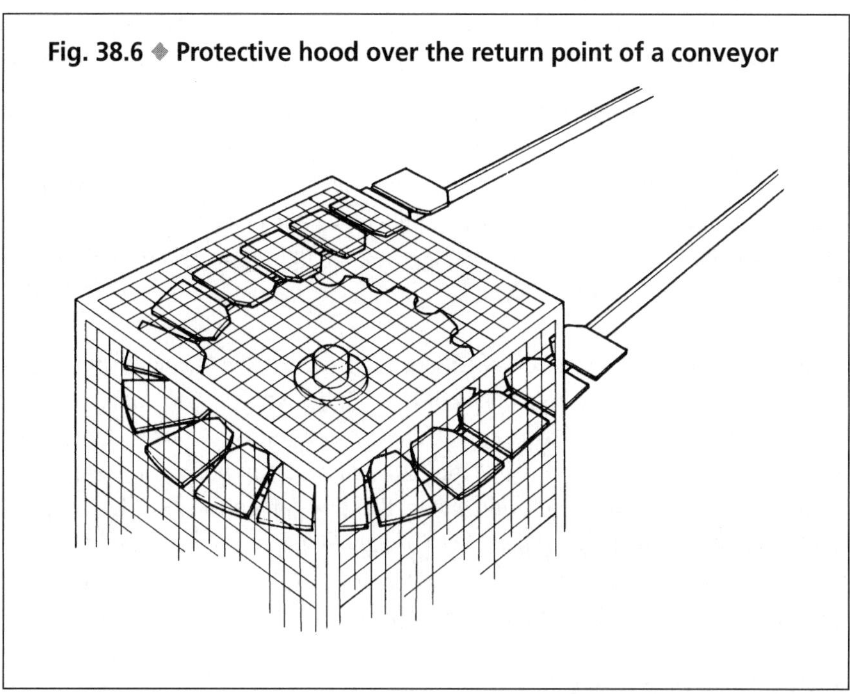

Source: BS 5667 Pt 18

Hazards from conveyor chain with exposed sharp edges: Before chain edges become sufficiently sharp, they should be replaced.

General aspects of conveyor guarding

Types of guard

Fixed guards: Fixed guards should be used wherever practicable and must be securely fixed in position when the conveyor is in motion or likely to be put in motion. It should not be possible to open or remove these guards without the aid of a specific tool. Moreover, the fastener should be captive to the guard. Fixed guards may be an integral part of the conveyor or freestanding from the floor and securely fixed to the floor of the conveyor structure. Guards should not allow space for a person to be trapped between the guard and the conveyor.

Interlocked guards: An interlocked guard should be so connected to the machine controls that:

(a) until the guard is closed the machine cannot operate;

(b) either the guard remains locked closed until the dangerous movement has ceased or, where overrun is insufficient to create danger, opening the guard disengages the drive. The interlocking system should be fail-safe.

Interlocked guards should be provided for dangerous parts of a machine where the operator needs frequent access. ('A part of machinery is dangerous if it is a possible cause of injury to anybody acting in a way in which a human being may be reasonably expected to act in circumstances which may reasonably be expected to occur' – per du Parcq J in *Walker* v. *Bletchley-Flettons Ltd* [1937] 1 AER 170.)

Tunnel guards: A tunnel guard is a form of distance guard preventing access to a danger point by reason of the relationship of the guard opening dimensions to length of tunnel. Where tunnel guards are used, the guards should be so designed and fitted that the relationship between the opening and the distance from the opening to the danger point complies with BS 5304: *Safeguarding of Machinery*, cl. 10, Fig. 5, 'Openings in Fixed Guards'. (*See* Chapter 32.) Clearance between the sides of the opening to the tunnel guard and the items being conveyed should be not less than 50 mm.

Other safety aspects

(a) Where practicable, arrangements should be made for lubrication with the guards in position, e.g. through suitably located small openings which do not allow access to danger points.

(b) To minimise the risk of conveyed items jamming or falling from conveyors, the radius of all bends should be maximised at the design stage.

(c) All fixed support members, including guide rails, should be free from sharp edges.

(d) Where conveyors rise to more than 1 metre above floor or walkway level, suitable rails or side members should be provided to a sufficient height above the conveyor to contain the top item of the load being conveyed.

Emergency devices

(a) Where a conveyor is greater than 20 metres in length, an emergency stop (trip) wire should be provided. (The alternative is a series of emergency stop buttons.)

(b) Emergency stop buttons, which must be easily identifiable and designated as such, should be provided. An emergency stop button, however, is not a substitute for effective guarding. It is a device for cutting off the power in order to stop the conveyor. The position and number of stop buttons should be determined by the following criteria –

(i) plant layout and product flow associated with the unit as a whole;

(ii) the operator positions about the plant: no point on the conveyor should be more than 10 metres from an emergency stop button;

(iii) at any point on a conveyor where an emergency stop button is not visible, e.g. where a conveyor passes through a wall, a further stop button should be provided.

Emergency stop buttons should be palm- or mushroom-shaped and coloured red. They should remain in the 'off' position until reset. Releasing the emergency stop button should not cause equipment to restart.

Elevators

Most elevators operate in a fixed position. In certain industries, however, mobile elevators, which can be moved from one point to another, are used for loading and unloading tasks.

Fixed elevators

Fixed elevators may be of the vertical or the adjustable angle type. Vertical elevators may take the form of:

(a) bucket elevators for transferring loose materials, such as grain; or

(b) bar elevators, on which items are placed or hung, e.g. sacked or boxed goods.

The elevator may be enclosed in a fixed shaft or hoistway and is generally continuous in operation, often being linked with a horizontal conveyor prior to and/or after elevation. One of the greatest hazards with elevating loose materials is dust explosions, and in flour mills, for instance, all elevator heads must be fitted with explosion reliefs. To prevent the spread of fire between floors in mills using bucket elevators in particular, all hoistways and floor openings must be fire-proofed with fire-resistant materials giving a notional period of fire resistance of 30 minutes.

Adjustable angle elevators are commonly used for loading and unloading the holds of ships, particularly where loose materials such as grain, metal ores and coal are involved.

Both types of elevator normally run in either direction. A fixed guard should be installed at the base of the elevator to prevent direct access to the moving flights and in-running nips formed between the chain and sprocket.

Mobile elevators

Mobile elevators are used for loading and unloading commodities such as sacked goods, regular shaped containers, and luggage into and out of aircraft. They may be of the bucket or bar type. A trap is created at the in-running nip between the elevator chains and sprockets, and both ends of the elevator should incorporate fixed guards. A further hazard is that such elevators can be run at variable speeds. Hand, arm and shoulder injuries have been sustained by operators because they have been unable to keep up with the speed of the elevator. Adequate supervision and control over the operation of these elevators is, therefore, crucial to the prevention of accidents.

Mobile handling equipment

A wide range of equipment is used in storage and handling operation. Although such equipment is necessary for speedy movement of goods, its use has frequently led to accidents. The basic requirements for the safe operation of mobile handling equipment, such as fork-lift trucks, are careful selection and use of the right equipment in the right place by the right people. With any mechanical handling task, the selection of the appropriate equipment for the material to be handled is important. The type and layout of the storage system, type, weight and shape of the materials to be handled, construction and layout of buildings and operational areas, as well as the potential for accidents, must be considered. Before acquiring mechanical handling equipment, manufacturers and/or suppliers should be consulted to ensure that the equipment selected matches the performance requirements. Where new storage systems are being developed, the manufacturer and/or supplier should again be consulted.

Classification of mobile mechanical handling equipment

Pedestrian-operated stacking trucks

These are of two types, manually operated and power-operated stackers (*see* Fig. 38.7). A manually operated stacker is normally restricted in operation to moving post pallets or heavy machinery. It has a manual shift with

Fig. 38.7 ◆ Pedestrian-operated stacking trucks (a) Manually operated stacker; (b) Power-operated stacker

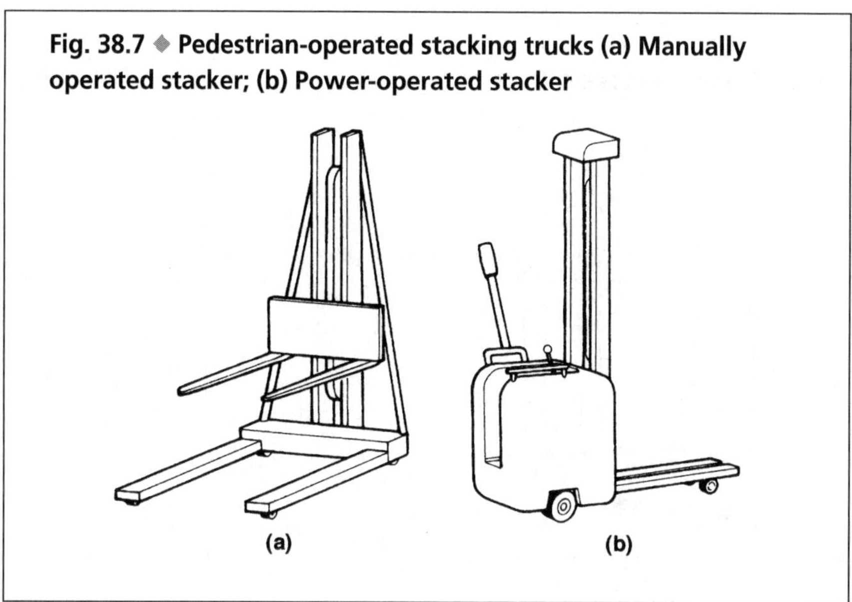

G

hydraulically operated lift, and cannot pick up directly from the floor. It has a capacity of 0.25 to 0.5 tonne, with a maximum lift of approximately 1.5 metres.

Power-operated stackers can be pedestrian- or rider-controlled, with power operation vertically and horizontally. They can pick up pallets from the floor. This type of stacker has a capacity of 0.5 to 1 tonne and a maximum lift of approximately three metres.

Reach trucks

A reach truck is a fork-lift truck that enables the load to be retracted within the wheel base, minimising overall working length and allowing reduced aisle widths. There are two separate forms, the moving mast reach truck and the pantograph reach truck (*see* Fig. 38.8). The former is rider-operated. Forward-mounted load wheels enable the fork carriage to move within the wheel base, so that forks can reach to pick up or deposit the load. The mast, forks and load move together. This truck has a capacity of 0.5 to 3 tonnes, with a maximum lift of 10 metres.

A pantograph reach truck is a rider-operated truck in which reach movement is by pantograph mechanism, whereby the fork and load can move away from the static mast. This truck has a capacity of 0.5 to 2.5 tonnes and a maximum lift of 10 metres.

Fig. 38.8 ◆ Reach trucks (a) Moving mast reach truck; (b) Pantograph reach truck

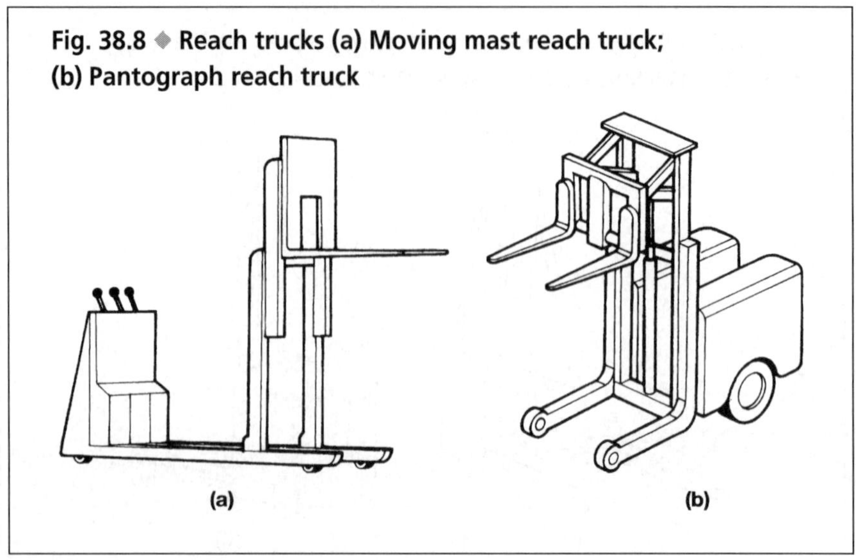

Counterbalance fork trucks

Such trucks are battery, petrol, diesel or gas powered. They carry the load in front which is counterbalanced to the weight of the vehicle over the rear wheels. They take three specific forms, namely lightweight pedestrian-controlled trucks, lightweight rider-controlled trucks and heavyweight rider-controlled trucks *(see* Fig. 38.9). The lightweight pedestrian-controlled type is normally a three-wheeled vehicle, and is used mainly where stacking rather than transfer is important. Such trucks provide a greater load-carrying capacity than the pedestrian stacker. They have a capacity of 0.5 to 1 tonne and maximum lift of approximately 3 metres.

The lightweight rider controlled truck is similar to the pedestrian controlled truck, except that the operator sits inside the truck. The handling rate is higher. Such trucks have a capacity of 0.5 to 1.25 tonnes and a maximum lift of 6 metres.

The heavyweight version is a four-wheeled truck, the high counterbalance weight over the rear wheels giving it a high load capacity. Many attachments, such as cradles, are available to suit different loads. They have a capacity from 1 to 9 tonnes and maximum lift from 6 to 12 metres.

Narrow aisle trucks

This type of truck differs from a reach truck in that the base of the truck does not turn within the working aisle to deposit or retrieve its load.

Fig. 38.9 ◆ Counterbalance fork trucks (a) Lightweight pedestrian-controlled; (b) Lightweight rider controlled; (c) Heavyweight rider-controlled

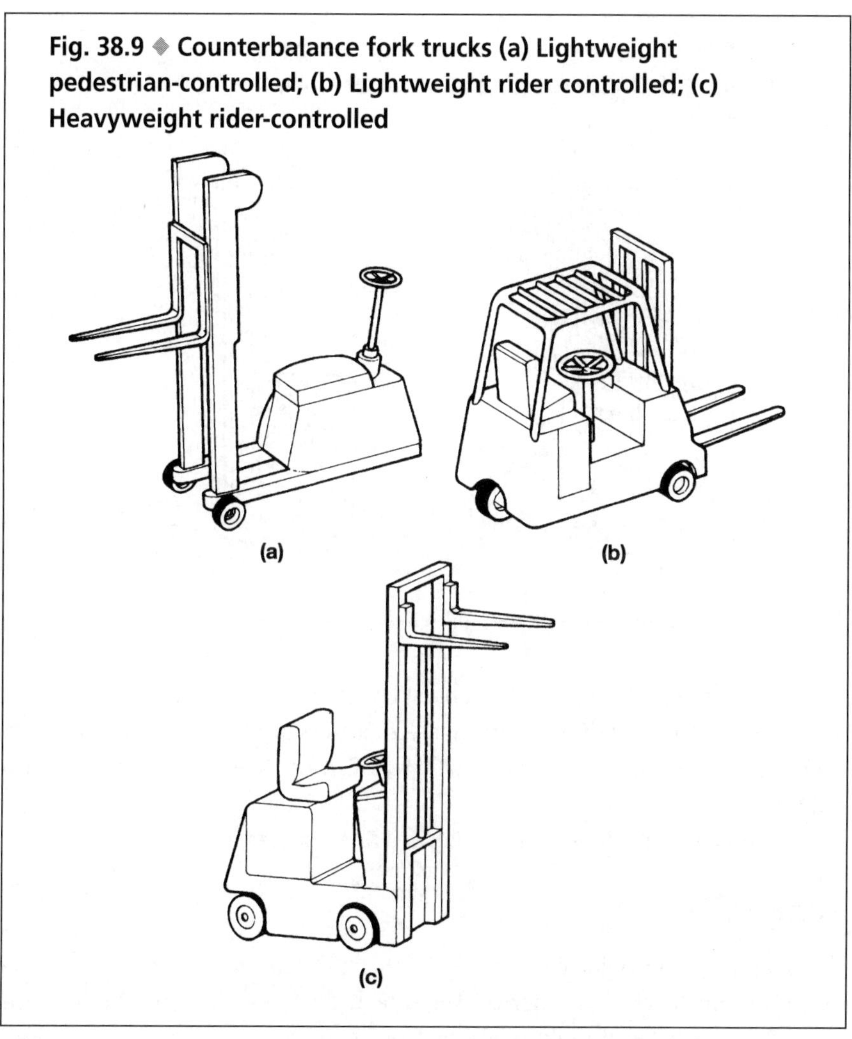

This enables the aisle width to be kept to a minimum. This type of truck takes two forms, namely side loaders and counterbalance rotating load turret trucks (*see* Fig. 38.10). Side loaders are ideal for long runs down narrow aisles. They are, however, only capable of stacking down one side of the aisle at a time and a large turning circle is needed at each end of the aisle in order to serve both faces of a racking system. Reach trucks have a capacity of up to 1.5 tonnes and a maximum lift to 9 to 12 metres.

The counterbalance rotating load turret truck has a rigid mast with telescopic sections. It incorporates a head which rotates through 180 degrees,

Fig. 38.10 ◆ Narrow aisle trucks (a) Side loader; (b) Counterbalance rotating load turret truck

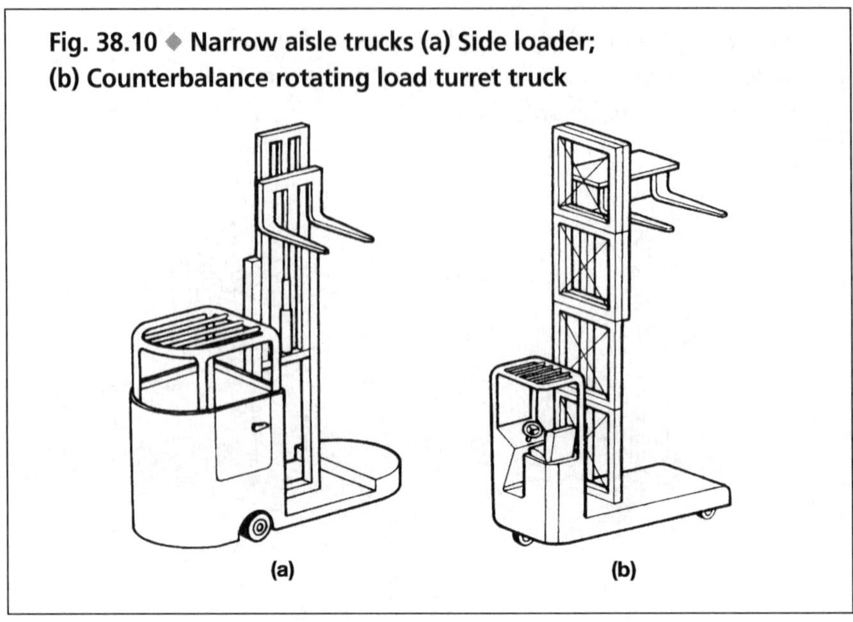

enabling it to slide sideways to deposit or retrieve a load. This type of truck can serve both faces of a racking system, and is guided by tracks or rails at floor level. They have a capacity of 1 to 1.5 tonnes and a maximum lift of 12 metres.

Both types of narrow aisle truck are rider-operated.

Order pickers

This device is derived from the fork-lift truck, incorporating a protected working platform permanently fixed to the lift forks. Thus the operator can pick goods from racking above floor level or place them in a racking system. The truck is operated from the picking platform and incorporates side shift, rotating mast and other purpose-added features. Order pickers allow maximum utilisation of racked storage areas owing to the narrowness of the aisles within which they can operate. Order pickers operate on a conventional basis or can be purpose designed for a specific task (*see* Fig. 38.11). Conventional order pickers operate on the same basis as a fork-lift truck with a cage fitted for the operator. The cage incorporates a small platform for the placement of goods picked from the racking. They have a capacity of 0.5 to 1 tonne with a maximum lift of 6 to 9 metres.

Order pickers can be designed for a multitude of storage tasks. The

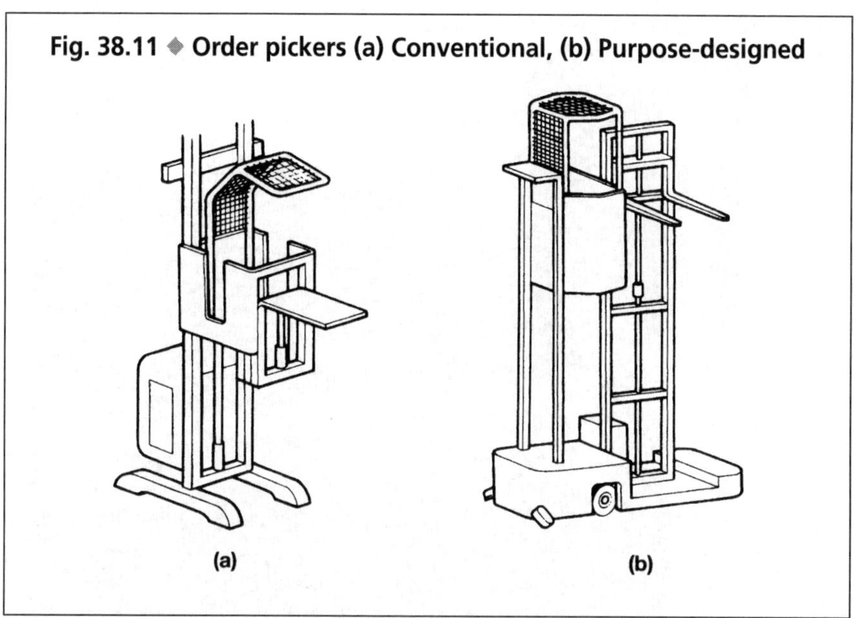

Fig. 38.11 ◆ Order pickers (a) Conventional, (b) Purpose-designed

heavy-duty type can incorporate traversing load masts or dual mast with independent load/operator control, and generally operates along rails or tracks within narrow aisles. With purpose-designed order pickers, capacity and maximum lift would be specified at the ordering stage.

Safe operation of mobile handling equipment

Equipment

The operation of mobile handling equipment results in many industrial accidents. The following rules should be applied to such operations:

(a) Untrained and/or unauthorised personnel should not drive or operate powered mechanical handling equipment.

(b) Rider trucks left unattended should have the forks lowered and be immobilised by –

   (i) leaving the controls in the neutral position;

   (ii) shutting off the power;

   (iii) applying the brakes;

   (iv) removing the key or connector plug.

(c) The maximum rated load capacity of the equipment, as stated on the manufacturer's identification plate, should never be exceeded (*see* Fig. 38.12a).

Fig. 38.12a ◆ Defects in fork-lift trucks
Serious distortion of the forks of a hand fork-lift truck due to overloading

(d) On no account should passengers be carried, unless in a properly constructed cage or platform. (*See* 'Use of fork-lift trucks as working platforms' later in this chapter.)

(e) When powered industrial trucks are used on public highways, they must comply with the Road Traffic Acts and be fitted with lights, brakes, steering, etc.

(f) The keys to the truck should be kept in a secure place when the equipment is not in use. Keys should be issued to authorised operators only and be retained by such persons until the end of the work period, when they should be returned to the manager responsible for the operation.

(g) A clearly defined maintenance programme, based on the manufacturer's recommendations for inspection, maintenance and servicing, should be operated. Repairs and maintenance should be carried out only by trained and experienced staff. Drivers should be trained to undertake simple periodic maintenance checks, and there should be a formal procedure for reporting defects identified in such checks and during normal operation. A typical daily check by the operator would include an examination and/or test of lights, including warning beacon, horn, tyres, brakes, steering, tilting, lifting and manipulation systems, operator controls, fluid levels, security of the overhead cage/guard and load backrest, as well as the integrity of hydraulic pipes, pipe joints and connections.

(h) Weekly maintenance on a truck should include all the operator checks mentioned above, together with an operational examination of steering gear, lifting gear, battery, mast, forks, attachments, and any chains or ropes used in the lifting mechanism (*see* Fig. 38.12b).

Fig. 38.12b ◆ Defects in fork-lift trucks

The fractured bracket (arrowed) attached the steering base to the fork-lift truck and affected the ability of the driver to steer the truck effectively

Reproduced by courtesy of National Vulcan Engineering Insurance Group Ltd

G

(i) Mobile handling and lifting equipment should also be subject to a six-monthly and an annual examination. In the case of the six-monthly examination (or 1,000 running hours) all parts of the truck should be examined by either a trained fitter or a representative of the manufacturer and a certificate issued by the examiner to the effect that the truck is safe to use. Lifting chains should be inspected on an annual basis and certificated in accordance with statutory requirements.

Operating area

Layout and maintenance of operating areas for mobile handling equipment are important in ensuring safe operation. The following points are relevant:

(a) Floors and roadways should be of adequate load-bearing capacity as well as being smooth surfaced and level. Moreover, a designer or consultant engineer of a factory or other workplace who fails to provide for this requirement can be sued for negligence (*Greaves & Company (Contractors) Ltd* v. *Baynham Meikle & Partners* [1975] 1 WLR 1095). Where 'sleeping sentries' ('sleeping policemen') are installed to slow down vehicular traffic, a bypass for mobile handling equipment should be provided. Furthermore, the edges of loading bays should be protected when not in use and 'tiger striped' to warn the driver, particularly when operating at night. Sharp bends and overhead obstructions, such as electric cables and pipework, should be eliminated where possible.

(b) Ramps should be installed to prevent displacement of the load at gutters, changes in floor level, etc.

(c) Gradients should not exceed 10 per cent and there should be a smooth gradual change of gradient at the bottom and top of the slope.

(d) Bridge plates should incorporate an adequate safety margin to support loaded equipment. These should be clearly and permanently marked with the maximum permissible load, secured to prevent accidental movement, and surfaced with a high-friction finish.

(e) Aisles should be of adequate width and overhead clearance to facilitate turning and safe movement, and should be kept clear at all times.

(f) Lighting should be adequate with a minimum overall illuminance level of 100 lux. If a permanent level of 100 lux is not available, auxiliary lighting should be installed on the equipment and so arranged to avoid glare which could affect other operators.

(g) Adequate general vehicle parking facilities should be provided away from the main operating areas and preferably in a secure compound.

(h) Actual layout of operating areas is crucial to the prevention of accidents. Doorways and overhead structures low enough to form an obstacle should have suitable warning notices displayed above them. Clear direction signs, marked barriers, electrically operated warning devices and convex mirrors should be used in order to prevent pedestrians coming into direct contact with trucks. Additionally, instructions to drivers to sound the horn and restrict speed should be posted at prominent positions. Separate routes designated crossing places and barriers, clearly marked at frequent intervals, should be arranged so as to restrict access by pedestrians to operational areas. Where pedestrians and handling equipment use the same access between parts of a building, a separate pedestrian access door should be available; alternatively, tubular steel barriers should be installed one metre from the side of the opening to provide a pedestrian passageway at the side. Windows or ports should be installed in rubber doors through which trucks frequently pass. Columns, pipework, racking, exposed electrical conduits and items of plant should be protected by impact-absorbing barriers.

(i) In truck battery-charging bays ventilation should be sufficient to prevent accumulations of hydrogen gas. Although not a statutory requirement, smoking should be forbidden and other sources of ignition eliminated. Notices prohibiting smoking, the use of naked lights and other sources of ignition should be displayed. Moreover, before disconnecting the truck battery from the charger, the current should be switched off to reduce the risk of sparking.

(j) Refuelling areas for petrol-driven trucks should ideally be located outside the building.

The operators

To ensure the safe operation of mobile handling equipment, it is essential that operators be responsible persons and physically fit for the job. They should be trained, and there should be an effective system for documentation of authorised operators, e.g. permits to drive. In particular:

(a) A high level of supervision and control should be exercised over all product- and goods-handling activities.

(b) Operators of mobile handling equipment should be physically and mentally fit, intelligent, mature and reliable. Handicapped persons need not be excluded, but medical advice should be sought as to suitability for specific tasks.

(c) Training should be given to operators and supervisors, and to managers responsible for areas where mobile handling is in operation, particularly to cover emergency situations.

(d) Operator training should be undertaken by trainers who are experienced in the specific tasks to be undertaken by trainees. Such training comprises three specific parts –

   (i) acquisition of the basic skills and knowledge required to operate the equipment safely and to undertake the required daily equipment checks;

   (ii) specific job training in a 'safe' working area to develop operational skills;

   (iii) familiarisation training under close supervision in the workplace. Training should be provided for all operators, even if they have been trained by a former employer. Supplementary or refresher training should be undertaken:

      (i) when there has been a significant change in operational layout;
      (ii) on transfer to a new operational area;
      (iii) when new or different equipment is introduced; or
      (iv) when there may have been a lapse in operator standards.

(e) Trainees should be tested. On passing a truck driving test, the operator should receive a 'permit to drive' (a form of written authority) for the class of truck on which he has qualified. Management should not allow persons to operate any mechanical handling equipment without this written authority. Additionally, a record should be maintained of all authorised operators and the serial number of the permit to drive issued. The date of training, and that of the refresher training which will be required in the future, should also be recorded.

(f) Operators should be provided with safety footwear and a safety helmet and, where appropriate, hearing protection, together with protective clothing to suit weather and/or temperature conditions. For example, for work in cold stores or on external loading, donkey jackets and gloves should be provided.

(g) Supervisors are responsible for ensuring that all operators are trained and working safely, that they carry out periodic checks and that there is a system for reporting deficiencies. (*See* Chapters 2 and 3 for the legal position relating to supervision/management.)

The HSC Approved Code of Practice and Supplementary Guidance *Rider operated lift trucks – operator training* provides excellent advice on this matter.

Use of fork-lift trucks as working platforms

There is now widespread industrial use of fork-lift trucks as a means of elevating workers and contractors to undertake tasks at high level, e.g. painting, cleaning, maintenance, repairs. Although, in principle, the use of a fork-lift truck affords considerable advantages for this type of work, nevertheless its primary function is the carriage and manipulation of materials and not as a means of support for a working platform. Therefore, if trucks are to be used for such purposes, certain safeguards are essential. Where practicable, the truck should be specifically designed for this purpose. In most cases, however, this is not the case and consequently working platforms are usually fitted to the forks of trucks. A platform designed for use on one particular truck should never be employed on any other type of fork-lift truck.

Where trucks are specially designed for or are regularly used with working platforms, movement of the platform should be controlled by the person on the platform. When trucks are only occasionally used with working platforms, either full platform controls or a platform-mounted emergency stop control should be provided.

G

Precautions with working platforms on fork-lift trucks

(a) The manufacturer's opinion as to the suitability of a truck for use in connection with a specific working platform should always be obtained.

(b) The weight of the platform and total superimposed load thereon should be not more than half of the truck manufacturer's rated capacity at the rated load centre distance of the truck at maximum lift height. A plate should be affixed to the platform indicating the maximum superimposed load and minimum rating of the truck on which it may be used.

(c) The platform should be secured to the forks and either the edges fenced to a minimum height of one metre either by guard rails comprising top rails, intermediate rails and toe board, or a steel mesh enclosure of similar height should be constructed.

(d) A locking device should be fitted to ensure that the mast remains vertical.

(e) Where controls are located on the platform, they should be of the 'dead man's handle' type, whereby the actuating lever or switch must be held or pressed continuously to effect motion of the platform. Preferably, controls should be positioned midway across the platform and at the rear to keep the operator away from the edges of the platform whilst it is in motion. This recommendation does not preclude provision of emergency controls at floor level which may be desirable to lower the platform in the event of breakdown or emergency. When fitted, such controls should be located and designed so as to prevent accidental or unauthorised operation.

(f) A prominent notice should be affixed to the platform with the instruction 'ENSURE THAT PARKING BRAKE IS APPLIED BEFORE ELEVATING PLATFORM'.

(g) On all machines designed specifically for, or likely to be used for, access purposes there should be a minimum of two suspension ropes or chains.

(h) No person should remain in the elevated working position when the truck is moved from one point to another.

(i) The appliance should only be used on well-maintained and level floors. (*See* reference earlier to the case of *Greaves & Company (Contractors) Ltd* v. *Baynham Meikle & Partners*.)

(j) All trapping points should be screened or guarded to protect persons carried on the platform, e.g. where a chain passes over sprockets, or where there is a crushing or shearing action between parts of the mast or its actuating mechanism.

Goods vehicles

Rail vehicles

Rail movement in depots and industrial complexes is a major hazard, frequently owing to poor visibility. The danger is greatly increased where long rafts or wagons are being shunted, especially by one-man diesel locomotives. Here the number of crossing places should be kept to a minimum and they should be clearly defined by signs and barriers. 'STOP, LOOK AND LISTEN' signs should be displayed well ahead of level crossings located in rail depots. Each crossing should be well lit at night and during bad weather. Shunting locomotives should be fitted with flashing beacons which operate during movement. Further guidance on rail vehicle operations is provided in the Locomotives in Sidings in Factories Regulations 1906.

Motor vehicles

Many of the requirements for industrial powered trucks apply to the operation of motor vehicles within the boundaries of a factory or industrial complex. The careful layout of loading bays, approaches, and other areas will minimise those hazards which can exist when vehicles need to manoeuvre in awkward or confined spaces.

Although goods vehicles visiting factory premises are not 'machinery' for the purposes of the Factories Act 1961, sec 14(1) (per Viscount Dilhorne in *British Railways Board* v. *Liptrot* [1967] 2 AER 1072), they come within the general requirements for a safe system of work, as laid down in HSWA, sec 2(2) and at common law. Consequently, drivers of such vehicles, whether company employees or those of another company delivering to the premises, must not reverse vehicles, unless guidance is given by an authorised person. In particular, where there is extensive vehicle manoeuvring in a loading area, marshalling stewards should be employed to ensure the safe reversing, loading and unloading of vehicles (*see* Chapter 13).

39 Dangerous substances

Many dangerous substances are used in industry, commerce, agriculture, research activities, hospitals and teaching establishments. The majority take the form of chemical compounds, but there are also naturally occurring substances, such as asbestos, heavy metals and siliceous dusts which can have adverse effects (*see* Chapter 19).

Generally, dangerous substances are of two types. They can directly affect the individual by entry into the body through the lungs, skin or mouth, or they can have a secondary effect through his coming into contact with them, e.g. when a highly flammable substance is ignited. Much depends upon the form – dust, liquid, aerosol, etc. – taken by the substance and its method of use. (*See further* Chapter 18 'Toxicology and health'.)

Classification of hazardous substances (supply requirements)

Hazardous substances are classified according to Schedule 1 of the Chemicals (Hazard Information and Packaging for Supply) (CHIP 2) Regulations 1994 as shown in Table 39.1.

Table 39.1 ◆ Part I Categories of Danger

Category of danger	*Property (See Note 1)*	*Symbol letter*
PHYSICO-CHEMICAL PROPERTIES		
Explosive	Solid, liquid, pasty or gelatinous substances and preparations which may react exothermically without atmospheric oxygen thereby quickly evolving gases, and which under defined test conditions detonate, quickly deflagrate or upon heating explode when partially confined.	E
Oxidising	Substances and preparations which give rise to a exothermic reaction in contact with other substances, particularly flammable substances.	O
Extremely flammable	Liquid substances and preparations having an extremely low flashpoint and a low boiling point and gaseous substances and preparations which are flammable in contact with air at ambient temperature and pressure.	F+
Highly flammable	The following substances and preparations, namely (a) substances and preparations which may become hot and finally catch fire in contact with air at ambient temperature without any application of energy; (b) solid substances and preparations which may readily catch fire after brief contact with a source of ignition and which continue to burn or to be consumed after removal of the source of ignition; (c) liquid substances and preparations having a very low flashpoint; (d) substances and preparations which, in contact with water or damp air, evolve highly flammable gases in dangerous quantities. (*See* Note 2)	F
Flammable	Liquid substances and preparations having a low flashpoint	None
HEALTH EFFECTS		
Very toxic	Substances and preparations which in *very low quantities* can cause death or acute or chronic damage to health when inhaled, swallowed or absorbed via the skin.	T+

Table 39.1 ◆ continued

Category of danger	*Property (See Note 1)*	*Symbol letter*
Toxic	Substances and preparations which in low *quantities* can cause death or acute or chronic damage to health when inhaled, swallowed or absorbed via the skin.	T
Harmful	Substances and preparations which may cause death or acute or chronic damage to health when inhaled, swallowed or absorbed via the skin.	Xn
Corrosive	Substances and preparations which may, on contact, with living tissues, *destroy* them.	C
Irritant	Non-corrosive substances and preparations which through immediate, prolonged or repeated contact with the skin or mucous membrane, may cause *inflammation.*	Xi
Sensitising	Substances and preparations which, if they are inhaled or if they penetrate the skin, are capable of eliciting a reaction by *hypersensitisation* such that on further exposure to the substance or preparation, characteristic adverse effects are produced.	
Sensitising by inhalation		Xn
Sensitising by skin contact		Xi
Carcino-genic (*See* Note 3)	Substances and preparations which, if they are inhaled or ingested or if they penetrate the skin, may induce *cancer* or increase its incidence.	
Category 1		T
Category 2		T
Category 3		Xn
Mutagenic (*See* Note 3)	Substances and preparations which, if they are inhaled or ingested or if they penetrate the skin, may induce *heritable genetic defects* or increase their incidence.	
Category 1		T
Category 2		T
Category 3		Xn

Table 39.1 ◆ continued

Category of danger	*Property (See Note 1)*	*Symbol letter*
Toxic for reproduction (See Note 3	Substances and preparations which, if they are inhaled or ingested or if they penetrate the skin, may produce or increase the incidence of *non-heritable* adverse effects in the progeny and/or an impairment of male or female reproductive functions or capacity.	
Category 1		T
Category 2		T
Category 3		Xn
Dangerous for the environment (*See* Note 4)	Substances which, were they to enter into the environment, would or might present an immediate or delayed danger for one or more components of the environment.	N

Notes

1. As further described in the *approved classification and labelling guide*.
2. Preparations packed in *aerosol dispensers* shall be classified as *flammable* in accordance with the additional criteria set out in Part II of this Schedule.
3. The categories are specified in the approved classification and labelling guide.
4. (a) In certain cases specified in the *approved supply list* and in the *approved classification and labelling guide* substances classified as dangerous for the environment do not require to be labelled with the symbol for this category of danger.
   (b) This category of danger does not apply to preparations.

Substance hazardous to health means any substance (including any preparation) which is

(a) a substance which is listed in Part 1 of the approved supply list as dangerous for supply within the meaning of Chemicals (Hazard Information and Packaging for Supply) Regulations 1994 and for which an indication of danger specified for the substance in Part V of that list is *very toxic, toxic, harmful, corrosive or irritant*;

(b) a substance specified in Schedule 1 (which lists substances assigned maximum exposure limits) or for which the Health and Safety Commission has approved an occupational exposure standard:

(c) a biological agent;

(d) dust of any kind, when present at a substantial concentration in air;

(e) a substance, not being a substance mentioned in sub-paragraphs (a) to (d) above, which creates a hazard to health of any person which is comparable with the hazards created by substances mentioned in those sub-paragraphs.

Physical state of dangerous substances

The form taken by a dangerous substance – e.g. liquid, gas, dust, mist, vapour, etc. – is a contributory factor to its potential for harm. Dangerous substances take many forms, the most common being as follows.

Dusts

These are solid airborne particles, often created by operations such as grinding, crushing, milling, sanding and demolition. Two of the principal harmful dusts encountered in industry are asbestos and silica (*see* Chapter 22).

Fumes

Fumes are solid particles which usually form an oxide in contact with air. They are created by industrial processes which involve the heating and melting of metals, such as welding, smelting and arc air gouging. A common fume danger is lead poisoning associated with the inhalation of lead fume.

Smoke

Smoke is the product of incomplete combustion, mainly of organic materials, and may include fine particles of carbon in the form of ash, soot and grit that are visibly suspended in air.

Mists

A mist is a finely dispersed liquid suspended in air. Mists are mainly created by spraying, foaming, pickling and electro-plating. Dangers arise most frequently from acid mists produced in industrial treatment processes.

Gases

These are formless fluids usually produced by chemical processes involving combustion or by the interaction of chemical substances. A gas will normally seek to fill the space completely into which it is liberated. One of the classic hazardous gases encountered in industry is carbon monoxide. Certain gases such as acetylene, hydrogen and methane are particularly flammable.

Vapours

A vapour is the gaseous form of a material normally encountered in a solid or liquid state at normal room temperature and pressure. Typical examples are solvents, such as trichlorethylene, which release vapours when the container is opened. Other liquids produce a vapour on heating, the amount of vapour being directly related to the boiling point of that particular liquid. A vapour contains very minute droplets of the liquid. However, in the case of a *fog*, the liquid droplets are much larger.

Solids

Certain substances in solid form can cause injury. Classic examples are cullet (broken glass), silica, asbestos and lead.

Liquids

Numerous dangerous substances are produced in liquid form including caustic and acid-based detergents, solvents and fuels.

The Control of Substances Hazardous to Health (COSHH) Regulations 1999

The COSHH Regulations 1999 apply to all forms of workplace and work activity involving the use of dangerous substances. The Regulations are supported by a number of ACOPs:

Control of substances hazardous to health

Control of substances hazardous to health in fumigation operations

Control of carcinogenic substances

Control of biological agents

together with the booklet produced by the HSE entitled *COSHH Assessments* and a number of HSE pamphlets entitled 'Introducing COSHH', 'Introducing Assessment' and 'Hazard and Risk Explained'.

It has been established that industry uses some 40,000 different substances, and each one could be hazardous to health due its inherent properties, i.e. toxic, harmful, corrosive or irritant. The Regulations apply to all substances which come within this classification under the CPLR. They also apply to all forms of business and workplace operation –

factories, offices, shops, schools, hospitals, etc. – irrespective of size, but the following industries will need to pay special attention:

(a) major manufacturers of chemicals and bulk users of chemicals;
(b) users of substances in situations most likely to involve high operator exposure levels, for instance in spraying activities;
(c) users of substances which may be highly toxic or present a special health risk, for example in chemical and bacteriological laboratories;
(d) the well-known 'dusty' trades and industries, for example, ceramics, metal manufacturing and finishing, foundries and quarrying operations;
(e) companies operating processes which generate substances hazardous to health in substantial amounts, such as milling and grinding, or welding and cutting operations, for example, engineering workshops.

The Regulations are based on good occupational hygiene practice and place particular emphasis on:

(a) the assessment of health risks from substances used at work;
(b) the provision, use and maintenance of certain physical controls to prevent liberation of dangerous substances into the working environment;
(c) monitoring of airborne contamination;
(d) in certain cases, health and/or medical surveillance of staff who may be exposed on a regular basis.

Definitions

Substance hazardous to health means any substance (including any preparation) which is:

(a) a substance which is listed in Part 1 of the approved supply list as dangerous for supply within the meaning of the Chemicals (Hazard Information and Packaging for Supply) Regulations 1994 and for which an indication of danger specified for the substance in Part V of that list is very toxic, toxic, harmful, corrosive or irritant;
(b) a substance specified in Schedule 1 (which lists substances assigned maximum exposure limits) or for which the HSC has approved an occupational exposure standard (See current HSE Guidance Note EH 40 *Occupational exposure limits*);
(c) a biological agent;
(d) dust of any kind when present at a substantial concentration in air;

(e) a substance, not being a substance mentioned in sub-paragraphs (a) to (d) above, which creates a hazard to the health of any person which is comparable with the hazards created by substances mentioned in those sub-paragraphs.

Assessment of health risks

The basis for implementing the Regulations is the assessment of health risks associated with exposure to hazardous substances being used in the workplace. Regulation 6 requires that an employer shall not carry on any work which is liable to expose any employees to any substance hazardous to health unless he has made a *suitable and sufficient assessment* of the risks created by that work to the health of those employees and of the steps that need to be taken to meet the requirements of the Regulations. Furthermore, the above assessment must be reviewed forthwith if there is reason to suspect that the assessment is no longer valid, or there has been a significant change in the work to which the assessment relates, and, where as a result of the review, changes in the assessment are required, those changes shall be made.

G

On this basis the following information should be available on completion of the health risk assessment for each substance:

(a) the risk involved (toxic, harmful, carcinogenic, etc.) including the route of entry of that substance into the body (for example, by inhalation);

(b) the physical and systems control measures necessary to control health risks;

(c) health and/or medical surveillance procedures, where necessary;

(d) environmental monitoring procedures necessary in the workplace.

Sec 6 of HSWA places specific duties on manufacturers and suppliers of substances used at work. Through the subsequent amendment of this section by the Consumer Protection Act 1987, such persons must provide adequate information to the user for him to take the appropriate precautions with regard to storage, use and disposal of such substances. This information should form the starting point of the health risk assessment for a particular substance with the objective of identifying the precautions necessary, including the development and implementation of safe systems of work, training needs of supervisors and operators and emergency procedures in the event of major spillages, gross body contamination or eye contact.

The employer must keep a written record of the assessment in all but the simplest cases, and inform his employees of the results of the assessment. Assessments must be 'suitable and sufficient'. This means that the detail and expertise used should be commensurate with the nature and degree of the risk to health and the relative complexity of the process in which the substances are used.

Prevention/control of exposure

Emphasis is on prevention of exposure to dangerous substances but, where this is not reasonably practicable, there must be adequate control. This applies whether the route of entry is through inhalation, ingestion or through the skin, or if the substance is dangerous when in contact with the skin.

In the case of substances hazardous through inhalation, which have been assigned Maximum Exposure Limits (MELs), exposure must never exceed these limits and should be reduced to below these limits to the greatest extent that is reasonably practicable. (*See* Schedule 1 to the Regulations – List of Substances Assigned Maximum Exposure Limits.)

Where there is exposure to a substance for which an Occupational Exposure Standard (OES) has been approved, the control of exposure shall, so far as the inhalation of the substance is concerned, only be treated as adequate if, firstly, the OES is not exceeded or secondly, where the OES is exceeded, the employer identifies the reasons for the standard being exceeded and takes appropriate action, as soon as is reasonably practicable, to remedy the situation.

In particular, where reasonably practicable, the prevention or adequate control of exposure of employees to a dangerous substance shall be secured by means other than by the provision of personal protective equipment, for example, respiratory protective equipment, hand and arm protection (Reg 7(2)). The significance of this subsection must be appreciated by employers. It will no longer be sufficient to issue personal protective equipment as a means of preventing ill-health at work. Other control and prevention strategies must be pursued, such as the use of LEV systems, substitution of less hazardous substances, change of process to reduce operator risks, enclosure of plant, segregation (by distance, age, sex or time) in limited situations, together with improvements in support strategies, particularly general and specific cleaning procedures, welfare amenity provisions and training of staff. The use of personal protective equipment must be seen as an additional precaution and not the sole precaution as is so frequently the case. (*See further* Chapter 26 'Prevention and control strategies in occupational hygiene'.) All forms of personal protective equipment must conform to specific standards, e.g. British Standards, must be of an approved type or comply with a standard approved by the HSE.

Once control measures have been installed, the employer must ensure they are properly used. Employees have a duty under the Regulations to make full and proper use of these control measures and must report any defect to their employer (Reg 8).

Maintenance, etc. of control measures

The employer must provide a formal maintenance programme to ensure that all control measures required under Reg 7 remain effective and continue to operate as originally intended. Engineering controls, e.g. LEV systems, must be thoroughly examined and tested at regular intervals and, in the case of LEV systems, the intervals between examination and testing must not exceed 14 months. Shorter intervals apply in the case of processes specified in Schedule 3 to the Regulations. Moreover, under Reg 9, respiratory protective equipment must be thoroughly examined and, if necessary, tested at suitable intervals. Suitable records of examination and test, and of any repairs or modifications found necessary, required by Reg 9 must be kept for at least five years.

Monitoring exposure at the workplace

Under Reg 10 the exposure of employees, and, so far as is reasonably practicable, other persons on the site, to hazardous substances should be monitored, where found necessary, by a suitable procedure. Schedule 4 specifies the time intervals at which the monitoring of exposure to certain specified substances, e.g. vinyl chloride monomer (VCM) and from other processes, must be carried out. Monitoring records must be kept in a suitable form for at least 30 years, where they refer to identifiable individuals, and five years in other cases.

Monitoring is necessary when failure or defects in the control measures could result in serious risk to health, or where it is necessary to ensure that an MEL or OES is not exceeded.

Health surveillance

'Health surveillance' is simply defined as 'including biological monitoring' (*See* Chapter 18.) Where a health risk assessment indicates a need for health surveillance of exposed employees, or those liable to be exposed to a dangerous substance, the employer must ensure that such employees are kept under suitable health surveillance (Reg 11). Health surveillance is necessary where:

(a) an employee may be exposed to one of the substances and is engaged in a process specified in Schedule 5, e.g. VCM manufacture, production reclamation, storage, discharge, transport, use or polymerisation, or in the manufacture of potassium or sodium chromate or dichromate; or

(b) the exposure of the employee to a substance hazardous to health is such that an identifiable disease or adverse health effect may be related to the exposure, there is reasonable likelihood that the disease or effect may occur under the particular conditions of his work and there are valid techniques for detecting indications of the disease or effect.

In such cases, detailed health records containing approved particulars must be kept for at least 30 years since the date of the last entry. These records must be offered to the HSE if the employer ceases to trade. An employer must:

(a) obey any directives made by an Employment Medical Adviser (EMA) or Appointed Doctor, i.e. doctors who have been specially appointed by EMAs, with respect to health records which relate to any employee engaged in work specified in Schedule 5 to the Regulations;

(b) allow employees covered by this Regulation to present themselves for health surveillance procedures during working hours and at his expense;

(c) allow employees access to their health records;

(d) allow EMAs and Appointed Doctors access to inspect the workplace and the records kept for the purposes of the Regulations, should the EMA or Appointed Doctor require it.

Information, instruction and training

Regs 12(1) and 12(2) are essential to the effective application of Regs 7, 8, and 9. They require an employer whose work may expose his employees or, so far as is reasonably practicable, any other person on the premises where the work is carried on, to substances hazardous to health to provide them with such information, instruction and training as is suitable and sufficient for them to know:

(a) the risk to health created by such exposure;

(b) the precautions which should be taken.

Information provided by the employer should include, in particular, details of:

(a) the nature and degree of risk involved, the consequences of exposure and any factors which might increase the risk:

(b) the control methods adopted, the reason for these, and their proper use;

(c) why and where personal protective equipment (PPE) is necessary;

(d) monitoring procedures, including arrangements for access to results and notification if an MEL is exceeded;

(e) the role of health surveillance, and access to individual health records and the results of collective health surveillance.

Instruction given to employees and others on the premises should enable them to know:

(a) what to do when dealing with substances hazardous to health, including the precautions that must be taken and when they should be taken;

(b) what cleaning, storage and disposal procedures are required, and why and when they are to be carried out;

(c) what emergency procedures to follow.

Persons on the premises who may have direct or indirect contact with hazardous substances should receive sufficient training to enable them to effectively apply and use:

(a) the various methods of control;

(b) personal protective equipment;

(c) emergency measures installed.

Reg 12(3) requires an employer to ensure that any person who does any work to ensure compliance with the Regulations, including carrying out any task in connection with their duties under them, has sufficient information, instruction and training to do the job safely. This implies a certain level of competence and may entail the use of external consultants and/or expertise, but the employer will still need to ensure that the persons engaged in the work receive sufficient information about the circumstances of the work.

Prohibited substances

Reg 4 ensures that prohibitions contained in the legislation repealed and revoked by the COSHH Regulations are carried forward. The full list of prohibited substances and the circumstances under which they are prohibited is contained in Schedule 2.

Under Reg 4(2) the importation into the United Kingdom, other than from another Member State, of the following substances and articles is prohibited, namely:

(a) 2-naphthylamine, benzidine, 4-aminodipheny, 4-nitrodiphenyl, their salts and any substance containing any of those compounds in a total concentration equal to or greater than 0.1 per cent by mass;

(b) matches made with white phosphorus.

Any contravention of this paragraph shall be punishable under the Customs and Excise Management Act 1979 and not as a contravention of a health and safety regulation.

The use of the above four substances and their salts is prohibited under all circumstances if their concentrations exceed 0.1 per cent. The importation of any of these compounds and of any substances containing them in concentrations greater than 0.1 per cent is banned, together with matches made with white phosphorous. Moreover, where such substances are imported, such an act is a contravention of the Customs and Excise Management Act 1979 and not of any health and safety regulations.

Defences under the regulations

Reg 16 is concerned with two important defences under the Regulations. In any alleged breach of the Regulations an employer can offer the defence that he took '*all reasonable precautions*' and exercised '*all due diligence*' to ensure compliance with the Regulations. However, an employer cannot delegate this responsibility to a third party and, therefore, it is no defence to show that an employee or other party was at fault.

To rely on this defence, an employer must establish that, on the balance of probabilities, he had taken *all* precautions that were reasonable and exercised *all* due diligence to ensure that these precautions were implemented in order to avoid such a contravention. It is unlikely that an employer could rely on the Reg 16 defence if:

(a) precautions were available which had not been taken; or

(b) he had not provided sufficient information, instruction and training together with adequate supervision, to ensure that the precautions were effective.

Thus a stated company policy on the safe use of hazardous substances, company code of practice, job safety instructions, displayed notices and other forms of documentation and propaganda to bring the risks to the attention of employees is insufficient without evidence of formal information, instruction, training and supervision.

Managing COSHH

There are a number of simple and practical steps (Fig. 39.2) which prudent management and health and safety specialists should take to ensure compliance with the Regulations. A step-by-step approach is detailed below.

1. Prepare an inventory of all the chemicals used in production, maintenance, cleaning, laboratory analysis, etc.
2. Identify the point of use for each substance.
3. Instigate a rationalisation programme with the principal objective of reducing the number of substances, rationalising the approach to purchase and ensuring the transmission of information on these substances to all users.
4. Obtain information (Safety Data Sheets) from the manufacturers or suppliers of each substance identified.
5. Commence the programme of health risk assessment starting with the more dangerous substances, identifying the prevention or control strategies in each case.
6. Monitor progress in the development of the programme, identifying exposures where medical and/or health surveillance may be necessary.
7. Institute a medical and/or health surveillance programme where appropriate.
8. Develop and commence a training programme aimed at informing users of the risk involved, the personal precautions necessary and the systems for controlling or preventing exposure.
9. Prepare and circulate a company Statement of Policy on the Use of Substances Hazardous to Health. Such a document should incorporate a statement of intent, organisation and arrangements (including individual responsibilities of all concerned) for implementing the policy and the procedures for monitoring compliance with the policy. The document should be signed and dated by the company chairman, chief executive or managing director, and incorporated as an Appendix to the Company Statement of Health and Safety Policy.
10. Install appropriate records and documentation on each assessed

substance, perhaps by means of a company Hazardous Substances Directory, which should be circulated to all managers, made available to operators, particularly during their training, and supported by job safety instructions and propaganda, such as posters, warning notices, accident bulletins and progress reports in the implementation of the programme.

The main duties of the employer are summarised below (*see* Fig. 39.2 COSHH Action Plan).

Fig. 39.2 ◆ COSHH Action Plan

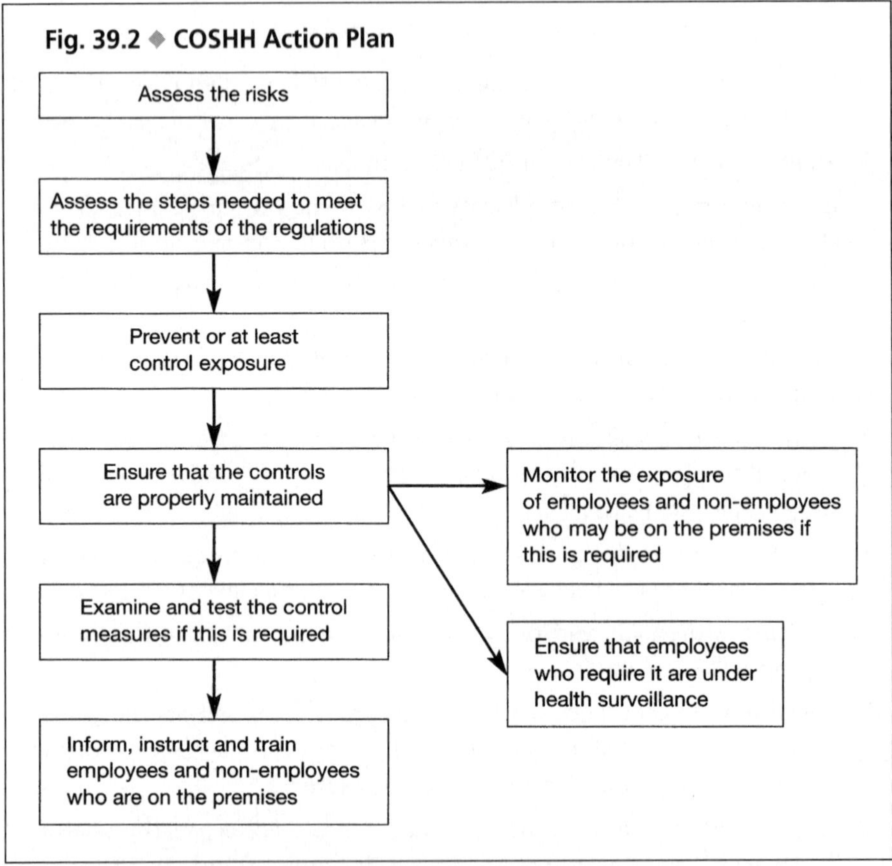
Assess the risks
Assess the steps needed to meet the requirements of the regulations
Prevent or at least control exposure
Ensure that the controls are properly maintained
Examine and test the control measures if this is required
Inform, instruct and train employees and non-employees who are on the premises
Monitor the exposure of employees and non-employees who may be on the premises if this is required
Ensure that employees who require it are under health surveillance

Highly Flammable Liquids and Liquefied Petroleum Gases Regulations 1972

These Regulations lay down requirements for the following matters in relation to highly flammable liquids etc. used in a factory, i.e.:

(a) manner of storage;
(b) marking of storage accommodation and vessels;
(c) precautions to be observed for the prevention of fire and explosion;
(d) provision of fire-fighting apparatus;
(e) securing means of escape in the event of fire.

'Highly flammable liquids' include both liquefied flammable gas, although not aqueous ammonia, and liquefied petroleum gases and therefore includes any liquid, liquid solution, emulsion or suspension which:

(a) gives off a flammable vapour at a temperature less than 32°C when tested in the manner set out in Schedule 1 to the Regulations (Closed cup flashpoint determination method);
(b) supports combustion when tested in a manner set out in Schedule 2 (Combustibility test).

'Liquefied flammable gas' is any substance which would be a flammable gas at a temperature of 20°C and a pressure of 760 mm of mercury, but which is in a liquid form as a result of the application of pressure or refrigeration or both.

'Liquefied petroleum gas' covers both commercial butane and commercial propane, and any mixture of them.

Application of the Regulations

The Regulations apply to factories and impose duties on the occupier of the premises or, in some cases, the owner of the substances.

Storage of highly flammable liquids – general

When not in use or being conveyed, all highly flammable liquids (HFLs) should be stored in a safe manner. All HFLs should be stored in one of the following ways:

(a) in suitable fixed storage tanks in a safe position; or
(b) in suitable closed vessels kept in a safe position in the open air and, where necessary, protected against direct sunlight; or
(c) in suitable closed vessels kept in a store room which either is in a safe position or is of fire-resistant structure; or
(d) in the case of a workroom where the aggregate quantity of HFL stored

does not exceed 50 litres, in suitable closed vessels kept in a suitably placed cupboard or bin which is a fire-resistant structure.

Other storage precautions

1. Bund walls

Storage tanks should be provided with a bund wall enclosure which is capable of containing 110 per cent of the capacity of the largest tank within the bund.

2. Ground beneath vessels

The ground beneath storage vessels should be impervious to liquid and be so sloped that any minor spillage will not remain beneath the vessels, but will run away to the sides of the enclosure.

3. Bulk storage

Bulk storage tanks should not be located inside buildings or on the roof of a building. Underground tanks should not be sited under the floors of process buildings.

4. Drum storage

The area to be utilised for drum storage should be surrounded with a sill capable of containing the maximum spillage from the largest drum in store.

Marking of store rooms and containers

Every store room, cupboard, bin, tank and vessel used for storing HFLs should be clearly and boldly marked 'Highly Flammable' or 'Flashpoint below 32°C' or 'Flashpoint in the range of 22°C to 32°C'.

Specific provisions for the storage of liquefied petroleum gas

All liquefied petroleum gas (LPG) must be stored in one of the following ways:

(a) in suitable underground reservoirs or in suitable fixed storage tanks located in a safe position, either underground or in the open air; or

(b) in suitable movable storage tanks/vessels kept in a safe position in the open air; or

(c) in pipelines or pumps forming part of an enclosed system; or

(d) in suitable cylinders kept in safe positions in the open air or, where this is not reasonably practicable, in a store room constructed of non-combustible material, having adequate ventilation, being in a safe position, of fire-resistant structure, and being used solely for the storage of LPG and/or acetylene cylinders.

LPG cylinders must be kept in a store until they are required for use, and any expended cylinder must be returned to store as soon as is reasonably practicable. This should ensure that only the minimum amount of LPG is kept in any workplace.

Marking of store rooms and containers

Every tank, cylinder, store room, etc. used for the storage of LPG should be clearly and boldly marked 'Highly Flammable – LPG'.

Precautions against spills and leaks (all HFLS)

Where HFLs are to be conveyed within a factory, a totally enclosed piped system should be used, where reasonably practicable. Where not reasonably practicable, a system using closed non-spill containers will be acceptable.

Portable vessels, when emptied, should be removed to a safe place without delay.

Where in any process or, operation any HFL is liable to leak or be spilt, all reasonably practicable steps should be taken to ensure that any such HFL should be contained or immediately drained off to a suitable container, or to a safe place, or rendered harmless.

Precautions against escaping vapours

No means likely to ignite vapour from any HFL should be present where a dangerous concentration of vapours from HFL may be present.

Where any HFL is being utilised in the workplace, reasonably practicable steps should be taken so as to minimise the risk of escape of HFL vapours into the general workplace atmosphere. Where such escape cannot be avoided, then the safe dispersal of HFL vapours should be effected, so far as is reasonably practicable.

Relaxation of fire resistance specifications in certain circumstances

In cases where either explosion pressure relief or adequate natural ventilation are required in a fire-resistant structure, a relaxation of the specification of a fire-resistant structure is allowable.

Fire escapes and fire certificates

There must be adequate and safe means of escape in case of fire from every room in which any HFL is manufactured, used or manipulated. This Regulation does not apply where there is storage only. Fire certificates are generally necessary where:

(a) HFLs are manufactured;

(b) LPG is stored;

(c) liquefied flammable gas is stored.

Prevention of build-up of deposits

Whenever, as a result of any process or, operation involving any HFL, a deposit of any solid waste residue liable to give rise to a risk of fire is liable to occur on any surface:

(a) steps must be taken to prevent the occurrence of all such deposits, so far as is reasonably practicable;

(b) where any such deposits occur, effective steps must be taken to remove all such residues, as often as necessary, to prevent danger.

Smoking controls

No person may smoke in any place in which any HFL is present and where the circumstances are such that smoking will give rise to the risk of fire.

Provision of fire-fighting equipment

Appropriate fire-fighting equipment should be made readily available for use in all factories where HFL is manufactured, used or manipulated.

Duties of employees

It is the duty of every employee to comply with them and co-operate in carrying them out.

If an employee discovers any defect in plant, equipment or appliance, it is his duty to report the defect without delay to the occupier, manager or other responsible person.

Handling and storage of dangerous substances

General precautions

Prior to handling and storing dangerous substances, it is imperative to consult sources of hazard data (*see* later in this chapter, 'Essential data'). Basic safety rules apply, however, with all dangerous substances and these are outlined below. Some are statutory requirements, whereas with others current best practice is outlined:

(a) Meticulous standards of housekeeping are necessary for any activity involving the handling and storage of dangerous substances.

(b) Smoking and the consumption of food or drink should be prohibited in any area in which substances are used or stored, e.g. laboratory, bulk chemical store (*see* particularly the statutory prohibitions set out in Chapter 5).

(c) Staff must be reminded regularly of the need for good personal hygiene, in particular washing hands after handling any chemical-based substance.

(d) The minimum quantities should be stored in the work area. Extra bulk storage should be provided separately well away from the work area.

(e) Containers should be clearly and accurately labelled.

(f) Chemical substances should always be transported with care and carriers used for Winchesters and other large containers.

(g) Fume cupboards should have a minimum face velocity of approximately 0.4 m/sec when measured with the sash opening set at 300 mm maximum, and performance should be checked frequently in accordance with the COSHH Regulations.

(h) Staff should wear protective clothing and equipment whenever handling or using dangerous substances (*see* Chapter 24).

(i) Any injury should be treated promptly, particularly skin wounds.

(j) Responsibility should be identified at senior management level, and written procedures published and used in the training and supervision of staff.

Precautions with specific substances

Flammable liquids

(a) All containers should be of the self-closing type. Caps should be replaced after dispensing. Liquids should be dispensed in a drip tray.

(b) Containers should be stored in a well-ventilated fire-protected area.

(c) Fire appliances should be located in a readily accessible position and staff trained in their use.

(d) Flammable liquids should be transported in closed containers of metal construction. (Some plastics may, however, be acceptable for this purpose.)

Carcinogens, poisons, etc.

(a) Staff must wear the appropriate protective clothing and equipment.

(b) First aid treatment, including the appropriate antidote, must be known and readily available.

(c) Substances producing fumes must be handled in fume cupboards.

(d) Substances should be transported in sealed and labelled containers.

Radioactive materials

(a) Materials should be clearly identified by the appropriate warning sign.

(b) Materials should never be handled with bare hands.

(c) Materials should only be moved under the direct supervision of an authorised person, and transported in sealed or other appropriate containers.

(d) The level of radioactivity should be checked before any radioactive source is approached.

(e) Staff must be trained to know the hazards involved and the precautions necessary (*see* Chapter 23).

Solids

(a) In the case of dusts and particulate materials, e.g. powders –
   (i) respiratory protection should be worn unless control measures are adequate;
   (ii) atmospheric concentrations should be measured and related to the current hygiene standard in order to determine the degree of danger present;
   (iii) there should be a complete ban on smoking where the solid is flammable.

(b) In the case of all solids, the nature of the substance to be handled must be ascertained.

Bulk storage of dangerous chemical substances

In the design and use of bulk storage facilities, the following aspects need attention:

(a) the range and quantities of chemical substances to be stored;

(b) dependent on (a) above, the degree of segregation by distance of –
   (i) the store from any other building;
   (ii) certain chemical substances within the store from other chemical substances stored.

Segregation

The aim in segregating stored chemicals should be:

(a) to facilitate emergency access and escape in the event of fire or other emergency;

(b) to separate incompatible chemicals to prevent their mixture, e.g. by spillage, or wetting during cleaning activities;

(c) to separate process areas, which normally contain relatively small quantities, from storage areas containing larger amounts;

(d) to prevent rapid fire spread, or the evolution of smoke and gases, which can be produced in a fire;

(e) to isolate oxidising agents which, when heated, will enhance a fire, perhaps to explosive condition;

(f) to isolate those substances which decompose explosively when heated;

(g) to minimise toxic hazards arising from loss of containment through spillage, seepage or package deterioration;

(h) to minimise risk of physical damage, e.g. by fork-lift trucks, to containers;

(i) to separate materials where the appropriate fire-fighting medium for one may be ineffective for, or cause an adverse reaction with, another.

Segregation distances: Reference should be made to Table 39.2 which specifies relative segregation distances for various forms of loose package. The hazard warning symbols for different forms of chemical substance classified in Table 39.2 are shown in Fig. 39.1. Table 39.3 details the categories of separation.

Table 39.2 ◆ The segregation of chemical substances

Hazard warning symbol	*Chemical substances*	*Alkali corrosives*	*Acid corrosives*	*Potential oxidising substances*	*Peroxides*	*Chlorine release agents*	*Flammable solids*	*Flammable liquids*	*Flammable gases*	*Non-flammable gases*	*Low-hazard products–human contact possible*	*Poisons*
A/E	Alkali corrosives		2	1	2	x	1	1	1	x	1	x
A/E	Acid corrosives	2		2	2	2	1	1	1	x	1	x
B/E	Potential oxidising substances	1	2		1	1	1	2	2	x	2	1
B/G	Peroxides	2	2	1		2	2	3	3	2	2	1
F/E	Chlorine release agents	x	2	1	2		2	2	2	x	1	2
F/C	Flammable solids	1	1	1	2	2		1	1	x	2	1
F	Flammable gases	1	1	2	3	2	1	2		x	2	1
C	Non-flammable gases	x	x	x	2	x	x	2	x		x	1
	Low-hazard products – human contact possible	1	1	2	2	1	2	2	2	x		2
D	Poisons	x	x	1	1	2	1	1	1	1	2	
	Food ingredients	3	3	3	3	3	3	3	3	1	1	3
	Miscellaneous dangerous substances	No general segregation requirements can be provided; consult individual data sheets										

Table 39.3 ◆ Categories of separation

Category of separation	*Requirement*
x	No specific separation required
1	Keep away from . . .
2	Keep well separated from . . .
3	Separate by fire-resistant partition or in a separate location

Structural requirements

Chemical storage may take two forms, viz. an open area or a purpose-built store. Open storage is not recommended but, when it is unavoidable, it should comprise a secure area fenced to a height of two metres with a lockable access point.

Purpose-built stores should be of the detached single-storey brick-built type or constructed in other suitable materials, e.g. concrete panels, with a sloping roof of weather-proof construction, the structure to have a notional period of fire resistance of at least one hour. Other features include:

(a) permanent ventilation by high- and low-level air bricks set in all elevations, except in those forming a boundary wall; low-level air bricks should be sited above door sill level;

(b) access doors constructed from material with at least one-hour notional period of fire resistance; doorways should provide access for fork-lift trucks with ramps on each side of the door sill; separate pedestrian access, which also serves as a secondary means of escape, should be provided;

(c) an impervious chemical-resistant finish to walls, floors and other parts;

(d) artificial lighting by sealed bulkhead or fluorescent fittings, to provide an overall illuminance level of 300 lux.

General requirements

In the cases of both open and closed storage, the following are required:

(a) provision of adequate space, with physical separation and containment for incompatible substances (*see* Table 39.2), each area to be marked with the permitted contents, the hazards and the necessary precautions, and incorporating an area for the storage of empty containers;

Fig. 39.1 ◆ Hazard warning symbols, as required by the CHIP 2 Regulations

FIG. 39.1 Hazard warning symbols, as required by the CHIP 2 Regulations

CATEGORY OF DANGER	SYMBOL LETTER	INDICATION OF DANGER	SYMBOL
Explosive	E	Explosive	
Oxidising	O	Oxidising	
Extremely Flammable	F+	Extremely Flammable	
Highly Flammable	F	Highly Flammable	
Flammable			
Very Toxic (Very Poisonous)	T+	Very Toxic	
Toxic	T	Toxic	
Harmful	Xn	Harmful	 HARMFUL
Corrosive	C	Corrosive	

Fig. 39.1 ◆ continued

FIG. 39.1 Continued.

CATEGORY OF DANGER	SYMBOL LETTER	INDICATION OF DANGER	SYMBOL
Irritant	XI	Irritant	 IRRITANT
Carcinogenic Category 1 & 2 Category 3 (May Cause Cancer)	T Xn	Toxic Harmful	 HARMFUL
Mutagenic Category 1 Category 2 & 3 (May Cause Mutation)	T Xn	Toxic Harmful	 HARMFUL
Teratogenic Category 1 Category 2 (May Cause Monstrosities)	T Xn	Toxic Harmful	 HARMFUL
Dangerous for the environment	N	Dangerous for the environment	

(b) fire separation of individual areas sufficient to prevent fire spreading;

(c) provision of the following equipment in a protected area outside the store –

  (i) fire appliances – dry powder and/or foam extinguishers;

  (ii) fixed hose reel appliance;

  (iii) emergency shower and eyewash station with water-heating facility to prevent freezing;

  (iv) personal protective equipment, i.e. safety helmet with visor, impervious gloves, disposable chemical-resistant overall, with storage facilities for same;

  (v) respirator and breathing apparatus in a marked enclosure;

(d) a total prohibition on the use of naked flames and smoking – appropriate warning signs should be displayed;

(e) prohibition on the use of the store for storage of other goods or for any other purpose;

(f) provision of racking or pallets to enable goods to be sorted clear of the floor.

Storage system

The system for storage must be simple to operate and compatible with existing legal requirements for classification and labelling.

Dangerous substances commonly encountered in industry

Within this category can be included materials such as cement, resins, coal tar pitch derivatives and fibreglass. Safety aspects of compressed gases are also considered.

Portland cement

Cement is used universally in the construction industry and other industries. Inhalation of cement dust may cause fibrosis, but the principal hazard is its propensity for causing dermatitis, through excessive contact or its specific sensitising effect. Cement, on contact with water, emits heat, resulting in burns to persons in contact with it.

Fibreglass

This material comes within the range of man-made mineral fibres. Continual contact with fibreglass causes dermatitis owing to the irritant effect of the fibres.

Coal tar pitch derivatives

This group includes pitch, creosote and tar produced in the distillation of crude hydrocarbon oils. Excessive contact may result in pitch warts, skin cancer and dermatitis. These substances are used widely in construction as constituents of roofing felt, bituminous paints, and asphalt for road surfacing, and in the form of creosote, a commonly used preservative and water repellant for timber.

Resins

Epoxy and polyester resins are used in adhesives, paints and sealants. Principal hazards are skin contact and inhalation of fumes. Other resins such as acrylic, phenolic and polyurethane resins have similar contact and inhalation risks. They should be used with considerable care and not in unventilated or badly ventilated areas.

Compressed gases

Compressed gases have numerous uses both commercially and domestically. In addition to their inherent flammable, toxic or corrosive properties, they are potentially dangerous as a consequence of their physical properties. Leakage from cylinders into an open room or workshop can give rise to dangerous concentrations resulting in fire, explosion, gassing incidents or oxygen depletion. The relative force with which the contents of a cylinder can be ejected can result in death, physical injury, damage to property and plant, and even the propulsion of the cylinder, like a rocket, across a working area.

Precautions with cylinders

Storage: One of the principal causes of accidents is incorrect storage of cylinders, resulting in incompatible reactions taking place between different gases leaking into the storage area. For this reason alone, the following precautions should be taken in the design of storage facilities:

(a) Cylinders should be stored outside the main buildings in a purpose-built store.

(b) The store should be a detached structure and well segregated from buildings frequently occupied.

(c) Storage should incorporate separate compartments for individual gases or groups of compatible gases.

(d) The structure should be of weather-proof lightweight construction e.g. single skin brick or lightweight concrete block walls and partitions providings a notional period of fire resistance of at least one hour – and should have a sloping roof.

(e) High- and low-level ventilation by air bricks should be located in the end and back walls.

(f) Each compartment should be provided with a lockable welded mesh door of mesh size within the range XM21 to XM26 or XM41 to XM43 (*see* BS 405:1945).

Only compatible gases should be stored in individual storage compartments (*see* Fig. 39.3). A typical arrangement is:

Compartment A – arcton, freon, argon, nitrogen, helium, hydrogen

Compartment B – propane, butane

Compartment C – oxygen, compressed air

Compartment D – ammonia

Compartment E – chlorine (This compartment should incorporate a lockable louvred door.)

Other storage design features: These include:

(a) concrete floor with slight fall to facilitate trolley loading; no drainage should be incorporated in the floor;

(b) total prohibition of the installation of electrical or heating equipment;

(c) marking of the store with approved signs indicating the contents and prohibiting smoking and the use of naked flames;

(d) racking, incorporating restraining chains, for cylinder storage; racking is not necessary where cylinders are of freestanding design;

(e) water supply through a 20 mm hose reel installation adjacent to the store, but well away from the chlorine cylinder storage compartment;

(f) artificial lighting to a minimum illuminance level of 150 lux, remote from the store but directed into the store and door openings;

(g) fire appliances, e.g. dry powder, of adequate size, located at a specific point.

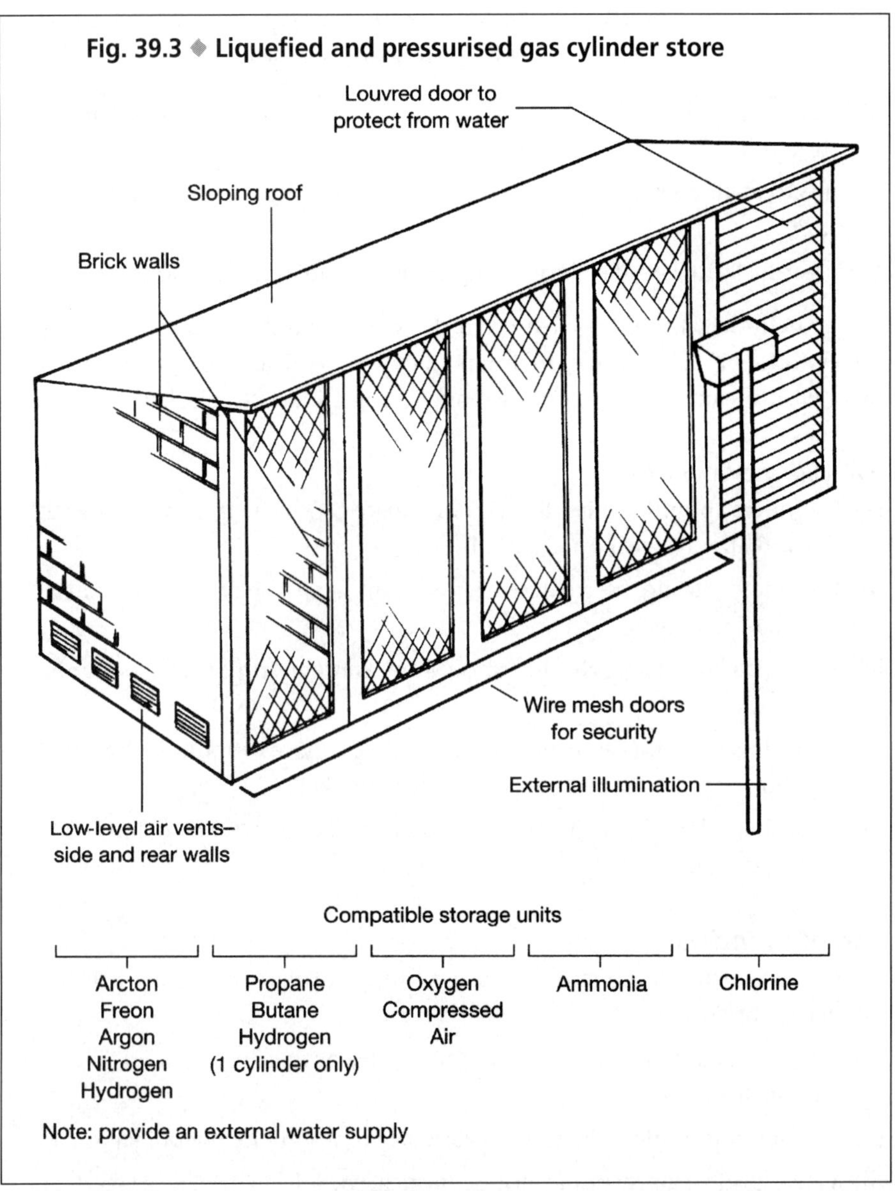

Fig. 39.3 ◆ Liquefied and pressurised gas cylinder store

Storage buildings should be so spaced that the distance between any aperture in the wall and the nearest point of any other building, boundary or ignition source is:

(a) not less than 3 metres for storage up to 1,000 kg;

(b) not less than 6 metres for storage over 1,000 kg.

Where the above standard is not practicable, it may be possible to locate the store against an existing building wall. However, there should be:

(a) no opening lights in the wall above the store, and any fixed lights should be glazed with georgian wired glass;

(b) no adjacent apertures – i.e. windows, doors – within three metres of the store;

(c) no ventilation intake above or immediately adjacent to the store.

Empty cylinders should be treated in the same manner as full or partially full cylinders, and the valves maintained closed. 'Full' and 'empty' cylinder racks within each compartment should be conspicuously marked.

Handling of cylinders

(a) Cylinders should not be dropped or allowed to come into contact with one another or with any hard object.

(b) Cylinders should be treated with extreme care; they are a potential source of energy.

(c) When transported, cylinders should be strapped in properly designed trolleys.

(d) Cylinders used for lecturing and training purposes should be handled in a fume cupboard and stood in a suitable rack.

(e) Improperly labelled cylinders should not be accepted. The colour code on the cylinder is only a secondary guide.

Use of cylinders

(a) Cylinders should be stored and used in an upright position with the valve uppermost.

(b) A regulator should be used to maintain the outlet pressure at a correct and uniform value.

(c) Cylinders should *not* be used as rollers for moving heavy objects.

(d) Valves and fittings should not be lubricated.

(e) Cylinders and valves should be kept clean.

(f) Cylinders with damaged threads or valves should immediately be labelled '*defective*' and returned to the supplier.

(g) When exchanging cylinders, the valves should be closed before the connections are transferred.

(h) After remaking a connection, the valve should be opened carefully in order to detect any leakage. In the event of leakage, the cylinder should be moved into the open air.

(i) Cylinder keys should not be extended to give greater leverage as valve spindles may be damaged.

(j) Leak detectors and alarms may be necessary when very dangerous gases are stored.

(k) Only the appropriate regulator should be used for each type of cylinder. Regulators should be examined at six-monthly intervals and labelled for use with one specific gas.

Dangerous gases – special precautions

These gases, especially flammable gases, should be housed in a suitably ventilated compartment outside the building and the gas piped in to the working area. Cylinders must:

(a) be fitted with automatic shut-off valves operable from inside the working area, e.g. laboratory;

(b) be fitted with flash-back arrestors in the line;

(c) have cylinder keys captive to the cylinder by non-ferrous chains;

(d) have their lines examined for leaks on commencement of work and on change of cylinder.

When cylinders are not in use:

(a) cylinder and bench valves must be closed tightly;

(b) protective caps screwed down over the valves.

Disposal

Procedures for the safe disposal of dangerous substances depend largely on the type and quantity of material involved. In all cases the local authority must be consulted since disposal facilities vary considerably in different areas of the country. The following must be considered prior to disposal of dangerous substances or their by-products.

a) Only trained and authorised staff should be permitted to dispose of dangerous substances: where large quantities are concerned, disposal should be carried out in conjunction with the local authority.

(b) Where contractors provided a service, it is necessary to know –

(i) the location of the disposal site;

(ii) the procedure for disposal;

(iii) whether formal licensing is required for the disposal;

(iv) the mode of transport of the dangerous substances from the premises to the disposal site.

The CHIP Regulations

The Chemicals (Hazard Information and Packaging for Supply) Regulations 1994 cover many important aspects with regard to the classification, labelling and packaging of chemicals. The following definitions are particularly significant:

Aerosol dispenser means an article which consists of a non-reusable receptacle containing a gas compressed, liquefied or dissolved under pressure, with or without liquid, paste or powder and fitted with a release device allowing the contents to be ejected as solid or liquid particles in suspension in a gas, as a foam, paste or powder or in a liquid state.

Category of danger means in relation to a substance or preparation dangerous for supply, one of the categories of danger specified in column 1 or Part I of Schedule 1.

Classification means, in relation to a substance or preparation dangerous for supply, classification in accordance with regulation 5 (classification for supply).

Indication of danger means, in relation to a substance or preparation dangerous for supply, one or more of the indications of danger referred to in column 1 of Schedule 2 and:

(a) in the case of a substance dangerous for supply *listed* in Part I of the approved supply list, it is one or more indications of danger for that substance specified by a symbol-letter in column 3 of Part V of that list; or

(b) in the case of a substance dangerous for supply *not so listed* or a *preparation dangerous for supply*, it is one or more indications of danger determined in accordance with the classification of that substance or preparation under Regulation 5 and the approved classification and labelling guide.

Package means, in relation to a substance or preparation dangerous for supply, the package in which the substance or preparation is supplied and which is liable to be individually handled during the course of the supply and includes the receptacle containing the substance or preparation and any other packaging associated with it and any pallet or other device which enables more than one receptacle containing a substance or preparation dangerous for supply to be handled as a unit, but *does not include*:

(a) a freight container (other than a tank container), a skip, a vehicle or other article of transport equipment; or

(b) in the case of supply by way of retail sale, any wrapping such as a paper or plastic bag in to which the package is placed when it is presented to the purchaser.

Packaging means, in relation to a substance or preparation dangerous for supply, as the context may require, the receptacle, or any components, materials or wrappings associated with the receptacle for the purpose of enabling it to perform its containment function or both.

Poisons advisory centre means a body approved for the time being for the purposes of Reg 14 (notification of constituents of certain preparations dangerous for supply) by the Secretary of State for Health in consultation with the Secretaries of State for Scotland and Wales, the HSC and such other persons or bodies as appear to the Secretary of State for Health to be appropriate.

Preparations means mixtures or solutions of two or more substances.

Preparation dangerous for supply means a preparation which is in one or more of the categories of danger specified in column 1 of Schedule 1.

Receptacle means, in relation to a substance or preparation dangerous for supply, a vessel, or the innermost layer of packaging, which is in contact with the substance and which is liable to be individually handled when the substance is used and includes any closure or fastener.

Risk phrase means, in relation to a substance or preparation dangerous for supply, a phrase listed in Part III of the approved supply list and in these Regulations specific risk phrases may be designated by the letter 'R' followed by a distinguishing number or combination of numbers but the risk phrase shall be quoted in full on any label or safety data sheet in which the risk phrase is required to be shown.

Safety phrase means, in relation to a substance or preparation dangerous for supply, a phrase listed in Part IV of the approved supply list and in these Regulations specific safety phrases may be designated by the letter 'S' followed by a distinguishing number or combination of numbers, but the safety phrase shall be quoted in full on any label or safety data sheet in which the safety phrase is required to be shown.

Substances means chemical elements and their compounds in the natural state or obtained by any production process, including any additive necessary to preserve the stability of the product and any impurity deriving from the process used, but excluding any solvent which may be separated without affecting the stability of the substance or changing its composition.

Substance dangerous for supply means:

(a) a substance listed in Part I of the approved supply list; or

(b) any other substance which is in one or more of the categories of danger specified in column 1 of Schedule 1.

Supplier means a person who supplies a substance or preparation dangerous for supply, and in the case of a substance which is imported (whether or not from a member State) includes the importer established in Great Britain of that substance or preparation.

Supply in relation to a substance or preparation:

(a) means, subject to paragraph (b) or (c) below, supply of that substance or preparation, whether as principal or agent for another, in the course or for use at work, by way of –
   (i) sale or offer for sale;
   (ii) commercial sample; or
   (iii) transfer from a factory, warehouse or other place of work and its curtilage to another place of work, whether or not in the same ownership;

(b) for the purposes of sub-paragraphs (a) and (b) of Reg 16(2), (HSE as enforcement agency), except in relation to Regs 7 (advertisements) and 12 (child-resistant fastenings and warning devices), in any case for which by virtue of those sub-paragraphs the enforcing authority for these Regulations is the Royal Pharmaceutical Society or the local weights and measures authority, has the meaning assigned to it by sec 46 of the Consumer Protection Act 1987 and also includes offer to supply and expose for supply; or

(c) in relation to Regs 7 (advertisements) and 12 (child-resistant fastenings and warning devices) shall have the meaning assigned to it by Regs 7(2) and 12(12) respectively.

Application of the Regulations

These Regulations apply to any substance or preparation which is dangerous for supply *except*:

(a) radioactive substances or preparations;

(b) animal feeds;

(c) cosmetic products;

(d) medicines and medicinal products;

(e) controlled drugs;

(f) substances or preparations which contain disease-producing micro-organisms;

(g) substances or preparations taken as samples under any enactment;

(h) munitions, which produce explosion or pyrotechnic effect;

(i) foods;

(j) a substance or preparation which is under customs control;

(k) a substance which is intended for export to a country which is not a member State;

(l) pesticides;

(m) a substance or preparation transferred within a factory, warehouse or other place of work;

(n) a substance to which Reg 7 of the Notification of New Substances Regulations 1993 applies;

(o) substances, preparations and mixtures in the form of wastes.

The Approved Supply List

This is the list entitled *Information Approved for the Classification and Labelling of Substances and Preparations Dangerous for Supply* approved by the HSC comprising Parts I to IV, together with such notes and explanatory material as are requisite for the use of the list.

Classification of Substances and Preparations Dangerous for Supply (Reg 5)

1. A supplier shall not supply a substance or preparation dangerous for supply unless it has been classified in accordance with the following paragraphs of this regulation.

2. In the case of a *substance which is listed in the approved supply list*, the classification shall be that specified in the entry for that substance in column 2 of Part V of that list.

3. In the case of a *substance which is a new substance* within the meaning of Reg 2(1) of the Notification of New Substances Regulations 1993 and which has been notified in accordance with Reg 4 or 6(1) or (2) of those Regulations, the substance shall be classified in conformity with that notification.

4. In the case of *any other substance dangerous for supply*, after an investigation to become aware of relevant and accessible data which may exist, the substance shall be classified by placing it in one or more of the *categories of danger* specified in column 1 of Part I of Schedule 1 corresponding to the properties specified in the entry opposite thereto in column 2 and by assigning appropriate *risk phrases* by the use of the criteria set out in the *approved classification and labelling guide*.

5. Subject to paragraph 6, *a preparation to which these Regulations apply* shall be classified as dangerous for supply in accordance with Schedule 3 by the use of the criteria set out in the approved classification and labelling guide.

6. A preparation which is intended for use as a pesticide (other than a pesticide which has been approved under the Food and Environment Protection Act 1985) shall be classified as dangerous for supply in accordance with Schedule 4.

Safety data sheets for substances and preparations dangerous for supply (Reg 6)

1. Subject to paragraphs 2 and 5, the supplier of a substance or preparation dangerous for supply shall provide the recipient of that substance or preparation with a safety data sheet containing information under the headings specified in Schedule 5 to enable the recipient of that substance or preparation to take the necessary measures relating to the protection of health and safety at work and relating to the protection of the environment and the safety data sheet shall clearly show its date of first publication or latest revision, as the case may be.

2. In this regulation, *supply* shall *not* include supply by way of:

(a) offer for sale;

(b) transfer from a factory, warehouse or another place of work and its curtilage to another place of work in the same ownership; or

(c) returning substances or preparations to the person who supplied them, provided that the properties of that substance or preparation remain unchanged.

3. The supplier shall keep the safety data sheet up to date and revise it forthwith if any significant new information becomes available regarding safety or risks to human health or the protection of the environment in relation to the substance or preparation concerned and the revised safety data sheet shall be clearly marked with the word '*revision*'.

4. Except in the circumstances to which paragraph 5 relates, the safety data sheet shall be provided free of charge no later than the date on which the substance or preparation is first supplied to the recipient and where the safety data sheet has been revised in accordance with paragraph 3, a copy of the revised safety data sheet shall be provided free of charge to all recipients who have received the substance or preparation in the last 12 months and the changes in it shall be brought to their notice.
5. Safety data sheets need *not* be provided with substances or preparations dangerous for supply sold to *the general public* in circumstances to which Reg 16(2)(a) or (b) applies (relating to supply from a shop, etc.) if sufficient information is furnished to enable users to take the necessary measures as regards the protection of health and safety, except that safety data sheets shall be provided free of charge at the request of persons who intend the substance or preparation to be used at work, but in those circumstances paragraph 4 (insofar as it relates to the subsequent provision of revised data sheets) shall not apply to such requests.
6. The particulars required to be given in the safety data sheets shall be in English, except that where a substance or preparation is intended to be supplied to a recipient in another Member State, the safety data sheet may be in the official language of that State.

Advertisements for substances dangerous for supply (Reg 7)

1. A person who supplies or offers to supply a substance dangerous for supply shall ensure that the substance is not advertised unless mention is made in the advertisement of the hazard of hazards presented by the substance.
2. In this regulation the word '*supply*' has the same meaning as in sec 46 of the Consumer Protection Act 1987.

Packaging of substances and preparations dangerous for supply (Reg 8)

The supplier of a substance or preparation which is dangerous for supply shall not supply any such substance or preparation unless it is in a package which is suitable for that purpose, and in particular, unless:

(a) the receptacle containing the substance or preparation and any associated packaging are *designed, constructed, maintained and closed* so as to prevent any of the contents of the receptacle from escaping when subjected to the stresses and strains of normal handling, except that this sub-paragraph shall not prevent the fitting of a suitable safety device;

(b) the receptacle and any associated packaging, in so far as they are likely to come into contact with the substance or preparation, are made of *materials* which are neither liable to be adversely affected by that substance nor liable in conjunction with that substance to form any other substance which is itself a risk to the health or safety of any person;

(c) where the receptacle is fitted with a *replaceable closure*, that closure is designed so that the receptacle can be repeatedly reclosed without its contents escaping.

Labelling of substances and preparations dangerous for supply (Reg 9)

1. Subject to Regs 9 and 10 of the Carriage of Dangerous Goods by Road and Rail (Classification, Packaging and Labelling) Regulations 1994 (which allow combined carriage and supply labelling in certain circumstances) and paragraphs 5 to 9, a supplier shall not supply a substance or preparation which is dangerous for supply unless the particulars specified in paragraph 2 relating to a substance or paragraph 3 relating to a preparation, as the case may be, are clearly shown in accordance with the requirements of Reg 11 (methods of marking or labelling of packages):

(a) on the *receptacle* containing the substance or preparation;

(b) if that receptacle is inside one or more layers of packaging, on any such layer which is likely to be to the outermost layer of packaging during the supply or use of the substance or preparation, unless such packaging permits the particulars shown on the receptacle or other packaging to be clearly seen.

2. The *particulars* required under paragraph 1 in relation to a *substance dangerous for supply* shall be:

(a) the *name, full address and telephone number* of a person in a member State who is responsible for *supplying* the substance, whether it be its manufacturer, importer or distributor;

(b) the *name of the substance*, being the name or one of the names for the substance listed in Part I of the approved supply list, or if it is not so listed an internationally recognised name;

(c) the following particulars ascertained in accordance with Part I of Schedule 6, namely –

(i) the *indication or indications of danger* and the corresponding *symbol or symbols*;

(ii) the *risk phrases* (set out in full);

(iii) the *safety phrases* (set out in full);

(iv) the *EEC number* (if any) and, in the case of a substance dangerous for supply which is listed in Part I of the approved supply list, the words '*EEC label*'.

3. The *particulars* required under paragraph 1 in relation to a *preparation which is, or (where sub-paragraph (d) below applies) may be dangerous for supply* shall be:

(a) the *name and full address and telephone number* of a person in a member State who is responsible for *supplying* the preparation, whether he be its manufacturer, importer or distributor;

G

(b) the *trade name or other designation* of the preparation;

(c) the following particulars ascertained in accordance with Part I of Schedule 6, namely –

(i) identification of the *constituents* of the preparation which result in the preparation being classified as dangerous for supply;

(ii) the *indication or indications of danger* and the corresponding *symbol or symbols*;

(iii) the *risk phrases* (set out in full);

(iv) the *safety phrases* (set out in full);

(v) in the case of a *pesticide,* the *modified* information specified in paragraph 5 of part I of Schedule 6;

(vi) in the case of a *preparation intended for sale to the general public,* the *nominal quantity* (nominal mass or nominal volume); and

(d) where required by paragraph 5(5) of Part I of Schedule 3, the *words specified* in that paragraph.

4. Where the Executive receives a notification of a derogation provided for by paragraph 3(1) of Part I of Schedule 6, it shall forthwith inform the European Commission thereof.

5. Indications such as '*non-toxic*' or '*non-harmful*' or any other statement indicating that the substance or preparation is not dangerous for supply shall not appear on the package.

6. Except for the outermost packaging of a package in which a substance or preparation is transferred, labelling in accordance with this regulation shall not be required where a substance or preparation dangerous for supply is supplied by way of transfer from a factory, warehouse or other place of work and its curtilage to another place of work if, at that other place of work, it is not subject to any form of manipulation, treatment or processing which results in the substance or preparation dangerous for supply being exposed or, for any purpose other than labelling in accordance with these Regulations, results in any receptacle containing the substance or preparation being removed from its outer packaging.
7. Except in the case of a substance or preparation dangerous for supply for which the indication of danger is required to be *explosive, very toxic or toxic* or which is classified as *sensitising,* labelling under this regulation shall not be required for such small quantities of that substance or preparation that there is no reason to fear danger to persons handling that substance or preparation or to other persons.
8. Where, in the case of a substance or preparation dangerous for supply, other than a *pesticide,* the package in which the substance or preparation is supplied does not contain more than *125 millilitres* of the substance or preparation, the *risk phrases* required by paragraph 2(c)(ii) or 3(c)(iii), and the *safety phrases* required by paragraph 2(c)(iii) or 3(c)(iv), as the case may be, need not be shown if the substance or preparation is classified only in one or more of the categories of danger, *highly flammable, flammable, oxidising or irritant* or in the case of *substances not intended to be supplied to the public, harmful.*
9. Where, because of the *size of the label,* it is not reasonably practicable to provide the *safety phrases* required under paragraph 2(c)(iii) or 3(c)(iv), as the case may be, on the label, that information may be given on a separate label or on a sheet acompanying the package.

Particular labelling requirements for certain preparations (Reg 10)

1. In the case of preparations to which Part II of Schedule 6 applies the appropriate provisions of that Part of the Schedule shall have effect to regulate the labelling of such preparations even if the preparations referred to in Part IIB of that Schedule would not otherwise be dangerous for supply.
2. In the case of preparations packaged in *aerosol dispensers,* the *flammability criteria* set out in Part II of Schedule I shall have effect for the

classification and labelling of those preparations for supply in place of the categories of danger *'extremely llammable'*, *'highly flammable'* or *'flammable'* set out in Part I of that Schedule, and where a dispenser contains a substance so classified, that dispenser shall be labelled in accordance with the provisions of paragraph 2 of the said Part II.

Methods of marking or labelling packages (Reg 11)

1. Any package which is required to be labelled in accordance with Regs 9 and 10 may carry the particulars required to be on the label clearly and indelibly marked on a part of that package reserved for that purpose and, unless the context otherwise requires, any reference in these Regulations to a label includes a reference to that part of the package so reserved.

2. Subject to paragraph 7, any label required to be carried on a package shall be *securely fixed* to the package with its entire surface in contact with it and the label shall be clearly and indelibly printed.

3. The *colour and nature of the marking* shall be such that the symbol (if any) and wording stand out from the background so as to be *readily noticeable* and the wording shall be of such a size and spacing as to be *easily read*.

G

4. The package shall be so labelled that the particulars can be *read horizontally* when the package is set down normally.

5. Subject to paragraph 7, the *dimensions* of the label required under Reg 9 shall be as follows

Capacity of package	Dimensions of label
(a) not exceeding 3 litres	if possible at least 52 × 74 mm
(b) exceeding 3 litres but not exceeding 50 litres	at least 74 × 105 mm
(c) exceeding 50 litres but not 500 litres	at least 108 × 148 mm
(d) exceeding 500 litres	at least 148 × 210 mm

6. Any symbol required to be shown in accordance with Reg 9(2)(c)(i) or 9(3)(c)(ii) and specified in column 3 of Schedule 2 shall be printed in *black on an orange-yellow background* and its size (including the orange-yellow background) shall be at least equal to an area of *one tenth* of that of a label which complies with paragraph 5 and shall not in any case be less than 100 sq mm.

7. If the package is an *awkward shape* or *so small* that it is unsuitable to attach a label complying with paragraphs 2 and 5, the label shall be so attached in some other appropriate manner.

8. The *particulars* required to be shown on the label shall be in English, except that where a substance or preparation is intended to be supplied to a recipient in another Member State, the label may be in the official language of the State.

Child-resistant fastenings and tactile warning devices (Reg 12)

1. The British and International Standards referred to in this regulation are further described in Schedule 7.

2. This regulation shall not apply in relation to a *pesticide*.

3. Subject to paragraph 5, a person shall not supply a substance or preparation referred to in paragraph 4 in a receptacle of any size fitted with a *replaceable closure* unless the packaging complies with the requirements of BS EN 28317 or ISO 8317.

4. Paragraph 3 shall apply to:

(a) substances and preparations dangerous for supply which are required to be labelled with the indications of danger '*very toxic*', '*toxic*' or '*corrosive*';

(b) preparations containing *methanol* in a concentration equal to or more than *3 per cent by weight*;

(c) preparations containing *dichloromethane* in a concentration equal to or more than *1 per cent by weight*;

(d) liquid preparations having a kinematic viscosity measured by rotative viscometry in accordance with BS2782 method 730B or ISO 3291 of less than 7 × 10 m s at 40°C and containing *aliphatic or aromatic hydrocarbons* or both in a total concentration equal to or more than *10 per cent by weight,* except where such a preparation is supplied in an *aerosol dispenser*.

5. Paragraph 3 shall not apply if the person supplying it can show that it is *obvious* that the packaging in which the substance or preparation is supplied is *sufficiently safe for children* because they cannot obtain access to the contents without the help of a tool.

6. If the packaging in which the substance or preparation is supplied was approved on or before 31 May 1993 by the British Standards Institution as complying with the requirements of the British Standards Specification 6652:1989 it shall be treated in all respects as complying with the requirements of BS EN 28317.

7. A person shall not supply a preparation dangerous for supply if the packaging in which the preparation is supplied has:

(a) a *shape or designation or both* likely to attract or arouse the *active curiosity of children* or to *mislead consumers;*

(b) a *presentation or designation or both* used for *human or animal foodstuffs, medicinal or cosmetic products.*

8. A person shall not supply a substance or preparation referred to in paragraph 9 in a receptacle of any size, unless the packaging carries a *tactile warning* of danger in accordance with BS7280 or EN Standard 272.

9. Paragraph 8 shall apply to substances and preparations dangerous for supply which are required to be labelled with the indication of danger *'very toxic', 'toxic', 'corrosive', 'harmful', 'extremely flammable' or 'highly flammable'.*

10. A duly authorised officer of the enforcing authority, for the purpose of ascertaining whether there has been a concentration of paragraph 3 may require the person supplying a substance or preparation to which that paragraph applies to provide him with a *certificate* from a qualified test house stating that:

(a) the closure is such that it is not necessary to test to BS EN 28317 or ISO 8317; or

(b) the closure has been tested and found to conform to that standard.

11. For the purpose of paragraph 10 a *qualified test house* means a laboratory that conforms to BS7501 or EN 45 000.

12. In this regulation, *'supply'* means offer for sale, or otherwise make available to the general public.

Retention of classification data for substances and preparations dangerous for supply (Reg 13)

A person who classifies a substance in accordance with Reg 5(4) or a preparation dangerous for supply shall keep a record of the information used for the purposes of classifying for at least three years after the date on which the substance or preparation was supplied by him for the last time and shall make the record or a copy of it available to the appropriate enforcing authority referred to in Reg 16(2) at its request.

Notification of the constituents of certain preparations dangerous for supply to the poisons advisory centre (Reg 14)

1. This regulation shall apply to any preparation which is classified on the basis of one or more of its health effects referred to in column 1 of Schedule I.

2. Subject to Reg 17 (transitional provisions), the supplier of a preparation to which this regulation applies shall, if it was first supplied before these regulations came into force (or, if it was first supplied after that date, before first supplying it), notify the *poisons advisory centre* of the information required to be in the safety data sheet prepared for the purposes of Reg 6 relating to the preparation.

3. The supplier shall ensure that the information supplied to the poisons advisory centre in pursuance of paragraph 2 is kept up to date.

4. The poisons advisory centre shall only disclose any information sent to it in pursuance of paragraph 2 or 3 on request by, or by a person working under the direction of, a registered medical practitioner in connection with the medical treatment of a person who may have been affected by the preparation.

Exemption certificates (Reg 15)

1. Subject to paragraph 2 and to any of the provisions imposed by the Community in respect of the free movement of dangerous substances and preparations, the Executive may by a certificate in writing exempt any person or class of persons, substance or preparation to which these Regulations apply, or class of such substances or preparations, from all or any of the requirements or prohibitions imposed by or under these Regulations and any such exemption may be granted subject to conditions and to a limit of time and may be revoked at any time by a certificate in writing.

2. The Executive shall *not* grant any such exemption unless, having regard to the circumstances of the case, and in particular to:

(a) the *conditions,* if any, which it proposes to attach to the exemption;

(b) any *requirements* imposed by or under any *enactments* which apply to the case,

it is satisfied that the health or safety of persons who are likely to be affected by the exemption will not be prejudiced in consequence of it.

Enforcement, civil liability and defence (Reg 16)

1. Insofar as any provisions of Reg 50 is made under sect 2 of the European Communities Act 1972

(a) subject to paragraph 2, the provisions of the Health and Safety at Work etc. Act 1974, which relate to the approval of *codes of practice* and their use in *criminal proceedings*, to enforcement and to offences shall apply to that provision as if that provision had been made under sec 15 of that Act;

(b) *a breach of duty imposed by that provision shall confer a right of action in civil proceedings*, insofar as that breach of duty causes damage.

2. Notwithstanding Reg 3 of the Health and Safety (Enforcing Authority) Regulations 1989, the enforcing authority for these Regulations shall be the Executive, except that:

(a) where a substance or preparation dangerous for supply is supplied in or from premises which are registered under sec 75 of the Medicines Act 1968 the enforcing authority shall be the *Royal Pharmaceutical Society*;

(b) where a substance or preparation dangerous for supply is supplied otherwise than as in sub-paragraph (a) above –

(i) in or from any *shop, mobile vehicle, market stall or other retail outlet*; or

(ii) otherwise to members of the public, including by way of *free sample, prize or mail order*, the enforcing authority shall be the local *weights and measures authority;*

(c) for Regs 7 and 12 the enforcing authority shall be the local weights and measures authority.

3. In every case where by virtue of paragraph 2 these Regulations are enforced by the *Royal Pharmaceutical Society* or the local *weights and measures authority*, they shall be enforced as if they were safety regulations made under sec 11 of the Consumer Protection Act 1987 and the provisions of sec 12 of that Act shall apply to these regulations and as if the *maximum period of imprisonment* on summary conviction specified in subsection (5) thereof were three months instead of six months.

4. In any proceedings for an *offence* under Regulations, it shall be a *defence* for any person to prove that he took *all reasonable precautions* and exercised *all due diligence* to avoid the commission of that offence.

Safety data sheets

Under the CHIP 2 Regulations obligatory information under the following headings must be provided in a safety sheet:

1. Identification of a substance/preparation.
2. Composition/information on ingredients.
3. Hazards identification.
4. First aid measures.
5. Fire-fighting measures.
6. Accidental release measures.
7. Handling and storage.
8. Exposure controls/Personal protection.
9. Physical and chemical properties.
10. Stability and reactivity.
11. Toxicological information.
12. Ecological information.
13. Disposal considerations.
14. Transport information.
15. Regulatory information.
16. Other information.

Monitoring

An important feature of preventing hazards to health associated with dangerous substances is the frequent assessment of systems and procedures relating to their identification, storage and handling, and of protection arrangements and the training requirements for operators. This monitoring exercise may take the form of a dangerous substances audit, and a typical example is shown in Fig. 39.4. The audit should be undertaken on either a weekly or a monthly basis dependent upon the range and quantities of dangerous substances being used and stored.

Fig. 39.4 ◆ Dangerous substances audit

Item	*Yes/No*	*Action*
1. Information and identification 1.1 Is an up-to-date list of all dangerous substances held on site readily available? 1.2 Are safety data sheets available for all dangerous substances on site? 1.3 Is this information adequate? 1.4 Are all packages and containers correctly labelled? 1.5 Are 'ready-use' containers suitable for that purpose and suitably marked?		
2. Storage 2.1 Are stores and external storage areas satisfactory in respect of construction, layout, security and control? 2.2 Are dangerous substances correctly segregated? 2.3 Are cleaning and housekeeping levels satisfactory? 2.4 Are all issues to staff controlled?		
3. Protection 3.1 Are the necessary warning signs posted in appropriate areas? 3.2 Is suitable personal protective equipment – available – serviceable – used? 3.3 Are emergency eye wash facilities and showers – available – suitably located – serviceable? 3.4 Are the above facilities provided with frost protection? 3.5 Are adequate and suitable first aid materials – available – suitably located? 3.6 Are the appropriate fire appliances – available		

G

Fig. 39.4 ◆ continued

Item	*Yes/No*	*Action*
– suitably located – serviceable – accessible? 3.7 Is a supply of neutralising compound readily available in the event of spillage?		
4. Procedures 4.1 Are written safe-handling procedures prepared and available for all dangerous substances? 4.2 Is there a specific procedure for dealing with spillages? 4.3 Is there a routine inspection procedure for – personal protective equipment – emergency showers and eye wash facilities – first aid boxes – fire appliances – chemical dosing to plant – neutralising compounds?		
5. Training 5.1 Are staff trained in – safe-handling procedures – use of fire appliances – dealing with spillages – use and care of personal protective equipment? 5.2 Are first aiders trained to deal with injuries associated with dangerous substances? 5.3 Are training records maintained?		

Signed .... Date ....

The COMAH Regulations

The Control of Major Accident Hazards (COMAH) Regulations 1999 replace the former Control of Industrial Major Accident Hazards (CIMAH) Regulations 1984. They are concerned with the prevention of major accidents, i.e. uncontrolled, unexpected or unplanned events, involving one or more of defined dangerous substances. A major accident is broadly defined as one having the potential to cause serious danger to people or the environment. In this context, serious danger means risk of death, physical injury or ill health whether acute (immediate) or chronic (delayed) to persons on the site or outside it, including the general public.

COMAH applies to all sites where dangerous substances listed in Schedule 1 to the regulations:

(a) are present at or above a threshold quantity;
(b) where any dangerous substance at the threshold quantity is generated either intentionally or unintentionally, for instance, as a result of an accident or malfunction in a process;
(c) nuclear and licensed explosives sites, where dangerous substances are stored or used at or above the threshold quantities.

Notification and marking of sites

The Dangerous Substances (Notification and Marking of Sites) Regulations 1990 cover the on-site storage of all dangerous substances, other than those substances specified in the Regulations. With the sole exception of petrol filling stations, operators of all sites having 'at any time a total aggregate quantity of dangerous substances of 25 tonnes or more' must erect suitable access warning signs. They are also required to notify the local authority of the quantities and range of dangerous substances stored. Full guidance for site operators, including the requirements for the erection of access signs and for notification are incorporated in the HSE publication *Notification and marking of sites – The Dangerous Substances (Notification and Marking of Sites) Regulations 1990: Guidance on the Regulations*.

Transport of dangerous substances

Introduction

Under the HSWA, employers have a duty to protect members of the public from hazards arising from their activities. This duty applies particularly in the case of the transport of dangerous substances by road and rail.

Specific legislation

A wide range of specific legislation applies to transport activities involving the carriage of dangerous substances, namely:

1. Radioactive Substances (Carriage by Road)(Great Britain) Regulations 1974.
2. Radioactive Material (Road Transport) Act 1991.
3. Freight Containers (Safety Convention) Regulations 1984.
4. Road Traffic (Carriage of Explosives) Regulations 1989.
5. Packaging of Explosives for Carriage Regulations 1991.
6. Road Traffic (Carriage of Dangerous Substances in Packages etc.) Regulations 1992.
7. Road Traffic (Carriage of Dangerous Substances in Road Tankers and Tank Containers) Regulations 1992.
8. Road Traffic (Training of Drivers Carrying Dangerous Goods) Regulations 1992.
9. Carriage of Dangerous Goods by Road and Rail (Classification, Packaging and Labelling) Regulations 1994.
10. Carriage of Dangerous Goods by Rail Regulations 1994.
11. Dangerous Substances in Harbour Areas Regulations 1987.
12. Placing on the Market and Supervision of Transfers of Explosives Regulations 1993.

Carriage of Dangerous Goods by Road and Rail (Classification, Packaging and Labelling) Regulations 1994

The objective of these regulations is to ensure that the rules in the UK meet the UN Recommendations on the Transport of Dangerous Goods. The regulations apply to all dangerous goods except explosives, radioactive materials and other items listed in Reg 3. The regulations are accompanied by the Approved Carriage List and the Approved Methods for the Classification and Packaging of Dangerous Goods for Carriage.

Duties under these regulations are placed on the consignor of dangerous goods who must classify, package and label dangerous goods in accordance with the requirements set out in the regulations.

Road Traffic (Carriage of Dangerous Substances in Road Tankers and Tank Containers) Regulations 1992

These regulations seek to control the risks arising from the transport of dangerous substances in road tankers and tank containers. Under these regulations the following definitions apply.

Road tanker means a goods vehicle which has a tank which is an integral part of the vehicle or is attached to the frame of the vehicle and is not intended to be removed from it.

Tank container means a tank, whether or not divided into separate compartments, with a total capacity of more than 450 litres and includes a *tube container and a tank swap body.*

Tank swap body means a tank specially designed for carriage by rail and road only and is without stacking capability.

Vehicles and tanks must be properly designed and of adequate strength and construction to convey dangerous substances by road. The operator is responsible for:

(a) ensuring that the regulations on the conveyance of certain substances are enforced and that tanks are not overfilled;

(b) giving written information to drivers on the hazards associated with their loads and the necessary emergency procedures.

The driver must ensure:

(a) the safe parking and supervision of the vehicle when not in use when prescribed substances are carried;

(b) that precautions to prevent fire or explosion are observed.

Hazard warning panels

All road tankers must display three hazard warning panels, one at the rear and one on each site.

All tank containers must display four hazard warning panels, one at each end and one on each side.

Hazard warning panels must be weather resistant, rigidly fixed and indelibly marked, and must contain the following information:

(a) *Emergency Action Code.*

(b) *Substance identification number.*

(c) *Telephone number for specialist advice.*

(d) *Warning classification sign for the substance carried.*

The regulations are enforced by the HSE, except when on a public road or in a public place, where enforcement becomes a police responsibility.

These regulations impose specific requirements for unloading petrol at petrol service stations and other premises licensed to keep petrol.

Road Traffic (Training of Drivers Carrying Dangerous Goods) Regulations 1992

Under these regulations (as amended) and the Carriage of Dangerous Goods by Rail Regulations 1994 there is a general duty to on the 'operators' of such vehicles and trains to instruct and train drivers in a range of safety procedures.

The regulations require that drivers of road tankers of more than 3,000 litres or maximum permissible weight of 3.5 tonnes, and of all vehicles carrying tank containers or explosives must hold a certificate of driver training from a Department of Transport approved training school. This was extended to vehicles of 3.5 tonnes on 1 January 1995.

Carriage of Dangerous Goods (Classification, Packaging and Labelling) and the Use of Transportable Gas Receptacles Regulations 1996

These regulations implement a number of European Directives covering the classification, packaging and labelling of dangerous goods under the European Agreement concerning the International Carriage of Dangerous Goods by Road (ADR). They cover gas cylinders used for the carriage of specified dangerous goods of volume capacity up to 5,000 litres in seamless cylinders and other cylinders up to 1,000 litres. (Most types of aerosol containers are excluded.) The regulations place specific requirements on manufacturers with regard to cylinders, first, manufactured before 1 January 1999 and, second, those manufactured after this date who must meet the standards and requirements of the Approved Requirements which accompany the regulations.

Transportable pressure receptacles

Part III of the regulations deals with transportable pressure receptacles (TPRs) intended to contain any dangerous goods. TPRs must meet specific requirements with regard to design, manufacture, modification and repair.

In particular, they must be safe and suitable for the purpose, and meet the design, construction and quality assurance requirements of the Approved Requirements. It is an offence for any person to fill or use a TPR which has been modified, repaired or damaged in such a way as to be dangerous, unless it has been examined and tested by an approved person in accordance with procedures laid down in the Approved Requirements. Manufacturers must manufacture receptacles in accordance with the Approved Requirements. No person shall import, supply or own a TPR containing dangerous goods unless:

(a) the receptacle conforms with the appropriate design and construction requirements of the Approved Requirements; or

(b) the receptacle is an EEC-type cylinder, i.e. there is an EEC Verification Certificate in force and it bears all the marks and inscriptions required by the Pressure Vessels Framework Directive and the separate Directive relating to that type of cylinder.

TPRs must be marked by a competent authority or an approved person following an initial examination and test and where further examination and test is needed under the Approved Requirements. For the purpose of the regulations, an approved person is a person approved by the HSE who has been issued with a certificate of approval by same, or a person approved by a competent authority other than the HSE for the performance of particular functions in relation to TPRs under the ADR. Further requirements relate to procedures when filling TPRs and the keeping of records by manufacturers.

40 Specific processes and activities

This chapter examines a number of typical processes and activities, many of which are ancillary to the principal activities of industry and commerce. These activities include catering, laboratory work, office work, workshop activities and welding operations.

Workshops

Many factors need to be considered to ensure safe working in engineering, vehicle maintenance and other types of workshop. A number of these factors are outlined below.

Structural features

Floors should be sound and kept clean, with adequate floor drainage where wet processes are undertaken. Vehicle inspection pits should be provided with safe access and egress, intrinsic flame-proof lighting, and a suitable cover, such as boards, when the pit is not in use. Moreover, elevated storage areas must be adequately lit, provided with safety rails and toe boards and a permanent safe means of access. In addition, there should be adequate external lighting.

Environmental features

The form of heating provided should be capable of maintaining a comfortable working temperature, e.g. 18°C, and heating appliances should not emit fumes or gases. Both general and specific lighting at workbenches should be to HSE Guidance Note *Lighting at work* recommended levels and an emergency lighting system should be installed. Moreover, ventilation requirements should accommodate the possibility of fumes from welding, engine testing and vehicle painting. Noise-producing activities, such as panel beating, should ideally be separated from the main workshop.

Machinery and equipment

Various hazards associated with the operation of woodworking machinery, abrasive wheels, lathes, lifting tackle and equipment, drills and drilling machines, gas and electric welding, metal-cutting guillotines, vehicle lifts, air compressors, compressed air equipment, pressure grease guns, jacking equipment and tyre inflation equipment should be readily appreciated by all staff, who should be adequately trained and supervised in their use.

Hand tools

Many accidents are caused by the misuse of hand tools or the use of defective hand tools. Examples would be hammers with split shafts, cold chisels with mushroomed heads, files with defective handles, screwdrivers with worn blades. Hand tools should be examined on a regular basis and defective tools rejected.

Electricity

Hand tools should be of the low-voltage type with efficient switches, earthing and double insulation. Flexes and connections should be frequently examined for wear and damage. The practice of overloading sockets should be prohibited. Flexes, leads, connections and earth clamps to battery-charging equipment and portable welding sets, together with hand-held inspection lamps, need regular examination. In the latter case, hand lamp bulbs should be fitted with a cage and operate at low voltage.

Vehicles

Lorry tilt cabs should be fitted with self-locking stays to prevent the cab from falling back into position whilst fitters are working on the engine beneath the cab. Where stays are not fitted, cabs should be maintained in the forward position by use of chains, wedges and props.

Laboratories

The principal hazard in laboratories is fire. This may result from spontaneous combustion, incompatible reactions, evolution of flammable gases during experiments or the use of defective electrical equipment. Electrical faults, the main cause of fire, can result in equipment overheating. Thermostats should be supported by thermal cut-outs and all electrical equipment checked on a regular basis. Other hazards include the risk of explosion or implosion, skin contact with strong acids, alkalis and organic compounds, incorrect labelling of reagents and poor storage of flammable liquids, unstable solids and compressed and liquefied gases. Spillages of flammable, toxic and corrosive substances can result in the evolution of gases and an enhanced fire risk. Neutral absorbing materials should always be available.

Many injuries involving glassware are caused by breakage of glass vessels under pressure or through inserting glass tubing into corks. Vessels under pressure should be guarded with mesh or tape and corks properly bored. Where experiments are carried out at pressures greater than atmospheric, using glass vessels, solid metal or wire mesh screens should be used to surround the area of the experiment.

Where possible, autoclaves should be separated from the rest of the laboratory area. They should be fitted with pressure bursting discs and subject to annual examination.

Special precautions are needed in the case of compressed gases (*see* Chapter 39). Centrifuges should be fitted with an interlocking device and brake so that the lid cannot be opened until the moving parts have come to rest. Good waste disposal provision is also important. Special waste solvent drums should be provided along with metal bins with close-fitting lids for other substances. Waste containers should be clearly marked for specific types of chemical waste. Above all, the practice of mouth pipetting should be prohibited and replaced by the use of purpose-designed pipette fillers.

Principles of laboratory safety

Safe design

Ideally all laboratories should be of single-storey construction and separate from other buildings, built in non-combustible materials and with floors impervious to chemical substances. Where attached to other buildings, they should be separated by fire-resisting construction with a minimum of half an hour notional period of fire resistance, and stairways should be encased in fire-resisting walls. Heating should be automatic with a complete prohibition on all types of portable fire.

Housekeeping

Good housekeeping is a key factor in safe laboratory practice. Many hazards can be eliminated by meticulous attention to detail, including environmental hygiene, and tidiness of work sections, benches and storage areas.

General conduct

A high standard of discipline and operational conduct is essential. Young laboratory workers should be trained in safe procedures and subject to regular supervision.

Personal hygiene

Personnel should be trained in the elements of sound personal hygiene, particularly hand washing after handling chemical compounds. There should be a total prohibition on eating, drinking and smoking in laboratories.

Personal protective equipment

The need to use the equipment provided – e.g. eye protection, respirators, overalls, gloves and visors – should be stressed. Cleaning facilities should be provided for such equipment.

First aid

A high standard of first aid provision is necessary. Eye baths, drenches and an emergency shower are necessary where large quantities of chemical substances are used and stored.

Staff training

All new staff should receive induction training in the hazards present. Everyone should know the relevant flashpoints, ignition temperatures and other hazardous properties of materials, and appreciate the need to separate incompatible and mutually reactive materials. Above all, staff should be trained in fire protection procedures.

Offices

The principal hazard in offices is fire. Fire hazards are created as a result of defective wiring and sockets, overloading of electrical circuits and the use of freestanding heating appliances. Much office machinery is now electrically operated and many offices are simply not provided with sufficient power outlets to meet the demands of an increased electrical load. As a result, it is not uncommon to see the use of multipoint adaptors and extension leads and the wiring of more than one appliance into a 13 amp plug. All these various forms of electrical abuse and misuse increase the potential for fire. Moreover, materials used in offices are highly flammable, in particular spirit-based cleaning fluids, floor polishes and packing materials. Smoking by office staff greatly increases the fire risk through contact with waste paper and packing materials. Indeed, a high proportion of office fires have resulted from a cigarette end left smouldering on the edge of a desk at the termination of work.

Fire prevention measures in offices should incorporate the following elements:

(a) a total ban on the use of freestanding heating appliances, particularly radiant type electric fires, oil heaters and gas-fired appliances: the central heating system should cope with temperature variations from winter to summer;

(b) electrical circuits should be examined by a competent electrical engineer every ten years, such examination to take account of current loading levels, the need for modifications to the system and electrical hazards which may exist;

(c) flammable substances should be stored in lockable metal cabinets;

(d) control over storage of waste paper and packing materials;

(e) a physical check of the premises prior to closing to ensure that all cigarettes and other ignition sources have been extinguished;

(f) a quarterly test of the fire alarm;

(g) an annual fire drill;

(h) annual servicing of fire appliances;

(i) the training of personnel in the correct use of fire appliances.

Accidents in offices

Equipment and materials used in offices present a wide range of hazards. The introduction of equipment such as refuse balers, photocopiers and guillotines has increased risks to staff. Accidents caused by staff tripping over partly projecting drawers of a filing cabinet or trailing flex to an electric typewriter are common, and even the humble drawing pin can inflict injury if left carelessly on a desk or seat. Moreover, many accidents are caused by human error or lack of perception, for instance reading while walking along a corridor or up stairs, restricted vision whilst carrying bulky items or inattention to obstructions such as cleaners' equipment or tea trolleys.

The standard of housekeeping in many offices is poor. The practice of storing materials such as stationery on staircases and landings, in sanitation areas and basement boiler houses, is all too common, resulting in falls, contact accidents and an increased fire risk. Many offices are poorly lit, particularly in staircase and landing areas, with the attendant risk of falls and contact accidents.

Visual display units (VDUs)/display screen equipment

Many office staff operate equipment incorporating VDUs. In the past, some units have suffered from poor ergonomic design leading to increased stress on the operator and accompanied by complaints of visual fatigue, postural fatigue, headaches, neck strain and nervous conditions. Therefore, the design of workstation, equipment and general environment, together with the occupational health aspects of VDU operation, need consideration.

Health and Safety (Display Screen Equipment) Regulations 1992

These Regulations came into operation on 1 January 1993. New workstations established on or after that date must meet the requirements laid down in the Schedule to the Regulations prior to use. Workstations in operation prior to 1 January 1993 must comply with the Schedule to the Regulations. A number of important definitions are incorporated in Reg 1 – Interpretation.

Reg 1 – Interpretation

Display screen equipment means any alphanumeric or graphic display screen, regardless of the display process involved.

Operator means a self-employed person who habitually uses display screen equipment as a significant part of his normal work.

Use means use for or in connection with work.

User means an employee who habitually uses display screen equipment as a significant part of his normal work.

Workstation means an assembly comprising:

(a) display screen equipment (whether provided with software determining the interface between the equipment and its operator or user, a keyboard or any other input device);

(b) any optional accessories to the display screen equipment;

(c) any disk drive, telephone, modem, printer, document holder, work chair, work desk, work surface or other item peripheral to the display screen equipment;

(d) the immediate work environment around the display screen equipment.

Exemptions from the Regulations

The Regulations do *not* apply to:

(a) drivers' cabs or control cabs for vehicles or machinery;

(b) display screen equipment on board a means of transport;

(c) display screen equipment mainly intended for public operation;

(d) portable systems not in prolonged use;

(e) calculators, cash registers or any equipment having a small data or measurement display required for direct use of the equipment; or

(f) window typewriters.

Reg 2 – Analysis of workstations to assess and reduce risks

1. Every employer shall perform a suitable and sufficient analysis of those workstations which:

  (a) (regardless of who has provided them) are used for the purposes of his undertaking by users; or

  (b) have been provided by him and are used for the purposes of his undertaking by operators, for the purpose of assessing the health and safety risks to which those persons are exposed in consequence of that use.

2. Any assessment made by an employer shall be reviewed by him if:

  (a) there is reason to suspect that it is no longer valid; or

  (b) there has been a significant change in the matters to which it relates; and where as a result of any such review changes in an assessment are required, the employer concerned shall make them.

3. The employer shall reduce the risks identified in consequence of the above assessment to the lowest extent reasonably practicable.

Reg 3 – Requirements for workstations

1. Every employer shall ensure that any workstation first put into service on or after 1 January 1993 meets the requirements laid down in the Schedule to these Regulations.
2. Every employer shall ensure that any workstation first put into service on or before 31 December 1992 meets the requirements laid down in the Schedule.

Reg 4 – Daily work routine

Every employer shall so plan the activities of users at work in his undertaking that their daily work on display screen equipment is periodically interrupted by breaks or changes of activity as reduce their workload at that equipment.

Reg 5 – Eyes and eyesight

1. Where a person:
   (a) is already a user on the date of coming into force of these Regulations; or
   (b) is an employee who does not habitually use display screen equipment as a significant part of his normal work but is to become a user in the undertaking in which he is already employed, his employer shall ensure that he is provided at his request with an appropriate eye and eyesight test, any such test to be carried out by a competent person.
2. Any eye and eyesight test carried out shall:
   (a) in any case to which sub-para (a) above applies, be carried out as soon as practicable after being requested by the user concerned;
   (b) in any case to which sub-para (b) above applies, be carried out before the employee concerned becomes a user.
3. At regular intervals after an employee has been provided with an eye and eyesight test, his employer shall ensure that he is provided with a further eye and eyesight test of an appropriate nature, any such test to be carried out by a competent person.
4. Where a user experiences visual difficulties which may reasonably be

considered to be caused by work on display screen equipment, his employer shall ensure that he is provided at his request with an appropriate eye and eyesight test, any such test to be carried out by a competent person as soon as practicable after being requested as aforesaid.

5. Every employer shall ensure that each user employed by him is provided with special corrective appliances appropriate for the work being done by the user concerned where:

(a) normal corrective appliances cannot be used;

(b) the result of any eye and eyesight test which the user has been given in accordance with this regulation shows such provision to be necessary.

6. Nothing in paragraph 3 above shall require an employer to provide any employee with an eye and eyesight test against that employee's will.

Reg 6 – Provision of training

1. Where a person:

(a) is already a user on the date of coming into force of these Regulations; or

(b) is an employee who does not habitually use display screen equipment as a significant part of his normal work but is to become a user in the undertaking in which he is already employed, his employer shall ensure that he is provided with adequate health and safety training in the use of any workstation upon which he may be required to work.

2. Every employer shall ensure that each user at work in his undertaking is provided with adequate health and safety training whenever the organisation of any workstation in that undertaking upon which he may be required to work is substantially modified.

Reg 6 – Provision of information

1. Every employer shall ensure that operators and users at work in his undertaking are provided with adequate information about:

(a) all aspects of health and safety relating to their workstations;

(b) such measures taken by him in compliance with his duties under Regs 2 and 3 as relate to them and their work.

2. Every employer shall ensure that users at work in his undertaking are provided with adequate information about such measures taken by him in compliance with his duties under Regs 4 and 6(2) as relate to them and their work.

3. Every employer shall ensure that users employed by him are provided with adequate information about such measures taken by him in compliance with his duties under Regs 5 and 6(1) as relate to them and their work.

The Schedule

(which sets out the minimum requirements for workstations which are contained in the Annex to Council Directive 90/270/EEC on the minimum safety and health requirements for work with display screen equipment)

1. Extent to which employers must ensure that workstations meet the requirements laid down in this schedule

An employer shall ensure that a workstation meets the requirements laid down in this Schedule to the extent that:

(a) those requirements relate to a component which is present in the workstation concerned;

(b) those requirements have effect with a view to securing the health, safety and welfare of persons at work;

(c) the inherent characteristics of a given task do not make compliance with those requirements inappropriate as respects the workstation concerned.

2. Equipment

General comment

The use as such of the equipment must not be a source of risk for operators or users (*see* Figs 40.1 and 40.2).

Display screen

The characters on the screen shall be well defined and clearly formed, of adequate size and with adequate spacing between the characters and lines.

The image on the screen should be stable, with no flickering or other

Fig. 40.1 ◆ Subjects dealt with in the schedule

1. Adequate lighting
2. Adequate contrast, no glare or distracting reflections
3. Distracting noise minimised
4. Leg room clearances to allow postural changes
5. Window covering
6. Software: appropriate to task, adapted to user, provides feedback on system status, no undisclosed monitoring
7. Screen: stable image, adjustable, readable, glare/reflection free
8. Keyboard: usable, adjustable, detachable, legible
9. Work surface: allow flexible arrangements, spacious, glare free
10. Work chair: adjustable
11. Footrest

Fig. 40.2 ◆ Seating and posture for typical office tasks

1. Seat back adjustability
2. Good lumbar support
3. Seat height adjustability
4. No excess pressure on underside of thighs and back of knees
5. Foot support if needed
6. Space for postural change, no obstacles under desk
7. Forearms approximately horizontal
8. Minimal extension, flexion or deviation of wrists
9. Screen height and angle should allow comfortable head position
10. Space in front of keyboard to support hands/wrists during pauses in keying

forms of instability. The brightness and the contrast between the characters and the background shall be easily adjustable by the user, and also easily adjustable to ambient conditions.

The screen must swivel and tilt easily and freely to suit the needs of the operator or user. It shall be possible to use a separate base for the screen or an adjustable table. The screen shall be free of reflective glare and reflections liable to cause discomfort to the user.

Keyboard

The keyboard shall be tiltable and separate from the screen so as to allow the operator or user to find a comfortable working position avoiding fatigue in the arms or hands.

The space in front of the keyboard shall be sufficient to provide support for the hands and arms of the operator or user. The keyboard shall have a matt surface to avoid reflective glare.

The arrangement of the keyboard and the characteristics of the keys shall be such as to facilitate the use of the keyboard. The symbols on the keys shall be adequately contrasted and legible from the design working position.

Work desk or work surface

The work desk or work surface shall have a sufficiently large, low-reflectance surface and allow a flexible arrangement of the screen, keyboard, documents and related equipment.

The document holder shall be stable and adjustable and shall be positioned so as to minimise the need for uncomfortable head and eye movements.

There shall be adequate space for operators or users to find a comfortable position.

Work chair

The work chair shall be stable and allow the user easy freedom of movement and a comfortable position. The seat shall be adjustable in height. The seat back shall be adjustable in both height and tilt. A footrest shall be made available to any user who wishes one.

3. Environment

Space requirements

The workstation shall be dimensioned and designed so as to provide sufficient space for the user to change position and vary movements.

Lighting

Any room lighting or task lighting provided shall ensure satisfactory lighting conditions and an appropriate contrast between the screen and the background environment, taking into account the type of work and the vision requirements of the operator or user.

Possible disturbing glare and reflections on the screen or other equipment shall be prevented by co-ordinating workplace and workstation layout with the positioning and technical characteristics of the artificial light sources.

Reflections and glare

Workstations shall be so designed that sources of light, such as windows and other openings, transparent or translucid walls, and brightly coloured fixtures or walls cause no direct glare and no distracting reflections on the screen.

Windows shall be fitted with a suitable system of adjustable covering to attenuate the daylight that falls on the workstation.

Noise

Noise emitted by equipment belonging to any workstation shall be taken into account when a workstation is being equipped, with a view in particular to ensuring that attention is not distracted and speech is not disturbed.

Heat

Equipment belonging to any workstation shall not produce excess heat which could cause discomfort to operators or users.

Radiation

All radiation with the exception of the visible part of the electromagnetic spectrum shall be reduced to negligible levels from the point of view of the protection of operators' or users' health and safety.

Humidity

An adequate level of humidity shall be established and maintained.

4. Interface between computer and operator/user

In designing, selecting, commissioning and modifying software, and in designing tasks using display screen equipment, the employer shall take into account the following principles:

(a) software must be suitable for the task;

(b) software must be easy to use and, where appropriate, adaptable to the user's level of knowledge or experience: no quantitative or qualitative checking facility may be used without the knowledge of the operators or users;
(c) systems must provide feedback to operators or users on the performance of those systems;
(d) systems must display information in a format and at a pace which are adapted to operators or users:
(e) the principles of software ergonomics must be applied, in particular to human data processing.

The Regulations are accompanied by a comprehensive guide issued by the HSE.

Catering operations

The principal injuries associated with catering operations are:

(a) scalds to hands, forearms, feet, legs and trunk through contact with boiling water, hot fats and hot liquids;
(b) cuts to hands from knives, bottles, slicing machinery and from the opening of cans;
(c) burns to hands and forearms from ovens, hotplates, ovenware, plates and hot liquids;
(d) bruising, abrasions and fractures from slips and falls caused by greasy floors and obstructions;
(e) back injuries associated with incorrect lifting and carrying techniques.

Good standards of safety in catering originate from the design of the kitchen area. The kitchen should be large enough to allow for safe movement, floors should have a non-slip finish with adequate floor drainage, racking should be provided for storage of equipment and utensils, and a high level of cleaning and housekeeping maintained.

Machinery and plant, such as meat slicers, ovens, bains-marie, bowl mixers and dish-washing machines all present specific hazards. Such items should be frequently inspected and maintained in sound working order, with the appropriate guards and safety devices fitted. Hand tool accidents are common in kitchens, particularly through the use of knives, cleavers, saws and can openers. Staff should be trained in their correct use.

Lighting and ventilation level should be of a high standard. The following illuminance levels are recommended:

Food preparation areas	500 lux
Storage and ancillary areas	300 lux
External storage areas	50 lux

This maintenance of good levels of illumination encourages safe working practices and sound standards of food hygiene. Ventilation levels should be within the range of 12 to 20 air changes per hour, with local exhaust ventilation over ovens and steam-producing appliances.

Kitchens also represent a considerable fire hazard, particularly where housekeeping standards may be poor. Fire appliances should be provided which are capable of dealing with electrical and fat fires, and staff should be trained in the correct use of such appliances. Fire exits should be clearly marked and kept unobstructed. In addition to regular fire drills being undertaken, there should be an effective fire alarm system.

Dry-cleaning processes

Dry cleaning involves the use of solvents, such as trichlorethylene and perchloroethylene along with other solvent-based products which are both flammable and toxic. Under the Dry Cleaning Special Regulations 1949 flammable liquids with flashpoints below 32°C must not be used, except as 'spotting agents' for the removal of stains and marks by hand. Here the liquid should be contained in a spotting bottle or container of not more than 570 ml capacity.

The various stages of the dry-cleaning process can result in vapours entering the working area, particularly when garments are removed from dry-cleaning plant. An effective exhaust ventilation system is, therefore, necessary to maintain concentrations in air well below dangerous levels. Before garments are handled or pressed, all residual solvent in them should have vaporised. Some dry-cleaning plants use diatomaceous earth as a filtering medium for the solvent. This is later removed from the plant as a sludge. Care is necessary to ensure that the sludge is stored safely in lockable bins in a secure area outside the workroom prior to collection. Work at solvent recovery stills should be undertaken only by staff trained in the use of respirators.

Welding operations

The two main forms of welding are gas welding and electric arc welding. Whilst the hazards peculiar to each form of welding are considered later, the following hazards are common to both forms.

Fire and explosion

Arcs, flames, sparks and metal spatter are sources of ignition which will readily ignite waste and other flammable materials in close proximity to the welding operation. Welding on systems or vessels under pressure can result in explosion. Welders should, therefore, ensure that welding arcs and flames do not come into contact with flammable materials. Moreover, care should be taken to ensure that welding does not take place in areas where flammable gases and vapours may be present. This is particularly appropriate in painting degreasing areas which should always be purged with an inert gas prior to welding commencing.

Burns

Welders should be provided with protective clothing to protect them from burns, i.e. face shields and helmets, gauntlets and aprons. Any newly welded work should be segregated from the workforce by barriers or screens, along with the display of warning notices.

Toxic fumes and gases

Inhalation of welding fumes and gases can lead to the condition known as 'welder's lung' or siderosis. Metallic fumes in the form of oxides can be evolved according to the nature of the base metals and electrodes in use. This is also true of fumes and dusts from flux coatings. The action of heat and ultraviolet leads to the evolution of ozone, carbon monoxide and oxides of nitrogen. Heavy particulate matters in the form of respirable dusts can be created as smoke and metal spatter. Many of the gases, vapours and dusts evolved during the welding process are invisible, colourless and odourless, and so considerable care should be taken during welding in confined spaces or unventilated areas. The operation of a permit to work system is, of course, necessary with such operations. (*See also* Chapter 22.)

Precautions during welding operations

(a) Welding workshops should be provided with effective mechanical ventilation capable of achieving six to ten air changes per hour, together with local exhaust ventilation in a designated welding area.

(b) Portable extraction and filtration units should be used where welding is undertaken *in situ* on production machinery and plant.

(c) Environmental monitoring should be undertaken in welding workshops wherever there is evidence of dust and fume accumulation.

(d) Welding in confined spaces, particularly, can present a noise hazard. Hearing protection should be provided and worn. (*See* Chapter 21.)

Gas welding

Fuel gases commonly used are acetylene and propane, both of which are inflammable and form mixtures with air or oxygen. Any leakage of fuel gas is potentially hazardous, as ignition may lead to rapid or explosive combustion, particularly in confined spaces or unventilated areas. Being heavier than air, propane can accumulate at floor level and will readily ignite. Acetylene, an unstable gas, can decompose explosively when subjected to heat or shock. This can occur in the absence of oxygen and under pressure.

A further hazard associated with gas welding is oxygen enrichment. Most welding and cutting operations use oxygen to support combustion of the fuel gas. Accidental leakage of oxygen has, therefore, considerable hazard potential. Oxygen enrichment will cause a change in ignition characteristics of all combustible materials, including those considered non-combustible. Any oxygen leakage in confined or unventilated areas is a matter for immediate concern. Oxygen should, therefore, never be used to purge or 'sweeten' the atmosphere of a confined space or vessel interior. Accidental leakage should be avoided by frequent inspections of hoses, valves and regulators.

Electric arc welding

Hazards can arise from poor standards of maintenance and/or repair of equipment, improper use, and use of unsuitable materials, e.g. insulation tape, to effect repairs to equipment and connections. Other dangers arise in the use of portable welding sets as a result of inadequate power supply, absence of isolating switches in the power supply circuit, the need to remake earth connections for each job, and strain or damage to terminals and connections of the welding set. A system for frequent examination, maintenance and repair of equipment is, therefore, essential. Such a system should ensure the following:

(a) The equipment rating is adequate for the job.

(b) The equipment is installed in accordance with the latest IEE *Regulations for Electrical Installations*, relevant British Standards and manufacturers' instructions.

(c) Isolation switches are readily accessible.

(d) The set is frequently examined by a competent electrician.

(e) All mains and secondary cables, terminals and cable connectors are of adequate size and construction for the maximum welding current.

(f) Terminal and live components are adequately protected.

(g) There is a separate earthing conductor in addition to the welding current return cable.

(h) Earthing circuits are of adequate capacity.

(i) Any damage to the insulation of cables, electrode holders, torches, etc., is repaired immediately or the item replaced.

(j) The amount of trailing cable is minimised to avoid impact damage and the danger from tripping.

(k) There are no exposed metal parts in clothing and protective equipment.

(l) Accidental arcing is avoided.

(m) Correct equipment is worn so that skin is protected, e.g. visor, gloves, apron, safety boots.

(n) Extra care is taken when working in wet, hot or damp conditions, in confined spaces or areas where access is difficult, and when working at heights.

(o) Records of equipment examinations and subsequent repairs, replacements, etc., are kept in the general register.

Another hazard associated with arc welding is that from ultraviolet radiation. This can have an acute effect on the eye, causing burning of the conjunctivae with attendant irritation and a painful feeling of grittiness ('arc eye'). Chronic effects can include permanent vision damage or, in extreme cases, blindness following prolonged exposure. The effect of ultraviolet radiation on degreasing solvents can be phosgene evolution. Phosgene is a highly toxic gas.

Helmets and shields should be kept in good condition and fitted with the correct grade of filter. Non-reflecting welding screens, e.g. matt green canvas, should always be placed around welding areas, and reflected glare should be reduced where possible by the use of non-reflective surfaces for wall finishes in welding workshops. Notices should be displayed in welding areas giving warning of arc flash, and welders should be instructed to warn other people present prior to striking an arc. Moreover, degreasing solvents should be excluded from welding areas. (*See also* Chapter 36.)

Work in compressed air

The risks to people working in compressed air in caissons, in particular the long-term effects of decompression sickness are well known. The Work in Compressed Air Regulations 1996 lay down requirements and prohibitions with respect to the health, safety and welfare of persons who work in compressed air.

G

The term *work in compressed* air means work within any working chamber, airlock or decompression chamber which (in each case) is used for the compression or decompression of persons, including a medical lock used solely for treatment purposes, the pressure of which exceeds 0.15 bar.

The regulations apply to construction work (within the meaning of the Construction (Design and Management) Regulations 1994 and have effect in addition to any applicable provisions of the Construction (Health, Safety and Welfare) Regulations 1996. They do not apply to diving operations within the meaning of the Diving Operations at Work Regulations 1981. The regulations:

(a) provide for the appointment of a competent contractor (the compressed air contractor) to execute or supervise the work in compressed air included in any project;

(b) require specified information to be notified in writing to the HSE and to specified hospitals and other bodies before work in compressed air is commenced and for further notification of the termaination or suspension of such work;

(c) require work in compressed air to be carried out only in accordance with a safe system of work and under adequate supervision;

(d) impose requirements with regard to the provision, use and maintenance of adequate and suitable plant and equipment;

(e) provide that a contract medical adviser be appointed to advise the compressed air contractor on matters relating to the health of persons who work in compressed air;

(f) impose a requirement on employers for adequate medical surveillance to be carried out in respect of such employees who work in compressed air;

(g) require compression and decompression to be carried out safely and in accordance with any procedures approved by the HSE.

Maintenance operations

According to the HSE Report *Deadly Maintenance* (1985) 326 people died during a period of approximately three years whilst engaged in maintenance operations. This report concluded that 83 per cent of these accidents could have been prevented by taking reasonably practicable precautions and 70 per cent of the fatalities could have been saved by positive management action. The seven principal activities featured in the accidents analysed by the HSE were plant and machinery maintenance, roofwork, vehicle repair work, painting, maintenance of electricity, gas and water systems, building maintenance, and window or industrial cleaning. In many cases, risks are further complicated by the presence of contractors, either undertaking work specifically or assisting maintenance staff in specific activities. The need for careful regulation of contractors (*see* Chapter 37 'Construction safety') must therefore be considered. Moreover, the risk is greatly increased because maintenance work is often carried out in relatively inaccessible areas which are difficult to supervise: also the work is frequently carried out under pressure in order to reduce production losses, and sometimes tends to be of an emergency nature, particularly following plant failure. As a result, in many cases, insufficient time is allocated to planning and organising safe systems of work.

Principal hazards

The principal hazards associated with maintenance operations can be classified thus:

1. *Mechanical* – machinery traps, entanglement, contact, ejection; reciprocating traps, shearing traps, in-running nips; uncovenanted strokes; unexpected start-up.

2. *Electrical* – electrocution, shock, burns.
3. *Pressure* – unexpected pressure releases, explosion.
4. *Physical* – extremes of temperature, noise, vibration, dust and fume.
5. *Chemical* – gases, fogs, mists, fumes, etc. prejudicial to health.
6. *Structural* – obstructions, floor openings.
7. *Access* – work at heights, confined spaces.

Two principal areas for consideration are mechanical maintenance procedures and the arrangements for safe access to process plant and buildings.

1. Mechanical maintenance

Plant and machinery maintenance accounted for 30 per cent of fatalities covered in the Report. The principal hazards were associated with maintenance work on conveyors, elevators, cranes, lifts, hoppers and process mixers, storage tanks, chemical, gas and oil process plant and degreasing plant, typical accidents including crushing by moving machinery, falls, burns, asphyxiation and electrocution. The principal causes of these accidents were failure to develop or implement safe systems of work, lack of effective supervision, a failure to isolate machinery and plant effectively, and mechanical faults. Employers should adopt the use of the procedures outlined below:

(a) Plan work in advance and visit the site of activity when systems of work are being developed.

(b) Adopt a formal approach using written safe systems of work, method statements, limitation of access or permit to work systems as appropriate.

(c) Plan specific operations using method statements and ensure that any contractors tendering for work submit method statements in advance of a contract being awarded.

(d) Use physical means of isolating or locking off plant; guard against risk of inadvertent movement by the use of props and chocks, ensure that any stored energy is dissipated whilst working upon plant where there is hydraulic or pneumatic power available.

(e) Systems of working should incorporate two-man working for high-risk operations; in addition the use of remote television monitoring, radios and lone worker alarms should be provided to assist in monitoring safety in maintenance work.

(f) Ensure maintenance considerations occupy a prominent position in design briefs for process plant and associated machinery.

(g) Use only competent contractors from an approved list.

(h) Ensure that maintenance operations, procedures and associated software, such as permits to work, are frequently audited.

2. Provision of safe access to process plant and buildings

Deadly maintenance records that almost 50 per cent of deaths during maintenance work involve falls, most of which occurred from roofs, through fragile roofing material or from fixed plant. The failure to provide safe systems of work and advance planning was a significant cause of these accidents.

The following procedures should, therefore, be covered for this type of work:

(a) Plan work and organise specific safety requirements and details well in advance of the date of commencement.

(b) Reach agreement with operators on procedures for safe access.

(c) Integrate safety requirements in the planning of specific high risk tasks.

(d) Develop written job safety instructions covering safe systems of work and obtain method statements where appropriate.

(e) Access needs should be considered at the drawing board stage of the exercise and incorporated in plans.

(f) Ensure a full exchange of information between occupier, contractors and subcontractors with regard to the above points.

(g) Monitor these operations, ensuring frequent examinations of fixed and temporary access equipment.

Gas safety legislation

Gas Appliances (Safety) Regulations 1992

Under these regulations gas appliances and fittings must comply with Schedule 3 which specifies that:

(a) Appliances and fittings must be safe when properly used and present no danger to persons.

(b) There must be comprehensive technical instructions for the installer when sold.

(c) All necessary information and instructions must be provided for the user.

(d) All materials must be capable of withstanding stresses imposed during foreseeable use.

(e) The manufacturer or supplier must guarantee that correct materials are used.

(f) Appliances must be constructed to ensure safe use.

(g) Appliances must have a valid EC-type examination or certificate.

It is a specific offence to put at risk the health and safety of domestic animals or property by breach of these regulations.

Registration of gas installers

The Gas Safety (Installation and Use) Regulations 1994 require any employer of, or self-employed, persons who work on gas fittings in domestic and commercial premises to be members of a body approved by the HSE, and such an approved body must operate within criteria prescribed by the HSE within the framework of the Regulations. The Council for Registered Gas Installers (CORGI) was approved for this purpose in 1991.

Under the scheme, individual gas fitting operatives are required to have their competence assessed and must possess a certificate issued by the accredited certification body, which consumers can ask to see.

Gas installation businesses are required to demonstrate to the registration body that these certificates are held by all their employees engaged in gas work.

Inspections

CORGI are empowered to appoint inspectors to undertake a continuing programme of work/site inspections. Inspections are based on:

(a) the nature of the business;

(b) the number of operatives employed on gas work;

(c) the type of work carried out;

(d) prior history, e.g. record on complaints, enforcement, etc.

Inspections of registered businesses take place every three years as a minimum, the frequency of inspection being based on the risks assessed for the business.

Inspectors may also inspect the work of a registered business which is the subject of a complaint, carrying out a full and unbiased investigation of that complaint and promoting a fair and satisfactory resolution of it.

Inspectors may also take appropriate action to secure safety where, in the case of either pro-active inspection or as a result of complaint investigation, a dangerous or potentially dangerous installation is discovered.

Gas Safety (Management) Regulations 1996

These regulations incorporate most of the key recommendations set out in the HSC's report *Britain's Gas Supply: A Safety Framework*, which dealt comprehensively with the safety implications of liberalising the domestic gas supply market.

The regulations are designed to maintain existing safety standards by ensuring that the management of the flow of gas through pipeline systems is properly controlled. Each gas transporter is not allowed to operate until their safety case has been accepted by the HSE.

The regulations also require gas transporters to appoint a Network Emergency Co-ordinator (NEC), who has powers to co-ordinate action in emergency circumstances where there may be a total or partial failure of supply, so as to ensure continued safe operation of the network. The NEC is required to prepare a safety case and have it accepted by the HSE.

Gas transporter's safety case

The key elements are:

(a) day-to-day management of their part of the network to ensure continuity of supply at the right pressure and composition so that gas appliances continue to burn safely;
(b) arrangements for dealing with reports, from consumers and others, of gas leaks and suspected emissions of carbon monoxide;
(c) arrangements for investigating fire and explosion incidents: the investigation of carbon monoxide poisoning incidents is the responsibility of gas suppliers.

NEC's safety case

The key elements are:

(a) arrangements to monitor the network to identify any potential national gas supply emergency, and to co-ordinate action to prevent it;
(b) arrangements for directing gas transporters to secure a reduction in gas consumption where it is not possible to prevent a gas emergency developing;
(c) procedures to restore gas supply safely to consumers, following an emergency;
(d) arrangements for conducting emergency services.

Duty to co-operate

All other organisations in the liberalised market must co-operate with transporters and the NEC so that arrangements set out in safety cases can be delivered in practice.

National emergency telephone number

The regulations also require the establishment of a single national emergency telephone number to allow the public and consumers to report gas leaks or suspected emissions of carbon monoxide from appliances. British Gas plc provide this emergency reporting facility via a national 0800 freephone number.

Gas transporters must provide, or secure the provision of, around-the-clock emergency cover to deal with reports of gas leaks, etc. in areas covered by their licences and to make situations safe.

Bibliography and further reading

Chapters 1–5

Dewis, M., and Stranks J. (1988), *Tolley's Health and Safety at Work Handbook* (2nd edn), Tolley Publishing Co. Ltd, Croydon, and RoSPA, Birmingham.

Fife, Judge Ian and Machin, E.A. (1990), *Redgrave's Health and Safety in Factories*, Butterworths, London.

Health and Safety Commission (1985), Leaflet HSC 9, *Time off for Training of Safety Representatives*, HSE Enquiry Points, London and Sheffield.

Health and Safety Executive (1977), *Factories Act 1961: A Short Guide*, HMSO, London.

Health and Safety Executive (1988), *Safety Representatives and Safety Comrnittees* (The Brown Book), HMSO, London.

Health and Safety Executive (1989), *Our Health and Safety Policy Statement: Writing Your Health and Safety Policy Statement: Guide to Preparing a Safety Policy Statement for a Small Business*, HMSO, London.

Health and Safety Executive (1990), *A Guide to the Health and Safety at Work etc. Act 1974: Guidance on the Act*, HMSO, London.

Health and Safety Executive (1992), *The Health and Safety System in Great Britain*, HMSO. London.

Health and Safety Executive, *Forms Used in Conjunction with Legislation*, HSE Enquiry Points, London and Sheffield.

Health and Safety Executive (1996), *A Guide to Information, Instruction and Training: Common Provisions in Health and Safety Law*, HSE Information Centre, Sheffield.

Health and Safety Executive (1996), *A Guide to the Health and Safety (Consultation with Employees) Regulations 1996*, HSE Books, Sudbury.

Health and Safety Executive (1996), *Consulting Employees on Health and Safety*, HSE Books, Sudbury.

Health and Safety Executive (1996), *New and Expectant Mothers at Work: A Guide for Employers* (HS(G) 122), HSE Books, Sudbury.

Health and Safety Executive (1996), *Consulting Employees on Health and Safety: A Guide to the Law*, HSE Books, Sudbury.

Health and Safety Executive (1998), *Health and Safety Law – What You Should Know*, HMSO, London.

Health and Safety Commission (2000), *Management of Health and Safety at Work: Approved Code of Practice: Management of Health and Safety at Work Regulations 1999*, HMSO, London.

Peters, Roger and Gill, Tess (1995), *Health and Safety Liability and Litigation*, FT Law & Tax, London.

Secretary of State for Employment (1974), *Health and Safety at Work etc. Act 1974*, HMSO, London.

Secretary of State for Employment (1998), *Working Time Regulations 1998:* TSO, London.

Secretary of State for Employment (1998), *Children (Protection at Work) Regulations 1998*, TSO, London.
Stranks, Jeremy (2001), *Health and Safety Law* (4th edn), Prentice Hall, London.
Stranks, Jeremy (1997), *Health and Safety Law Pass Notes*, Rapid Results College Business Training.

Chapter 6

Bird, F.E., and Loftus, R.G. (1984), *Loss Control Management*, RoSPA, Birmingham.
British Standards Institute (1996), *BS 8800: Guide to Occupational Health and Safety Management Systems*: BSI, London.
Fletcher, J.A. and Douglas, H.M. (1971), *Total Loss Control*, Associated Business Programmes, London.
Health and Safety Commission (1977), *Safety Representatives and Safety Committees*, HMSO, London.
Health and Safety Executive (1980), *Effective Policies for Health and Safety*, HMSO, London.
Health and Safety Executive (1989), *Access to Occupational Health and Safety Information*, HSE Enquiry Points, London and Sheffield.
Health and Safety Executive (1989), *Essentials of Health and Safety at Work*, HMSO, London.
Health and Safety Executive/Industrial Society (1988), *The Management of Health and Safety*, Industrial Society, Birmingham.
Health and Safety Executive (1991), *Successful Health and Safety Management*, HMSO, London.
Health and Safety Executive (1992), *Five Steps to Successful Health and Safety Management: Special Help for Directors and Managers*, HSE Enquiry Points, London and Sheffield.
Health and Safety Executive, (1992), *Tips for a Safer Business*, HSE Enquiry Points, London and Sheffield.
Health and Safety Executive (1992), *Selecting a Health and Safety Consultancy*, HSE Enquiry Points, London and Sheffield.
Health and Safety Executive (1990), *Safety Pays*, HSE Enquiry Points, London and Sheffield.
Health and Safety Executive (1994), *Five Steps to Successful Health and Safety Management: Special Help for Directors and Managers* ((INDG) 132L), HSE Books, Sudbury.
Health and Safety Executive (1996), *Signpost to the Health and Safety (Safety Signs and Signals) Regulations 1996* (S.I. 1996: No. 341), HMSO London.
Health and Safety Executive (1998), *Working Alone in Safety*, HSE Books, Sudbury.
Institution of Occupational Safety and Health (1997), *Safety and Health . . . Part of Everybody's Working Life*, IOSH, Wigston.
Pirani, M., and Reynolds J. (1976), 'Gearing up for Safety', *Personal Management*, 8, 2, 25–9.
Secretary of State for Employment (1972), *Report of the Committee on Safety and Health at Work* (Robens Report) (Cmnd 5034), HMSO, London.
Secretary of State for Employment (1996), *Health and Safety (Safety Signs and Signals) Regulations 1996* (SI 1996: No. 341), HMSO, London.
Stranks, Jeremy (1994), *Management Systems for Safety*, Pitman Publishing, London.
Stranks, Jeremy (2001), *A Manager's Guide to Health and Safety at Work* (6th Edition), Kogan Page, London.
Stranks, Jeremy (1998): *Health and Safety at Work in* the UK, Blackhall Publishing, Dublin.
Stranks, Jeremy (1998): *Health and Safety at Work in Ireland*, Blackhall Publishing, Dublin.
Stranks, Jeremy (2000): *One Stop Health and Safety* (2nd edn):, ICSA Publishing, London.

Chapter 7

Health and Safety Executive (1996), *The Reporting of Injuries, Diseases and Dangerous Occurrences*, HMSO, London.

Health and Safety Executive (1996), *Guide to the Reporting of Injuries, Diseases and Dangerous Occurrences Regulations 1995*, HMSO, London.

Health and Safety Executive (1996), *Reporting an Injury or Dangerous Occurrence*, HSE Enquiry Points, London and Sheffield.

Health and Safety Executive (1996), *Reporting a Case of Disease*, HSE Enquiry Points, London and Sheffield.

Health and Safety Executive (1988), *Report that Accident – The Reporting of Injuries, Diseases and Dangerous Occurrences Regulations 1985*, HSE Enquiry Points, London and Sheffield.

Chapter 8

Chemical Industries Association Ltd (1975), *Safety Audits*, Chemical Industries Association, London.

Chemical Industries Association Ltd (1977), *A Guide to Hazard and Operability Studies*, Chemical Industries Association, London.

Chapter 9

Chemical Industries Association Ltd (1974), *Recommended Procedures for Handling Major Emergencies*, Chemical Industries Association, London.

Home Office and Scottish Home and Health Department (1977), *Guide to the Fire Precautions Act, 1971: No. 2 Factories*, HMSO, London.

Chapter 11

Bird, F.E. (1974), *Management Guide to Loss Control*, Institute Press, Atlanta.

Morgan, P., and Davies, N. (1981), 'The Cost of Occupational Accidents and Diseases in Great Britain', *Employment Gazette*, HMSO, London.

Chapter 12

Bird, F.E., and Loftus, R.G. (1984), *Loss Control Management*, RoSPA, Birmingham.

Department of Employment (1974), *Accidents in Factories: The Pattern of Causation and Scope for Prevention*, HMSO, London.

Health and Safety Executive (1985), *Watch your Step; Prevention of Slipping, Tripping and Falling Accidents at Work*, HSE Enquiry Points, London and Sheffield.

Health and Safety Executive (1989), *Standards Significant to Health and Safety at Work*, HSE Enquiry Points, London and Sheffield.

Health and Safety Executive (1994), *5 Steps to Risk Assessment*, HSE Books, Sudbury.

Health and Safety Commission (1997), *Health and Safety in Small Firms*, HSE Books, Sudbury.

Health and Safety Executive (1999), *Workplace Health, Safety and Welfare*, HSE Books, Sudbury.

Heinrich, H.W. (1931), *Unsafe Acts and Conditions* (1959 edn), McGraw-Hill, London.

Powell, P.I., Hale, M., Martin, J., and Simon, M. (1971), *2000 Accidents*, National Institute of Industrial Psychology, London.

Royal Society for the Prevention of Accidents (1971), *Factory Accidents: Their Causes and Prevention*, RoSPA, Birmingham.

Stranks, Jeremy (1996), *The Law and Practice of Risk Assessment*, Pitman Publishing, London.

Chapter 13

Department of Employment (1974), *Accidents in Factories: The Pattern of Causation and Scope for Prevention*, HMSO, London.

Health and Safety Executive (1986), *Guide to the Principles and Operation of Permit-to-Work Procedures as Applied in the UK Petroleum Industry*, HMSO, London.

Health and Safety Executive (1989), *Safe Systems of Work*, HSE Enquiry Points, London and Sheffield.

Health and Safety Executive (1990), *Working Alone in Safety*, HSE Enquiry Points, London and Sheffield.

Health and Safety Executive (1996), *Homeworking; Guidance for Employers and Employees on Health and Safety*, HSE Books, Sudbury.

Health and Safety Executive (1999), *Simple Guide to the Lifting Operations and Lifting Equipment Regulations 1998*, HSE Books, Sudbury.

Stevenson, A. (1980), *Planned Safety Management*, Alan Osborne & Associates, Cradley Heath.

Chapter 14

Health and Safety Executive (1978), *Road Transport in Factories*, Guidance Note GS9, HMSO, London.

Health and Safety Executive (1982), *Transport Kills*, HMSO, London.

Health and Safety Executive (1992), *Workplace Health, Safety and Welfare; Approved Code of Practice: Workplace (Health, Safety and Welfare) Regulations 1992*, HMSO, London.

Health and Safety Executive (1995), *Managing Vehicle Safety at the Workplace*, HSE Books, Sudbury.

Chapter 15

Electricity Council, *Better Office Lighting*, Electricity Council, London.

Grundy, J.W., and Rosenthal, S.G. (1978), *Vision and VDUs*, Association of Optical Practitioners, London.

Health and Safety Executive (1980), *Flame Arrestors and Explosion Reliefs*, HMSO, London.

Health and Safety Executive (1987), *Lighting at Work*, HMSO, London.

Health and Safety Executive (1992), *Workplace Health, Safety and Welfare: Approved Code of Practice: Workplace (Health, Safety and Welfare) Regulations 1992*, HMSO, London.

Lyons, S. (1984), *Management Guide to Modern Industrial Lighting*, Butterworths, Sevenoaks.

Chapter 16

Health and Safety Executive (1992), *Workplace Health, Safety and Welfare: Approved Code of Practice: Workplace (Health, Safety and Welfare) Regulations 1992*, HMSO, London.

Chapter 17

Alcock, P.A. (1983), *Food Poisoning*, H.K. Lewis, London.

Sprenger, R.A. (1986), *The Food Hygiene Handbook*, Institution of Environmental Health Officers, London.

Chapter 18

Health and Safety Executive, *Guidance Notes in the Environmental Hygiene and Medical Series*, HMSO, London.

Health and Safety Executive (1982), *A Guide to the Notification of New Substances Regulations*, 1982, HMSO, London.

Health and Safety Executive (1990), *Solvents and You,* HSE Information Centre, Sheffield.

Health and Safety Executive (1990), *Respiratory Sensitisers*, HSE Information Centre, Sheffield.

Health and Safety Executive (1998), *Occupational Exposure Limits*, Guidance Note EH40, HMSO, London.

Plunkett, E.R. (1976), *Handbook of Industrial Toxicology*, Heyden & Son, London.

Sax, N.I. (1979), *Dangerous Properties of Industrial Materials*, Reinhold, New York.

Chapter 19

Atherley, G.R.C. (1978), *Occupational Health and Safety Concepts*, Applied Science Publishers Ltd, London.

Department of Health and Social Security (1983), *Notes on the Diagnosis of Occupational Diseases,* HMSO, London.

Financial Times (1995), *Health in the Workplace*, Financial Times, London.

Gray, H. (1977), *Anatomy, Descriptive and Surgical*, Bounty Books, New York.

Health and Safety Commission (1992), *Vibration White Finger in Foundries; Advice for Employers*, HSE Enquiry Points, London and Sheffield.

Health and Safety Executive (1988), *Control of Asbestos at Work: The Control of Asbestos at Work Regulations* 1987, HMSO, London.

Health and Safety Executive (1988), *Work with Asbestos Insulation, Asbestos Coating and Asbestos Insulating Board: Revised Approved Code of Practice*, HMSO, London.

Health and Safety Executive (1992), *Work-related Upper Limb Disorders: A Guide to Prevention, HS(G)60*, HMSO, London.

Health and Safety Executive (1993), *Protecting Your Health at Work*, HSE Enquiry Points, London and Sheffield.

Health and Safety Executive (1994), *Upper Limb Disorders: Assessing the Risks* (IND(G)171(L)), HSE Books, Sudbury.

Health and Safety Executive (1994), *Work Related Upper Limb Disorders: A Guide to Prevention* (HS(G)60), HSE Books, Sudbury.

Health and Safety Executive (1996), *Good Health is Good Business: Employers' Guide*, HSE Information Centre, Sheffield.

Health and Safety Executive (1996), *Good Health is Good Business: Managing Health Risks in Manufacturing Industry*, HSE Books, Sudbury.

Health and Safety Executive (1996), *Hand-Arm Vibration: Advice on Vibration White Finger for Employees and the Self-Employed* (IND(G)126(L)), HSE Books, Sudbury.

Health and Safety Executive (1996), *Preventing Dermatitis at Work: Advice for Employers and Employees* (IND(G)233(L)), HSE Books, Sudbury.

Health and Safety Executive (1998), *Control of Lead at Work: Approved Code of Practice*, TSO, London.

Health and Safety Executive (1998), *Control of Lead at Work: Approved Code of Practice,:* HMSO, London.

Health and Safety Executive (1998), *Check It Out: Health Risks for Supermarket Cashiers*, HSE Books, Sudbury.

Health and Safety Executive (1998), *Managing Asbestos in Workplace Buildings*, HSE Books, Sudbury.
Hunter, D. (1976), *The Diseases of Occupations*, English Universities Press, London.
International Labour Organisation (1977), *Protection of Workers Against Noise and Vibration in the Working Environment*, ILO, Geneva.
Chalmers Mill, Wendy (1994), *RSI – Repetitive Strain Injury,* Thorsons, London.
Stranks, Jeremy (1995), *Occupational Health and Hygiene*, Pitman Publishing, London.
Taylor, W., and Palmear, P.L. (1975), *Vibration White Finger in Industry*, Academic Press, London.
Wingate, P. (1972), *The Penguin Medical Encyclopedia*, Penguin Books, Harmondsworth.

Chapter 20

Deacon, Steve (1996), *Health Surveillance at Work*, Technical Communications (Publishing), Hitchin.
Health and Safety Executive (1981), *Problem Drinkers at Work*, HMSO, London.
Health and Safety Executive (1982), *Guidelines for Occupational Health Services*, HMSO, London.
Health and Safety Executive (1982), *Pre-employment Health Screening*, Guidance Note MS20, HMSO, London.
Health and Safety Executive (1990), *Surveillance of People Exposed to Health Risks at Work*, HS(G)61, HMSO, London.
Health and Safety Executive (1990), *Health Surveillance under COSHH: Guidance for Employers*, HMSO, London.
Health and Safety Executive (1992), *COSHH and Peripatetic Workers*, HS(G)77, HMSO, London.
Health and Safety Executive (1992), *Health Surveillance of Occupational Skin Disease*, GN: MS 23, HMSO, London.
Health and Safety Executive (1992), *Drug Abuse at Work: A Guide for Employers HSE Enquiry Points*, HSE Enquiry Points, London and Sheffield.
International Labour Organisation (1984), *Occupational Health Services*, ILO, Geneva.
Royal College of Nursing (1975), *An Occupational Health Service*, Royal College of Nursing, London.
Schilling, R.S.F. (1975), *Occupational Health Practice*, Butterworths, London.

Chapter 21

Bilsom International, *In Defence of Hearing* (1992), Bilsom International, Henley-on-Thames.
Bruel & Kjaer, *Noise Control* (1978), Bruel & Kjaer, Naerum, Denmark.
Bruel & Kjaer (1981), *Measuring Vibration*, Bruel & Kjaer, Naerum, Denmark.
Bruel & Kjaer (1984), *Measuring Sound*, Bruel & Kjaer, Naerum, Denmark.
Burns, W. (1973), *Noise and Man,* John Murray, London.
Burns, W., and Robinson, D.W. (1970), *Hearing and Noise in Industry*, HMSO, London.
Harland, I. (1972), *Woods' Practical Guide to Noise Control*, Woods Acoustics, Colchester.
Health and Safety Executive (1976), *Noise and the Worker*, HSW Booklet 26, HMSO, London.
Health and Safety Executive (1983), *100 Practical Applications of Noise Reduction Methods*, HMSO, London.

Health and Safety Executive (1989), Noise at Work: Noise Guide No. 1: *Legal Duties of Employers to Prevent Damage to Hearing.* Noise Guide No. 2: *Legal Duties of Designers, Manufacturers, Importers and Suppliers to Prevent Damage to Hearing. The Noise at Work Regulations 1989*, HMSO, London.

Health and Safety Executive (1990), Noise at Work: Noise Guide No. 3: *Equipment and Procedures for Noise Surveys.* Noise Guide No. 4: *Engineering Control of Noise.* Noise Guide No. 5: *Types and Selection of Personal Ear Protectors.* Noise Guide No 6: *Training for Competent Persons.* Noise Guide No. 7: *Procedures for Noise Testing Machinery.* Noise Guide No. 8: *Exemptions from Certain Requirements of the Noise at Work Regulations 1989*, HMSO, London.

Health and Safety Executive (1995), *Health Surveillance in Noisy Industries: Advice for Employers* (IND(G)193L), HSE Books, Sudbury.

Health and Safety Executive (1995), *Listen Up* (IND(G)122L), HSE Books, Sudbury.

Health and Safety Executive (1998), *Health Risks from Hand–arm Vibration*, HSE Books, Sudbury.

International Labour Organisation (1977), *Protection of Workers Against Noise and Vibration in the Working Environment*, ILO, Geneva.

Chapter 22

Atherley, G.R.C. (1978), *Occupational Health and Safety Concepts*, Applied Science Publishers, London.

Gill, F.S., and Ashton, I. (1982), *Monitoring for Health Hazards at Work*, RoSPA, Birmingham.

Hackett, W.J., and Robbins, G.P. (1979), *Safety Science for Technicians*, Longman, London.

Health and Safety Executive (1988), Guidance Note EH22, *Ventilation of the Workplace*, HMSO, London.

Health and Safety Executive (1984), Guidance Note EH44, *Dust in the Workplace: General Principles of Protection*, HMSO, London.

Schilling, R.S.F. (1975), *Occupational Health Practice*, Butterworths, London.

Chapter 23

Hackett, W.J., and Robbins, G.P. (1979), *Safety Science for Technicians*, Longman, London.

Harvey, B., *et al.* (1983 onwards), *Handbook of Occupational Hygiene – Ionising Radiations*, Kluwer Publishing, London.

Health and Safety Executive (1984), *Wear Your Film Badge*, HSE Enquiry Points, London and Sheffield.

National Radiological Protection Board (1981), *Living with Radiation*, HMSO, London.

Secretary of State for Employment (1999), *Ionising Radiations Regulations 1999*, TSO, London.

United Kingdom Atomic Energy Authority (1982), *Nuclear Facts*, UKAEA, London.

Chapter 24

Hamilton, M. (1983), *The Hand Book*, RoSPA, Birmingham.

Health and Safety Executive (1990), Guidance Note HS(G)53, *Respiratory Protective Equipment: A Practical Guide for Users*, HMSO, London.

Health and Safety Executive (1990), *Respiratory Protective Equipment*, HS(G)53, HMSO, London.

Health and Safety Executive (1992), *Personal Protective Equipment at Work; Guidance on the Personal Protective Equipment at Work Regulations 1992*, HMSO, London.

Health and Safety Executive (1992), *Respiratory Protective Equipment: Legislative Requirements and Lists of HSE Approved Standards and Type Approved Equipment*, HMSO, London.

Chapter 25

Atherley, G.R.C. (1978), *Occupational Health and Safety Concepts*, Applied Science Publishers, London.

Cullis, C.F., and Firth, J.G. (1981), *Detection and Measurement of Hazardous Gases*, Heinemann, London.

Harvey, B., *et al.* (1983 onwards), *The Handbook of Occupational Hygiene*, Kluwer Publishing, London.

Health and Safety Executive (1986), Guidance Note EH28, *Control of Lead: Air Sampling Techniques and Strategies*, HMSO, London.

Health and Safety Executive (1989), Guidance Note EH42, *Monitoring Strategies for Toxic Substances*, HMSO, London.

Health and Safety Executive, *Occupational Exposure Limits*, Guidance Note EH40, HMSO, London.

Jones, A.L., Hutcheson, D.M.W., and Dymott, S. (1981), *Occupational Hygiene*, Croom Helm Ltd, London.

Chapter 26

Harvey, B., *et al.* (1983 onwards), *Handbook of Occupational Hygiene*, Kluwer Publishing, London.

Health and Safety Executive (1987), Guidance Note HS(G)37, *Introduction to Local Exhaust Ventilation*, HMSO, London.

Health and Safety Executive (1988), Free leaflet SIR 16, *Low-volume High-velocity Extraction Systems*, HSE Enquiry Points, London and Sheffield.

Health and Safety Executive (1990), *The Maintenance, Examination and Testing of Local Exhaust Ventilation*, HMSO, London.

Schilling, R.S.F. (1975), *Occupational Health Practice*, Butterworths, London.

Chapter 27

Anderson, P.W. (1990), *Safety Manual for Mechanical Plant Construction: Lifting and Handling*, Kluwer Publishing, London.

Anderson, T.M. (1969), *Human Kinetics*, RoSPA, Birmingham.

Creber, F.L. (1967), *Safety for Industry*, RoSPA and Queen Anne Press, London.

Health and Safety Executive (1992), *Manual Handling: Guidance on the Manual Handling Operations Regulations 1992*, HMSO, London.

Health and Safety Executive (1992), *Lighten the Load: Guidance for Employees on Musculoskeletal Disorders*, HSE Enquiry Points, London and Sheffield.

Health and Safety Commission (1992), *Guidance on Manual Handling of Loads in the Health Services*, HMSO, London.

Health and Safety Executive (1993), *Getting to Grips with Manual Handling*: HSE Enquiry Points, London and Sheffield.

Health and Safety Executive (1994), *Manual Handling: Solutions You Can Handle*, HSE Books, Sudbury.

International Labour Organisation (1984), *I.L.O. Encyclopaedia: Lifting and Carrying*, ILO, Geneva.

Pheasant, S., and Stubbs, D. (1991), *Lifting and Handling: An Ergonomic Approach*, National Back Pain Association, Teddington.

Chapter 28

Health and Safety Commission (1990), *First Aid at Work – Health and Safety (First Aid) Regulations 1981 and Guidance – Approved Code of Practice*, HMSO, London.

Health and Safety Executive (1981), *First Aid at Work*, HSS Booklet HS(R)11, HMSO, London.

Health and Safety Executive (1997), *First Aid at Work: Your Questions Answered*, HSE Information Centre, Sheffield.

Secretary of State for Employment (1981), *Health and Safety (First Aid) Regulations 1981*, SI 1981 No. 917, HMSO, London.

Chapter 29

Health and Safety Executive (1989), Guidance Note HS(G)48, *Human Factors in Industrial Safety*, HMSO, London.

Health and Safety Executive/Industrial Society (1990), *The Management of Health and Safety*, Industrial Society, Birmingham.

Health and Safety Commission (1991), *Advisory Committee on Safety of Nuclear Installations; Study group on Human Factors; Second Report: Human Reliability Assessment – A Critical Overview*, HMSO, London.

Health and Safety Executive (1999), *Help On Work-related Stress: A Short Guide*, HSE Books, Sudbury.

Heinrich, H.W. (1931), *Unsafe Acts and Conditions*, McGraw-Hill, New York.

Maslow, A.H. (1954), *Motivation and Personality*, Harper, New York.

Powell, P.I., Hale, M., Martin, J., and Simon, M. (1971), *2000 Accidents*, National Institute of Industrial Psychology, London.

Stranks, Jeremy (1994), *Human Factors and Safety*, Pitman Publishing, London.

Stranks, Jeremy (1996), *People at Work: The Human Factors Approach to Health and Safety*, Technical Communications (Publishing), Hitchin.

Chapter 30

Bell, C.R. (1974), *Men at Work*, George Allen & Unwin, London.

Brown, B.L., and Martin, J.T. (1977), *Human Aspects of Man-Machine Systems*, Open University Press, Milton Keynes.

Central Computer and Telecommunications Agency and the Council of Civil Service Unions (1988), *Ergonomic Factors Associated with the Use of Visual Display Units*, CCTA, London.

Edholm, O.G. (1967), *The Biology of Work*, World University Library, London.

Health and Safety Executive (1980), *Effective Policies for Health and Safety: A Review Drawn from the Experience of the Accident Prevention Advisory Unit of HM Factory Inspectorate*, HMSO, London.

Health and Safety Executive (1989), *Essentials of Health and Safety at Work*, HMSO, London.

Health and Safety Executive (1989), Guidance Note HS(G)48, *Human Factors in Industrial Safety*, HMSO, London.

Health and Safety Executive (1994), *If the Task Fits: Ergonomics at Work*, HSE Books, Sudbury.

McCormick, E.J. (1976), *Human Factors Engineering*, McGraw-Hill, New York.

Murrell, K.F.H. (1965), *Ergonomics, Man and his Working Environment*, Chapman & Hall, London.

National Electronics Council (1983), *Human Factors and Information Technology*, National Electronics Council, London.

Chapter 31

Cox, T. (1978), *Stress*, Macmillan Press, London.

Health and Safety Executive (1996), *Violence at Work, A guide for Employers*, HSE Books, Sudbury.

Humphrey, J. and Taylor, R. (1990), 'Stress and the Health and Safety Professional', *Health and Safety Practitioner*, (7/90), IOSH, Leicester.

Selye, H. (1936), *The Stress of Life*, revised 1976, McGraw-Hill, New York.

Warr, P.B. (1971), *Psychology at Work*, Penguin Books, Harmondsworth.

Chapter 32

Booth, R.T. (1976), *Machinery Guarding*, Technical File No. 36, *Engineering*.

British Standards Institution (1988), *Code of Practice No. 5304 (EN 292) 1988: Safeguarding of Machinery* (BS 5304), BSI, London.

Department of Employment (1967), *Drilling Machines: Guarding of Spindles and Attachments* (HSW 20), HMSO, London.

Department of Employment (1968), *Safety in the Use of Guillotines and Shears* (HSW 33), HMSO, London.

Department of Employment (1968), *Safety in the Use of Machinery in Bakeries* (HSW 9), HMSO, London.

Department of Employment (1969), *Safety in the Use of Woodworking Machines* (HSW 4), HMSO, London.

Department of Employment (1970), *Guarding of Cutters of Horizontal Milling Machines* (HSW 43), HMSO, London.

Department of Employment (1970), *Safety in the Use of Mechanical Power Presses* (HSW 14), HMSO, London.

Department of Employment (1971), *Safety Devices for Hand and Foot Operated Presses* (HSW 3), HMSO, London.

Department of Employment (1972), *Safety at Drop Forging Hammers* (HSW 12), HMSO, London.

Department of Employment and Productivity (1970), *Electrical Limit Switches and their Applications* (HSW 24), HMSO, London.

Engineering Industry Training Board (1977), *Instruction Manual: Mechanical Maintenance for Engineering Craftsmen*, EITB, Watford.

Health and Safety Executive (1981), *Microprocessors in Industry: Implications in the Use of Programmable Electronic Systems* (HSE Occasional Paper OP2), HMSO, London.

Health and Safety Executive (1998), *Work Equipment: Guidance on the Provision and Use of Work Equipment Regulations 1998*, TSO, London.

Health and Safety Executive (1998), *Buying New Machinery*, HSE Books, Sudbury.
Health and Safety Executive (1999), *Simple Guide to the Provision and Use of Work Equipment Regulations 1998*, HSE Books, Sudbury.
Stranks, Jeremy (1996), *Safety Technology*, Pitman Publishing, London.

Chapter 33

Chemical Industries Association, *Guide to the Storage and Use of Highly Flammable Liquids*, CIA, London.
Department of Employment (1975), *Dust Explosions in Factories* (HSW 22), HMSO, London.
Dewis, M., and Stranks, J. (1988), *Fire Prevention and Regulations Handbook*, Royal Society for the Prevention of Accidents, Birmingham.
Dewis, M., and Stranks, J. (1988), *Health and Safety at Work Handbook* (2nd edn), Tolley, Croydon.
Fire Protection Association, *Compendium of Fire Safety Data* (vols 1 to 6), FPA, London.
Fire Protection Association (1982), *Fire, Safety and Security Planning in Industry and Commerce*, Fire Data Sheet MR2, FPA, London.
Health and Safety Executive (1992), *Assessment of Fire Hazards from Solid Materials and the Precautions Required for their Safe Storage and Use*, HS(G)64, HMSO, London.
Health and Safety Executive (1996), *Safe Working with Flammable Substances*, (IND(G)227L), HSE Information Centre, Sheffield.
Home Office (1977), *Guide to the Fire Precautions Act*, 1971, HMSO, London.
Home Office, *Manual of Firemanship* (Books 1 to 12), HMSO, London.
Lyons, W.A. (1981), *Action Against Fire*, Alan Osborne, London.
Secretary of State for Employment (1997), *The Fire Precautions (Workplace) Regulations 1997* (SI 1997: No. 1840), HMSO, London.
Underdown, G.W. (1979), *Practical Fire Precautions*, Gower Press, Farnborough.

Chapter 34

Anderson, P.W.P. (1990), *Safety Manual for Mechanical Plant Construction*, Kluwer Publishing, London.
Dickie, D.E. (1981), *Crane Handbook*, Butterworths, London.
Dickie, D.E. (1981), *Lifting Tackle Manual*, Butterworths, London.
Health and Safety Executive, *Precautions in the Working of Lifts* (SHW 276), HMSO, London.
Health and Safety Executive (1981), *Safety in Working with Lift Trucks*, Booklet HS(G)6, HMSO, London.
Health and Safety Executive (1988), *Rider Operated Lift Trucks – Operator Training: Approved Code of Practice and Supplementary Guidance*, HMSO, London.
Health and Safety Executive (1992), *A Guide to the Lifting Plant and Equipment (Records of Test and Examination etc.) Regulations 1992*, HMSO, London.
National Joint Industrial Council for the Flour Milling Industry (1956), *Health and Safety Handbook*, NJICFMI, London.

Chapter 35

Ministry of Power (1958), *The Efficient Use of Fuel*, HMSO, London.
Secretary of State for Employment (2000), *Pressure Systems Safety Regulations 2000*, TSO, London.

Chapter 36

Beckingsale, A.A. (1976), *The Safe Use of Electricity*, RoSPA, Birmingham.

Health and Safety Executive (1980), *Electrical Testing: Safety in Electrical Testing* (HSE Booklet HS(G)13), HMSO, London.

Health and Safety Executive (1989), *Memorandum of Guidance on the Electricity at Work Regulations 1989*, Guidance on Regulations, HMSO, London.

Health and Safety Executive (1990), Free leaflet – *Guidance for Small Businesses on Electricity at Work*, HSE Enquiry Points, London and Sheffield.

Health and Safety Executive (1996), *Electrical Safety and You*, HSE Books.

Health and Safety Executive (1996), A *Guide to the Construction (Health, Safety and Welfare) Regulations 1996*, HSE Books, Sudbury.

Hughes, E. (1978), *Electrical Technology*, Longman, London.

Imperial College of Science and Technology (1976), *Safety Precautions in the Use of Electrical Equipment*, ICST, London.

Institution of Electrical Engineers (1989), *IEE Regulations for Electrical installations* (The 'Wiring Regulations'), IEE, Hitchin.

Secretary of State for Employment (1989), *The Electricity at Work Regulations 1989* SI No. 635, HMSO, London.

Secretary of State for Employment (1994) *Construction (Design and Management) Regulations 1994*, HMSO, London.

Secretary of State for Employment (1996) *Construction (Health, Safety and Welfare) Regulations 1996*, HMSO, London.

St John's Ambulance Association and Brigade (1972), *First Aid Manual*, SJAB, London.

Chapter 37

Anderson, P.W.P. (1984), *Safety Manual for Mechanical Plant Construction*, Kluwer Publishing, London.

Armstong, P.T. (1980), *Fundamentals of Construction Safety*, Hutchinson, London.

Health and Safety Executive (n.d.), *Roofwork: Prevention of Falls*, Guidance Note GS 10, HMSO, London.

Health and Safety Executive (1988), *Blackspot Construction: A Study of Five Years Fatal Accidents in the Building and Civil Engineering Industries*, HMSO, London.

Health and Safety Executive (1989), *Construction (Head Protection) Regulations 1989: Guidance on Regulations*, HMSO, London.

Health and Safety Executive (1989), *Avoiding Danger from Underground Services*, HS(G)58, HMSO, London.

Health and Safety Executive (1996), *Health and Safety in Construction* (HS(G)150), HSE Books, Sudbury

Health and Safety Executive (1996), *Managing Asbestos in Workplace Buildings* (IND(G)223(L)), HSE Information Centre, Sheffield.

Health and Safety Executive (1996), *Noise in Construction: Further Guidance on Noise at Work Regulations 1989*, HSE Information Centre, Sheffield.

Health and Safety Executive (1997), *Electrical Safety: Electrical Safety on Construction Sites*, HSE Books, Sudbury.

Health and Safety Executive (1997), *Guidance for Everyone in Construction Work: Health and Safety in Construction*, HSE Books, Sudbury.

Health and Safety Executive (1997), *Guidance on How to Protect the Public: Protecting the Public – Your Next Move*, HSE Books, Sudbury.

International Labour Organisation (1982), *Safety and Health in Building and Civil Engineering Work*, ILO, Geneva.

Royal Society for the Prevention of Accidents (1982), *Construction Regulations Handbook*, RoSPA, Birmingham.

Chapter 38

Health and Safety Executive (1980), *Safe Working with Lift Trucks* (HS(G)6), HMSO, London.

Health and Safety Executive (1988), *Rider Operated Lift Trucks – Operator Training: Approved Code of Practice and Supplementary Guidance*, HMSO, London.

Royal Society for the Prevention of Accidents (1975), *Training Manual for Power Truck Operators*, RoSPA, Birmingham.

Chapter 39

Atherley, G.R.C. (1978), *Occupational Health and Safety Concepts*, Applied Science Publishers, London.

Health and Safety Executive (1988), *Classification and Labelling of Substances Dangerous for Supply. Notification of New Substances Regulations* 1982. *Classification, Packaging and Labelling of Dangerous Substances Regulations* 1984: *Approved Code of Practice*, HMSO, London.

Health and Safety Executive (1988), *Control of Asbestos at Work: The Control of Asbestos at Work Regulations* 1987: *Approved Code of Practice*, HMSO, London.

Health and Safety Executive (1988), *COSHH Assessments: A Step by Step Guide to Assessment and the Skills Needed for it. Control of Substances Hazardous to Health Regulations* 1988, HMSO, London.

Health and Safety Executive (1988), Guidance booklet HS(G)27, *Substances for Use at Work: The Provision of Information* (2nd edn), HMSO, London.

Health and Safety Executive (1989), *COSHH: An Open Learning Course*, HMSO, London.

Health and Safety Executive (1989), Regulation booklet HS(R)19, *Guide to the Asbestos (Licensing) Regulations 1983*, HMSO, London.

Health and Safety Executive (1989), *Transport of Compressed Gases in Tube Trailers and Tube Containers: Dangerous Substances (Conveyance by Road in Road Tankers and Tank Containers) Regulations 1981: Approved Code of Practice*, HMSO, London.

Health and Safety Executive (1990), *Classification and Labelling of Dangerous Substances for Conveyance by Road in Tankers, Tank Containers and Packages (revision 1). Dangerous Substances Conveyance by Road in Road Tankers and Tank Containers Regulations 1981. Classification, Packaging and Labelling of Dangerous Substances* Regulations *1984. Road Traffic (Carriage of Dangerous Substances in Packages, etc.) Regulations 1986. Approved Code of Practice*, HMSO, London.

Health and Safety Executive (1990), *Notification and Marking of Sites: The Dangerous Substances (Notification and Marking of Sites) Regulations 1990: Guidance on the Regulations*, HMSO, London.

Health and Safety Executive (1993), *A Step by Step Guide to COSHH Assessment*, HS(G)97, HMSO, London.

Health and Safety Executive (1995), *The Complete Idiot's Guide to CHIP2*, HSE Books, Sudbury.

Hunter, D. (1978), *The Disease of Occupations*, English Universities Press, London.
Imperial College of Science and Technology (1977), *Safety in Chemical Laboratories and in the Use of Chemicals*, ICST, London.
Kletz, T.A. (1983), *HAZOP and HAZAN*, Institution of Chemical Engineers, Rugby.
Muir, G.D. (1974), *Hazards in the Chemical Laboratory*, Royal Institute of Chemistry, London.
Parkes, W.R. (1974), *Occupational Lung Disorders*, Butterworths, London.
Royal Institute of Chemistry (1976), *Code of Practice for Chemical Laboratories*, RIC, London.
Sax, N.I. (1979), *Dangerous Properties of Industrial Materials*, Reinhold, New York.
Schilling, R.S.F. (1981), *Occupational Health Practice*, Butterworths, London.
Secretary of State for Employment (1999), *Control of Substances Hazardous to Health Regulations 1999*, TSO, London.

Chapter 40

Health and Safety Commission (1982), *Principles of Good Laboratory Practice*, HMSO, London.
Health and Safety Executive (1979), Guidance booklet HS(G)5, *Hot Work: Welding and Cutting on Plant Containing Flammable Materials*, HMSO, London.
Health and Safety Executive (1983), *Principles of Good Laboratory Practice: Notification of New Substances Regulations 1982; Approved Code of Practice*, HMSO, London.
Health and Safety Executive (1983), *Visual Display Units*, HMSO, London.
Health and Safety Executive (1984), *Health Effects of VDUs: A Bibliography*, HMSO, London.
Health and Safety Executive (1987), *Dangerous Maintenance: A Study of Maintenance Accidents in the Chemical Industry and How to Prevent Them*, HMSO, London.
Health and Safety Executive (1987), Guidance booklet HS(G)35, *Catering Safety; Food Preparation Machinery*, HMSO, London.
Health and Safety Executive (1988), Guidance booklet HS(G)45, *Safety in Meat Preparation: Guidance for Butchers*, HMSO, London.
Health and Safety Executive (1990), *Health and Safety in Kitchens and Food Preparation Areas*, HMSO, London.
Health and Safety Executive (1991), *Health and Safety in Motor Vehicle Repair* HS(G)67, HMSO, London.
Health and Safety Executive (1992), *Display Screen Equipment Work: Guidance on the Health and Safety (Display Screen Eguipment) Regulations* 1992, HMSO, London.
Health and Safety Executive (1992), *Working with VDUs*, HSE Enquiry Points, London and Sheffield.
Health and Safety Executive (1993), *Electric Storage Batteries: Safe Charging and Use*, HSE Enquiry Points, London and Sheffield.
Health and Safety Executive (1994), *Officewise* (IND(G)173L), HSE Books, Sudbury.
Health and Safety Executive (1995), *Health and Safety in Engineering Workshops*, HSE Books, Sudbury.
Health and Safety Executive (1996), *Homeworking*, HSE Books, Sudbury.
Health and Safety Executive (1997), *Computer Control: A Question of Safety* (IND(G)243L), HSE Books, Sudbury.
Health and Safety Executive (1997), *Gas Appliances: Get Them Checked: Keep Them Safe* (IND(G)238L), HSE Information Centre, Sheffield.

Road Transport Industry Training Board (1975), *Basic Safety Training in Motor Vehicle Workshops, Stores and Forecourts*, RTITB, Wembley.
Royal Society for the Prevention of Accidents (1972), *Safety in Offices and Shops*, RoSPA, Birmingham.
Royal Society for the Prevention of Accidents (1976), *Catering Care*, RoSPA, Birmingham.
Saunders, Roger (1995), *The Office Safety Handbook*, Pitman Publishing, London.
Secretary of State for Employment (1996) *Work in Compressed Air Regulations 1996*, HMSO, London.

Index